Lecture Notes in Computer Science 16878

The series Lecture Notes in Computer Science (LNCS), including its subseries Lecture Notes in Artificial Intelligence (LNAI) and Lecture Notes in Bioinformatics (LNBI), has established itself as a medium for the publication of new developments in computer science and information technology research, teaching, and education.

LNCS enjoys close cooperation with the computer science R & D community, the series counts many renowned academics among its volume editors and paper authors, and collaborates with prestigious societies. Its mission is to serve this international community by providing an invaluable service, mainly focused on the publication of conference and workshop proceedings and postproceedings. LNCS commenced publication in 1973.

Guang Yang · Ehsan Adeli · Marleen de Bruijne ·
Bartłomiej W. Papież · Stefanie Speidel ·
Pallavi Tiwari · Guoyan Zheng ·
Mohammad Yaqub · Qi Dou · Islem Rekik
Editors

Medical Image Computing and Computer Assisted Intervention – MICCAI 2026

29th International Conference
Strasbourg, France, September 27 – October 1, 2026
Proceedings, Part I

Springer

Editors
Guang Yang
Imperial College London
London, UK

Marleen de Bruijne
Erasmus MC
Rotterdam, The Netherlands

Department of Computer Science
University of Copenhagen
Copenhagen, Denmark

Stefanie Speidel
TU Dresden/German Cancer Research Center
Dresden, Sachsen, Germany

Guoyan Zheng
Shanghai Jiao Tong University
Shanghai, China

Qi Dou
The Chinese University of Hong Kong
Hong Kong, Hong Kong

Ehsan Adeli
Stanford University
Palo Alto, CA, USA

Bartłomiej W. Papież
University of Oxford
Oxford, UK

Pallavi Tiwari
University of Wisconsin-Madison
Madison, WI, USA

Mohammad Yaqub
Mohamed bin Zayed University of Artificial Intelligence
Abu Dhabi, United Arab Emirates

Islem Rekik
Imperial College London
London, UK

ISSN 0302-9743 ISSN 1611-3349 (electronic)
Lecture Notes in Computer Science
ISBN 978-3-032-38058-6 ISBN 978-3-032-38059-3 (eBook)
https://doi.org/10.1007/978-3-032-38059-3

This Springer imprint is published by the registered company Springer Nature Switzerland AG
The registered company address is: Gewerbestrasse 11, 6330 Cham, Switzerland

Preface

We are delighted to present the proceedings of the 29th International Conference on Medical Image Computing and Computer Assisted Intervention (MICCAI 2026), held in Strasbourg, France on September 27 – October 1, 2026. Following the missed opportunity to gather in Strasbourg in 2021, it was especially joyful to finally welcome the MICCAI community to this wonderful city. Beyond its remarkable cultural heritage, Strasbourg is a vibrant European capital shaped by the exchange of ideas across borders and disciplines. It has also fostered a thriving interdisciplinary research ecosystem in which artificial intelligence, medical imaging, engineering, and clinical sciences come together, perfectly reflecting the collaborative spirit and foundations of MICCAI.

The conference featured a broad programme of satellite events on September 27 and October 1, comprising 112 workshops, tutorials, and challenges. This year also marked an important milestone for the Clinical Translation in Medical Image Computing and Computer Assisted Intervention conference (CLINICCAI), whose fifth edition was held on September 29. For the first time, CLINICCAI was organized in close collaboration with the MICCAI Organizing Committee rather than as an independent event, further reinforcing MICCAI's commitment to clinical translation and engagement. Open science remained another important area of focus, with the conference hosting its second Open Data Session on September 30.

MICCAI 2026 set a new record for submissions, with a 25% increase compared with 2025. We received 4,601 full submissions, of which 4,402 advanced to full review. After review, 1,167 papers were accepted for presentation, representing an acceptance rate of approximately 27%. One hundred and ninety-nine papers were desk-rejected because they did not comply with the submission guidelines, including papers with unverifiable references and papers that violated the double-submission rules. The final accepted papers comprise the 19 volumes of the Lecture Notes in Computer Science (LNCS) proceedings, as follows:

- Part I, LNCS Volume 16878: Adapting Foundation Models for Medical Imaging: Large Vision and Language Models (I)
- Part II, LNCS Volume 16879: Adapting Foundation Models for Medical Imaging: Large Vision and Language Models (II)
- Part III, LNCS Volume 16880: Efficient Learning Strategies in Medical Imaging
- Part IV, LNCS Volume 16881: Domain Adaptation and Generalization in Medical Imaging
- Part V, LNCS Volume 16882: Trustworthy, Explainable, and Responsible Medical AI
- Part VI, LNCS Volume 16883: Medical Image Segmentation (I)
- Part VII, LNCS Volume 16884: Medical Image Segmentation (II)
- Part VIII, LNCS Volume 16885: Computer-Aided Diagnosis (I)
- Part IX, LNCS Volume 16886: Computer-Aided Diagnosis (II)
- Part X, LNCS Volume 16887: Longitudinal Modeling, Staging, and Prognosis

- Part XI, LNCS Volume 16888: Image Reconstruction and Image Enhancement
- Part XII, LNCS Volume 16889: Medical Image Registration; Shape Analysis
- Part XIII, LNCS Volume 16890: Medical Image Synthesis and Super-resolution
- Part XIV, LNCS Volume 16891: Computational Pathology; Microscopic Imaging
- Part XV, LNCS Volume 16892: Surgical Data Science and Scene Understanding
- Part XVI, LNCS Volume 16893: Image-Guided Surgery and Intervention
- Part XVII, LNCS Volume 16894: Neuroimaging and Brain-Computer Interfaces
- Part XVIII, LNCS Volume 16895: Benchmarking Datasets and Methods; Clinical Applications
- Part XIX, LNCS Volume 16896: Emerging Techniques and Applications

The papers for the proceedings were selected through a rigorous double-blind peer-review process in which submissions received three reviews each on average. The MICCAI 2026 Program Committee (PC) consisted of 244 Area Chairs (ACs) and over 3,900 scientific reviewers, representing 71 countries across all major continents. About a quarter (23%) of the ACs self-identified as women. Each area chair managed the review of roughly 33 papers, and the average workload was four papers per reviewer. Area chairs and reviewers were selected following public calls to the community and were vetted by the program chairs. As in the past, the Microsoft Conference Management Toolkit (CMT) was used for paper management and organization of the overall review process. The Toronto Paper Matching System (TPMS) was again employed to help ensure that papers were evaluated by area chairs and reviewers with the appropriate expertise. While a special effort was made this year to improve the quality of paper matches by reworking both the TPMS formulae and the CMT categorization of papers and PC member profiles, we recognize the ongoing challenge of securing high-quality reviews in the face of ever-increasing submission numbers.

Based on the double-blind reviews, area chairs' recommendations, and program chairs' decisions, 9% (376) of papers were provisionally accepted without rebuttal, 51% (2255) were rejected, and 40% (1771) proceeded to the rebuttal stage. Final decisions were made based on all available information, including paper screening, adherence to the author submission guidelines, reviews, rebuttals, 2–3 independent meta-reviews, communications during the review process, and, in some cases, additional assessment by the program chairs. A dedicated team was formed to enforce the dual-submission policy. The program committee scrutinized papers flagged during review, manuscripts of special interest, potential cases of plagiarism, and appeals from authors. The committee was particularly attentive to indications of Large Language Model (LLM) misuse in both submissions and reviews, an increasingly vexing issue not only at MICCAI but also in science more generally.

The 1,167 accepted papers that constituted this year's scientific program originated from 48 countries. Almost 80% of these papers came from the following five countries, as identified by the first author's affiliation: China, including Hong Kong SAR and Macao SAR (547); the USA (141); South Korea (95); the UK (68); and Germany (63). The next five countries, in descending order of the number of accepted papers, were Canada (29), France (29), Australia (28), India (24), and the United Arab Emirates (18), highlighting MICCAI's continuing efforts to broaden its global impact.

To build the program of 153 podium presentations, accepted papers were first clustered by technical theme. For each theme, oral presentation candidates were chosen from the papers that received overwhelming enthusiasm from the area chairs and reviewers. These candidates were then narrowed down to the final list by ensuring that the program represented the range of technical contributions and clinical applications covered by the papers and reflected both geographic and demographic inclusivity among its presenters. Among the 153 presentations, 99 were selected as long oral presentations and 54 as spotlight presentations (short talks). The distinction between oral and spotlight presentations was not meant to reflect a difference in technical quality or contribution. These assignments were made primarily based on programming considerations, thematic balance, session structure, and time constraints. All of these papers were recognized by the reviewers and area chairs as having strong potential for clinical translation, making significant contributions to both the technical and clinical communities, or advancing the field through the release of open-source code and publicly available datasets.

MICCAI 2026 was organized around two overarching themes. The first, From Concept to Clinical Impact, emphasized the translational trajectory of MICCAI methods from foundational research to robust deployable systems that improve clinical decision-making and patient care. To support this theme, we strengthened engagement with industry, healthcare providers, and technology partners through dedicated panels, translational showcases, and structured mentoring activities in close collaboration with the MICCAI Student Board and leading experts across academia and industry.

The second theme, Connected Communities and Global Collaboration, reflected MICCAI's long-standing role as an international forum that brings together diverse research communities across continents. It focused on strengthening collaboration across established and emerging ecosystems, fostering inclusivity, and improving access across career stages and regions. Targeted mentoring initiatives supported early-career researchers in scientific writing, submission strategy, and rebuttal preparation, while additional programs broadened participation from underrepresented and resource-constrained communities.

MICCAI 2026 also emphasized the translation of research into real-world healthcare deployment. Dedicated sessions addressed clinical challenges, data governance, and responsible AI, including privacy, regulatory constraints, and reproducibility. A key focus was strengthening collaboration between academia, hospitals, industry, and startups to accelerate clinical translation.

In addition, MICCAI 2026 promoted an innovation-oriented ecosystem connecting researchers, clinicians, and industry stakeholders. This included an expanded Startup Village, industry showcases, mentoring sessions, and entrepreneurship activities that supported technology transfer and new collaborations. Complementary workshops, hackathons, and thematic schools extended engagement beyond the conference. Through these efforts, MICCAI 2026 reinforced its role as the leading global forum for medical image computing and computer-assisted intervention while accelerating clinical impact and strengthening collaboration across the healthcare AI ecosystem.

To further the mission of enhancing inclusion at MICCAI, we offered travel grants to candidates from lower- and middle-income countries as well as from countries historically underrepresented in the MICCAI community, including those from Africa, Asia, and Latin America. We also provided partial support to additional participants through the MICCAI Society Registration Grants. Funding was made possible by generous support from the MICCAI Society, donations from individual groups within the MICCAI community, and financial support from non-commercial sponsors. Support for US-based students and early-career investigators was provided by the National Institute of Biomedical Imaging and Bioengineering of the National Institutes of Health (NIH) under Award Number R13EB035922. The content expressed in written conference materials or publications and by speakers and moderators does not necessarily reflect the official policies of the NIH; nor does mention of trade names, commercial practices, or organizations imply endorsement by the U.S. Government.

A meeting of MICCAI's scale and impact is only possible through the collective dedication and tireless efforts of many individuals, most notably the members of this year's organizing committees. While every member played a vital role, we would like to especially recognize several key contributors. The Workshops Committee (Xiu Wang, Sophia Bano, Shadi Albarqouni, and Chunfeng Lian), Tutorials Committee (Andrea Lara and Anees Kazi), and Challenges Committee (Olivier Colliot, Moi Hoon Yap, Ulas Bagci, Annika Reinke, Gorkem Durak, and Charlotte Godard) reviewed a total of 191 satellite event proposals and assembled a diverse, high-quality program. We are particularly grateful to our Satellite Program Committee (Spyridon Bakas, Ghada Zamzmi, and Xiaoxiao Li) for skillfully coordinating the 112 accepted satellite events into a cohesive and well-structured program. The Panel Committee, headed by Tanveer Syeda-Mahmood, established a strong panel that facilitated in-depth discussions of our conference themes. The Student Committee, under the leadership of Camila Gonzalez and Numan Saeed, introduced a new and impactful program that further enriched the participation and engagement of students and early-career researchers.

We also appreciate the exemplary work of our Outreach & Sponsorship Team: Angelica Aviles Rivero, Rachid Jennane, Mohammad Yaqub, Nassir Navab, Lina Chato, Mohammed Almasni, Ilkay Oksuz, Muhammad Rushdi, Shereen Fouad, Alaa Alkaysi, and Salma Hassan, for their efforts in securing financial support for the conference despite the challenging economic climate and helping to further expand MICCAI's reach into underrepresented regions. Joseph Chin-Chi Kuo, Farouk Dako, Shahrukh Hashmi, Rafat Damseh, and Michael B. Wallace, together with the CLINICCAI Program Committee, ensured the continued success of CLINICCAI as a distinctive flagship of MICCAI. Martijn P.A. Starmans, Apostolia Tsirikoglou, Lidia Garrucho Moras, and Kaouther Mouheb provided distinguished leadership in advancing the Open Data Initiative.

A tremendous amount of work was carried out by the Local Organizing Committee, chaired by Caroline Essert, whose commitment and meticulous attention to every detail were critical to the smooth execution of the meeting. We are also grateful to the team of Student Volunteers, led by Jonas Mehtali, for their enthusiasm and dedication in organizing the student activities that complemented the scientific program and helped build a vibrant and cohesive student community. We are deeply grateful to our local supporting institutions: the University of Strasbourg, ICube Institute, ITI HealthTech, and ENACT Cluster, for their generous contributions to and support for the conference. We also wish to extend our heartfelt thanks to the City of Strasbourg and the team at the Convention Centre of Strasbourg, led by Ramona Fischer and Anne Knaus, whose dedication, professionalism, and unwavering support made it possible to bring MICCAI 2026 to France.

We would also like to express our sincere appreciation to Geng Chen, our Publication Chair, for his outstanding efforts in overseeing the preparation and production of the conference proceedings. His professionalism, efficiency, and meticulous attention to detail ensured the timely and high-quality publication of the scientific program. Our heartfelt thanks also go to our Keynote Chairs: Min Xu, Michael Brady, and Nassir Navab, for assembling an exceptional keynote program that brought together internationally renowned leaders whose inspiring talks helped shape the scientific discussions and vision of MICCAI 2026. We are deeply grateful to our Equity, Diversity & Inclusion Chairs: Celia Cintas, Esther Puyol-Antón, Aamna Al Shehhi, and Jihad Zahir, for their dedication to fostering a welcoming, inclusive, and diverse conference environment. Their initiatives and thoughtful leadership strengthened our commitment to ensuring that MICCAI remains an open, supportive, and accessible community for researchers from all backgrounds. In addition, we sincerely thank our Career Development and Mentorship Chairs: Antonio Porras, Baiying Lei, and Anees Kazi, for their commitment to supporting the professional growth of our community. Through the organization of mentorship activities and career development initiatives, they created valuable opportunities for students, early-career researchers, and emerging leaders to connect, learn, and thrive within the MICCAI community.

We would also like to express our sincere gratitude to the MICCAI Society Board of Directors and, in particular, to its President, Marius George Linguraru, and former President, Caroline Essert, for their support and constructive feedback. We gratefully acknowledge the contributions of the MICCAI Society secretariat personnel: Janette Wallace and Johanne Langford, who oversaw logistics and budgets, and Diana Cunningham and John Baxter, who provided timely conference announcements in the MICCAI Society newsletters and MICCAI Morning issues. The proficient on-site organization of the conference, including budget management and event operations, was the commendable work of our Professional Conference Organizer, Dekon Congress & Tourism, led by Mehmet Eldegez.

The proceedings of MICCAI 2026 are the fruit of the extraordinary work of our entire community. We would like to give special recognition to Kitty Wong, the MICCAI Submission System Manager, and express our deepest gratitude for her unwavering dedication and exceptional support throughout every stage of the conference organization. From managing the submission and review system to coordinating countless operational details behind the scenes, Kitty has consistently gone above and beyond to ensure that every aspect of the process runs smoothly and efficiently. Her professionalism, responsiveness, meticulous attention to detail, and tireless commitment have been instrumental in making the review process and conference organization a success. Kitty's dedication, patience, and positive spirit are invaluable assets to the MICCAI community. We have been truly fortunate to work alongside her, and the conference would simply not have been possible without her extraordinary efforts. On behalf of the entire MICCAI community, we extend our heartfelt appreciation and sincere thanks for everything she has done.

We are especially grateful to the members of the Program Committee team for their dedication and tremendous efforts throughout the conference preparations, from planning and paper review to decision-making and handling countless emails. It was certainly challenging, but also immensely rewarding! Our warmest congratulations and sincere thanks go to all our ACs and reviewers for their outstanding work and for meeting every deadline. Most importantly, we extend our deepest appreciation to all the authors for the time, creativity, and effort they invested in preparing their manuscripts and supporting the conference through their submissions. The success of MICCAI 2026 is a direct reflection of the exceptional quality of the work presented at the meeting. Heartfelt congratulations to all our authors and contributors on their outstanding achievements!

In closing, we once again thank every individual and team involved in organizing MICCAI 2026. Their inspiring dedication, professionalism, enthusiasm, and collaborative spirit made this year's conference a truly outstanding and memorable occasion. We hope you thoroughly enjoyed MICCAI 2026 and your time with us in Strasbourg. We now look forward with great excitement to reuniting with the MICCAI community next year in Auckland, New Zealand, for MICCAI 2027!

September 2026

Guang Yang
Ehsan Adeli
Marleen de Bruijne
Bartłomiej W. Papież
Stefanie Speidel
Pallavi Tiwari
Guoyan Zheng
MICCAI 2026 Program Chairs

Mohammad Yaqub
Islem Rekik
Qi Dou
MICCAI 2026 General Chairs

Organization

General Chairs

Mohammad Yaqub	Mohamed bin Zayed University of Artificial Intelligence, UAE
Islem Rekik	Imperial College London, UK
Qi Dou	Chinese University of Hong Kong, China

Program Chairs

Guang Yang (Lead)	Imperial College London, UK
Ehsan Adeli	Stanford University, USA
Marleen de Bruijne	Erasmus MC, Netherlands and University of Copenhagen, Denmark
Bartłomiej W. Papież	University of Oxford, UK
Stefanie Speidel	TU Dresden & German Cancer Research Center, Germany
Pallavi Tiwari	University of Wisconsin–Madison, USA
Guoyan Zheng	Shanghai Jiao Tong University, China

Satellite Event Committee

Spyridon Bakas (Lead)	Indiana University, USA
Xiaoxiao Li	University of British Columbia, Canada
Ghada Zamzmi	HeartFlow, USA

Workshop Committee

Xiu Wang (Lead)	University of Sydney, Australia
Sophia Bano	University College London, UK
Shadi Albarqouni	University of Bonn, Germany
Chunfeng Lian	Xi'an Jiaotong University, China

Challenges Committee

Olivier Colliot (Lead)	French National Centre for Scientific Research, France
Moi Hoon Yap	Manchester Metropolitan University, UK
Ulas Bagci	Northwestern University, USA
Annika Reinke	Heidelberg University, Germany
Charlotte Godard	Paris Brain Institute, France
Gorkem Durak	Northwestern University, USA
Lena Maier-Hein (Advisor)	Heidelberg University, Germany

Tutorials Committee

Andrea Lara (Lead)	Universidad Galileo, Guatemala
Anees Kazi	Harvard Medical School, USA

MICCAI-Clinical (CLINICCAI) Committee

Joseph Chin-Chi Kuo (Lead)	China Medical University Hospital, China
Farouk Dako	University of Pennsylvania, USA
Shahrukh Hashmi	Department of Health, Abu Dhabi, UAE
Rafat Damseh	United Arab Emirates University, UAE
Michael B. Wallace	Mayo Clinic, USA
Tausifa Jan Saleem	Mohamed bin Zayed University of Artificial Intelligence, UAE

Publication Committee

Geng Chen	Northwestern Polytechnical University, China

Keynote Committee

Min Xu (Lead)	Carnegie Mellon University, USA
Michael Brady	Oxford Clinical Diagnostic Centre, UK
Nassir Navab	Technical University of Munich, Germany

Sponsorship Committee

Angelica Aviles Rivero (Lead)	Tsinghua University, China
Rachid Jennane	University of Orléans, France
Nassir Navab	Technical University of Munich, Germany
Mohammad Yaqub	Mohamed bin Zayed University of Artificial Intelligence, UAE

Outreach Committee

Lina Chato (Lead)	University of South Dakota, USA
Mohammed Almasni	Sejong University, South Korea
Mohamed Mabrok	Qatar University, Qatar
Ilkay Oksuz	Istanbul Technical University, Turkey
Muhammad Rushdi	Cairo University, Egypt
Shereen Fouad	Aston University, UK
Alaa Alkaysi	University of Technology-Iraq, Iraq

Open Data Committee

Martijn P. A. Starmans (Lead)	Erasmus MC, Netherlands
Apostolia Tsirikoglou	Karolinska Institutet, Sweden
Kaouther Mouheb	Erasmus MC, Netherlands
Lidia Garrucho Moras	Universitat de Barcelona, Spain

Students Committee

Camila Gonzalez (Lead)	Stanford University, USA
Numan Saeed	Mohamed bin Zayed University of Artificial Intelligence, UAE

Equity, Diversity and Inclusion Committee

Celia Cintas (Lead)	IBM, USA
Esther Puyol-Antón	King's College London and Heartflow, UK
Aamna Al Shehhi	Khalifa University, UAE

Jihad Zahir	Cadi Ayyad University, Morocco
Baoru Huang	Imperial College London, UK

Career Development and Mentorship Committee

Antonio Porras (Lead)	University of Chicago, USA
Baiying Lei	Shenzhen University, China
Anees Kazi	Harvard Medical School, USA

Local Organizers

Caroline Essert (Lead)	University of Strasbourg, France
Aamna Al Shehhi	Khalifa University, UAE
Omar Alkadi	University of Jordan, Jordan
Shadab Khan	ADIA Lab, UAE
Mai Ahmed Shaaban	Mohamed bin Zayed University of Artificial Intelligence, UAE
Ibrahim Almakky	Mohamed bin Zayed University of Artificial Intelligence, UAE
Numan Saeed	Mohamed bin Zayed University of Artificial Intelligence, UAE
Shahad Hardan	Mohamed bin Zayed University of Artificial Intelligence, UAE
Salma Hassan	Mohamed bin Zayed University of Artificial Intelligence, UAE
Mostafa Salem	Mohamed bin Zayed University of Artificial Intelligence, UAE
Tausifa Jan Saleem	Mohamed bin Zayed University of Artificial Intelligence, UAE
Dwarikanath Mahapatra	Khalifa University, UAE
Yutong Xie	Mohamed bin Zayed University of Artificial Intelligence, UAE
Asif Hanif	Mohamed bin Zayed University of Artificial Intelligence, UAE
Mohammad Yaqub	Mohamed bin Zayed University of Artificial Intelligence, UAE

Program Committee (Area Chairs)

Sharib Ali	University of Leeds, UK
Syed Anwar	Children's National Hospital, USA
Pablo Arbelaez	Universidad de los Andes, Colombia
Ulas Bagci	Northwestern University, USA
Abhirup Banerjee	University of Oxford, UK
Joao Barreto	University of Coimbra, Portugal
Neslihan Bayramoglu	University of Oulu, Finland
Cosmin Bercea	Technical University of Munich and Helmholtz AI and Helmholtz Center Munich, Germany
Alaa Bessadok	Université de Carthage, Tunisia
Binod Bhattarai	University of Aberdeen, UK
Benjamin Billot	Inria, France
Ryoma Bise	Kyushu University, Japan
Federico Bolelli	Università degli Studi di Modena e Reggio Emilia, Italy
Katharina Breininger	Center for AI and Data Science, University of Würzburg, Germany
Esther Bron	Erasmus MC, Netherlands
Mariano Cabezas	Universitat de Girona, Spain
Peng Cao	Northeastern University, China
Cheng Chen	University of Hong Kong, China
Hao Chen	Hong Kong University of Science and Technology, China
Jianan Chen	University College London, UK
Xin Chen	University of Nottingham, UK
Zhaolin Chen	Monash University, Australia
Zhen Chen	Hong Kong Polytechnic University, China
Jun Cheng	Institute for Infocomm Research, A*STAR, Singapore
Li Cheng	University of Alberta, Canada
Toby Collins	IRCAD, France
Kang Dang	Xi'an Jiaotong-Liverpool University, China
Behdad Dashtbozorg	Netherlands Cancer Institute, Netherlands
Elena De Momi	Politecnico di Milano, Italy
Ruining Deng	Weill Cornell Medicine, USA
Christian Desrosiers	École de technologie supérieure, Canada
Blake Dewey	Johns Hopkins University, USA
Jose Dolz	ETS Montreal, Canada
Reuben Dorent	Inria, France
Niharika D'Souza	IBM Research, USA

Dmitry V. Dylov	Computational Imaging Laboratory, Russia
Roy Eagleson	Western University, Canada
Mostafa El Habib Daho	University of Western Brittany, France
Huihui Fang	Nanyang Technological University, Singapore
Azade Farshad	ELLIS Institute Finland and Aalto University, Finland and Technical University of Munich, Germany
Huazhu Fu	IAIC, A*STAR, Singapore
Jin Kyu Gahm	Pusan National University, South Korea
Yue Gao	Tsinghua University, China
Zhifan Gao	Sun Yat-sen University, China
Zongyuan Ge	Monash University, Australia
Guido Gerig	New York University, USA
Elodie Germani	Université de Rennes, France
Ricardo Gonzales	Harvard Medical School, USA
Camila Gonzalez	Medical University of Vienna, Austria
Karthik Gopinath	Heartflow, USA
Pietro Gori	Télécom Paris, France
Pengfei Guo	NVIDIA, USA
Xueqi Guo	Siemens Healthineers, USA
Prashnna Gyawali	West Virginia University, USA
Ilker Hacihaliloglu	University of British Columbia, Canada
Xiangmin Han	Tsinghua University, China
Yuting He	Case Western Reserve University, USA
Monica Hernandez	University of Zaragoza, Spain
Yi Hong	Shanghai Jiao Tong University, China
Benjamin Hou	Imperial College London, UK
Dan Hu	University of North Carolina at Chapel Hill, USA
Xiaoling Hu	Harvard Medical School, USA
Baoru Huang	University of Liverpool, UK
Junzhou Huang	University of Texas at Arlington, USA
Yuankai Huo	Vanderbilt University, USA
Jana Hutter	King's College London, UK
Abdullah-Al-Zubaer Imran	University of Kentucky, USA
Mohammad Jafari	Amazon, Canada
Amir Jamaludin	University of Oxford, UK
Won-Ki Jeong	Korea University, South Korea
Meirui Jiang	Ant Group, China
Tingting Jiang	Peking University, China
Zhifan Jiang	Children's National Hospital, USA
Zhongliang Jiang	University of Hong Kong, China
Dakai Jin	Alibaba USA Inc., USA

Qiangguo Jin	Northwestern Polytechnical University, China
Samuel Kadoury	Polytechnique Montréal, Canada
Bernhard Kainz	Imperial College London, UK and Friedrich-Alexander-Universität Erlangen-Nürnberg, Germany
Tina Kapur	Brigham and Women's Hospital, USA
Youngwook Kee	Stony Brook University, USA
Benjamin Killeen	Technical University of Munich, Germany
Minjeong Kim	University of North Carolina at Greensboro, USA
Seong Tae Kim	Kyung Hee University, South Korea
Won Hwa Kim	POSTECH, South Korea
Sila Kurugol	Boston Children's Hospital and Harvard Medical School, USA
Christian Ledig	University of Bamberg, Germany
Soochahn Lee	Kookmin University, South Korea
Baiying Lei	Shenzhen University, China
Gang Li	University of North Carolina at Chapel Hill, USA
Hongwei Li	Technical University of Munich, Germany
Lei Li	National University of Singapore, Singapore
Yuexiang Li	City University of Macau, China
Zeju Li	Fudan University, China
Zhen Li	Chinese University of Hong Kong, Shenzhen, China
Chunfeng Lian	Xi'an Jiaotong University, China
Hongxiang Lin	Zhejiang Lab, China
Daochang Liu	University of Western Australia, Australia
Peirong Liu	Johns Hopkins University, USA
Xiaofeng Liu	Yale University, USA
Hervé Lombaert	Polytechnique Montréal, Canada
Le Lu	Ant Group, USA
Gongning Luo	Harbin Institute of Technology, China
Jax Luo	Cleveland Clinic, USA
Luyang Luo	Harvard University, USA
Xiangde Luo	Stanford University, USA
Dwarikanath Mahapatra	Khalifa University, UAE
Diana Mateus	Centrale Nantes, France
Evangelos Mazomenos	University College London, UK
Deval Mehta	Monash University, Australia
Qingjie Meng	University of Birmingham, UK
Yanda Meng	King Abdullah University of Science and Technology, Saudi Arabia
Raffaele Mineo	University of Catania, Italy

Deepak Mishra	IIT Jodhpur, India
Sara Moccia	Università degli Studi G. d'Annunzio Chieti–Pescara, Italy
Tony C.W. Mok	DAMO Academy, Alibaba Group, China
Rodrigo Moreno	KTH Royal Institute of Technology, Sweden
Daniel Moyer	Vanderbilt University, USA
Karthik Nandakumar	Michigan State University, USA
Hien Nguyen	University of Houston, USA
Haomiao Ni	University of Memphis, USA
Marc Niethammer	University of California, San Diego, USA
Jack Noble	Vanderbilt University, USA
Masahiro Oda	Nagoya University, Japan
Felipe Orihuela-Espina	University of Birmingham, UK
Cheng Ouyang	University of Oxford, UK
Jiazhen Pan	Technical University of Munich, Germany
Sang Hyun Park	POSTECH, South Korea
Magdalini Paschali	Stanford University, USA
Hieu Pham	VinUniversity, Vietnam
Szymon Płotka	Jagiellonian University, Poland
Prateek Prasanna	Stony Brook University, USA
Mengyun Qiao	University College London, UK
Yuchuan Qiao	Fudan University, China
Jing Qin	Hong Kong Polytechnic University, China
Yulei Qin	Tencent, China
Wu Qiu	Huazhong University of Science and Technology, China
Xihe Qiu	Shanghai University of Engineering Science, China
Liangqiong Qu	University of Hong Kong, China
Linhao Qu	Harvard Medical School, Harvard University, USA
Daniele Ravì	University of Messina, Italy
Mirabela Rusu	Stanford University, USA
Numan Saeed	Mohamed Bin Zayed University of Artificial Intelligence, UAE
Duygu Sarikaya	University of Leeds, UK
Cian Scannell	Eindhoven University of Technology, Netherlands
Thomas Schultz	University of Bonn, Germany
Hongming Shan	Fudan University, China
Harshita Sharma	Microsoft Research, UK
Feng Shi	United Imaging Intelligence, China
Keewon Shin	University of Ulsan, South Korea

Julio Silva-Rodríguez	ETH Zurich, Switzerland
Vivek Singh	Siemens Healthineers, USA
Kevin Smith	KTH Royal Institute of Technology, Sweden
Yang Song	University of New South Wales, Australia
Youyi Song	China Pharmaceutical University, China
Sahar Soussa	Nile University, Egypt
Rachel Sparks	King's College London, UK
Vinkle Srivastav	IIT Madras, India
Ruisheng Su	Eindhoven University of Technology, Netherlands
Carole Sudre	University College London, UK
Yao Sui	Peking University, China
Kyung Sung	University of California, Los Angeles, USA
Tanveer Syeda-Mahmood	IBM Research, USA
Tao Tan	Macao Polytechnic University, China
Xiaoying Tang	Southern University of Science and Technology, China
Yucheng Tang	NVIDIA, USA
Zhenyu Tang	Beihang University, China
Qian Tao	Delft University of Technology, Netherlands
Austin Tapp	Children's National Hospital, USA
Lin Tian	Harvard Medical School and Massachusetts General Hospital, USA
Martin Urschler	Medical University of Graz, Austria
Maria Vakalopoulou	CentraleSupélec, France
María del C. Valdés Hernández	University of Edinburgh, UK
Fons van der Sommen	Eindhoven University of Technology, Netherlands
Francisco Vasconcelos	University College London, UK
Harini Veeraraghavan	Memorial Sloan Kettering Cancer Center, USA
Archana Venkataraman	Johns Hopkins University, USA
Sulaiman Vesal	Stanford University, USA
Christian Wachinger	Technical University of Munich, Germany
Changmiao Wang	Shenzhen Research Institute of Big Data, China
Chengjia Wang	Heriot-Watt University, UK
Hong Wang	Xi'an Jiaotong University, China
Manning Wang	Fudan University, China
Puyang Wang	Ant Group, USA
Shujun Wang	Hong Kong Polytechnic University, China
Shuo Wang	Fudan University, China
Xiaosong Wang	Shanghai AI Laboratory, China
Yaqi Wang	Hangzhou Dianzi University, China
Donglai Wei	Boston College, USA
Thomas Wendler	University Hospital Augsburg, Germany

Bryan Williams	Lancaster University, UK
Ken C.L. Wong	IBM Research – Almaden Research Center, USA
Jonghye Woo	Massachusetts General Hospital and Harvard Medical School, USA
Ye Wu	Nanjing University of Science and Technology, China
Yicheng Wu	Imperial College London, UK
Yong Xia	Northwestern Polytechnical University, China
Yiming Xiao	Concordia University, Canada
Fuyong Xing	University of Colorado, Denver, USA
Xiaohan Xing	Stanford University, USA
Xuanang Xu	Yale University, USA
Yan Xu	Beihang University, China
Yanwu Xu	South China University of Technology, China
Yanyu Xu	Shandong University, China
Zhe Xu	Columbia University, USA
Wufeng Xue	Shenzhen University, China
Ke Yan	Alibaba DAMO Academy, China
Zengqiang Yan	Huazhong University of Science and Technology, China
Jiancheng Yang	ELLIS Institute Finland and Aalto University, Finland
Qiushi Yang	City University of Hong Kong, China
Yanwu Yang	University of Tübingen and Max Planck Institute of Intelligent Systems, Germany
Jiawen Yao	DAMO Academy, Alibaba Group, China
Qingsong Yao	Stanford University, USA
Dong Hye Ye	Georgia State University, USA
Zhaozheng Yin	Stony Brook University, USA
Chenyu You	Stony Brook University, USA
Lequan Yu	University of Hong Kong, China
Fatemeh Zabihollahy	University of Toronto, Canada
Fan Zhang	University of Electronic Science and Technology of China, China
Jianpeng Zhang	Alibaba DAMO Academy, China
Jiong Zhang	Nanjing University, China
Lichi Zhang	Shanghai Jiao Tong University, China
Xiaofan Zhang	Shanghai Jiao Tong University, China
Yi Zhang	Sichuan University, China
Yizhe Zhang	Nanjing University of Science and Technology, China
Fenqiang Zhao	DAMO Academy, Alibaba Group, China
He Zhao	University of Liverpool, UK

Lin Zhao	New Jersey Institute of Technology, USA
Qingyu Zhao	Weill Cornell Medicine, USA
Rongchang Zhao	Central South University, China
Yitian Zhao	Cixi Institute of Biomedical Engineering, Ningbo Institute of Industrial Technology, Chinese Academy of Sciences, China
Yefeng Zheng	Westlake University, China
Yushan Zheng	Beihang University, China
Zichun Zhong	Wayne State University, USA
Bo Zhou	Northwestern University, USA
Luping Zhou	University of Sydney, Australia
Qin Zhou	East China University of Science and Technology, China
Tianfei Zhou	Beijing Institute of Technology, China
Yan-Jie Zhou	DAMO Academy, Alibaba Group, China
Yi Zhou	Southeast University, China
Yuyin Zhou	University of California, Santa Cruz, USA
Zongwei Zhou	Johns Hopkins University, USA
Meilu Zhu	University of Hong Kong, China
Veronika Zimmer	Technical University of Munich and Siemens Healthineers, Germany
Ke Zou	Sichuan University, China
Lianrui Zuo	Vanderbilt University, USA

Outstanding Area Chairs

Carole Sudre	University College London, UK
Lianrui Zuo	Vanderbilt University, USA
Mostafa El Habib Daho	University of Western Brittany, France

Outstanding Reviewers

Yunlong He	Télécom Paris, France
Julius Hense	TU Berlin, Germany
Arthur Longuefosse	RIKEN, Japan
Johanna Müller	Friedrich-Alexander-Universität Erlangen-Nürnberg, Germany
Aminu Musa	Federal University Dutse, Nigeria
Andreas Østvik	Norwegian University of Science and Technology, Norway

Nathan Painchaud	CREATIS, France
Luca Parolari	University of Padua, Italy
Mostafa Salem	Mohamed bin Zayed University of Artificial Intelligence, UAE
Weirong Wang	Hangzhou Institute for Advanced Study, University of Chinese Academy of Sciences, China

Honorable Mentions (Reviewers)

Marwan Abbas	Mayo Clinic, USA
Jaesin Ahn	Kyungpook National University, South Korea
Michel Antunes	Smith & Nephew, Portugal
Alberto Arrigoni	Istituto di Ricerche Farmacologiche Mario Negri IRCCS, Italy
Shreeram Athreya	University of California, Los Angeles, USA
Johannes Brandt	Technische Universität München, Germany
Salvatore Calcagno	University of Catania, Italy
Liane Canas	King's College London, UK
Lara Cavinato	Politecnico di Milano, Italy
Satrajit Chakrabarty	GE HealthCare, USA
Dongdong Chen	Anhui University, China
Weihao Cheng	Hangzhou Dianzi University, China
Olivier Colliot	CNRS, France
Pierrick Coupé	CNRS and University of Bordeaux, France
Jürgen Dammers	Forschungszentrum Jülich, Germany
Sanuwani Dayarathna	Monash University, Australia
Lei Deng	Central South University, China
Wenting Duan	University of Lincoln, UK
Ertunc Erdil	ETH Zurich, Switzerland
Yannik Frisch	University of Heidelberg, Germany
Boxun Fu	Xidian University, China
Adrian Galdran	Universitat Pompeu Fabra, Spain
Fei Gao	University of Science and Technology of China, China
Haifan Gong	University of Pennsylvania, USA
Alvaro Gonzalez-Jimenez	University Hospital of Basel, Switzerland
Valerio Guarrasi	Università Campus Bio-Medico di Roma, Italy
Diandian Guo	Chinese University of Hong Kong, China
Arnela Hadzic	Medical University of Graz, Austria

Salma Hassan	Mohamed Bin Zayed University of Artificial Intelligence, UAE
Dongdong He	Harbin Institute of Technology, China
Da-Woon Heo	Korea University, South Korea
Mariko Isogawa	Keio University, Japan
Ji-Hoon Jeong	Sungkyunkwan University, South Korea
Zhuoyun Jiang	University of Shanghai for Science and Technology, China
Shuo Jiang	Hangzhou Dianzi University, China
Amelia Jiménez-Sánchez	Universitat de Barcelona, Spain
Erwan Kerrien	Inria, France
Abbas Khan	Queen Mary University of London, UK
Young-Gon Kim	Seoul National University Hospital, South Korea
Youngwoo Kim	Kumoh National Institute of Technology, South Korea
Ender Konukoglu	ETH Zurich, Switzerland
Naveen Kumar Kummari	Mohamed bin Zayed University of Artificial Intelligence, UAE
Frank Langbein	Cardiff University, UK
François Lecomte-Denis	Inria, France
Won Hee Lee	Kyung Hee University, South Korea
Huiqi Li	Beijing Institute of Technology, China
Jun Li	Technical University of Munich, Germany
Miao Li	Tsinghua University, China
Ruizhe Li	University of Nottingham, UK
Simone Lionetti	Lucerne University of Applied Sciences and Arts, Switzerland
Feng Liu	University of Queensland, Australia
Andrea Loddo	University of Cagliari, Italy
Angela Lombardi	Politecnico di Bari, Italy
Luca Lumetti	University of Modena and Reggio Emilia, Italy
Huoling Luo	Shenzhen University of Information Technology, China
Ngoc Ly	University of Science-VNUHCM, Vietnam
Jamie McClelland	University College London, UK
Martin Menten	Technical University of Munich, Germany
Klaus Mueller	Stony Brook University, USA
Favour Nerrise	Stanford University, USA
Ahmad P. Tafti	University of Pittsburgh, USA
Yongsheng Pan	Northwestern Polytechnical University, China
Mario Parreno Centeno	Oulu University, Finland
Kelly Payette	University Children's Hospital Zurich, Switzerland

João Pedrosa	INESCTEC, Portugal
Aloys Portafaix	Polytechnique Montréal, Canada
Liangce Qi	Changchun University of Science and Technology, China
Lin Qi	Northeastern University, China
Feiwei Qin	Hangzhou Dianzi University, China
Sandro Queirós	University of Minho, Portugal
Samuel Remedios	Johns Hopkins University, USA
Anne-Marie Rickmann	Yale University, USA
Ranga Rodrigo	University of Moratuwa, Sri Lanka
Binesh Sadanandan	University of New Haven and Medtronic, USA
Carlos Santiago	Universidade Lisboa, Portugal
Philipp Seeböck	Medical University of Vienna, Austria
Chen Shen	Xi'an Jiaotong University, China
Linlin Shen	Shenzhen University, China
Jie Shu	North China University of Technology, China
Md. Saiful Bari Siddiqui	BRAC University, Bangladesh
Weronika Smolak-Dyżewska	Jagiellonian University, Poland
Qi Song	University of Science and Technology of China, China
Yiran Song	University of Minnesota, USA
Ziga Spiclin	University of Ljubljana, Slovenia
Emma Stanley	Imperial College London, UK
Nil Stolt-Ansó	Technical University of Munich, Germany
Zhichao Sun	Wuhan University, China
Jeremy Tan	Aalborg University, Denmark
Kaibo Tang	University of North Carolina at Chapel Hill, USA
Zihao Tang	University of Sydney, Australia
Vasiliki Tassopoulou	University of Pennsylvania, USA
Eneko Uruñuela	University of Calgary, Canada
Muhammad Usman	Stanford University, USA
Danush Kumar Venkatesh	National Centre for Tumor Diseases, Germany
Tom Vercauteren	King's College London, UK
Huitao Wang	Peking University, China
Meng Wang	National University of Singapore, Singapore
Ruiqi Wang	Washington University in St. Louis, USA
Ruiqi Wu	Southeast University, China
Taoyu Wu	Xi'an Jiaotong-Liverpool University, China
Yu-Te Wu	National Yang Ming Chiao Tung University, Taiwan
Fengbo Xie	North University of China, China
Jiayi Xin	University of Pennsylvania, USA

Yoon Seok Yang	SUNY Korea, South Korea
Chunzhi Yi	Harbin Institute of Technology, China
Kyungho Yoon	Yonsei University, South Korea
Han Yu	Nanjing University of Posts and Telecommunications, China
Zheng Yu	University of Adelaide, Australia
Mariam Zabihi	University College London, UK
Stefan Zachow	Zuse Institute Berlin, Germany
Qing Zhang	East China Normal University, China
Wanjun Zhang	Henan University, China
Shilun Zhao	ShanghaiTech University, China
Lujia Zhong	University of Southern California, USA
Jiayuan Zhu	University of Oxford, UK

Mentorship Program (Mentors)

Ekaterina Antipushina	Applied AI Institute, Russia
Stephen Aylward	NVIDIA, USA
Rotimi-Williams Bello	Tshwane University of Technology, South Africa
Mariano Cabezas Grebol	Macquarie University, Australia
Zhen Chen	Hong Kong Polytechnic University, China
Li Cheng	University of Alberta, Canada
Kathleen Curran	University College Dublin, Ireland
Behdad Dashtbozorg	Netherlands Cancer Institute, Netherlands
Kanchana Devanathan	Vanderbilt University, USA
Qi Dou	Chinese University of Hong Kong, China
Niharika Dsouza	IBM Research, USA
Mohammad Eslami	Harvard Medical School, USA
Caroline Essert	University of Strasbourg, France
Ruogu Fang	University of Florida, USA
Huazhu Fu	Institute of Advanced Intelligence and Computing, A*STAR, Singapore
Zhifan Gao	Sun Yat-sen University, China
Zulqarnain Gilani	Edith Cowan University and University of Western Australia, Australia
Polat Goktas	Sabanci University, Turkey
Leo Joskowicz	Hebrew University of Jerusalem, Israel
Bernhard Kainz	Friedrich-Alexander-Universität Erlangen-Nürnberg, Germany and Imperial College London, UK
Neerav Karani	QuantCo, USA

Karim Lekadir	Icrea and Universitat de Barcelona, Spain
Xiaoxiao Li	University of British Columbia, Canada
Guiqiu Liao	University of Pennsylvania, USA
Marius George Linguraru	Children's National Hospital, USA
Peirong Liu	Johns Hopkins University, USA
Siladittya Manna	Hong Kong Baptist University, China
Tommaso Mansi	Janssen R&D Data Science, USA
Qingjie Meng	University of Birmingham, UK
Nassir Navab	Technical University of Munich, Germany
Roseline Ogundokun	Redeemer's University, Nigeria
Mohammad Rizwan	Hammond Chemicals, UK
Divya Saleela	University of Southampton, UK
Julia Schnabel	Computational Imaging and AI in Medicine, Germany
Bella Specktor Fadida	University of Haifa, Israel
Kyung Sung	University of California, Los Angeles, USA
Mathias Unberath	Johns Hopkins University, USA
Archana Venkataraman	Boston University, USA
Shandong Wu	University of Pittsburgh, USA
Fuyong Xing	University of Colorado Denver, USA
Jiancheng (JC) Yang	ELLIS Institute Finland and Aalto University, Finland
Le Zhang	University of Birmingham, UK
Rongchang Zhao	Central South University, China
S. Kevin Zhou	Institute of Computing Technology, China
Lei Zhu	Hong Kong University of Science and Technology, Guangzhou, China
Maria A. Zuluaga	EURECOM, France

Additional Reviewers

Abdullah
Marwan Abbas
Syed Farhan Abbas
Hamid Abbasi
Gabriele Abbate
Zeinab Abboud
Nassib Abdallah
Lamiaa Abdel-Hamid
Abeer Abdelhamid
Kumar Abhishek
Farhan Fuad Abir
Purang Abolmaesumi
Mohammad Mahdi Abootorabi
Mansour Abtahi
Jumana Abu-Khalaf
Titipat Achakulvisut
Ashok Vardhan Addala
Hamidreza Aftabi
Akshay Agarwal
Oktay Agcaoglu
Aarush Aggarwal
Zahra Aghaloo

Shashank Agnihotri
Antonio Agudo
Marco Agus
Sakib Ahammed
Artur Aharonyan
Raha Ahmadi
Seyed-Ahmad Ahmadi
Sheraz Ahmed
Euijoon Ahn
Jaesin Ahn
Suhyun Ahn
Antti Ahola
Happiness Aigbogun
Aroof Aimen
Naren Akash
Elif Ecem Akbaba
Younes Akbari
Faranak Akbarifar
Emre Akbas
Muhammad Usman Akram
Mumu Aktar
Tanya Akumu
Dawood Al Chanti
Wafa Al Ghallabi
Mugahed A. Al-antari
Khaoula Alaoui Belghiti
Abbas Alatrany
Baidaa Al-Bander
Chadi Albitar
Abdalrahman Alblwi
Julia Alekseenko
Daniel Alexander
Felicia Alfano
Saghir Alfasly
Walid Al-Haidri
Hassan Alhajj
Danish Ali
Hazrat Ali
Kausar Ali
Mansoor Ali
Adamu Ali-Gombe
Farouq Aliyu
Omar Al-Kadi
Mohammad Alkhatib
Ahmed Alksas
Erwin Alles
Mohammed Al-masni
Ahmed Al-Qurri
Arwa Alrubaian
Mohammad Alsharid
Pablo Alvarez
Charlems Alvarez-Jimenez
Jennifer Alvén
Jakob Ambsdorf
Ankit Amrutkar
Dayeong An
Delin An
Haill An
Jinseo An
Jonghoon An
Sion An
Rishi Ananth
Dimitrios Anastasiou
Wang Anbang
Gustavo Andrade-Miranda
Vincent Andrearczyk
Naga Surya Sandeep Angara
Fatai Anifowose
Anna Anikina
André Anjos
Faizanuddin Ansari
Ekaterina Antipushina
Michela Antonelli
Michel Antunes
Kian Anvari Hamedani
Yunke Ao
Yoshimitsu Aoki
Erchan Aptoula
Madhu Aravindan
Ignacio Arganda-Carreras
Tomas Arias-Vergara
Mikel Ariz
Mehak Arora
Ahmad Arrabi
Alberto Arrigoni
Muhammad Waqas Arshad
Sai Susmitha Arvapalli
Mohammad Asadi
Yuta Asano
John Ashburner

Tajamul Ashraf
Pooya Ashtari
Mehdi Astaraki
Arda Atalık
Chinmayee Athalye
Thomas Athey
Shreeram Athreya
Indriyati Atmosukarto
Mohammad Atwany
Benjamin Aubert
Marc Aubreville
Mehmet Yigit Avci
Paolo Avesani
Jordina Aviles Verdera
Navchetan Awasthi
Suyash Awate
Ishak Ayad
Orhun Utku Aydin
Ilhan Aytutuldu
Lev Ayzenberg
Puria Azadi Moghadam
Faiza Azhar
Mohammad Saber Azimi
Abu Zahid Bin Aziz
Iyad Ba Gari
Hyeon-Min Bae
Jongduk Baek
Seung Baek
Seunghun Baek
Priyanka Bagade
Negin Baghbanzadeh
Ali Bahari Malayeri
Mohammed Baharoon
Harrison Bai
Jieyun Bai
Long Bai
Wenjia Bai
Xiangzhi Bai
Xiaoyu Bai
Xuguang Bai
Yunhao Bai
Ujjwal Baid
Chandrajit Bajaj
Ghassen Baklouti
Shreyan Bakshi
Yoganand Balagurunathan
Guha Balakrishnan
Shidin Balakrishnan
Shafa Balaram
Abdourahmane Balde
Vishnu Vardhan Baligodugula
Yutong Ban
Imon Banerjee
Oishi Banerjee
Soumyanil Banerjee
Chandan Ganesh Bangalore Yogananda
David Bani-Harouni
Fangxun Bao
Rina Bao
Shunxing Bao
D M Bappy
Catarina Barata
Payam Barnaghi
Juan Barragan Noguera
Hritam Basak
Soumen Basu
Tanmay Basu
Deepti R Bathula
Dheerendranath Battalapalli
Matthew Baugh
Christoph Baur
Simon Baur
Nourhan Bayasi
Mahdi Bayat
Siming Bayer
Roza Bayrak
Clément Beitone
Soufiane Belharbi
Giovanni Bellitto
Rotimi-Williams Bello
Dafna Ben Bashat
Andreas Bender
Sean Benson
Jamal Bentahar
Mariana Bento
Jose Bernal
Olivier Bernard
Gabriel Bernardino
Richard Bethlehem
Can Deniz Bezek

Riddhish Bhalodia
Rahul Bhardwaj
Nithya Bhasker
Divya Bhatia
Deblina Bhattacharjee
Sourav Bhattacharjee
Ning Bi
Yuan Bi
Diego Biagini
Cheng Bian
Wanyu Bian
Zhangxing Bian
Khawaja Bilal
Harald Binder
Nur Suriza Syazwany
Binti Ahmad Nizam
Kyriaki-Margarita Bintsi
Kimberley Bird
Manuel Birlo
George Biros
Alceu Bissoto
Soma Biswas
David Black
Matthew Blackledge
Nathan Blake
Yuntian Bo
Sudhakavya Bodapati Venkata
Lea Bogensperger
Patrick Bolan
Christopher Boland
Vanessa Borst
Marc Boubnovski Martell
Said Boumaraf
Anis Bourou
Alexandre Bousse
Arnaud Boutillon
Nicolas Boutry
Maria Boyko
Rowan Bradbury
Abdelbasset Brahim
Johannes Brandt
Kit Bransby
Laura Brattain
Diego Bravo
Ali Braytee
Keno Bressem
Christopher Bridge
Ricardo Brioso
Tom Brosch
James Brown
Ekaterina Brui
David Brundage
Michal Brzus
Zhenyu Bu
Lukas Buess
Katja Bühler
Doanh Cao Bui
Nhat-Tan Bui
Phuoc-Nguyen Bui
Christina Bukas
Junghyun Bum
Valay Bundele
Arianna Bunnell
Ninon Burgos
Nikolay Burlutskiy
Yunsu Byeon
Ji Young Byun
Alessandro Cacciatore
Francisco Caetano
Noa Cahan
Chengfei Cai
Jiushen Cai
Kunyan Cai
Linghan Cai
Lishan Cai
Tongan Cai
Weidong Cai
Zemin Cai
Zhuotong Cai
Zongyou Cai
Salvatore Calcagno
Ross Callaghan
Lilian Calvet
Duy-Cat Can
Pere Canals
Liane Canas
Antonio Candito
Bo Cao
Chao Cao
Guogang Cao

Jianfeng Cao
Kangyang Cao
Rui Cao
Shuang Cao
Tian Cao
Yankun Cao
Zhenjie Cao
Daniel Capellán-Martín
Aaron Carass
Ionut Cardei
Paula Cárdenas
Luca Carlini
Owen Carmichael
Diedre Carmo
Patrick Carnahan
Gustavo Carneiro
Anna Caroli
Hector Carrion
Jacob Carse
Adrià Casamitjana
Alessandro Casella
Eduardo Castañeda
Philippe C. Cattin
Thiago Cavalcanti
Lara Cavinato
Alberto Mario Ceballos-Arroyo
Stefano Cerri
Dong-Kyu Chae
Han Joo Chae
Shurong Chai
Krishna Chaitanya
Satrajit Chakrabarty
Souradeep Chakraborty
Chi Kwan Chan
Trevor Chan
Wei Xuan Chan
Weng Ian Chan
Wing Chi Chan
Yi Hao Chan
Deeph Chana
Dibaloke Chanda
Shekhar Chandra
Heyou Chang
Qing Chang
Yao-Jen Chang
Yue Changpeng
Pradyumna Chari
Masoud Charkhabi
Rajdeep Chatterjee
Sudhanya Chatterjee
Pratik Chattopadhyay
Soumitri Chattopadhyay
Gustavo Ramon Chau Loo Kung
Muhammad Faizyab Ali Chaudhary
Joohi Chauhan
Joël Chavas
Levy Chaves
Tongtong Che
Yiming Che
Bohao Chen
Bolin Chen
Boyu Chen
Changfang Chen
Chu Chen
Chuan Chen
Cunjian Chen
Dehua Chen
Dongdong Chen
Duoyou Chen
Fuqiang Chen
Geng Chen
Guannan Chen
Hairui Chen
Han Chen
Hongbo Chen
Hsiang-Ting Chen
Huan-Yu Chen
Hung-Hsuan Chen
Jiabao Chen
Jiangning Chen
Jiayi Chen
Jiazhou Chen
Jie Chen
Jieneng Chen
Jixiang Chen
Jun Chen
Junxiang Chen
Junzhou Chen
Kai Chen
Kanghao Chen

Kevin Chen
Li Chen
Lifang Chen
Ling Chen
Lingyu Chen
Lixuan Chen
Ma Chen
Minheng Chen
Peikai Chen
Peilin Chen
Pingyi Chen
Qi Chen
Qiang Chen
Qingchao Chen
Shaobin Chen
Shengchao Chen
Shengjia Chen
Shijing Chen
Tao Chen
Tong Chen
Wanwen Chen
Wanyi Chen
Wei Chen
Weijie Chen
Wenting Chen
Wenxuan Chen
Xiang Chen
Xiaohua Chen
Xiaoyang Chen
Xin Chen
Xizi Chen
Xueli Chen
Yaowen Chen
Yaqian Chen
Yaxi Chen
Yifei Chen
Yihong Chen
Yiliang Chen
Yiming Chen
Yinda Chen
Yingyu Chen
Yinran Chen
Yixiong Chen
Yong Chen
Yu Chen
Yuanyuan Chen
Yufei Chen
Yuhang Chen
Zhanghao Chen
Zhaojin Chen
Zhaomin Chen
Zhaozheng Chen
Zhennong Chen
Zhenqin Chen
Zhi Chen
Zhihao Chen
Zhong Chen
Zifan Chen
Zihao Chen
Zijiao Chen
Ziyang Chen
Chang-Chieh Cheng
Chun-Wun Cheng
Chunyang Cheng
Jiale Cheng
Jian Cheng
Jianhong Cheng
Jing Cheng
Junfu Cheng
Junlong Cheng
Kun Cheng
Pujin Cheng
Teng Cheng
Weihao Cheng
Xinxing Cheng
Yanqi Cheng
Zihan Cheng
Daeyoung Cheon
Farida Cheriet
Teja Krishna Cherukuri
James Chung-Wai Cheung
Ka Chun Cheung
Matt Cheung
Anju Chhetri
Mark Chiew
Philip Chikontwe
Ching-Hao Chiu
Hyuna Cho
Jaejin Cho
Jaeyoung Cho

Minji Cho
Sue Min Cho
Yeong-Jun Cho
Tomáš Chobola
Changyong Choi
Gyuho Choi
Hanseung Choi
Hyoungshin Choi
Jang-Hwan Choi
Jin Woo Choi
Kyu Sung Choi
Sang-Il Choi
Woohyeok Choi
Wookjin Choi
Yangsean Choi
Yujin Choi
Kyobin Choo
Omar Choudhry
Argyrios Christodoulidis
Dimosthenis Chrysochoou
Lingyang Chu
Yuetan Chu
Artur Chudzik
Il Yong Chun
Albert Chung
Minyoung Chung
Pau-Choo Chung
Woojin Chung
Yunsung Chung
Nikhil Churamani
Konrad Ciecierski
Ozkan Cigdem
Catalin Ciobirca
Allison Clement
Dana Cobzas
Paulo Jorge Coelho
Olivier Colliot
Chao Cong
Cong Cong
William Consagra
Pierre-Henri Conze
Tim Cootes
Hélène Corbaz
Valentina Corbetta
Baris Coskunuzer
Stephane Cotin
Pierrick Coupé
Francesco Cremonesi
Alessandro Crimi
Ines Alejandro Cruz Guerrero
Angel Cruz-Roa
Beilei Cui
Hengfei Cui
Jian Cui
Jianan Cui
Jiaqi Cui
Mengmeng Cui
Ruize Cui
Wenhui Cui
Xiaoxiao Cui
Zhe Cui
Sasa Cukovic
Tolga Cukur
Kelvin Cunha
Ariel Curiale
Yudha Raja Singam D.
Mohamad Dabboussi
Lavsen Dahal
Rema Daher
Di Dai
Guowei Dai
Ling Dai
Linrui Dai
Wei Dai
Weicheng Dai
Connor Daly
Salma Dammak
Jürgen Dammers
Rafat Damseh
Corentin Dancette
Iman Dandapath
Danilo Danese
Min Dang
Trung Dang
Meghal Dani
Bilel Daoud
Salman Ul Hassan Dar
Eleonora D'Arnese
Abhijit Das
Badhan Kumar Das

Debanjan Das
Dhritiman Das
Debashis Das Chakladar
Muhammed Furkan Dasdelen
Soma Dasgupta
Christos Davatzikos
Muhammad Dawood
Tareen Dawood
Sanuwani Dayarathna
Akshay Daydar
Laura Daza
Ezequiel de la Rosa
Swakshar Deb
Charles Delahunt
Herve Delingette
Başar Demir
Ugur Demir
Omer Demirel
Botao Deng
Chao Deng
Lei Deng
Liang-Jian Deng
Qingxin Deng
Xiang Deng
Yingjiao Deng
Yu Deng
Zhipeng Deng
Zhiwei Deng
Zhongying Deng
Elena Denisova
Cem Deniz
Felix Denzinger
Adrien Depeursinge
Kanchana Devanathan
Alina Devkota
Dipayan Dewan
Samrat Kumar Dey
Swarnava Dey
Manish Dhakal
Sobhan Dhara
Jennifer Dhont
Mariachiara Di Cosmo
Théo Di Piazza
Francesco Di Salvo
Roberto Di Via
Yueqin Diao
Jouke Dijkstra
Avtantil Dimitriadis
Sadia Din
Guanqun Ding
Han Ding
Kexin Ding
Qiaoqiao Ding
Rizhi Ding
Saisai Ding
Tianyi Ding
Xiaohuan Ding
Xinpeng Ding
Ying Ding
Kerol Djoumessi
Ha Thanh Do
Hyungrok Do
Daniel Dobkin
Sebastian Doerrich
Mischa Dombrowski
Aimei Dong
Di Dong
Fangfang Dong
Haoyu Dong
Hexin Dong
Mengjin Dong
Min Dong
Shoubin Dong
Siyuan Dong
Xiuyu Dong
Yafei Dong
Yihong Dong
Zhengyuan Dong
Zijian Dong
Jonathan Donnelly
Gianfranco Doretto
Zolnamar Dorjsembe
Haoran Dou
Shuguang Dou
Yimeng Dou
Yulong Dou
Hassen Drira
Anghong Du
Chenhe Du
Dongyang Du

Hao Du
Honghui Du
Ji Du
Jiawei Du
Jie Du
Lei Du
Penghui Du
Shaoyi Du
Siyi Du
Tianming Du
Yao Du
Ye Du
Yuetian Du
Yuexi Du
Dingna Duan
Jinming Duan
Peiyu Duan
Wenting Duan
Yaofei Duan
Kauê Duarte
Thomas Duboudin
David Dueñas Gaviria
Benoit Dufumier
Felix Dülmer
Andrea Dunn Beltran
Fatih Duran
Nicholas Durr
Tapas Dutta
Nicha Dvornek
Jagan Mohan Reddy Dwarampudi
Witold Dyrka
Obed Dzikunu
Oleh Dzyubachyk
Chamira Edussooriya
Aniek Eijpe
Favour Ekong
Nafie El Amrani
Sidaty El Hadramy
Marawan Elbatel
Abdussalam Elhanashi
Mohamed Elmanna
Mohamed Elsharkawy
Mohamed Elshrif
Daniel Elson
Eyad Elyan
Noha El-Zehiry
Taha Emre
Sandy Engelhardt
Ertunc Erdil
Marius Erdt
Ayhan Can Erdur
Franciskus Xaverius Erick
Alina Ermilova
Ilker Ersoy
Dennis Eschweiler
Meltem Esengönül
Mohammad Eslami
Nazila Esmaeili
Parhom Esmaeili
Elyar Esmaeilzadeh
Shahrooz Faghihroohi
Alexandre Falcao
Chao Fan
Di Fan
Jianan Fan
Jiansong Fan
Kailong Fan
Liang Fan
Mengchen Fan
Wenrui Fan
Xiaoya Fan
Xiaoyao Fan
Xin Fan
Xinqi Fan
Yubo Fan
Yuheng Fan
Zhiting Fan
Zipei Fan
Fen Fang
Hao Fang
Songke Fang
Wei Fang
Xi Fang
Yingying Fang
Yuqi Fang
Zhenghan Fang
Aly Farag
Somayeh Farahani
Manel Farhat
Andreia Faria

Aiman Farooq
Hasan Farooq
Muhammad Ali Farooq
Muhammad Umar Farooq
Syed Hasib Akhter Faruqui
Giuseppe Fasano
Jamil Fayyad
Botond Fazekas
Edoardo Fazzari
Maximilian Fehrentz
Manman Fei
Yuchen Fei
Paula Feldman
Lina Felsner
Chaolu Feng
Jing Feng
Jun Feng
Ming Feng
Siwei Feng
Siyang Feng
Yidan Feng
Yushi Feng
Zehui Feng
Zetian Feng
Zishun Feng
Philipp Feodorovici
Aasa Feragen
Abdesselam Ferdi
Tharindu Fernando
André Ferreira
Jean Feydy
Leo Fillioux
Marten Finck
Maria Chiara Fiorentino
Alessio Fiorin
Paul Fischer
Stefan Fischer
Roman Flepp
Kyriakos Flouris
Stefano Fogarollo
Bernhard Föllmer
A Fong
Fahimeh Fooladgar
Nils Daniel Forkert
Jean-Rassaire Fouefack
Joris Fournel
Kevin François-Bouaou
Moti Freiman
Michael Friebe
Christoph Friedrich
Paul Friedrich
Yannik Frisch
Baochen Fu
Boxun Fu
Jia Fu
Jingru Fu
Qifan Fu
Rong Fu
Suzhong Fu
Xiangling Fu
Xueming Fu
Yibing Fu
Yingji Fu
Yu Fu
Yuxuan Fu
Zeyu Fu
Moritz Fuchs
Hiroshi Fujita
Florian Fürnrohr
Ryo Furukawa
Michael Gadermayr
Fasil Gadjimuradov
Emilien Gadoux
Miruna-Alexandra Gafencu
Laurent Gajny
Efstathios Galanakis
Adrian Galdran
Gurur Gamgam
Haitao Gan
Yu Gan
Iyyakutti Iyappan Ganapathi
Vaibhav Gandhi
Aniruddha Ganguly
Chang Gao
Cong Gao
Fei Gao
Rui Gao
Shangqi Gao
Shuo Gao
Weiwei Gao

Xiaoru Gao
Xinbo Gao
Xingyu Gao
Yajun Gao
Yang Gao
Yao Gao
Yaozong Gao
Yi Gao
Yibo Gao
Yifan Gao
Yuan Gao
Yurui Gao
Zeyu Gao
Zheyao Gao
Zhirui Gao
Zhuangzhi Gao
Mengxi Gao
Utpal Garain
Alvaro Garcia-Tornel Garcia-Camba
Gautam Gare
Gaurav Garg
Eleftherios Garyfallidis
Sergios Gatidis
Ioannis Gatopoulos
Romane Gauriau
Rongjun Ge
Ruiquan Ge
Yuyan Ge
Haixiao Geng
Konstantinos Georgas
Ajo George
Yasmeen George
Beerend Gerats
Negin Ghamsarian
Yazan Gharaibeh
Mohammed Amine Gharsallaoui
Ghazal Ghazaei
Madhusudan Ghosh
Shantanu Ghosh
Yongha Gi
Debora Gil
Simon Ging
Rémi Giraud
Tristan Glatard
Ben Glocker
Patrick Godau
Bharti Goel
Wesley Gohn
Polina Golland
Alberto Gomez
Catalina Gomez
Santiago Gómez Hernández
Estibaliz Gómez-de-Mariscal
Haifan Gong
Wei Gong
Xun Gong
Md Raihan Goni
German Gonzalez
Cristina González
Ines Gonzalez Pepe
Jesus Gonzalez Riveros
Alvaro Gonzalez-Jimenez
Sharath Gopal
Shekoufeh Gorgi Zadeh
Mikhail Gorodnichev
Subrahmanyam Gorthi
Michael Götz
Shuiping Gou
Maged Goubran
Darshana Govind
Shreyank Gowda
Aman Goyal
Puneet Goyal
Robert Graf
Costantino Grana
Eric Granger
Jason Granstedt
Vicente Grau
Thomas Grenier
Clément Grisi
Fabian Gröger
Michal Grzeszczyk
Fei Gu
Feng Gu
Jinchen Gu
Leslie Gu
Qiangqiang Gu
Ran Gu
Xue Gu
Yannian Gu

Yi Gu
Yuanyuan Gu
Yun Gu
Haoyue Guan
Qingji Guan
Valerio Guarrasi
Luis Guayacan
Houssem-Eddine Gueziri
Luying Gui
Halil Gulluk
Malitha Gunawardhana
Ayse Gungor
Manuel Günther
Changlu Guo
Diandian Guo
Fei Guo
Fumin Guo
Guangyu Guo
Heng Guo
Hongbo Guo
Jia Guo
Jiayu Guo
Jixiang Guo
Jun Guo
Lijun Guo
Wenzhangzhi Guo
Xiaojuan Guo
Yanfei Guo
Yong Guo
Zaiyang Guo
Zuwei Guo
Ankur Gupta
Gagan Gupta
Sukrit Gupta
Kali Gürkahraman
Boris Gutman
Carmen Guzman-Garcia
Seunghyun Gwak
Sung Min Ha
Stathis Hadjidemetriou
Arnela Hadzic
Ibrahim Hadzic
Ida Häggström
James Hahn
Justin Haldar
Verena Jasmin Hallitschke
Ghassan Hamarneh
Emadeldeen Hamdan
Abdullah Hamdi
Donghee Han
Hu Han
Hua Han
Hyunkyung Han
Ji Yong Han
Jingxin Han
Jinyoung Han
John Han
Kangfu Han
Luyi Han
Seokmin Han
Xian-Hua Han
Xiao Han
Xiaodong Han
Xu Han
Yoseob Han
Yuqiang Han
Jonny Hancox
Asif Hanif
Muhammad Abdullah Hanif
Hartmut Häntze
Degan Hao
Huaying Hao
Jilei Hao
Jing Hao
Qin Hao
Shuang Hao
Teqi Hao
Xinyu Hao
Yingqi Hao
Nazim Haouchine
Shahad Hardan
Mohamed Harmanani
Mahade Hasan
Mahmudul Hasan
Sam Hashemi
Sara Hashemi
Anam Hashmi
Marta Hasny
Ghulam Mubashar Hassan
Salma Hassan

Charles Hatt
Hideaki Hayashi
Mansoor Hayat
Savannah Hays
Along He
Chloe He
Dinglun He
Dongdong He
Hangzhou He
Hongliang He
Huiguang He
Lei He
Linchao He
Mengshen He
Mengxian He
Runlong He
Sheng He
Xinzi He
Xuelei He
Yimeng He
Ying He
Yongjun He
Yunlong He
Zhenqi He
Zhibin He
Ziyang He
Moein Heidari
Jonas Hein
Mattias Heinrich
Jon Heiselman
Halil Helvaci
Anja Hennemuth
Julius Hense
Chanyeong Heo
Da-Woon Heo
Juampablo Heras Rivera
Lisa Herzog
Risa Higashita
Adam Hilbert
Meva Himmetoglu
Carlos Hinojosa
Florian Hinterwimmer
Rebecca Hisey
Md Hasibul Husain Hisham
David Ho
Edmond S. L. Ho
Ming-Yang Ho
Ngoc Huynh Ho
Matthew Holden
Felix Holm
Gregory Holste
Charmgil Hong
Jaeseong Hong
Pa Hong
Qingqi Hong
Sungrae Hong
Xiaoming Hong
Yang Hong
Nicolas Honnorat
Hidekata Hontani
Amir Hooshiar
Md. Enamul Hoq
Mobarak Hoque
Md Belayat Hossain
Tonmoy Hossain
Md Azam Hossain
Golriz Hosseinimanesh
Junlin Hou
Qingshan Hou
Rui Hou
Wenguang Hou
Xiangjian Hou
Yucun Hou
Prantik Howlader
Cheyu Hsu
Chih-Chung Hsu
William Hsu
Anjun Hu
Chuanfei Hu
Haigen Hu
Hao Hu
Huawen Hu
Jinlong Hu
Jisu Hu
Junjie Hu
Kai Hu
Liwei Hu
Ming Hu
Qiang Hu
Renjiu Hu

Shiang Hu
Shishuai Hu
Taiyuan Hu
Wei Hu
Xi Yuan Hu
Xingcan Hu
Yang Hu
Yibo Hu
Yuqi Hu
Zhangfeng Hu
Zhehao Hu
Zhisen Hu
Shengyi Hua
Zheang Huai
Bangyan Huang
Bin Huang
Bingsheng Huang
Cheng Huang
Chun-Rong Huang
Dianye Huang
Guantian Huang
Haolin Huang
Jiahao Huang
Jiahui Huang
Jiaju Huang
Jiaxing Huang
Jinghan Huang
Kai Huang
Kaiwen Huang
Kaizhu Huang
Le Huang
Luojie Huang
Peixiang Huang
Ruobing Huang
Shiqi Huang
Shuo Huang
Siyi Huang
Tianqi Huang
Wei Huang
Weimin Huang
Weitian Huang
Wenhui Huang
Wenli Huang
Wenqi Huang
Xingyu Huang
Xinrui Huang
Yan Huang
Yanyan Huang
Yaping Huang
Yi Huang
Yigeng Huang
Yi-Jie Huang
Yijin Huang
Yiming Huang
Ying Huang
Yongqiang Huang
Yuhao Huang
Yunzhi Huang
Zhe Huang
Zhen Huang
Zheng Huang
Zhenxing Huang
Zhi-An Huang
Zixuan Huang
Arnaud Huaulmé
Charmayne Hughes
Michael Hughes
Edward Hui
Hui Hui
Jiayu Huo
Mohammad Arafat Hussain
Muhammad Ishfaq Hussain
Tanveer Hussain
Yasir Hussain
Seong Jae Hwang
Sunil Hwang
Yechan Hwang
Yunseob Hwang
Bulat Ibragimov
Kamil Ibragimov
Mahmoud Ibrahim
Ilknur Icke
Azeez Idris
Mudassir Iftikhar
Kay Igwe
Efe Ilicak
Alfredo Illanes
Abdul-Mojeed Ilyas
Talha Ilyas
Hyeonseung Im

Sunghoon Im
Baris Imre
Veera Venkata Raghunath Indugu
Radu Tudor Ionescu
Asim Iqbal
Amanda Isaac
Md Saiful Islam
Mariko Isogawa
Koichi Ito
Hayato Itoh
Krithika Iyer
Athira Jacob
Mathews Jacob
Kshitij Jadhav
Mohit Jain
Pankaj Jain
Bushra Jalil
Mahtab Jamali
Amoon Jamzad
Beakcheol Jang
Boa Jang
Han Jang
Hyensoo Jang
Ikbeom Jang
Pierre Jannin
Vincent Jaouen
Ronnachai Jaroensri
Tim Jaspers
Alexander Jaus
Nivetha Jayakumar
Umarani Jayaraman
Niyonteze Jean De Dieu
Hana Jebril
Ahmadreza Jeddi
Duhee Jeon
Yongjun Jeon
Young Seok Jeon
Hwihun Jeong
Ji-Hoon Jeong
Jiwoong Jeong
Minjae Jeong
Rathinaraja Jeyaraj
Ranjeet Ranjan Jha
Changkai Ji
Jianhang Ji
Zexin Ji
Dengqiang Jia
Fucang Jia
Junhao Jia
Qi Jia
Tong Jia
Xibin Jia
Zhenge Jia
Bailiang Jian
Jia-Xian Jian
Benxiang Jiang
Bo Jiang
Dengrong Jiang
Hao Jiang
Haojun Jiang
Haotian Jiang
Hongxu Jiang
Juntao Jiang
Lihong Jiang
Ning Jiang
Peilin Jiang
Shancheng Jiang
Shuo Jiang
Wei Jiang
Weili Jiang
Wenbo Jiang
Wenchao Jiang
Wenhao Jiang
Xi Jiang
Xiajun Jiang
Xiaoyi Jiang
Yishan Jiang
Yiwen Jiang
Yufeng Jiang
Yuming Jiang
Yuncheng Jiang
Yuxian Jiang
Zhuoyun Jiang
Yining Jiao
Betshrine Rachel Jibinsingh
Amelia Jiménez-Sánchez
Kudirat Jimoh
Lingbo Jin
Qiyu Jin
Qunchao Jin

Rui Jin
Tao Jin
Yili Jin
Yueming Jin
Ze Jin
Bin Jing
Peiyuan Jing
Sanghyun Jo
Matjaz Jogan
Craig Jones
Gihun Joo
Anand Joshi
Shantanu Joshi
Leo Joskowicz
Samuel Joutard
Ce Ju
Lie Ju
Guil Jung
Heechul Jung
Jay Hwasung Jung
Juho Jung
Julip Jung
Kyu-Hwan Jung
Seung-Won Jung
Wonsik Jung
Sreekanth K Manikandan
Mounir Kaaniche
Twaha Kabika
Ashad Kabir
Krish Kabra
Takeaki Kadota
Ali Kafaei Zad Tehrani
Sevgi Kafali
Joshua Kaggie
Fredrik Kahl
Johannes Kaiser
John Kalafut
Niveditha Kalavakonda
Joanna Kaleta
John Kalkhof
Grigorios Kalliatakis
Odysseus Kaloidas
Panagiotis Kalozoumis
Tae-Eui Kam
Sreeram Kamabattula
Sharif Amit Kamran
Shichao Kan
Bhavik Kanekar
Dong Un Kang
June Kang
Junghwa Kang
Luoyao Kang
Ming Kang
Mingon Kang
Myeongkyun Kang
Qingbo Kang
Solha Kang
Praveenkumar Kanithi
Arunkumar Kannan
Madhavi Kannan
Dmitrii Kaplun
Saarthak Kapse
Siddhartha Kapuria
Aupendu Kar
Mokshagna Sai Teja Karanam
Amirhossein Kardoost
Yasmin Kassim
Tushar Kataria
Alexander Katzmann
Gurwinder Kaur
Eiryo Kawakami
Ali Kazemi
Mohammad Mahdi Kazemi Esfeh
Jing Ke
Xiao Ke
Razmig Kechichian
Tanzila Kehkashan
Matthias Keicher
Noam Keidar
Ayse Keles
Connah Kendrick
Roshan Kenia
Erwan Kerrien
Sara Ketabi
Nabeel Khalid
Sirvan Khalighi
Abbas Khan
Abdul Rehman Khan
Misaal Khan
Muhammad Irfan Khan

Muneeb Khan
Naimul Khan
Saba Khan
Shadab Khan
Sheheryar Khan
Sulaiman Khan
Ufaq Khan
Ziaullah Khan
Pulkit Khandelwal
Mahdieh Khanmohammadi
Rezkellah Noureddine Khiati
Julien Khlaut
Prathamesh Pradeep Khole
Hee Guan Khor
Mahapara Khurshid
Behnam Kiani Kalejahi
Tomohiro Kikuchi
Boah Kim
Bryan Sangwoo Kim
Byeong Soo Kim
Byung Hyung Kim
Byung-Hoon Kim
Chanyoung Kim
Daeseung Kim
Dohyun Kim
Harim Kim
Heejong Kim
Ho Heon Kim
Hongin Kim
Hye Ryun Kim
Hyeongsub Kim
Hyunmin Kim
Jeongin Kim
Jiehyun Kim
Jin Kim
Jinman Kim
Jonghun Kim
Jongwoo Kim
June Sic Kim
Junmo Kim
Junyeong Kim
Ka Young Kim
Kiduk Kim
Kyeong Ho Kim
Kyungsu Kim
Kyuri Kim
Mansu Kim
Minjun Kim
Minwoo Kim
Namkug Kim
Sanghyuk Kim
Sangwook Kim
Seongjun Kim
Sihwan Kim
So Hyun Kim
Songsoo Kim
Soopil Kim
Suhyeon Kim
Sunghun Kim
Taehun Kim
Taek Min Kim
Young-Gon Kim
Young-Min Kim
Youngwoo Kim
Yunsoo Kim
Alvin Kimbowa
Scott Kinder
Kaleab Kinfu
Andrew King
Miranda Kirby
Yannick Kirchhoff
Maxim Kireev
Tristan Kirscher
Gabriel Kiss
Itaru Kitahara
Yoshiro Kitamura
Klaus-Rudolf Kladny
Jens Kleesiek
Tobias Klinder
Vladimir Klinshov
Hannah Kniesel
Hanbin Ko
Seonghyeon Ko
Kazuma Kobayashi
Erich Kobler
Lisa Koch
Ryoichi Koga
Tatsuki Koike
Taha Koleilat
Satoshi Kondo

Praneeta R Konduri
Fanwei Kong
Jindi Kong
Ming Kong
Yan Kong
Youyong Kong
Anastasia Konik
Mirko Konstantin
Ender Konukoglu
Aishik Konwer
Nicholas Konz
Myoung-Wan Koo
Ivica Kopriva
Yilmaz Korkmaz
Kivanc Kose
Niklas Koser
Shyam Reddy Kotha
Konstantinos Koukoutegos
Deepika Koundal
Adam Krawczyk
Zuzanna Krawczyk-Borysiak
Julian Kreimeier
Anithapriya Krishnan
Deepa Krishnaswamy
Felix Krones
Frithjof Kruggel
Alexander Krull
Henry Krumb
Dongyang Kuang
Hulin Kuang
Dariusz Kucharski
Thomas Kuestner
Thijs Kuipers
Harshita Kukreja
Joshua Kulasingham
Aleksandra Kulbaka
Pranav Kulkarni
Abhay Kumar
Ashwin Kumar
Harshit Kumar
Nishant Kumar
Sandeep Kumar
Snehil Kumar
Nagendra Kumar
Naveen Kumar Kummari
Shinjini Kundu
Soumya Kundu
Manuela Kunz
Po-Chih Kuo
Tahsin Kurc
Nikhil Cherian Kurian
Semen Kurkin
Vinod Kurmi
Rafsanjany Kushol
Jin Tae Kwak
Hyeokjin Kwon
Jii Kwon
Jong Kwon
Junmo Kwon
Sunyoung Kwon
Jan Kybic
Agnieszka Lach
Chih Lai
Songning Lai
Taotao Lai
Hala Lamdouar
Hengrong Lan
Qizhen Lan
Tianzhong Lan
Yinyu Lan
Bennett Landman
Frank Langbein
Georg Langs
Qicheng Lao
Othmane Laousy
Andrea Lara
Bianca Lassen-Schmidt
Patricia Lasserre
Sarah Latus
Wai Tak Lau
Shlomi Laufer
Max-Heinrich Laves
Aonghus Lawlor
Scindhiya Laxmi
Duc Tai Le
Loic Le Folgoc
François Lecomte-Denis
Maria Ledesma-Carbayo
Brian Lee
Bumshik Lee

Dongheon Lee
Euijong Lee
Eungjoo Lee
Gyeongmin Lee
Hansang Lee
Hoi Leong Lee
Hong Joo Lee
Hyeonjun Lee
Hyun Gyu Lee
Hyungyung Lee
Hyunyeol Lee
Jaeho Lee
Jaesung Lee
Jangho Lee
Jiyang Lee
Jonghyun Lee
Jongmin Lee
Jong-Min Lee
Jooho Lee
Ju Hwan Lee
Jungwook Lee
Junhyeok Lee
Junseok Lee
Min Hun Lee
Minsik Lee
Moon Hwan Lee
Sanghyuck Lee
Sangmin Lee
Sangyoon Lee
Seoyoung Lee
Seungeun Lee
Sing Chun Lee
Won Hee Lee
Woonghee Lee
Youngjoon Lee
Yugyung Lee
Yurim Lee
Antoine Legouhy
Long Lei
Minglong Lei
Xuan Lei
Yiming Lei
Karoline Leiberg
Nick Lemke
Yilin Leng
Matthias Lenga
Dimitrios Lenis
Gabriel Lepetit-Aimon
Carson Leung
Meir Levi
Michael Levine
Ao Li
Bin Li
Binghua Li
Bo Li
Chao Li
Chaoyi Li
Cheng Li
Chengkun Li
Chengyin Li
Chentao Li
Chuanpu Li
Cynthia Xinran Li
Dawei Li
Fan Li
Feng Li
Frank Li
Guang Li
Guanya Li
Guoshi Li
Haifang Li
Haiqing Li
Haitao Li
Han Li
Hanyu Li
Hao Li
Haofeng Li
Haojin Li
Haoshen Li
Haoxiang Li
He Li
Heng Li
Hongming Li
Hongshuo Li
Hongyu Li
Hongzhao Li
Huiqi Li
Jiafeng Li
Jialin Li
Jialu Li

Jiapeng Li
Jiawen Li
Jihe Li
Jincheng Li
Jingming Li
Jinpeng Li
Jun Li
Juncheng Li
Junjian Li
Jupeng Li
Kaiwen Li
Kaiyan Li
Kang Li
Kehan Li
Kun Li
Lei Li
Lin Li
Ling Li
Miao Li
Min Li
Peng Li
Qi Li
Qiaoxin Li
Qingli Li
Qingya Li
Qiyue Li
Rui Li
Ruiyang Li
Ruizhe Li
Ruocheng Li
Sanqian Li
Shaowei Li
Shibo Li
Shiwei Li
Shuaibo Li
Shumeng Li
Shusheng Li
Sijia Li
Siqi Li
Tao Li
Teng Li
Tiansong Li
Weiling Li
Wen Li
Wenbo Li
Wenxuan Li
Wenxue Li
Xiang Li
Xiangchen Li
Xiangyu Li
Xiaomeng Li
Xiaoxiao Li
Xiao-Xin Li
Xiaoye Li
Xieji Li
Xing Li
Xinshan Li
Xinyue Li
Xue Li
Xueshen Li
Xuesong Li
Xueyang Li
Yang Li
Yaoqian Li
Yaxuan Li
Ye Li
Yicong Li
Yidi Li
Yijie Li
Yijin Li
Yingshu Li
Yingtai Li
Yinsheng Li
Yiping Li
Yiran Li
Yitong Li
Yonghong Li
Yuchen Li
Yudong Li
Yuheng Li
Yutong Li
Zhe Li
Zhen Li
Zhenghong Li
Zhenping Li
Zhiyuan Li
Zhjin Li
Zhuoan Li
Zi Li
Zihan Li

Ziyu Li
Zuoyong Li
Chenyu Lian
Sheng Lian
Dong Liang
Huanwen Liang
Jianwen Liang
Junjie Liang
Kewei Liang
Li Liang
Wei Liang
Xiaokun Liang
Xinglong Liang
Yixiong Liang
Yongjie Liang
Yuxuan Liang
Zhichao Liang
Guiqiu Liao
Wei Liao
Weicheng Liao
Xiangyun Liao
Zehui Liao
Andrzej Liebert
Gilbert Lim
Min Hyuk Lim
Michelle Espranita Liman
Bowen Lin
Che Lin
Fengming Lin
Fuchun Lin
Haoneng Lin
Jhih-Yuan Lin
Jianxin Lin
Junjie Lin
Nan Lin
Rizhong Lin
Runjia Lin
Tianyu Lin
Tong Lin
Weiping Lin
Wenjun Lin
Xian Lin
Yi Lin
Ying-Chih Lin
Yiqun Lin
Yiwei Lin
Yiyang Lin
Youn-Long Lin
Zehui Lin
Haibin Ling
Simone Lionetti
A.J. Lisowska
Ciro Listone
Mattia Litrico
Baodi Liu
Chang Liu
Che Liu
Chen Liu
Cheng Liu
Chenxi Liu
Chenyu Liu
Chunwei Liu
Congcong Liu
Dichao Liu
Fangyi Liu
Feihong Liu
Feng Liu
Fengbei Liu
Fenglin Liu
Han Liu
Hao Liu
Hengjie Liu
Hengyan Liu
Hong Liu
Hongying Liu
Huan Liu
Hui Liu
Jiahe Liu
Jiameng Liu
Jianlei Liu
Jiaxiang Liu
Jiaxuan Liu
Jie Liu
Jin Liu
Jinchao Liu
Jingsong Liu
Jingxin Liu
Jingyu Liu
Jinyuan Liu
Jiquan Liu

Jiyao Liu
Juan Liu
Jun Liu
Junqi Liu
Kang Liu
Lei Liu
Li Liu
Linjie Liu
Mengjun Liu
Mengting Liu
Mianxin Liu
Mingsi Liu
Mingtian Liu
Mingxin Liu
Mingxuan Liu
Mingyuan Liu
Peng Liu
Quanying Liu
Ruishan Liu
Runhao Liu
Shangwang Liu
Shengfeng Liu
Shengyuan Liu
Shu Liu
Shuting Liu
Sidong Liu
Tiantian Liu
Ting Liu
Tingwei Liu
Wei Liu
Wenjian Liu
Wenzheng Liu
Xiabi Liu
Xianjing Liu
Xiaohua Liu
Xiaoyue Liu
Xinyu Liu
Xixi Liu
Xuan Liu
Xueli Liu
Xueyu Liu
Yajun Liu
Yan Liu
Yanbei Liu
Yang Liu
Yanzhen Liu
Yihao Liu
Yikang Liu
Yiming Liu
Yingcheng Liu
Yiqing Liu
Yu Liu
Yuanye Liu
Yuhang Liu
Yunbi Liu
Yuxi Liu
Yuxiao Liu
Yuxuan Liu
Zeyu Liu
Zhe Liu
Zhentao Liu
Zhenyu Liu
Zhi Liu
Zhiyuan Liu
Zhu Liu
Ziteng Liu
Ziwen Liu
Zuozhu Liu
Andrea Loddo
Ojaswita Lokre
Andreas Lolos
Angela Lombardi
Bining Long
Jiake Long
Lifan Long
Ziyang Long
Arthur Longuefosse
Renaud Lopes
Miguel López-Pérez
Marco Lorenzi
Meng Lou
Wei Lou
Yiwei Lou
Marta Lovino
Chenggang Lu
Chenyang Lu
Chunyao Lu
Donghuan Lu
Hengfa Lu
Jiacheng Lu

Jingfeng Lu
Li Lu
Lin Lu
Shuai Lu
Wenjing Lu
Yao Lu
Yifan Lu
Yuhuan Lu
Zhengda Lu
Zhicheng Lu
Zhihe Lu
Zhixiang Lu
Zilin Lu
Ziqian Lu
Haijing Luan
Xiao Luan
Luca Lumetti
Chen Luo
Guibo Luo
Haozhe Luo
Huoling Luo
Jiajie Luo
Ma Luo
Mingyuan Luo
Qinxuan Luo
Tao Luo
Xiaoling Luo
Xin Luo
Xinzhe Luo
Xuexiong Luo
Ye Luo
Yiwen Luo
Yong Luo
Yu Luo
Zhiming Luo
Zihao Luo
Marie Luong
Xingguo Lv
Buntheng Ly
Ngoc Ly
Son Ly
Donghang Lyu
Ilwoo Lyu
Jun Lyu
Mengye Lyu
Qing Lyu
Yanjun Lyu
Yiheng Lyu
Yilin Lyu
Vivek M
Meharban M S
Sharath M S
Deeksha M Shama
Changsheng Ma
Da Ma
Gengchen Ma
He Ma
Hehuan Ma
Jiaxin Ma
Jun Ma
Junbo Ma
Lei Ma
Ling Ma
Liyan Ma
Qi Ma
Qiang Ma
Shixing Ma
Tai Ma
Tengfei Ma
Wei Ma
Xiangyuan Ma
Xiaodong Ma
Xinke Ma
Xinyu Ma
Xiyao Ma
Yuefeng Ma
Yutao Ma
Zihan Ma
Lennart Maack
Mohamed Mabrok
Julia Machnio
Frederik Maes
Nooshin Maghsoodi
Amirreza Mahbod
Mohammed Mahmoud
Md Ishtyaq Mahmud
Amulya Kumar Mahto
Andreas Maier
Klaus H. Maier-Hein
Boris Mailhe

Aamir Malik
Hashmat Shadab Malik
Norberto Malpica
Richa Malviya Dutta
Michail Mamalakis
Jan Emily Mangulabnan
Merry Mani
Ali Mansouri
Yiji Mao
Arooba Maqsood
Vincenzo Marcianò
Jan Margeta
Umar Marikkar
Brett Marinelli
Zdravko Marinov
Jose Diogo Marques dos Santos
Pablo Márquez Neila
Carsten Marr
Yassine Marrakchi
Nicolas Martin
Thomas Martinetz
Fabio Martínez
Maria Martini
Queralt Martín-Saladich
Pierre Marza
Axel Masquelin
Yoshitaka Masutani
Tejas Sudharshan Mathai
Raji Mathew
Prateek Mathur
Sasan Matinfar
Hiroki Matsuzaki
Petr Matula
Lev Matveev
Dimitrios Mavroeidis
Perla Mayo
Badhan Mazumder
Ashery Mbilinyi
Jamie McClelland
Niccolo McConnell
Yasamin Medghalchi
Pau Medrano-Gracia
Mostafa Mehdipour Ghazi
Raghav Mehta
Jonas Mehtali
Liye Mei
Siyuan Mei
Erik Meijering
M S Mekala
Alaa Melek
Alexandre Mello
Cai Meng
Hui Meng
Jia Meng
Mingyuan Meng
Nan Meng
Runqi Meng
Yan Meng
Zhu Meng
Zijie Meng
Martin Menten
Engelbert Mephu Nguifo
Talha Meraj
Dorit Merhof
Weiming Mi
Dong Miao
Juzheng Miao
Yao Miao
Filippo Michelotti
Leo Milecki
Dongbo Min
Hyun-Seok Min
Zhe Min
Wenlong Ming
Duy Minh Ho Nguyen
Davneet Minhas
Giulio Minore
Daniel Minsky
Qazaleh Mirsharif
Ghazaleh Mirzaee
Leo Misera
Rashika Mishra
Suraj Mishra
Virendra Mishra
Sampa Misra
Tadashi Miyamoto
Tadahaya Mizuno
Marc Modat
C. Krishna Mohan
Sovesh Mohapatra

Ramtin Mojtahedi
Marc Molina van den Bosch
Hendrik Möller
Joyanta Jyoti Mondal
Sumita Mondal
Sidratul Montaha
Javier A. Montoya-Zegarra
Jong Hak Moon
Ashkan Moradi
Saed Moradi
José Morano
Laurin Mordhorst
Alejandra Moreno Tarazona
Carlos Moreno-Garcia
Kensaku Mori
Henrique Morimitsu
Matteo Moro
Ken'ichi Morooka
Aliasghar Mortazi
Mayar Mostafa
Lei Mou
Lichao Mou
Tian Mou
Simmi Mourya
Samir Moustafa
Nan Mu
Wei Mu
Klaus Mueller
Pritam Mukherjee
Anirban Mukhopadhyay
Henning Müller
Johanna Müller
Philip Müller
Sarah Müller
Marion Mundt
Nusrat Munia
Asbjørn Munk
Aminu Musa
Toufiq Musah
Przemyslaw Musialski
Vit Musil
Saif Musluh
Yasmine Mustafa
Pushkar Mutha
Inye Na
Shahabedin Nabavi
Saad Nadeem
Roman Naeem
Manas Nag
Hajime Nagahara
Reza Naghne
Md Jaber Al Nahian
Toshiya Nakaguchi
Sahil Nalawade
Woo-Jeoung Nam
Khashayar Namdar
Nandakishor Nandakishor
Vittoria Nardone
André Nascimento
Sarah Nassar
Yassine Nasser
Rodrigo Nava
Deepak Ranjan Nayak
Herve Nicolas Nbonsou Tegang
Giacomo Nebbia
Sven Nebelung
Fnu Neha
Omid Nejatimanzari
Iman Nematollahi
Alberto Neri
Favour Nerrise
Sara Neves Silva
Jacob Newman
Alasdair Newson
Cuong Ngo
Binh Nguyen
Huy Nguyen
Khoa Nguyen
Tan Nguyen
Tan-Cong Nguyen
Tri-Thien Nguyen
Trong-Thuan Nguyen
Phi Nguyễn
Thanh Nguyen-Duc
Chang Ni
Jun Ni
Ke Ni
Shuo Ni
Hannes Nickisch
Jingxin Nie

Weizhi Nie
Xinyu Nie
Maximilian Nielsen
Andreas Nienkötter
Aditya Nigam
Dmitry Nikolaev
Emmanouil Nikolakakis
Guochen Ning
Junzhi Ning
Wenbo Ning
Xiaolin Ning
Zeeshan Nisar
Kazuya Nishimura
Sijie Niu
Usma Niyaz
Vincent Noblet
Shahryar Noei
Jeonghyun Noh
Junhyug Noh
Wonjong Noh
Fuad Noman
Anass Nouri
Alexey Novikov
Franciszek Nowak
Hanung Nugroho
Felix Nützel
Chinedu Nwoye
Emilio Ochoa
Gilberto Ochoa-Ruiz
Samuel Ofosu Mensah
Marzieh Oghbaie
Roseline Ogundokun
Hyun-Jic Oh
Ji-Hun Oh
Kwanseok Oh
Sang-Il Oh
Seong Je Oh
Sewook Oh
Ajibola Oladokun
Oluwole Olajide
Laura Oliva Maza
Sara Oliveira
Antoine Olivier
Juan Olmos
Blessing Olorunfemi
Jimena Olveres
Doruk Oner
John Onofrey
Eliza Orasanu
Alexei Ossadtchi
Andreas Østvik
Yoshito Otake
Yafei Ou
Mehdi Ounissi
Nora Ouzir
Utku Ozbulak
Gökhan Özbulak
Caner Ozer
Kutsev Ozyoruk
Andre Pacheco
Suraj Pai
Nathan Painchaud
Eduardo Pais Pooch
Ashkan Pakzad
Aditya Pal
Basudha Pal
Subhash Pal
Balamurugan Palaniappan
Christoph Palm
Naveen Paluru
Viswanath Pamulakanty Sudarshan
Fei Pan
Haoyu Pan
Meihong Pan
Qingtao Pan
Wentao Pan
Xiang Pan
Xinlei Pan
Yongsheng Pan
Eleftheria Panagiotaki
Manish Pandey
Egor Panfilov
Fengqian Pang
Haowen Pang
Kaifeng Pang
Xintao Pang
Yunkui Pang
Hsing-Kuo Pao
Tasos Papastylianou
Ganesh Paramasivam

Jay Paranjape
Abhijeet Parida
Sourav Parija
Chae Jung Park
Chunsu Park
Doohyun Park
Doyoung Park
Heejung Park
Hogun Park
Ho-Min Park
Hyejin Park
Hyoung Suk Park
Hyunjin Park
Jonggwon Park
Joon Sik Park
Joonhyuk Park
Juyoun Park
Robin Park
Sung-Hong Park
Wonjung Park
Luca Parolari
Mario Parreno Centeno
Srivibha Parthasarathy
Akshay Paruchuri
Andrea Passerini
Srivathsa Pasumarthi
Oankar Patil
Shubham Patil
Arijit Patra
Mayank Patwari
Angshuman Paul
Riti Paul
Ruben Pauwels
Kelly Payette
Nicolo Pecco
João Pedrosa
Jialun Pei
Yuchen Pei
Yuru Pei
Himashi Peiris
Chantal Pellegrini
Akila Pemasiri
Jiachuan Peng
Jie Peng
Liangrui Peng
Linkai Peng
Tingying Peng
Xiaoling Peng
Yanjun Peng
Yige Peng
Yunsong Peng
Zelin Peng
Zhihao Peng
Matteo Pennisi
Jacob Peoples
Sergio Pereira
Alessandro Perelli
Alejandra Perez
Paula Andrea Pérez-Toro
Matthias Perkonigg
Daniele Perlo
Terry Peters
Eike Petersen
Simon Pezold
Elisabeth Pfaehler
Micha Pfeiffer
Duy Khoa Pham
Dzung Pham
Trong Thang Pham
Van-Nguyen Pham
Tanmoy Sarkar Pias
Tomasz Pieciak
Omar Piña-Ramirez
Thomas Pinetz
Stefan Ploner
Gasper Podobnik
Eleonora Poeta
Kilian Pohl
Sandesh Pokhrel
Isabella Poles
Robert Policelli
Giuseppe Pontillo
Aloys Portafaix
Michael Potter
Nakul Poudel
Pranav Poudel
Lighittha PR
Chinmay Prabhakar
John Prada
Raphael Prevost

Federica Proietto Salanitri
Yao Pu
Sergi Pujades
Vineet Punyamoorty
Elodie Puybareau
Melika Qahqaie
Ashwaq Qasem
Adnan Qayyum
Kehan Qi
Liangce Qi
Lin Qi
Quan Qi
Shile Qi
Xiaoran Qi
Yaolei Qi
Chen Qian
Xusheng Qian
Yan Qiang
Maoying Qiao
Yanyuan Qiao
Chendong Qin
Fangbo Qin
Feiwei Qin
Genggeng Qin
Jiaolong Qin
Yi Qin
Linbo Qing
Jie Qiu
Junlai Qiu
Liang Qiu
Shi Qiu
Shihan Qiu
Song Qiu
Xingyu Qiu
Xinmei Qiu
Zhike Qiu
Tran Minh Quan
Ningfeng Que
Sandro Queirós
Willmer Rafell Quiñones Robles
Sharon Quispe
Neeraja R.
Sinyoung Ra
Akm Shahariar Azad Rabby
Hamid R. Rabiee
Daniel Racoceanu
Alicja Rączkowska
Lavdie Rada
Dharini Raghavan
Chandravardhan Singh Raghaw
Matthew Ragoza
Maqsudur Rahman
Md Mostafijur Rahman
Muhammad Mahboob Ur Rahman
Sowad Rahman
Tarek Rahman
Wahyu Rahmaniar
Shyam Nandan Rai
Sudarshan Rajagopalan
Ajitha Rajan
Santhiya Rajan
Jagath Rajapakse
João Ramalhinho
Jayroop Ramesh
Ruisheng Ran
Amin Ranem
Aneesh Rangnekar
Asha Rani
Rakesh Ranjan
Amog Rao
Carolina Raposo
Essam Rashed
Hatem Rashwan
Sebastian Rassmann
Sparsh Rastogi
Lakmali Nadeesha Kumari Rathukohe Mudiyanselage
Mahesh Raveendranatha Panicker
Keerthi Sravan Ravi
Sadhana Ravikumar
Vikram Ravindra
Yogesh Rawat
Manahil Raza
Imran Razzak
Divya D Reddy
Mourad Rehali
Abdul Rehman
Paolo Remagnino
Samuel Remedios
Moritz Rempe

Ge Ren
Qin Ren
Sijie Ren
Siyang Ren
Tianyi Ren
Yudan Ren
Yuyang Ren
Zehua Ren
Zhihang Ren
David Restrepo
Mauricio Reyes
Abel Reyes-Angulo
Hadrien Reynaud
Hayley Reynolds
Thiago Rezende
Eman Rezk
Helena Ribeiro Torres
Anne-Marie Rickmann
Bastian Rieck
Brandalyn Riedel
Isabel Rio-Torto
Jens Rittscher
Duhyun Ro
Youngmin Ro
Charlotte Robert
Harry Robertshaw
Robert Robinson
Jessica Rodgers
Ranga Rodrigo
Antonio Rodriguez-Sanchez
Harry Rogers
Maximilian Rokuss
Eduardo Romero
Ali Roozbehi
Timo Ropinski
Pooneh Roshanitabrizi
Paul Rosin
Holger Roth
Rahimeh Rouhi
Gideon Rouwendaal
Saikat Roy
Dan Ruan
Josué Ruano-Balseca
Mathieu Rubeaux
Nejung Rue
Filippo Ruffini
Danny Ruijters
Josef Lorenz Rumberger
Hosuk Ryou
Hoe Sung Ryu
Kanghyun Ryu
Andrzej Rzepiela
J. S.
Muralikrishna S. N.
Sowmya S Sundaram
Armin Saadat
Sari Saba-Sadiya
Amirhossein Sabour
Maziar Sabouri
Bhuvan Sachdeva
Ario Sadafi
Binesh Sadanandan
Amir Reza Sadri
Fahad Saeed
Shaheer Saeed
Faezeh Safari
Aon Safdar
Monjoy Saha
Pramit Saha
Farhang Sahba
Heba Sailem
Simone Saitta
Md Sajid
Shadman Sakib
Soorena Salari
Divya Saleela
Afsah Saleem
Tausifa Jan Saleem
Md Sirajus Salekin
Mostafa Salem
Emmanuelle Salin
Samuele Salti
Md Samiullah
Karen Sanchez
Alejandro Sanchez Guinea
Ricardo Sanchez-Matilla
Yudi Sang
Sara Sangalli
Daniel Santak Wolf
Caio Santana

Ruggiero Santeramo
Ainkaran Santhirasekaram
Carlos Santiago
Alice Santilli
Nishchal Sapkota
Parham Saremi
Rachana Sathish
Sandra Emi Sato Imuro
Olivier Saut
Shier Nee Saw
Christian Schiffer
Robin Schirrmeister
Alexander Schlaefer
Adam Schmidt
Julia Schnabel
Till Schnabel
Daniel Scholz
Daan Schouten
Sebastian Schröder
Peter Schüffler
Johannes Schuiki
Nico Schulthess
Jeanette Schulz-Menger
Assaf Schuster
Davide Scorza
Michaël Sdika
Thalia Seale
Kyle See
Philipp Seeböck
Lalithkumar Seenivasan
Matthias Seibold
Alexander Seitel
Yang Sejung
Alexander Selivanov
Raghavendra Selvan
Jothiraj Selvaraj
Ilia Semenkov
Ahmet Sen
Sourya Sengupta
Thilo Sentker
Chang Wook Seo
Hyewon Seo
Sumin Seo
Yuri Seo
Maxime Sermesant
Manuel Serna-Aguilera
Giuseppe Serra
Carmen Serrano
Xavier Sevillano
Marvin Seyfarth
Qingrui Sha
Tianming Sha
Mai A. Shaaban
Nazmus Shakib Shadin
Ahmed Shaffie
Hina Shah
Jay Shah
Keyur Devendra Shah
Nisarg Shah
Nagur Shareef Shaik
Nazim Shaikh
S. Shailja
Amit Shakya
Mahesh Shakya
Nina Shamsi
Niyas Shamsudeen Kutty
Caifeng Shan
Jiwei Shan
Shanshan Shan
Yixin Shan
Dattesh Dayanand Shanbhag
Jiang Shang
Yeqin Shao
Maxim Sharaev
Lalith Sharan
Ajay Sharma
Saurav Sharma
Vanshali Sharma
Varsha Sharma
Furqan Shaukat
Mona Sheikh Zeinoddin
Ao Shen
Chen Shen
Gang Shen
Hualei Shen
Linlin Shen
Wei Shen
Xiang-Jun Shen
Xinke Shen
Yi Shen

Yiqing Shen
Zhenrong Shen
Zhiqiang Shen
Hongwei Sheng
Yi Sheng
John Shepherd
Bo Shi
Enze Shi
Hang Shi
Haoyuan Shi
Hongjian Shi
Jian Shi
Jiayang Shi
Jun Shi
Kangdi Shi
Peng Shi
Pengcheng Shi
Ruohua Shi
Shuo Shi
Xiaoshuang Shi
Xiaoyu Shi
Yaying Shi
Yi Shi
Yinghuan Shi
Yiyu Shi
Zhan Shi
Shu-Fu Shih
Takuro Shimaya
Jaeseung Shin
Jitae Shin
Seung Yeon Shin
Rakesh Shiradkar
Tamir Shor
David Shorthouse
Hongchao Shu
Jie Shu
Yucheng Shu
Ananya Shukla
Ankita Shukla
Md Rezowan Hossain Ferdous Shuvo
Weixin Si
Rosa Sicilia
Md. Saiful Bari Siddiqui
Sidi Mohamed Sid'El Moctar
Alberto Signoroni
Wilson Silva
Margarida Silveira
Jaeyoon Sim
Ivor Simpson
Zachary Sims
Amritpal Singh
Azad Singh
Dinesh Singh
Karanpartap Singh
Aniruddha Sinha
Ayushi Sinha
Vasilis Siomos
Raviteja Sista
Ssharvien Kumar Sivakumar
Elena Sizikova
Luka Škrlj
Gregory Slabaugh
Dane Smith
Weronika Smolak-Dyżewska
Tiffany Y So
Rajath Soans
Roger Soberanis-Mukul
Paolo Soda
Maina Sogabe
Maëlys Solal
Elanchezhian Somasundaram
Moo Hyun Son
Seungyeon Son
Anthony Song
Haojie Song
Hong Song
Hyunwoo Song
Pei-Cheng Song
Qi Song
Ruoxian Song
Shuang Song
Sifan Song
Tao Song
Xiao Song
Yantao Song
Yiran Song
Yu Song
Yucheng Song
Zhiwei Song
Zhiying Song

Zhiyun Song
Tananun Songdechakraiwut
Kristine Sørensen
Mazen Soufi
Raissa Souza
Roberto Souza
Georgia Sovatzidi
Michal Španěl
Bella Specktor Fadida
William Speier
Ziga Spiclin
Jon Sporring
Przemysław Spurek
Pradeeba Sridar
Anuj Srivastava
Lawrence Staib
Emma Stanley
Martijn Starmans
Johannes Stegmaier
Jan Steinbrener
Nil Stolt-Ansó
Skylar Stolte
Duncan Stothers
Joshua Stough
Robin Strand
Eric Strobl
Jérémie Stym-Popper
Martin Styner
Bo Su
Fei Su
Jiawei Su
Juntao Su
Weifeng Su
Yao Su
Barathi Subramanian
Hongyiel Suh
Julian Suk
K. M. Arefeen Sultan
Muhammad Ahmad Sultan
Nurjahan Sultana
Waqas Sultani
Chao Sun
Diya Sun
Dongbiao Sun
Guanqun Sun
Jiajun Sun
Jian Sun
Jianqi Sun
Jiarui Sun
Jindong Sun
Kaicong Sun
Minghao Sun
Roger Sun
Rui Sun
Ruiqing Sun
Shangquan Sun
Shoukun Sun
Shuifa Sun
Song Sun
Susu Sun
Xia Sun
Xiaowu Sun
Xiaoxue Sun
Xuebin Sun
Yan Sun
Yifei Sun
Yihua Sun
Yiran Sun
Yongheng Sun
Zhaoyi Sun
Zhe Sun
Zhenglong Sun
Zhichao Sun
Zhongxiang Sun
Zijun Sun
Vaanathi Sundaresan
Nak-Jun Sung
Elifnur Sunger
Hanna Suominen
Akara Supratak
Kadambari Suryavanshi
Krunal Suthar
Mohammod Naimul Islam Suvon
Zaneta Swiderska-Chadaj
Tomasz Szczepański
Shabana Tabassum
Sairam Tabibu
Ahmad P. Tafti
Behrad Taghibeyglou
Pablo Tahoces

Riccardo Taiello
Razane Tajeddine
Divyanshu Tak
Eichi Takaya
Saori Takeyama
Shumpei Takezaki
Hugues Talbot
Roger Tam
Utkarsh Tambe
Jose Tamez-Peña
Jeremy Tan
Kaiwen Tan
Shan Tan
Shuangyi Tan
Weimin Tan
Wenjun Tan
Xiaoyu Tan
Arthur Tang
Feilong Tang
Fenghe Tang
Haifeng Tang
Hao Tang
He Tang
Huadong Tang
Jianbo Tang
Jin Tang
Jinshan Tang
Kaibo Tang
Lingzhi Tang
Ming Tang
Shiqin Tang
Yilin Tang
Yucheng Tang
Yuxian Tang
Zhenzhou Tang
Zihao Tang
Michael Tanzer
Tianli Tao
Wenzheng Tao
Yitian Tao
Mickael Tardy
Sansiri Tarnpradab
Giacomo Tarroni
Virginia Tasso
Vasiliki Tassopoulou
Wallapak Tavanapong
Thomas Tavolara
Soroosh Tayebi Arasteh
Gbègninougbo Aurel Davy Tchokponhoue
Kartikay Tehlan
Lin Teng
Atsushi Teramoto
Rafeek Thahakoya
Antoine Theberge
Sarina Thomas
Peter Thompson
Jacob Thrasher
Sai Natarajan Thrissur Ranganathan
Cristiana Tiago
Jiyu Tian
Qingyao Tian
Qiyuan Tian
Xuanyu Tian
Yibin Tian
Yun Tian
Vijay Srinivas Tida
Zumnan Timbi
Chee-Ming Ting
Fung Fung Ting
Aleksei Tiulpin
Yuta Tokuoka
Santiago Toledo-Cortés
Paulina Tomaszewska
Lyuyang Tong
Shao Tong
Akif Tosun
Nicolas Toussaint
Henry Tregidgo
Georgios Triantafyllou
Eleftherios Trivizakis
Jakob Troidl
Daniel Truhn
Tuan Truong
Liyun Tu
Puxun Tu
Jason Turuwhenua
Aayush Tyagi
Mrinal Tyagi
Georgios Tziritas

Kwang-Hyun Uhm
Constantin Ulrich
Ilkay Ulusoy
Rao Umer
Tamas Ungi
Saritha Unnikrishnan
Eneko Uruñuela
Fatmatulzehra Uslu
Muhammad Usman
Hristina Uzunova
Sami Väänänen
Anil Kumar Vadathya
Theodoros Vagenas
Puru Vaish
Vanya Valindria
Juan Miguel Valverde
Gijs van Tulder
Henk van Voorst
Jef Vandemeulebroucke
Joris Vankerschaver
Juan Vargas Garcia
Bino Varghese
Hooman Vaseli
Rajat Vashistha
Michael Vasilakakis
Neil Vaughan
S. Swaroop Vedula
Jai Prakash Veerla
Danush Kumar Venkatesh
Vikram Venkatraghavan
Tom Vercauteren
Manisha Verma
Mitko Veta
Pierre-Frederic Villard
Pk Vinod
Thanaporn Viriyasaranon
Christiaan Viviers
Hung Vo
Wolf-Dieter Vogl
Anh Vu
Truong Vu
Milos Vukadinovic
Kushal Vyas
Assefa Wahd
Victor Wåhlstrand
Gordon Waiter
Zuzanna Wakefield-Skórniewska
Christian Wallraven
Ricky Walsh
Cheng Wan
Kaiwen Wan
Liang Wan
Shouhong Wan
Xiaohua Wan
Zhiyu Wan
Zishuo Wan
Alan Wang
An Wang
Baizhi Wang
Bin Wang
Bo Wang
Bomin Wang
Changfei Wang
Changjian Wang
Chaoqun Wang
Cheng Wang
Chengkai Wang
Chenglong Wang
Chengyan Wang
Chenhui Wang
Chong Wang
Chunjiang Wang
Chunliang Wang
Cong Wang
Congcong Wang
Dekui Wang
Di Wang
Dongang Wang
Fangyijie Wang
Fanwen Wang
Fuying Wang
Guangzhi Wang
Guoan Wang
Guotai Wang
Haifeng Wang
Hao Wang
Haolin Wang
Haoyu Wang
Hongpeng Wang
Hongqiu Wang

Huanqin Wang
Huayu Wang
Huitao Wang
Janet Wang
Jiacheng Wang
Jiahao Wang
Jian Wang
Jiazhen Wang
Jiazheng Wang
Jihang Wang
Jinfeng Wang
Jinhao Wang
Jixin Wang
Jiyao Wang
Juan Wang
Jueqi Wang
Jui-Kai Wang
Junyi Wang
Junze Wang
Kai Wang
Kailing Wang
Kaini Wang
Lehan Wang
Lei Wang
Li Rong Wang
Liansheng Wang
Lifang Wang
Lihui Wang
Lin Wang
Linwei Wang
Lipo Wang
Lituan Wang
Lu Wang
Luo Wang
Luoyu Wang
Meili Wang
Meng Wang
Mengxiao Wang
Mingliang Wang
Peng Wang
Pingping Wang
Qian Wang
Qianqian Wang
Qijie Wang
Qingying Wang
Qiuli Wang
Rongfang Wang
Rongsheng Wang
Rui Wang
Ruiqi Wang
Ruixuan Wang
Runmin Wang
Sen Wang
Shangfei Wang
Shiru Wang
Shuangyi Wang
Shuhui Wang
Shuihua Wang
Shuo Wang
Sihan Wang
Tao Wang
Tianyu Wang
Tony Wang
Wei Wang
Weichung Wang
Weirong Wang
Xi Wang
Xin Wang
Xingce Wang
Xingyue Wang
Xinkai Wang
Xinyu Wang
Yan Wang
Yi Wang
Yifan Wang
Yifeng Wang
Yihui Wang
Yinghui Wang
Yingni Wang
Yipei Wang
Yiqing Wang
Yiying Wang
Yonghao Wang
Yu Wang
Yuli Wang
Yulin Wang
Yuqi Wang
Zesong Wang
Zhewei Wang
Zhihui Wang

Zhisong Wang
Zhiyan Wang
Zhonghua Wang
Zhou Wang
Zi Wang
Zichen Wang
Zihao Wang
Zipei Wang
Ziyi Wang
Ziyue Wang
Zizhou Wang
Zuhui Wang
Talha Waqas
Tyler Ward
Simon Warfield
Valentine Wargnier Dauchelle
William Warner
Azmine Toushik Wasi
Saad Wazir
Michael Wehrli
Chen Wei
Dong Wei
Guangshun Wei
Hao Wei
Jia Wei
Jie Wei
Linda Wei
Qijie Wei
Ruofeng Wei
Shuwen Wei
Xinjian Wei
Yongbin Wei
Zechen Wei
Ziquan Wei
Jonas Weidner
Michael Wels
Xuyun Wen
Yannuo Wen
Zheyu Wen
Cui Wenju
Philipp Wesp
Kithmin Wickremasinghe
Brady Williamson
Matthias Wilms
Paul Wilson
Eric Wisotzky
Adam Wittek
Marek Wodzinski
Ivo Wolf
Julia Wolleb
Jelmer Wolterink
Dong-Ok Won
Ji Hye Won
Jongjun Won
Bryan Wong
Chun Kit Wong
Joshua Wong
Kitty Wong
Anna Woodard
Mckell Woodland
Stefan Wörz
Beining Wu
Chao Wu
Chengchuan Wu
Chengyu Wu
Chenwei Wu
Cong Wu
Dan Wu
Donghan Wu
Feng Wu
Fuping Wu
Fuzhi Wu
Guoqing Wu
Han Wu
Hao Wu
Honghan Wu
Ji Wu
Jiamin Wu
Jianghao Wu
Jie Ying Wu
Jinlin Wu
Junde Wu
Junyang Wu
Lei Wu
Luohong Wu
Mengqi Wu
Ming Wu
Nian Wu
Qian Wu
Qing Wu

Rong Wu
Ruifan Wu
Ruiqi Wu
Ruoyou Wu
Shaoju Wu
Shaoxuan Wu
Shibin Wu
Shuang Wu
Shuqiong Wu
Taoyu Wu
Victoria Wu
Wei Wu
Weiyi Wu
Wenjun Wu
Wenxia Wu
Wenxuan Wu
Xiangcen Wu
Xiangjun Wu
Xiao Wu
Xinyao Wu
Yanlin Wu
Yihang Wu
Yinhao Wu
Yinzhe Wu
Yongfei Wu
Yonghuang Wu
Yujia Wu
Yunheng Wu
Yu-Te Wu
Zhaolong Wu
Zhengwang Wu
Zhijian Wu
Zhiying Wu
Zhuoyu Wu
Zihao Wu
Zijian Wu
Joris Wuts
Julian Wyatt
Magdalena Wysocki
Haoran Xi
Nan Xi
Pengcheng Xi
Xiaoming Xi
Chao Xia
Chen Xia
Jianan Xia
Mingze Xia
Qing Xia
Tian Xia
Weihao Xia
Weiran Xia
Yihao Xia
Jinhai Xiang
Lei Xiang
Nan Xiang
Tiange Xiang
Tianqi Xiang
Bin Xiao
Deqiang Xiao
Furen Xiao
Jingjing Xiao
Li Xiao
Pan Xiao
Ruolin Xiao
Sa Xiao
Bin Xie
Fengbo Xie
Guo-Sen Xie
Hai Xie
Hao Xie
Housheng Xie
Jianyang Xie
Kangxian Xie
Kunzi Xie
Long Xie
Shiyu Xie
Sihong Xie
Xingyu Xie
Yining Xie
Yiting Xie
Yutong Xie
Zhaoheng Xie
Jack Xin
Jiayi Xin
Changda Xing
Fangxu Xing
Jiarui Xing
Shuwei Xing
Ying Xing
Yucheng Xing

Yuxiang Xing
Zhaohu Xing
Conghao Xiong
Hao Xiong
Honglin Xiong
Huimin Xiong
Sijie Xiong
Xiangyu Xiong
Xinyu Xiong
Yujian Xiong
Zicheng Xiong
Chenchen Xu
Chenfan Xu
Chuankai Xu
Dexuan Xu
Di Xu
Dunyuan Xu
Fangqiang Xu
Gelei Xu
Geng-Xin Xu
Guoping Xu
Hao Xu
Hongming Xu
Jialang Xu
Jiangchang Xu
Jianwei Xu
Jiaqi Xu
Jiashu Xu
Jiaxing Xu
Jinjing Xu
Jun Xu
Junlin Xu
Lei Xu
Lijian Xu
Lina Xu
Mengya Xu
Min Xu
Pengcheng Xu
Qing Xu
Rui Xu
Ruru Xu
Weixin Xu
Xiaowei Xu
Xinran Xu
Xiuyuan Xu
Xuan Xu
Yiwen Xu
Yongchao Xu
Yunqiu Xu
Zhuowei Xu
Ziang Xu
Zibo Xu
Zikang Xu
Ziyue Xu
Cheng Xue
Hui Xue
Jie Xue
Nian Xue
Peng Xue
Rundong Xue
Xiaofei Xue
Yunzhe Xue
Siyi Xun
Yue Xun
Goshiro Yamamoto
Yu Yamaoka
Amine Yamlahi
Bingjie Yan
Chongzhe Yan
Fang Yan
Fanzhe Yan
Jingwen Yan
Junyu Yan
Rui Yan
Siyuan Yan
Tianyi Yan
Yan Yan
Yige Yan
Zhiling Yan
Zhonghao Yan
Zipei Yan
Baoyao Yang
Chen Yang
Chunna Yang
Dong Yang
Fan Yang
Guang Yang
Guangqian Yang
Guangzhengao Yang
Guanyu Yang

Hongjia Yang
Hongxi Yang
Hongxu Yang
Hsin-Jung Yang
Huanqi Yang
Hyeonsik Yang
Hyung-Jeong Yang
Jiachen Yang
Jianhua Yang
Jiaqi Yang
Jiewen Yang
Jijiang Yang
Jin Yang
Junlin Yang
Kaiwen Yang
Kaixiang Yang
Ke Yang
Lei Yang
Litao Yang
Liutao Yang
Long Yang
Peng Yang
Qian Yang
Qianye Yang
Shu Yang
Shuo Yang
Shuojue Yang
Sicheng Yang
Sooyoung Yang
Su Yang
Wanqi Yang
Wei Yang
Xiaochen Yang
Xiaofei Yang
Xiaofeng Yang
Xikai Yang
Xin Yang
Xinnian Yang
Xinyi Yang
Xuan Yang
Yiguang Yang
Yijun Yang
Yingyu Yang
Yoon Seok Yang
Yun Yang
Yuxuan Yang
Zefan Yang
Zekang Yang
Zhaocan Yang
Zhenhong Yang
Zhidong Yang
Zhiwei Yang
Zhiwen Yang
Zijiang Yang
Zixin Yang
Ziyuan Yang
Zongbao Yang
Berrin Yanikoglu
Chenglin Yao
Haoran Yao
Linlin Yao
Tianyuan Yao
Xiaohui Yao
Zhao Yao
Milad Yazdani
Chenfei Ye
Chengze Ye
Chuyang Ye
Jiayu Ye
Johan Ye
Jong Chul Ye
Kangqing Ye
Meng Ye
Shuchang Ye
Xuesong Ye
Xujiong Ye
Yiwen Ye
Yubo Ye
Zhaoyi Ye
Yousef Yeganeh
Varduhi Yeghiazaryan
Chun-Hsiao Yeh
Chunzhi Yi
Huahui Yi
Jirong Yi
Mun Yi
Murong Yi
Weixi Yi
Xin Yi
Yuhao Yi

Zhenyu Yi
Rüveyda Yilmaz
Wen-Wai Yim
Chong Yin
Jiahui Yin
Ming Yin
Pengshuai Yin
Xuefei Yin
Yi Yin
Yunfei Yin
Zihan Yin
Haochao Ying
Qichao Ying
Jay Yoo
Min Kyoon Yoo
Seok Bong Yoo
Yongseon Yoo
Youngjin Yoo
Dan Yoon
Jihun Yoon
Kyungho Yoon
Siyeop Yoon
Susik Yoon
Weiqiu You
Xin You
Khalid Youssef
Andrew Yu
Boxiao Yu
Chaohui Yu
Chengjin Yu
Dahui Yu
De Yu
Fanyang Yu
Han Yu
Jiejiang Yu
Jin-Gang Yu
Jinhua Yu
Mingyang Yu
Qinji Yu
Qinkai Yu
Tingting Yu
Xiangchun Yu
Xiaopeng Yu
Xiaowei Yu
Yang Yu
Yanxuan Yu
Yi Yu
Zhen Yu
Zheng Yu
Ziqi Yu
Cheng Yuan
Gang Yuan
Haoyue Yuan
Jie Yuan
Kun Yuan
Manman Yuan
Mingze Yuan
Xiaohan Yuan
Xinrui Yuan
Yuchen Yuan
Zhouhang Yuan
Novanto Yudistira
Jiangbei Yue
Jiaxin Yue
Peiyan Yue
Boxiang Yun
Juseung Yun
Seong Dae Yun
Ka Yung
Chang Yuwen
Elizabeth Yuwono
Mariam Zabihi
Stefan Zachow
Anas Zafar
Abdelrahman Zaian
Nayyar Zaidi
Fahim Ahmed Zaman
Omar Zamzam
Alexey Zaytsev
Rachid Zeghlache
Ramy Zeineldin
Dong Zeng
Guangjian Zeng
Hongye Zeng
Hui Zeng
Jiajun Zeng
Na Zeng
Pinxian Zeng
Qi Zeng
Qingjie Zeng

Qingyuan Zeng
Shuang Zeng
Tianyi Zeng
Xinyi Zeng
Zhitao Zeng
Ruyi Zha
Xinbei Zha
Yongping Zhai
Aiying Zhang
Baochang Zhang
Bingbing Zhang
Chen Zhang
Chensheng Zhang
Chenxi Zhang
Chuanhao Zhang
Chujie Zhang
Dan Zhang
Dexun Zhang
Dingzhong Zhang
Dong Zhang
Fan Zhang
Feifei Zhang
Francis Xiatian Zhang
Gengwei Zhang
Gongyu Zhang
Guoning Zhang
Haixian Zhang
Han Zhang
Hang Zhang
Hanrong Zhang
Hanwen Zhang
Hanxiao Zhang
Hanyuan Zhang
Hao Zhang
Haochen Zhang
Haofeng Zhang
Haosen Zhang
Haoyue Zhang
Hongyuan Zhang
Huahui Zhang
Hui Zhang
Huihuang Zhang
Jiajin Zhang
Jianhai Zhang
Jianjia Zhang
Jiansong Zhang
Jianwei Zhang
Jiaqing Zhang
Jiazhen Zhang
Jing Zhang
Jingyang Zhang
Jintan Zhang
Jinwei Zhang
Kai Zhang
Ke Zhang
Kun Zhang
Kunyu Zhang
Le Zhang
Lei Zhang
Li Zhang
Lihe Zhang
Lin Zhang
Lingtong Zhang
Ling-Xiao Zhang
Liping Zhang
Lu Zhang
Miaomiao Zhang
Minghui Zhang
Minqing Zhang
Molin Zhang
Nan Zhang
Peng Zhang
Pingping Zhang
Qi Zhang
Qianni Zhang
Qibin Zhang
Qing Zhang
Rongzhao Zhang
Ruicheng Zhang
Ruipeng Zhang
Shipeng Zhang
Shu Zhang
Shuai Zhang
Shuaitong Zhang
Shugang Zhang
Song Zhang
Songyao Zhang
Teng Zhang
Tengda Zhang
Tianqiao Zhang

Tianyu Zhang
Ting Zhang
Tongxu Zhang
Wanjun Zhang
Wei Zhang
Weihang Zhang
Weitong Zhang
Weiwen Zhang
Weiyi Zhang
Wengfeng Zhang
Wenhua Zhang
Wenhui Zhang
Wenjie Zhang
Xiao Zhang
Xiaodong Zhang
Xiaoman Zhang
Xiaoming Zhang
Xiaorui Zhang
Xin Zhang
Xingpeng Zhang
Xuhong Zhang
Yanan Zhang
Yanchao Zhang
Yanteng Zhang
Ye Zhang
Yi Zhang
Yichi Zhang
Yijia Zhang
Yikun Zhang
Yilan Zhang
Yinglin Zhang
Yingying Zhang
You Zhang
Yu Zhang
Yuan Zhang
Yuanwang Zhang
Yubo Zhang
Yudong Zhang
Yue Zhang
Yuelin Zhang
Yuhan Zhang
Yukun Zhang
Yundi Zhang
Yunyan Zhang
Yuyao Zhang
Zeng Zhang
Zeyu Zhang
Zheng Zhang
Zhenxuan Zhang
Zheyuan Zhang
Zhuoneng Zhang
Zichen Zhang
Zijie Zhang
Zijun Zhang
Ziyu Zhang
Zuyu Zhang
Beidi Zhao
Bin Zhao
Bokai Zhao
Chen Zhao
Chongyue Zhao
Fengjun Zhao
Jianhui Zhao
Jichao Zhao
Jun Zhao
Junyong Zhao
Lei Zhao
Li Zhao
Meng Zhao
Na Zhao
Puyang Zhao
Qi Zhao
Shilun Zhao
Weibing Zhao
Weiqin Zhao
Wenbo Zhao
Wenjie Zhao
Wenzhao Zhao
Xiaowei Zhao
Xingzhong Zhao
Xinkai Zhao
Yang Zhao
Yawu Zhao
Yi Zhao
Yidong Zhao
Yijun Zhao
Yue Zhao
Yujing Zhao
Zhicheng Zhao
Zhihao Zhao

Zihao Zhao
Ziwei Zhao
Zixu Zhao
Bo Zhao
Boyun Zheng
Fudan Zheng
Hao Zheng
Jialun Zheng
Jieyu Zheng
Minghang Zheng
Qingqing Zheng
Shenhai Zheng
Shuai Zheng
Yalin Zheng
Yifan Zheng
Yijiang Zheng
Ying Zheng
Yubin Zheng
Aocheng Zhong
Chunlin Zhong
Gaoyan Zhong
Haiqin Zhong
Jian Zhong
Jia-Xing Zhong
Jiayang Zhong
Liang Zhong
Lujia Zhong
Tao Zhong
Wenqi Zhong
Xinliu Zhong
Xinyun Zhong
Yuan Zhong
Feixiang Zhou
Guangquan Zhou
Hangqi Zhou
Hao Zhou
Haoyin Zhou
Huajun Zhou
Jun Zhou
Kang Zhou
Lei Zhou
Linkuan Zhou
Longxi Zhou
Meihua Zhou
Meng Zhou
Mo Zhou
Nan Zhou
Qian Zhou
Qifeng Zhou
Quan Zhou
Ran Zhou
Rong Zhou
Ruixi Zhou
Shihao Zhou
Shuo Zhou
Sijin Zhou
Tao Zhou
Wei Zhou
Wu Zhou
Xiaobing Zhou
Xinrui Zhou
Xueying Zhou
Yao Zhou
Yaxuan Zhou
Youjia Zhou
Yuan Zhou
Yuanpin Zhou
Yubo Zhou
Yue Zhou
Yufei Zhou
Zhen Zhou
Zhengbo Zhou
Zhengchao Zhou
Zhenghua Zhou
Zhenhuan Zhou
Zijian Zhou
Zirui Zhou
Ziteng Zhou
Ziyu Zhou
Chunzheng Zhu
Congcong Zhu
Hancan Zhu
Hengde Zhu
Heqin Zhu
Hongjie Zhu
Jianfei Zhu
Jiayuan Zhu
Jinghua Zhu
Junchao Zhu
Lei Zhu

Lijing Zhu
Mu Zhu
Ning Zhu
Pengli Zhu
Shenghao Zhu
Simeng Zhu
Xiaohui Zhu
Xiliang Zhu
Xun Zhu
Yi Zhu
Yinghao Zhu
Yinheng Zhu
Yitao Zhu
Yuanyi Zhu
Yuanzhuo Zhu
Yuemin Zhu
Yunzheng Zhu
Zhanshi Zhu
Zhengjie Zhu
Zhiqin Zhu
Zhu Zhu
Zirui Zhu
Luoting Zhuang
Mingrui Zhuang
Yipeng Zhuang
Zhemin Zhuang
Li Zhuo
Yupeng Zhuo
Muhammad Zia Ur Rehman
Arif Hassan Zidan
Alexander Ziller
David Zimmerer
Kpêtchéhoué Merveille Santi Zinsou
Lilla Zöllei
Fangrong Zong
Dongmian Zou
Sarah Zou
Shihao Zou
Xinrui Zou
Yunmin Zou
Ziyun Zou
Maria A. Zuluaga
Wenbin Zuo
Gerald Zwettler

Contents

Adapting Foundation Models for Medical Imaging: Large Vision and Language Models (I)

An Artifact-Based Agent Framework for Adaptive and Reproducible Medical Image Processing

Lianrui Zuo[1(✉)], Yihao Liu[1], Gaurav Rudravaram[1], Karthik Ramadass[1,2], Aravind R. Krishnan[1], Michael D. Phillips[1], Yelena G. Bodien[6], Mayur B. Patel[6], Paula Trujillo[7], Yency Forero Martinez[8], Stephen A. Deppen[9,13], Eric L. Grogan[9,13], Fabien Maldonado[9], Kevin McGann[9], Hudson M. Holmes[9,10], Laurie E. Cutting[4,5], Yuankai Huo[2], and Bennett A. Landman[1,2,3,11,12]

[1] Department of Electrical and Computer Engineering, Vanderbilt University, Nashville, TN, USA
lianrui.zuo@vanderbilt.edu
[2] Department of Computer Science, Vanderbilt University, Nashville, TN, USA
[3] Department of Biomedical Engineering, Vanderbilt University, Nashville, TN, USA
[4] Peabody College, Vanderbilt University, Nashville, TN, USA
[5] Vanderbilt Brain Institute, Vanderbilt University, Nashville, TN, USA
[6] Department of Surgery, Vanderbilt University Medical Center, Nashville, TN, USA
[7] Department of Neurology, Vanderbilt University Medical Center, Nashville, TN, USA
[8] Department of Medicine, Vanderbilt University Medical Center, Nashville, TN, USA
[9] Department of Thoracic Surgery, Vanderbilt University Medical Center, Nashville, TN, USA
[10] Department of Biomedical Informatics, Vanderbilt University Medical Center, Nashville, TN, USA
[11] Department of Radiology and Radiological Sciences, Vanderbilt University Medical Center, Nashville, TN, USA
[12] Vanderbilt University Institute of Imaging Science, Nashville, TN, USA
[13] Veteran Affairs Tennessee Valley Healthcare System, Nashville, TN, USA

Abstract. Medical imaging research is increasingly shifting from controlled benchmark evaluation toward real-world clinical deployment. In such settings, applying analytical methods extends beyond model design and requires dataset-aware workflow configuration and provenance tracking. Two requirements therefore become central: **adaptability**, the ability to configure workflows according to dataset-specific conditions and evolving analytical goals; and **reproducibility**, the guarantee that all transformations and decisions are explicitly recorded and re-executable. Here, we present an artifact-based agent framework that introduces a semantic layer to augment medical image processing. The framework formalizes intermediate and final outputs through an artifact contract, enabling structured interrogation of workflow state and goal-conditioned assembly of configurations from a modular rule library. Execution is del-

G. Yang et al. (Eds.): MICCAI 2026, LNCS 16878, pp. 3–13, 2027.
https://doi.org/10.1007/978-3-032-38059-3_1

egated to a workflow executor to preserve deterministic computational graph construction and provenance tracking, while the agent operates locally to comply with most privacy constraints. We evaluate the framework on real-world clinical CT and MRI cohorts, demonstrating adaptive configuration synthesis, deterministic reproducibility across repeated executions, and artifact-grounded semantic querying. These results show that adaptive workflow configuration can be achieved without compromising reproducibility in heterogeneous clinical environments. Our code is available at https://github.com/MASILab/medimg-agent.

Keywords: CT · MRI · Clinical translation · Agent

1 Introduction

Methodological progress in medical imaging has been driven by advances in model architectures and learning paradigms, typically evaluated on curated benchmark datasets or structured research cohorts [3,6,9]. Public initiatives and grand challenges have further accelerated innovation by providing standardized datasets and controlled evaluation protocols [1,8,18–20]. These settings enable rigorous comparison of methods under well-defined experimental conditions.

In contrast, when the goal becomes applying existing analytical methods to a new research or clinical dataset, the operational landscape changes substantially. Imaging data originate in hospital and research picture archiving and communication systems (PACS), where heterogeneous file organization, mixed modalities, nested archives, and acquisition planning images are common (Fig. 1). For example, a single subject folder very often contains raw DICOM slices mixed with localizer images, reformatted series, and non-imaging documents, without standardized naming or directory structure. Transforming such data into analyzable form requires multi-stage curation, format conversion, quality control, and preprocessing. These steps often demand considerable manual and cohort-specific effort. As a result, workflow assembly becomes an expertise-driven process that is difficult to generalize and reproduce.

These challenges are further compounded in clinical environments governed by privacy regulations and data use agreements. In many settings, raw images or metadata cannot be transmitted to external settings. Ironically, the scenarios that require the greatest interactive effort for dataset adaptation are precisely those in which cloud-based models may be constrained or require additional regulatory and institutional approvals. Intelligent support systems must therefore be designed to operate within such constraints while preserving deterministic execution and traceable provenance.

To this end, two key requirements emerge. First is adaptability: the ability to reason about heterogeneous datasets and evolving analytical goals to configure appropriate processing workflows. Second is reproducibility: the guarantee that all transformations and decisions are explicitly represented and deterministically re-executable [11,14]. Existing workflow engines [2,4] ensure deterministic execution once a workflow is defined, and standardization initiatives [12] improve

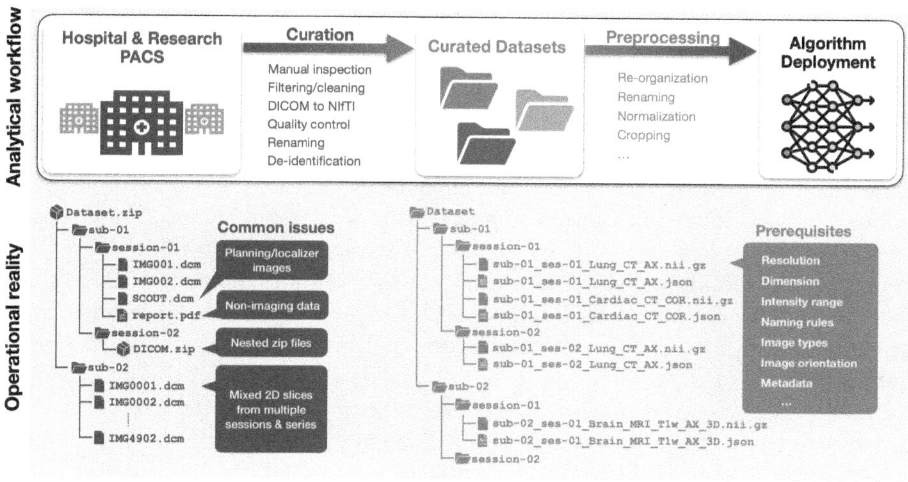

Fig. 1. The typical analytical workflow and operational reality in imaging research. Imaging data originate in hospital and research PACS, where heterogeneous file structures, mixed modalities, and nested archives are common. Extensive curation and processing are often required before most analytical algorithms can be deployed.

structural consistency. However, these tools assume pre-specified configurations and do not provide mechanisms for dataset-aware adaptation or semantic inspection of intermediate workflow state.

To bridge this gap, we propose an artifact contract-based agent framework that introduces a semantic layer above deterministic workflow execution. The framework defines an artifact contract that encodes all workflow outputs and states (intermediate and final) into structured, queryable records with explicit provenance. A semantic query module operates strictly over these contract-compliant artifacts to support dataset-level inspection. For analytical objectives, a workflow assembly module synthesizes goal-conditioned configurations from a modular rule library, enabling dataset-aware adaptation while preserving deterministic execution through an underlying workflow executor. The entire system is deployed fully locally to comply with regulations in many clinical environments. The contributions are threefold:

- An artifact-based contract that formalizes workflow state for structured and auditable reasoning;
- A constrained agent layer enabling goal-conditioned workflow assembly and artifact-grounded semantic querying;
- Empirical validation across clinical CT and MRI cohorts demonstrating adaptive configuration synthesis with deterministic reproducibility.

2 Methods

2.1 Problem Setup

Dataset, Workflow, and Artifact Contract. Let $\mathcal{D} = \{d_i\}_{i=1}^{N}$ denote a dataset consisting of N subjects, where each d_i may include multiple longitudinal imaging sessions, file types, and associated metadata. Let $\mathcal{S} = \{s_k\}_{k=1}^{K}$ denote a library of modular processing rules that a user can choose. Each rule s_k specifies a processing step and its dependencies, such as file format conversion or a segmentation algorithm. A workflow configuration is defined as $C = (\pi, \theta)$, where $\pi \subseteq \mathcal{S}$ represents an ordered subset of selected rules together with their dependency graph, and θ denotes rule-level parameters. Given dataset $\mathcal{D}$ and configuration C, execution is performed by a deterministic workflow executor $P(\mathcal{D}, C) = \mathcal{A}$, where $\mathcal{A} = \{a_j\}_{j=1}^{M}$ denotes the set of artifacts produced during execution. Artifacts comprise intermediate and final workflow outputs such as transformed images, derived measurements (e.g., classification labels), quality control indicators, and log records. Conventional workflow engines guarantee reproducibility under fixed configuration C, but assume that configurations are manually specified and that interpretation of workflow state occurs externally. They do not provide structured mechanisms for adaptive configuration to new datasets or semantic interrogation of intermediate state.

Agentic Systems. Recent advances in large language models (LLMs) have enabled agentic systems that perform goal-directed reasoning through iterative planning and tool invocation [10,13,16]. In a general agent architecture, a planner selects actions from a tool set conditioned on current state and user objectives, executes those actions, and updates memory based on resulting observations [13]. In the context of medical image processing, however, unrestricted tool invocation is insufficient: execution must remain deterministic and reproducible.

Within our formulation, the dataset $\mathcal{D}$ and artifact set $\mathcal{A}$ define the observable state; the rule library $\mathcal{S}$ defines the available tools [13]; and the configuration $C = (\pi, \theta)$ represents a structured plan over $\mathcal{S}$. We therefore seek a constrained agentic system where planning operates over defined workflow components, memory is grounded in contract-compliant artifacts $\mathcal{A}$ [10], and execution is delegated to the workflow executor $P(\mathcal{D}, C)$. This enables adaptability through configuration synthesis while preserving reproducibility through deterministic execution.

2.2 Proposed Framework

Overview We introduce a constrained artifact-based agent layer (Fig. 2) that operates above deterministic workflow execution. Each artifact $a_j \in \mathcal{A}$ is defined under an artifact contract as a tuple $a_j = (t_j, \phi_j, \psi_j)$, where t_j denotes artifact type, ϕ_j denotes structured attributes (e.g., modality, resolution, measurements), and ψ_j encodes provenance information including generating rule and dependency relationships. This contract provides a formally defined, machine-readable state representation over which semantic reasoning is grounded. On top of $\mathcal{A}$, the agent layer implements two constrained functions:

Workflow Assembly: $\Gamma_{\text{assemble}}(g, \mathcal{A}, \mathcal{S}) = C$, which synthesizes a configuration C by selecting and composing rules from $\mathcal{S}$ conditioned on the current artifact state and analytical goal g; **Semantic Querying:** $\Gamma_{\text{query}}(q, \mathcal{A}) = r$, which returns a response r derived strictly from contract-compliant artifacts to address the user's query q. These operators form a constrained agentic loop where planning and querying are performed over structured state, while execution remains delegated to the deterministic workflow executor $P(\mathcal{D}, C)$ (Fig. 2).

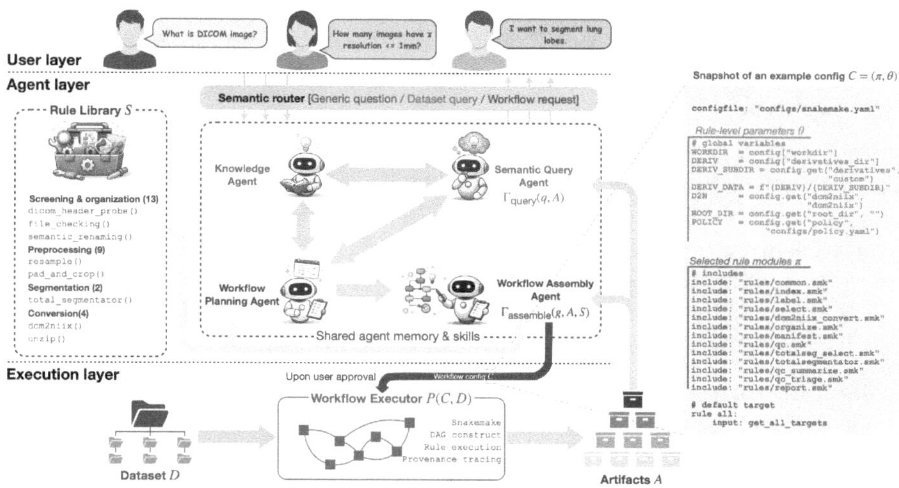

Fig. 2. Overview of the proposed agentic system. User requests are routed to either (1) Knowledge Agent (generic questions), (2) a Semantic Query Agent operating on artifacts $\mathcal{A}$, or (3) a workflow pathway for analytical goals. The Workflow Planning and Assembly Agents construct a configuration C from rule library $\mathcal{S}$ and artifacts $\mathcal{A}$, which is executed by a workflow manager to transform dataset $\mathcal{D}$ into artifacts $\mathcal{A}$.

Adaptation via Workflow Assembly. To deploy a workflow on a new dataset $\mathcal{D}_{\text{new}}$, the system first performs an inspection skill to identify available modalities, file organization, and prior processing status. This inspection summarizes observable dataset properties and becomes part of the artifact state.

Given an initial analytical request g_0, the planning component evaluates whether the objective is well-defined with respect to the current artifact state and available rule library $\mathcal{S}$. Because users may not fully know dataset characteristics or available tools, the agent may iteratively refine the request into a structured and executable goal g by clarifying required inputs, outputs, and processing assumptions. Once a feasible goal g is established, Γ_{assemble} selects rules from $\mathcal{S}$ whose declared inputs are satisfied by existing artifacts and whose outputs fulfill the objective. Dependency constraints are enforced during selection, and rule-level parameters are assigned to produce an explicit configuration $C = (\pi, \theta)$, as shown in Fig. 2. The assembled workflow and directed acyclic graph (DAG) are then presented to the user for approval before execution. Adaptation therefore occurs through goal planning and configuration synthesis, while computation remains deterministic.

Semantic Query. The semantic query function enables inspection of workflow state without re-accessing raw files. Natural-language queries are interpreted as constraints over artifact attributes ϕ_j and provenance ψ_j. For example, a query about available T1-weighted scans with specific resolution is translated into filters over modality and voxel spacing fields stored in artifacts. All responses are generated exclusively from contract-compliant artifacts and reference explicit ϕ_j and ψ_j. Because the language model operates only over structured artifact fields, responses are grounded in recorded workflow state rather than inferred from unstructured data.

Reproducibility through Workflow Executor. Once a configuration C is generated, execution is delegated to a workflow executor (`Snakemake` [4] in our implementation). The executor constructs a DAG from declared rule inputs and outputs, ensuring that dependencies are resolved prior to execution. Given fixed dataset $\mathcal{D}$, configuration C, and computational environment, execution is deterministic. Intermediate and final artifacts are produced only when declared inputs are satisfied, and rule-level provenance is recorded automatically.

Implementation Details. The agent operates locally via `Ollama` using the DeepSeek-R1 (14B) model in inference mode. Workflow rules are organized as modular `.smk` components and registered in a rule catalog that explicitly declares inputs, outputs, and parameter interfaces. Artifacts are stored as structured `.json` or `.csv` records consistent with the artifact contract. Workflow execution is performed using Snakemake (v7.x), which constructs the DAG from the generated configuration and executes tasks with file-based dependency tracking. Experiments were conducted on a Linux workstation equipped with an NVIDIA A6000 GPU for downstream model inference.

3 Experiments and Results

We evaluate whether the proposed framework (i) preserves deterministic execution under repeated runs, (ii) assembles goal-conditioned workflows across varying cohorts, and (iii) supports artifact-grounded semantic query without re-running. Experiments were conducted on three cohorts including both research and clinical settings (Table 1): `NLST` (research CT; moderate heterogeneity), `LungCaTrial` (clinical trial CT; high heterogeneity), and `BrainICU` (research MRI; moderate heterogeneity). Across cohorts, we test our method under multiple analytical goals including data conversion/curation and downstream analysis (i.e., lung lobe and nodule segmentation [15], cancer risk estimation [7], harmonization [21], and brain segmentation [17]), resulting in 9 different workflows.

Reproducibility. For each dataset and analytical goal, we invoke the agent twice after clearing memory and intermediate states, then compare the resulting

Table 1. Dataset characteristics and quantitative evaluation of reproducibility and adaptability. For each dataset and analytical goal, we report the number of rules in the domain-expert ground-truth configuration, reproducibility via DAG equivalence across repeated executions, and adaptability metrics†.

Dataset Characterization						Assembly		Reprod.	Adaptability		
Dataset	Modality	Cohort	Heteroge.	#subj	#ses	Analytical goal	#Rules	DAG	IRM(%)↑	PL↓	FO(%)↑
`NLST` [5]	CT	Research	Mid.	200	592	convert+curate	11	✓	100	1	98.8
						lung lobe seg	14	✓	85.7	4	97.9
						lung nodule seg	14	✓	100	3	97.4
`LungCaTrial`	CT	Clinical	High	636	2,950	convert+curate	12	✓	100	2	99.0
						lung nodule seg	15	✓	100	4	90.7
						lung cancer risk	17	✓	88.2	6	91.1
`BrainICU`	MRI	Research	Mid.	113	135	convert+curate	17	✓	94.1	5	100
						harmonization	19	✓	100	8	97.7
						brain seg	20	✓	100	8	96.3

Example: Goal Planning (`NLST`, lung nodule seg)

```
User > I want to segment the lung nodules in this dataset.
Agent> Current status: DICOMs have been converted to NIFTI through dcm2niix (registered rule
0.1.3a). Needs confirmation: 1. Kernel. I see multiple kernels of the same session. Do you want
to segment them all? 2. Resolution. Do you want to choose the best resolution? ...
User > 1. Segment them all. 2. Choose the best resolution (in z-dimension) ···
Agent> This is an updated workflow with rules ... Assumptions made: [omit to save space]. If
you approve, I will assemble and execute.
```

†**Abbreviations: IRM:** Initial Rule Matching. Percentage of rules matched between agent's first selection and ground truth; **PL:** Number of planning iterations (chats from user) before a workflow is assembled by the agent; **FO:** Percentage of sessions whose final outputs satisfy the intended goal against manually curated results.

workflow graphs. Reproducibility is measured by DAG equivalence between the two runs. Across all settings, the induced DAGs were identical (Table 1), indicating that, under fixed goals and environment, the planning-to-configuration pathway yields consistent executable workflows and deterministic execution graphs.

Adaptability. We measure how well the agent adapts workflows to each cohort and analytical goal using three metrics. First, Initial Rule Matching (IRM) quantifies overlap between the agent's initial proposed rule set and a domain-expert configuration, reflecting the quality of dataset-aware planning immediately after inspection. Second, Planning Iterations (PL) counts the number of user-agent exchanges before finalizing a configuration, capturing interaction cost. Third, Final Output (FO) reports the percentage of sessions whose outputs satisfy the intended goal against manually curated results.

As shown in Table 1, the agent's initial plans are close to expert-defined pipelines, with IRM> 85% across all cohorts and analytical goals. Additionally, iterative planning resolves ambiguities in user intent (e.g., which series to target), as shown in Table 1 example. This is reflected by modest PL values (a workflow is accurately assembled after 1 to 8 rounds of chats between agent and user) even in the clinical cohorts, while maintaining high session-level success in most tasks. We then evaluated all failure cases: mismatches are concentrated in cases of missing/corrupted data and ambiguous series selection (e.g., multiple candidate series satisfying similar constraints), which typically require manual intervention even in conventional workflows.

Table 2. Semantic query evaluation (`deepseek-r1:14b` and `qwen2.5:14b`) and ablation experiment with/without artifact contract. Accuracy denotes exact-match percentage over 20 queries per category.

Dataset	Model	Contract?	Status	Filter	Provenance
NLST	DeepSeek	✗	95%	20%	80%
	Qwen	✓	100%	90%	100%
	DeepSeek	✓	100%	95%	100%
LungCaTrial	DeepSeek	✗	90%	15%	80%
	Qwen	✓	100%	85%	100%
	DeepSeek	✓	100%	90%	100%
BrainICU	DeepSeek	✗	95%	10%	75%
	Qwen	✓	95%	90%	100%
	DeepSeek	✓	100%	90%	100%

Example (Filter/counting)

```
User > How many NIfTI images were
acquired using a Siemens scanner and
have slice thickness greater than 1 mm?
Agent> 23. File paths can be retrieved
from artifact records in
data_inventory.csv and qa_report.csv.
```

Semantic Query. We evaluate semantic querying across three categories: status, filtering/counting, and provenance. For each category, 20 representative questions are evaluated using exact-match accuracy against ground truth. We used two local LLMs in this experiment (Table 2). Status queries are generally answered correctly, as they simply require checking whether certain files exist (e.g., "Has lung nodule segmentation been executed for subject X?"). Filtering/counting queries are more demanding because they require structured attribute reasoning, such as checking specific voxel resolution or scanner manufacturer. Provenance queries require tracing rule-level dependencies, for example identifying which preprocessing steps generated a particular result. The proposed method accurately answers all status and provenance queries, and only made one to two mistakes in dataset query questions. Upon investigating error cases, we found that most failures were caused by incomplete or corrupted DICOM headers, leading to missing or inaccurate artifact attributes ϕ_j. For example, body part was inferred from DICOM headers, and this process can be unreliable. These errors therefore reflect limitations of upstream metadata rather than ungrounded generation. Incorporating additional skills, such as image-based body-part regression, may further improve robustness.

Ablation Study. To isolate the effect of artifact contract, we perform an ablation in which the agent is denied access to the artifact registry and instead receives only directory listings and raw filenames, mimicking a conventional manual scripting workflow. The same query set is evaluated under this condition.

After removing the artifact contract, the largest degradation occurs in filtering/counting queries. For example, when asked to identify sessions with voxel resolution below a specified threshold, filename-based reasoning fails because resolution is not encoded in directory names; it is only available in structured artifact fields extracted during preprocessing. Provenance queries also degrade without artifact access. When asked which rule generated a particular segmentation output, the grounded system retrieves the recorded rule identifier and dependency chain from ψ_j, whereas the filename-only condition lacks explicit traceability and must rely on heuristic inference. These results demonstrate the importance of contract-compliant artifacts. The artifact contract therefore serves not only as a reproducibility mechanism but also as a grounding interface that reduces inference over unstructured file organization.

4 Discussion and Conclusion

We presented an artifact contract-based agent framework for adaptive and reproducible medical image processing across heterogeneous research and clinical datasets. The system separates semantic planning from deterministic execution: workflow configurations are synthesized at the semantic layer, while execution remains reproducible through a workflow manager with explicit provenance tracking. Experiments show that configurations can be generated across diverse CT and MRI cohorts, and that repeated runs produce identical DAGs. In our ablation study, we show that the artifact contract improves the semantic query accuracy. The main limitation arises from incomplete artifact attributes. In our experiments, most query errors were caused by missing or corrupted metadata, leading to inaccurate artifact fields. This highlights the dependence of semantic query on upstream data quality. Similarly, adaptability is bounded by the coverage of the rule catalog; unsupported analytical objectives require additional rules. Additionally, a more detailed analysis on the computational overhead and time cost during user–agent communication is valuable for the clinical application of the proposed framework. Nevertheless, the framework is designed for extension. New rules and skills can be added without altering the overall architecture, enabling progressive enrichment of the artifact schema and planning capabilities. By grounding interaction in structured workflow state and preserving deterministic execution, the proposed system provides a practical and extensible foundation for adaptive medical image analysis in real-world settings.

Acknowledgments. We sincerely thank Adam M. Saunders and Michael E. Kim for sharing their experience with `Snakemake`, which contributed to the development of this work. This research was funded by the National Cancer Institute (NCI) grant R01-CA253923-04 and R01-CA253923-04S1. This work was also supported by the following awards: National Science Foundation CAREER1452485; U01CA196405; UL1-RR024975-01 from the National Center for Research Resources and UL1-TR000445-06 from the National Center for Advancing Translational Sciences; the Martineau Innovation Fund grant through the Vanderbilt-Ingram Cancer Center Thoracic Working Group; and NCI Early Detection Research Network grant 2U01-CA152662. Additional support was provided by U01-CA152662, SNF300329 (Helmsley), NSF NAIRR Pilot 240468, and P50-HD103537. During the development of this work, we used generative AI to create code segments from task descriptions and to assist with debugging, editing, and code autocompletion. In addition, generative AI technologies were used to help structure sentences and perform grammatical checks. We emphasize that the conceptualization, ideation, and all prompts provided to the AI originated entirely from the authors' intellectual and creative contributions. The authors take full responsibility for reviewing and validating all AI-generated content included in this work.

Disclosure of Interest. The authors have no competing interests to declare that are relevant to the content of this article.

References

1. Chen, J., et al.: Beyond the LUMIR challenge: the pathway to foundational registration models. arXiv preprint arXiv:2505.24160 (2025)
2. Di Tommaso, P., Chatzou, M., Floden, E.W., Barja, P.P., Palumbo, E., Notredame, C.: Nextflow enables reproducible computational workflows. Nat. Biotechnol. **35**(4), 316–319 (2017)
3. Ferrucci, L., Resnick, S.M., Deal, J.A.: Baltimore Longitudinal Study of Aging (BLSA). Hearing Loss Rehabilitation and Higher-Order Auditory and Cognitive Processing, p. 116 (2023)
4. Köster, J., Rahmann, S.: Snakemake–a scalable bioinformatics workflow engine. Bioinformatics **28**(19), 2520–2522 (2012)
5. Kramer, B.S., Berg, C.D., Aberle, D.R., Prorok, P.C.: Lung cancer screening with low-dose helical CT: results from the National Lung Screening Trial (NLST) (2011)
6. LaMontagne, P.J., et al.: OASIS-3: longitudinal neuroimaging, clinical, and cognitive dataset for normal aging and Alzheimer disease. medrxiv, pp. 2019–12 (2019)
7. Li, T.Z., et al.: Time-distance vision transformers in lung cancer diagnosis from longitudinal computed tomography. In: Medical Imaging 2023: Image Processing, vol. 12464, pp. 229–238. SPIE (2023)
8. Menze, B.H., et al.: The multimodal brain tumor image segmentation benchmark (BRATS). IEEE Trans. Med. Imaging **34**(10), 1993–2024 (2014)
9. Naseer, I., Akram, S., Masood, T., Jaffar, A., Khan, M.A., Mosavi, A.: Performance analysis of state-of-the-art CNN architectures for LUNA16. Sensors **22**(12), 4426 (2022)
10. Park, J.S., O'Brien, J., Cai, C.J., Morris, M.R., Liang, P., Bernstein, M.S.: Generative agents: interactive simulacra of human behavior. In: Proceedings of the 36th Annual ACM Symposium on User Interface Software and Technology, pp. 1–22 (2023)
11. Peng, R.D.: Reproducible research in computational science. Science **334**(6060), 1226–1227 (2011)
12. Poldrack, R.A., et al.: The past, present, and future of the brain imaging data structure (BIDS). Imaging Neurosci. **2**, imag–2 (2024)
13. Schick, T., et al.: Toolformer: language models can teach themselves to use tools. Adv. Neural. Inf. Process. Syst. **36**, 68539–68551 (2023)
14. Stodden, V., et al.: Enhancing reproducibility for computational methods. Science **354**(6317), 1240–1241 (2016)
15. Wasserthal, J., et al.: TotalSegmentator: robust segmentation of 104 anatomic structures in CT images. Radiol. Artif. Intell. **5**(5), e230024 (2023)
16. Yao, S., et al.: React: synergizing reasoning and acting in language models. In: The Eleventh International Conference on Learning Representations (2022)

17. Yu, X., et al.: Unest: local spatial representation learning with hierarchical transformer for efficient medical segmentation. Med. Image Anal. **90**, 102939 (2023)
18. Zhang, J., et al.: Harmonization-enriched domain adaptation with light fine-tuning for multiple sclerosis lesion segmentation. In: Medical Imaging 2024: Clinical and Biomedical Imaging, vol. 12930, pp. 633–639. SPIE (2024)
19. Zhang, J., et al.: Uniself: a unified network with instance normalization and self-ensembled lesion fusion for multiple sclerosis lesion segmentation. Med. Image Anal. 103954 (2026)
20. Zuo, L., et al.: Synthesizing realistic brain MR images with noise control. In: Burgos, N., Svoboda, D., Wolterink, J.M., Zhao, C. (eds.) SASHIMI 2020. LNCS, vol. 12417, pp. 21–31. Springer, Cham (2020). https://doi.org/10.1007/978-3-030-59520-3_3
21. Zuo, L., et al.: HACA3: a unified approach for multi-site MR image harmonization. Comput. Med. Imaging Graph. **109**, 102285 (2023)

APEX-SAM: Anatomy-Aware Prompting with Expert Retrieval for Training-Free Medical Image Segmentation

Zhihao Mao, Bangpu Chen, Qi Lei, Jiaqi Tan, and Kun Sun(✉)

School of Computer Science, China University of Geosciences (Wuhan), Wuhan, China
{chenbangpu,kevinlei,sunkun}@cug.edu.cn

Abstract. Training-free cross-domain few-shot medical image segmentation aims to segment unseen anatomies without parameter updates, addressing the high cost of dense annotation and domain-specific fine-tuning in clinical practice. Existing support-driven prompting methods face three limitations: support exemplars are randomly selected without quality assurance, geometric alignment is poorly modeled, and multi-modal prompt capabilities remain underexploited. We present APEX-SAM, a retrieval-augmented framework with three innovations. Quality-aware expert retrieval (QAR) builds a dual-stream DINO/SigLIP expert bank with diversity-aware selection to ensure support-query compatibility. Anatomy-aware prompt mining (APM) performs style-aligned geometric matching and anatomy-guided point sampling from morphological priors. Hybrid multi-modal fusion (HMF) fuses three SAM branches (point, text, and box prompts) via training-free feature-consensus weighting. Experiments on three cross-domain benchmarks confirm state-of-the-art performance among training-free methods, with ablations validating each component's contribution.

Keywords: Cross-domain segmentation · Few-shot learning · Retrieval-augmented prompting · SAM · Training-free

1 Introduction

Medical image segmentation is central to diagnosis, intervention planning, and longitudinal assessment in routine clinical workflows [2]. Accurate organ and lesion delineation enables quantitative biomarker extraction, treatment response monitoring, and surgical planning. However, its main practical bottleneck is expensive dense annotation, especially for rare diseases and newly adopted protocols [22]. When combined with domain gaps across modalities (e.g., MRI vs. CT), scanners, and institutions, this makes cross-domain few-shot segmentation both clinically urgent and technically difficult.

Z. Mao and B. Chen—Equal contribution.
Project page: https://trump0412.github.io/APEX-SAM/.

G. Yang et al. (Eds.): MICCAI 2026, LNCS 16878, pp. 14–23, 2027.
https://doi.org/10.1007/978-3-032-38059-3_2

Recent methods fall into two paradigms: those requiring training and those that are training-free. Training-based methods achieve strong accuracy via prototype matching, meta-learning, and domain adaptation [5,6,10,15–17]. These methods typically learn transferable representations from large labeled source datasets and adapt them to new targets via iterative optimization. However, they rely heavily on source domain supervision and repeated fine-tuning for each new target domain. Moreover, robustness often degrades under large clinical distribution shifts, and deployment requires substantial computational resources and domain expertise.

Within the training-free branch, support-driven prompting has become a promising direction for fast adaptation. Representative methods such as MAUP, SynPo, and OSAM build prompts from support-query correspondence without parameter updates, which is particularly attractive for high-frequency clinical environments where rapid deployment and minimal infrastructure are essential [11,18,25]. These methods leverage pre-trained foundation models (e.g., SAM [8]) and construct prompts from a small set of support exemplars at test time, enabling immediate inference on unseen anatomies without retraining.

Current support-driven pipelines still face three unresolved bottlenecks: inadequate support quality assurance, poor support-query alignment modeling, and underused multi-modal prompt signals [4,20,23,25]. These issues directly limit stability when support and query are weakly aligned in real cross-domain settings. First, support samples are often selected randomly without explicit quality control or diversity constraints. Second, direct mask transfer ignores geometric misalignment between domains, and existing point-prompting strategies do not leverage anatomical structure priors. Third, although modern foundation models such as SAM3 [4] support rich multi-modal prompts, current methods rely on single-modal prompts and lack principled fusion strategies.

Medical images offer rich exploitable priors that are often overlooked by existing methods. For instance, 3D scans exhibit strong inter-slice morphological continuity due to organ shape consistency, and cross-patient data share semantic consistency in anatomical structures despite appearance variations. Together with recent advances in promptable segmentation and retrieval-augmented generation, this opens a practical path for training-free inference without retraining [4]. The key insight is to unify quality-aware retrieval, geometry-guided adaptation, and multi-modal prompt orchestration in a single inference pipeline that exploits both data priors and model capabilities.

To this end, we propose **APEX-SAM**, a training-free retrieval-augmented framework for cross-domain few-shot medical image segmentation. Our contributions are threefold:

- **QAR**: a dual-stream DINO [13]/SigLIP [21] expert bank with diversity-aware quality scoring and two-level retrieval.
- **APM**: wavelet style normalisation, semantic gating, orientation-aware geometric alignment, and anatomy-guided prompt generation.
- **HMF**: point, text, and box prompts as three independent SAM branches, fused by reliability-weighted feature consensus.

APEX-SAM outperforms the strongest training-free baseline by an average of 21.1 pp mean Dice across three cross-domain benchmarks. Code and project materials are available on the project page.

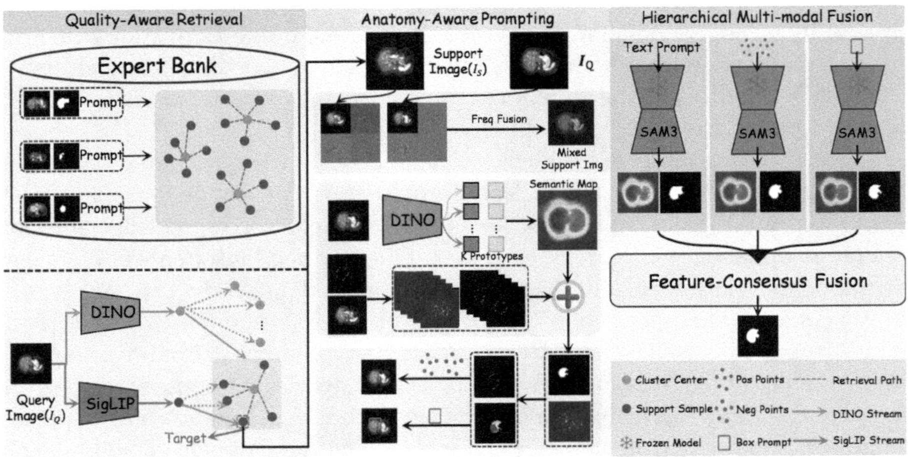

Fig. 1. Overview of APEX-SAM. QAR (left) builds a hierarchical expert bank with DINO/SigLIP dual-stream indexing. APM (middle) performs style alignment, semantic gating, orientation-aware geometric alignment, and anatomy-aware point sampling. HMF (right) fuses three SAM3 prompt branches via feature-consensus weighting.

2 Method

Problem Setting. We study training-free cross-domain few-shot medical image segmentation with a frozen foundation model [4,8]. Given a query image $\mathbf{I}_q$ and a textual target description t_q, the goal is to predict a binary segmentation mask without any gradient update. Formally, we require a mapping

$$\hat{\mathbf{M}}_q = \mathcal{F}(\mathbf{I}_q, t_q, \mathcal{D}), \tag{1}$$

where $\mathcal{D}$ is an external expert bank and $\mathcal{F}$ is entirely executed at inference time. Each entry $e_i = (I_i, M_i, t_i, z_i^D, z_i^S, q_i)$ stores the support image, mask, text label, a DINO structural key z_i^D, a SigLIP multi-modal key z_i^S, and a quality score q_i. Our method instantiates three principles as **QAR**, **APM**, and **HMF**: retrieval before prompting, geometry-guided alignment before transfer, and multi-branch prompt fusion. Figure 1 provides an end-to-end overview.

2.1 Quality-Aware Expert Retrieval (QAR)

Expert Bank Construction. Naively pooling all annotated images creates a redundant bank dominated by easy, similar cases; we instead cluster supports by DINO [13] structural keys $\{z_i^D\}$ and greedily select entries per cluster using a quality-coverage-diversity score. The SigLIP [21] retrieval key for each candidate is formed as

$$z_i^S = \mathrm{Norm}(\alpha\, v_i + (1-\alpha)\, u_i)\,, \tag{2}$$

where $v_i = \mathrm{SigLIP}_{\mathrm{img}}(I_i)$ captures visual appearance and $u_i = \mathrm{SigLIP}_{\mathrm{txt}}(t_i)$ encodes the anatomy label, ensuring cross-modal alignment in the retrieval index. Inside each cluster k with selected subset $\mathcal{S}_k$ and centroid μ_k^S, a candidate i is scored by

$$\pi(i \mid \mathcal{S}_k) = q_i + \cos\bigl(z_i^S, \mu_k^S\bigr) + \min_{j\in\mathcal{S}_k}\bigl(1 - \cos\bigl(z_i^S, z_j^S\bigr)\bigr)\,, \tag{3}$$

where the three equally-weighted terms reward annotation quality, cluster representativeness, and diversity relative to already-selected supports, respectively. This ensures the bank is compact yet covers the intra-class appearance variation arising from different imaging protocols and scanner vendors.

Two-Level Retrieval at Inference. Given a query, QAR routes it to the top-L clusters by DINO structural similarity r_k. QAR then re-ranks the candidate supports by multi-modal similarity

$$s_{qi} = \cos\bigl(z_q^S,\, z_i^S\bigr)\,, \quad \mathcal{S}_q = \mathrm{TopM}(\{s_{qi}\})\,, \tag{4}$$

which fuses structural and semantic cues into a single retrieval score; the two-level index keeps cost sublinear in bank size while multi-modal keys prevent structural matches from overriding semantic misalignment. For append-only bank growth—built offline from source-domain ground-truth masks—a new entry is inserted only when its maximum cosine similarity to existing cluster members falls below β, with the cluster centre updated online in $\mathcal{O}(1)$ time; at inference the append path uses the predicted mask and SAM confidence as quality proxy, requiring no target-domain annotation.

2.2 Anatomy-Aware Prompt Mining (APM)

Cross-Domain Style Normalisation. Cross-modality intensity mismatches corrupt feature comparisons, so we apply a one-level Haar DWT and swap the support's low-frequency sub-band with the query's:

$$\tilde{I}_s = \mathrm{IDWT}(LL_q,\, LH_s,\, HL_s,\, HH_s)\,. \tag{5}$$

This preserves organ-boundary detail in the high-frequency sub-bands while removing global intensity shift.

Semantic Gating. We cluster support foreground features into $K_r = 8$ DINO prototypes $\{c_k\}$ and build a binary gate

$$G(x) = \mathbb{1}\bigl[\max_k \cos\bigl(F_q(x),\, c_k\bigr) > \eta\bigr], \tag{6}$$

masking out query pixels with low semantic affinity to the support, suppressing false-positive activations before geometric alignment.

Orientation-Aware Geometric Alignment. Even after style normalisation, organ scale, orientation, and boundary curvature may still differ across patients. We find the best-fitting transform T^* by minimising a directional chamfer cost over the support boundary:

$$T^* = \arg\min_{T \in \mathcal{T}} \frac{1}{|B_s|} \sum_{(x,\theta)\in B_s} \mathrm{DT}_{k(\theta+r)}(T(x)), \tag{7}$$

where $\mathrm{DT}_{k(\cdot)}$ is the oriented distance transform binned by edge direction. Unlike isotropic Chamfer matching, the directional penalty yields finer shape correspondence on thin structures such as myocardium and kidney cortex. The aligned pre-mask is then $M_{\text{pre}} = T^*(M_s) \cap G$, confining warped support shape to semantically gated query regions.

Prompt Generation. From M_{pre} we extract a bounding box b, positive points at Voronoi cell centroids [1] of the foreground for spatial diversity, and negative points at the lowest-affinity pixel in each boundary sector—placing negative cues in the anatomically adjacent tissue most prone to false-positive activation.

2.3 Hybrid Multi-modal Fusion (HMF)

Three-Branch Inference. We exploit the multi-prompt interface by running SAM once per modality: Branch A uses geometry-only points $\{P^+, P^-\}$; Branch B supplies the anatomy text label t_s; Branch C uses the bounding box b from M_{pre}. Each branch produces last-layer logits M^b and feature maps F^b ($b \in \{po, tp, bx\}$).

Training-Free Reliability-Weighted Fusion. Rather than treating all branches equally, we score each by its cosine agreement with the per-pixel mean feature $\bar{F}$: $q_b = |\Omega|^{-1} \sum_x \cos(F^b(x), \bar{F}(x))$, then normalised into weights w_b via softmax with temperature γ. The final prediction is

$$\hat{M}_q = \sigma\left(\sum_{b\in\{po,tp,bx\}} w_b \cdot \mathrm{logit}\big(\mathrm{clip}(M^b, \epsilon, 1-\epsilon)\big) \right), \tag{8}$$

where ϵ clips probabilities away from $\{0, 1\}$ to prevent log-odds divergence, and no learned parameters are required.

3 Experiments

3.1 Experimental Setup

We evaluate APEX-SAM on Abd-MRI, Abd-CT, and Card-MRI [7,9,26] under a cross-domain 5-shot protocol: a unified expert bank is built offline from all three source-domain splits (patient-level leave-one-out); at inference, retrieval is label-filtered to entries matching the queried anatomy, and no target-domain labels are used. Dice is reported as the per-class mean over 1000 episodes (Fig. 2).

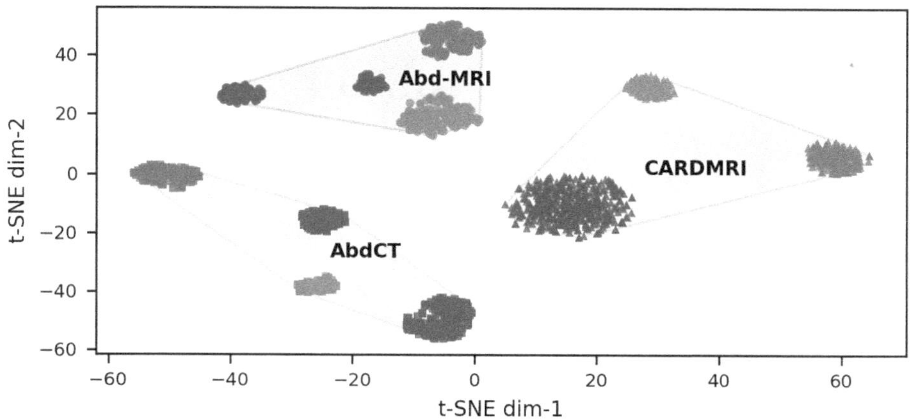

Fig. 2. Dataset visualization. Samples and DINO-feature t-SNE (one point per slice) for Abd-MRI, Abd-CT, and Card-MRI, illustrating cross-domain appearance variation.

Table 1. Quantitative comparison (Dice %) on Abd-MRI and Abd-CT. Best in **bold**, second underlined.

Method	Ref.	Abd-MRI					Abd-CT				
		Liver	LK	RK	Spleen	Mean	Liver	LK	RK	Spleen	Mean
PANet	ICCV'19	39.24	26.47	37.35	26.79	32.46	40.29	30.61	26.66	30.21	31.94
SSL-ALP	TMI'22	70.74	55.49	67.43	58.39	63.01	71.38	34.48	32.32	51.67	47.46
RPT	MICCAI'23	49.22	42.45	47.14	48.84	46.91	65.87	40.07	35.97	51.22	48.28
PATNet	ECCV'22	57.01	50.23	53.01	51.63	52.97	75.94	46.62	42.68	63.94	57.29
IFA	CVPR'24	50.22	35.99	34.00	42.21	40.61	46.62	25.13	26.56	24.85	30.79
FAMNet	AAAI'25	73.01	57.28	<u>74.68</u>	58.21	65.79	73.57	57.79	61.89	<u>65.78</u>	64.75
MAUP	MICCAI'25	<u>78.16</u>	<u>58.23</u>	72.34	<u>59.65</u>	<u>67.09</u>	<u>78.25</u>	<u>59.41</u>	<u>71.80</u>	60.38	<u>67.46</u>
APEX-SAM	Ours	**95.10**	**96.88**	**96.01**	**95.23**	**95.81**	**93.47**	**90.28**	**91.83**	**92.06**	**91.91**

Our implementation uses DINO/SigLIP retrieval keys, SAM3 decoding, and rigid/similarity geometric alignment; q_i is mean SAM mask-confidence over foreground pixels; query/support slices are patient-disjoint. Hyperparameters: $L = 3$, $M = 5$, $\alpha = 0.6$, $\eta = 0.94$, $\gamma = 0.1$, $N_v = 6$, $N_s = 8$.

3.2 Main Results

APEX-SAM achieves consistent gains over all baselines on three benchmarks; Tables 1 and 2 compare against PANet [17], SSL-ALP [14], RPT [24], PATNet [19], IFA [12], FAMNet [3], and MAUP [25].

Table 2. Quantitative comparison (Dice %) on Card-MRI. Best in **bold**, second underlined.

Method	Ref.	LV-BP	LV-MYO	RV	Mean
PANet	ICCV'19	51.42	25.75	25.75	36.66
SSL-ALP	TMI'22	83.47	22.73	66.21	57.47
RPT	MICCAI'23	60.84	42.28	57.30	53.47
PATNet	ECCV'22	65.35	50.63	68.34	61.44
IFA	CVPR'24	50.43	31.32	30.74	37.50
FAMNet	AAAI'25	86.64	51.82	76.26	71.58
MAUP	MICCAI'25	88.36	52.74	78.29	73.13
APEX-SAM	Ours	**92.75**	**68.41**	**88.23**	**83.13**

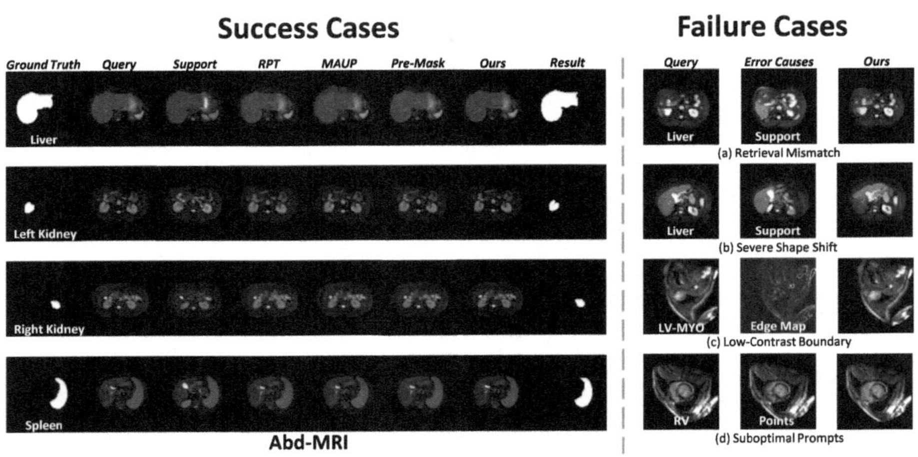

Fig. 3. Qualitative and failure cases. Left: successful cross-domain masks. Right: representative failure modes.

Table 1 and Table 2 report per-class Dice on all three benchmarks; abdominal gains are larger because cross-modality appearance divergence (MRI vs. CT) degrades direct feature matching, while cardiac structures such as LV-MYO have thin, irregular morphology that limits the benefit of geometric alignment.

Figure 3 visualises successful cross-domain masks alongside four failure modes: low-contrast boundaries (27%), severe shape shift (35%), retrieval mismatch (23%), and suboptimal prompts (15%).

3.3 Ablation Study

Module-wise gains are +7.8 pp for QAR, +6.1 pp for APM, and +5.5 pp for HMF; the thresholded append-only memory update contributes an additional +4.0 pp by continuously enriching the bank with high-confidence inference samples (Table 3).

Table 3. Ablation on core designs and memory update strategy (Dice %).

Configuration	Enabled Modules			Memory Rule	Mean Dice
	QAR	APM	HMF		
Prompt-only baseline	✗	✗	✗	–	72.4
+ QAR	✓	✗	✗	Fixed	80.2
+ QAR + APM	✓	✓	✗	Fixed	86.3
+ QAR + APM + HMF	✓	✓	✓	Fixed	91.8
+ **Full + thresholded append-only (ours)**	✓	✓	✓	Thresholded append	**95.81**

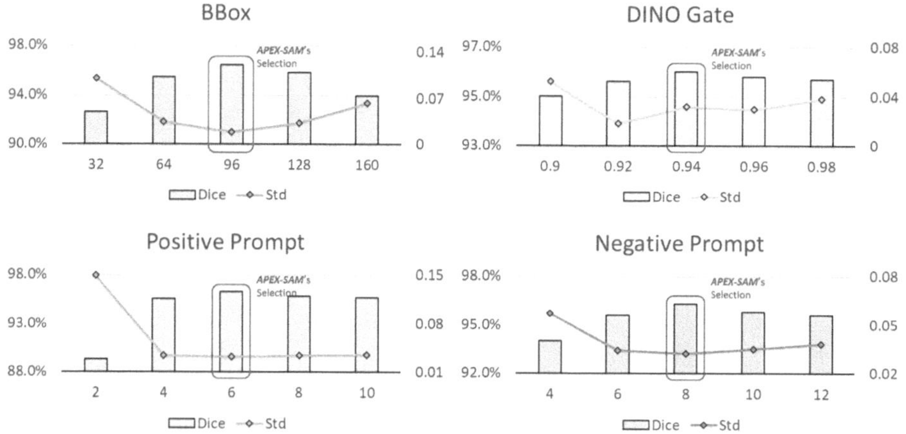

Fig. 4. Hyperparameter sensitivity. Mean Dice (bars, left axis) and standard deviation (lines, right axis) on Abd-MRI; each parameter is swept independently. Red boxes mark the selected values. (Color figure online)

3.4 Hyperparameter Sensitivity

Figure 4 sweeps key hyperparameters on Abd-MRI; Dice remains stable within ±one step of each selected value, confirming robust deployment across varied clinical scenarios.

4 Conclusion

APEX-SAM is a training-free framework for cross-domain few-shot medical image segmentation that chains quality-aware expert retrieval, anatomy-aware prompt mining, and hybrid multi-modal fusion without any gradient update. Experiments on three benchmarks confirm that reliable support selection, geometry-aware alignment, and complementary prompt fusion are key to robust cross-domain adaptation, while the append-only bank and frozen parameters reduce deployment barriers and facilitate clinical adoption without infrastructure overhead. The current implementation still assumes a known target anatomy label and invokes multiple SAM prompt branches, which increases inference cost relative to single-pass prompting. Future work will explore multi-class and 3D extensions with uncertainty-aware retrieval.

Acknowledgments. This work was supported by the National Natural Science Foundation of China (No. 62176242) and Hubei Provincial Natural Science Foundation of China (No. 2025AFB832).

Disclosure of Interests. The authors have no competing interests to declare that are relevant to the content of this article.

References

1. Aurenhammer, F.: Voronoi diagrams: a survey of a fundamental geometric data structure. ACM Comput. Surv. (CSUR) **23**(3), 345–405 (1991)
2. Azad, R., et al.: Medical image segmentation review: the success of u-net. IEEE Trans. Pattern Anal. Mach. Intell. **46**(12), 10076–10095 (2024)
3. Bo, Y., Zhu, Y., Li, L., Zhang, H.: Famnet: frequency-aware matching network for cross-domain few-shot medical image segmentation. In: AAAI Conference on Artificial Intelligence (2025)
4. Carion, N., et al.: SAM 3: segment anything with concepts. In: International Conference on Learning Representations (2026). https://openreview.net/forum?id=r35clVtGzw
5. Cheng, Z., Wang, S., Xin, T., Zhou, T., Zhang, H., Shao, L.: Few-shot medical image segmentation via generating multiple representative descriptors. IEEE Trans. Med. Imaging **43**(6), 2202–2214 (2024)
6. Cheng, Z., Zhu, Y., Wang, S., Xin, T., Zhang, H.: Cascaded alternating refinement transformer for few-shot medical image segmentation. ACM Trans. Intell. Syst. Technol. (2025)
7. Kavur, A.E., et al.: Chaos challenge-combined (CT-MR) healthy abdominal organ segmentation. Med. Image Anal. **69**, 101950 (2021)
8. Kirillov, A., et al.: Segment anything. In: Proceedings of the IEEE/CVF International Conference on Computer Vision, pp. 4015–4026 (2023)
9. Landman, B., Xu, Z., Igelsias, J., Styner, M., Langerak, T., Klein, A.: MICCAI multi-atlas labeling beyond the cranial vault–workshop and challenge. In: Proceedings of the MICCAI—Workshop Challenge, vol. 5, p. 12 (2015)
10. Lei, S., Zhang, X., He, J., Chen, F., Du, B., Lu, C.T.: Cross-domain few-shot semantic segmentation. In: European Conference on Computer Vision, pp. 73–90 (2022)
11. Liu, Y., et al.: Synpo: boosting training-free few-shot medical segmentation via high-quality negative prompts (2025). https://arxiv.org/abs/2506.15153
12. Nie, J., et al.: Cross-domain few-shot segmentation via iterative support-query correspondence mining. In: IEEE/CVF Conference on Computer Vision and Pattern Recognition, pp. 3380–3390 (2024)
13. Oquab, M., et al.: Dinov2: learning robust visual features without supervision. arXiv preprint arXiv:2304.07193 (2023)
14. Ouyang, C., Biffi, C., Chen, C., Kart, T., Qiu, H., Rueckert, D.: Self-supervised learning for few-shot medical image segmentation. IEEE Trans. Med. Imaging **41**(7), 1837–1848 (2022)

15. Snell, J., Swersky, K., Zemel, R.: Prototypical networks for few-shot learning. In: Advances in Neural Information Processing Systems, p. 30 (2017)
16. Tang, S., et al.: Few-shot medical image segmentation with high-fidelity prototypes. Med. Image Anal. **100**, 103412 (2025)
17. Wang, K., Liew, J.H., Zou, Y., Zhou, D., Feng, J.: Panet: few-shot image semantic segmentation with prototype alignment. In: proceedings of the IEEE/CVF International Conference on Computer Vision, pp. 9197–9206 (2019)
18. Wang, R., Yang, Z., Song, Y.: Osam-fundus: a training-free, one-shot segmentation framework for optic disc and cup in fundus images. Biomed. Signal Process. Control **100**, 107069 (2025)
19. Wu, H., Xiao, F., Liang, C.: Dual contrastive learning with anatomical auxiliary supervision for few-shot medical image segmentation. In: European Conference on Computer Vision, pp. 417–434 (2022)
20. Yang, Y., Fang, X., Li, X., Han, Y., Yu, Z.: CDSG-SAM: a cross-domain self-generating prompt few-shot brain tumor segmentation pipeline based on SAM. Biomed. Signal Process. Control **100**, 106936 (2025)
21. Zhai, X., Mustafa, B., Kolesnikov, A., Beyer, L.: Sigmoid loss for language image pre-training. In: Proceedings of the IEEE/CVF International Conference on Computer Vision, pp. 11975–11986 (2023)
22. Zhang, S., Metaxas, D.: On the challenges and perspectives of foundation models for medical image analysis. Med. Image Anal. **91**, 102996 (2024)
23. Zhu, Y., Li, M., Ye, Q., Wang, S., Xin, T., Zhang, H.: Robustemd: domain robust matching for cross-domain few-shot medical image segmentation (2025). https://arxiv.org/abs/2410.01110
24. Zhu, Y., Wang, S., Xin, T., Zhang, H.: Few-shot medical image segmentation via a region-enhanced prototypical transformer. In: International Conference on Medical Image Computing and Computer-Assisted Intervention, pp. 271–280. Springer, Cham (2023)
25. Zhu, Y., Zhang, H.: MAUP: training-free multi-center adaptive uncertainty-aware prompting for cross-domain few-shot medical image segmentation. In: Proceedings of Medical Image Computing and Computer Assisted Intervention - MICCAI 2025. LNCS, vol. 15966. Springer, Cham (2025)
26. Zhuang, X.: Multivariate mixture model for myocardial segmentation combining multi-source images. IEEE Trans. Pattern Anal. Mach. Intell. **41**(12), 2933–2946 (2018)

Are Foundational Models Less Biased Than Specialized Models? An Ultrasound Study

Joris Fournel[1(✉)], Jakob Ambsdorf[2], Benjamin Jønch Jürgensen[1], Christopher Boland[1], Aya Elgebaly[1], Kamil Mikolaj[1], Paraskevas Pegios[1], Emilie Pi Fogtmann Sejer[3], Martin Tolsgaard[3], Anders Nymark Christensen[1], Mads Nielsen[2], and Aasa Feragen[1]

[1] Technical University of Denmark, Kongens Lyngby, Denmark
jorisfournel1@gmail.com
[2] University of Copenhagen, Copenhagen, Denmark
[3] Region Hovedstaden Hospital, Copenhagen, Denmark

Abstract. Foundation models (FMs) pre-trained via self-supervised learning (SSL) on large, diverse datasets are widely expected to offer superior generalization—and, by extension, reduced bias—compared to classically supervised (CS) models. Yet whether this holds in practice, and whether the choice of pretraining data matters, remains largely unresolved. We present a controlled comparison between SSL-pretrained FMs and a CS baseline fine-tuned on identical data across two prenatal ultrasound tasks: spontaneous preterm birth (sPTB) prediction and fetal scan weight estimation. Using visual radar-based bias profiles spanning a comprehensive set of clinical and acquisition subgroups, we show that FMs reproduce the bias profile of their CS counterpart—best- and worst-performing subgroups largely coincide, and mean performance gaps are comparable in magnitude. Pretraining data strategy does influence bias: oversampling cervical images during pretraining outperformed scaling to 13 million images for sPTB in both global performance and subgroup fairness, while naïve oversampling of rare devices degraded overall performance. These results challenge the assumption that large-scale diverse pretraining uniformly reduces bias, and suggest that data alignment with the downstream task matters more than scale.

Keywords: Foundational Models · Generalization · AI Fairness

1 Introduction

Foundation models—pre-trained via self-supervised learning (SSL) on large, diverse, unlabeled datasets—offer a compelling response to the chronic shortage of annotated data in medical imaging [3,16,23]. By transferring broad visual representations through lightweight fine-tuning [1,2,4], they are particularly attractive in healthcare, where annotation costs, disease rarity, and privacy constraints severely limit labeled data availability.

G. Yang et al. (Eds.): MICCAI 2026, LNCS 16878, pp. 24–33, 2027.
https://doi.org/10.1007/978-3-032-38059-3_3

Yet the clinical deployment of medical AI remains fundamentally constrained by bias. Models trained on unrepresentative data produce systematic performance disparities across patient subgroups [6,12,13,19,20]. On this topic, regulatory pressure is mounting [5], indicating that bias is not a peripheral concern, but a barrier to deployment.

A natural hypothesis is that SSL pre-training on large, diverse data should reduce susceptibility to the spurious correlations that drive these disparities in classically supervised (CS) models [8,22]. However, foundation models (FMs) appear to reproduce many of the same biases as their purely-supervised counterparts [10,11,21], and Glocker *et al.* [7] found FMs to be *more* biased—though their model was pre-trained in a supervised manner, leaving the SSL case unresolved. Whether SSL pre-training at scale can mitigate fine-tuning biases, and whether pretraining strategy matters, remain open questions of direct practical consequence given the enormous cost of building foundation models.

This study provides a controlled, direct comparison between SSL-pretrained foundation models and a "traditional", purely-supervised baseline fine-tuned on identical data across two prenatal ultrasound tasks.

Our **research questions** are as follows:

1. Are foundation models less biased than classical, purely-supervised models?
2. What is the effect of the pretraining data regime on the foundation model's bias?

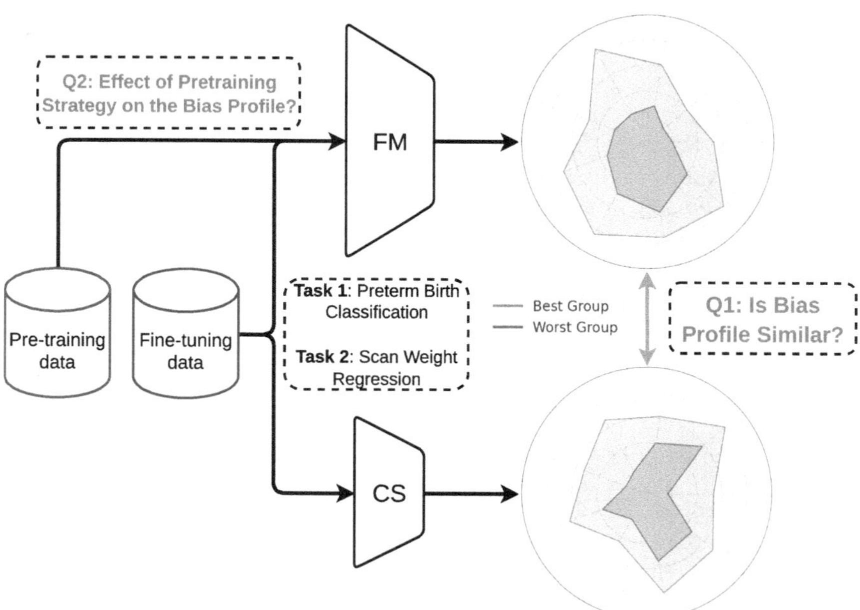

Fig. 1. Study Design. CS: classically supervised; FM: Foundation Model.

2 Method

A meaningful comparison between FMs and CS models requires: (i) a large unlabeled dataset for SSL pretraining; (ii) an identical, labeled fine-tuning set for the FM and CS models; and (iii) a CS baseline trained on that fine-tuning data alone. This work evaluates multiple FM variants with different pretraining corpora on the same test set across two tasks (Fig. 1).

2.1 Bias Quantification

The test set is stratified by clinical and acquisition factors. For each factor, subgroups with a minimum sample size are retained, and the task metric is computed for each subgroup.

Visual Analysis. Radar plots display the best- and worst-performing subgroup metric on each axis (one per factor), forming two polygons. Overlaying plots from two models enables direct comparison.

Quantitative Metrics. Four complementary measures characterize a model's bias profile:

- *Absolute gap*: $|\text{worst} - \text{best}|$, averaged across factors.
- *Relative gap*: $|\text{worst} - \text{best}| / \text{best} \times 100\,\%$, averaged across factors.
- *Improved/worsened subgroups*: number of subgroups with a statistically significant performance gain or loss relative to a reference model, assessed by 1000-fold bootstrap resampling and Mann–Whitney test for classification and independent two-sample t-test in regression.
- *Bias similarity index*: a DSC value is computed between the two gap regions (area of the best/worse gaps) using the radar plots of two models, after normalizing so that the mean best-group metric is equal across models. This index is purely indicative to associate some familiar metric value of overlapping next to the two visuals of the bias profiles.

3 Experiment

3.1 Datasets

All dataset came from the Country X national fetal ultrasound database:

1. **FM Pretraining (PT) Data**: 13 million fetal ultrasound images from the same national cohort as described in a prior published work [Anon.].
2. **Spontaneous Preterm Birth Prediction (sPTB)**: 7862 transvaginal ultrasound images corresponding to n=3931 preterm births and n=3931 term births. The dataset was divided in 10 splits with 6290/786/786 samples for training/validation/test sets. The performance metric was the sensitivity at the threshold corresponding to a specificity of 85.

3. **Scan Weight Prediction**: 433,096 ultrasound images derived from 94,538 examinations conducted on 65,752 patients. The fetal weight ground truth at scan time was derived from the birth weight after using the Marsál [14] growth curve. The data was divided on a patient basis between training, validation, and test sets (85/5/10%). The performance metric was the Mean Relative Error (MRE) (%).

The finetuning datasets were stratified over the following studied factors: ethnicity, body-to-mass index (BMI), maternal hormonal treatment, conception method, cervical length, gestational age, birth weight, birth year, birth hospital, ultrasound device, pixel spacing. This resulted in a total of 55 and 38 subgroups for the sPTB and scan weight datasets, respectively.

3.2 Models

In order to ensure a fair comparison, we implement qualitative and validated CS baselines for each task:

1. **CS model for sPTB**: The CNN baseline, described in [18], predicts the probability of preterm birth from a cervical scan.
2. **CS model for Scan Weight**: CNN baseline as in [15] predicting the fetal scan weight from a set of head, abdomen, and femur standard planes.
3. **FM**: a ViT-B foundation model pretrained from scratch using a DINOv2 [17] self-supervised learning objective on fetal ultrasound images. All FM parts are finetuned: encoder, decoder, up to the final classification layer.

3.3 Pretraining Data Strategies

We evaluate four FM variants with the following pretraining regimes:

1. **FM2M**: Base FM pretrained on randomly selected 2 million ultrasound images out of the 13 million available.
2. **FM13M**: Base FM pretrained on all 13 million ultrasound images.
3. **FM2M E10**: Base FM pretrained on 2 million ultrasound images with a rare ultrasound device (Voluson E10) oversampled (500k).
4. **FM2M Cervix**: Base FM pretrained on 2 million ultrasound images with cervical scans oversampled, based on 120k images identified as cervical scans in the 13M PT dataset by a ResNet18 [9] trained to classify anatomical planes. These were oversampled five times (600k), and the remaining 1.4 million with a random sample from the rest of the PT data.

4 Results

We divide the analysis of our results along our two research questions: First, how different are the biases of the FM and CS model? Second, what is the impact of the pretraining data regime on bias?

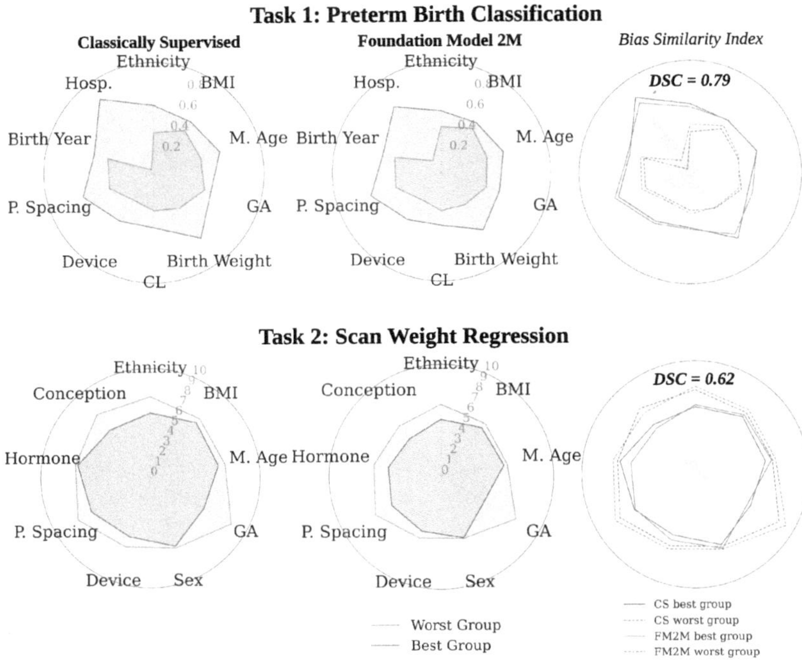

Fig. 2. Is the Foundation Model Less Biased Than Its Classically Supervised Counterpart? The blue radar gives the performance metric of the best subgroups; the red radar of the worst subgroups. Classification metric: sensitivity at 85% specificity (higher is better); regression metric: mean relative error (lower is better); S; Bias Similarity Index: DSC between the two gap areas (in the center) after scaling so that the mean metric of the best groups is the same for the two models. (Color figure online)

4.1 Does the Bias Profile of a FM Resemble That of a CS Model?

The bias profiles of the FM2M and CS models, for both the sPTB classification and scan weight regression tasks, show strong similarity in shape (Fig. 2). This is reflected in the overall relative gap between the best- and worst-performing groups across the CS and FM2M models for each task, at 37.3% and 33.8% for sPTB classification and 14.9% and 17.8% in scan weight prediction (Table 1). For the preterm birth prediction task, FM2M also failed to substantially improve overall performance compared to the CS model (0.516 sensitivity at 85% specificity for CS, 0.518 for FM2M).

Notably, the majority of the best- and worst-performing subgroups were the same across the FM and CS models (14 out of 20 in the classification task, 9 out of 18 in the regression task). There was also no consistent improvement in fine-grained subgroup performance with FM2M compared to the CS model. For the sPTB classification task, 17 subgroups (out of 55) improved significantly, while 17 worsened significantly (Table 2). These findings challenge the hypothesis that large-scale SSL pretraining reduces bias in downstream tasks, and suggest instead that bias is primarily a product of the fine-tuning data.

Table 1. Global performance and mean subgroup performance shift (best vs. worst subgroup, averaged over all bias factors) for all methods. sPTB: sensitivity at 85% specificity; Scan Weight: mean relative error (%). Rel. Gap: |worst − best|/ max ×100; Abs. Gap: |worst − best|. Models are highlighted **best** and <u>second-best</u>. In the case of a tie, both are marked as **best**. (**): different from CS baseline with $p < .001$.

Model	sPTB			Scan Weight		
	↑Sens.	Rel. gap	Abs. gap	↓MRE (%)	Rel. gap	Abs. gap
CS	51.6	37.3%	0.25	6.7±5.4	14.9%	<u>1.14</u>
FM2M	51.8	**33.8%**	**0.21**	**5.9±4.9****	17.8%	1.18
FM13M	<u>53.1</u>**	<u>33.9</u>%	0.22	<u>6.0±5.1</u>**	<u>14.8</u>%	**1.00**
FM2M E10	50.8**	34.4%	**0.21**	9.7±7.3**	**14.4%**	1.61
FM2M Cervix	**54.7****	34.5%	0.23	-	-	-

4.2 Does Pretraining Data Strategy Influence the Bias Profile of Foundation Models?

Here, we review the results associated to each pretraining strategy.

Scaling PT Data (FM13): The model pretrained with 13 million ultrasound images did result in a better bias profile in terms of improved/worsened subgroups (sPTB, Table 2) and in terms of relative gap (Scan Weight, Table 1). In sPTB, the effect of scaling was noticeable on certain ethnic and device subgroups, but subtle overall (Fig. 3).

Oversampling Rare Device (FM2M E10): Oversampling the Voluson E10 resulted in an improvement (0.31 to 0.36) on this particular device in sPTB prediction. But it also led to a decrease in overall performance, especially on the regression task, where errors increased from 5.9% to 9.7%.

Oversampling Cervical Scans (FM2M Cervix): This strategy significantly improved both the global performance and subgroup performance of the foundation model for sPTB prediction, which uses a cervical scan as input. 39 subgroups improved, while only 7 worsened, compared to 17 and 17, respectively, for FM2M (Table 2). In particular, some worrying disparities in the CS model were mitigated (much more strongly than in FM13M), as evidenced by the noticeable drop in performance for babies born in 2016–2017 and after 2017 (Fig. 4). However, in terms of the absolute and relative gaps between the best and worst subgroups, the bias profile was not improved compared to FM2M.

These results suggest that the benefit of SSL pretraining depends critically on the relevance of the pretraining data to the downstream task. Scaling and naive oversampling offer limited or even negative returns.

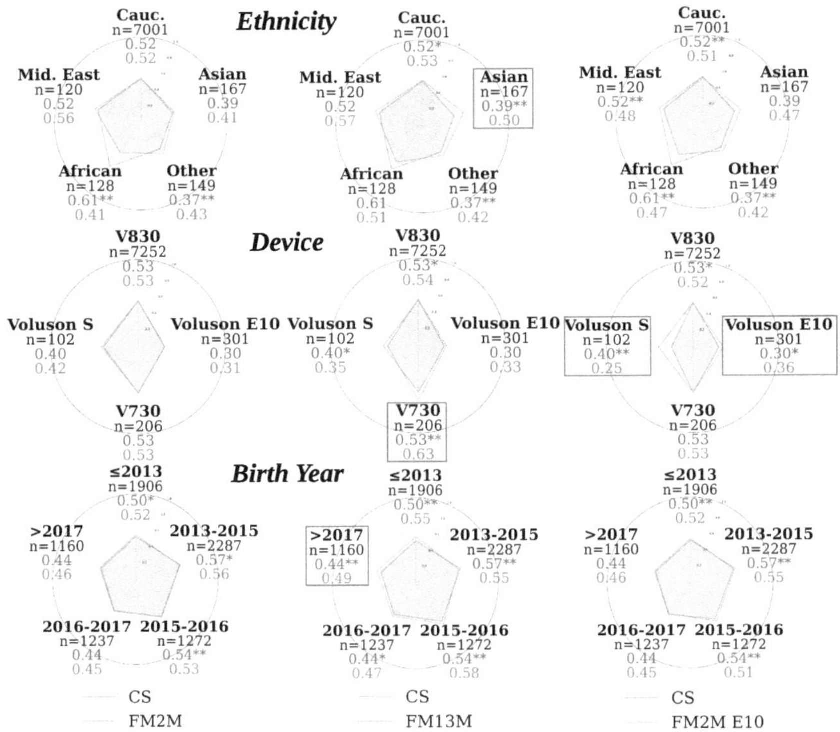

Fig. 3. Can the Pretraining Strategy Change the FM's sPTB Bias Profile? Sensitivity values are displayed; stars indicate significance of difference: (*) or (**): p < .05 or .001.

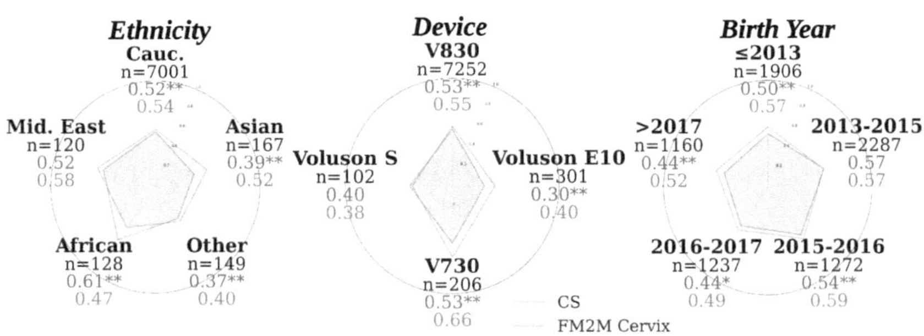

Fig. 4. Effect on sPTB Bias of Oversampling Cervical Scans in Pretraining. Sensitivity values are displayed; stars indicate significance of difference: (*) or (**): p < .05 or .001.

Table 2. Number of subgroups (out of 55) with a statistically significant sensitivity gain or loss (at 85% specificity) relative to CS. Models are highlighted **best** and <u>second-best</u>.

Transition	Improved	Worsened
CS → FM2M	17	17
CS → FM13M	<u>30</u>	<u>11</u>
CS → FM2M E10	16	26
CS → FM2M Cervix	**39**	**8**

5 Discussion

Our **main findings** are:

1. Foundation models reproduce the bias profile of specialized models.
2. The pretraining data strategy can *moderately* influence the bias profile, in particular on smaller subgroups.
3. Oversampling task-related images outperformed scaling to 13 million images on sPTB prediction, both in global performance and subgroup fairness.

FMs mirrored the bias profile of specialized models. The best- and worst-performing subgroups largely coincided, and mean performance gaps were comparable in magnitude. These findings align with prior work [10,21]. But to our knowledge, this is the first study to systematically compare foundation and CS models by a direct, subgroup-level performance analysis across a comprehensive set of factors. This contrasts with Glocker *et al.* [7], whose notion of bias mainly focused on the encoding of demographic attributes in latent representations rather than performance disparities across numerous subgroups.

Pretraining strategy did matter, and its effect depended on its alignment with the downstream task. This is supported by the differential benefit of FM2M across tasks. For example, fetal biometric scans (head, abdomen, femur), used as input for scan weight prediction, were abundant in the pretraining corpus, which may explain why FM2M improved performance on that task without any targeted curation. Cervical scans, by contrast, are rare in standard prenatal ultrasound collections, and generic pretraining provided little benefit for sPTB prediction. Oversampling cervical images during pretraining outperformed scaling to 13 million images, challenging the assumption that large-scale, diverse pretraining uniformly benefits any downstream task.

This study had limitations. We only investigated ViT-B models (90 million parameters), while some foundation models as DINOv2/v3, CLIP, are much larger (up to 1B parameters). Also, we focused on ultrasound imaging and two obstetric tasks. But such comparisons are inherently resource-intensive, requiring simultaneous access to huge unlabeled data pools, large labeled fine-tuning datasets, and the infrastructure to train multiple foundation models—making the scale achieved here (13 million PT images, 94k and 8k labeled samples) a strong baseline. We also ignored possible interaction effects between clinical/acquisition factors. Such intersectional analysis could reveal some additional and more subtle differences between models. But given enough factors, two very similar bias profiles already tell us that the two bias behaviors correspond in their major directions, which was what we wanted to know.

Our findings have direct practical implications. FMs carry substantially higher computational costs than CS models; if their bias profiles are not demonstrably superior, the return on investment is questionable in high-data regimes. Certain configurations (e.g., rare-device oversampling) even degraded performance, underscoring that naïve pretraining data augmentation can be counterproductive.

In **conclusion**, SSL pretraining alone does not guarantee reduced bias relative to classically supervised models. Aligning pretraining data with the downstream task may be more impactful than indiscriminate scaling, and future work should rigorously compare this strategy with fine-tuning data curation.

Acknowledgments. The work was partly funded by the Novo Nordisk foundation through the project NNF0087102.

Disclosure of Interests. ANC, MT, AF, and MN hold shares in Prenaital ApS.

References

1. Alayrac, J.B., et al.: Flamingo: a visual language model for few-shot learning. Adv. Neural. Inf. Process. Syst. **35**, 23716–23736 (2022)
2. Balestriero, R., et al.: A cookbook of self-supervised learning. arXiv preprint arXiv:2304.12210 (2023)
3. Bommasani, R., et al.: On the opportunities and risks of foundation models. arXiv preprint arXiv:2108.07258 (2021)
4. Devlin, J., Chang, M.W., Lee, K., Toutanova, K.: BERT: pre-training of deep bidirectional transformers for language understanding. In: Proceedings of the 2019 Conference of the North American Chapter of the Association for Computational Linguistics: Human Language Technologies (Volume 1: Long and Short Papers), pp. 4171–4186 (2019)
5. FDA: Artificial intelligence-enabled device software functions: Lifecycle management and marketing submission recommendations (2025). https://www.fda.gov/regulatory-information/search-fda-guidance-documents/artificial-intelligence-enabled-device-software-functions-lifecycle-management-and-marketing
6. Geirhos, R., et al.: Shortcut learning in deep neural networks. Nat. Mach. Intell. **2**(11), 665–673 (2020)
7. Glocker, B., Jones, C., Roschewitz, M., Winzeck, S.: Risk of bias in chest radiography deep learning foundation models. Radiol. Artif. Intell. **5**(6), e230060 (2023). https://doi.org/10.1148/ryai.230060. https://pubs.rsna.org/doi/full/10.1148/ryai.230060
8. Goyal, P., et al.: Vision models are more robust and fair when pretrained on uncurated images without supervision. arXiv preprint arXiv:2202.08360 (2022)
9. He, K., Zhang, X., Ren, S., Sun, J.: Deep residual learning for image recognition (2015). https://arxiv.org/abs/1512.03385
10. Jin, R., et al.: Fairmedfm: fairness benchmarking for medical imaging foundation models. Adv. Neural Inf. Process. Syst. **37**, 111318–111357 (2024)

11. Khan, M.O., Afzal, M.M., Mirza, S., Fang, Y.: How fair are medical imaging foundation models? In: Hegselmann, S., et al. (eds.) Proceedings of the 3rd Machine Learning for Health Symposium. Proceedings of Machine Learning Research, vol. 225, pp. 217–231. PMLR (2023). https://proceedings.mlr.press/v225/khan23a.html
12. Lee, T., Puyol-Antón, E., Ruijsink, B., Shi, M., King, A.P.: A systematic study of race and sex bias in CNN-based cardiac MR segmentation. In: International Workshop on Statistical Atlases and Computational Models of the Heart, pp. 233–244. Springer, Cham (2022)
13. Lin, M., et al.: Shortcut learning in medical image segmentation. In: International Conference on Medical Image Computing and Computer-Assisted Intervention, pp. 623–633. Springer, Cham (2024)
14. Maršál, K., Persson, P.H., Larsen, T., Lilja, H., Selbing, A., Sultan, B.: Intrauterine growth curves based on ultrasonically estimated foetal weights. Acta Paediatr. **85**(7), 843–848 (1996)
15. Mikołaj, K.W., et al.: Predicting abnormal fetal growth using deep learning. NPJ Digit. Med. **8**(1) (2025). https://doi.org/10.1038/s41746-025-01704-0
16. Moor, M., et al.: Foundation models for generalist medical artificial intelligence. Nature **616**(7956), 259–265 (2023)
17. Oquab, M., et al.: Dinov2: learning robust visual features without supervision. arXiv preprint arXiv:2304.07193 (2023)
18. Pegios, P., et al.: Leveraging shape and spatial information for spontaneous preterm birth prediction, pp. 57–67. Springer, Cham (2023). https://doi.org/10.1007/978-3-031-44521-7_6
19. Puyol-Antón, E., et al.: Fairness in cardiac magnetic resonance imaging: assessing sex and racial bias in deep learning-based segmentation. Front. Cardiovasc. Med. **9**, 859310 (2022)
20. Seyyed-Kalantari, L., Zhang, H., McDermott, M.B., Chen, I.Y., Ghassemi, M.: Underdiagnosis bias of artificial intelligence algorithms applied to chest radiographs in under-served patient populations. Nat. Med. **27**(12), 2176–2182 (2021)
21. Yang, Y., et al.: Demographic bias of expert-level vision-language foundation models in medical imaging. Sci. Adv. **11**(13), eadq0305 (2025)
22. Yfantidou, S., Spathis, D., Constantinides, M., Vakali, A., Quercia, D., Kawsar, F.: Using self-supervised learning can improve model fairness. In: Proceedings of the 30th ACM SIGKDD Conference on Knowledge Discovery and Data Mining, pp. 3942–3953 (2024)
23. Zhang, S., Metaxas, D.: On the challenges and perspectives of foundation models for medical image analysis. Med. Image Anal. **91**, 102996 (2024)

ASTAR: Automated Induction of STAndardized Radiology Reporting Templates from Large-Scale Clinical Free-Text Corpora

Xinfeng Zhang[1], Mingxuan Liu[1], Yifei Chen[1], Juncheng Zhu[2], Kasidit Anmahapong[1], Yiming Huang[3], Yuan Zhang[4], Hongjia Yang[1], Yi Liao[2], Gang Ning[2], Haibo Qu[2], and Qiyuan Tian[1](✉)

[1] Tsinghua University, Beijing, China
qiyuantian@tsinghua.edu.cn
[2] Sichuan University, Chengdu, China
[3] University of California San Diego, San Diego, USA
[4] Southeast University, Nanjing, China

Abstract. Structured reporting converts free-text radiology narratives into queryable data keys, facilitating cohort assembly, longitudinal tracking, and training label generation for medical AI. The prevailing paradigm follows a two-stage pipeline: (1) constructing a reporting template, (2) extracting information to populate it. While the extraction stage has benefited from advances in large language models (LLMs), template construction remains a manual bottleneck relying on labor-intensive expert consensus that is static, difficult to scale, and may fail to capture real-world reporting diversity. We address this limitation with `ASTAR`, an LLM-based framework for Automated induction of STAndardized radiology Reporting templates from large-scale clinical free-text corpora. Extensive experiments on 4,215 fetal brain MRI reports from multiple centers demonstrate that, in this reporting scenario, the `ASTAR`-induced template surpasses two expert-curated templates across template coverage, information fidelity, diagnostic fidelity, and expert-rated usability, reducing template development from weeks of committee deliberation to hours of automated processing. Code: https://github.com/birthlab/ASTAR.

Keywords: Radiology report · LLM · Structured information extraction

1 Introduction

Radiology departments worldwide generate clinical reports at an ever-growing scale. In the United States alone, over 16.1 million CT examinations were

X. Zhang and M. Liu—Equal contribution.

G. Yang et al. (Eds.): MICCAI 2026, LNCS 16878, pp. 34–44, 2027.
https://doi.org/10.1007/978-3-032-38059-3_4

reported during a nine-month period in 2020 [11], and imaging utilization across all modalities is projected to rise 17–27% by 2055 [7]. Each examination produces a detailed free-text narrative synthesizing categorical descriptors, quantitative measurements, and diagnostic impressions, thereby forming one of the richest data repositories in modern healthcare [5,17]. However, the utility of these reports is constrained because findings remain archived as unstructured text with wide stylistic, terminological, and structural variations across radiologists, institutions, and languages [15,16]. Such heterogeneity hinders systematic case retrieval, cohort assembly, longitudinal tracking, and training-label generation for medical AI [6,17]. Radiology societies therefore increasingly advocate for structured reporting [4,20], converting narratives into queryable keys that support rapid cohort identification and high-quality ground-truth generation [1,2].

Nevertheless, achieving these benefits at scale requires automated tools, since manual structuring of the massive and continuously growing volume of narrative reports is highly impractical. Therefore, recent studies increasingly employ natural language processing (NLP) and large language models (LLMs) for automated information extraction [12,23]. Specifically, the prevailing paradigm follows a two-stage pipeline: (1) a reporting template (i.e., a target template defining the set of keys and their permissible values) is constructed; and (2) relevant information is extracted from free-text reports to populate this template. For example, BURExtract-Llama [6] extracts clinical concepts from breast ultrasound reports with an average F1 score of 84.6%.

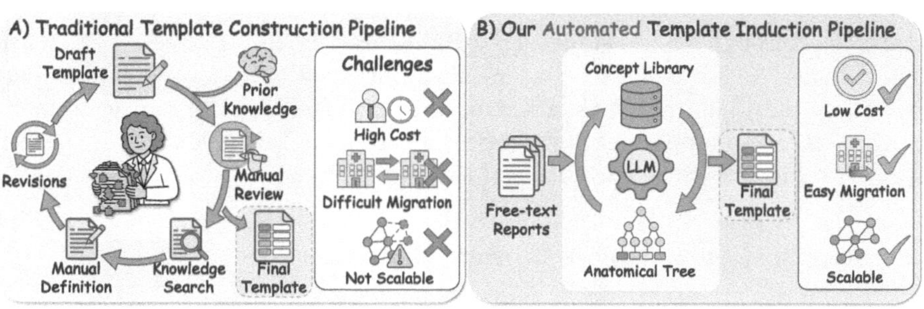

Fig. 1. (A) Traditional template construction pipeline. (B) Our automated template induction pipeline (ASTAR).

However, while the extraction stage has benefited substantially from recent advances in NLP and LLMs, the template construction stage, as an upstream prerequisite for all extraction methods, remains a manual bottleneck, since existing templates are typically produced through labor-intensive expert consensus (Fig. 1A). For example, the ESPR template [20] was derived from multidisciplinary panel discussions, and the JSON template used by FetalExtract-LLM [15] was compiled from ESPR guidelines [20], MeSH [14], and a medical dictionary. Beyond this scalability limitation, expert-designed templates are inherently static and may fail to capture the full spectrum of linguistic diversity and evolving clinical descriptions present in massive historical report corpora [4]. As a result, rare but clinically meaningful findings risk being absent from the predefined template, while institution-specific reporting conventions cannot be accommodated by a one-size-fits-all structure.

To the best of our knowledge, no automated method exists for standardized radiology reporting templates construction. To fill this gap, we propose ASTAR (Automated induction of STAndardized radiology Reporting templates), an LLM-based framework (Fig. 1B) that automatically mines unified reporting structures from large-scale radiology report corpora. We demonstrate its utility on 4,215 fetal brain MRI reports from three centers, where the ASTAR-induced template achieved stable information retention with progressive consolidation of long-tail keys as corpus size increases, offering a scalable, data-driven complement to labor-intensive expert consensus.

2 Method

2.1 Overview of ASTAR

Given a corpus of N de-identified free-text radiology reports $\mathcal{D} = \{r_1, \ldots, r_N\}$, the goal of ASTAR is to automatically induce a hierarchical reporting template $\mathcal{T} = (\mathcal{V}, \mathcal{E}, \{f_v\}_{v \in \mathcal{V}_{\text{leaf}}})$, where internal nodes represent anatomical or semantic groupings and each leaf node $v \in \mathcal{V}_{\text{leaf}}$ carries a field descriptor $f_v = (k_v, \tau_v, \Omega_v)$ specifying a standardized field name k_v, a value type $\tau_v \in \{\texttt{categorical}, \texttt{numerical}, \texttt{free-text}\}$, and a type-dependent value specification Ω_v. ASTAR obtains $\mathcal{T}$ through three stages (Fig. 2): $\mathcal{D} \xrightarrow{\text{I}} \Phi \xrightarrow{\text{II}} T_1 \xrightarrow{\text{III}} \mathcal{T}^*$, where Stage I builds a de-duplicated concept slots library $\Phi = \{\phi_1, \ldots, \phi_L\}$ with each $\phi_j = (k_j, \tau_j, \Omega_j)$ sharing the same structure as f_v; Stage II organizes Φ into a hierarchical tree, mapping each surviving slot to exactly one leaf ($f_v \leftarrow \phi_j$); and Stage III refines the tree into the final template $\mathcal{T}^*$.

2.2 Stage I: Concept Slots Library Construction

In stage I, raw narratives are converted into a de-duplicated library of canonical clinical concept slots in two steps. (1) Span Extraction. An LLM (Qwen-Max [21]) in JSON-mode segments each report r_i into spans $\mathcal{S}_i$, each represented as a 7-tuple $(a_{\text{struct}}, a_{\text{sub}}, a_{\text{attr}}, a_{\text{type}}, a_{\text{val}}, a_{\text{role}}, a_{\text{sec}})$ encoding anatomical location (structure / substructure), attribute identity (name / type), a canonicalized value (numerics abstracted as ⟨NUM⟩), clinical role (finding | measurement | diagnosis), and report section (description | impression). Spans from all reports are pooled and de-duplicated by the composite key $(a_{\text{val}}, a_{\text{role}})$ into a global span set $\mathcal{S}$. (2) Clustering and Slot Induction. Each span is embedded via Qwen3-Embedding-8B [21] and grouped by K-means into C clusters, with the optimal number of clusters C^* chosen by maximizing the silhouette score [19] over $C \in [50, 300]$. Because raw clusters may still mix heterogeneous concepts (e.g., *ventricular width* and *ventricular morphology*), the LLM sub-partitions each cluster into semantically coherent subgroups and induces a canonical slot $\phi = (k, \tau, \Omega)$ per subgroup, where k is a standardized key name, $\tau \in \{\texttt{categorical}, \texttt{numerical}, \texttt{free-text}\}$ the value type, and Ω the permissible value set. Slots sharing identical standardized key names (k) are then globally merged by the LLM, yielding the unified concept slots library $\Phi = \{\phi_1, \ldots, \phi_L\}$.

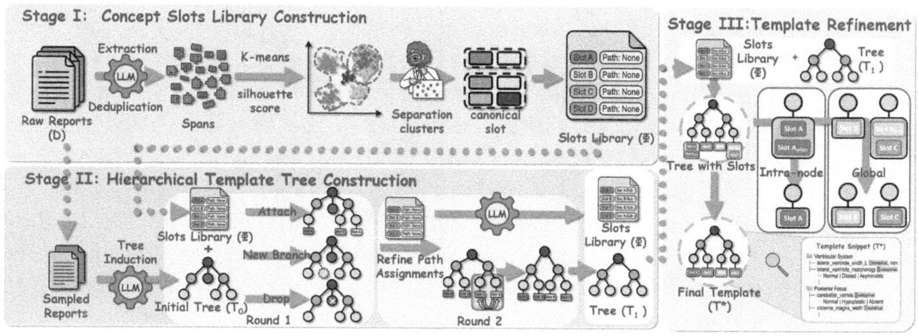

Fig. 2. Overview of ASTAR. (1) Stage I: Concept Slots Library Construction. (2) Stage II: Hierarchical Template Tree Construction. (3) Stage III: Template Refinement.

2.3 Stage II: Hierarchical Template Tree Construction

In Stage II, given the slots library Φ, the flat set of slots is organized into a clinically meaningful hierarchy through a two-round LLM-driven process. **Round 1: Template Construction and Initial Assignment.** An LLM (Qwen-Max) first generates an initial Template $\mathcal{T}_0$ from a representative subset of reports, with internal nodes organized by major anatomical regions. Each slot $\phi_j \in \Phi$ is then presented to the LLM together with its descriptive statistics and the current template. The LLM decides to `attach` it under an existing node, `propose` a new branch, or `drop` it. Proposed branches are incorporated into the template incrementally. **Round 2: Global Path Refinement.** Because Round 1 slots are assigned against an evolving template, early slots never see branches proposed later. A second pass therefore re-evaluates all surviving slots in batches on the now-complete tree, taking each slot's Round 1 path as reference, and outputs a revised path, yielding $\mathcal{T}_1$.

2.4 Stage III: Template Refinement

The final stage performs two refinement operations on $\mathcal{T}_1$ to ensure global consistency and output the final template $\mathcal{T}^*$. (1) Intra-node Semantic Reorganization. Within each node, the LLM consolidates synonymous permissible values (e.g., [*"normal"*, *"no abnormality"*, *"unremarkable"*] → [*"Normal"*]) and harmonizes value type assignments across sibling keys, reducing value-level redundancy while preserving clinical granularity. (2) Global Concept Harmonization. A tree-wide pass identifies cross-node redundancies, i.e., keys whose names differ across subtrees but refer to the same concept, and standardizes terminology throughout the tree (e.g., unifying *"cerebral hemisphere"* vs. *"brain hemisphere"* across subtrees).

2.5 Evaluation Protocol for Radiology Reporting Templates

Because there is no widely used protocol for evaluating radiology reporting templates, we introduce a three-level protocol to fill this gap: (1) Template quality assesses whether the template is comprehensive and well-structured; (2) structuring fidelity measures whether populating the template preserves the information in the original report; and (3) radiologist judgment captures whether clinicians find the template usable and clinically valuable. Specifically,

(1) Template quality, computed on a held-out test dataset, quantifies how completely $\mathcal{T}^*$ captures the clinical concepts present in unseen reports, measured by *case-level* (macro-averaged over reports) and *key-level* (micro-averaged over all keys) coverage. For each test report r_i, concept keys K_i are extracted by concatenating entity and attribute keys of each span. An LLM (Qwen-Max) judges whether each key has a semantic match in the template key set $\mathcal{F}(\mathcal{T}^*)$. The two metrics are defined as:

$$\mathrm{Cov_{case}} = \frac{1}{|\mathcal{D}_{\mathrm{test}}|}\sum_i \frac{\sum_{k\in K_i}\mathrm{match}(k,\mathcal{F})}{|K_i|}, \quad \mathrm{Cov_{key}} = \frac{\sum_i\sum_{k\in K_i}\mathrm{match}(k,\mathcal{F})}{\sum_i |K_i|}. \tag{1}$$

(2) Structuring fidelity, also computed on the held-out test dataset, measures whether populating the template preserves the information in the original report, assessed via *information fidelity* and *diagnostic fidelity.* Specifically, (a) *information fidelity*, evaluated through a four-step extract–reconstruct round trip, quantifies textual information preservation. *(i)* An LLM (Gemini-2.5-Pro-Preview [8]), chosen independently of the template-construction LLM to prevent extraction bias, populates the template from each report's `imaging description`; *(ii)* extracted key–value pairs are mapped into the full hierarchy and re-flattened for cross-template consistency; *(iii)* another LLM (Qwen-Max) reconstructs a free-text report solely from the structured report, guided by a fixed style-reference text; *(iv)* the reconstruction $\hat{r}_i$ is compared with the original r_i using ROUGE-1/2/L (F1) [13], chrF [18], $\mathrm{BERTScore}_R$ (F1) (with RoBERTa-wwm-ext [10]), and $\mathrm{BERTScore}_M$ (F1) [22] (with MacBERT [9]). As a sensitivity check for model-dependent evaluation, we repeated the round-trip reconstruction with GPT and observed the same relative ranking among the three templates. (b) *Diagnostic Fidelity*, evaluated through an infer-then-judge protocol, tests whether the structured report retains sufficient clinical information for diagnosis. Qwen-Max receives only the structured output, infers a diagnosis d_i^{struct} under strict information closure, and an LLM-as-Judge protocol (with Qwen-Max) is used to compare it against the ground-truth impression d_i^{gt} (i.e., the `diagnostic impression` in the original free-text report) using three metrics (all normalized to $[0,1]$): *Primary Diagnosis Accuracy* (PDA; 0–5 integer, macro correctness), *Key Finding Preservation* (KFP; 0–1 continuous, abnormal-finding retention), and *Clinical Actionability* (CA; 0–5 integer, decision-support sufficiency).

(3) Radiologist judgment directly evaluates the template via structured clinician ratings. Two radiologists from different countries, ensuring cross-national diversity in clinical conventions and mitigating institution-specific bias, rated all three templates on 28 Likert-scale items (1–5) across three dimensions. (a) *Template Structure* (12 items) evaluates the schema design across six sub-dimensions: coverage & completeness (COMP, ABNO), structure & granularity (LOGI, GRAN) terminology & clarity (NAME, ENUM), safety & error prevention (NOMS, NOFN, UNCE), adaptability (ADAP), and deployment readiness (ADOP, SIGN); (b) *System Usability* (10 items) follows the standard scale [3], comprising FREQ (willingness to use), COMP (complexity†), EASE (ease of use), TECH (technical support needed†), INTE (integration), INCO (inconsistency†), LEAR (learnability), CUMB (cumbersomeness†), CONF (confidence), and PREL (prior learning needed†), where † marks negatively worded items reverse-scored as $6-x_{\text{raw}}$; (c) *Clinical Impact* (6 items) assesses downstream utility: NOMS (omission reduction), CONS (consistency), ORDE (workflow alignment), NAEX (not-assessed expressibility), COMM (communication efficiency), and OVR (overall satisfaction).

Table 1. Template quality (coverage) and structuring fidelity ((information fidelity & diagnostic fidelity)) across templates evaluated on ID (n = 100) and OoD (n = 115) test sets. Bold indicates best results. The three templates differ in key count (ESPR: 159, FetalExtract: 51, ASTAR: 88), and the results suggest that key relevance and organization, rather than key count alone, contribute to the observed differences.

Category	Metric	ESPR [20]	FetalExtract [15]	ASTAR
Structure	Number of keys	159	51	**88**
ID test dataset (n=100)				
Coverage	Case-level	45.41%	36.34%	**81.34%**
	Key-level	45.76%	37.82%	**82.42%**
Information Fidelity	ROUGE-1 (F1)	0.651±0.139	0.409±0.107	**0.749±0.079**
	ROUGE-2 (F1)	0.513±0.167	0.275±0.102	**0.689±0.104**
	ROUGE-L (F1)	0.562±0.124	0.397±0.103	**0.694±0.100**
	chrF	0.576±0.076	0.539±0.093	**0.578±0.075**
	BERTScore_M (F1)	0.873±0.016	0.865±0.020	**0.877±0.017**
	BERTScore_R (F1)	0.892±0.016	0.882±0.021	**0.895±0.016**
Diagnostic Fidelity	PDA	0.879±0.119	0.858±0.158	**0.953±0.090**
	KFP	0.882±0.134	0.844±0.226	**0.958±0.093**
	CA	0.876±0.104	0.858±0.147	**0.953±0.085**
OoD test dataset (n=115)				
Coverage	Case-level	54.32%	37.22%	**66.12%**
	Key-level	55.48%	37.09%	**66.40%**
Information Fidelity	ROUGE-1 (F1)	0.482±0.228	0.316±0.215	**0.717±0.264**
	ROUGE-2 (F1)	0.347±0.233	0.219±0.227	**0.590±0.318**
	ROUGE-L (F1)	0.394±0.175	0.305±0.219	**0.503±0.192**
	chrF	0.322±0.094	0.269±0.123	**0.376±0.073**
	BERTScore_M (F1)	0.843±0.024	0.843±0.031	**0.850±0.020**
	BERTScore_R (F1)	0.853±0.029	0.851±0.033	**0.862±0.023**
Diagnostic Fidelity	PDA	0.790±0.189	0.802±0.143	**0.847±0.125**
	KFP	0.733±0.206	0.739±0.182	**0.811±0.154**
	CA	0.751±0.137	0.757±0.134	**0.847±0.107**

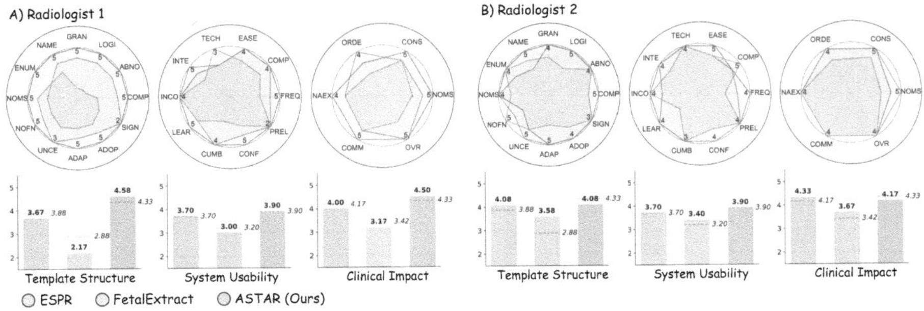

Fig. 3. Two radiologists from different countries rated all templates on 28 Likert-scale items (1–5, higher is better). **(A)** Radiologist 1; **(B)** Radiologist 2. **Top row:** per-item radar charts for Template Structure (12 items), System Usability (10 items), and Clinical Impact (6 items); numbers on each axis denote the maximum score among the three templates for that item. **Bottom row:** dimension-level means; dashed lines show cross-radiologist averages.

3 Experiments and Results

3.1 Data Collection

A total of 4,100 de-identified fetal brain MRI free-text reports were retrospectively collected from West China Second University Hospital, covering examinations performed between October 2015 and December 2024 on one 3.0 T system (Siemens MAGNETOM Skyra) and two 1.5 T systems (Philips Achieva and United Imaging uMR 570). Each report comprises a narrative *imaging description* section and a *diagnostic impression* section. Of these, 4,000 reports were used for template construction with `ASTAR` and 100 were held out as an in-distribution (ID) test dataset. To assess cross-institutional generalizability, an additional out-of-distribution (OoD) test dataset of 115 de-identified fetal brain MRI reports was collected from Sichuan Provincial Woman's and Children's Hospital and Chengdu Women's and Children's Central Hospital.

3.2 Comparison with Expert-Designed Templates

The ASTAR-induced template was compared against two expert-designed templates. (1) ESPR Template [20]: a standardized fetal MRI reporting template released in 2024 by the Fetal Task Force of the European Society of Paediatric Radiology. (2) FetalExtract Template [15]: a JSON template compiled from ESPR guidelines, MeSH, and a medical dictionary, used for structured information extraction with FetalExtract-LLM.

Across all metrics, `ASTAR` consistently outperforms both expert-designed templates on both ID and OoD datasets, achieving higher coverage with fewer keys, superior information and diagnostic fidelity, and the highest expert-rated usability (Table 1, Fig. 3). (1) Template Quality: On the ID test dataset, `ASTAR` achieves case-/key-level coverage of 81.34%/82.42%, outperforming ESPR and

FetalExtract. ASTAR is data-induced, higher ID coverage is expected and should be interpreted together with OoD and radiologist evaluations. This advantage persists on the OoD set, though the gap narrows due to expected domain shift. (2) Structuring Fidelity: ASTAR consistently achieves the highest scores across all six metrics for *information fidelity* on both datasets. Specifically, on the ID dataset, ASTAR attained ROUGE-L 0.694 $\pm$ 0.100, substantially outperforming ESPR and FetalExtract; BERTScore$_{R}$ reached 0.895 $\pm$ 0.016 vs. 0.892 $\pm$ 0.016 and 0.882 $\pm$ 0.021. On the OoD test dataset, ASTAR maintains a clear lead (ROUGE-L 0.503 $\pm$ 0.192 vs. 0.394 $\pm$ 0.175 and 0.305 $\pm$ 0.219), with larger standard deviations reflecting greater cross-institutional stylistic heterogeneity. For *diagnostic fidelity*, ASTAR achieves the highest scores across all three dimensions on the ID set: PDA 0.953 $\pm$ 0.090, KFP 0.958 $\pm$ 0.093, and CA 0.953 $\pm$ 0.085, outperforming ESPR (PDA/KFP/CA: 0.879/0.882/0.876) and FetalExtract (0.858/0.844/0.858). On the OoD set, ASTAR retains its advantage (PDA 0.847, KFP 0.811, CA 0.847), with consistently high KFP (>0.81) confirming that the automatically induced template preserves critical abnormal findings (e.g., ventriculomegaly, posterior fossa anomalies) essential for clinical decision-making. (3) Radiologist Judgment: ASTAR receives the highest scores across *Template Structure* 4.33/5.00 (vs. ESPR 3.88, FetalExtract 2.88), *System Usability* 3.90 (vs. 3.70, 3.20), and *Clinical Impact* 4.33 (vs. 4.17, 3.42).

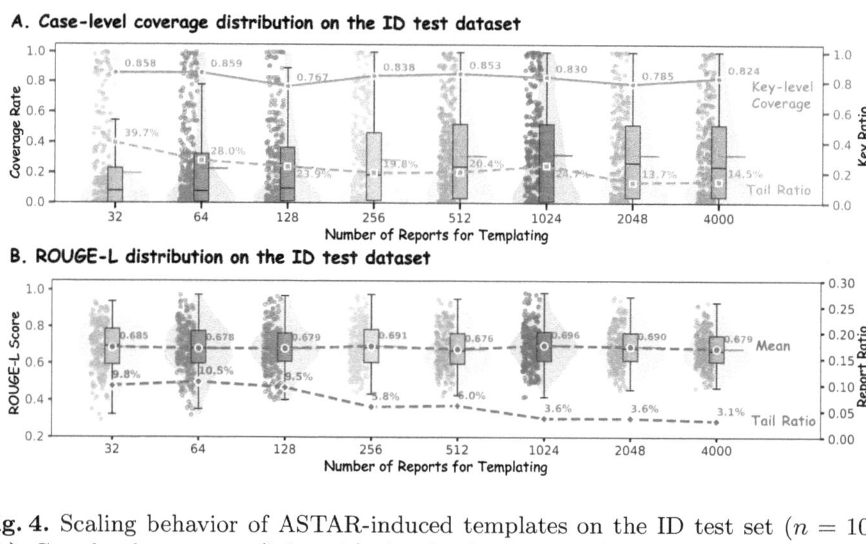

Fig. 4. Scaling behavior of ASTAR-induced templates on the ID test set (n = 100). **(A)** Case-level coverage (left axis), key-level coverage (orange, left axis) and key coverage tail ratio (gold, right axis). Coverage values are ratios bounded in [0,1]. **(B)** ROUGE-L distributions (left axis), mean ROUGE-L (blue, left axis) and report-level tail ratio (red dashed, right axis). (Color figure online)

3.3 Scaling Behavior of ASTAR

To investigate how corpus size affects template quality, we constructed ASTAR templates from subsets of $n = 32$ to $n = 4000$ reports in the training dataset and evaluated each on the ID test dataset (Fig. 4). Key-level coverage remains consistently high (0.767–0.859), while the tail ratio (case-level coverage < 0.01) decreases from 39.7% at $n = 32$ to 14.5% at $n = 2048$ before stabilizing, confirming that ASTAR progressively consolidates rare-but-valid concepts into coherent keys (Fig. 4A). Additionally, mean ROUGE-L remains stable across all corpus sizes (0.676–0.696), while the report-level tail ratio (case-level ROUGE-L $<$ 0.5) drops from 9.8–10.5% at $n \leq 64$ to 3.1% at $n = 4000$, indicating that additional reports primarily improve coverage of structurally atypical or linguistically rare cases rather than boosting average fidelity (Fig. 4B).

4 Conclusion

We presented ASTAR, the LLM-based framework that automatically induces standardized radiology reporting templates from large-scale free-text corpora. Evaluated on 4,215 multi-center fetal brain MRI reports, the ASTAR-induced template outperforms two expert-designed baseline templates in template quality, structuring fidelity, and radiologist-rated usability, reducing template development from weeks of expert consensus to hours of automated processing. Future work will extend the evaluation beyond fetal brain MRI to broader radiology reporting scenarios.

Acknowledgments. This work was supported by the National Natural Science Foundation of China (grant number 82572198), the Natural Science Foundation of Sichuan Province (grant number 2025ZNSFSC1768), the Scientific Research Project of Sichuan Medical Association (grant number 2024HR130), the Science and Technology Department of Sichuan Province, China (grant number 25SYSX0255), the Tsinghua University Startup Fund, and the Tsinghua University Dushi Program (grant numbers 20241080026, 20251080056).

Disclosure of Interests. The authors have no competing interests to declare.

References

1. Bai, X., Liu, M., Chen, Y., Yang, H., Tian, Q.: Chest-OMDL: organ-specific multidisease detection and localization in chest computed tomography using weakly supervised deep learning from free-text radiology report. In: Medical Imaging with Deep Learning (2025). https://openreview.net/forum?id=ns6nq592HX
2. Bai, X., et al.: Exact: an explainable anomaly-aware vision foundation model for analysis of 3D chest CT (2026). https://arxiv.org/abs/2604.24146
3. Brooke, J.: SUS: a 'quick and dirty' usability scale. In: Jordan, P.W., Thomas, B., Weerdmeester, B.A., McClelland, I.L. (eds.) Usability Evaluation in Industry, pp. 189–194. Taylor & Francis, London (1996)

4. Busch, F., et al.: Large language models for structured reporting in radiology: past, present, and future. Eur. Radiol. **35**(5), 2589–2602 (2024). https://doi.org/10.1007/s00330-024-11107-6
5. Chen, H., et al.: MMLNB: multi-modal learning for neuroblastoma subtyping classification assisted with textual description generation (2025). https://arxiv.org/abs/2503.12927
6. Chen, Y., et al.: BURExtract-Llama: an LLM for clinical concept extraction in breast ultrasound reports. In: Proceedings of the 1st International Workshop on Multimedia Computing for Health and Medicine, pp. 53–58. ACM, Melbourne, VIC, Australia (2024). https://doi.org/10.1145/3688868.3689200
7. Christensen, E.W., Drake, A.R., Parikh, J.R., Rubin, E.M., Rula, E.Y.: Projected US imaging utilization, 2025 to 2055. J. Am. Coll. Radiol. **22**(2), 151–158 (2025). https://doi.org/10.1016/j.jacr.2024.10.017
8. Comanici, G., Bieber, E., Schaekermann, M.: Gemini 2.5: pushing the frontier with advanced reasoning, multimodality, long context, and next generation agentic capabilities (2025)
9. Cui, Y., Che, W., Liu, T., Qin, B., Wang, S., Hu, G.: Revisiting pre-trained models for Chinese natural language processing. In: Findings of the Association for Computational Linguistics: EMNLP 2020, pp. 657–668. Association for Computational Linguistics (2020). https://doi.org/10.18653/v1/2020.findings-emnlp.58
10. Cui, Y., Che, W., Liu, T., Qin, B., Yang, Z.: Pre-training with whole word masking for Chinese BERT. IEEE Trans. Audio Speech Lang. Process. (2021). https://doi.org/10.1109/TASLP.2021.3124365. https://ieeexplore.ieee.org/document/9599397
11. Davenport, M.S., Fruscello, T., Chatfield, M., Weinstein, S., Sensakovic, W.F., Larson, D.B.: CT volumes from 2,398 radiology practices in the united states: a real-time indicator of the effect of COVID-19 on routine care, January to September 2020. J. Am. Coll. Radiol. **18**(3), 380–387 (2021). https://doi.org/10.1016/j.jacr.2020.10.010
12. Fytas, P., Breger, A., Selby, I., Baker, S., Shahipasand, S., Korhonen, A.: Can rule-based insights enhance LLMs for radiology report classification? Introducing the RadPrompt methodology. In: Demner-Fushman, D., Ananiadou, S., Miwa, M., Roberts, K., Tsujii, J. (eds.) Proceedings of the 23rd Workshop on Biomedical Natural Language Processing, pp. 212–235. Association for Computational Linguistics, Bangkok, Thailand (2024). https://doi.org/10.18653/v1/2024.bionlp-1.17
13. Lin, C.Y.: ROUGE: a package for automatic evaluation of summaries. In: Text Summarization Branches Out, pp. 74–81. Association for Computational Linguistics, Barcelona, Spain (2004). https://aclanthology.org/W04-1013/
14. Lipscomb, C.E.: Medical subject headings (mesh). Bull. Med. Libr. Assoc. **88**(3), 265 (2000)
15. Liu, M., et al.: FetalExtract-LLM: structured information extraction from free-text fetal MRI reports based on privacy-ensuring open-weights large language models. In: Link-Sourani, D., Abaci Turk, E., Bastiaansen, W., Hutter, J., Melbourne, A., Licandro, R. (eds.) Perinatal, Preterm and Paediatric Image Analysis, pp. 119–129. Springer, Cham (2026). https://doi.org/10.1007/978-3-032-05997-0_11
16. Nowak, S., et al.: Transformer-based structuring of free-text radiology report databases. Eur. Radiol. **33**(6), 4228–4236 (2023). https://doi.org/10.1007/s00330-023-09526-y
17. Nowak, S., et al.: Privacy-ensuring open-weights large language models are competitive with closed-weights GPT-4o in extracting chest radiography findings from free-text reports. Radiology **314**(1), e240895 (2025). https://doi.org/10.1148/radiol.240895

18. Popović, M.: chrF: character n-gram F-score for automatic MT evaluation. In: Proceedings of the Tenth Workshop on Statistical Machine Translation, pp. 392–395. Association for Computational Linguistics, Lisbon, Portugal (2015)
19. Rousseeuw, P.J.: Silhouettes: a graphical aid to the interpretation and validation of cluster analysis. J. Comput. Appl. Math. **20**, 53–65 (1987). https://doi.org/10.1016/0377-0427(87)90125-7
20. Sofia, C., Aertsen, M., Garel, C., Cassart, M.: Standardised and structured reporting in fetal magnetic resonance imaging: recommendations from the Fetal Task Force of the European Society of Paediatric Radiology. Pediatr. Radiol. **54**(10), 1566–1578 (2024). https://doi.org/10.1007/s00247-024-06010-7
21. Yang, A., et al.: Qwen3 technical report. arXiv preprint arXiv:2505.09388 (2025)
22. Zhang, T., Kishore, V., Wu, F., Weinberger, K.Q., Artzi, Y.: Bertscore: evaluating text generation with BERT (2020). https://arxiv.org/abs/1904.09675
23. Zhang, Y., et al.: Content generation models in computational pathology: a comprehensive survey on methods, applications, and challenges. IEEE Rev. Biomed. Eng. **19**, 374–395 (2026). https://doi.org/10.1109/RBME.2025.3619086

AutoLand: A Data-Efficient Automated Agentic Workflow for Universal Medical Landmark Detection

Ziyi Li[1], Jiarui Li[1], Yulong Shi[1], Xue Jiang[2], Shouliang Qi[1], Ting Liu[2(✉)], Lin Qi[1(✉)], and Wei Qian[1]

[1] Northeastern University, Shenyang, China
qilin@bmie.neu.edu.cn
[2] The First Hospital of China Medical University, Shenyang, China
20082180@cmu.edu.cn

Abstract. Medical landmark detection varies significantly across modalities and clinical protocols, posing unique challenges for developing universally applicable models. Conventional approaches typically entail training bespoke architectures per dataset or relying on general-purpose heuristics with fixed rules. However, these methods suffer from limited scalability and lack the semantic reasoning required to adapt to complex anatomical variations. To address these limitations, we propose AutoLand, a data-efficient autonomous workflow for landmark detection in medical images, leveraging a Vision Language Model (VLM) as a medical analysis agent. Specifically, AutoLand integrates a Physics Fingerprint Extraction module to extract precise physical characteristics, ensuring the agent aligns with modality-specific data properties. Subsequently, a Medical Deep Learning Library is queried during reasoning to retrieve specialized network architectures and optimization protocols, enabling the formulation of expert-level training strategies with only a few samples. This workflow enhances robustness and adaptability while reducing manual intervention. Experiments on three datasets show that AutoLand significantly outperforms state-of-the-art methods. Moreover, ablation studies confirm that our agentic paradigm effectively overcomes the limitations of fixed baselines through semantic reasoning. The source code is available at https://github.com/ChiZai123456/AutoLand.

Keywords: Medical landmark detection · Vision language model · Automated agentic workflow · Data-Efficient learning

1 Introduction

Medical landmark detection aims to localize anatomically meaningful key points in medical images and serves as a fundamental step in many clinical applications, such as treatment planning and surgical interventions [25]. Manual landmark annotation is time-consuming and subject to inter-observer variability, motivating automated methods.

G. Yang et al. (Eds.): MICCAI 2026, LNCS 16878, pp. 45–54, 2027.
https://doi.org/10.1007/978-3-032-38059-3_5

Current approaches can be categorized into task-specific methods [15,17,23] and general-purpose methods [5,6,9,11,24,26]. Task-specific methods achieve strong performance by designing bespoke architectures for a particular modality or protocol, but they often require non-trivial redesign of the model when transferred to new scenarios. In contrast, general-purpose methods attempt to infer configurations from dataset statistics or heuristic rules, yet such fixed heuristics lack semantic reasoning, making them suboptimal under complex anatomical variations or few-sample scenarios.

To address these limitations, we introduce AutoLand, which leverages a Vision Language Model (VLM) as a medical analysis agent to formulate dataset-specific training strategies automatically. By interpreting both the visual context and physics fingerprints of the input data, the agent can adapt its decisions to modality-specific properties. Moreover, during reasoning, it queries a specialized medical Deep Learning Library to retrieve valid network architectures and training strategies, enabling robust strategy instantiation without manual tuning.

The contributions of our work are summarized as follows: (1) We propose AutoLand, a data-efficient automated agentic workflow that covers model configuration, preprocessing, training, and validation for landmark detection on new datasets. (2) We leverage the agent's semantic reasoning capability to reduce data dependency, generating effective training strategies from only a few annotated samples. (3) We demonstrate robust generalizability of the proposed workflow and validate it across diverse anatomical targets and imaging modalities.

2 Method

Figure 1 illustrates our data-efficient autonomous landmark detection workflow named AutoLand, which employs a VLM as a medical analysis agent. By leveraging the agent's semantic reasoning capabilities to interpret the physics fingerprints and visual contexts of input datasets, our workflow can formulate the training strategy autonomously. Consequently, our approach achieves universal landmark detection across diverse datasets without requiring manual tuning or redesigning network architecture.

2.1 Physics Fingerprint Extraction

The first step of our approach is to enable the medical analysis agent to understand the input medical images. However, feeding raw medical images to the agent directly limits information fidelity due to resolution constraints. Meanwhile, the determination of training strategies relies on precise physical parameters of input medical images, which are obscured in the visual input of agent [21]. To address these challenges, we extract physics parameters as fingerprints from the input medical images and feed them to the agent alongside the raw images. This ensures that the agent can perceive the dataset comprehensively, combining visual context with precise physics fingerprints, thereby supporting data-efficient strategy reasoning under limited annotations.

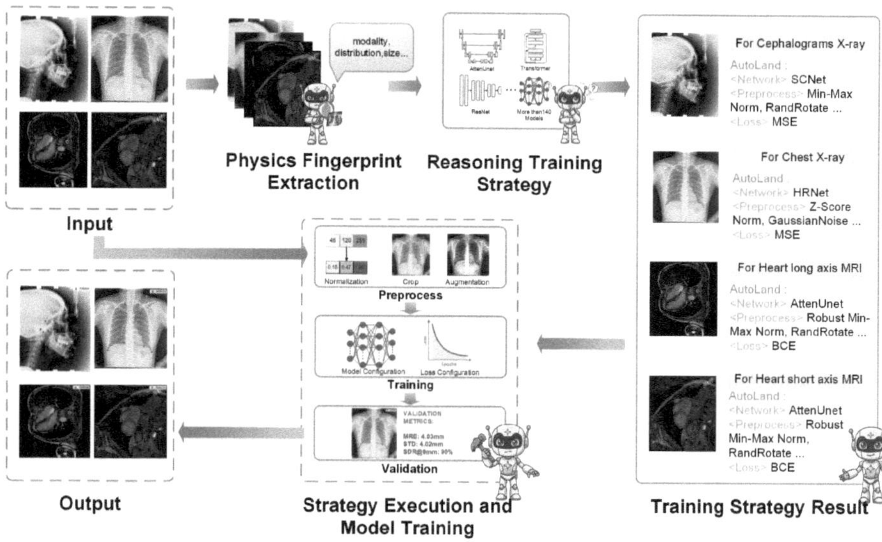

Fig. 1. Overview of our approach. Physics fingerprints are extracted from input data to represent dataset characteristics. Then, the medical analysis agent generates optimal training strategies. Finally, the automated strategy execution module builds the workflow for landmark detection.

Specifically, we define the input medical images as $\mathbf{D} = \{I_n\}_{n=1}^{K}$, where I_n denotes the n-th medical image and K denotes the number of input images. For each image, we define an extraction function as $E(I_n) = \{\mathbf{v}_n, \mathbf{s}_n, \mu_n, m_n\}$, where $\mathbf{v}_n$ denotes voxel spacing, $\mathbf{s}_n$ denotes spatial dimensions, μ_n denotes mean intensity, and m_n denotes max intensity. To eliminate the impact of outliers, we define the median voxel spacing $\tilde{\mathbf{v}}$, median spatial dimension $\tilde{\mathbf{s}}$, and mean intensity $\bar{\mu}$ as $\tilde{\mathbf{v}} = \text{median}(\{\mathbf{v}_n\}_{n=1}^{K})$, $\tilde{\mathbf{s}} = \text{median}(\{\mathbf{s}_n\}_{n=1}^{K})$, and $\bar{\mu} = \frac{1}{K}\sum_{n=1}^{K}\mu_n$. Then, to infer the bit depth of the input images, we define $\omega_{bit} = 16$ for high bit-depth images if $m_n > 255$, and $\omega_{bit} = 8$ for low bit-depth images otherwise. Finally, the physics fingerprints can be described as $\mathbf{f}_{phy} = [\tilde{\mathbf{v}}, \tilde{\mathbf{s}}, \bar{\mu}, \omega_{bit}]$.

2.2 Reasoning Training Strategy from Medical DL Library

The second step of our approach is to enable the medical analysis agent to formulate executable training strategies for different datasets. However, general-purpose agents lack specialized knowledge about domain-specific medical deep learning methods and training protocols. Consequently, the agent may generate strategies that are theoretically reasonable but actually invalid, leading to inefficient trial-and-error under limited annotations [13,18]. To address these challenges, we constructed a specialized medical Deep Learning Library (DL Library) as $\mathbf{M}_{lib}$, which integrates over 140 network architectures from MONAI and preprocessing protocols from SOTA methods [2–4,7,9,16]. Serving as a knowledge base, this DL Library enables the agent to determine training strategies for specific datasets, improving data efficiency by constraining reasoning to executable options.

Specifically, to effectively utilize the DL Library for formulating training strategies, we employ a multi-modal prompt strategy. We define $\mathbf{V}$ as visual tokens encoded from the input image. Then, we define a tokenization mapping function $\Phi(\cdot)$ that serializes textual descriptions into a sequence of text tokens. We apply the mapping function $\Phi(\cdot)$ to transform the physics fingerprint $\mathbf{f}_{phy}$ and the DL Library $\mathbf{M}_{lib}$ into their respective text tokens: $\mathbf{T}_{phy} = \Phi(\mathbf{f}_{phy})$ and $\mathbf{T}_{lib} = \Phi(\mathbf{M}_{lib})$, where $\mathbf{T}_{phy}$ denotes text tokens that describe the physics fingerprint and $\mathbf{T}_{lib}$ denotes text tokens that describe the DL Library. Finally, the multi-modal prompt $\mathbf{P}$ can be described as

$$\mathbf{P} = \begin{bmatrix} \texttt{<System Role>} & : \mathbf{I}_{sys} \\ \texttt{<Visual Input>} & : \mathbf{V} \\ \texttt{<Physics Fingerprint>} & : \mathbf{T}_{phy} \\ \texttt{<Medical DL Library>} & : \mathbf{T}_{lib} \\ \texttt{<Instruction>} & : \mathbf{I}_{task} \end{bmatrix}, \tag{1}$$

Under the guidance of this prompt, the agent aligns the input images with the physics fingerprints and performs reasoning based on the DL Library to formulate the training strategy. The training strategy $\mathbf{S}$ is defined as

$$\mathbf{S} = \begin{bmatrix} \texttt{<Net Arch>} & : f_{net} \in \mathbf{M}_{lib} \\ \texttt{<Pre-process>} & : \rho_{trans} \\ \texttt{<Loss Func>} & : \mathcal{L} \\ \texttt{<Hyper-param>} & : \mathbf{h} \end{bmatrix}, \tag{2}$$

where f_{net} denotes the selected network architecture from the medical DL Library, ρ_{trans} denotes the preprocessing protocol, $\mathcal{L}$ denotes the loss function, and $\mathbf{h}$ denotes the optimization hyperparameters such as learning rate and batch size.

2.3 Strategy Execution and Model Training

The final step of our approach is to instantiate the training strategy into an executable training routine. In this section, we present model training as preprocessing, training, and validation.

In the preprocessing phase, we instantiate the preprocessing stage based on the ρ_{trans} in $\mathbf{S}$. Specifically, ρ_{trans} contains transformation identifiers such as "normalization", "crop", and "augmentation". We map these identifiers to an ordered sequence of executable transformations. In the training phase, we instantiate the network architecture by mapping the identifier f_{net} to its corresponding class definition in the medical DL Library. The executable model is constructed as $\mathbf{M} = \text{Instantiate}(f_{net}, \mathbf{M}_{lib})$. The model parameters are optimized according to the hyperparameters $\mathbf{h}$ and the loss function $\mathcal{L}$. To address invalid strategies generated by the agent, we design a fallback mechanism that defaults to a standard UNet with basic preprocessing upon instantiation failure. In the validation phase, we evaluate the instantiated model on a hold-out validation set to verify the effectiveness of the instantiated strategy. The instantiated model performs inference to predict landmark coordinates, and the validation results are used to assess generalization.

3 Experiments

3.1 Datasets and Evaluation Metrics

We validated our approach on multimodal medical image datasets to demonstrate its effectiveness and generalizability. The Cephalometric X-ray Dataset from the ISBI 2015 Challenge includes 400 images from 400 patients [20], with 19 expert-annotated landmarks per image. The Chest X-ray Dataset from a public Kaggle repository includes 279 images from 279 patients [19], with 6 expert-annotated landmarks per image. The Cardiac MRI dataset consists of 5,293 LAX and 15,024 SAX slices collected from 144 patients at a hospital, with 2 expert-annotated landmarks per image. This study was approved by the local Institutional Review Board (IRB). To evaluate the data-efficient performance of our workflow in few-sample scenarios, we partitioned all datasets into training, validation, and test sets in a 1:1:8 ratio at the patient level for all experiments.

We used standard metrics to evaluate the detection performance: Mean Radial Error (MRE) [16] to quantify accuracy, Standard Deviation (STD) [22] to assess stability, and Success Detection Rate (SDR) [12] at thresholds of 1–10 mm to measure the percentage of successful detections.

3.2 Implementations

All experiments leveraged a single RTX 4090 GPU system with 24 GB memory, running PyTorch 2.9.1, Python 3.10, Ubuntu 22.04, and CUDA 12.8. We used Qwen/Qwen3-VL-235B-A22B-Thinking as the medical analysis agent throughout all experiments.

3.3 Comparisons with SOTA

Table 1 and Fig. 2 present the quantitative comparison and visual results, respectively. As shown, conventional methods such as Unet [1], Shape-Model [8], and RFRV [14] exhibit limited performance due to their fixed network architectures and lack of priors. Advanced methods such as NFDP [7] attempt to improve performance by leveraging specifically designed normalizing flow models. NFDP achieves competitive performance on the Chest X-rays dataset but suffers significant degradation on the Cardiac MRI dataset, indicating poor generalization across modalities. Similarly, universal methods like GU2Net [27] and UniverDetect [16] typically require extensive supervision, leading to suboptimal adaptation in our few-sample scenarios. For instance, both GU2Net and UniverDetect

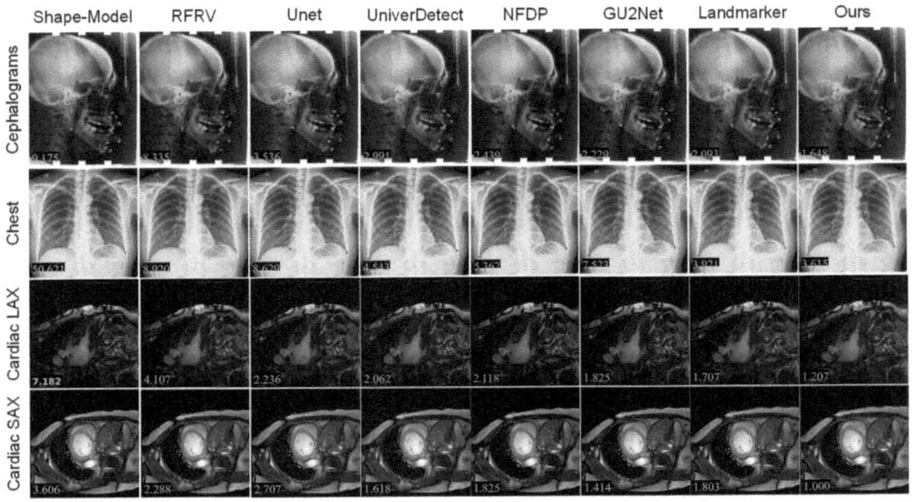

Fig. 2. Visualization of the comparative experiments. The predicted landmarks are red dots, and the ground-truth are green dots. The MRE for each image is displayed in the bottom-left corner. (Color figure online)

lag significantly behind our method, yielding errors of 2.90 mm and 3.70 mm compared to our results, respectively. These inconsistencies highlight the limitations of static training strategies.

In contrast, our approach achieves the best performance across different datasets by using agents to tailor customized training strategies aligned with domain specificity. Notably, our method reduces the MRE to 1.05 mm compared to 2.06 mm for the second-best GU2Net in the Cardiac MRI dataset, highlighting its efficacy in MRI data. Furthermore, our approach attains a 75.82% SDR within 2 mm in the Cephalograms dataset, exceeding the 45.29% achieved by Landmarker [10]. This performance is attributed to the agent's capability to formulate specific training strategies for different datasets based on physics fingerprints rather than relying on fixed protocols. While UniverDetect aims for universality, its performance plateau implies potential difficulties in adapting to the distinct physical properties of MRI versus X-ray. Conversely, our approach addresses these limitations and achieves the best performance across all datasets.

3.4 Ablation Studies

To analyze the influence of the Physics Fingerprint and the DL Library on the agent's strategy reasoning, we conduct an ablation study by removing each component. We summarize the resulting strategy configurations in Table 2 and report the quantitative performance in Table 3.

The Physics Fingerprint is essential for enabling the agent to perceive modality-specific physical characteristics and to derive appropriate preprocessing policies from fewer samples. When the fingerprint is removed, the agent tends to

Table 1. Comparisons of our approach against SOTAs. All dataset was partitioned into training, validation, and test in 1:1:8 to simulate few-sample scenarios. The best results are red, and the second-best results are blue.

Models	Cephalograms					Chest X-ray				
	MRE↓ (mm)	STD↓ (mm)	SDR_d(%)↑ 2 mm	3 mm	4 mm	MRE↓ (mm)	STD↓ (mm)	SDR_d(%)↑ 3 mm	6 mm	9 mm
Shape-Model[8]	5.96	17.55	36.18	55.65	69.07	16.91	21.72	11.39	31.23	41.77
RFRV[14]	5.73	11.96	38.91	56.17	70.31	14.3	22.34	16.93	36.41	49.97
Unet[1]	5.05	12.59	32.86	51.82	65.24	13.81	22.05	18.88	38.96	53.61
UniverDetect[16]	3.7	7.76	41.26	58.52	70.21	7.71	25.61	47.05	78.92	88.23
NFDP[7]	3.66	4.95	47.34	55.26	70.5	5.57	6.08	56.61	78.36	86.73
GU2Net[27]	2.9	3.99	46.94	64.21	76.21	5.61	22.62	50.81	80.08	87.39
Landmarker[10]	2.55	3.67	45.29	70.36	85.58	7.75	20.94	43.92	72.35	84.14
Ours	1.41	1.36	75.82	93.51	98.34	4.03	4.02	59.04	81.12	90.56
Models	**Cardiac LAX**					**Cardiac SAX**				
	MRE↓ (mm)	STD↓ (mm)	SDR_d(%)↑ 2 mm	2.5 mm	3 mm	MRE↓ (mm)	STD↓ (mm)	SDR_d(%)↑ 2 mm	2.5 mm	3 mm
Shape-Model[8]	14.86	18.3	20.82	32.01	40.17	15.71	17.11	20.11	30.17	39.81
RFRV[14]	4.16	4.11	51.34	65.13	75.31	4.11	4.6	51.45	65.03	74.9
Unet[1]	6.31	17.2	33.9	43.75	51.79	7.11	15.84	31.36	42.14	51.34
UniverDetect[16]	3.08	7.34	56.92	70.48	79.68	3.27	5.6	56.11	69.77	79.11
NFDP[7]	6.11	7.5	35.17	45.61	52.37	5.12	11.01	44.2	57.56	66.39
GU2Net[27]	2.06	2.11	84.12	89.98	92.62	2.64	3.31	82.38	89.98	91.11
Landmarker[10]	2.18	2.75	82.34	88.9	91.92	3.17	10.16	66.84	77.45	84.2
Ours	1.05	1.97	94.49	97.08	98.54	1.18	2.48	92.52	94.89	96.13

adopt conservative, less adaptive normalization choices, which limit robustness to intensity variability. This effect is consistently reflected in performance drops across datasets. For example, on the Cardiac LAX dataset, removing the fingerprint increases MRE from 1.05 mm to 2.56 mm and reduces SDR at 2 mm from 94.49% to 43.25%. On the Chest X-ray dataset, the same ablation increases MRE from 4.03 mm to 6.44 mm and reduces SDR at 3 mm from 59.04% to 21.48%. These results indicate that physical fingerprints play a key role in guiding the agent toward preprocessing strategies that align with each dataset.

The DL Library is critical for constraining the agent's reasoning within valid medical deep learning practices, especially for network selection and optimization design. When the DL Library is removed, the agent falls back to generic configurations that are easier to justify but less suited to the anatomical and optimization characteristics of each dataset. This ablation yields the most severe degradation among all variants. On the Cephalogram dataset, removing the DL Library increases MRE from 1.41 mm to 3.89 mm and reduces SDR at 2 mm from 75.82% to 25.05%. On the Cardiac LAX dataset, the same ablation increases MRE from 1.05 mm to 3.26 mm, indicating that domain knowledge is indispensable for identifying effective optimization targets and loss formulations under limited supervision.

Overall, the ablation results confirm the complementary roles of the two components. The Physics Fingerprint improves data perception and steers preprocessing toward modality-aware decisions, while the DL Library provides domain priors that guide strategy reasoning for architecture and optimization. Together,

Table 2. Ablation study on the reasoning training strategy. The table summarizes the specific training strategy formulated by the agent under different Variants.

Dataset	Variant	Network	Preprocessing	Optimization & Loss
Cephalogram	w/o fingerprint	SCNet	Resize, Rot, Standard Min-Max Norm, CLAHE	AdamW, MSE, Cosine, lr = 1×10^{-3}
	w/o DL Library	UNet	Resize, Rot, Flip, Robust Min-Max Norm, Bright, CLAHE	Adam, MSE, lr = 1×10^{-4}
	Ours(AutoLand)	SCNet	Resize, Rot, Flip, Robust Min-Max Norm, Bright, CLAHE	AdamW, MSE, Cosine, lr = 1×10^{-3}
Chest	w/o fingerprint	HRNet	Resize, Rot, Standard Min-Max Norm, CLAHE	AdamW, MSE, Cosine, lr = 1×10^{-3}
	w/o DL Library	UNet	Resize, Flip, Rot, Z-Score Norm, CLAHE	Adam, MSE, lr = 1×10^{-4}
	Ours(AutoLand)	HRNet	Resize, Rot, Flip, Z-Score Norm, Gaussian Noise, CLAHE	AdamW, MSE, Cosine, lr = 1×10^{-3}
Cardiac	w/o fingerprint	AttenUnet	Resize, Rot, Standard Min-Max, CLAHE	AdamW, BCE, Cosine, lr = 5×10^{-4}
	w/o DL Library	UNet	Resize, Rot, Flip, Robust Min-Max Norm, CLAHE	Adam, MSE, lr = 1×10^{-4}
	Ours(AutoLand)	AttenUnet	Resize, Rot, Flip, Robust Min-Max Norm, CLAHE	AdamW, BCE, Cosine, lr = 5×10^{-4}

Table 3. Quantitative ablation results.

Variant	Cephalograms					Chest X-ray				
	MRE↓ (mm)	STD↓ (mm)	SDR_d(%)↑ 2 mm	3 mm	4 mm	MRE↓ (mm)	STD↓ (mm)	SDR_d(%)↑ 3 mm	6 mm	9 mm
w/o fingerprint	3.07	1.91	32.1	55.84	74.36	6.44	7.31	21.48	56.82	80.92
w/o DL Library	3.89	2.23	25.05	46.82	66.55	7.4	12.35	20.68	54.21	76.1
Ours(AutoLand)	1.41	1.36	75.82	93.51	98.34	4.03	4.02	59.04	81.12	90.56
Variant	Cardiac LAX					Cardiac SAX				
	MRE↓ (mm)	STD↓ (mm)	SDR_d(%)↑ 2 mm	2.5 mm	3 mm	MRE↓ (mm)	STD↓ (mm)	SDR_d(%)↑ 2 mm	2.5 mm	3 mm
w/o fingerprint	2.56	4.43	43.25	67.79	81.88	2.77	5.39	41.73	67.11	82.08
w/o DL Library	3.26	4.72	39.19	63.78	79.35	3.34	6.93	37.83	66.66	75.03
Ours(AutoLand)	1.05	1.97	94.49	97.08	98.54	1.18	2.48	92.52	94.89	96.13

they enable the agent to generate executable strategies that generalize across modalities in few-sample scenarios.

4 Conclusion

The proposed AutoLand presents a novel autonomous framework for data-efficient, agentic workflows in universal medical anatomical landmark detection. By coupling the Physics Fingerprint for data perception with the Medical DL Library for strategy reasoning, AutoLand demonstrates robust and adaptable detection across diverse anatomical targets and imaging modalities. Replacing static heuristics with dynamic agentic reasoning offers a promising path for reducing reliance on manual expertise when deploying models to new clinical scenarios. Future work will extend AutoLand to large-scale 3D volumetric datasets and incorporate interactive physician feedback to facilitate clinical integration.

Acknowledgements. This work was supported by the Key Research and Development Program of Liaoning Province (2024JH2/102500076) and Fundamental Research Funds for the Central Universities (N25BJD013).

Disclosure of Interests. The authors have no competing interests to declare that are relevant to the content of this article.

References

1. Cao, H., et al.: Swin-unet: unet-like pure transformer for medical image segmentation. In: European Conference on Computer Vision, pp. 205–218. Springer, Cham (2022)
2. Cardoso, M.J., et al.: MonAI: an open-source framework for deep learning in healthcare. arXiv preprint arXiv:2211.02701 (2022)
3. Choi, M., Jang, J.S.: Heatmap-based active shape model for landmark detection in lumbar x-ray images. J. Imaging Inform. Med. **38**(1), 291–308 (2025)
4. Goyal, S., et al.: HealAI: a healthcare LLM for effective medical documentation. In: Proceedings of the 17th ACM International Conference on Web Search and Data Mining, pp. 1167–1168 (2024)
5. He, Y., Yang, D., Roth, H., Zhao, C., Xu, D.: Dints: differentiable neural network topology search for 3D medical image segmentation. In: Proceedings of the IEEE/CVF Conference on Computer Vision and Pattern Recognition, pp. 5841–5850 (2021)
6. Huang, Z., et al.: Pele scores: pelvic x-ray landmark detection with pelvis extraction and enhancement. Int. J. Comput. Assist. Radiol. Surg. **19**(5), 939–950 (2024)
7. Huang, Z., et al.: Landmark localization from medical images with generative distribution prior. IEEE Trans. Med. Imaging **43**(7), 2679–2692 (2024)
8. Ibragimov, B., Likar, B., Pernuš, F., Vrtovec, T.: Shape representation for efficient landmark-based segmentation in 3-D. IEEE Trans. Med. Imaging **33**(4), 861–874 (2014)
9. Isensee, F., Jaeger, P.F., Kohl, S.A., Petersen, J., Maier-Hein, K.H.: nnu-net: a self-configuring method for deep learning-based biomedical image segmentation. Nat. Methods **18**(2), 203–211 (2021)
10. Jonkers, J., Duchateau, L., Van Wallendael, G., Van Hoecke, S.: landmarker: a toolkit for anatomical landmark localization in 2D/3D images. SoftwareX **30**, 102165 (2025)
11. Kevin, Z.: Learning to localize cross-anatomy landmarks in x-ray images with a universal model. BME Front. (2022)
12. Lee, M., Chung, M., Shin, Y.G.: Cephalometric landmark detection via global and local encoders and patch-wise attentions. Neurocomputing **470**, 182–189 (2022)
13. Li, C., et al.: Llava-med: training a large language-and-vision assistant for biomedicine in one day. Adv. Neural. Inf. Process. Syst. **36**, 28541–28564 (2023)
14. Lindner, C., Bromiley, P.A., Ionita, M.C., Cootes, T.F.: Robust and accurate shape model matching using random forest regression-voting. IEEE Trans. Pattern Anal. Mach. Intell. **37**(9), 1862–1874 (2014)

15. Liu, C., Xie, H., Zhang, S., Mao, Z., Sun, J., Zhang, Y.: Misshapen pelvis landmark detection with local-global feature learning for diagnosing developmental dysplasia of the hip. IEEE Trans. Med. Imaging **39**(12), 3944–3954 (2020)
16. Lu, C., Yang, G., Qiao, X., Chen, W., Zeng, Q.: Univerdetect: universal landmark detection method for multidomain x-ray images. Neurocomputing **600**, 128157 (2024)
17. Nie, X., Feng, J., Zhang, J., Yan, S.: Single-stage multi-person pose machines. In: Proceedings of the IEEE/CVF International Conference on Computer Vision, pp. 6951–6960 (2019)
18. Pal, A., Umapathi, L.K., Sankarasubbu, M.: Med-halt: medical domain hallucination test for large language models. arXiv preprint arXiv:2307.15343 (2023)
19. Pandey, N.: Chest x-ray masks and labels (2018). https://www.kaggle.com/datasets/nikhilpandey360/chest-xray-masks-and-labels
20. Rakshit, S.: ISBI challenge dataset (2015) (2018). https://www.kaggle.com/datasets/soumikrakshit/isbi-challenge-dataset
21. Simpson, A.L., et al.: A large annotated medical image dataset for the development and evaluation of segmentation algorithms. arXiv preprint arXiv:1902.09063 (2019)
22. Wang, C.W., et al.: A benchmark for comparison of dental radiography analysis algorithms. Med. Image Anal. **31**, 63–76 (2016)
23. Wang, L., Xu, Q., Leung, S., Chung, J., Chen, B., Li, S.: Accurate automated cobb angles estimation using multi-view extrapolation net. Med. Image Anal. **58**, 101542 (2019)
24. Yan, X., Jiang, W., Shi, Y., Zhuo, C.: MS-NAS: multi-scale neural architecture search for medical image segmentation. In: International Conference on Medical Image Computing and Computer-Assisted Intervention, pp. 388–397. Springer, Cham (2020)
25. Zhou, S.K., et al.: A review of deep learning in medical imaging: imaging traits, technology trends, case studies with progress highlights, and future promises. Proc. IEEE **109**(5), 820–838 (2021)
26. Zhu, H., Quan, Q., Yao, Q., Liu, Z., Zhou, S.K.: Uod: universal one-shot detection of anatomical landmarks. In: International Conference on Medical Image Computing and Computer-Assisted Intervention, pp. 24–34. Springer, Cham (2023)
27. Zhu, H., Yao, Q., Xiao, L., Zhou, S.K.: You only learn once: universal anatomical landmark detection. In: International Conference on Medical Image Computing and Computer-Assisted Intervention, pp. 85–95. Springer, Cham (2021)

CARE: Clinically Aligned Retrieval Evidence for Consistent Radiology Report Generation

Jintao Zhu[1], Yi Gu[2], and Jiaxin Ma[1,3](✉)

[1] Zhejiang CAS Angels Biotechnology Co. Ltd., Jiaxing, China
[2] Nara Institute of Science and Technology, Ikoma, Japan
[3] OMRON SINIC X Corporation, Bunkyō, Japan
jiaxin.ma@sinicx.com

Abstract. Automatic radiology report generation has the potential to assist radiologists by providing preliminary clinical descriptions. Despite recent progress in medical vision-language models, generated reports often contain factual inconsistencies and repetitive templated findings, indicating weak grounding to image evidence. Retrieval-augmented generation (RAG) improves grounding by conditioning on prior reports; however, existing end-to-end optimization suffers from representation inconsistency, as actively updated query representations diverge from stale historical keys. This mismatch destabilizes the retrieval space and disrupts supervision. We propose CARE (Clinically Aligned Retrieval Evidence), a multimodal framework introducing temporally coherent retrieval supervision. CARE maintains a shared memory queue populated by a slowly evolving momentum key encoder to ensure a stable representation space. We further introduce a Disease-Aware Alignment objective that enforces diagnostic agreement between input images and retrieved reports to guide generation. Trained on IU-Xray and evaluated on the MIMIC-CXR-RRG dataset, CARE significantly improves report quality. It nearly doubles clinical entity extraction performance in training-free adaptation settings and substantially reduces repetitive report collapse across multiple backbones. The source code is available at: https://github.com/Levii-zhu/CARE.

Keywords: Radiology report generation · Retrieval-augmented generation · Vision-language models · Chest X-ray

1 Introduction

Automated radiology report generation aims to produce clinically grounded diagnostic descriptions from medical images, alleviating radiologists' workload and reducing diagnostic variability [2,10]. A high-quality report must provide structured and precise interpretations of visual findings to support downstream therapeutic decisions [12]. Recent progress in multimodal large language models (MLLMs) has demonstrated strong biomedical visual understanding [15,21].

G. Yang et al. (Eds.): MICCAI 2026, LNCS 16878, pp. 55–65, 2027.
https://doi.org/10.1007/978-3-032-38059-3_6

Despite this, directly applying these models to radiology reporting remains challenging: generated reports often contain factual inconsistencies or unsupported clinical findings, or fall back to repetitive templates lacking patient-specific clinical details [9,24,33]. Unlike generic image captioning, radiology reporting requires strict alignment between visual evidence and pathological interpretation [13], making even minor factual errors clinically problematic.

Retrieval-Augmented Generation (RAG) introduces external clinical evidence by conditioning report generation on prior cases [4]. While traditional systems employ a frozen retriever, end-to-end RAG jointly optimizes the retriever with the generation objective [14]. However, as the query encoder is updated through backpropagation, its representations may gradually diverge from stale historical keys stored in the retrieval queue, creating inconsistent retrieval supervision. This mismatch can weaken reliable image–pathology associations and encourage the generator to rely on frequent textual patterns rather than retrieved visual evidence. Such shortcut learning [5] is particularly problematic under cross-dataset shifts, where diagnostic statements should remain grounded in the input image. Unlike conventional momentum-contrastive learning, where queued keys primarily serve as negative samples, our method uses the queue as a source of retrieved clinical evidence for report generation. Therefore, temporal drift affects not only contrastive supervision but also the evidence conditioned on by the generator.

Motivated by these challenges, we propose CARE (Clinically Aligned Retrieval Evidence), a retrieval-grounded training framework designed to maintain coherent retrieval supervision during optimization. CARE adopts a MoCo-inspired architecture [6] with a gradient-updated query encoder and a slowly evolving momentum encoder, ensuring that historical keys in the memory queue reside in a smoothly evolving representation space. In addition, we introduce a Disease-Aware Alignment objective using CheXbert [27]. Unlike prior disease-matching retrieval constraints [26], CARE couples diagnostic compatibility with differentiable retrieval weights, encouraging the generator to condition on clinically compatible evidence rather than superficially similar text.

We train on the IU-Xray dataset [3] and evaluate out-of-domain generalization via training-free adaptation on the MIMIC-CXR benchmark [11]. **The contributions** of this work are threefold: 1) We propose CARE, a retrieval-grounded framework that maintains temporally coherent retrieval supervision through a momentum-updated key encoder and a shared memory queue. Unlike conventional contrastive queues, the stored representations are associated with reports used as generation evidence, making retrieval consistency important for both representation learning and report generation. 2) We introduce a differentiable disease-aware evidence alignment objective that couples retrieval weighting with diagnostic compatibility, encouraging the model to select clinically compatible retrieved reports rather than merely textually similar evidence. 3) Extensive evaluations demonstrate superior clinical efficacy on IU-Xray and robust out-of-domain generalization on MIMIC-CXR. Empirically, CARE nearly doubles the

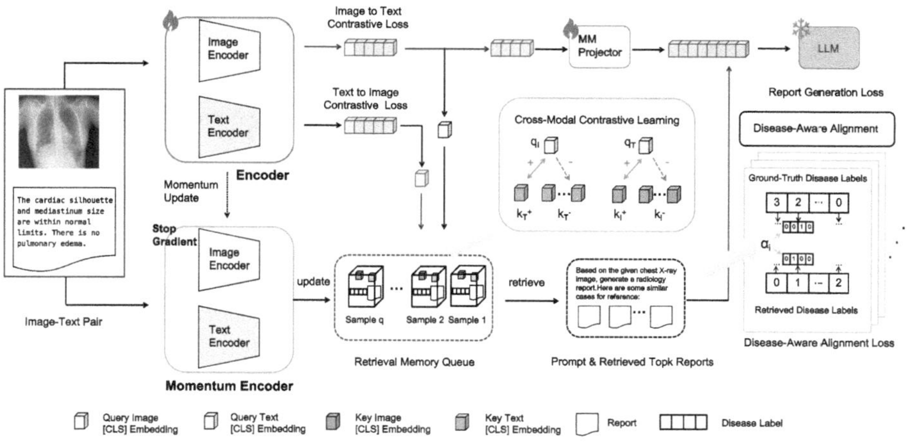

Fig. 1. Overview of the CARE framework. A chest X-ray image is processed by the query encoder to extract query features q_I and q_T, while the momentum key encoder (updated via exponential moving average) provides temporally coherent key representations k_I and k_T to populate a shared Retrieval Memory Queue. We enforce a Disease-Aware Alignment Loss using soft retrieval weights α_i to ensure the retrieved top-K reports are diagnostically consistent with the visual input. Finally, the projected image features H_{proj} and the retrieved clinical evidence R are fed into a frozen LLM to autoregressively generate the diagnostic report.

clinical entity extraction performance of its base model in training-free adaptation settings and remains competitive with fully supervised models.

2 Method

Consistent Memory Queue via Momentum Encoders. To prevent the retrieval representation space from becoming inconsistent during active end-to-end optimization, CARE adopts a momentum-driven memory architecture [6]. As illustrated in Fig. 1, a gradient-updated query encoder f_q extracts query features $q_I \in \mathbb{R}^d$ and $q_T \in \mathbb{R}^d$ from an input image I and its corresponding report T. Concurrently, a slowly evolving momentum encoder f_k processes the same inputs to produce key features k_I and k_T. We maintain a First-In-First-Out (FIFO) dynamic memory queue $\mathcal{Q}$ of size M. For each historical sample, $\mathcal{Q}$ stores the momentum-encoded keys (k_I, k_T), the ground-truth disease label vector y, and the raw report text r. Crucially, the momentum encoder parameters θ_k are updated strictly via an exponential moving average (EMA) of the query parameters ($\theta_k \leftarrow m\theta_k + (1-m)\theta_q$). This mathematical constraint decouples the memory keys from abrupt backpropagation updates, ensuring that the historical representations in $\mathcal{Q}$ reside in a smoothly evolving, consistent space.

Contrastive Retrieval Learning. Building upon recent successes in cross-modal representation learning (e.g., ALBEF [17]) and medical vision-language pre-training [30,35], we optimize the retrieval space using the coherent keys established in $\mathcal{Q}$. Specifically, we apply a symmetric InfoNCE loss [22] to contrast the active queries against their corresponding positive keys, while pushing away negative keys sampled from both the dynamic memory queue and the current mini-batch. The objective is defined as:

$$\mathcal{L}_{contrast} = -\frac{1}{2} \sum_{v \in \{I,T\}} \log \frac{\exp(q_v \cdot k_{\bar{v}}^{+} / \tau_{nce})}{\sum_{k \in \{k_{\bar{v}}^{+}\} \cup \mathcal{Q}_{\bar{v}} \cup \mathcal{B}_{\bar{v}}} \exp(q_v \cdot k / \tau_{nce})}, \tag{1}$$

where $v \in \{I, T\}$ denotes the query modality, and $\bar{v}$ denotes its cross-modal counterpart. For a given query q_v, $k_{\bar{v}}^{+}$ represents the positive key feature from the paired sample. The denominator computes the similarity over the positive key and a comprehensive set of negative keys k, drawn from the union of the cross-modal memory queue $\mathcal{Q}_{\bar{v}}$ and the current mini-batch $\mathcal{B}_{\bar{v}}$. By leveraging the momentum queue, this contrastive objective effectively decouples the negative sample size from the mini-batch limit while maintaining representation consistency.

Disease-Aware Clinical Evidence Alignment. For enforcing diagnostic consistency, we retrieve the top-K candidates from $\mathcal{Q}$ based on cosine similarity $s_i = \text{sim}(q_I, k_{T_i})$ and assign soft weights via softmax relaxation as $\alpha_i = \exp(s_i/\tau_{align}) / \sum_{j=1}^{K} \exp(s_j/\tau_{align})$, where τ_{align} is a temperature hyperparameter controlling the sharpness of the retrieval weight distribution. Let y and y_{T_i} denote the clinical state vectors [27] of the query and the i-th retrieved sample. The alignment objective minimizes the expected diagnostic penalty computed by Mean Squared Error (MSE) as $\mathcal{L}_{align} = \sum_{i=1}^{K} \alpha_i \cdot \text{MSE}(\text{OneHot}(y), \text{OneHot}(y_{T_i}))$. Crucially, gradients propagate through α_i to the query encoder, coupling retrieval weighting with visual representation learning.

Retrieval-Grounded Report Generation and Optimization. The frozen LLM is conditioned on the retrieved reports $R = \{r_1, \ldots, r_K\}$ and the image patch tokens H_p. Specifically, H_p is mapped to the LLM's text embedding space via a trainable Multimodal (MM) Projector [19], yielding the projected features H_{proj}. For seamless integration of visual and textual evidence, we construct the multimodal input using a fixed instruction prompt: *"Based on the given chest X-ray image, generate a radiology report. Here are some similar cases for reference: [R]"*, where [R] represents the concatenated text of the top-K retrieved reports. The generation loss is computed as the standard autoregressive negative log-likelihood $\mathcal{L}_{gen} = -\sum_t \log P(w_t \mid w_{<t}, H_{proj}, R)$, where w_t denotes the t-th token of the ground-truth report and $w_{<t}$ represents all preceding tokens. During training, the LLM remains frozen, and the momentum encoders are updated strictly via exponential moving average. Only the query encoders and the MM Projector are updated by minimizing the joint objective. The three objectives respectively regularize representation consistency, clinical compatibility, and generative fidelity, combined as $\mathcal{L}_{total} = \mathcal{L}_{contrast} + \mathcal{L}_{align} + \mathcal{L}_{gen}$.

3 Experiments

Datasets and Evaluation Metrics. CARE is evaluated under two settings: *in-domain* performance and *out-of-domain* generalization. IU-Xray [3] serves as the primary training corpus and is evaluated using the official patient-level split. For out-of-domain assessment, we adopt the MIMIC-CXR-RRG benchmark [33,34], a standardized evaluation subset derived from the official MIMIC-CXR test set [11]. This cross-dataset setting introduces substantial variations in disease prevalence, institutional reporting styles, and imaging protocols. To assess generalization without prior target-domain supervised training, we employ the LLaVA-Med [15] backbone, whose pre-training data excludes MIMIC-CXR. All parameters optimized on IU-Xray remain frozen, and adaptation to the target reporting style is achieved solely by populating the external memory queue with text-only reports from MIMIC-CXR. We report standard natural language generation (NLG) metrics (BLEU, METEOR, ROUGE-L) [1,18,25], together with clinically grounded evaluations, including RadGraph F1 and Exact Reward (RG-F_1, RG_{ER}) [8], as well as CheXbert-based micro- and macro-F1 over five CheXpert categories [7,27], denoted ^{5}Mi-F_1 and ^{5}Ma-F_1, respectively. We also report RadCliQ0 ($CliQ_0$; $\downarrow$), a composite metric reflecting radiologist preferences [31].

Implementation Details. We instantiate CARE on LLaVA-Med [15] and LLaVA-Rad [32]. For fair comparison, we freeze the core LLM and optimize only the query encoder, multimodal projector, and momentum retrieval components. Models are trained on 8 NVIDIA A100 GPUs for 20 epochs using AdamW [20] (lr$=5 \times 10^{-5}$, batch size$=32$, bf16). Retrieval parameters follow the configuration of $K = 5$, $m = 0.999$, $|\mathcal{Q}| = 1000$, and $\tau_{nce} = \tau_{align} = 0.07$. Open-source baselines are locally re-evaluated using official weights unless otherwise noted.

In-Domain Performance. Evaluation on the IU-Xray dataset demonstrates the effectiveness of CARE across both the general-domain LLaVA-Med backbone and the medically optimized LLaVA-Rad backbone. As shown in Table 1, CARE consistently outperforms the corresponding SFT variants on most evaluation metrics, while remaining competitive with the compared baselines(XrayGPT, R2GenGPT, and EVCAP [16,28,29]). An interesting observation is that the improvements are not always reflected in RG-F_1 and $CliQ_0$, two clinically important report-level metrics. In several cases, CARE achieves comparable or slightly lower scores than the corresponding SFT models despite outperforming them on most other metrics. We hypothesize that this behavior is related to differences in report diversity. Standard SFT tends to imitate dominant normal-style reporting templates, whereas retrieval-augmented CARE generates more diverse reports conditioned on retrieved evidence. To examine this effect, Fig. 2 reports the cumulative number of unique N-grams generated across the test set. For both LLaVA-Med and LLaVA-Rad, CARE exhibits substantially higher vocabulary diversity, while the corresponding SFT models produce markedly flatter

Table 1. Performance evaluation on the IU-Xray dataset. Models are grouped by their exposure to the training set.

Dataset: IU-Xray									
Model	Clinical Metrics					Lexical Metrics			
	^{5}Mi-F$_1$	^{5}Ma-F$_1$	RG-F$_1$	RG$_{ER}$	CliQ$_0$ (↓)	B-1	B-4	MTR	R-L
Zero- or Few-Shot Models									
MedFlamingo [21] (Few-shot)	5.1	3.2	5.0	5.1	4.86	14.2	1.3	8.8	10.0
LLaVA-Med [15]	3.9	1.0	6.4	7.0	4.51	13.4	0.7	12.4	10.1
Supervised Models (Either Pretrained or Finetuned on IU-Xray)									
XrayGPT [28]	12.2	12.3	30.4	32.4	3.02	32.3	5.6	23.9	24.1
R2GenGPT [29]	3.9	3.5	24.8	24.8	3.27	38.2	12.8	**30.9**	24.6
EVCAP [16]	11.2	5.4	20.2	32.1	3.72	16.8	3.3	25.2	16.5
LLaVA-Med (SFT) [15]	1.0	0.3	30.1	25.3	3.18	25.7	8.6	19.6	27.9
CARE (LLaVA-Med)	20.4	15.4	27.5	29.5	3.17	30.3	6.9	24.7	24.0
LLaVA-Rad [32]	27.9	23.9	28.5	30.0	3.03	37.0	8.9	25.3	27.3
LLaVA-Rad (SFT) [32]	34.4	23.2	**38.0**	34.0	**2.55**	39.1	13.3	28.0	**34.4**
CARE (LLaVA-Rad)	**35.1**	**26.9**	32.2	**35.4**	2.75	**40.7**	**14.6**	30.3	31.9

diversity curves, indicating stronger reliance on recurrent report templates. We further argue that this trade-off is generally favorable from a clinical perspective. Although template-driven generation can maintain competitive RG-F$_1$ and CliQ$_0$ scores, excessive reliance on normal-style templates may impair abnormality recognition. This effect is particularly evident for LLaVA-Med, whose SFT version exhibits extremely low diversity and even achieves lower ^{5}Mi-F$_1$ and ^{5}Ma-F$_1$ scores than the original backbone, suggesting degraded abnormality-label agreement after fine-tuning. A similar but less severe trend can be observed for LLaVA-Rad, where the SFT version attains a lower ^{5}Ma-F$_1$ than the original model. In contrast, CARE avoids this performance regression. These findings suggest that CARE achieves a better balance between report-level similarity and pathology-level correctness, mitigating the mode collapse commonly observed in standard SFT.

Out-of-Domain Generalization. Evaluation of whether CARE captures invariant diagnostic semantics—rather than overfitting to dataset-specific patterns—is conducted via out-of-domain assessment on the MIMIC-CXR-RRG benchmark [11] (Table 2). For training-free adaptation, all IU-Xray optimized parameters remain frozen. Leveraging RAG's plug-and-play nature, report-style adaptation is achieved simply by populating the memory queue with 500 randomly sampled text-only reports from MIMIC-CXR-RRG. As inference relies solely on cross-modal retrieval, this setting requires no target-domain images or labels and evaluates adaptation under the absence of target-domain supervision. Unlike specialized medical baselines (e.g., LLaVA-Rad [32], XrayGPT [28], and

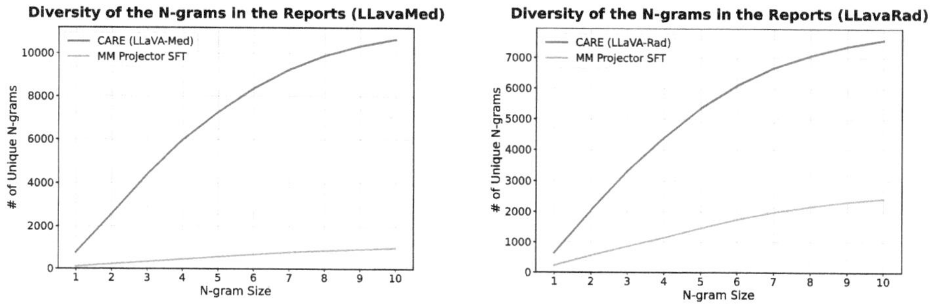

Fig. 2. Cumulative count of unique N-grams generated across the test set. CARE maintains substantially higher vocabulary diversity.

Table 2. Out-of-domain generalization on the MIMIC-CXR-RRG benchmark. Baseline results for GPT-4V, MedFlamingo, and LLaVA are reported from [32]. Best and second-best results under the training-free adaptation setting are highlighted in bold and underlined, respectively.

Dataset: MIMIC-CXR-RRG									
Model	Clinical Metrics					Lexical Metrics			
	[5]Mi-F_1	[5]Ma-F_1	RG-F_1	RG$_{ER}$	CliQ$_0$ (↓)	B-1	B-4	MTR	R-L
Supervised Models (Trained on MIMIC-CXR)									
XrayGPT [28]	39.5	30.0	19.7	18.1	3.66	30.3	5.9	19.1	21.9
R2GenGPT [29]	22.7	19.2	17.1	17.6	3.68	32.3	7.8	23.3	20.5
LLaVA-Rad [32]	55.7	46.6	21.2	20.7	3.41	33.4	9.7	23.0	25.1
Zero- or Few-Shot Models									
GPT-4V [23]	25.8	19.6	**13.2**	–	–	16.4	1.9	-	13.2
MedFlamingo [21] (Few-shot)	21.1	12.1	9.2	7.1	4.37	10.6	1.3	9.6	14.4
LLaVA [19]	23.4	17.5	4.5	–	–	21.0	1.3	-	13.8
LLaVA-Med [15]	35.2	17.5	7.0	6.3	4.26	23.2	1.2	12.0	15.6
CARE (LLaVA-Med)	**37.7**	**30.4**	13.0	**13.2**	**3.88**	**31.4**	**4.3**	**17.3**	**17.8**

R2GenGPT [29]) that inherently include MIMIC-CXR in their training corpora, the LLaVA-Med [15] backbone was never exposed to MIMIC-CXR data. Evaluated under this training-free domain adaptation setting, CARE demonstrates robust generalization, nearly doubling the clinical entity extraction performance of the LLaVA-Med baseline (RG-F1 [8] improving from 7.0 to 13.0). Furthermore, as qualitatively demonstrated in Fig. 3, CARE effectively overcomes the memorization of dataset-specific lexical priors, successfully identifying subtle clinical pathologies while baseline models suffer from severe mode collapse.

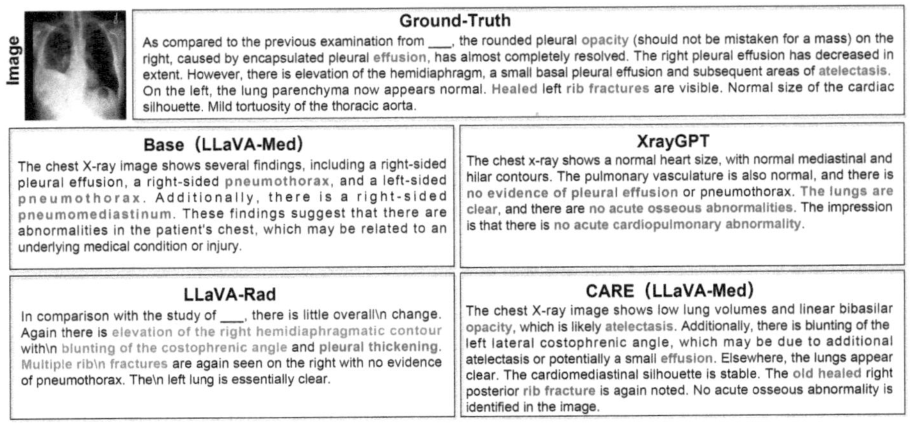

Fig. 3. Qualitative comparison under the training-free adaptation setting. Unlike baselines that suffer from mode collapse (e.g., outputting rigid "clear lungs" templates) or unsupported clinical findings (e.g., inventing pneumothorax), CARE effectively aligns with genuine clinical pathology, correctly identifying atelectasis, small effusion, and the subtle "old healed" attribute of the rib fracture.

Table 3. Ablation study evaluating the step-wise contributions of CARE components using the LLaVA-Med [15] backbone on the IU-Xray dataset.

Dataset: IU-Xray									
Ablation Setting	Clinical Metrics					Lexical Metrics			
	^{5}Mi-F$_1$	^{5}Ma-F$_1$	RG-F$_1$	RG$_{ER}$	CliQ$_0$ (↓)	B-1	B-4	MTR	R-L
Backbone: LLaVA-Med									
Pre-trained Base	3.9	1.0	6.4	7.0	4.51	13.4	0.7	12.4	10.1
+ In-batch CRL	13.8	8.5	25.4	28.9	3.44	22.0	3.8	23.5	19.9
+ SFT (MM Projector) [19]	14.6	12.2	26.3	29.0	3.20	28.1	5.5	24.3	22.8
+ Disease-Aware Alignment	14.9	12.4	26.9	29.3	3.22	28.2	6.6	24.6	23.1
→ Momentum Memory (**CARE**)	**20.4**	**15.4**	**27.5**	**29.5**	**3.17**	**30.3**	**6.9**	**24.7**	**24.0**

Ablation Study. We conduct an ablation study on the LLaVA-Med [15] backbone to analyze the incremental contributions of CARE components (Table 3). Starting from the *Pre-trained Base*, we first introduce retrieval-augmented generation optimized via In-batch Contrastive Retrieval Learning (+ In-batch CRL), where training relies solely on in-batch negatives without a memory queue. The retriever is trained within the batch but remains fixed at inference. We then incorporate Supervised Fine-Tuning (+ SFT) to optimize the multimodal projector for end-to-end report generation, which substantially improves clinical extraction (e.g., RG-F$_1$ reaching 26.3). However, in-batch sampling limits the diversity of retrieval supervision and does not provide a temporally consistent representation space for dynamic optimization. Adding the Disease-Aware Alignment objective further enforces diagnostic compatibility between images and retrieved reports, yielding consistent gains. Finally, replacing in-batch sampling with a momentum-driven memory queue (→ Momentum Memory, **CARE**) introduces a dynamically maintained yet temporally stable retrieval space. This design enlarges the negative pool beyond batch constraints while ensuring smooth representation evolution during joint training. As a result, CARE achieves the best overall clinical performance, demonstrating the importance of coherent and stable retrieval supervision.

4 Conclusion

We introduced CARE, a retrieval-grounded framework mitigating representation inconsistency for end-to-end radiology report generation. By integrating a momentum memory queue and a Disease-Aware Alignment Objective, CARE encourages reports to be conditioned on clinically compatible visual evidence rather than dataset-specific priors. Evaluations show that CARE reduces shortcut learning [5] and template collapse while improving clinical entity extraction in training-free adaptation settings. CARE still lacks explicit filtering for severely erroneous retrieved evidence and has not been evaluated under adversarially corrupted retrieval evidence. It also lacks entity- or region-level grounding and expert radiologist evaluation. Future work will address these limitations through retrieval verification, fine-grained grounding, and parameter-efficient LLM adaptation.

Disclosure of Interests. The authors have no competing interests to declare that are relevant to the content of this article.

References

1. Banerjee, S., Lavie, A.: METEOR: an automatic metric for MT evaluation with improved correlation with human judgments. In: Proceedings of the ACL Workshop on Intrinsic and Extrinsic Evaluation Measures for Machine Translation and/or Summarization, pp. 65–72 (2005)
2. Chen, Z., Song, Y., Chang, T.H., Wan, X.: Generating radiology reports via memory-driven transformer. In: Proceedings of the 2020 Conference on Empirical Methods in Natural Language Processing (EMNLP), pp. 1439–1449 (2020)
3. Demner-Fushman, D., Kohli, M.D., Rosenman, M.B., et al.: Preparing a collection of radiology examinations for distribution and retrieval. J. Am. Med. Inf. Assoc. **23**(2), 304–310 (2016)
4. Endo, M., Krishnan, R., Krishna, V., Ng, A.Y., Rajpurkar, P.: Retrieval-based chest x-ray report generation using a pre-trained contrastive language-image model. In: Proceedings of Machine Learning for Health, vol. 158, pp. 209–219 (2021)
5. Geirhos, R., Jacobsen, J.H., Michaelis, C., et al.: Shortcut learning in deep neural networks. Nat. Mach. Intell. **2**(11), 665–673 (2020)
6. He, K., Fan, H., Wu, Y., Xie, S., Girshick, R.: Momentum contrast for unsupervised visual representation learning. In: Proceedings of the IEEE/CVF Conference on Computer Vision and Pattern Recognition, pp. 9729–9738 (2020)

7. Irvin, J., Rajpurkar, P., Ko, M., Yu, Y.: CheXpert: a large chest radiograph dataset with uncertainty labels and expert comparison. In: Proceedings of the AAAI Conference on Artificial Intelligence, vol. 33, pp. 590–597 (2019)
8. Jain, S., Agrawal, A., Saporta, A., Truong, S.Q.H.: RadGraph: extracting clinical entities and relations from radiology reports. In: Proceedings of the Neural Information Processing Systems Track on Datasets and Benchmarks, vol. 1 (2021)
9. Ji, Z., Lee, N., Frieske, R., Yu, T., et al.: Survey of hallucination in natural language generation. ACM Comput. Surv. **55**(12), 248:1–248:38 (2023)
10. Jing, B., Xie, P., Xing, E.: On the automatic generation of medical imaging reports. In: Proceedings of the 56th Annual Meeting of the Association for Computational Linguistics, vol. 1: Long Papers, pp. 2577–2586 (2018)
11. Johnson, A.E.W., Pollard, T.J., Berkowitz, S.J., et al.: MIMIC-CXR, a de-identified publicly available database of chest radiographs with free-text reports. Sci. Data **6**(1), 317 (2019)
12. Kahn, C.E., Jr., Langlotz, C.P., Burnside, E.S., et al.: Toward best practices in radiology reporting. Radiology **252**(3), 852–856 (2009)
13. Langlotz, C.P.: The Radiology Report: A Guide to Thoughtful Communication for Radiologists and Other Medical Professionals. CreateSpace Independent Publishing Platform (2015)
14. Lewis, P., Perez, E., Piktus, A., Petroni, F.: Retrieval-augmented generation for knowledge-intensive NLP tasks. Adv. Neural Inf. Process. Syst. **33**, 9459–9474 (2020)
15. Li, C., Wong, C., Zhang, S., Usuyama, N.: LLaVA-med: training a large language-and-vision assistant for biomedicine in one day. Adv. Neural Inf. Process. Syst. **36**, 28541–28564 (2023)
16. Li, J., Vo, D.M., Sugimoto, A., Nakayama, H.: EVCap: retrieval-augmented image captioning with external visual-name memory for open-world comprehension. In: Proceedings of the IEEE/CVF Conference on Computer Vision and Pattern Recognition, pp. 13733–13742 (2024)
17. Li, J., Selvaraju, R., Gotmare, A., Joty, S.: Align before fuse: vision and language representation learning with momentum distillation. Adv. Neural Inf. Process. Syst. **34**, 9694–9705 (2021)
18. Lin, C.Y.: ROUGE: a package for automatic evaluation of summaries. In: Text Summarization Branches Out, pp. 74–81 (2004)
19. Liu, H., Li, C., Wu, Q., Lee, Y.J.: Visual instruction tuning. Adv. Neural Inf. Process. Syst. **36**, 34892–34916 (2023)
20. Loshchilov, I., Hutter, F.: Decoupled weight decay regularization. In: International Conference on Learning Representations (2019)
21. Moor, M., Huang, Q., Wu, S., Yasunaga, M.: Med-Flamingo: a multimodal medical few-shot learner. In: Proceedings of the 3rd Machine Learning for Health Symposium, vol. 225, pp. 353–367 (2023)
22. van den Oord, A., Li, Y., Vinyals, O.: Representation learning with contrastive predictive coding. arXiv preprint arXiv:1807.03748 (2018)
23. OpenAI: GPT-4V(ision) system card, OpenAI. Technical report (2023)
24. Pal, A., Umapathi, L.K., Sankarasubbu, M.: Med-HALT: medical domain hallucination test for large language models. arXiv preprint arXiv:2307.15343 (2023)
25. Papineni, K., Roukos, S., Ward, T., Zhu, W.J.: BLEU: a method for automatic evaluation of machine translation. In: Proceedings of the 40th Annual Meeting of the Association for Computational Linguistics, pp. 311–318 (2002)

26. Park, S.J., Heo, K.S., Shin, D.H., Son, Y.H., Oh, J.H., Kam, T.E.: Dart: disease-aware image-text alignment and self-correcting re-alignment for trustworthy radiology report generation. In: Proceedings of the IEEE/CVF Conference on Computer Vision and Pattern Recognition, pp. 15580–15589 (2025)
27. Smit, A., Jain, S., Rajpurkar, P., Pareek, A.: CheXbert: combining automatic labelers and expert annotations for accurate radiology report labeling using BERT. In: Proceedings of the 2020 Conference on Empirical Methods in Natural Language Processing (EMNLP), pp. 1500–1519 (2020)
28. Thawkar, O., Shaker, A., Mullappilly, S.S.: XrayGPT: chest radiographs summarization using medical vision-language models. arXiv preprint arXiv:2306.07971 (2023)
29. Wang, Z., Liu, L., Wang, L., Zhou, L.: R2GenGPT: radiology report generation with frozen LLMs. Meta-Radiol. **1**(3), 100033 (2023)
30. Wang, Z., Wu, Z., Agarwal, D., Sun, J.: MedCLIP: contrastive learning from unpaired medical images and text. In: Proceedings of the 2022 Conference on Empirical Methods in Natural Language Processing, pp. 3876–3887 (2022)
31. Yu, F., Endo, M., Krishnan, R., Pan, I., et al.: Evaluating progress in automatic chest x-ray radiology report generation. Patterns **4**(9), 100802 (2023)
32. Zambrano Chaves, J.M., Huang, S.C., Xu, Y.: Towards a clinically accessible radiology foundation model: open-access and lightweight, with automated evaluation. arXiv preprint arXiv:2403.08002 (2024)
33. Zhang, X., Meng, Z., Lever, J., Ho, E.S.L.: CCD: mitigating hallucinations in radiology mllms via clinical contrastive decoding. arXiv preprint arXiv:2509.23379 (2025)
34. Zhang, X., Meng, Z., Lever, J., Ho, E.S.L.: Libra: leveraging temporal images for biomedical radiology analysis. In: Findings of the Association for Computational Linguistics: ACL 2025, pp. 17275–17303 (2025)
35. Zhang, Y., Jiang, H., Miura, Y., Manning, C.D., Langlotz, C.P.: Contrastive learning of medical visual representations from paired images and text. In: Proceedings of the 7th Machine Learning for Healthcare Conference, vol. 182, pp. 2–25 (2022)

CARE: Confidence-Aware Reasoning for Reliable Medical VQA

Yuetian Du[1], Yucheng Wang[1], Zhenyuan Chen[1], Luyuan Chen[1], Rongyu Zhang[1], Jinjian Zhang[2], Wei Zhou[2], Zhijie Xu[3], Ming Kong[1], Zhan Zhou[1], Jie Liu[4(✉)], and Qiang Zhu[1(✉)]

[1] Zhejiang University, Hangzhou, China
zhuq@zju.edu.cn
[2] Ant Group, Hangzhou, China
[3] University of Michigan, Ann Arbor, China
[4] City University of Hong Kong, Hong Kong, China
22421227@zju.edu.cn

Abstract. Reinforcement Fine-Tuning (RFT) has enabled medical Multimodal Large Language Models (MLLMs) to produce Chain-of-Thought (CoT) reasoning for visual question answering, yet these models suffer from *confidence miscalibration*—a systematic gap between expressed certainty and actual diagnostic accuracy that undermines clinical trust. We propose **CARE**, a **C**onfidence-**A**ware medical **RE**asoning framework that jointly optimizes accuracy and calibration through a dual-stage pipeline. First, a scalable Medical-CoT synthesis provides structured cold-start data for Supervised Fine-Tuning. Second, Group Relative Policy Optimization (GRPO) with a novel **Confidence-Aware Reward (CAR)** mechanism ties the model's confidence to diagnostic correctness within the reward signal. Across three Medical VQA benchmarks, **CARE** achieves the highest diagnostic accuracy while obtaining the lowest Expected Calibration Error and Hallucination Rate, establishing a foundation for trustworthy clinical decision support. Our code is available at https://github.com/anotherbricki/CARE.

Keywords: Reinforcement Fine-Tuning · Medical VQA · Confidence Calibration

1 Introduction

Medical Multimodal Large Language Models (MLLMs) are increasingly applied to clinical decision support tasks such as medical visual question answering (VQA) [12]. Typically built through Supervised Fine-Tuning (SFT) on curated clinical instruction datasets, these models learn a direct input-output mapping

Y. Du and Y. Wang—Equal contribution.

G. Yang et al. (Eds.): MICCAI 2026, LNCS 16878, pp. 66–75, 2027.
https://doi.org/10.1007/978-3-032-38059-3_7

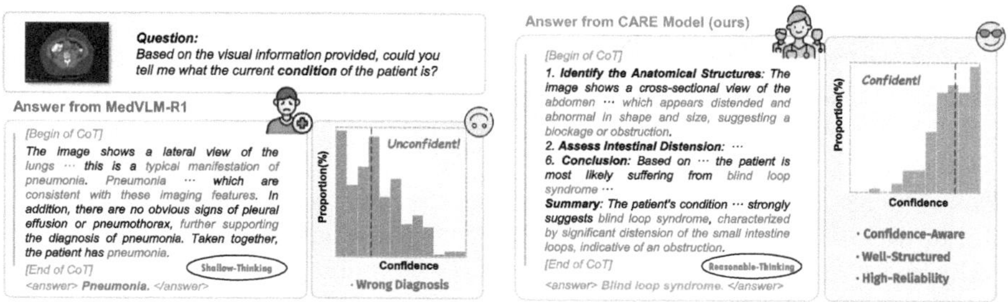

Fig. 1. *Comparative Analysis of Diagnostic Reasoning Paths.* Existing medical reasoning MLLMs such as MedVLM-R1 exhibit confidence miscalibration, where expressed certainty fails to reflect actual diagnostic accuracy. **CARE** explicitly aligns subjective confidence with diagnostic correctness, producing verifiable and interpretable reasoning trajectories for clinical decision-making.

that produces predictions without transparent reasoning processes. In clinical practice, where diagnostic errors carry severe consequences, physicians are unlikely to trust model predictions that lack interpretable reasoning chains. This highlights an urgent need for medical MLLMs that can perform transparent, step-by-step clinical reasoning.

Reinforcement Fine-Tuning (RFT), where RL algorithms such as Group Relative Policy Optimization (GRPO) [16] optimize models against verifiable reward signals, has recently enabled reasoning-focused models that generate extended Chain-of-Thought (CoT) outputs. This paradigm has been adopted in the medical domain, producing early medical reasoning models [9,15] for Medical VQA. While these models improve reasoning transparency, a critical issue persists: *confidence miscalibration* [5,20,21]. As illustrated in Fig. 1, existing models exhibit a systematic misalignment between expressed confidence and actual diagnostic accuracy: they may appear underconfident on correct diagnoses or assign unwarranted certainty to incorrect ones. In clinical deployment, such miscalibration undermines trust, as a model that cannot reliably reflect its own uncertainty provides no safe basis for decision-making.

Related Work. Reinforcement Fine-Tuning (RFT) has emerged as a new paradigm for improving reasoning in large models [13,22], surpassing traditional SFT-based CoT approaches. Unlike SFT, which passively fits existing data distributions, RFT enables models to receive direct feedback from verifiable outcomes and dynamically refine their reasoning paths [2]. Pioneered in mathematical reasoning [16] and popularized by DeepSeek-R1 [3], RFT has recently been extended to multimodal medical applications, producing models such as Med-R1 [9], MedVLM-R1 [15], and others [4,8,18,19] for clinical VQA. However, existing RFT frameworks optimize primarily for answer correctness through verifiable rewards [10], without explicitly addressing the alignment between model confidence and prediction accuracy. Unlike methods that elicit verbal confidence or simply reward high confidence, **CARE** uses confidence as a correctness-conditioned calibration signal within GRPO. Medical-CoT synthesis mainly provides a verified cold start that stabilizes reasoning format and answer extraction for subsequent confidence-aware RL.

To address this gap, we propose **CARE**—a **C**onfidence-**A**ware medical **RE**asoning framework that jointly optimizes diagnostic accuracy and confidence calibration within a unified two-stage RFT pipeline. Our main contributions are as follows:

- **Confidence-Aware Reinforcement Fine-Tuning:** We integrate a novel Confidence-Aware Reward (**CAR**) into the GRPO framework, explicitly aligning the model's expressed confidence with diagnostic accuracy during RL optimization.
- **Scalable Medical-CoT Data Construction:** We design an automated synthesis pipeline to construct high-quality medical reasoning data for multiple clinical scenarios, providing structured diagnostic trajectories with verifiable conclusions.
- **Consistent Improvements in Accuracy and Calibration:** Evaluations across multiple Medical VQA benchmarks demonstrate that **CARE** achieves superior diagnostic accuracy while substantially reducing confidence miscalibration, establishing a reliable foundation for clinical decision support.

2 Method

2.1 Overview

As illustrated in Fig. 2, the **CARE** framework consists of two phases: (1) *Medical-CoT Data Synthesis*, which constructs scalable, well-structured diagnostic reasoning paths from existing clinical datasets; and (2) *Two-Stage Optimization*, transitioning from SFT-based domain adaptation to GRPO reinforcement learning driven by a novel **Confidence-Aware Reward (CAR)** mechanism. This design jointly optimizes diagnostic accuracy and confidence calibration, mitigating the risk of confidence miscalibration in clinical judgments.

2.2 Scalable Medical-CoT Data Synthesis

To facilitate structured reasoning without relying solely on prohibitive expert annotations, we design an automated, reverse-thinking synthesis pipeline. Let $\mathcal{D}_{\text{VQA}} = \{(V_i, Q_i, Y_i)\}_{i=1}^{N}$ denote a standard Medical-VQA dataset comprising visual contexts V, text queries Q, and ground truth diagnoses Y. We employ a capable base MLLM, denoted as π_θ, to retroactively generate intermediate reasoning trajectories $\mathcal{T}$:

$$\mathcal{T}_i \sim \pi_\theta(V_i, Q_i, Y_i). \tag{1}$$

To ensure the clinical validity and logical coherence of $\mathcal{T}_i$, we enforce a strict structural template requiring explicit phase identifiers (e.g., visual analysis, differential diagnosis) and a conclusive summary. Crucially, rather than relying on

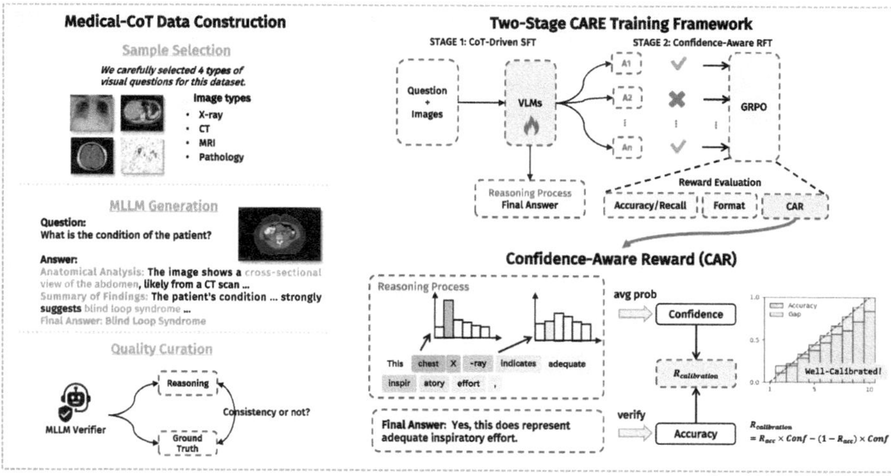

Fig. 2. *Overview of the **CARE** Framework.* The pipeline consists of two phases: *(1) Medical-CoT Data Construction* for synthesizing structured diagnostic reasoning paths, and *(2) Two-Stage Optimization*, comprising SFT cold-start and GRPO with the proposed **Confidence-Aware Reward (CAR)** mechanism to align confidence with diagnostic accuracy.

raw generative outputs, we implement a rigorous verification mechanism. An auxiliary verifier (e.g., GPT-4o) evaluates each trajectory against the ground truth Y_i. A generated trajectory is admitted into the final training corpus $\mathcal{D}_{\mathrm{CoT}}$ if and only if its reasoning chain logically deduces a conclusion that perfectly aligns with Y_i. This objective filtering acts as a scalable proxy for quality assurance, ensuring the model learns from goal-oriented, logically sound diagnostic paths.

2.3 Two-Stage Optimization of CARE

Leveraging the curated $\mathcal{D}_{\mathrm{CoT}}$, **CARE** is trained in two stages: SFT cold-start for domain adaptation, followed by confidence-aware GRPO-based reinforcement learning.

Phase I: SFT Cold Start. We conduct standard supervised fine-tuning on $\mathcal{D}_{\mathrm{CoT}}$ to establish the model's capacity for structured clinical reasoning. For each tuple $(V, Q, \mathcal{T}, Y) \in \mathcal{D}_{\mathrm{CoT}}$, the model maximizes the likelihood of the target sequence $S = [\mathcal{T}; Y]$:

$$\mathcal{L}_{\mathrm{SFT}}(\theta) = -\mathbb{E}_{(V,Q,S)\sim\mathcal{D}_{\mathrm{CoT}}}\left[\sum_{t=1}^{|S|} \log \pi_\theta(s_t | V, Q, s_{<t})\right]. \quad (2)$$

This phase yields a reference policy π_{ref} that follows the structured reasoning format, serving as initialization for the RL phase.

Phase II: GRPO-Based RL. We then optimize π_θ using GRPO, which eliminates the need for a separate value network. For each query (V, Q), the policy samples K candidate outputs $\{o_1, \ldots, o_K\}$. The objective maximizes group-relative advantages while constraining divergence from π_{ref}:

$$J_{\text{GRPO}}(\theta) = \mathbb{E}\left[\frac{1}{K}\sum_{i=1}^{K} \min\left(\rho_i, \text{clip}(\rho_i, 1-\epsilon, 1+\epsilon)\right) A_i - \beta \mathbb{D}_{\text{KL}}(\pi_\theta \| \pi_{\text{ref}})\right], \quad (3)$$

where $\rho_i = \frac{\pi_\theta(o_i|V,Q)}{\pi_\theta^{\text{old}}(o_i|V,Q)}$ is the importance weight, ϵ is the clip ratio, and β controls the KL penalty. The advantage A_i is computed by normalizing the composite reward R_i within the sampled group:

$$A_i = \frac{R_i - \mu(R_{1:K})}{\sigma(R_{1:K})}. \quad (4)$$

2.4 Confidence-Aware Reward (CAR) Formulation

Existing RFT frameworks rely on binary format or accuracy rewards, leaving confidence miscalibration unaddressed. We introduce the **Confidence-Aware Reward (CAR)** to close this gap by directly incorporating confidence alignment into the reward signal. The composite reward $R_i = R_{\text{form}} + R_{\text{out}} + R_{\text{calib}}$ consists of three components:

Format and Output Rewards ($R_{\text{form}}, R_{\text{out}}$). $R_{\text{form}} \in \{0, 1\}$ enforces the use of designated <think> and <answer> delimiters. R_{out} evaluates diagnostic correctness: for closed-ended tasks, it is an exact-match indicator $I(Y \subseteq a_i)$; for open-ended tasks, it uses a recall-based metric to assess the coverage of ground truth Y within the predicted answer a_i.

Calibration Reward (R_{calib}). This is the core component that distinguishes **CARE** from standard RFT. We use the model's answer-level predictive confidence as a calibration signal, rather than treating token probability as a complete epistemic uncertainty estimate. Specifically, for each output o_i, let $a_i = \{t_1, \ldots, t_{|a_i|}\}$ denote only the tokens inside the <answer> span, excluding reasoning tokens, formatting tokens, and special tokens. We compute:

$$C(a_i) = \frac{1}{|a_i|}\sum_{j=1}^{|a_i|} \pi_\theta(t_j \mid V, Q, t_{<j}). \quad (5)$$

The calibration reward ties this answer-level confidence to diagnostic correctness:

$$R_{\text{calib}}(o_i, Y) = R_{\text{out}} \cdot C(a_i) - \lambda(1 - R_{\text{out}}) \cdot C(a_i), \quad (6)$$

where λ controls the penalty on overconfident incorrect predictions. When the prediction is correct, higher answer confidence is rewarded; when it is incorrect, high confidence is penalized. This correctness-conditioned design differs from simply encouraging high confidence, and directly optimizes the confidence-accuracy alignment measured by ECE.

3 Experiments

3.1 Experimental Setup

Datasets. We evaluate CARE on three widely adopted Medical VQA benchmarks:

- **VQA-RAD** [11]: Focused on radiology, comprising 315 clinician-annotated images and 3,515 query-answer pairs.
- **SLAKE** [14]: A semantically-labeled, knowledge-enhanced dataset featuring 642 images and 14,000 bilingual pairs.
- **PathVQA** [7]: Dedicated to pathology, containing 4,998 images and 32,799 question-answer pairs.

Baselines. We evaluate CARE against state-of-the-art reasoning-focused Medical MLLMs, categorized by parameter scale: (1) *Compact-Scale Models ($\leq$ 3B):* Including Med-R1-3B [9] and MedVLM-R1-2B [15]. (2) *Standard-Scale Models (7B–8B):* Including Lingshu-7B [19], MedVLThinker-7B [8], Fleming-VL-8B [17], and MedMO-8B [4].

Evaluation Metrics. We assess model performance along three dimensions:

- *Diagnostic Accuracy:* We report Accuracy for closed-ended questions and Recall for open-ended questions, following standard Medical-VQA evaluation protocols.
- *Confidence Calibration:* We adopt Expected Calibration Error (ECE) [6] to measure the alignment between model confidence and actual accuracy. Confidence scores are partitioned into M equally spaced bins B_m, and ECE computes the weighted deviation:

$$\mathrm{ECE} = \sum_{m=1}^{M} \frac{|B_m|}{N} |\mathrm{acc}(B_m) - \mathrm{conf}(B_m)|, \tag{7}$$

  where N is the total number of samples. Lower ECE indicates better calibration.
- *Hallucination Rate:* We employ Lingshu-32B [19] as a VLM judge to evaluate reasoning trajectories under a fixed prompt and rubric. The judge is run with deterministic decoding, and all methods are evaluated with identical inputs to ensure fairness. Each sample receives a normalized hallucination score r_i, and the overall HR is computed as:

$$\mathrm{HR} = \frac{C}{\mathbb{E}\left[(1-r_i)^2\, t_i\right]}, \tag{8}$$

  where t_i is the output token count and C is a global normalization constant ensuring HR $\in [0, 1]$. The squared term $(1 - r_i)^2$ applies a quadratic penalty that disproportionately suppresses samples with high hallucination rates, amplifying the contribution of heavily hallucinated outputs to the overall score.

3.2 Implementation Details

CARE is built upon Qwen2.5-VL-7B-Instruct [1]. The same model architecture serves as both the CoT data synthesizer in Sect. 2.2 and the policy model for subsequent training, ensuring consistency between the generated reasoning format and the optimization target. Training is conducted via full-parameter fine-tuning on 6× NVIDIA A100 GPUs. During the *SFT Cold Start*, we use the AdamW optimizer with a learning rate of 1×10^{-5} and cosine annealing. In the *GRPO RL Phase*, we generate $K = 4$ rollouts per query with a batch size of 2 and `bfloat16` mixed precision. In the calibration reward of **CAR**, the penalty coefficient is set to $\lambda = 0.5$.

Table 1. Main results on Medical VQA benchmarks. The best and second-best results in each column are highlighted in **bold** and <u>underlined</u>, respectively. The same applies to Table 2 (**ACC** = accuracy, **ECE** = expected calibration error, **HR** = hallucination rate).

Model	VQA-RAD			SLAKE			PathVQA		
	ACC ↑	ECE ↓	HR ↓	ACC ↑	ECE ↓	HR ↓	ACC ↑	ECE ↓	HR ↓
Compact-Scale Models (≤ 3B)									
Med-R1-3B	0.513	0.451	0.195	0.596	0.369	0.257	0.387	0.584	0.155
MedVLM-R1-2B	0.532	0.448	0.200	0.504	0.474	0.237	0.387	0.587	0.168
Standard-Scale Models (7B–8B)									
MedMO-8B	0.647	<u>0.264</u>	0.365	0.816	0.277	0.368	0.563	<u>0.383</u>	0.223
Lingshu-7B	<u>0.679</u>	0.375	0.083	<u>0.831</u>	0.322	0.105	0.619	0.522	0.098
MedVLThinker-7B	0.637	0.413	<u>0.071</u>	0.678	0.424	0.099	<u>0.652</u>	0.569	0.082
Fleming-VL-8B	0.668	0.333	0.073	0.819	<u>0.181</u>	<u>0.089</u>	0.629	0.455	<u>0.073</u>
CARE-7B (Ours)	**0.767**	**0.202**	**0.048**	**0.873**	**0.115**	**0.070**	**0.689**	**0.290**	**0.059**

3.3 Experimental Results and In-Depth Analysis

CARE Simultaneously Achieves the Best Accuracy, Calibration, and Lowest Hallucination Across All Benchmarks. As shown in Table 1, existing medical reasoning models consistently exhibit trade-offs among these three dimensions: strong accuracy often comes with poor calibration or high hallucination, and vice versa. No baseline ranks first across all three metrics on any single benchmark. **CARE** is the only model to do so consistently, achieving the highest accuracy (e.g., 0.873 on SLAKE, 0.767 on VQA-RAD) while simultaneously obtaining the lowest ECE and HR. The substantial ECE reduction (e.g., 0.115 on SLAKE, a 36% relative improvement over the second-best Fleming-VL-8B) directly confirms the effectiveness of **CAR** in aligning confidence with accuracy. The consistently lowest HR (e.g., 0.048 on VQA-RAD) further validates our reverse-thinking data synthesis: answer-grounded reasoning paths reduce fabrication at the source by providing structured, goal-oriented CoT trajectories.

Table 2. Ablation study results for different training stages.

Training Stages	VQA-RAD				SLAKE				PathVQA			
	Open ↑	ECE ↓	Closed ↑	ECE ↓	Open ↑	ECE ↓	Closed ↑	ECE ↓	Open ↑	ECE ↓	Closed ↑	ECE ↓
Training-Free	0.483	0.485	0.658	0.341	0.513	0.456	0.690	0.287	0.150	0.823	0.664	0.326
SFT	0.520	0.468	0.629	0.359	0.803	0.188	0.767	0.225	<u>0.349</u>	<u>0.640</u>	0.857	0.133
RL	<u>0.563</u>	<u>0.407</u>	**0.864**	**0.096**	<u>0.819</u>	<u>0.176</u>	**0.861**	**0.119**	0.166	0.727	**0.955**	**0.020**
SFT+RL	**0.620**	**0.363**	<u>0.658</u>	<u>0.323</u>	**0.881**	**0.112**	<u>0.779</u>	<u>0.209</u>	**0.421**	**0.561**	<u>0.863</u>	<u>0.121</u>

Different Question Types Favor Different Training Configurations. Table 2 reveals a clear pattern: for closed-ended questions, RL alone achieves the best accuracy and ECE across all three benchmarks (e.g., 0.864 ACC and 0.096 ECE on VQA-RAD), as the constrained answer space allows RL exploration to converge efficiently without SFT initialization. For open-ended questions, however, SFT+RL consistently dominates (e.g., 0.881 on SLAKE Open, 0.421 on PathVQA Open), since generating free-form diagnostic reasoning requires the structured CoT foundation established during SFT. Notably, SFT alone already yields substantial open-ended gains (e.g., SLAKE Open rises from 0.513 to 0.803), confirming that the CoT cold-start is the primary driver for open-ended performance, while RL further refines calibration on top of it. The results also reflect the complementary roles of the two stages: Medical-CoT provides a stable reasoning and answer-extraction format, especially for open-ended questions, while the RL stage further optimizes answer correctness and calibration through CAR.

CAR Improves Confidence-Accuracy Alignment Rather than Merely Increasing Confidence. Figure 3 compares the normalized confidence distributions before and after **CAR** alignment. The post-training distribution shifts toward higher-confidence regions, but this shift should be interpreted together with the simultaneous ECE reductions in Table 1. Since ECE directly measures the gap between confidence and empirical accuracy, the improved ECE indicates that the confidence shift is better aligned with diagnostic correctness rather than being a simple increase in likelihood. This supports the design of **CAR**, which rewards confident correct predictions while penalizing overconfident incorrect ones.

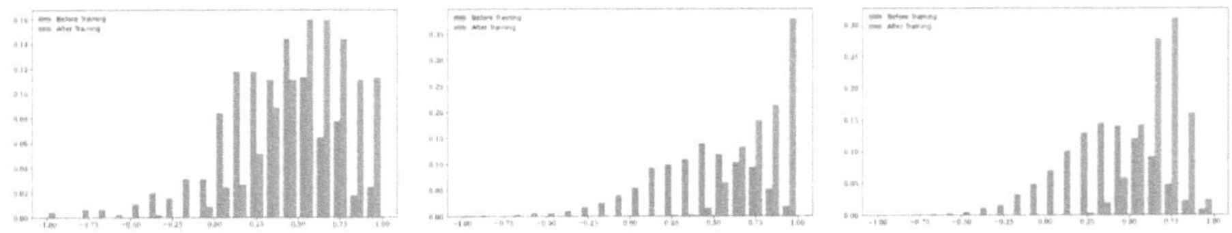

Fig. 3. *Confidence distributional shift.* Red and green bars represent the normalized confidence scores pre- and post-training, respectively. After training, the mean shifts rightward with increased magnitude, reflecting improved confidence-accuracy alignment. (Color figure online)

4 Conclusion

We present **CARE**, a framework that addresses confidence miscalibration in RFT-based medical reasoning models. By combining a scalable Medical-CoT synthesis pipeline with a Confidence-Aware Reward (**CAR**) mechanism within GRPO, **CARE** explicitly aligns the model's expressed confidence with diagnostic accuracy during RL optimization. Experiments across three Medical VQA benchmarks show that **CARE** achieves the best diagnostic accuracy while simultaneously obtaining the lowest ECE and Hallucination Rate, demonstrating that accuracy and calibration can be jointly improved rather than traded off. These results suggest that incorporating confidence signals into the reward design is an effective path toward reliable medical AI systems. We sincerely hope this work encourages further exploration of calibration-aware training objectives in safety-critical domains beyond medical VQA.

Acknowledgments. This work was supported by the National Natural Science Foundation of China under Grant 42394060 and 42394064, Ant Group Research Fund, and the Zhejiang University - Jolly Pharmaceutical Joint R&D Center for Intelligent Empowerment in Food and Medicine.

Disclosure of Interests. The authors have no competing interests to declare.

References

1. Bai, S., et al.: Qwen2.5-vl technical report (2025). https://arxiv.org/abs/2502.13923
2. Chu, T., et al.: Sft memorizes, rl generalizes: a comparative study of foundation model post-training (2025). https://arxiv.org/abs/2501.17161
3. DeepSeek-AI, et al.: Deepseek-r1: incentivizing reasoning capability in llms via reinforcement learning (2025). https://arxiv.org/abs/2501.12948
4. Deria, A., Kumar, K., Dukre, A.M., Segal, E., Khan, S., Razzak, I.: Medmo: grounding and understanding multimodal large language model for medical images (2026). https://arxiv.org/abs/2602.06965

5. Du, Y., et al.: Confidence calibration for multimodal llms: an empirical study through medical vqa. In: proceedings of Medical Image Computing and Computer Assisted Intervention - MICCAI 2025. LNCS, vol. 15965, Springer, Cham (2025). https://doi.org/10.1007/978-3-032-04978-0_9
6. Guo, C., Pleiss, G., Sun, Y., et al.: On calibration of modern neural networks. In: International Conference on Machine Learning, pp. 1321–1330. PMLR (2017)
7. He, X., Zhang, Y., Mou, L., Xing, E., Xie, P.: Pathvqa: 30000+ questions for medical visual question answering (2020). https://arxiv.org/abs/2003.10286
8. Huang, X., Wu, J., Liu, H., Tang, X., Zhou, Y.: Medvlthinker: simple baselines for multimodal medical reasoning (2025). https://arxiv.org/abs/2508.02669
9. Lai, Y., Zhong, J., Li, M., Zhao, S., Yang, X.: Med-r1: reinforcement learning for generalizable medical reasoning in vision-language models (2025). https://arxiv.org/abs/2503.13939
10. Lambert, N., et al.: Tulu 3: pushing frontiers in open language model post-training (2025). https://arxiv.org/abs/2411.15124
11. Lau, J., Gayen, S., Ben Abacha, A.: A dataset of clinically generated visual questions and answers about radiology images. Sci. Data **5**, 180251 (2018)
12. Li, C., Wong, C., Zhang, S.: LLaVA-Med: training a large language-and-vision assistant for biomedicine in one day. Adv. Neural Inf. Process. Syst. **36** (2024)
13. Li, Z.Z., et al.: From system 1 to system 2: a survey of reasoning large language models (2025). https://arxiv.org/abs/2502.17419
14. Liu, B., Zhan, L.M., Xu, L.: SLAKE: a semantically-labeled knowledge-enhanced dataset for medical visual question answering. In: IEEE 18th International Symposium on Biomedical Imaging, pp. 1650–1654 (2021)
15. Pan, J., et al.: MedVLM-R1: incentivizing medical reasoning capability of vision-language models (VLMs) via reinforcement learning. arXiv preprint arXiv:2502.19634 (2025)
16. Shao, Z., et al.: Deepseekmath: pushing the limits of mathematical reasoning in open language models (2024). https://arxiv.org/abs/2402.03300
17. Shu, Y., Liu, C., Chen, R., Li, D., Dai, B.: Fleming-vl: towards universal medical visual reasoning with multimodal llms (2025). https://arxiv.org/abs/2511.00916
18. Sun, H., et al.: Chiron-o1: igniting multimodal large language models towards generalizable medical reasoning via mentor-intern collaborative search (2025). https://arxiv.org/abs/2506.16962
19. Team, L., et al.: Lingshu: a generalist foundation model for unified multimodal medical understanding and reasoning (2025). https://arxiv.org/abs/2506.07044
20. Tian, K., Mitchell, E., Zhou, A., et al.: Just ask for calibration: strategies for eliciting calibrated confidence scores from language models fine-tuned with human feedback. arXiv preprint arXiv:2305.14975 (2023)
21. Xiong, M., Hu, Z., Lu, X., et al.: Can LLMs express their uncertainty? An empirical evaluation of confidence elicitation in LLMs. arXiv preprint arXiv:2306.13063 (2023)
22. Xu, F., et al.: Towards large reasoning models: a survey of reinforced reasoning with large language models (2025). https://arxiv.org/abs/2501.09686

ConVL: Interpretable Concept-Guided Vision-Language MIL for Survival Analysis in Whole Slide Images

Junjian Li, Hulin Kuang(✉), Jin Liu, Hailin Yue, Mengshen He, Xinyu Li, and Jianxin Wang(✉)

Hunan Provincial Key Lab on Bioinformatics, School of Computer Science and Engineering, Central South University, Changsha 410083, Hunan, China
hulinkuang@csu.edu.cn, jxwang@mail.csu.edu.cn

Abstract. Whole Slide Images (WSIs) pose significant challenges for survival modeling due to their complex tissue heterogeneity. While Multiple Instance Learning (MIL) shows promise, it lacks semantic transparency, failing to identify the underlying pathological concepts driving predictions. Moreover, although vision-language models introduce semantic priors, they rely on static text prompts that remain invariant during inference across all patients, failing to capture the morphological variations of phenotypes. To bridge this gap, we propose ConVL, a novel, interpretable concept-guided vision-language MIL framework that explicitly grounds survival prediction in explicit pathological knowledge. ConVL introduces three novel mechanisms: (1) Instance Conditioned Prompt Learning, which dynamically adapts concept prototypes to slide-specific visual contexts, enhancing robustness to tissue heterogeneity; (2) a Synergistic Dual-Stream Mechanism combining data-driven visual features with a knowledge-driven Concept Stream, utilizing Concept-Guided Cross Attention to quantify pathological attributes; and (3) a Distribution Alignment Loss that enforces consistency between the visual feature manifold and the concept-based prediction manifold. We evaluate ConVL across nine real-world cancer cohorts from TCGA. ConVL not only achieves state-of-the-art survival prediction performance but also provides fine-grained, concept-driven interpretability. Code is available at https://github.com/junjianli106/ConVL.

Keywords: Computational Pathology · Whole Slide Images · Survival Analysis · Vision-Language Model

1 Introduction

Deep learning has shown strong potential in a broad range of medical image analysis tasks [2,8,9,23,25,26]. Whole Slide Images (WSIs) are the clinical gold standard for cancer diagnosis, encapsulating rich prognostic information that spans cellular morphology, tissue architecture, and the tumor microenvironment. However, the gigapixel resolution of WSIs, coupled with the scarcity of fine-grained

G. Yang et al. (Eds.): MICCAI 2026, LNCS 16878, pp. 76–86, 2027.
https://doi.org/10.1007/978-3-032-38059-3_8

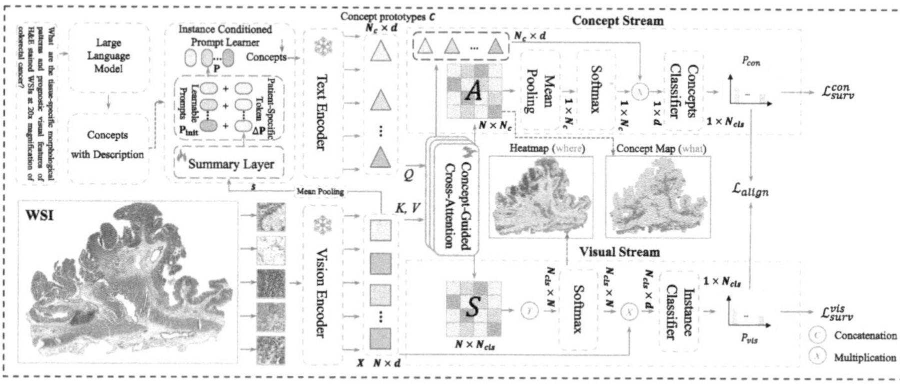

Fig. 1. Overview of the ConVL for WSI survival prediction. ConVL integrates Instance Conditioned Prompt Learning for dynamic concept adaptation, a Synergistic Dual-Stream Mechanism for coupling visual context with semantic evidence, and a Distribution Alignment Loss for dual-branch consistency.

annotations and significant intra-class heterogeneity, poses severe challenges for computational survival modeling. Tumors of the same subtype often exhibit substantial morphological variations across different patients, which complicates the extraction of robust prognostic features [21].

Multiple Instance Learning (MIL) has become the dominant paradigm for WSI analysis by aggregating features from local patches into slide-level representations [1,3–5,7,10,12,16,19,20,24]. Attention-based MIL models [1,22] focus on localizing discriminative regions, **answering where the model looks**, while Transformer-based approaches [16,17] capture broader contextual dependencies. Despite these advances, existing MIL methods face a critical interpretability bottleneck. They rely on purely visual statistical correlations without explicit semantic guidance. Consequently, they can highlight high-attention regions but fail to explain the underlying pathological rationale. **They cannot answer what morphological phenotype is present.** This absence of semantic anchors makes it difficult to capture consistent prognostic patterns across morphological variations, limiting model generalization when facing highly heterogeneous samples.

Recent advances in Vision-Language Models (VLMs) have inspired prompt-based WSI analysis [13,15]. For instance, VLSA [11] utilizes language-encoded prognostic priors to guide MIL aggregation. However, these methods primarily rely on learned static prompts that remain invariant during inference across all patients. Such static representations fail to adapt to the significant morphological heterogeneity inherent in cancer tissue, where the visual manifestation of the same concept (e.g., 'tumor infiltration') can vary drastically between slides. Consequently, applying such static prompts often fails to accurately capture patient-specific pathological nuances, inevitably leading to feature misalignment.

To address these limitations, we propose ConVL, an concept-guided vision-language MIL framework that deeply couples pathological domain knowledge with visual features. Specifically, ConVL employs a dual-branch architecture integrated with three novel mechanisms. First, Instance Conditioned Prompt Learning generates patient-specific context offsets to adaptively refine concept prototypes to mitigate tissue heterogeneity. Second, a Synergistic Dual-Stream Mechanism combining data-driven visual features with a knowledge-driven Concept Stream. The visual branch functions as a data-driven feature aggregator to capture latent prognostic cues, while the concept branch explicitly assesses the histological burden of pathological phenotypes. Third, a Distribution Alignment Loss ensures consistent prediction patterns between the visual and concept branches. Experiments on nine real-world TCGA cancer cohorts demonstrate that ConVL not only delivers competitive prognostic performance but also provides intelligible explanations by explicitly associating tissue regions with specific pathological concepts.

2 Methodology

2.1 Overview of ConVL

To address the semantic gap and tissue heterogeneity in WSI-based survival modeling, we propose ConVL, an interpretable concept-guided vision-language MIL framework. As shown in Fig. 1, ConVL employs a synergistic dual-branch architecture, comprising a Visual Branch that aggregates visual features using standard attention, and a Concept Branch that grounds predictions in learnable pathological prototypes to achieve semantic evidence localization.

2.2 WSI Preprocessing and Concept Construction

For each patient, the WSIs are segmented to remove background and tiled into non-overlapping patches at 20× magnification, forming a bag of instances $\mathbf{X} = \{x_n\}_{n=1}^{N} \in \mathbb{R}^{N \times d}$. Each patch x_n is a d-dimensional feature embedding extracted via the image encoder of a pre-trained pathology-specific VLM (e.g., CONCH).

To establish a prognostic semantic prior, we prompt a Large Language Model (e.g., Gemini 3 Pro) with a domain-specific inquiry: 'What are the tissue-specific morphological patterns and prognostic visual features of H&E stained WSIs at 20× magnification for [Cancer Type]?' This process yields a set of pathological concepts describing histological evidence (e.g., 'solid sheets of tumor cells'). These texts serve as the foundational semantic descriptions for our prompt learning module. The full list of concepts is available in our public repository.

2.3 Instance Conditioned Prompt Learning

Prior prompt-based MIL methods (e.g., VLSA) rely on fixed and static concept prototypes shared across all patients. While providing a robust semantic prior, these static prompts struggle with severe tissue heterogeneity, as visual manifestations of the same pathological concept vary substantially across slides. Consequently, fixed prototypes fail to capture patient-specific morphological variations, leading to domain misalignment during inference. To address this, we introduce Instance Conditioned Prompt Learning (ICPL), which dynamically refines concept prototypes using global slide-level context. By generating tailored concepts for each WSI, even slides from the same diagnostic class retrieve distinct concept prototypes. This enables fine-grained adaptation, effectively shifting the prompt paradigm from static priors to patient-specific knowledge.

Specifically, we first introduce a set of Learnable Prompt Vectors $\mathbf{P}_{init} \in \mathbb{R}^{L\times d}$. To inject patient-specific information, we then derive a global contextual representation $\mathbf{s} \in \mathbb{R}^d$ by aggregating all instance embeddings within a bag. We employ a Summary Layer $\mathcal{M}$ (acting as a transformation bridge) to map the visual summary $\mathbf{s} = \frac{1}{N}\sum_{i=1}^{N} x_i$ into the prompt embedding space. The transformation is formulated as:

$$\varDelta\mathbf{P} = \text{Reshape}(\mathbf{W}_2(\sigma(\mathbf{W}_1\mathbf{s}))), \tag{1}$$

where $\mathbf{W}_1 \in \mathbb{R}^{\frac{d}{2}\times d}$ and $\mathbf{W}_2 \in \mathbb{R}^{(L\cdot d)\times\frac{d}{2}}$ are learnable projection matrices, σ denotes the ReLU activation, and $\varDelta\mathbf{P} \in \mathbb{R}^{L\times d}$ represents the patient-specific tokens. These tokens serve as dynamic offsets that capture the unique morphological characteristics of the current patient. The prompt prototypes are then formed by:

$$\mathbf{P} = \mathbf{P}_{init} + \varDelta\mathbf{P}. \tag{2}$$

Finally, each text concept T_i is tokenized and mapped into a sequence of word embeddings e_i via the embedding layer of the VLM text encoder. We then concatenate the dynamic prompt prototypes P with the concept embeddings e_i, formulating the input text sequence as $[P, e_i]$. Here, each concept description provides the pathological semantic content, while the patient-specific tokens generated from the WSI-level summary inject slide-level visual context. The concatenated sequence is then fed into the subsequent Transformer layers of the VLM text encoder, producing a dynamic concept prototype $C_i \in \mathbb{R}^d$ that is explicitly grounded in the specific visual context of the patient.

2.4 Synergistic Dual-Stream Survival Prediction

To leverage both holistic visual context and fine-grained semantic evidence, ConVL employs a synergistic dual-stream architecture. We introduce a Concept-Guided Cross-Attention (CGA) module to align the bag of instance embeddings $\mathbf{X} \in \mathbb{R}^{N\times d}$ with patient-specific concept prototypes $\mathbf{C} \in \mathbb{R}^{N_c\times d}$. Formally, we treat $\mathbf{C}$ as queries and $\mathbf{X}$ as keys and values; the CGA module computes a conceptâĂŞinstance affinity matrix $\mathbf{A} \in \mathbb{R}^{N\times N_c}$ via scaled dot-product attention. Each entry $\mathbf{A}_{j,i}$ quantifies the activation of the i-th concept at the j-th patch. From $\mathbf{A}$, we obtain patch-level class logits $\mathbf{S} \in \mathbb{R}^{N\times N_{cls}}$ by a linear projection over the concept dimension (N_{cls} is number of survival risk bins).

Visual Stream. Unlike standard MIL approaches that derive attention weights solely from visual features, this stream leverages the concept-guided patch-class logits $\mathbf{S}$ to direct instance aggregation. Specifically, per-class attention over the patches is obtained by applying a Softmax function along the instance dimension, denoted as $\boldsymbol{\Theta} = \text{Softmax}(\mathbf{S}^\top)$. The bag-level visual representation $\mathbf{h}_{vis}$ is then generated by weighting the instance embeddings $\mathbf{X}$ with $\boldsymbol{\Theta}$, followed by mean pooling across the class dimension:

$$\mathbf{h}_{vis} = \text{Mean}\big(\boldsymbol{\Theta}\mathbf{X}\big) = \text{Mean}\big(\text{Softmax}(\mathbf{S}^\top)\mathbf{X}\big), \quad \text{and} \quad P_{vis} = f_{cls}^{vis}(\mathbf{h}_{vis}). \tag{3}$$

Here, the Mean operator computes the average across the N_{cls} class-conditioned aggregations. Thus, $\mathbf{h}_{vis}$ aggregates patch features weighted by class-wise semantic priors, capturing holistic prognostic cues in relevant regions beyond the predefined concept set.

Concept Stream. This stream interprets the WSI by evaluating concept prevalence. Specifically, we apply global average pooling to the affinity matrix $\mathbf{A}$ across the instance dimension, yielding a concept distribution $\mathbf{v} \in \mathbb{R}^{N_c}$. Here, each element $v_i = \frac{1}{N}\sum_{j=1}^{N} \mathbf{A}_{j,i}$ explicitly quantifies the global burden of the i-th concept across the entire tissue. The slide-level concept representation is then formulated by weighting the concept prototype embeddings with this distribution:

$$\mathbf{h}_{con} = \mathbf{Z}^\top \mathbf{v}, \quad \text{and} \quad P_{con} = f_{cls}^{con}(\mathbf{h}_{con}), \tag{4}$$

where $\mathbf{Z} \in \mathbb{R}^{N_c \times d}$ denotes the concept prototype embeddings. Consequently, $\mathbf{h}_{con}$ characterizes the WSI as a weighted combination of clinical concepts, effectively grounding the final survival prediction in explicit phenotypic evidence.

2.5 Distribution Alignment Loss

To ensure semantic consistency between modalities, we introduce a Distribution Alignment Loss (DAL) as a mutual regularization constraint. This mechanism explicitly aligns the interpretable concept-based predictions with the holistic visual representations, enforcing a robust consensus between the two distinct feature manifolds. Formally, since the model predicts survival probabilities across discrete time intervals, we compute the alignment loss using the symmetric Kullback-Leibler (KL) divergence between the predictive distributions P_{vis} and P_{con}:

$$\mathcal{L}_{align} = \text{KL}(P_{vis}||P_{con}) + \text{KL}(P_{con}||P_{vis}). \tag{5}$$

The overall training objective is formulated as a weighted sum of the task-specific losses and the alignment constraint:

$$\mathcal{L} = \alpha\mathcal{L}_{surv}^{vis} + (1-\alpha)\mathcal{L}_{surv}^{con} + \beta\mathcal{L}_{align}, \tag{6}$$

where $\mathcal{L}_{surv}^{vis}$ and $\mathcal{L}_{surv}^{con}$ denote the discrete-time Negative Log-Likelihood (NLL) losses for the visual and concept branches, respectively. The hyperparameter β

controls the strength of the alignment regularization. Crucially, $\mathcal{L}_{align}$ enforces a bidirectional consensus: it constrains the explainable concept predictions to remain faithful to the robust visual evidence, while simultaneously regularizing the visual stream with structured semantic priors. This synergy effectively mitigates modality-specific overfitting and ensures that the final prognosis is supported by consistent cross-modal reasoning.

3 Experiments

3.1 Datasets and Implementation Details

We evaluate ConVL on nine TCGA cohorts for survival prediction: BLCA, BRCA, CO&RE, LUSC, KIRC, SKCM, LIHC, SARC, and UCEC. We compare ConVL against representative MIL baselines and state-of-the-art survival models, including ABMIL [1], TransMIL [16], ILRA [18], PseMix [12], RRT-MIL [17], WiKG [3], ACMIL [22], and MiCo [6], as well as recent VLM-based MIL methods such as TopMIL [14] and VLSA [11]. For feature extraction, we follow CONCH [13] to generate 448×448 patches at 20× magnification. We evaluate our method using a 4-fold cross-validation scheme, randomly splitting the data into 60% training, 15% validation, and 25% test sets for each fold. For fair comparison, all baselines utilized the same CONCH features network and data splits. We train for 200 epochs with batch size 1, learning rate 2e-4, and Ranger optimizer. Early stopping is applied with patience 8 based on validation loss. Following prior work [6], the number of survival risk bins N_{cls} is set to 4. Furthermore, we set the alignment weight $\beta = 0.1$ and prompt length $L = 16$, while the number of concepts N_c is tailored per dataset. All experiments are conducted on a single 3080Ti GPU.

Table 1. Survival prediction performance results (C-index) of different methods. Best results are highlighted in **bold**, and second-best results are underlined.

Methods	BLCA	BRCA	CO&RE	LUSC	KIRC	SKCM	LIHC	SARC	UCEC	Average
ABMIL (18' ICML)	$0.575_{0.068}$	$0.563_{0.079}$	$0.650_{0.038}$	$0.556_{0.030}$	$0.703_{0.041}$	$0.581_{0.036}$	$0.682_{0.034}$	$0.565_{0.082}$	$0.717_{0.103}$	0.621
TransMIL (21' NeurIPS)	$0.564_{0.070}$	$0.607_{0.071}$	$0.668_{0.083}$	$0.517_{0.026}$	$0.711_{0.047}$	$0.547_{0.033}$	$0.685_{0.046}$	$0.595_{0.065}$	$0.710_{0.066}$	0.623
ILRA (23' ICLR)	$0.597_{0.026}$	$0.590_{0.081}$	$0.645_{0.078}$	$0.541_{0.067}$	$0.698_{0.046}$	$0.524_{0.044}$	$0.660_{0.032}$	$0.525_{0.054}$	$0.712_{0.067}$	0.610
PseMix (24' TMI)	$0.531_{0.055}$	$0.561_{0.116}$	$0.669_{0.049}$	$0.509_{0.057}$	$0.719_{0.041}$	$0.584_{0.028}$	$0.670_{0.017}$	$0.513_{0.089}$	$0.686_{0.062}$	0.605
RRTMIL (24' CVPR)	$0.579_{0.018}$	$0.577_{0.120}$	$0.698_{0.007}$	$0.520_{0.037}$	$0.689_{0.041}$	$0.581_{0.029}$	$0.679_{0.050}$	$0.550_{0.041}$	$0.675_{0.080}$	0.616
WiKG (24' CVPR)	$0.599_{0.031}$	$0.534_{0.064}$	$0.670_{0.068}$	$0.544_{0.018}$	$0.718_{0.044}$	$0.576_{0.022}$	$0.642_{0.029}$	$0.575_{0.032}$	$0.707_{0.085}$	0.618
ACMIL (24' ECCV)	$0.590_{0.047}$	$\underline{0.613_{0.052}}$	$0.676_{0.042}$	$0.543_{0.014}$	$0.714_{0.055}$	$0.570_{0.027}$	$0.688_{0.043}$	$0.484_{0.026}$	$\underline{0.731_{0.066}}$	0.623
MiCo (25' MICCAI)	$0.600_{0.055}$	$0.607_{0.072}$	$\mathbf{0.714}_{0.069}$	$0.552_{0.048}$	$\underline{0.738}_{0.046}$	$0.574_{0.022}$	$\mathbf{0.690}_{0.031}$	$0.529_{0.069}$	$0.722_{0.087}$	$\underline{0.636}$
TopMIL (23' NeurIPS)	$0.604_{0.045}$	$0.532_{0.046}$	$0.696_{0.040}$	$\underline{0.572_{0.023}}$	$0.733_{0.046}$	$\underline{0.601_{0.021}}$	$0.675_{0.029}$	$0.575_{0.020}$	$0.720_{0.092}$	0.634
VLSA (25' ICLR)	$\underline{0.625_{0.020}}$	$0.581_{0.051}$	$0.683_{0.055}$	$0.531_{0.026}$	$0.712_{0.057}$	$0.547_{0.029}$	$0.657_{0.028}$	$\underline{0.596_{0.015}}$	$0.714_{0.075}$	0.627
ConVL (Ours)	$\mathbf{0.629}_{0.021}$	$\mathbf{0.618}_{0.076}$	$\underline{0.709_{0.051}}$	$\mathbf{0.579}_{0.024}$	$\mathbf{0.746}_{0.052}$	$\mathbf{0.612}_{0.025}$	$\underline{0.688_{0.037}}$	$\mathbf{0.608}_{0.059}$	$\mathbf{0.734}_{0.080}$	**0.658**

3.2 Results and Discussion

Comparison with State-of-the-Art Methods. As summarized in Table 1, ConVL establishes a new state-of-the-art benchmark in WSI-based survival prediction, achieving the highest average C-index of 0.658. This performance represents a comprehensive improvement of 2.2% over the strongest competitor, MiCo (0.636), and outperforms recent baselines such as VLSA (0.627). Notably, ConVL demonstrates remarkable robustness and generalizability, ranking first in seven out of the nine evaluated cohorts. The performance gains are particularly pronounced in the SARC and SKCM cohorts, where our ConVL outperforms the second-best approaches by margins of 1.2% and 1.1%, respectively. Furthermore, even in cohorts where ConVL does not secure the top rank (i.e., CO&RE and LIHC), it remains highly competitive, trailing the leading methods by negligible margins of 0.005 and 0.002.

Table 2. Ablation study for ConVL.

Methods	BLCA	BRCA	CO&RE	LUSC	KIRC	SKCM	LIHC	SARC	UCEC	Average
w/o DAL	$\underline{0.617_{0.054}}$	$\underline{0.617_{0.072}}$	$0.670_{0.080}$	$0.554_{0.017}$	$\underline{0.742_{0.045}}$	$\underline{0.609_{0.024}}$	$0.682_{0.041}$	$\underline{0.564_{0.073}}$	$0.710_{0.077}$	$\underline{0.641}$
w/o CGA	$0.599_{0.018}$	$0.590_{0.080}$	$\underline{0.681_{0.081}}$	$\underline{0.576_{0.028}}$	$0.733_{0.041}$	$0.602_{0.020}$	$\mathbf{0.697}_{0.031}$	$0.550_{0.091}$	$\underline{0.723_{0.075}}$	0.639
w/o ICPL	$0.550_{0.118}$	$0.607_{0.076}$	$0.680_{0.040}$	$0.554_{0.018}$	$0.735_{0.038}$	$0.606_{0.022}$	$0.685_{0.026}$	$0.545_{0.025}$	$0.717_{0.066}$	0.631
ConVL	$\mathbf{0.629}_{0.021}$	$\mathbf{0.618}_{0.076}$	$\mathbf{0.709}_{0.051}$	$\mathbf{0.579}_{0.024}$	$\mathbf{0.746}_{0.052}$	$\mathbf{0.612}_{0.025}$	$\underline{0.688_{0.037}}$	$\mathbf{0.608}_{0.059}$	$\mathbf{0.734}_{0.080}$	**0.658**

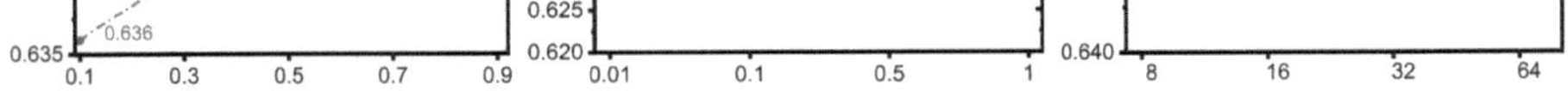

Fig. 2. Hyperparameter sensitivity analysis on the average C-Index across nine datasets. We illustrate the impact of (a) the concept fusion weight α, (b) the distribution alignment weight β, and (c) the prompt length L. The red markers denote the mean C-Index. (Color figure online)

Ablation Study. To validate the contribution of each component, we conduct an ablation study as shown in Table 2. The results demonstrate that all proposed modules are essential for optimal performance. The most significant degradation occurs when replacing the Instance Conditioned Prompt Learning (w/o ICPL) with static prompts, which causes the average C-index to drop from 0.658 to

0.631. This confirms the critical role of our adaptive mechanism in handling tissue heterogeneity. Similarly, substituting the Concept-Guided Cross-Attention with standard Gated Attention (w/o CGA) or removing the Distribution Alignment Loss (w/o DAL) leads to performance declines to 0.639 and 0.641, respectively. These findings verify that semantic-aware evidence retrieval and dual-branch consistency are indispensable for robust survival prediction.

Hyperparameter Analysis. As illustrated in Fig. 2, we conduct sensitivity analyses on three key hyperparameters. The model achieves optimal performance with a distribution alignment weight of $\beta = 0.1$, a prototype count of $L = 16$, and a balanced concept fusion weight of $\alpha = 0.5$. Deviations from these values lead to performance degradation, which validates the necessity of moderate alignment regularization, sufficient context capacity, and the equal contribution of visual and textual modalities.

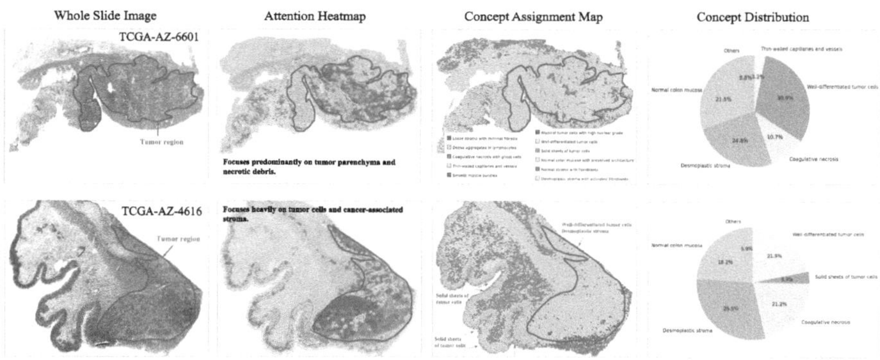

Fig. 3. Interpretability Analysis. Visual comparison of ConVL against standard attention mechanisms. Columns from left to right display the Original WSI, standard Attention Heatmap, Concept Assignment Map, and the quantified Concept Distribution pie chart. The visualizations contrast a Low-Risk patient (top row) characterized by well-differentiated cells with a High-Risk patient (bottom row) driven by malignant desmoplastic stroma.

Interpretability Analysis. To validate the semantic grounding of ConVL, Fig. 3 compares the original WSIs and standard Attention Heatmaps alongside our Concept Assignment Maps and slide-level Concept Distributions. Unlike standard heatmaps that only indicate where the model looks, these visualizations explicitly reveal what phenotypes drive the prediction, with the distribution pie charts quantifying the overall morphological burden across the entire WSI. In the High-Risk case (bottom row), the model identifies tumor cells intermingled with *Desmoplastic stroma* (light green). The concept distribution (Column

4) corroborates this by assigning the highest importance (29.5%) to this malignant stroma. As confirmed by an expert pathologist, this signifies a desmoplastic reaction at the invasive front, which is a hallmark of aggressive tumors and a primary high-risk indicator. Conversely, the Low-Risk case (top row) is dominated by *Well-differentiated tumor cells* (30.9%) and *Normal colon mucosa* (21.5%). Here, the desmoplastic stroma burden is notably reduced, outweighed by preserved architectures and localized *Coagulative necrosis* (10.7%). This reflects a confined, expansile growth pattern rather than a highly infiltrative one. Furthermore, this localized necrosis resembles dirty necrosis, a morphological feature often associated with high microsatellite instability and favorable outcomes.

4 Conclusion

In this paper, we present ConVL, a novel, interpretable concept-guided vision-language framework designed to overcome severe intra-class heterogeneity and the lack of semantic transparency in WSI-based survival analysis. To shift from static semantic priors to patient-specific knowledge, we introduce Instance Conditioned Prompt Learning, which dynamically adapts pathological concepts to unique slide-level contexts. This is coupled with a Concept-Guided Cross Attention module to explicitly retrieve and quantify discriminative semantic evidence. Furthermore, a Distribution Alignment Loss ensures robust consensus between visual and concept-driven prediction manifolds. Extensive evaluations across nine TCGA cancer cohorts demonstrate that ConVL not only establishes a new SOTA in prognostic accuracy but also delivers fine-grained, clinically actionable interpretability.

Acknowledgement. This work was supported in part by the National Natural Science Foundation of China (Nos. U25A20449 and U24A20256), and by the Key R&D Program of Xinjiang Uygur Autonomous Region under Grant No. 2024B03039-3. This work was also partially carried out using computing resources at the High Performance Computing Center of Central South University.

Disclosure of Interests. The authors have no competing interests to declare that are relevant to the content of this article.

References

1. Ilse, M., Tomczak, J., Welling, M.: Attention-based deep multiple instance learning. In: International Conference on Machine Learning, pp. 2127–2136. PMLR (2018)
2. Kui, X., Xiao, Q., Li, Q., Ji, Z., Zhang, J., Zou, B.: Flip distribution alignment vae for multi-phase mri synthesis. In: International Conference on Medical Image Computing and Computer-Assisted Intervention, pp. 208–218. Springer, Heidelberg (2025). DOI: https://doi.org/10.1007/978-3-032-05185-1_21
3. Li, J., et al.: Dynamic graph representation with knowledge-aware attention for histopathology whole slide image analysis. In: Proceedings of the IEEE/CVF Conference on Computer Vision and Pattern Recognition, pp. 11323–11332 (2024)

4. Li, J., Kuang, H., Liu, J., Yue, H., He, M., Wang, J.: Universal-to-specific: dynamic knowledge-guided multiple instance learning for few-shot whole slide image classification. In: Proceedings of the IEEE/CVF Conference on Computer Vision and Pattern Recognition, pp. 26614–26623 (2026)
5. Li, J., Kuang, H., Liu, J., Yue, H., Wang, J.: Ca2cl: cluster-aware adversarial contrastive learning for pathological image analysis. IEEE J. Biomed. Health Inf. (2025)
6. Li, J., Liu, J., Kuang, H., Yue, H., He, M., Wang, J.: Mico: multiple instance learning with context-aware clustering for whole slide image analysis. In: International Conference on Medical Image Computing and Computer-Assisted Intervention, pp. 376–385. Springer, Heidelberg (2025). https://doi.org/10.1007/978-3-032-04927-8_36
7. Li, J., et al.: Darc: deep adaptive regularized clustering for histopathological image classification. Med. Image Anal. **80**, 102521 (2022)
8. Lin, J., Zhu, C., Kneuertz, P.J., Bai, Y., Xue, Y.: When models learn to ask why: adaptive causal reasoning for trustworthy medical vision-language models. In: Proceedings of the IEEE/CVF Conference on Computer Vision and Pattern Recognition, pp. 5556–5568 (2026)
9. Liu, M., et al.: Data-and knowledge-driven multimodal learning in computational pathology: a comprehensive survey. EngMedicine **3**(2), 100132 (2026)
10. Liu, M., et al.: Murrenet: modeling holistic multimodal interactions between histopathology and genomic profiles for survival prediction. In: International Conference on Medical Image Computing and Computer-Assisted Intervention, pp. 396–406. Springer, Heidelberg (2025). https://doi.org/10.1007/978-3-032-05182-0_39
11. Liu, P., Ji, L., Gou, J., Fu, B., Ye, M.: Interpretable vision-language survival analysis with ordinal inductive bias for computational pathology. In: The Thirteenth International Conference on Learning Representations (2025)
12. Liu, P., Ji, L., Zhang, X., Ye, F.: Pseudo-bag mixup augmentation for multiple instance learning-based whole slide image classification. IEEE Trans. Med. Imaging **43**(5), 1841–1852 (2024)
13. Lu, M.Y., et al.: A visual-language foundation model for computational pathology. Nat. Med. **30**, 863–874 (2024)
14. Qu, L., Fu, K., Wang, M., Song, Z., et al.: The rise of ai language pathologists: exploring two-level prompt learning for few-shot weakly-supervised whole slide image classification. Adv. Neural. Inf. Process. Syst. **36**, 67551–67564 (2023)
15. Radford, A., et al.: Learning transferable visual models from natural language supervision. In: International Conference on Machine Learning, pp. 8748–8763. PmLR (2021)
16. Shao, Z., et al.: Transmil: transformer based correlated multiple instance learning for whole slide image classification. Adv. Neural. Inf. Process. Syst. **34**, 2136–2147 (2021)
17. Tang, W., Zhou, F., Huang, S., Zhu, X., Zhang, Y., Liu, B.: Feature re-embedding: towards foundation model-level performance in computational pathology. In: Proceedings of the IEEE/CVF Conference on Computer Vision and Pattern Recognition, pp. 11343–11352 (2024)
18. Xiang, J., Zhang, J.: Exploring low-rank property in multiple instance learning for whole slide image classification. In: The Eleventh International Conference on Learning Representations (2023)
19. Zhang, D., et al.: Stadis: stability distance to detecting out-of-distribution data in computational pathology. Med. Image Anal. 103774 (2025)

20. Zhang, D., et al.: Care: a molecular-guided foundation model with adaptive region modeling for whole slide image analysis. In: Proceedings of the IEEE/CVF Conference on Computer Vision and Pattern Recognition, pp. 21078–21088 (2026)
21. Zhang, Q., et al.: Integrated multiomic analysis reveals comprehensive tumour heterogeneity and novel immunophenotypic classification in hepatocellular carcinomas. Gut **68**(11) (2019)
22. Zhang, Y., Li, H., Sun, Y., Zheng, S., Zhu, C., Yang, L.: Attention-challenging multiple instance learning for whole slide image classification. In: European Conference on Computer Vision, pp. 125–143. Springer, Heidelberg (2024). DOI: https://doi.org/10.1007/978-3-031-73668-1_8
23. Zhu, C., Lin, Y., Chen, S., Wang, Y., Lin, J.: Medeyes: learning dynamic visual focus for medical progressive diagnosis. In: Proceedings of the AAAI Conference on Artificial Intelligence, vol. 40, pp. 13916–13924 (2026)
24. Zhu, C., Lin, Y., Shao, J., Lin, J., Wang, Y.: Pathology-aware prototype evolution via llm-driven semantic disambiguation for multicenter diabetic retinopathy diagnosis. In: Proceedings of the 33rd ACM International Conference on Multimedia, pp. 9196–9205 (2025)
25. Zhu, C., et al.: fmri2ges: co-speech gesture reconstruction from fmri signal with dual brain decoding alignment. IEEE Trans. Circuits Syst. Video Technol. (2025)
26. Zhu, C., Zeng, J., Jiang, J., Lin, J., Wang, Y.: Medsynapse-v: bridging visual perception and clinical intuition via latent memory evolution. arXiv preprint arXiv:2604.26283 (2026)

Counterfactual Anatomy-Guided Spatial-Temporal Decoding for Annotation-Free Hallucination Mitigation in Medical VLMs

Yifan Lu[1], Adinath Dukre[1], Abhijit Das[1], Ziyun Zou[1], Haolin Yang[1], Yutong Xie[1], and Imran Razzak[1,2(✉)]

[1] Mohamed bin Zayed University of Artificial Intelligence, Abu Dhabi, UAE
[2] MedOS, Abu Dhabi, UAE
Imran.Razzak@mbzuai.ac.ae

Abstract. Medical vision-language models (Med-VLMs) have demonstrated strong performance on medical visual question answering, yet they remain prone to hallucination, generating clinically unsupported statements that are insufficiently grounded in image evidence. Mitigation methods applied during decoding offer a practical solution, but they typically lack anatomical awareness or rely heavily on ground truth annotations, which limits their applicability. We propose Counterfactual Anatomy-guided Spatial-Temporal decoding (CAST), a framework that operates entirely during inference and requires no manual annotations for anatomically grounded hallucination mitigation. CAST automatically discovers anatomical regions relevant to the given query through broad medical segmentation. It then selects a compact, causally informative area using counterfactual intervention based on the drop in answer likelihood under occlusion. Guided by this chosen region, CAST performs a unified contrastive decoding process, combining classifier-free guidance to correct spatial attention with stepwise temporal contrast to regulate generation dynamics. Experiments on the SLAKE and MIMIC-CXR datasets across three Med-VLMs demonstrate that CAST consistently outperforms strong baselines and surpasses decoding strategies reliant on ground truth. Our results indicate that compact, automatically selected regions provide highly effective contrastive guidance without expert annotations, offering a practical and generalizable solution for improving spatial grounding and reducing hallucinations.

Keywords: Medical Vision-Language Models · Hallucination Mitigation · Contrastive Decoding · Causal Discovery

1 Introduction

Medical vision-language models (Med-VLMs) have recently demonstrated impressive performance on medical visual question answering (Med-VQA) by

G. Yang et al. (Eds.): MICCAI 2026, LNCS 16878, pp. 87–97, 2027.
https://doi.org/10.1007/978-3-032-38059-3_9

jointly reasoning over images and clinical language [5,11,24]. Despite their strong average accuracy, these models remain prone to hallucination, generating fluent but clinically unsupported statements that lack sufficient grounding in the image [6,14,20]. In medical settings, such errors can have serious consequences, as a single hallucinated finding may mislead clinical interpretation or subsequent decision-making [3,19,26]. Improving faithfulness without expensive retraining has therefore become a central challenge for the reliable deployment of Med-VLMs [17,23,27].

Decoding-time methods are attractive because they are plug-and-play and do not require additional supervision [7,8,10]. Visual Contrastive Decoding (VCD) [15] mitigates hallucinations by contrasting predictions from the original image and a noise-perturbed version, encouraging tokens that remain stable under visual corruption. Decoding by Contrasting Layers (DoLA) [4] improves factuality by contrasting logits from early and late transformer layers, suppressing tokens driven by memorized knowledge rather than contextual evidence. OPERA [9] penalizes over-confident attention patterns and reallocates attention during generation to reduce over-trust in spurious visual cues. Although effective for general-purpose VLMs [13], these methods are largely anatomy-agnostic and apply global corrections that do not explicitly account for the localized nature of medical evidence. Anatomical Region-Guided Contrastive Decoding (ARCD) [18] addresses this limitation by introducing spatial grounding through classifier-free guidance (CFG) conditioned on an anatomical region-of-interest (ROI). At each decoding step, ARCD contrasts a full-image branch with an ROI-suppressed branch, steering generation toward tokens supported by the diagnostic region. However, ARCD depends on ground-truth (GT) ROI annotations, which are costly, scarce, and dataset-specific, limiting real-world applicability.

We investigate whether region-guided contrastive decoding can be made annotation-free without sacrificing effectiveness by automatically discovering question-relevant ROIs for guidance. Empirically, we find that such automatically selected regions not only remove the dependency on expert annotations but can also provide stronger guidance in practice. This is partly because the effectiveness of ROI-guided CFG depends on mask granularity: ground-truth bounding boxes in Med-VQA benchmarks are often coarse and high-coverage, and suppressing them removes most visual information from the unconditional branch, weakening the region-specific contrast. In contrast, compact and precise masks preserve contextual cues while selectively removing relevant evidence, yielding a cleaner contrastive signal during decoding.

Motivated by this observation, we propose **C**ounterfactual **A**natomy-guided **S**patial-**T**emporal (**CAST**) decoding, an annotation-free framework for hallucination mitigation in Med-VLMs. CAST consists of three stages. First, in the proposal stage, we generate diverse anatomical region candidates using a medical segmentation model through efficient multi-concept inference. Second, in the selection stage, we identify the most question-relevant and CFG-compatible region via a counterfactual intervention criterion that measures the likelihood drop under regional occlusion while discouraging overly large masks. Third, in the decoding stage, we introduce a spatial-temporal contrastive strategy that corrects where the model attends via anatomy-guided CFG and regulates what it predicts via step-wise temporal contrast. The entire pipeline operates at inference time and remains fully plug-and-play. Across three Med-VLMs on SLAKE and MIMIC-CXR, CAST outperforms strong decoding-time baselines and GT-dependent region-guided methods, achieving effective spatial grounding without expert annotations.

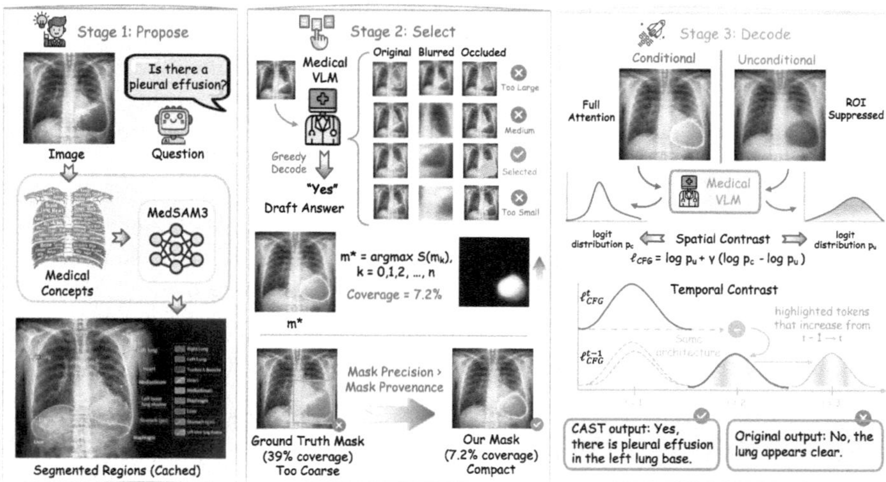

Fig. 1. Overview of the proposed CAST framework. It (1) proposes anatomical regions via multi-concept medical segmentation, (2) selects a compact question-relevant ROI using counterfactual likelihood drop under occlusion, and (3) applies spatial-temporal contrastive decoding: anatomy-guided CFG for grounding and step-wise temporal contrast to suppress hallucinations.

2 Method

CAST is an annotation-free, inference-time framework for mitigating hallucinations in Med-VLMs via automatic ROI discovery and spatial-temporal contrastive decoding. As illustrated in Fig. 1, we first generate candidate anatomical masks with multi-concept medical segmentation and filter them for CFG compatibility. Given a question, we select the ROI that most affects the model's likelihood under regional occlusion while favoring compactness. Finally, CAST performs anatomy-guided CFG for region grounding and a step-wise temporal contrast to damp language-driven momentum. The entire pipeline is fully plug-and-play, requiring no retraining or expert annotations.

2.1 Zero-Shot Anatomical Region Proposal

The first stage of CAST aims to extract a diverse set of anatomically coherent region candidates directly from the input image $\mathbf{x}$. To achieve this without dataset-specific annotations, we leverage the robust medical priors embedded in a text-guided medical segmentation model, MedSAM3 [21]. Rather than sequentially querying individual anatomical structures, we employ an efficient multi-concept inference strategy. We construct a comprehensive vocabulary of anatomical concepts (spanning thoracic, abdominal, musculoskeletal, and vascular systems) and encode them into a unified prompt. The model's DETR-style [1] decoder processes this joint embedding to activate diverse parallel queries, generating segmentation masks for multiple structures in a single forward pass. To ensure the proposals are suitable for localized decoding, the raw masks undergo a compactness-aware filtering pipeline. Masks are filtered by area coverage to remove pixel-level noise and excessively global masks. We then apply spatial non-maximum suppression based on intersection over union to eliminate near-duplicates [25], followed by morphological filtering to discard degenerate shapes. This results in a spatially diverse and anatomically valid set of K candidate regions $[m_1, \ldots, m_K]$, extracted with minimal computational overhead.

2.2 Counterfactual Region Selection

Given the K candidates, CAST must autonomously identify the single region m^* that is most causally responsible for the model's clinical reasoning regarding the current query q. We frame this as a causal discovery problem, evaluating each region via counterfactual intervention. We first generate a draft sequence $\hat{y} = (y_1, \ldots, y_T)$ utilizing greedy decoding on the factual, unoccluded image $\mathbf{x}$. To isolate the causal contribution of a candidate region m_k, we define a counterfactual intervention do(occlude m_k) by selectively corrupting the pixel space strictly within the mask boundaries. To prevent the model from over-indexing on occlusion artifacts, we utilize two complementary operators, $o \in \{\text{blur}, \text{mean}\}$, which independently disrupt fine-grained texture and local contrast statistics. The causal effect (CE) of region m_k is quantified as the drop in the predictive log-likelihood of the draft answer when the region is removed:

$$\mathrm{CE}_o(m_k) = \log p(\hat{y} \mid \mathbf{x}, q) - \log p(\hat{y} \mid \mathbf{x}_{\mathrm{occ}}^{(o,k)}, q), \tag{1}$$

where $\mathbf{x}_{\mathrm{occ}}^{(o,k)}$ represents the image under occlusion mode o. Effective contrastive guidance requires spatial precision; a region that spans the entire image provides no localized contrastive signal. Therefore, our selection objective explicitly balances causal relevance with mask compactness and cross-modal consistency:

$$S(m_k) = \frac{1}{2} \sum_o \mathrm{CE}_o(m_k) - \lambda \cdot \mathrm{area}(m_k) - \mu \cdot \mathrm{Var}\big[\mathrm{CE}_{\mathrm{blur}}(m_k), \mathrm{CE}_{\mathrm{mean}}(m_k)\big]. \tag{2}$$

Here, the area penalty λ enforces mask compactness for localized contrast, while μ penalizes variance to filter out occlusion-specific artifacts. The highest-scoring mask $m^* = \arg\max_k S(m_k)$ is then selected as the optimal diagnostic anchor.

2.3 Unified Spatial-Temporal Contrastive Decoding

With the optimal region m^* identified, CAST introduces a lightweight, sequential decoding strategy to simultaneously correct two orthogonal hallucination axes: spatial grounding (*where* the model attends) and temporal generation (*what* it predicts). At each decoding step t, a dual-branch forward pass is constructed. The unconditional branch heavily suppresses attention to visual tokens within m^*, forcing reliance on surrounding context and language priors, while the conditional branch processes the full image natively. Spatially guided logits are computed as:

$$\boldsymbol{\ell}_{\mathrm{CFG}}^{(t)} = \log p_u^{(t)} + \gamma \cdot \left(\log p_c^{(t)} - \log p_u^{(t)}\right), \tag{3}$$

where γ is the guidance scale, and p_c and p_u are conditional and unconditional token probabilities. This amplifies tokens dependent on the targeted anatomical region, reducing reliance on spurious background correlations. Following spatial grounding, CAST applies an orthogonal temporal correction across consecutive steps:

$$\boldsymbol{\ell}_{\mathrm{final}}^{(t)} = w_m \cdot \boldsymbol{\ell}_{\mathrm{CFG}}^{(t)} - w_p \cdot \boldsymbol{\ell}_{\mathrm{CFG}}^{(t-1)}, \tag{4}$$

where w_m and w_p are weighting hyper-parameters. The theoretical foundation for this temporal contrast is that the prior step's distribution, $\boldsymbol{\ell}_{\mathrm{CFG}}^{(t-1)}$, heavily encodes the default trajectory driven by the model's internal language momentum. By subtracting this prior distribution from the current active prediction $\boldsymbol{\ell}_{\mathrm{CFG}}^{(t)}$, CAST selectively amplifies tokens that are newly activated by the current visual evidence, while suppressing tokens that are merely passively inherited from previous steps. Because the spatial contrast modulates the source of visual attention and the temporal contrast regulates habitual token transitions, the two mechanisms operate harmoniously without interference.

3 Experiments

3.1 Experimental Setup

Datasets and Baselines. We evaluate on SLAKE [22] and MIMIC-CXR [12]. SLAKE contains 1,061 multi-modal (CT/MRI/X-ray) QA pairs with open- and closed-ended questions and bounding boxes. MIMIC-CXR includes 2,134 chest X-ray QA pairs, primarily closed-ended with pathology annotations. Following prior work [18], we report Accuracy (closed), Recall (open), and weighted Overall performance. CAST is compared against the standard decoding baseline and four decoding-time methods: VCD [15], DoLA [4], OPERA [9], and ARCD [18].

Implementation Details. To demonstrate the plug-and-play versatility of CAST, we apply it to three Med-VLMs: Phi-3.5V-Med [18], HuatuoGPT-Vision-7B [2], and LLaVA-Med-7B [16]. MedSAM3 processes images at a resolution of 1008×1008. We apply NMS with an IoU threshold of 0.9 to ensure spatial diversity, retaining up to K=16 candidate regions per question for counterfactual scoring. For the selection objective, both the area penalty factor λ and the consistency coefficient μ are set to 0.1. In the decoding stage, the spatial CFG employs a guidance scale of γ=1.5. The attention modulation weights for regions inside and outside the ROI (α and β, respectively) are empirically determined via grid search over the set $\{0.01, 0.1, 0.5, 1.0, 2.0, 3.0\}$. For the step-wise temporal contrast, we set the active momentum weight w_m=1.0 and the prior-step penalty w_p=0.5. All experiments are executed on a single NVIDIA A100 40GB GPU.

Table 1. Comprehensive comparison on SLAKE and MIMIC benchmarks. Best in **bold**, second underlined within each model block.

Method	SLAKE			MIMIC		
	Open	Closed	Overall	Open	Closed	Overall
Phi-3.5V-Med	38.58	61.42	50.00	14.29	58.12	47.24
+ VCD	40.16	61.42	50.79	15.87	59.16	48.43
+ DoLA	**47.24**	59.84	53.54	19.05	56.54	47.24
+ OPERA	36.22	65.35	50.79	14.29	61.78	50.00
+ ARCD	44.54	65.87	52.90	28.85	60.44	52.61
+ **CAST (Ours)**	44.90	**81.49**	**59.25**	**33.03**	**67.85**	**59.22**
HuatuoGPT-V-7B	47.94	75.48	58.74	29.84	62.68	54.54
+ VCD	**49.73**	75.48	59.83	31.05	63.61	55.54
+ DoLA	43.97	78.61	57.55	29.64	65.30	56.46
+ OPERA	46.32	74.04	57.19	**34.74**	63.99	56.74
+ ARCD	48.33	75.00	58.79	30.97	63.12	55.15
+ **CAST (Ours)**	47.56	**81.49**	**60.86**	31.31	**67.29**	**58.37**
LLaVA-Med-7B	39.62	62.50	48.59	22.52	53.71	45.98
+ VCD	40.00	62.98	49.01	22.77	53.71	46.04
+ DoLA	32.75	66.35	45.92	21.34	56.14	47.51
+ OPERA	**40.59**	65.38	50.31	26.17	54.52	47.49
+ ARCD	37.95	62.02	47.39	22.88	53.77	46.11
+ **CAST (Ours)**	36.49	**73.32**	**50.93**	**27.16**	**59.69**	**51.63**

3.2 Experimental Results and Analysis

Comparison with Decoding-Time Methods. Table 1 reports the comprehensive comparison across three Med-VLMs on SLAKE and MIMIC-CXR. CAST consistently improves overall performance over strong decoding-time baselines (VCD, DoLA, OPERA) and GT-dependent region-guided decoding across all models. Most notably, CAST demonstrates substantial and consistent improvements in Closed-ended question accuracy, which directly reflects a model's ability to remain factually grounded and resist hallucination. For instance, when applied to Phi-3.5V-Med on SLAKE, CAST elevates closed-ended accuracy from 61.42 to a striking 81.49. On MIMIC, CAST improves both open and closed metrics simultaneously, indicating that spatial-temporal contrast benefits both factual grounding and short-form diagnostic responses. While anatomy-agnostic baselines such as DoLA, OPERA, and VCD occasionally yield isolated gains on Open-ended queries for specific models, they frequently exhibit volatile performance, trading closed-ended accuracy for open-ended recall. In contrast, CAST maintains highly competitive open-ended performance while decisively dominating closed-ended factual reasoning. Notably, CAST consistently outperforms ARCD, this empirical superiority strongly validates the core theoretical hypothesis introduced in our methodology: automatically discovered, dynamically selected, and highly compact region masks provide a cleaner, more targeted contrastive signal than coarse GT annotations. By explicitly optimizing for causal relevance and spatial compactness, CAST achieves superior spatial grounding and hallucination mitigation entirely free of expert annotation dependency.

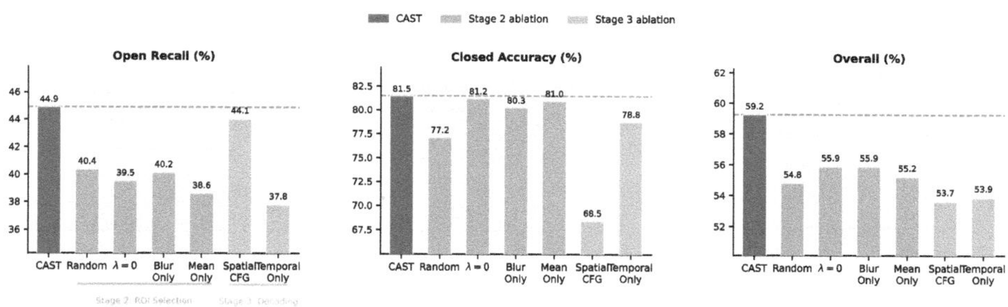

Fig. 2. Ablation study on SLAKE (Phi-3.5V-Med).

Ablation Study. Figure 2 ablates the core components of CAST on SLAKE, demonstrating that both our unified decoding mechanisms and counterfactual selection criteria are essential. Isolating spatial CFG or temporal contrast drops overall accuracy from 59.25% to 53.66% and 53.91%, respectively, confirming their orthogonal roles: spatial CFG anchors where the model attends to rescue closed-question accuracy (from 68.51% to 81.49%), while temporal contrast regulates what it predicts to sustain open-question recall (44.90% vs. 37.83%). Beyond decoding, counterfactual region selection provides a critical signal; replacing it with random mask selection degrades overall accuracy by 4.4%. Within the scoring function itself, removing the area penalty (λ=0) causes a severe 3.4% drop, cementing that mask compactness is strictly required for effective CFG. Finally, relying on a single occlusion mode (blur or mean) lowers accuracy by 3.3% and 4.0%, highlighting the necessity of combining dual occlusions to average out modality-specific artifacts.

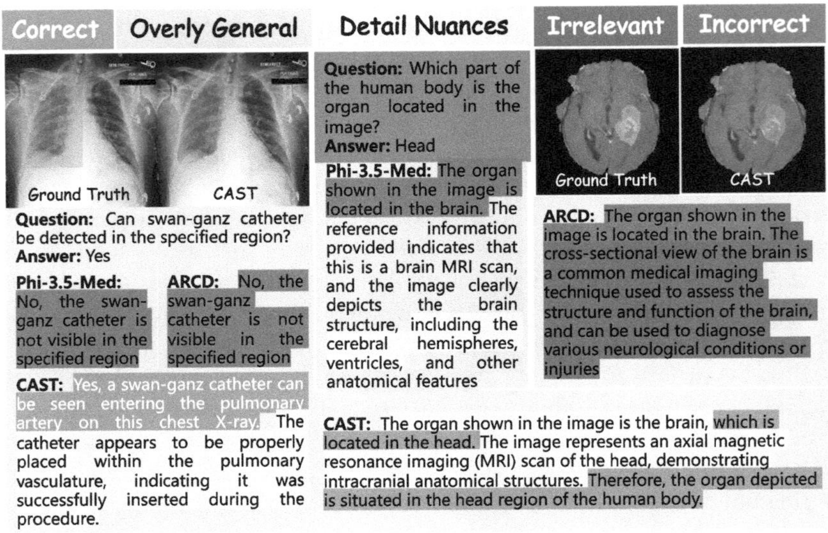

Fig. 3. Qualitative comparison of model-generated responses.

Qualitative Analysis. Figure 3 shows two failure cases where CAST overcomes limitations of coarse GT masks. In Case 1 (MIMIC X-ray), GT-ARCD fails to detect a subtle Swan-Ganz catheter because its bounding box spans the entire right lung, diluting the contrastive signal. CAST instead isolates a compact 10.6% region around the catheter tip and correctly confirms its presence. In Case 2 (SLAKE MRI), the model must identify the anatomical region (head) rather than the highly salient organ (brain). Both the Baseline and GT-ARCD incorrectly output "brain" because the broad GT segmentation reinforces organ identity over spatial location, whereas CAST selects a highly compact mask highlighting specific craniofacial landmarks to correctly guide the response. Together, these cases concretely demonstrate that dynamically selected, compact masks provide superior contrastive guidance compared to static GT annotations, which can actively mislead or dilute critical visual evidence.

Mask Compactness Analysis. To understand CAST's advantage over GT-based methods, we analyze mask coverage and performance on HuatuoGPT-V-7B (SLAKE). GT bounding boxes are substantially larger than CAST-selected masks (39.41% vs. 7.05% mean coverage), with some spanning the entire image. When GT coverage exceeds 40%, closed-question accuracy drops noticeably. Per-sample analysis shows that CAST mainly improves cases with large GT masks, while GT performs competitively only when annotations are compact. Notably, GT slightly reduces closed accuracy compared to NULL decoding, whereas CAST significantly improves it. These results indicate that oversized masks dilute the localized contrast required by classifier-free guidance, while compact, causally selected regions provide sharper and more effective signals.

Computational Efficiency. As a plug-and-play inference-time framework, CAST introduces minimal overhead compared to standard decoding method. In Stage 1, MedSAM3 proposals are precomputed and cached, effectively eliminating per-question segmentation costs; even in an online setting, this overhead remains marginal relative to VLM decoding. In Stage 2, counterfactual ROI scoring operates via a lightweight teacher-forced evaluation over a short sequence ($T_{score} = 32$ tokens), which is roughly an order of magnitude cheaper than a full autoregressive generation pass. Although Stage 3 employs dual-branch spatial-temporal contrast, the overall wall-clock latency of CAST remains nearly identical to that of DoLA, placing it squarely within the practical operating range of accepted inference-time methods. Furthermore, the K candidate regions are independent and can be seamlessly parallelized at the engineering level for further efficiency gains.

4 Conclusion

In this paper, we presented CAST, a plug-and-play decoding framework for mitigating hallucinations in Med-VLMs. By formulating spatial grounding as a causal discovery problem, CAST automatically identifies compact and question-relevant anatomical regions through counterfactual intervention. We show that these dynamically selected regions address the compactness limitations of coarse ground-truth annotations and provide a sharper and more informative contrastive signal. Combined with a unified spatial-temporal decoding strategy that jointly refines visual attention and step-wise generation dynamics, CAST consistently achieves state-of-the-art performance across multiple Med-VLMs on SLAKE and MIMIC-CXR. Overall, CAST offers a practical inference-time approach to improving factual reliability without requiring expert annotations or model retraining.

Disclosure of Interests. The authors have no competing interests to declare that are relevant to the content of this article.

References

1. Carion, N., Massa, F., Synnaeve, G., Usunier, N., Kirillov, A., Zagoruyko, S.: End-to-end object detection with transformers. In: European Conference on Computer Vision, pp. 213–229. Springer (2020)
2. Chen, J., et al.: Towards injecting medical visual knowledge into multimodal LLMs at scale. In: Proceedings of the 2024 Conference on Empirical Methods in Natural Language Processing, pp. 7346–7370 (2024)
3. Chen, Y., Yang, Z., Huang, Y., Yang, X., Yeo, S., Zhang, Y.: Trustworthy disentangled framework for multi-label medical image classification with multimodal refinement. In: 2025 IEEE International Conference on Bioinformatics and Biomedicine (BIBM), pp. 839–846. IEEE (2025)

4. Chuang, Y.S., Xie, Y., Luo, H., Kim, Y., Glass, J.R., He, P.: Dola: decoding by contrasting layers improves factuality in large language models. In: International Conference on Learning Representations, vol. 2024, pp. 54158–54183 (2024)
5. Deria, A., Kumar, K., Dukre, A.M., Segal, E., Khan, S., Razzak, I.: Medmo: grounding and understanding multimodal large language model for medical images. arXiv preprint arXiv:2602.06965 (2026)
6. Dutta, N., Bose, K., Syailendra, E., Chu, L., Gupta, P.: Vision-language models in diagnostic imaging: review of technical advances, clinical validation, and practical deployment. Int. J. Med. Inf., 106227 (2025)
7. Favero, A., et al.: Multi-modal hallucination control by visual information grounding. In: Proceedings of the IEEE/CVF Conference on Computer Vision and Pattern Recognition, pp. 14303–14312 (2024)
8. Ghosh, S., et al.: Visual description grounding reduces hallucinations and boosts reasoning in LVLMs. In: International Conference on Learning Representations, vol. 2025, pp. 66510–66547 (2025)
9. Huang, Q., et al.: Opera: alleviating hallucination in multi-modal large language models via over-trust penalty and retrospection-allocation. In: Proceedings of the IEEE/CVF Conference on Computer Vision and Pattern Recognition, pp. 13418–13427 (2024)
10. Ji, Y., et al.: Specvlm: enhancing speculative decoding of video LLMs via verifier-guided token pruning. In: Proceedings of the 2025 Conference on Empirical Methods in Natural Language Processing, pp. 7216–7230 (2025)
11. Jiang, S., et al.: Hulu-med: a transparent generalist model towards holistic medical vision-language understanding. arXiv preprint arXiv:2510.08668 (2025)
12. Johnson, A.E., et al.: Mimic-CXR, a de-identified publicly available database of chest radiographs with free-text reports. Sci. Data **6**(1), 317 (2019)
13. Kang, J., Shu, H., Li, W., Zhai, Y., Chen, X.: Vispec: accelerating vision-language models with vision-aware speculative decoding. Adv. Neural. Inf. Process. Syst. **38**, 115511–115532 (2026)
14. Khanal, B., et al.: Hallucination-aware multimodal benchmark for gastrointestinal image analysis with large vision-language models. In: International Conference on Medical Image Computing and Computer-Assisted Intervention, pp. 235–245. Springer (2025)
15. Leng, S., et al.: Mitigating object hallucinations in large vision-language models through visual contrastive decoding. In: Proceedings of the IEEE/CVF Conference on Computer Vision and Pattern Recognition, pp. 13872–13882 (2024)
16. Li, C., et al.: Llava-med: training a large language-and-vision assistant for biomedicine in one day. Adv. Neural. Inf. Process. Syst. **36**, 28541–28564 (2023)
17. Liang, X., et al.: Chexpo: preference optimization for chest x-ray VLMs with counterfactual rationale. In: Proceedings of the 33rd ACM International Conference on Multimedia, pp. 2606–2615 (2025)
18. Liang, X., et al.: Anatomical region-guided contrastive decoding: a plug-and-play strategy for mitigating hallucinations in medical VLMs. In: Proceedings of the AAAI Conference on Artificial Intelligence. vol. 40, pp. 6871–6879 (2026)
19. Liang, X., et al.: Uncertainty-driven expert control: enhancing the reliability of medical vision-language models. In: Proceedings of the IEEE/CVF International Conference on Computer Vision, pp. 21144–21154 (2025)
20. Liao, Z., Hu, S., Zou, K., Fu, H., Zhen, L., Xia, Y.: Vision-amplified semantic entropy for hallucination detection in medical visual question answering. In: International Conference on Medical Image Computing and Computer-Assisted Intervention, pp. 669–679. Springer (2025)

21. Liu, A., et al.: Medsam3: delving into segment anything with medical concepts. arXiv preprint arXiv:2511.19046 (2025)
22. Liu, B., Zhan, L.M., Xu, L., Ma, L., Yang, Y., Wu, X.M.: Slake: a semantically-labeled knowledge-enhanced dataset for medical visual question answering. In: 2021 IEEE 18th International Symposium on Biomedical Imaging (ISBI), pp. 1650–1654. IEEE (2021)
23. Nath, V., et al.: Vila-m3: enhancing vision-language models with medical expert knowledge. In: Proceedings of the Computer Vision and Pattern Recognition Conference, pp. 14788–14798 (2025)
24. Shu, Y., Liu, C., Chen, R., Li, D., Dai, B.: Fleming-vl: towards universal medical visual reasoning with multimodal LLMs. arXiv preprint arXiv:2511.00916 (2025)
25. Si, K.S., et al.: Accelerating non-maximum suppression: a graph theory perspective. Adv. Neural. Inf. Process. Syst. **37**, 121992–122028 (2024)
26. Silva-Rodríguez, J., Ben Ayed, I., Dolz, J.: Trustworthy few-shot transfer of medical VLMs through split conformal prediction. In: International Conference on Medical Image Computing and Computer-Assisted Intervention, pp. 658–668. Springer (2025)
27. Zhao, Y., et al.: Med-VLM: enhancing medical image segmentation accuracy through vision-language model. In: Proceedings of the IEEE/CVF International Conference on Computer Vision, pp. 7283–7293 (2025)

CRIMSON: A Clinically-Grounded LLM-Based Metric for Generative Radiology Report Evaluation

Mohammed Baharoon[1], Thibault Heintz[2], Siavash Raissi[1], Mahmoud Alabbad[3], Mona Alhammad[4], Hassan AlOmaish[5], Sung Eun Kim[1], Oishi Banerjee[1], and Pranav Rajpurkar[1](✉)

[1] Department of Biomedical Informatics, Harvard Medical School, Boston, MA, USA
pranav_rajpurkar@hms.harvard.edu
[2] Department of Radiation Oncology, Mass General Brigham, Boston, MA, USA
[3] King Fahad Hospital, Al-Ahsa Health Cluster, Al Hofuf, Saudi Arabia
[4] Ras-Tanura General Hospital, Ministry of Health, Ras-Tanura, Eastern Province, Saudi Arabia
[5] Department of Medical Imaging, King Abdulaziz Medical City, Ministry of National Guard, Riyadh, Saudi Arabia
https://github.com/rajpurkarlab/CRIMSON

Abstract. We introduce **CRIMSON**, a clinically grounded evaluation framework for chest X-ray report generation that assesses reports based on diagnostic correctness, contextual relevance, and patient safety. Unlike prior metrics, CRIMSON incorporates full clinical context, including patient age, indication, and guideline-based decision rules, and prevents normal or clinically insignificant findings from exerting disproportionate influence on the overall score. The framework categorizes errors into a comprehensive taxonomy covering false findings, missing findings, and eight attribute-level errors (e.g., location, severity, measurement, and diagnostic overinterpretation). Each finding is assigned a clinical significance level (urgent, actionable non-urgent, non-actionable, or expected/benign), based on a guideline developed in collaboration with attending cardiothoracic radiologists, enabling severity-aware weighting that prioritizes clinically consequential mistakes over benign discrepancies. CRIMSON is validated through strong alignment with clinically significant error counts annotated by six board-certified radiologists in ReXVal (Kendall's τ = 0.61-0.71; Pearson's r = 0.71-0.84), and through two additional benchmarks that we introduce. In *RadJudge*, a targeted suite of clinically challenging pass-fail scenarios, CRIMSON shows consistent agreement with expert judgment. In *RadPref*, a larger radiologist preference benchmark of over 100 pairwise cases with structured error categorization, severity modeling, and 1–5 overall quality ratings from three cardiothoracic radiologists, CRIMSON achieves the strongest alignment with radiologist preferences. We release the metric, the evaluation benchmarks, RadJudge and RadPref, and a fine-tuned MedGemma

M. Baharoon, T. Heintz and S. Raissi—These authors contributed equally.

G. Yang et al. (Eds.): MICCAI 2026, LNCS 16878, pp. 98–107, 2027.
https://doi.org/10.1007/978-3-032-38059-3_10

model to enable reproducible evaluation of report generation, all available at https://github.com/rajpurkarlab/CRIMSON.

Keywords: Report Evaluation · Radiology Report Generation

1 Introduction

Automated radiology report generation has advanced rapidly with the emergence of large vision-language models, yet reliable evaluation remains a fundamental challenge [6,17,21]. Recent radiology-specific metrics have moved beyond surface-level text similarity and instead assess factual correctness through structured error counting and finding-level comparison [3,4,9,15,22,23]. These approaches represent important progress toward clinically meaningful evaluation by explicitly detecting hallucinations and omissions.

Despite these advances, current metrics largely treat detected errors as either uniformly important or binary (significant vs. not significant), and evaluate findings in relative isolation from broader clinical context. In practice, the clinical consequences of errors vary substantially. For instance, failing to report a life-threatening pneumothorax is categorically different from missing age-related aortic calcification. Moreover, the relevance and interpretation of findings depend on a patient's age and indication. Existing evaluation frameworks do not explicitly encode clinical severity as a fine-grained spectrum; instead, they collapse findings into coarse categories, lack sufficient clinical context to determine true significance, or rely on LLM judgments without structured in-context guidelines [4,9,15]. Consequently, these frameworks conflate minor, clinically inconsequential discrepancies with omissions that directly impact patient safety.

To address these limitations, we introduce CRIMSON, a clinically grounded LLM-based evaluation framework designed to align automated assessment with real-world radiologic reasoning. CRIMSON evaluates reports at the level of individual findings while incorporating full clinical context, including patient age and indication. The framework models a comprehensive taxonomy of errors—false findings, missing findings, attribute-level errors (e.g., location, severity, measurement) and significance labels (urgent, actionable non-urgent, non-actionable, or expected/benign)—defined according to a rubric developed with cardiothoracic radiologists. The severity labels determine weights within a principled scoring formulation that prioritizes clinically consequential errors over benign discrepancies and supports partial credit for partially correct findings under fine-grained attribute rules. We validate CRIMSON through alignment with radiologist-annotated clinically significant error counts in ReXVal [17,18] (PhysioNet Credentialed Health Data License 1.5.0). We also introduce and validate the metric using RadJudge, a targeted ranking test suite of clinically challenging scenarios, and RadPref, a comprehensive radiologist preference benchmark, demonstrating improved agreement with expert judgment compared to existing metrics.

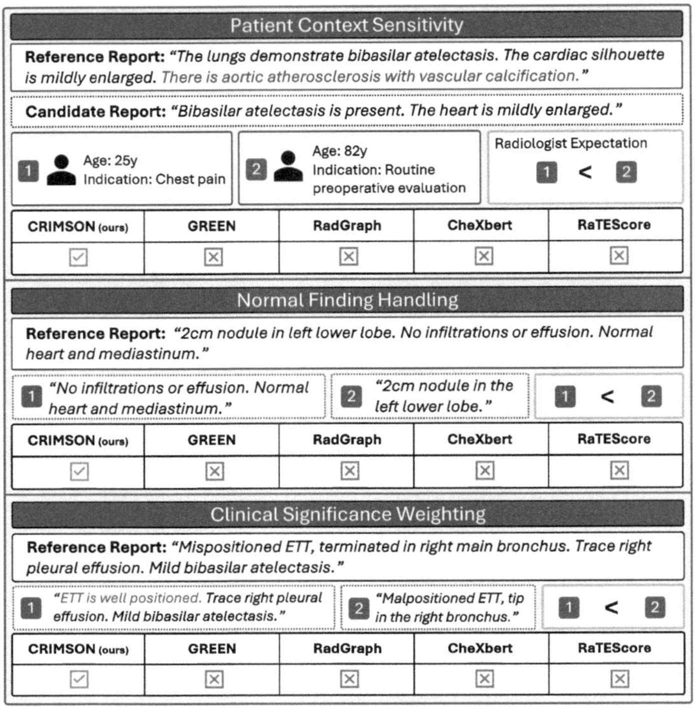

Fig. 1. Representative RadJudge cases illustrating core design principles of CRIMSON. **Top:** Patient context sensitivity. The clinical impact of an omission (e.g., aortic atherosclerosis) varies by age and indication, and CRIMSON adjusts severity accordingly. **Middle:** Normal finding handling. CRIMSON does not reward mentioning normal findings, preventing score inflation. **Bottom:** Clinical significance weighting. Errors are weighted by consequence, prioritizing clinically important findings. In each case, CRIMSON aligns with radiologist expectations, whereas prior metrics fail.

2 CRIMSON: Clinically-Grounded Report Evaluation

CRIMSON evaluates a candidate radiology report against a reference report by identifying and categorizing discrepancies at the finding level, weighing each error by its clinical significance, and computing a normalized score that reflects clinical preferences. GPT-5.2 [12] is used as the backbone throughout this pipeline. The framework operates in three stages: (1) finding extraction and clinical significance assignment, (2) error detection and classification, and (3) severity-aware score computation.

2.1 Finding Extraction and Clinical Significance Assignment

Given a reference report R_{ref} and a candidate report R_{cand}, CRIMSON first extracts all abnormal findings from each report. Normal findings are excluded from evaluation because including them can introduce spurious variability due to stylistic differences across radiologists [1,2]. Each extracted finding f is assigned a clinical significance weight $w(f)$ based on standard radiological practice and a structured severity framework adapted from [14] with input from attending cardiothoracic radiologists. The clinical significance weight $w(f)$ is defined as $w(f) = 1.0$ if the finding is urgent, 0.5 if actionable but not urgent, 0.25 if not actionable and not urgent, and 0.0 if expected or benign. Findings classified as *urgent* correspond to abnormalities requiring immediate intervention or indicating life-threatening conditions, such as a tension pneumothorax. *Actionable non-urgent* findings are those that alter patient management but are not immediately critical, including nodules, moderate pleural effusions, or consolidations. *Not actionable* are findings that represent minimal clinical impact but remain worth documenting, such as a cervical rib or appropriately positioned support devices. *Expected benign* findings include expected or age-appropriate changes with no impact on care, such as degenerative spine changes or a tortuous aorta. We separate non-actionable from expected benign findings because reporting of expected benign findings varies substantially by radiologist style; penalizing based on them introduces unnecessary randomness. Clinical significance assignment incorporates patient context when available, including age and indication. For example, aortic calcification in a 75-year-old patient is classified as expected benign, whereas the same finding in a 25-year-old patient may be considered actionable non-urgent due to atypical early onset.

2.2 Error Taxonomy and Classification

CRIMSON characterizes discrepancies through three primary error categories: false findings, missing findings, and attribute errors. False findings are abnormal findings present in R_{cand} but absent from R_{ref}, representing a hallucination. Missing findings are abnormal findings present in R_{ref} but absent from R_{cand}, representing diagnostically meaningful omissions.

Findings that appear in both reports are considered "matched." For such findings, CRIMSON evaluates attribute-level correctness across eight dimensions: (1) anatomical location or laterality, (2) severity or extent, (3) morphological descriptors, (4) quantitative measurements, (5) certainty level, (6) diagnostic underinterpretation, (7) overinterpretation, and (8) temporal or comparison descriptors.

Each attribute error e is assigned a severity-based weight $w_{\text{attr}}(e)$, where $w_{\text{attr}}(e) = 0.5$ if the error is labeled as significant, and $w_{\text{attr}}(e) = 0.0$ if it is labeled as negligible. Significant attribute errors are those that could alter treatment decisions or patient management, whereas negligible errors correspond to clinically inconsequential differences. For example, incorrect lung laterality is considered significant, whereas positional differences within the same lobe, such as 'apical' vs 'lateral' are considered negligible. For pulmonary nodules smaller than 6 mm, measurement discrepancies exceeding 2 mm are considered significant; for nodules 6 mm or larger, discrepancies exceeding 4 mm are considered significant, reflecting established practice [8]. Changes in severity descriptors that affect urgency, such as "small" versus "large," are significant, whereas as "small" versus "tiny" are negligible. A single matched finding may contain multiple attribute errors, each evaluated independently.

2.3 Severity-Aware Scoring

The framework produces a score in the range of $(-1, 1]$ that can be easily interpreted in clinical workflows. The scale is grounded at 0, which corresponds to a normal candidate report. This reflects a practical assumption that a radiologist begins from a normal template and modifies the report by adding abnormal findings. A score greater than zero indicates that the candidate report contains more correct findings than errors after severity weighting. A score equal to zero indicates that the report is no more informative than submitting a normal template, except when the reference report is also normal, in which case a correct normal report receives a score of 1. A score less than zero indicates that the report contains more errors than correct findings, implying that a radiologist would likely spend more effort correcting it than editing a normal template. The upper bound of 1 represents a perfect report with no missed findings, no false positives, and no significant attribute errors. Negative scores approach -1 asymptotically because errors are theoretically unbounded: a candidate can always become worse by introducing additional false findings.

For each matched finding m_i, let its clinical significance weight be $w_i = w(m_i)$. Attribute-level penalties are aggregated as $E_{\text{attr},i} = \sum_j w_{\text{attr}}(e_{j,i})$, where $e_{j,i}$ referes to attribute error j for finding i, and total credit, C, across matched findings is $C = \sum_{i \in \text{matched}} w_i \cdot \frac{w_i}{w_i + E_{\text{attr},i}}$. Let $W_{\text{ref}} = \sum_{f \in R_{\text{ref}}} w(f)$ denote the total weighted clinical significance of the reference report, and let $E_{\text{false}} = \sum_{f \in \text{false}} w(f)$ denote the weighted sum of false positive findings. The raw score is defined as:

$$S = \begin{cases} \dfrac{C - E_{\text{false}}}{W_{\text{ref}}} & \text{if } W_{\text{ref}} > 0, \\ -E_{\text{false}} & \text{if } W_{\text{ref}} = 0 \text{ and } E_{\text{false}} > 0. \\ 1 & \text{if } W_{\text{ref}} = 0 \text{ and } E_{\text{false}} = 0, \end{cases} \tag{1}$$

To bound the negative range while preserving relative ordering, let $A = E_{\text{false}} - C$, which represents the excess weighted errors relative to correct findings. The final score is:

$$\text{CRIMSON} = \begin{cases} S & \text{if } S \geq 0 \\ -\dfrac{A}{1 + A} & \text{if } S < 0 \end{cases} \tag{2}$$

3 Results

We perform three complementary forms of validation: correlation with radiologist-annotated clinically significant error counts (Sect. 3.1), a radiologist-

RadJudge	CRIMSON	GREEN	RadGraph	CheXbert	RaTEScore	RadCliQ-v1	ROUGE-L	BLEU	BERTScore
1 False Finding Penalization	3	0	1	0	1	3	1	1	0
2 Patient Context Sensitivity	3	0	0	0	0	0	0	0	0
3 Normal Finding Handling	3	0	0	2	0	0	0	0	0
4 Paraphrase Robustness	3	1	1	2	0	0	0	1	0
5 Clinical Practicality	3	2	0	2	1	0	0	1	0
6 Location Error Handling	3	3	0	0	0	0	0	0	0
7 Measurement Error Sensitivity	3	1	1	0	1	2	1	1	2
8 Diagnostic Precision	3	1	0	0	3	1	0	0	3
9 Clinical Significance Weighting	3	0	0	2	2	2	0	0	1
10 Partial Credit Assignment	3	2	2	0	0	0	0	0	0
Total (/30)	30	10	5	8	8	8	2	4	6

Fig. 2. RadJudge results. For each case, metrics are evaluated based on whether their relative ranking of multiple candidate reports agrees with the expected ordering determined with agreement across three attending cardiothoracic radiologists. Each category contains three cases; entries are cases passed (out of 3), with totals out of 30.

guided pass-fail clinical judgment test on RadJudge (Sect. 3.2), and large-scale radiologist preference alignment on RadPref (Sect. 3.3).

3.1 Correlation with Radiologist-Annotated Significant Errors

We evaluated CRIMSON on 50 cases from ReXVal [18] annotated by six board-certified radiologists. We computed Kendall's τ and Pearson r correlations between automatic metric scores and radiologist-derived clinically significant error counts. As shown in Table 1, CRIMSON demonstrates strong alignment with these expert annotations. Furthermore, error counts (E) exhibited even stronger alignment, and severity-weighted errors (Weighted E) achieved the highest correlations overall, demonstrating that explicitly modeling clinical consequence further improves agreement with expert judgment.

3.2 Radiologist-Guided Clinical Judgment Test

We also developed RadJudge, a targeted pass-fail test suite reflecting real-world radiologist intuition. RadJudge comprises 30 curated cases across 10 clinically nuanced categories in which multiple candidate reports are compared and independently reviewed by three cardiothoracic radiologists, with agreement required to establish the reference preference. A metric passes if it ranks reports in accordance with expert judgment or assigns equivalent scores when radiologists deem them clinically indistinguishable, defined as differences within a threshold of

Table 1. Kendall τ and Pearson r correlations (95% CI) between automatic metrics and radiologist-derived clinically significant error counts ($n = 50$). Columns refer to different candidate reports on ReXVal, each of which was chosen to optimize a specific metric [18]. GREEN E and CRIMSON E denote the total (unweighted) error count, while CRIMSON Weighted E applies clinical severity-based weighting to errors. All other metrics were calculated using RadEval [16]. *CRIMSON results are averaged across 5 runs due to non-deterministic API outputs.

Metric	Correlation with Significant Error Counts			
	CheXbert		BERTScore	
	Kendall τ	Pearson r	Kendall τ	Pearson r
RadGraph [5]	0.41 [0.19,0.61]	0.59 [0.41,0.75]	0.54 [0.36,0.68]	0.65 [0.51,0.78]
BLEU [10]	0.49 [0.30,0.65]	0.60 [0.47,0.72]	0.36 [0.16,0.54]	0.48 [0.32,0.63]
BERTScore [19]	0.52 [0.35,0.67]	0.65 [0.52,0.78]	0.49 [0.30,0.66]	0.60 [0.44,0.74]
GREEN [9]	0.62 [0.46,0.75]	0.75 [0.64,0.86]	0.67 [0.54,0.78]	0.70 [0.59,0.80]
ROUGE-L [7]	0.58 [0.44,0.71]	0.71 [0.60,0.81]	0.54 [0.37,0.70]	0.62 [0.46,0.75]
CheXbert [13]	0.46 [0.26,0.63]	0.45 [0.18,0.70]	0.30 [0.08,0.51]	0.34 [0.09,0.60]
RaTEScore [23]	0.39 [0.17,0.57]	0.52 [0.31,0.69]	0.49 [0.32,0.65]	0.56 [0.37,0.73]
RadCliQ-v1 [17]	0.34 [0.21,0.46]	0.34 [0.19,0.52]	0.35 [0.21,0.48]	0.35 [0.22,0.53]
CRIMSON*	0.68 [0.54,0.79]	0.84 [0.76,0.90]	0.71 [0.60,0.80]	0.82 [0.74,0.89]
GREEN E	0.71 [0.59,0.81]	0.75 [0.65,0.86]	0.75 [0.63,0.85]	0.85 [0.75,0.92]
CRIMSON E	0.73 [0.61,0.83]	0.88 [0.79,0.94]	0.72 [0.62,0.80]	0.86 [0.77,0.93]
CRIMSON Weighted E	**0.78** [0.67,0.86]	**0.90** [0.85,0.95]	**0.80** [0.71,0.87]	**0.91** [0.88,0.95]
	RadGraph		BLEU	
	Kendall τ	Pearson r	Kendall τ	Pearson r
RadGraph	0.59 [0.46,0.71]	0.60 [0.50,0.72]	0.64 [0.50,0.75]	0.75 [0.65,0.84]
BLEU	0.13 [0.09,0.34]	0.23 [0.02,0.39]	0.52 [0.34,0.68]	0.67 [0.53,0.79]
BERTScore	0.46 [0.29,0.61]	0.54 [0.39,0.68]	0.58 [0.41,0.72]	0.72 [0.60,0.83]
GREEN	0.62 [0.48,0.74]	0.65 [0.53,0.77]	0.70 [0.55,0.83]	0.79 [0.68,0.89]
ROUGE-L	0.54 [0.38,0.67]	0.60 [0.49,0.70]	0.67 [0.52,0.79]	0.80 [0.70,0.87]
CheXbert	0.29 [0.08,0.48]	0.33 [0.08,0.55]	0.18 [0.05,0.40]	0.23 [0.04,0.47]
RaTEScore	0.57 [0.42,0.70]	0.62 [0.45,0.78]	0.54 [0.39,0.68]	0.67 [0.54,0.79]
RadCliQ-v1	0.12 [0.05,0.28]	0.06 [0.11,0.28]	0.28 [0.11,0.43]	0.16 [0.01,0.53]
CRIMSON*	0.61 [0.45,0.75]	0.71 [0.53,0.85]	0.67 [0.54,0.79]	0.81 [0.71,0.89]
GREEN E	0.71 [0.59,0.81]	0.75 [0.65,0.86]	**0.80** [0.71,0.88]	**0.88** [0.82,0.93]
CRIMSON E	0.73 [0.61,0.83]	0.86 [0.78,0.92]	0.74 [0.61,0.84]	0.87 [0.78,0.93]
CRIMSON Weighted E	**0.77** [0.67,0.85]	**0.86** [0.80,0.93]	0.78 [0.69,0.86]	**0.88** [0.82,0.93]

0.01. Three representative cases are shown in Fig. 1. The suite probes challenging scenarios, including urgent omissions versus benign hallucinations, context-dependent findings, diagnostic over- and under-interpretation, and situations

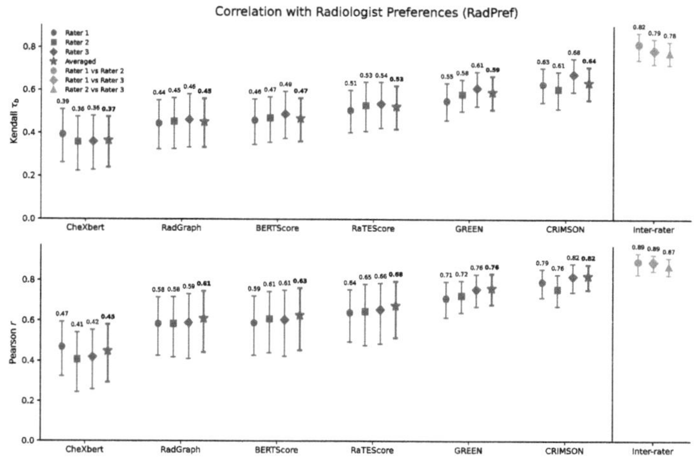

Fig. 3. Radiologist Preference Alignment (RadPref). Correlation between metric score and radiologist rating differences across 100 pairwise cases. Each point corresponds to a case comparing two candidate reports for the same reference report.

reflecting the clinical reality of imperfect reference reports that omit localization or age-expected benign findings.

As shown in Fig. 2, CRIMSON is the only metric that correctly solves all 30 out of 30 cases, consistently ranking candidate reports in accordance with expert radiologist judgment. In contrast, all prior metrics perform substantially worse, correctly resolving fewer than 35% of cases, highlighting their limited ability to capture nuanced clinical judgment.

3.3 Radiologist Preference Alignment

Correlation with error counts may not fully capture radiologist preference, as experts don't weigh different types of errors equally in overall judgment. To directly assess preference alignment, we introduce RadPref, a benchmark of 100 cases, each containing a reference report from ReXGradient-160K [20] (Non-Commercial Data Access and Use Agreement (v1, April 7, 2025)) and two candidate reports randomly generated using diverse regimes: report generation with MedGemma [11], randomly sampled reports, BERT similarity-matched reports, and LLM-based editing, addition, or removal of findings. Each candidate was rated on a 1–5 scale by three cardiothoracic radiologists based on overall clinical quality and correctness relative to the referene report. Scores were computed separately for both candidates.

Figure 3 shows that CRIMSON demonstrates the strongest alignment with radiologist pairwise preferences among all evaluated metrics. Across Kendall's τ_b and Pearson r, CRIMSON consistently outperforms prior approaches and approaches inter-rater radiologist agreement. These findings indicate that CRIMSON more faithfully reflects expert clinical preference in relative report quality.

4 Conclusion

We introduce CRIMSON, a clinically grounded and severity-aware framework for fine-grained radiology report evaluation that explicitly models patient context, diagnostic consequence, and structured attribute-level errors. By incorporating clinical significance weights and score normalization, CRIMSON aligns automated evaluation with real-world radiologist reasoning. Across correlation with radiologist-annotated significant error counts, RadJudge, and RadPref, CRIMSON consistently demonstrates stronger agreement with expert judgment than prior metrics, establishing a robust and interpretable foundation for clinically meaningful evaluation of report generation systems.

Acknowledgments. This research was supported in part by Nebius AI and the Harvard Medical School Dean's Innovation Award for Accelerating Foundation Model Research.

Disclosure of Interests. The authors declare no competing interests.

References

1. Baharoon, M., Ma, J., Fang, C., Toma, A., Wang, B.: Exploring the design space of 3d mllms for ct report generation. In: International Conference on Medical Image Computing and Computer-Assisted Intervention, pp. 237–246. Springer, Heidelberg (2025). https://doi.org/10.1007/978-3-032-04978-0_23
2. Baharoon, M., et al.: Radgame: an ai-powered platform for radiology education. arXiv preprint arXiv:2509.13270 (2025)
3. Chaves, J.M.Z., et al.: Towards a clinically accessible radiology foundation model: open-access and lightweight, with automated evaluation. arXiv preprint arXiv:2403.08002 (2024)
4. Huang, A., Banerjee, O., Wu, K., Reis, E.P., Rajpurkar, P.: Fineradscore: a radiology report line-by-line evaluation technique generating corrections with severity scores. arXiv preprint arXiv:2405.20613 (2024)
5. Jain, S., et al.: Radgraph: extracting clinical entities and relations from radiology reports. arXiv preprint arXiv:2106.14463 (2021)
6. Li, R., Li, J., Jian, B., Yuan, K., Zhu, Y.: Reevalmed: rethinking medical report evaluation by aligning metrics with real-world clinical judgment. In: Proceedings of the 2025 Conference on Empirical Methods in Natural Language Processing, pp. 11823–11837 (2025)
7. Lin, C.Y.: Rouge: a package for automatic evaluation of summaries. In: Text Summarization Branches Out, pp. 74–81 (2004)
8. MacMahon, H., et al.: Guidelines for management of incidental pulmonary nodules detected on ct images: from the fleischner society 2017. Radiology **284**(1), 228–243 (2017)
9. Ostmeier, S., et al.: Green: generative radiology report evaluation and error notation. In: Findings of the Association for Computational Linguistics: EMNLP 2024, pp. 374–390 (2024)
10. Papineni, K., Roukos, S., Ward, T., Zhu, W.J.: Bleu: a method for automatic evaluation of machine translation. In: Proceedings of the 40th Annual Meeting of the Association for Computational Linguistics, pp. 311–318 (2002)

11. Sellergren, A., et al.: Medgemma technical report. arXiv preprint arXiv:2507.05201 (2025)
12. Singh, A., et al.: Openai gpt-5 system card. arXiv preprint arXiv:2601.03267 (2025)
13. Smit, A., Jain, S., Rajpurkar, P., Pareek, A., Ng, A.Y., Lungren, M.P.: Chexbert: combining automatic labelers and expert annotations for accurate radiology report labeling using bert. arXiv preprint arXiv:2004.09167 (2020)
14. Tian, K., et al.: Refisco: report fix and score dataset for radiology report generation. PhysioNet (2023)
15. Wang, Z., Luo, X., Jiang, X., Li, D., Qiu, L.: Llm-radjudge: achieving radiologist-level evaluation for x-ray report generation. arXiv preprint arXiv:2404.00998 (2024)
16. Xu, J., et al.: Radeval: a framework for radiology text evaluation. In: Proceedings of the 2025 Conference on Empirical Methods in Natural Language Processing: System Demonstrations, pp. 546–557 (2025)
17. Yu, F., et al.: Evaluating progress in automatic chest x-ray radiology report generation. Patterns **4**(9) (2023)
18. Yu, F., et al.: Radiology report expert evaluation. Rexval dataset (2023)
19. Zhang, T., Kishore, V., Wu, F., Weinberger, K.Q., Artzi, Y.: Bertscore: evaluating text generation with bert. arXiv preprint arXiv:1904.09675 (2019)
20. Zhang, X., Acosta, J.N., Miller, J., Huang, O., Rajpurkar, P.: Rexgradient-160k: a large-scale publicly available dataset of chest radiographs with free-text reports. In: Machine Learning for Health (2025)
21. Zhang, X., et al.: Automated chest x-ray report generation remains unsolved. In: Biocomputing 2026: Proceedings of the Pacific Symposium, pp. 236–250. World Scientific (2025)
22. Zhang, Z., et al.: Gema-score: granular explainable multi-agent score for radiology report evaluation. arXiv preprint arXiv:2503.05347 (2025)
23. Zhao, W., Wu, C., Zhang, X., Zhang, Y., Wang, Y., Xie, W.: Ratescore: a metric for radiology report generation. arXiv preprint arXiv:2406.16845 (2024)

DEER: A Foundation Model with Dual-Branch CT-Enhanced Embeddings for Thoracoabdominal Radiographs

Yihua Sun[1,2], Jia Guo[2], Qijing Wang[2,3], Jingzhou Ouyang[3], Ge Li[3], Nan Zhang[4(✉)], Fang Chen[1(✉)], and Hongen Liao[1,2(✉)]

[1] School of Biomedical Engineering, Shanghai Jiao Tong University, Shanghai, China
{chen-fang,liaohongen}@sjtu.edu.cn
[2] School of Biomedical Engineering, Tsinghua Medicine, Tsinghua University, Beijing, China
[3] Zhongshi Kangkai Technology Co., Ltd., Beijing, China
[4] State Key Laboratory of Multimodal Artificial Intelligence Systems, Institute of Automation, Chinese Academy of Sciences, Beijing, China
zhangnan@ia.ac.cn

Abstract. The chest and abdomen are critical, anatomically complex regions with diverse pathologies, with digital radiography (DR) serving as the primary modality for screening and diagnosis. Developing an effective DR foundation model with strong disease discrimination through self-supervised learning remains a significant challenge. Moreover, prior research has largely overlooked abdominal DR and the potential of computed tomography (CT) localizers as a bridge to fine-grained CT knowledge. In this work, we pioneer the incorporation of large-scale, previously neglected CT localizers and CT reports to train a unified DR foundation model on both chest and abdominal regions. We propose a dual-branch contrastive pretraining approach using 1.1 million DR and 1 million CT localizers, each paired with radiologist reports. By integrating comprehensive knowledge from CT reports, we substantially enhance the model's capability in DR interpretation. Extensive evaluations on 10 downstream tasks demonstrate the superiority of our model. To the best of our knowledge, this is the first foundational pretraining framework that jointly leverages DR, CT localizers, and reports, scaling to an unprecedented 2.1 million image-report pairs. It offers valuable insights into CT-enhanced intelligent DR diagnosis and scalable multimodal medical foundation models, with demonstrated robustness across diverse clinical tasks. The model is available at https://github.com/sunyh1/DEER.

Keywords: Contrastive learning · Foundation model · Localizer · Radiograph · Thoracoabdominal region

Y. Sun, J. Guo and Q. Wang—Co-first authors.

G. Yang et al. (Eds.): MICCAI 2026, LNCS 16878, pp. 108–118, 2027.
https://doi.org/10.1007/978-3-032-38059-3_11

1 Introduction

The chest and abdomen are anatomically complex regions associated with a broad spectrum of pathologies. Digital radiography (DR) is the primary imaging modality for screening and diagnosis [18]. However, current research has predominantly focused on chest DR, while abdominal DR remains underexplored. This highlights the need for more generalizable foundation models for interpreting a wide range of pathologies across trunk DR images.

However, scaling up foundation models for robust disease interpretation remains challenging. Current methods include image-only self-supervised learning [13,21], which lacks the ability to extract disease-relevant semantics. Alternatively, image-label supervision [12] relies on labor-intensive manual annotations or on automatically generated labels that often oversimplify the rich information in radiology reports. In contrast, multimodal self-supervised approaches combining images and radiology reports [15] offer a promising solution. They allow foundation models to scale up disease interpretation capabilities directly from routine clinical data, without additional human labeling.

On the other hand, radiologists have limited ability to interpret diseases from projective 2D DR, which limits the diagnostic information in DR reports. In contrast, 3D computed tomography (CT) provides much richer disease insights. Therefore, leveraging CT to enhance DR-based models could be highly beneficial. Here, we identify the previously underexplored potential of CT localizers (LOC) as a bridge between the detailed disease information in CT reports and DR images, enhancing disease discriminative representations from DR.

In this study, we jointly leverage DR, LOC, and reports, scaling to an unprecedented 2.1 million image-report pairs for multimodal self-supervised pretraining. Building on this, we present a foundation model (Fig. 1a) with **D**ual-branch CT-**E**nhanced **E**mbeddings for thoracoabdominal **R**adiographs (**DEER**). Our key novelties and contributions are: 1) **Thoracoabdominal DR foundation model**: The model enables chest and abdominal disease interpretation and generalizes well to pelvic diseases. 2) **CT-enhanced DR embeddings**: We improve disease discriminative representations from DR using LOC and CT reports. 3) **Large-scale multimodal pretraining**: DEER is pretrained on 2.1 million image-report pairs, the largest multimodal radiograph dataset to date.

Evaluations on 10 downstream tasks demonstrate DEER's superiority. Our results highlight the potential of CT-enhanced embeddings for advancing DR intelligence. Notably, by leveraging routinely collected image-report pairs for self-supervised pretraining, DEER surpasses models trained with large-scale label supervision. This demonstrates the feasibility of scaling self-supervised learning to develop more generalizable and high-performing medical foundation models.

2 Methods

2.1 Pretraining Datasets

We use only image-report pairs from routine radiology workflows, without extra manual labels, enabling scalable self-supervised foundation model development.

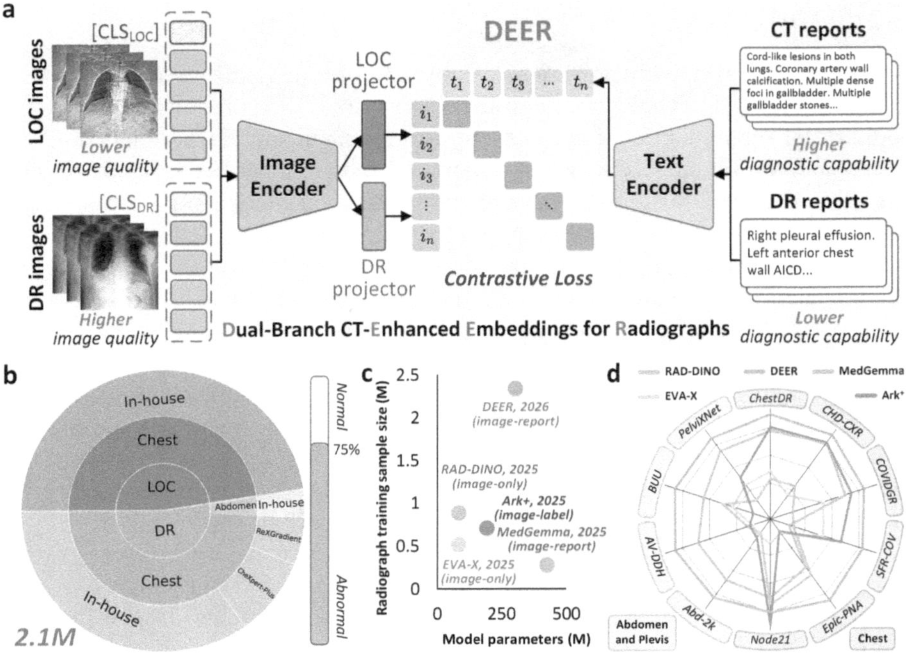

Fig. 1. (a) DEER trains on routinely collected clinical image-report data without additional manual annotations. We propose a dual-branch contrastive learning approach that uses CT localizers and CT reports to enhance DR image representations. (b) DEER jointly learns from both chest and abdominal DRs, as well as CT localizers. (c) DEER uses a larger model trained on more data than SOTA methods, enabling further foundation model scaling. (d) DEER achieves superior performance in downstream evaluations (fine-tuning), showing strong robustness across diverse clinical applications.

DR Image and Report. The **chest DR** primarily covers the lungs and may include part of the upper abdomen. Similarly, **abdominal DR** mainly encompasses the abdomen and typically includes the pelvis and part of the lower lungs. Therefore, jointly learning from chest and abdominal DR facilitates the integration of both overlapping and region-specific information across different DR types, enabling a more comprehensive understanding of trunk DR. We use public chest DR datasets for training: CheXpert-Plus [1] and ReXGradient [22]. In addition, we incorporate a large-scale private chest and abdominal DR dataset, resulting in a total of 1.1 million DR image-report pairs (Fig. 1b).

CT Localizer and Report. Radiologists face inherent limitations in interpreting diseases from 2D DR, leading to relatively limited diagnostic information in DR reports. To address this gap, we investigate the largely unexplored role of CT localizers as a bridge between diagnostic-rich CT reports and DR images to enhance disease-discriminative DR representations. We utilize 1 million private LOC images paired with CT reports to enhance DR training (Fig. 1b).

To the best of our knowledge, we construct the first multimodal dataset combining DR, LOC, and reports, totaling an unprecedented 2.1 million image-report pairs. We define a report-image pair as abnormal if the report contains any positive finding or impression. In our dataset, 75% of the pairs are abnormal (Fig. 1b), highlighting a diversity that enables robust foundation model development.

2.2 Model Framework

DR images exhibit higher image quality, yet DR reports provide relatively limited diagnostic information. In contrast, LOC has lower image quality, while CT reports offer richer and more diagnostically informative content. DEER is designed to leverage the strengths of both modalities, which combine high-quality DR image understanding with strong diagnostic capability (Fig. 1a).

Dual-Branch Visual Encoding. To incorporate LOC while preserving the ability to interpret high-quality DR images, we design a dual-branch visual encoder (E_I) based on a shared ViT [4] backbone (Fig. 1a). For DR and LOC input images (I_{DR}, I_{LOC}), we employ two independent classification tokens ([CLS$_{\mathrm{DR}}$], [CLS$_{\mathrm{LOC}}$]) and separate projection layers (f_{DR}, f_{LOC}) to extract visual features: $i_{mod} = f_{mod}\left(E_I\left([\mathrm{CLS}_{mod}], I_{mod}\right)\right)$, where $mod \in \{\mathrm{DR}, \mathrm{LOC}\}$. The resulting features i_{DR} and i_{LOC} are the extracted visual features, respectively.

Image-Report Contrastive Learning. As shown in Fig. 1a, to fully leverage radiology report knowledge for DR image representation learning, we adopt the contrastive language-image pretraining (CLIP) framework [5,14]. Specifically, both DR and CT reports (T) are encoded into textual feature representations t using CLIP's Transformer-based text encoder E_T: $t = E_T(T)$.

A contrastive learning objective is then applied to pull matched image-text pairs closer in the shared embedding space while pushing unmatched pairs farther apart. Given a batch of size N, let p, q denote the index of an element within the batch. The contrastive learning loss can be written as:

$$\mathcal{L}_{i2t} = -\frac{1}{N}\sum_{p}^{N} \log \frac{\exp\left(\langle i_p, t_p\rangle/\tau\right)}{\sum_q^N \exp\left(\langle i_p, t_q\rangle/\tau\right)}, \; \mathcal{L}_{t2i} = -\frac{1}{N}\sum_{q}^{N} \log \frac{\exp\left(\langle i_q, t_q\rangle/\tau\right)}{\sum_p^N \exp\left(\langle i_p, t_q\rangle/\tau\right)},$$

where $\langle \cdot, \cdot \rangle$ denotes the cosine similarity, and τ is the temperature parameter.

The final training loss is $\mathcal{L} = (\mathcal{L}_{i2t} + \mathcal{L}_{t2i})/2$. CT reports are introduced to enhance diagnostic discriminability and refine textual representations, thereby optimizing image features in the shared embedding space via contrastive loss.

3 Results

3.1 Implementation Details and Evaluation Metrics

Radiology reports were standardized for positive findings and impressions by directly prompting a large language model. DEER was pretrained using 2 × 96 GB RTX PRO 6000 GPUs. The model is implemented based on OpenCLIP [3,9]. We continued pretraining on our datasets from OpenAI's ViT-L/14-336 checkpoint [14] for 3 epochs, using an increased input resolution of 518 × 518. The private datasets were curated by Zhongshi Kangkai Technology Co., Ltd. following rigorous de-identification and anonymization protocols.

For evaluation, we fine-tune DEER and the following state-of-the-art (SOTA) approaches on downstream tasks: Ark$^+$ [12], MedGemma (we only evaluate the MedSigLIP encoder) [15], RAD-DINO [13], and EVA-X [21]. We evaluate the above models on the 10 downstream datasets, including chest, abdominal, pelvic, and spinal DR scans, as shown in Table 1. Model performances are assessed using both the area under the receiver operating characteristic curve (AUROC) and the area under the precision-recall curve (AUPRC). We follow the official splits and report the mean and std. over five runs. If the official split does not provide a validation set, we construct one from the training set. In the absence of an official train/test split, we adopt five-fold cross-validation and report the mean and std. across folds. Statistical significance is evaluated using t-test.

Table 1. 10 downstream datasets for comprehensive evaluation across anatomies.

Dataset	Anatomy	Labels
Abd-2k (in-house)	Abdomen	Intestinal obstruction, Dilated bowel, Aortic calcification, Kidney calculi, Urinary bladder calculi, Fracture, Hyperostosis, Scoliosis
AV-DDH [7]	Pelvis	Left/Right/Bilateral developmental dysplasia of the hip
BUU-LSPINE [10]	Spine	Left/Right laterolisthesis, Anterolisthesis, Retrolisthesis
ChestDR [19]	Chest	19 thoracic diseases
CHD-CXR [23]	Chest	Atrial septal defect, Ventricular septal, Patent ductus arteriosus
COVIDGR [17]	Chest	COVID-19 severity: Normal-PCR+, Mild, Moderate, Severe
Epic-PNA [8]	Chest	Pneumonia
Node21 [16]	Chest	Nodule
PelviXNet [2]	Pelvis	Hip and pelvic fracture
SFR-COVID-19 [11]	Chest	Typical, indeterminate, or atypical for COVID-19

3.2 Chest Disease Diagnosis

Reports and Labels Strengthen Model Performance. We conducted extensive experiments on chest DR images (Table 1) covering conventional pulmonary diseases (ChestDR), pediatric congenital heart diseases (CHD-CXR), COVID-19 (COVIDGR, SFR-COVID-19), pneumonia (Epic-PNA), and nodules (Node21). As shown in Fig. 2, models trained with image-report (DEER, MedGemma) or image-label (Ark^+) supervision consistently achieve higher performance across 6 downstream chest datasets. In contrast, image-only supervision (EVA-X, RAD-DINO) provides no explicit diagnostic guidance, making it difficult for models to disentangle and encode clinically meaningful semantics.

Image-Report Training Improves Generalization and Scalability. Notably, DEER, based on image-report pretraining, achieves leading performance on 5 out of 6 downstream chest datasets (Fig. 2). This training paradigm not only leverages image-report pairs routinely generated in clinical radiology practice without requiring additional human annotation, but also yields stronger generalization across diverse tasks. In contrast, image-label training requires either manual annotation or automated extraction from reports, which oversimplifies the rich findings information contained in the reports. Notably, the self-

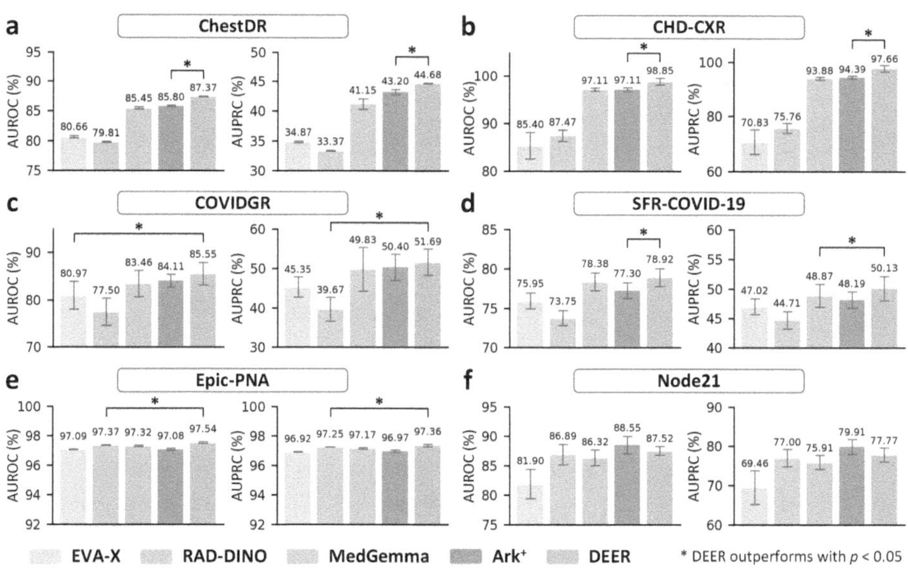

Fig. 2. DEER achieves superior performance in chest disease diagnosis. Image-report pretraining (DEER, MedGemma) learns more diagnostically discriminative representations, outperforming image-only training (EVA-X, RAD-DINO). Moreover, DEER, pretrained via self-supervised image-report learning, outperforms models trained with large-scale label supervision (Ark^+), demonstrating the feasibility of scaling self-supervised medical foundation models with superior diagnostic performance.

supervised image-report pretrained DEER outperforms Ark$^+$ trained on large-scale chest DR labels. This result highlights the effectiveness of self-supervised learning and its scalability toward larger medical foundation models.

3.3 Abdominal Disease Diagnosis

To evaluate the model's performance on the abdomen, we collected a dataset comprising 2,000 abdominal DR images across eight diseases (Abd-2k, Table 1). DEER demonstrates strong capability in abdominal DR analysis (Fig. 3a). Notably, this dataset exhibits a long-tailed, class-imbalanced distribution. Therefore, AUPRC better reflects the improvements of our model. In contrast, although abdominal diseases are clinically important, they remain largely under-explored in prior work. Existing DR foundation models show noticeable performance degradation when applied to abdominal cases (Fig. 3a).

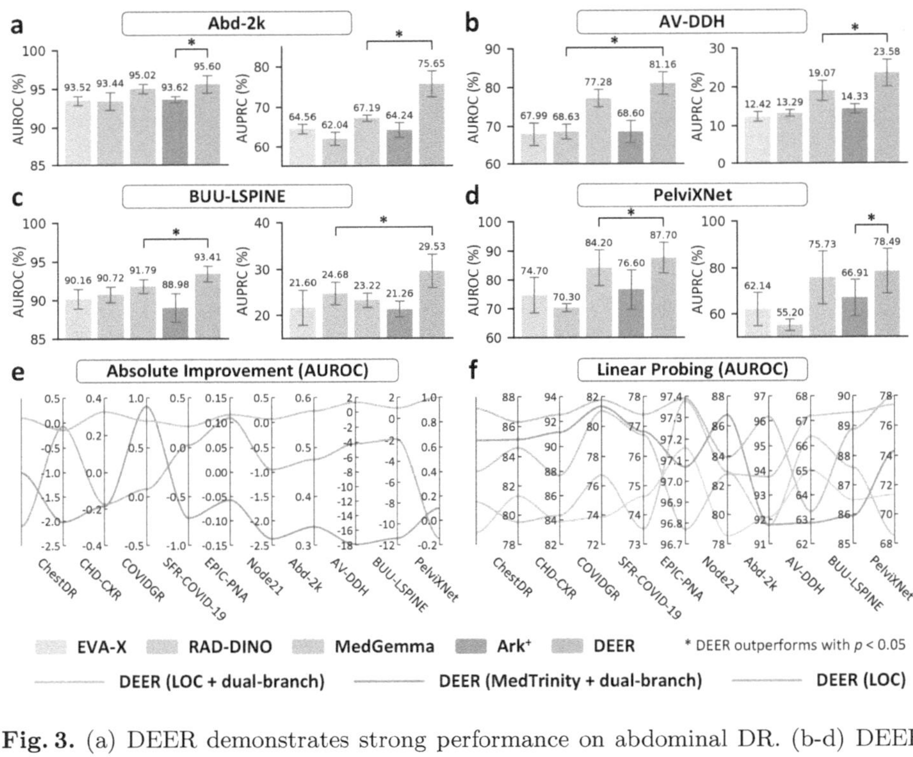

Fig. 3. (a) DEER demonstrates strong performance on abdominal DR. (b-d) DEER shows superior transferability on pelvic DR, while MedGemma performs better among baseline methods. This finding supports the advantage of image-report pretraining for improved generalization. (e) Incorporating LOC via dual-branch encoding further enhances performance, with greater gains in the abdomen and pelvis than in the chest. (f) DEER demonstrates superior linear probing performance on 7 out of 10 datasets.

3.4 Transferability to Pelvic and Spinal Scans

Chest and abdominal DR scans share overlapping anatomical regions, and abdominal scans frequently include the pelvic area. Learning from both types of DR therefore facilitates a more comprehensive understanding of trunk DR. We conducted extensive experiments on pelvic DR images (Table 1), including hip developmental dysplasia (AV-DDH), lumbar spondylolisthesis (BUU-LSPINE), and fractures (PelviXNet). As shown in Fig. 3c-d, DEER demonstrates strong transferability to pelvic and spinal scans, with consistently superior performance.

In contrast, current DR foundation models exhibit limited generalization to the pelvic region (Fig. 3c-d). Among the baseline methods, MedGemma achieves the best cross-region transfer performance, whereas label-supervised approaches (Ark^+) achieve relatively lower performance. This observation further supports the effectiveness of image-report self-supervised pretraining in building foundation models with stronger generalization.

3.5 CT Localizers Further Boost the Performance

We further enhanced the embeddings by incorporating LOC and CT reports through the dual-branch encoding framework, leading to additional performance gains (Fig. 3e). Without dual-branch encoding, the entanglement of DR and LOC information often degrades representation quality and downstream performance (Fig. 3e). Notably, the improvement in the abdomen and pelvis settings was greater than that observed for chest DR. This difference may be attributed to the broader anatomical coverage of LOC compared with conventional chest DR, which enriches the image distribution of abdominal and pelvic regions.

To further demonstrate that LOC contributes to effective training benefits, we sampled 1 million non-DR image-text pairs from MedTrinity-25M [20], including CT, magnetic resonance, ultrasound, etc. These samples were combined with the DR data to train the model. As shown in Fig. 3e, compared with an equal amount of LOC data, incorporating other data sources does not yield performance gains and may even substantially degrade model performance. This finding further suggests that LOC and CT reports are essential for improving DR foundation model training, rather than simply increasing data volume.

3.6 Qualitative *t*-SNE Visualizations

The *t*-SNE (Fig. 4a) shows that both the label-supervised model (Ark^+) and the image-report self-supervised model (DEER) achieve clearer feature separation on chest DR, particularly between "severe" (orange) and "normal-pcr+" (red) cases. In contrast, the image-only self-supervised model (EVA-X) produces more entangled features with overlapping diagnostic clusters. On abdominal tasks, DEER maintains stronger feature discriminability, while the current DR foundation model exhibits mixed and less structured representations (Fig. 4b). This demonstrates the feasibility of leveraging self-supervised image-text learning to train foundation models with strong diagnostic discriminative representations. Also,

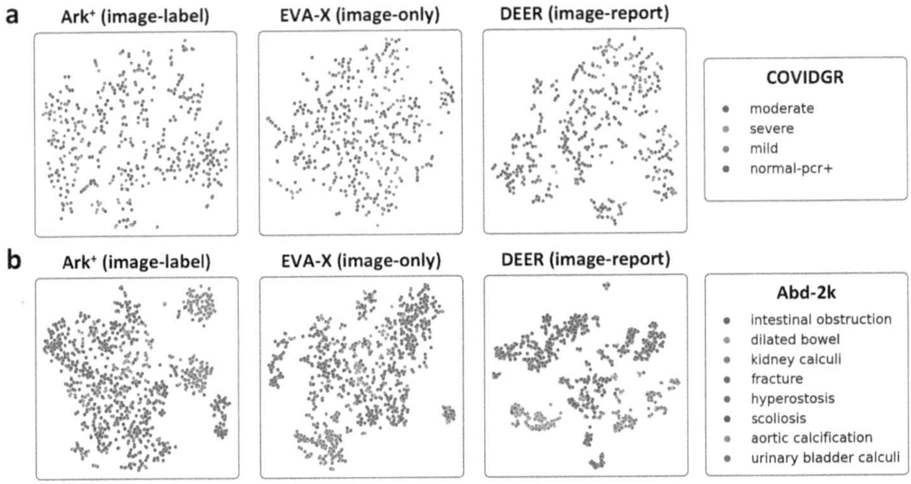

Fig. 4. (a) The label-supervised model (Ark$^+$) and the image-report model (DEER) show clearer feature separation on chest DR, whereas the image-only model yields more entangled clusters. (b) On abdominal tasks, DEER maintains stronger discriminability, while the current DR foundation model shows overlapping representations.

DEER achieves superior linear probing performance on 7 out of 10 datasets, further validating the effectiveness of the learned representations (Fig. 3f).

4 Discussion

DR is the primary tool for screening and diagnosis [18], especially in under-resourced regions [6]. DEER demonstrates strong performance across diverse chest, abdominal, and pelvic tasks, covering underexplored abdominal DR. This offers significant clinical value for intelligent DR diagnosis and could be deployed in remote areas. By leveraging previously overlooked CT localizers and reports, our image-text self-supervised approach outperforms large-scale label-supervised methods. These findings provide insights for scaling DR foundation models with multimodal CT integration. Future directions include evaluating intelligent DR models for complex diseases that typically require CT for diagnosis, as well as developing a full-body DR foundation model to enhance clinical coverage.

Acknowledgments. The authors acknowledge supports from Science and Technology Commission of Shanghai Municipality (24511104100), National Natural Science Foundation of China (82572314, U22A2051, 62271246, 62501594), National Key Research and Development Program of China (2022YFC2405200).

Disclosure of Interests. The authors declare no competing interests.

References

1. Chambon, P., Delbrouck, J.B., Sounack, T., Huang, S.C., et al.: Chexpert plus: augmenting a large chest x-ray dataset with text radiology reports, patient demographics and additional image formats (2024). https://arxiv.org/abs/2405.19538
2. Cheng, C.T., et al.: A scalable physician-level deep learning algorithm detects universal trauma on pelvic radiographs. Nat. Commun. **12**(1), 1066 (2021)
3. Cherti, M., Beaumont, R., Wightman, R., Wortsman, M.: Reproducible scaling laws for contrastive language-image learning. In: Proceedings of the IEEE/CVF Conference on Computer Vision and Pattern Recognition, pp. 2818–2829 (2023)
4. Dosovitskiy, A., Beyer, L., Kolesnikov, A., Weissenborn, D., Zhai, X., Unterthiner, T.: An image is worth 16×16 words: transformers for image recognition at scale. In: International Conference on Learning Representations (2021)
5. Du, J.: Ret-clip: A retinal image foundation model pre-trained with clinical diagnostic reports. In: Medical Image Computing and Computer Assisted Intervention - MICCAI 2024, pp. 709–719. Springer, Cham (2024). https://doi.org/10.1007/978-3-031-72390-2_66
6. Frija, G., Blažić, I., Frush, D.P., Hierath, M.: How to improve access to medical imaging in low- and middle-income countriesâĂŕ? eClinicalMedicine 38 (2021)
7. Haddad, B., AlHosanie, T., et al.: Acetabular-vision hip developmental dysplasia's (av-ddh). Mendeley Data (2025). https://doi.org/10.17632/4gvcb6gmh2.1
8. Hira, M.I.K., Bithee, M.M.A., Ahmed, S., Akter, L., Anonna, M.J.M.: A primary chest x-ray dataset of normal and pneumonia cases from epic Chittagong, Bangladesh. Mendeley Data V2 (2025). https://doi.org/10.17632/wndbd5r26y.2
9. Ilharco, G., et al.: OpenCLIP (2021). https://doi.org/10.5281/zenodo.5143773
10. Klinwichit, P., Yookwan, W., Limchareon, S.: Buu-lspine: a thai open lumbar spine dataset for spondylolisthesis detection. Appl. Sci. **13**(15) (2023)
11. Lakhani, P., Mongan, J., Singhal, C., Zhou, Q., et al.: The 2021 siim-fisabio-rsna machine learning covid-19 challenge: annotation and standard exam classification of covid-19 chest radiographs. J. Digit. Imaging **36**(1), 365–372 (2023)
12. Ma, D., Pang, J., Gotway, M.B., Liang, J.: A fully open AI foundation model applied to chest radiography. Nature **643**(8071), 488–498 (2025)
13. Pérez-García, F., Sharma, H., et al.: Exploring scalable medical image encoders beyond text supervision. Nat. Mach. Intell. **7**(1), 119–130 (2025)
14. Radford, A., Kim, J.W., Hallacy, C., Ramesh, A., Goh, G., Agarwal, S., et al.: Learning transferable visual models from natural language supervision. In: Proceedings of the 38th International Conference on Machine Learning. Proceedings of Machine Learning Research, vol. 139, pp. 8748–8763. PMLR (2021)
15. Sellergren, A., Kazemzadeh, S., Jaroensri, T., Kiraly, A., Traverse, M., Kohlberger, T., et al.: Medgemma technical report (2025). https://arxiv.org/abs/2507.05201
16. Sogancioglu, E., et al.: Node21 (2021). https://doi.org/10.5281/zenodo.5548363
17. Tabik, S., Gómez-Ríos, A., Martín-Rodríguez, J.L., et al.: Covidgr dataset and covid-sdnet methodology for predicting covid-19 based on chest x-ray images. IEEE J. Biomed. Health Inf. **24**(12), 3595–3605 (2020)
18. United Nations Scientific Committee on the Effects of Atomic Radiation: Sources, Effects and Risks of Ionizing Radiation, United Nations Scientific Committee on the Effects of Atomic Radiation (UNSCEAR) 2020/2021 Report. United Nations, 2021 edn. (2022)
19. Wang, D., Wang, X., et al.: A real-world dataset and benchmark for foundation model adaptation in medical image classification. Sci. Data **10**(1), 574 (2023)

20. Xie, Y., et al.: Medtrinity-25m: a large-scale multimodal dataset with multigranular annotations for medicine (2025). https://arxiv.org/abs/2408.02900
21. Yao, J., Wang, X., Song, Y.: EVA-X: a foundation model for general chest x-ray analysis with self-supervised learning. npj Dig. Med. **8**(1), 678 (2025)
22. Zhang, X., et al.: Rexgradient-160k: a large-scale publicly available dataset of chest radiographs with free-text reports (2025). https://arxiv.org/abs/2505.00228
23. Zhixin, L., Gang, L., Zhixian, J., Sibao, W., Silin, P.: Chd-cxr: a de-identified publicly available dataset of chest x-ray for congenital heart disease. Front. Cardiovasc. Med. **11** (2024). https://doi.org/10.3389/fcvm.2024.1351965

Defining Robust Ultrasound Quality Metrics via an Ultrasound Foundation Model

Ziyang Huang[1], Bingyan Li[1], Chen Ma[1], Tianyi Liu[2,3], Yihui Zhai[2,3], Hong Xu[2,3], Yi Guo[1], Zeju Li[1](✉), and Yuanyuan Wang[1](✉)

[1] College of Biomedical Engineering, Fudan University, Shanghai, China
{zejuli,yywang}@fudan.edu.cn

[2] Department of Nephrology, Children's Hospital of Fudan University, National Children's Medical Center, Shanghai, China

[3] Shanghai Kidney Development and Pediatric Kidney Disease Research Center, Shanghai, China

Abstract. Clinicians lack a principled framework to quantify diagnostic utility in ultrasound reconstructions. Existing standards like PSNR and VGG-LPIPS are inadequate, failing to account for modality-specific physics or the structural nuances of acoustic imaging. We close this gap with a TinyUSFM-based evaluation framework featuring two distinct metrics: TinyUSFM-uLPIPS, a full-reference perceptual distance based on multi-layer token relations, and TinyUSFM-NRQ, a deployable no-reference quality score utilizing clean-manifold modeling and worst-region aggregation to detect localized harmful artifacts. We demonstrate that the presented metrics have four unique advantages: 1) Task-linked quality, where TinyUSFM-uLPIPS achieves superior calibration with semantic task damage, accurately reflecting Dice-score drops in segmentation where VGG-based metrics fail; 2) Cross-organ comparability, maintaining stable scoring scales and consistent severity rankings across diverse anatomical sites and domain-shifted data; 3) PSNR-consistent sensitivity, with TinyUSFM-NRQ providing a reliable quality score without ground-truth images that remains consistent with traditional fidelity benchmarks (i.e. PSNR); and 4) Clinical utility, improving the prediction of expert preference from 47.2% to 72.8% accuracy and producing super-resolution reconstructions preferred by sonographers. By integrating these advantages into a unified assessment and optimization loop, this work establishes a modality-aligned standard that finally bridges the gap between algorithmic performance and diagnostic utility. Our code is available at https://github.com/sextant-fable/US-Metrics.

Keywords: Ultrasound · Image Quality Assessment · Perceptual Metric

Z. Huang, B. Li and C. Ma—Equal contribution.

G. Yang et al. (Eds.): MICCAI 2026, LNCS 16878, pp. 119–129, 2027.
https://doi.org/10.1007/978-3-032-38059-3_12

1 Introduction

Ultrasound is ubiquitous in clinical imaging due to its portability and safety. However, evaluating the perceptual and clinical usefulness of ultrasound reconstructions and generative outputs remains poorly served by commonly used image-quality metrics, especially across organs, devices, and acquisition settings.

Ultrasound appearance is shaped by modality-specific physics and sampling. Coherent interference produces speckle, specular responses are view dependent, attenuation and shadowing create structured dropouts, and scan conversion introduces anisotropic resolution and sampling artifacts [2,3,18]. Clinically meaningful failures are often localized: a small region of clutter or shadowing may erase a critical boundary even when global intensity changes slightly. In contrast, adjustments to global amplification or depth-dependent amplification alter pixel values across the image without degrading its diagnostic interpretability.

Existing metrics are poorly calibrated for ultrasound due to three primary limitations: 1) Fidelity vs. Clinical Utility: Pixel-level measures (PSNR and SSIM [20]) prioritize global intensity agreement, making them overly sensitive to benign brightness shifts while failing to detect localized, clinically harmful artifacts. 2) Domain Mismatch: Natural-image perceptual metrics (VGG-LPIPS and FID [6,24]) rely on photographic priors that misinterpret ultrasound-specific textures as noise and under-penalize structured occlusions. 3) Lack of Generalization: While generic no-reference IQA metrics such as BRISQUE [14] and NIQE [15], and ultrasound-specific learned IQA/QA methods including fetal ultrasound IQA [21,23], expert-agnostic ultrasound IQA [17], and KDE-based QA [8], have been proposed, most existing approaches are tied to restricted clini-

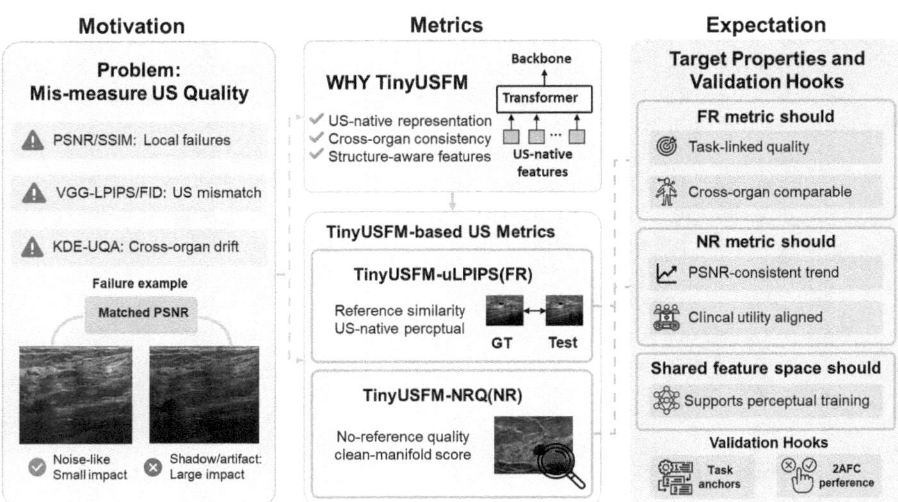

Fig. 1. Overview of the TinyUSFM-based ultrasound quality evaluation framework, including metric motivation (left), proposed full-reference (FR) and no-reference (NR) metrics (middle), and target properties with shared-feature downstream utility (right).

cal protocols, organ-specific assumptions, or task-specific labels. In contrast, our goal is to define both FR and NR metrics in a shared ultrasound foundation-model feature space and evaluate whether this space supports task-linked, cross-organ, and clinically aligned quality assessment.

As illustrated in Fig. 1, we turn to USFM [7] and its lightweight variant, TinyUSFM [11] to develop a principled, ultrasound-native evaluation framework that bridges the gap between image-quality metrics and diagnostic utility. By leveraging the foundation model's feature space, we introduce two metrics—the full-reference (FR) TinyUSFM-uLPIPS and the no-reference (NR) TinyUSFM-NRQ—which offer superior calibration with clinical tasks and stable performance across diverse anatomical sites. We validate both metrics using task-based semantic anchors and clinician two-alternative forced choice (2AFC) studies. Furthermore, we employ the same feature space as a perceptual loss for super-resolution, establishing a validated pipeline for clinical image enhancement.

2 Methodology

2.1 Ultrasound-Native Representations from TinyUSFM

For an input image x, let $f_\ell(x) \in \mathbb{R}^{(1+T_\ell)\times C_\ell}$ denote the TinyUSFM output at layer ℓ, where T_ℓ is the number of spatial patch tokens and C_ℓ is the channel dimension. After removing the global [CLS] token, we denote the remaining patch-token matrix by $F_\ell(x) \in \mathbb{R}^{T_\ell \times C_\ell}$. For TinyUSFM-NRQ, each layer is further summarized by average pooling its patch tokens to obtain $u_\ell(x) \in \mathbb{R}^{C_\ell}$, and the pooled features from layers $\mathcal{L}$ are concatenated and L_2-normalized to form a global descriptor $z(x)$. Thus, $F_\ell(x)$ is used for full-reference comparison, whereas $z(x)$ is used for no-reference quality modeling.

2.2 TinyUSFM-uLPIPS: a Full-Reference Quality Metric

Given a reference image x and a test image y, TinyUSFM-uLPIPS compares their TinyUSFM patch-token features across layers $\mathcal{L} = \{3, 5, 7, 11\}$, where $F_\ell(x), F_\ell(y) \in \mathbb{R}^{T_\ell \times C_\ell}$ denote the [CLS]-removed patch-token matrices at layer ℓ. In all experiments, we use $\mathcal{L} = \{3, 5, 7, 11\}$, $r = 3$, and $\tau = 20$.

Let $\hat{F}_\ell(x)$ and $\hat{F}_\ell(y)$ denote the row-wise L_2-normalized versions of $F_\ell(x)$ and $F_\ell(y)$. For each token location $q \in \{1, \ldots, T_\ell\}$, let $\mathcal{N}_r(q)$ denote its local neighborhood with radius r, and let $A_{\ell,q}(x) \in \mathbb{R}^{|\mathcal{N}_r(q)|\times C_\ell}$ be the matrix formed by stacking the normalized token vectors in that neighborhood. We then define

$$d_\ell^{\mathrm{s}}(x, y) = \frac{1}{T_\ell} \sum_{q=1}^{T_\ell} \mathrm{MeanAbs}\big(\mathrm{Softmax}\big(\tau A_{\ell,q}(x) A_{\ell,q}(x)^\top\big) - \mathrm{Softmax}\big(\tau A_{\ell,q}(y) A_{\ell,q}(y)^\top\big)\big), \tag{1}$$

where Softmax($\cdot$) is applied row-wise and τ is a temperature parameter.

Channel co-activation statistics are compared using Gram matrices computed from the normalized token matrices [5], with $G_\ell(x) = \frac{1}{T_\ell C_\ell}\hat{F}_\ell(x)^\top \hat{F}_\ell(x)$ and $d_\ell^{\mathrm{g}}(x,y) = \mathrm{MeanAbs}(G_\ell(x) - G_\ell(y))$. The per-layer distance is $d_\ell(x,y) = d_\ell^{\mathrm{s}}(x,y) + d_\ell^{\mathrm{g}}(x,y)$, and the final TinyUSFM-uLPIPS score is

$$\text{TinyUSFM-uLPIPS}(x,y) = \frac{1}{|\mathcal{L}|}\sum_{\ell\in\mathcal{L}} d_\ell(x,y). \tag{2}$$

2.3 TinyUSFM-NRQ: a No-Reference Quality Metric

TinyUSFM-NRQ scores a single ultrasound image without a reference by combining organ-aware clean-manifold likelihood with worst-region aggregation. This design models valid appearance variation across devices and settings while remaining sensitive to localized failures.

For an image x, we extract overlapping patches $\{x_i\}_{i=1}^{N_x}$ of size 224×224 with stride 112, where N_x is the number of extracted patches. Each patch x_i is encoded as $\hat{z}_i = z(x_i)$ using the global TinyUSFM descriptor from Sect. 2.1. For each organ o, clean patch descriptors are projected to a PCA space of dimension $d = 128$ and modeled by a diagonal-covariance Gaussian mixture model p_o with $K = 4$ components. If the organ label is available, patch x_i is scored by $s_i(x) = \log p_o(\hat{z}_i)$; otherwise, we use a uniform mixture over all O organ models, i.e., $s_i(x) = \log\left(\frac{1}{O}\sum_{o=1}^{O} p_o(\hat{z}_i)\right)$. To emphasize localized failures, we average the lowest-scoring patches, where $\kappa = \max(1, \mathrm{round}(0.15\,N_x))$ and $\mathcal{W}_\kappa(x)$ denotes the index set of the κ worst patches. We then define

$$\text{TinyUSFM-NRQ}(x) = \frac{1}{\kappa}\sum_{i\in\mathcal{W}_\kappa(x)} s_i(x), \tag{3}$$

where higher values indicate better quality.

2.4 TinyUSFM Features as a Perceptual Loss for Image Restoration

For SR optimization, we use a differentiable token-distance loss in the same TinyUSFM feature space: Let $\hat{F}_\ell(\cdot)$ denote the row-wise L_2-normalized [CLS]-removed token matrix at layer ℓ. Given a reconstruction $\hat{y}$ and its target image y, we define

$$\mathcal{L}_{\text{TinyPerc}}(\hat{y},y) = \frac{1}{|\mathcal{L}|}\sum_{\ell\in\mathcal{L}} \frac{1}{T_\ell}\sum_{t=1}^{T_\ell} \left\| \hat{F}_{\ell,t}(\hat{y}) - \hat{F}_{\ell,t}(y) \right\|_2^2. \tag{4}$$

For 512×512 images, the TinyUSFM perceptual loss is computed on sliding 224×224 windows and averaged across windows to avoid resizing artifacts.

3 Experiments and Results

3.1 Evaluation Protocol

Task-Linked Quality (FR): We assess metric ability to capture task-linked quality by using performance degradation of downstream models as an objective proxy for clinical utility. We systematically apply distortions using a binary search to ensure each reaches an identical PSNR-matched magnitude relative to the clean baseline. By evaluating a pre-trained model across these controlled degradations, we use the resulting drop in task performance (e.g., segmentation or classification accuracy) as a "ground truth" for how much a specific distortion actually interferes with diagnostic information. The rank correlation (Spearman's ρ and Kendall's τ) between TinyUSFM-uLPIPS and these accuracy drops serves as a primary indicator of whether the metric prioritizes task-relevant structural integrity over benign global fluctuations like intensity shifts.

Cross-Organ Comparability (FR): To evaluate cross-organ comparability, we assess the consistency of degradation rankings derived by TinyUSFM-uLPIPS across diverse anatomical sites using Kendall's W, ensuring that the relative severity of distortions is penalized uniformly regardless of the organ. We further analyze score-scale stability using the interquartile range (IQR), which measures the statistical spread of the middle 50% of the data. A lower IQR indicates that the metric produces more consistent and predictable scores for a given quality level, suggesting that the numerical scale remains stable and is not skewed by anatomical variations.

PSNR-Consistent Sensitivity (NR): To evaluate our NR metric for TinyUSFM-NRQ, we utilize a ground-truth ranking protocol. We synthesize six variants of each clean image by applying distortions of known, increasing severity. Since the true quality hierarchy is established by design, we assess the metric's effectiveness by measuring how closely its predicted scores correlate with these pre-defined ranks. High correlation indicates that the framework can accurately perceive diagnostic degradation without requiring a pristine reference image.

Clinician Utility (NR): To validate the metric's practical utility, we conducted a blinded 2AFC study. Clinicians evaluated PSNR-matched image pairs featuring different degradation types and selected the more clinically acceptable version in each case. These expert preferences were then compared against TinyUSFM-NRQ scores across 540 cross-degradation pairs to determine if the metric aligns with professional diagnostic standards when traditional pixel-fidelity measures fail to distinguish quality.

The evaluation covers 8 distortions categorized into four primary groups: noise and texture corruption, blur and resolution loss, ultrasound-specific artifacts (such as shadowing, specular clipping, and missing scanlines), and geometric deformation.

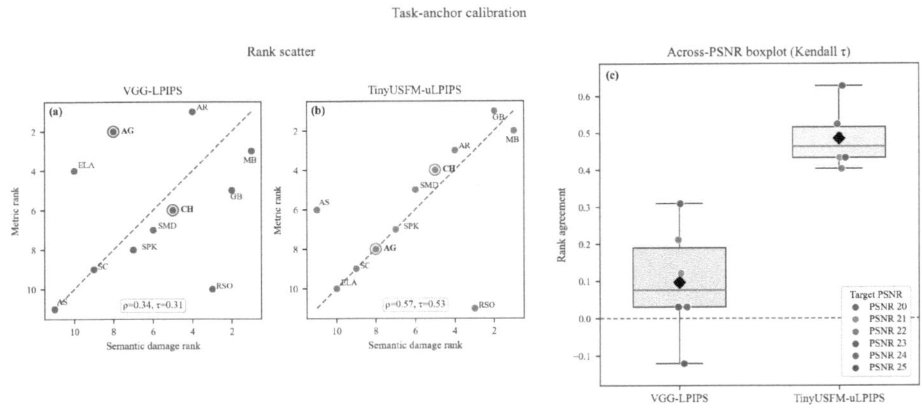

Fig. 2. Task-anchor calibration of FR metrics under PSNR-aligned evaluation. Left: representative DDTI rank–rank scatter at PSNR= 20 for VGG-LPIPS and TinyUSFM-uLPIPS. Right: Kendall's τ summarized across segmentation anchors and PSNR targets; colored dots indicate individual anchor–PSNR settings.

3.2 Full-Reference Evaluation (TinyUSFM-uLPIPS)

Experimental Setup: We use thyroid and kidney ultrasound images for Dice-based segmentation evaluation, including thyroid nodule segmentation from DDTI [16] and kidney capsule segmentation from the Open Kidney Ultrasound Data Set [19]. We degrade the images to matched PSNR levels (20–25) and run the same segmentation model to obtain Dice scores. We then compare VGG-LPIPS and TinyUSFM-uLPIPS on the degraded images and test which metric better reflects the true severity of task damage.

Task-Linked Quality: Figure 2 shows a representative DDTI rank-scatter example and Kendall's τ summaries across segmentation anchors and PSNR targets. TinyUSFM-uLPIPS aligns better with semantic damage than VGG-LPIPS on the thyroid and kidney ultrasound datasets, and it shows consistently higher and more stable Kendall's τ across PSNR targets (Fig. 2). At low PSNR, the difference is most clear because distortion effects are easier to separate; at higher PSNR, rank comparison becomes harder because Dice drops are smaller, but TinyUSFM-uLPIPS remains more stable than VGG-LPIPS.

Under matched PSNR, VGG-LPIPS shows a systematic bias: it tends to over-penalize noise-like degradations and under-penalize ultrasound-specific structural artifacts. For example, at PSNR= 20 on the thyroid dataset, VGG-LPIPS ranks AdditiveGaussian as more severe than ClutterHaze/ROIShadow Occlusion, even though the latter causes larger Dice drop, while TinyUSFM-uLPIPS gives the correct ordering.

Cross-Organ Comparability: TinyUSFM-uLPIPS produces more consistent cross-organ degradation rankings than VGG-LPIPS (Fig. 3(a)) and shows a

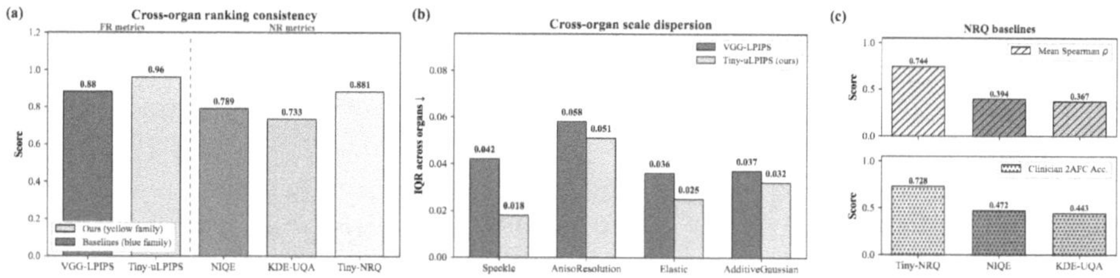

Fig. 3. Cross-organ robustness and NRQ baseline comparison under PSNR-aligned evaluation. (a) Cross-organ ranking consistency for FR and NRQ metrics (Kendall's W, PSNR 20–25 dB). (b) Cross-organ scale dispersion for FR metrics (IQR across organs) on representative degradations (aggregated over PSNR 20–25 dB). (c) NRQ baseline comparison: within-organ correlation (Spearman ρ) and clinician 2AFC agreement.

smaller inter-organ IQR on several representative degradations, such as speckle-related and resolution-related corruptions (Fig. 3(b)).

3.3 No-Reference Evaluation (TinyUSFM-NRQ)

Experimental Setup. We evaluate TinyUSFM-NRQ on public ultrasound datasets covering multiple organs and acquisition domains, including CUBS [13], UF1990 [4], STMUS [12], AUL [22], BUSI [1], MMOTU [25], DDTI [16], KidneyUS [19] and CAMUS [9]. Robustness analysis for TinyUSFM-NRQ and baselines is conducted on 9 organs following the shared protocol in Sect. 3.1, yielding (9×8) organ×distortion conditions. We compare TinyUSFM-NRQ with NIQE, BRISQUE, and KDE-UQA (Fig. 3(c)).

PSNR-Consistent Sensitivity. TinyUSFM-NRQ achieves a mean Spearman $\rho = 0.744$ over PSNR levels, indicating strong within-organ monotonicity. NIQE and KDE-UQA fail systematically on ultrasound-typical structural degradations, especially scanline missing and specular clipping. At PSNR= 25, these metrics become numerically ill-conditioned for structural artifacts, and several distortion families demonstrate near-zero or negative correlations with clinical quality across different organs. This makes them unreliable as a unified cross-organ quality control scale under heterogeneous ultrasound artifacts and is consistent with their lower monotonicity and weaker clinician alignment relative to TinyUSFM-NRQ (Fig. 3(c)).

Using the same PSNR-aligned degradation settings on unseen organs (stomach, multifidus muscle, and neck nerve), TinyUSFM-NRQ maintains strong monotonicity and reliable severity ranking under domain shift. Its degradation scoring trends are consistent with the in-distribution results, showing similar relative severity patterns across major artifact families.

Clinician Utility. TinyUSFM-NRQ predicts clinician preference with 72.8% accuracy (95% CI [0.689, 0.764], Wilson), significantly above chance (two-sided binomial test, $p<10^{-20}$). In embedded sanity checks (same degradation, different PSNR) and duplicate pairs, the readers achieved 100% accuracy and consistency.

3.4 Ultrasound Super-Resolution

Experimental Setup. We train the same SwinIR backbone [10] with identical splits under three objectives: $\mathcal{L}_1$, $\mathcal{L}_1 + \mathcal{L}_{\text{VGG}}$, and $\mathcal{L}_1 + \mathcal{L}_{\text{TinyUSFM}}$ (Sect. 2.4). Training uses a combined breast+echocardiography ultrasound dataset for 100 epochs (LR 10^{-4}) with $\lambda_{\text{percep}}/\lambda_1 = 0.1/1$ for both perceptual losses.

Perceptual-Clinical Alignment Loop. Compared with $\mathcal{L}_{\text{VGG}}$, using $\mathcal{L}_{\text{TinyUSFM}}$ as the perceptual loss improves PSNR and SSIM by 5.3% and 3.3%, respectively. Although $\mathcal{L}_1$ still gives the strongest PSNR, it produces over-smoothed reconstructions, while $\mathcal{L}_{\text{VGG}}$ reduces fidelity and introduces ultrasound-inconsistent artifacts. In contrast, $\mathcal{L}_{\text{TinyUSFM}}$ maintains competitive fidelity and better preserves diagnostically relevant acoustic texture with fewer conspicuous structural hallucinations. As shown in Fig. 4, the ultrasound-native perceptual metrics (TinyUSFM-uLPIPS and TinyUSFM-NRQ) also favor $\mathcal{L}_{\text{TinyUSFM}}$ reconstructions. These findings validate the use of ultrasound-native features for both model optimization and evaluation. By prioritizing acoustic-specific structures over pixel-level fidelity, the metric aligns more closely with clinical utility.

In a blinded clinician pairwise evaluation, expert sonographers compared reconstructions from the same case and selected the one more suitable for clinical diagnosis. Clinicians preferred $\mathcal{L}_{\text{TinyUSFM}}$ reconstructions over $\mathcal{L}_1$ in 87.5% of cases and over $\mathcal{L}_{\text{VGG}}$ in 100% of cases. Moreover, $\mathcal{L}_1$ was preferred over $\mathcal{L}_{\text{VGG}}$ in 97.5% of cases, indicating that visually sharper but hallucinated structures are clinically unacceptable in ultrasound. This is consistent with TinyUSFM-NRQ, which favors $\mathcal{L}_{\text{TinyUSFM}}$ reconstructions over the $\mathcal{L}_1$ and $\mathcal{L}_{\text{VGG}}$ baselines.

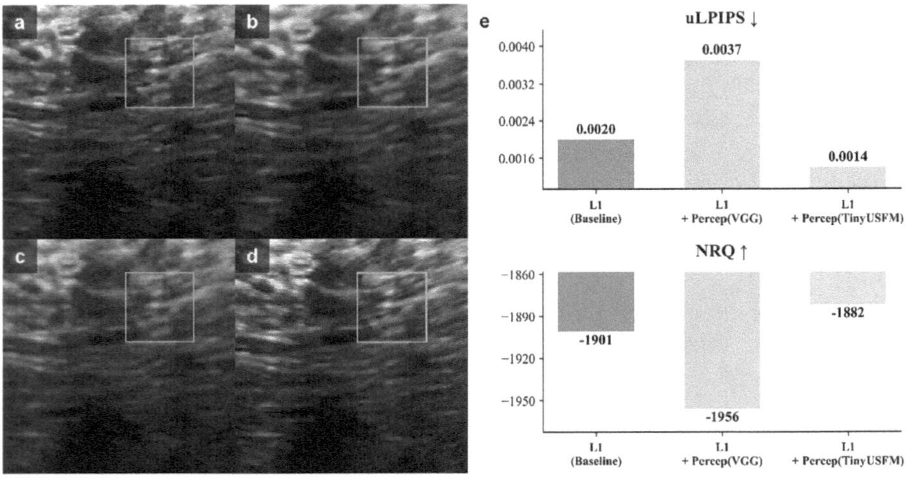

Fig. 4. SR comparison under three training objectives. (a–d) Representative breast ultrasound example (ground truth; $\mathcal{L}_1$; $\mathcal{L}_1 + \mathcal{L}_{\text{VGG}}$; $\mathcal{L}_1 + \mathcal{L}_{\text{TinyUSFM}}$). (e) Quantitative comparison using TinyUSFM-uLPIPS and TinyUSFM-NRQ.

Runtime analysis confirms a functional trade-off between the two modes. TinyUSFM-uLPIPS is optimized for high-fidelity offline research evaluation, whereas TinyUSFM-NRQ provides significantly lower latency and higher throughput. This efficiency makes the NRQ variant well-suited for real-time, no-reference quality control within active clinical and system pipelines.

4 Conclusion

We introduced a TinyUSFM-based ultrasound quality evaluation framework with two ultrasound-native metrics: TinyUSFM-uLPIPS for FR perceptual assessment and TinyUSFM-NRQ for NR quality scoring. Using a shared TinyUSFM feature space, the framework provides a modality-aligned alternative to pixel-fidelity metrics, natural-image perceptual metrics, and generic quality control baselines for ultrasound reconstruction and generative imaging. Future work will explore extending the framework to other ultrasound foundation models, higher-PSNR quality regimes, dynamic ultrasound, and diverse medical imaging modalities. The proposed metrics inherit the coverage and potential biases of the TinyUSFM backbone, and their reliability may be affected by rare pathologies, unseen devices, or under-represented acquisition settings. Overall, we believe this work provides the ultrasound community with a concrete answer to the question clinicians have always implicitly asked: Does this reconstruction show what I need to see?

Acknowledgements. This work was supported in part by the National Key R&D Program of China under Grant 2024YFF0507300 and Grant 2024YFF0507303, in part by the National Natural Science Foundation of China under Grant 62531004, in part by the Fudan University Medicine-Engineering Collaboration Program under Grant yg-2025-general-07, and in part by the Shanghai Municipal Education Commission Program on AI for Transforming Academic Research Paradigms (24RGZNB05).

Disclosure of Interests. The authors have no competing interests to declare.

References

1. Al-Dhabyani, W., Gomaa, M., Khaled, H., Fahmy, A.: Dataset of breast ultrasound images. Data Brief **28**, 104863 (2020). https://doi.org/10.1016/j.dib.2019.104863
2. Baad, M., Lu, Z.F., Reiser, I., Paushter, D.: Clinical significance of us artifacts. Radiographics **37**(5), 1408–1423 (2017). https://doi.org/10.1148/rg.2017160175
3. Bamber, J.C., Daft, C.: Adaptive filtering for reduction of speckle in ultrasonic pulse-echo images. Ultrasonics **24**(1), 41–44 (1986). https://doi.org/10.1016/0041-624X(86)90072-7
4. Cai, P., Yang, T., Xie, Q., Liu, P., Li, P.: A lightweight hybrid model for the automatic recognition of uterine fibroid ultrasound images based on deep learning. J. Clin. Ultrasound **52**(6), 753–762 (2024). https://doi.org/10.1002/jcu.23703
5. Gatys, L.A., Ecker, A.S., Bethge, M.: A neural algorithm of artistic style. arXiv preprint arXiv:1508.06576 (2015). https://arxiv.org/abs/1508.06576

6. Heusel, M., Ramsauer, H., Unterthiner, T., Nessler, B., Hochreiter, S.: Gans trained by a two time-scale update rule converge to a local nash equilibrium. In: Advances in Neural Information Processing Systems (NeurIPS) (2017). https://arxiv.org/abs/1706.08500
7. Jiao, J., et al.: Usfm: A universal ultrasound foundation model generalized to tasks and organs towards label efficient image analysis. Med. Image Anal. **96**, 103202 (2024). https://doi.org/10.1016/j.media.2024.103202
8. Kwon, J., Jiao, J., Self, A., Noble, J.A., Papageorghiou, A.: A kernel density estimation based quality metric for quality assessment of obstetric ultrasound video. In: Trustworthy Machine Learning for Healthcare (TML4H 2023). Lecture Notes in Computer Science, vol. 13932, pp. 134–146. Springer, Cham (2023). https://doi.org/10.1007/978-3-031-39539-0_12
9. Leclerc, S., Smistad, E., Pedrosa, J., Östvik, A., Grenier, T., Espinosa, F., et al.: Deep learning for segmentation using an open large-scale dataset in 2d echocardiography. IEEE Trans. Med. Imaging **38**(9), 2198–2210 (2019). https://doi.org/10.1109/TMI.2019.2900516
10. Liang, J., Cao, J., Sun, G., Zhang, K., Van Gool, L., Timofte, R.: Swinir: Image restoration using swin transformer. In: Proceedings of the IEEE/CVF International Conference on Computer Vision Workshops (ICCVW) (2021). https://arxiv.org/abs/2108.10257
11. Ma, C., et al.: Tinyusfm: Towards compact and efficient ultrasound foundation models. arXiv preprint arXiv:2510.19239 (2025). https://doi.org/10.48550/arXiv.2510.19239, https://arxiv.org/abs/2510.19239
12. Marzola, F., van Alfen, N., Doorduin, J., Meiburger, K.M.: Deep learning segmentation of transverse musculoskeletal ultrasound images for neuromuscular disease assessment. Comput. Biol. Med. **135**, 104623 (2021). https://doi.org/10.1016/j.compbiomed.2021.104623
13. Meiburger, K.M., Marzola, F., Zahnd, G., Faita, F., Loizou, C.P., Lainé, N., et al.: Carotid ultrasound boundary study (cubs): Technical considerations on an open multi-center analysis of computerized measurement systems for intima-media thickness measurement on common carotid artery longitudinal b-mode ultrasound scans. Comput. Biol. Med. **144**, 105333 (2022). https://doi.org/10.1016/j.compbiomed.2022.105333
14. Mittal, A., Moorthy, A.K., Bovik, A.C.: No-reference image quality assessment in the spatial domain. IEEE Trans. Image Process. **21**(12), 4695–4708 (2012). https://doi.org/10.1109/TIP.2012.2214050
15. Mittal, A., Soundararajan, R., Bovik, A.C.: Making a "completely blind" image quality analyzer. IEEE Signal Processing Letters **20**(3), 209–212 (2013). https://live.ece.utexas.edu/research/Quality/niqe_spl.pdf
16. Pedraza, L., Vargas, C., Narváez, F., Durán, O., Muñoz, E., Romero, E.: An open access thyroid ultrasound image database. In: Medical Imaging 2015: Computer-Aided Diagnosis. Proceedings of SPIE, vol. 9287 (2015). https://doi.org/10.1117/12.2073532
17. Raina, D., Ntentia, D., Chandrashekhara, S.H., Voyles, R., Saha, S.K.: Expert-agnostic ultrasound image quality assessment using deep variational clustering. arXiv preprint arXiv:2307.02462 (2023). https://arxiv.org/abs/2307.02462
18. Robinson, D.E., Knight, P.C.: Interpolation scan conversion in pulse-echo ultrasound. Ultrason. Imaging **4**(4), 297–310 (1982). https://doi.org/10.1177/016173468200400401

19. Singla, R., et al.: The open kidney ultrasound data set. In: Medical Image Computing and Computer Assisted Intervention – MICCAI 2023 (2023). https://doi.org/10.1007/978-3-031-44521-7_15
20. Wang, Z., Bovik, A.C., Sheikh, H.R., Simoncelli, E.P.: Image quality assessment: From error visibility to structural similarity. IEEE Trans. Image Process. **13**(4), 600–612 (2004). https://doi.org/10.1109/TIP.2003.819861
21. Wu, L., Cheng, J.Z., Li, S., Lei, B., Wang, T., Ni, D.: Fuiqa: Fetal ultrasound image quality assessment with deep convolutional networks. IEEE Trans. Cybern. **47**(5), 1336–1349 (2017). https://doi.org/10.1109/TCYB.2017.2671898
22. Xu, Y., et al.: Improving artificial intelligence pipeline for liver malignancy diagnosis using ultrasound images and video frames. Brief, Bioinf. **24**(1), bbac569 (2023). https://doi.org/10.1093/bib/bbac569
23. Zhang, B., Liu, H., Luo, H., Li, K.: Automatic quality assessment for 2d fetal sonographic standard plane based on multitask learning. Medicine **100**(4), e24427 (2021)
24. Zhang, R., Isola, P., Efros, A.A., Shechtman, E., Wang, O.: The unreasonable effectiveness of deep features as a perceptual metric. In: Proceedings of the IEEE/CVF Conference on Computer Vision and Pattern Recognition (CVPR), pp. 586–595 (2018). https://doi.org/10.1109/CVPR.2018.00068
25. Zhao, Q., et al.: Mmotu: A multi-modality ovarian tumor ultrasound image dataset for unsupervised cross-domain semantic segmentation. arXiv preprint arXiv:2207.06799 (2022). https://arxiv.org/abs/2207.06799

Deterministic Hallucination Detection in Medical VQA via Confidence-Evidence Bayesian Gain

Mohammad Asadi[1(✉)], Tahoura Nedaee[2], Jack W. O'Sullivan[3,4], Euan Ashley[3,4], and Ehsan Adeli[4,5,6]

[1] Department of Electrical Engineering, Stanford University, Stanford, CA, USA
masadi@stanford.edu
[2] Department of Biology, Stanford University, Stanford, CA, USA
[3] Division of Cardiology, Department of Medicine, Stanford University, Stanford, CA, USA
[4] Department of Biomedical Data Science, Stanford University, Stanford, CA, USA
[5] Department of Computer Science, Stanford University, Stanford, CA, USA
[6] Department of Psychiatry and Behavioral Sciences, Stanford University, Stanford, CA, USA

Abstract. Multimodal large language models (MLLMs) have shown strong potential for medical Visual Question Answering (VQA), yet they remain prone to hallucinations, defined as generating responses that contradict the input image, posing serious risks in clinical settings. Current hallucination detection methods, such as Semantic Entropy (SE) and Vision-Amplified Semantic Entropy (VASE), require 10 to 20 stochastic generations per sample together with an external natural language inference model for semantic clustering, making them computationally expensive and difficult to deploy in practice. We observe that hallucinated responses exhibit a distinctive signature directly in the model's own log-probabilities: inconsistent token-level confidence and weak sensitivity to visual evidence. Based on this observation, we propose **C**onfidence-**E**vidence **B**ayesian **G**ain (CEBaG), a deterministic answer-level hallucination detection method for short-form medical VQA that requires no stochastic sampling, no external models, and no task-specific hyperparameters. CEBaG combines two complementary signals: *token-level predictive variance*, which captures inconsistent confidence across response tokens, and *evidence magnitude*, which measures how much the image shifts per-token predictions relative to text-only inference. Evaluated across four medical MLLMs and three VQA benchmarks (16 experimental settings), CEBaG achieves the highest AUC in 13 of 16 settings and improves over VASE by 8 AUC points on average, while being fully deterministic and self-contained. https://github.com/masadi-99/CEBaG.

Keywords: Hallucination detection · Medical VQA · Multimodal LLMs

G. Yang et al. (Eds.): MICCAI 2026, LNCS 16878, pp. 130–140, 2027.
https://doi.org/10.1007/978-3-032-38059-3_13

1 Introduction

Medical Visual Question Answering (VQA), powered by multimodal large language models (MLLMs), enables the interpretation of medical images in response to natural language queries, supporting tasks from abnormality detection to differential diagnosis [18,24]. However, MLLMs are susceptible to *hallucinations*, defined as responses that misinterpret or contradict the input image, posing substantial risks including misdiagnoses, inappropriate treatments, and diminished clinician trust [8,19]. Despite efforts in data optimization [26], training [9], and decoding refinements [23], hallucinations remain a persistent challenge, highlighting the need for effective detection methods.

Recent approaches to hallucination detection include detector fine-tuning [5], cross-checking [3,26], visual evidence verification [25], and uncertainty estimation [1,4,10]. Among these, uncertainty estimation stands out for its simplicity, as it uses only the model itself without additional annotated data or external knowledge bases [8,19]. Semantic Entropy (SE) [4] quantifies uncertainty by computing entropy over semantically clustered responses. Vision-Amplified Semantic Entropy (VASE) [16] extends SE by contrasting semantic distributions from original and augmented images, amplifying the influence of visual input. Other contrastive methods such as VCD [12] and LCD [13] leverage output differences under perturbed inputs, while Semantic Entropy Probes [11] train a probe on internal activations, requiring task-specific data unlike our training-free approach. A common thread in these methods is their reliance on stochastic sampling, external models, or additional training, which raises the question of whether the uncertainty signal they seek is accessible through simpler means.

These methods also face a concrete efficiency bottleneck. SE requires M stochastic generations (typically M=10) with high-temperature sampling, followed by pairwise semantic equivalence checks via an external NLI model [6]. VASE doubles this to $2M$ generations plus $O(M^2)$ NLI comparisons per input, making it prohibitive for clinical deployment where real-time feedback is essential. We propose **Confidence-Evidence Bayesian Gain (CEBaG)**, which replaces stochastic sampling with deterministic log-probability analysis. CEBaG combines two complementary signals: *token-level predictive variance* (the standard deviation of per-token log-probabilities, elevated when the model fabricates content) and *evidence magnitude* (the absolute per-token log-probability shift when the image is present versus absent, capturing the strength of visual grounding). The resulting score, $\sigma \cdot (1 + |G|/L)$, is parameter-free, requires only one generation and two scoring passes, uses no external models, and is fully deterministic (Fig. 1). We position CEBaG specifically for answer-level hallucination detection in short-form medical VQA. Its two signals, the sequence-level evidence gain G and the global standard deviation of token log-probabilities, operate over the entire response and thus capture whether an answer, taken as a whole, is grounded in the image. This makes CEBaG well-suited to flagging holistic answer–image consistency in concise medical VQA.

Our contributions are: (1) We propose CEBaG, a deterministic answer-level hallucination detection method requiring three forward passes and no external models, compared to 20+ stochastic generations and an NLI model for SE/VASE. (2) We conduct an extensive evaluation on four MLLMs (MedGemma-4b, MedGemma-1.5-4b, HuatuoGPT-Vision-7B, LLaVA-Med-v1.5-7B) and three datasets (VQA-RAD, SLAKE, PathVQA), 16 settings in total. (3) CEBaG consistently outperforms SE, VASE, and RadFlag [22], achieving the best AUC in 13 of 16 settings and improving over VASE by 8 AUC points on average.

2 Method

Given a medical image x_v, a textual question x_q, and a medical MLLM f, the model generates a response $r = f(x_v, x_q)$ via greedy or low-temperature decoding. Our goal is to determine whether r is a hallucination (a response that contradicts the visual evidence in x_v) without access to ground-truth answers. For evaluation, we follow prior work [4,16] and use the GREEN model [20] to produce reference-based ground-truth labels.

2.1 Evidence Gain as Pointwise Mutual Information

The key quantity in CEBaG is the *evidence gain*: how much the model's belief in its own answer changes when it observes the image. Formally, we interpret the text-only probability $P(r \mid x_q)$ as the model's **language prior**, i.e., its belief about the answer based solely on the question and its pre-trained knowledge. The multimodal probability $P(r \mid x_v, x_q)$ represents the **posterior** belief after updating with the visual evidence x_v. We use "prior" and "posterior" in the information-theoretic sense: $P(r \mid x_q)$ represents the model's belief before observing visual evidence, and $P(r \mid x_v, x_q)$ its updated belief after conditioning on the image. This is distinct from Bayesian inference over model parameters; rather, G captures how the image updates the model's predictive distribution for a fixed set of parameters. The sequence log-probabilities are defined as:

$$\log P(r \mid x_v, x_q) = \sum_{j=1}^{L} \log P(r_j \mid r_{<j}, x_v, x_q), \tag{1}$$

$$\log P(r \mid x_q) = \sum_{j=1}^{L} \log P(r_j \mid r_{<j}, x_q). \tag{2}$$

The **evidence gain** G is the difference between the posterior and the prior:

$$G = \log P(r \mid x_v, x_q) - \log P(r \mid x_q). \tag{3}$$

By Bayes' rule, $P(r \mid x_v, x_q) = \frac{P(x_v|r,x_q)P(r|x_q)}{P(x_v|x_q)}$. Substituting this into Eq. (3):

$$G = \log \left(\frac{P(x_v \mid r, x_q) P(r \mid x_q)}{P(x_v \mid x_q)} \right) - \log P(r \mid x_q) = \log P(x_v \mid r, x_q) - \log P(x_v \mid x_q). \tag{4}$$

Since $\log P(x_v \mid x_q)$ is constant with respect to the response r, G is directly proportional to $\log P(x_v \mid r, x_q)$, the log-likelihood of the image given the response. Thus, G is the **Pointwise Mutual Information (PMI)** between the response and the image (conditioned on the question). A large positive G implies that the response makes the observed image highly probable (strong grounding), while a negative G suggests the response contradicts the visual evidence.

The **evidence magnitude** is a length-normalized visual influence measure:

$$E = \frac{|G|}{L}, \tag{5}$$

which captures the average absolute shift in belief per token. We use the magnitude $|G|$ because a strong visual influence (increasing or decreasing token probabilities) indicates attention to the image, whereas a lack of influence ($G \approx 0$) suggests the model is relying entirely on its language prior.

2.2 Confidence-Evidence Bayesian Gain

We propose CEBaG, a scoring function that combines the model's internal uncertainty with its sensitivity to visual evidence. We define the **token-level predictive variance**: $\sigma = \text{std}_{j=1}^{L}\big[\log P(r_j \mid r_{<j}, x_v, x_q)\big]$, which measures the consistency of the model's confidence across the generated sequence. High σ indicates "confidence spikes" on function words interspersed with low-confidence content tokens, a pattern characteristic of hallucination. The CEBaG score is defined as:

$$\text{CEBaG} = \sigma \cdot \big(1 + E\big). \tag{6}$$

This formulation can be interpreted as **posterior uncertainty weighted by visual information gain**. Hallucinations typically arise in two scenarios: (1) The model is uncertain about the content (σ is high); or (2) The visual evidence causes a large shift in belief (E is high) but fails to resolve the uncertainty (or contradicts the language prior).

The multiplicative term $(1 + E)$ amplifies the base uncertainty σ when the response is highly sensitive to the image, flagging cases where the model is "struggling" to reconcile visual evidence with its language prior.

This amplification is realized as a multiplicative *gain* on σ rather than an independent additive term. The gain is monotone in both signals and preserves σ as its base, such that the score reduces to σ when the image leaves the prediction unchanged (E=0), so visual evidence can only sharpen, never override, the uncertainty estimate. Because it multiplies rather than adds, it also requires no coefficient to reconcile the differing scales of σ and E.

Importantly, this formulation is **hyperparameter-free**, requiring no hyperparameter tuning or normalization, making it robust across different models and datasets. Unlike SE and VASE, which require configuring the number of generations M, sampling temperature T, and (for VASE) the amplification ratio α, CEBaG applies the same fixed formula across all models and datasets.

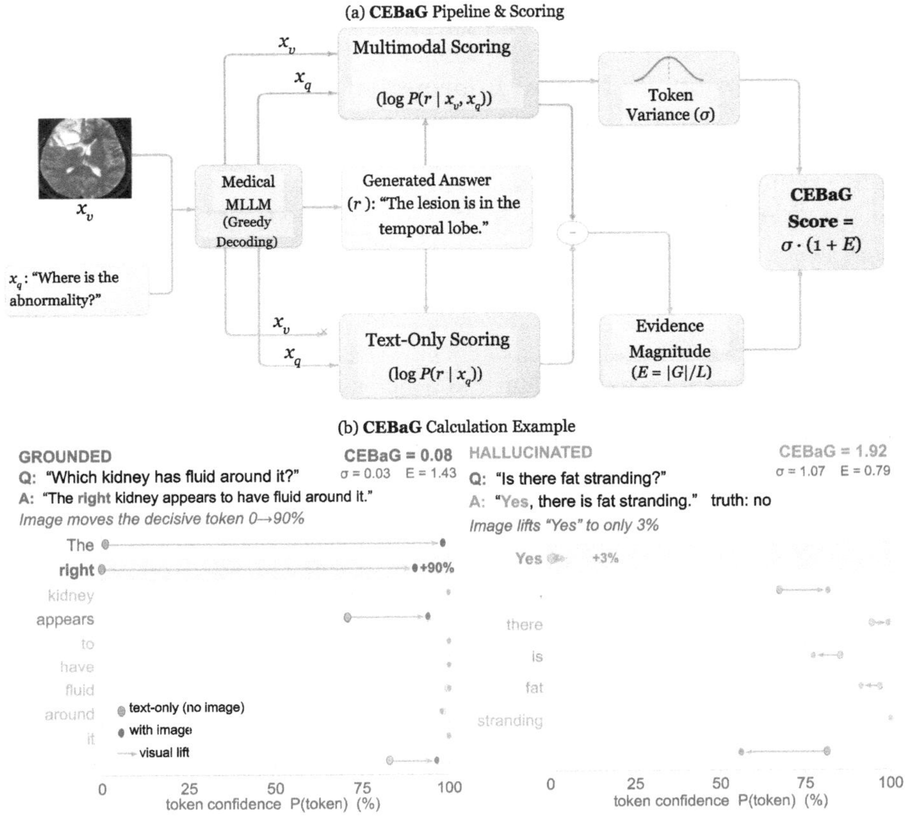

Fig. 1. Overview of CEBaG for hallucination detection in medical VQA. (a) The CEBaG pipeline: the model generates answer r, scored with and without the image to yield two signals: token-level predictive variance σ and evidence magnitude $E{=}|G|/L$. CEBaG combines these as $\sigma \cdot (1{+}E)$. (b) Per-token probabilities ($100\,P(\text{token})$) with vs. without the image, and CEBaG scores for a grounded and a hallucinated example.

3 Experiments

Datasets: We evaluate on three medical VQA benchmarks: *VQA-RAD* [14] (451 test samples, including 200 open-ended), a radiology VQA dataset covering chest X-rays, CT, and MRI; *SLAKE* [17], a bilingual medical VQA dataset with semantically labeled knowledge (1061 test samples); and *PathVQA* [7] (1000 test samples), covering pathology images. For VQA-RAD, we report performance on both open-ended questions and all questions (including yes/no), following [16].

Models: Four medical MLLMs spanning two architecture families: *MedGemma-4b-it* and *MedGemma-1.5-4b-it* [21] (encoder–decoder), and *HuatuoGPT-Vision-7B* [2] and *LLaVA-Med-v1.5-Mistral-7B* [15] (LLaVA-style decoder-only).

Ground Truth and Metrics: Following the evaluation protocol established by VASE [16], we use the GREEN model [20] to generate reference-based ground-truth labels: a response is labeled as hallucinated if its GREEN score falls below 1.0. We adopt this protocol to ensure direct comparability with prior work and use the same metrics as [16]: the Area Under the ROC Curve (AUC) and the Area Under the GREEN Curve (AUG).

Baselines. We compare CEBaG against four state-of-the-art hallucination detection methods: (1) *AvgProb* [10], which uses the average token probability in the sequence; (2) *Semantic Entropy (SE)* [4], which measures uncertainty over semantic clusters of generated answers; (3) *Vision-Amplified Semantic Entropy (VASE)* [16], which enhances SE by generating additional answers conditioned on visually augmented images; and (4) *RadFlag* [22], which checks for consistency between the generated answer and the findings in the radiology report. We adhere to the official implementations and hyperparameters for all baselines. For SE and VASE, we use 10 generations (plus 10 augmented for VASE) and a temperature of 1.0, as specified in their respective papers.

Implementation Details. For CEBaG, the generated answer r is produced via greedy decoding (temperature T=0.1). The two scoring passes (Eqs. 1–2) use teacher-forced forward passes; σ is computed from the same forward pass as $\log P(r \mid x_v, x_q)$ at no additional cost. For the text-only pass, visual input is removed at the interface level: for MedGemma, no image is included in the chat message and no pixel values are passed to the model; for LLaVA-style models (LLaVA-Med, HuatuoGPT), the `<image>` token is removed from the prompt template and the processor is called without image input. This ensures a clean text-only baseline across all architectures. All experiments were conducted on 8 NVIDIA H100 GPUs. Following the evaluation protocol of SE [4] and VASE [16], we report point estimates of AUC and AUG. Being deterministic, CEBaG avoids the baselines' sampling variance, giving exact per-answer scores.

3.1 Main Results

Table 1 presents the hallucination detection performance across all 16 experimental settings. CEBaG achieves the best AUC in **13 of 16** settings and ranks in the top two in 15 of 16. On average, CEBaG attains 67.9% AUC, outperforming VASE (59.6%), SE (57.4%), and RadFlag (57.8%) by substantial margins, with an improvement of +8.2 AUC points over the previous state-of-the-art VASE. Significantly, CEBaG achieves this without any hyperparameter tuning.

The gains are particularly pronounced on larger datasets: on PathVQA, CEBaG outperforms VASE by +10.5 (MedGemma), +6.5 (MedGemma-1.5), +28.8 (HuatuoGPT), and +17.2 (LLaVA-Med) AUC points. On SLAKE, the improvements are similarly large, with CEBaG surpassing VASE by +9.3 (MedGemma) and +7.1 (HuatuoGPT). On the smaller VQA-RAD datasets, the improvements remain consistent across most models, with CEBaG achieving the best AUC in 6 of 8 VQA-RAD settings. Furthermore, our method maintained

Table 1. Hallucination detection performance: AUC (%)↑ and AUG (%)↑ across four medical MLLMs and three VQA datasets. Best and second-best among the detection methods (SE, VASE, RadFlag, CEBaG) are in **bold** and underlined, respectively. CEBaG achieves the best AUC in 13 of 16 settings.

Dataset	AvgProb [10]		SE [4]		VASE [16]		RadFlag [22]		**CEBaG (Ours)**	
	AUC	AUG	AUC	AUG	AUC	AUG	AUC	AUG	AUC	AUG
MedGemma-4b-it [21]										
VQA-RAD (Open)	39.9	48.0	**61.9**	**65.3**	60.8	63.3	58.4	63.6	<u>61.0</u>	<u>64.8</u>
VQA-RAD (All)	42.9	54.6	52.2	58.0	<u>57.6</u>	<u>59.9</u>	54.7	58.9	**60.8**	**64.6**
SLAKE	35.3	52.9	54.9	64.2	<u>62.6</u>	<u>70.1</u>	57.0	66.2	**71.9**	**75.9**
PathVQA	46.4	45.1	58.0	48.9	<u>61.5</u>	<u>49.6</u>	56.8	47.4	**71.9**	**59.0**
MedGemma-1.5-4b-it [21]										
VQA-RAD (Open)	69.1	47.4	53.2	37.2	<u>60.1</u>	<u>40.3</u>	54.5	39.3	**72.9**	**47.2**
VQA-RAD (All)	63.6	56.2	53.4	48.6	<u>54.8</u>	<u>51.5</u>	54.3	46.3	**66.9**	**58.2**
SLAKE	57.4	52.0	<u>57.7</u>	51.9	**59.5**	<u>52.9</u>	57.2	**53.0**	57.5	52.0
PathVQA	65.9	48.9	56.8	40.9	<u>60.1</u>	<u>45.7</u>	56.3	41.6	**66.6**	**48.8**
HuatuoGPT-Vision-7B [2]										
VQA-RAD (Open)	43.7	48.1	<u>64.1</u>	<u>62.8</u>	61.7	59.6	59.9	58.4	**70.4**	**63.1**
VQA-RAD (All)	39.3	52.2	58.3	65.1	58.9	<u>66.1</u>	<u>60.0</u>	63.5	**67.7**	**67.6**
SLAKE	30.5	38.4	57.7	54.9	59.9	<u>58.0</u>	<u>60.6</u>	54.2	**67.0**	**64.2**
PathVQA	30.5	28.7	53.6	38.2	53.2	<u>39.5</u>	<u>55.5</u>	38.9	**82.0**	**55.6**
LLaVA-Med-v1.5-Mistral-7B [15]										
VQA-RAD (Open)	62.9	38.6	63.8	**40.4**	**68.4**	<u>40.2</u>	<u>65.8</u>	40.1	65.8	39.1
VQA-RAD (All)	58.8	48.7	<u>58.1</u>	**50.7**	57.6	49.3	56.8	48.1	**62.1**	<u>50.0</u>
SLAKE	59.9	50.0	57.9	50.2	57.4	<u>50.8</u>	<u>58.7</u>	50.7	**63.8**	**52.3**
PathVQA	75.1	52.0	57.4	41.3	<u>60.4</u>	<u>41.7</u>	58.8	39.7	**77.6**	**53.1**
Average	51.3	47.6	57.4	51.2	59.6	52.4	57.8	50.6	**67.9**	**57.2**

stable detection performance across GREEN thresholds between 0.4 and 0.8, confirming its insensitivity to specific ground-truth boundaries.

3.2 Efficiency Analysis

Table 2 compares the computational requirements and resulting average performance for each method. With the standard setting $M{=}10$, SE requires 10 autoregressive generations plus $O(M^2){=}O(100)$ pairwise entailment comparisons per sample. VASE doubles the generation cost to 20 passes plus the same entailment overhead. RadFlag also requires M generations and entailment checks. All three methods additionally require loading and running a DeBERTa-v2-xlarge-MNLI model alongside the target MLLM, and all involve hyperparameters (M, sampling temperature T, and for VASE, the vision-amplified ratio α).

Table 2. Computational cost and performance. Despite an order of magnitude fewer forward passes, CEBaG achieves the highest average AUC and AUG.

Method	Forward passes	External model	Deterministic	Hyperparams	$\overline{AUC}$	$\overline{AUG}$
AvgProb [10]	1 gen.	None	Yes	None	51.3	47.6
SE [4]	M gen. + NLI	DeBERTa-MNLI	No	M, T	57.4	51.2
VASE [16]	$2M$ gen. + NLI	DeBERTa-MNLI	No	M, T, α	59.6	52.4
RadFlag [22]	M gen. + NLI	DeBERTa-MNLI	No	M, T	57.8	50.6
CEBaG (Ours)	1 gen. + 2 evals	None	Yes	None	**67.9**	**57.2**

In contrast, CEBaG requires exactly **three forward passes**: one autoregressive generation and two teacher-forced scoring passes (with and without the image). No external model is needed, and no hyperparameters require tuning. The scoring passes are substantially cheaper than generation as they process fixed token sequences without autoregressive decoding. Despite its simplicity and efficiency, CEBaG achieves the highest average AUC (67.9%) and AUG (57.2%) across all settings, outperforming the computationally expensive baselines.

3.3 Ablation Study

We compare four variants: σ only (token-level variance); E only (evidence magnitude $|G|/L$); CEBaG, the parameter-free $\sigma \cdot (1 + E)$; and CEBaG$_\lambda$, an oracle that, per model-dataset pair, min-max normalizes G and a penalty term and selects the gain sign, penalty term, and weight $\lambda \in [-5, 5]$ maximizing AUC, an in-sample configuration unattainable without test labels.

Table 3 reports the ablation; the bottom row averages all 16 settings. 1- Token-level variance dominates: σ-only achieves 67.0% average AUC, above all baselines (SE 57.4%, VASE 59.6%, RadFlag 57.8%). 2- Evidence magnitude is complementary: while E only (59.0%) trails σ only, CEBaG (67.9%) outperforms it in more than half of the settings, with notable gains on MedGemma-1.5 VQA-RAD Open (+4.7) and MedGemma PathVQA (+4.1). In the rest, $(1+E)$ lowers AUC, usually marginally (−0.3 on HuatuoGPT VQA-RAD All, −1.4 on LLaVA-Med SLAKE). Because σ and E combine without normalization, their relative influence varies across models; CEBaG keeps this factor because it raises the average and needs no tuning. 3- The tuned upper bound does not transfer. CEBaG$_\lambda$ reaches 71.9% average AUC, 4.0 points above CEBaG, entirely from per-setting λ selection: the optimal λ is unstable, ranging from -5 to $+2.75$. The best single λ recovers under one of these four points, and the tuned value is unattainable without test-label access.

Table 3. Ablation study: AUC (%)↑ of CEBaG components. σ only uses token-level variance alone; E only uses absolute per-token gain alone; CEBaG combines both (parameter-free); CEBaG$_\lambda$ adds a tunable weight (upper bound).

Setting	AUC (%)				AUG (%)			
	σ	E	CEBaG	CEBaG$_\lambda$	σ	E	CEBaG	CEBaG$_\lambda$
MedGemma-4b-it								
SLAKE	71.1	51.6	71.9	**73.5**	74.8	64.9	**75.9**	74.6
PathVQA	67.8	63.4	71.9	**77.5**	55.6	57.0	**59.0**	61.7
MedGemma-1.5-4b-it								
VQA-RAD (Open)	68.2	63.9	72.9	**75.7**	44.7	40.8	47.2	**47.0**
PathVQA	57.9	70.8	66.6	**77.8**	42.6	52.2	48.8	**56.3**
HuatuoGPT-Vision-7B								
PathVQA	80.6	67.3	**82.0**	82.5	55.3	48.3	**55.6**	55.5
VQA-RAD (All)	68.0	53.8	67.7	**68.6**	**67.5**	59.6	67.6	66.9
LLaVA-Med-v1.5-Mistral-7B								
VQA-RAD (All)	62.9	52.0	62.1	**66.2**	50.8	45.6	50.0	**54.0**
SLAKE	65.2	51.8	63.8	**68.8**	52.7	46.8	52.3	**55.3**
Average (all 16)	67.0	59.0	67.9	**71.9**	53.8	50.5	57.2	**58.8**

4 Conclusion

We introduced CEBaG, a deterministic answer-level hallucination detection method for medical VQA combining token-level predictive variance and evidence magnitude. Unlike sampling-based approaches (SE, VASE) that need 10–20 stochastic generations and an external NLI model, CEBaG uses only three forward passes. Across four medical MLLMs and 16 settings, it achieves the best AUC in 13 and surpasses state-of-the-art VASE by 8 AUC points on average at much lower cost. Its efficiency, simplicity, and zero-configuration design make it practical for real-time clinical deployment; its main limitation is requiring white-box access to log-probabilities. Future work can extend CEBaG to black-box settings via output perturbation and to multi-turn clinical dialogue. Because both signals aggregate over the full response, CEBaG operates at the whole-answer level; extending the per-token gain into a span-level signal for long-form outputs like report generation is a natural next direction. GREEN-generated labels, used for comparability with prior work, also make validation on a clinician-annotated subset an important next step.

Acknowledgment. This work was partially supported by the NIH Grants AG089169, AG084471, Stanford HAI Hoffman-Yee Award & GCP Credits, and UIT. M. Asadi is supported by the Amazon AI PhD and the Stanford HAI Graduate Fellowships.

Disclosure of Interests. The authors have no competing interests to declare.

References

1. Chen, C., et al.: INSIDE: LLMs' internal states retain the power of hallucination detection. In: ICLR (2024)
2. Chen, J., et al.: HuatuoGPT-Vision, towards injecting medical visual knowledge into multimodal LLMs at scale. arXiv:2406.19280 (2024)
3. Cohen, R., Hamri, M., Geva, M., Globerson, A.: LM vs LM: Detecting factual errors via cross examination. In: EMNLP, pp. 12621–12640 (2023)
4. Farquhar, S., Kossen, J., Kuhn, L., Gal, Y.: Detecting hallucinations in large language models using semantic entropy. Nature **630**(8017), 625–630 (2024)
5. Gunjal, A., Yin, J., Bas, E.: Detecting and preventing hallucinations in large vision language models. In: AAAI, vol. 38, pp. 18135–18143 (2024)
6. He, P., Liu, X., Gao, J., Chen, W.: DeBERTa: Decoding-enhanced BERT with disentangled attention. In: ICLR (2021)
7. He, X., Zhang, Y., Mou, L., Xing, E., Xie, P.: PathVQA: 30000+ questions for medical visual question answering. arXiv:2003.10286 (2020)
8. Huang, L., et al.: A survey on hallucination in large language models: Principles, taxonomy, challenges, and open questions. ACM Trans. Inf. Syst. (2025)
9. Jiang, C., et al.: Hallucination augmented contrastive learning for multimodal large language model. In: CVPR, pp. 27036–27046 (2024)
10. Li, Q., Geng, J., Lyu, C., Zhu, D., Panov, M., Karray, F.: Reference-free hallucination detection for large vision-language models. In: Findings of the Association for Computational Linguistics: EMNLP 2024, pp. 4542–4551 (2024)
11. Kossen, J., Han, J., Razzak, M., Schut, L., Malik, S., Gal, Y.: Semantic entropy probes: Robust and cheap hallucination detection in LLMs. arXiv:2406.15927 (2024)
12. Leng, S., et al.: Mitigating object hallucinations in large vision-language models through visual contrastive decoding. In: CVPR (2024)
13. Manevich, A., Tsarfaty, R.: Mitigating hallucinations in large vision-language models via language-contrastive decoding. In: ACL Findings, pp. 6008–6022 (2024)
14. Lau, J.J., Gayen, S., Ben Abacha, A., Demner-Fushman, D.: A dataset of clinically generated visual questions and answers about radiology images. Sci. Data **5**(1), 180251 (2018)
15. Li, C., et al.: LLaVA-Med: Training a large language-and-vision assistant for biomedicine in one day. NeurIPS **36** (2023)
16. Liao, Z., Hu, S., Zou, K., Fu, H., Zhen, L., Xia, Y.: Vision-amplified semantic entropy for hallucination detection in medical visual question answering. In: MICCAI (2025)
17. Liu, B., Zhan, L.M., Xu, L., Ma, L., Yang, Y., Wu, X.M.: SLAKE: A semantically-labeled knowledge-enhanced dataset for medical visual question answering. In: ISBI, pp. 1650–1654 (2021)
18. Liu, B., et al.: GEMeX: A large-scale, groundable, and explainable medical VQA benchmark for chest X-ray diagnosis. arXiv:2411.16778 (2024)
19. Liu, H., et al.: A survey on hallucination in large vision-language models. arXiv:2402.00253 (2024)
20. Ostmeier, S., et al.: GREEN: Generative radiology report evaluation and error notation. arXiv:2405.03595 (2024)
21. Sellergren, A., et al.: MedGemma technical report. arXiv:2507.05201 (2025)
22. Zhang, S., Sambara, S., Banerjee, O., Acosta, J., Fahrner, J., Rajpurkar, P.: RadFlag: A black-box hallucination detection method for medical vision language models. arXiv:2411.00299 (2024)

23. Wang, X., Pan, J., Ding, L., Biemann, C.: Mitigating hallucinations in large vision-language models with instruction contrastive decoding. In: ACL, pp. 15840–15853 (2024)
24. Xiao, H., et al.: A comprehensive survey of large language models and multimodal large language models in medicine. arXiv:2405.08603 (2024)
25. Yin, S., et al.: Woodpecker: Hallucination correction for multimodal large language models. SCIENCE CHINA Inf. Sci. **67**(12), 220105 (2024)
26. Yu, Q., et al.: HalluciDoctor: Mitigating hallucinatory toxicity in visual instruction data. In: CVPR, pp. 12944–12953 (2024)

EchoVQA: Enabling Conversational Assistance for Point-of-Care Cardiac Ultrasound

Filippos Bellos[1(✉)], Yutong Li[1], Jessie N. Dong[2], Zaiyang Guo[2], Emily Mackay[2], Yayuan Li[1], Yannis Avrithis[3], Alison Pouch[2], and Jason J. Corso[1]

[1] University of Michigan, Michigan, USA
{fbellos,jjcorso}@umich.edu
[2] University of Pennsylvania, Philadelphia, USA
[3] Athens, Greece

Abstract. Point-of-care transthoracic echocardiography enables cardiac assessment in virtually any clinical setting, yet its diagnostic utility remains constrained by the expertise required for image acquisition and interpretation. Visual question answering (VQA) offers a promising paradigm for bridging this expertise gap through interactive clinical assistance, but existing echocardiography VQA datasets are limited in scale, restricted to high-quality images, and only cover a few views. We introduce EchoVQA, the first large-scale VQA dataset for echocardiography, comprising 14,299 images and 74,819 question-answer pairs. The dataset integrates public sources (EchoNet-Dynamic, CAMUS) with our own point-of-care acquisitions from two handheld probes (Lumify, Clarius), spanning diverse views and including both high-quality and suboptimal images. Uniquely, EchoVQA includes acquisition guidance questions to help users optimize transducer positioning toward a diagnostic apical 4-chamber view for left ventricular ejection fraction estimation—a challenging task for novice operators in point-of-care settings. We further develop a parameter-efficient method based on multimodal learnable prompts achieving state-of-the-art performance on most benchmarks, including EchoVQA, with significantly fewer trainable parameters than existing state-of-the-art approaches. Data and code: https://huggingface.co/datasets/filbel/echoVQA.

Keywords: Point-of-Care Echocardiography · Visual Question Answering · Multimodal Large Language Models · VQA Benchmark

1 Introduction

Transthoracic echocardiography (TTE) is the first-line imaging modality for assessing cardiac function due to its portability, lack of ionizing radiation, and

Y. Li, J. N. Dong and Z. Guo—Contributed equally.

Y. Avrithis—Independent Scientist.

G. Yang et al. (Eds.): MICCAI 2026, LNCS 16878, pp. 141–151, 2027.
https://doi.org/10.1007/978-3-032-38059-3_14

ease of rapid acquisition [2]. Point-of-care (POC) TTE platforms—comprising handheld transducers connected to smartphones or tablets—have expanded cardiac imaging access to resource-limited settings [9]. However, interpreting echocardiographic images remains highly operator-dependent, requiring simultaneous identification of standard views, assessment of image quality, evaluation of chamber visibility, and determination of suitability for downstream measurements such as left ventricular ejection fraction (LVEF). LVEF is a key metric of systolic function commonly assessed from the apical 4-chamber (A4C) view, yet optimizing transducer positioning for high-quality A4C acquisition is challenging for novice users [1,17]. In POC settings, trained sonographers are often unavailable, and clinicians must acquire and interpret images with limited expertise and computational resources.

VQA has emerged as a powerful paradigm bridging vision and language processing to provide interactive assistance for medical image analysis [13]. Several medical benchmark datasets have been introduced: VQA-RAD [10] provides 3,515 QA pairs across radiology images, SLAKE [14] offers bilingual physician-verified annotations, PathVQA [5] contributes 32,799 pairs for pathology, and recent large-scale efforts such as MedTrinity-25M [26] provide millions of image-text pairs across diverse modalities. However, none specifically target echocardiography. To our knowledge, the only existing echocardiography VQA benchmark [23] provides just 622 QA pairs from high-quality hospital acquisitions, insufficient for training and lacking acquisition guidance for POC settings.

Enabling conversational assistance for medical imaging requires not only comprehensive datasets but also capable models. Multimodal large language models (MLLMs) are a promising solution. LLaVA-Med [12] adapts general-domain vision-language models to biomedicine through curriculum learning on figure-caption pairs followed by instruction-tuning. LLaVA-Tri [26] further improves performance through pretraining on MedTrinity-25M's multigranular annotations. However, these approaches require training 7–8B parameters, posing significant computational demands. Parameter-efficient fine-tuning (PEFT) methods aim to reduce this burden—PeFoMed [4] fine-tunes an MLLM with LoRA using only 33M trainable parameters—yet underperforms fully fine-tuned approaches.

In this work, we address these challenges with two contributions. First, we introduce EchoVQA, the largest VQA dataset for echocardiography, comprising 14,299 images and 74,819 question-answer pairs, featuring multi-turn conversations covering view classification, image quality assessment, chamber visibility, LVEF estimation feasibility, and acquisition guidance. Second, we propose a parameter-efficient method for adapting multimodal large language models to medical VQA, combining visual adaption prompts with text prompt learning [16]—previously unexplored in multimodal medical settings—and instance-level prompt optimization. Our method achieves state-of-the-art performance on most Med-VQA benchmarks with significantly fewer trainable parameters than LLaVA-Med (7B) and LLaVA-Tri (8B), while substantially outperforming PeFoMed.

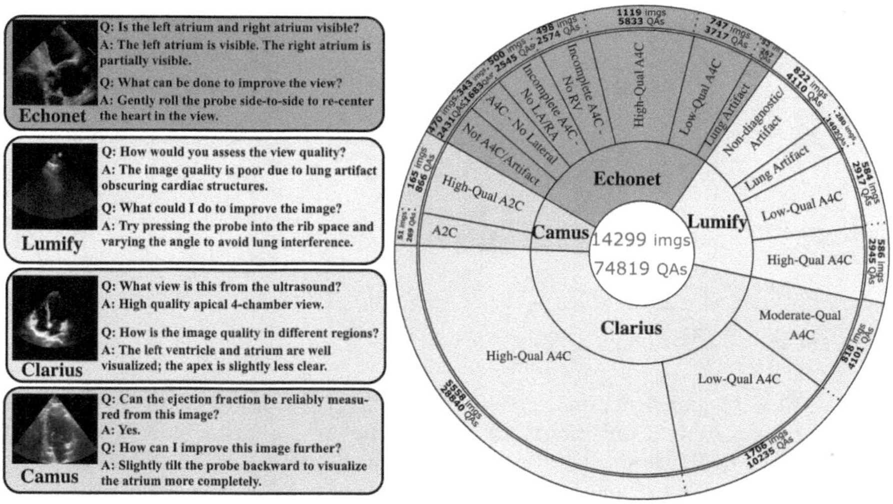

Fig. 1. EchoVQA dataset overview. **Left**: Example QA pairs from each data source demonstrating the different question types. **Right**: Distribution of images and QA pairs across data sources (inner ring) and quality categories (outer ring). The dataset comprises 14,299 images and 74,819 QA pairs spanning public (EchoNet-Dynamic, CAMUS) and our POC (Lumify, Clarius) acquisitions.

2 EchoVQA Dataset

2.1 Data Collection and Pre-processing

We collect echocardiographic images from two complementary sources: public datasets and our own POC acquisitions.

Public Data. Public datasets such as EchoNet-Dynamic [19] and CAMUS [11] are acquired by trained sonographers in hospital echo labs using cart-based systems. Our inspection reveals substantial variation within EchoNet-Dynamic, including non-standard views (e.g. A4C missing structures), lung artifacts, etc.) (Fig. 1). We explicitly examine and categorize these variations to enable targeted question-answer generation.

Point-of-Care Data. To complement public data with realistic POC acquisitions, typically performed with handheld probes by non-specialist clinicians, we collect additional scans using the Philips Lumify and Clarius PAL HD3, under institutional IRB approval. Using the Lumify, two operators performed scans on 3 subjects, and using the Clarius, scans were collected from 9 subjects (12 total; 7F/5M; age 23.5–40.6; 3 White, 9 Asian).These acquisitions capture the full spectrum of image quality a novice sonographer would encounter, stratified into five categories reflecting the progression toward a diagnostic A4C view

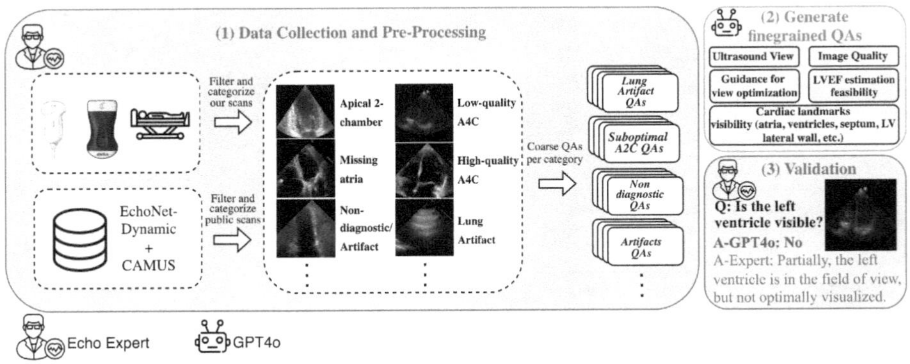

Fig. 2. EchoVQA construction pipeline. (1) Echocardiography experts categorize images from our POC acquisitions and public datasets into view and quality-based categories, then define coarse QA templates for each category. (2) GPT-4o generates diverse, image-specific QA pairs conditioned on category labels and expert templates, covering cardiac landmark visibility, LVEF estimation feasibility, ultrasound view classification, acquisition guidance, and image quality. (3) Experts review all generated QAs, correcting inaccurate or poorly grounded responses to ensure clinical accuracy.

for LVEF estimation: (1) *high-quality A4C*—all four chambers clearly visible, LVEF readily measurable; (2) *moderate-quality A4C*—partial chamber visibility with medium image quality, LVEF generally measurable; (3) *low-quality A4C*—recognizable A4C but compromised by suboptimal visualization, missing structures, off-axis orientation, or poor image settings, LVEF typically not feasible; (4) *lung artifact*—rib shadows or pleural interfaces obscuring cardiac structures; and (5) *non-diagnostic/artifacts*—inadequate for clinical interpretation due to severe artifacts or cardiac structures not in view. The specific nuances within each category are captured in the corresponding QA pairs. We adopt image-level rather than video-level QA because corrective guidance is frame-specific—differing across frames of a single clip—and composes with downstream video LVEF models; we target A4C as the canonical and hardest-to-acquire POC view for LVEF.

2.2 Question-Answer Generation

We develop a human-in-the-loop annotation pipeline that ensures clinical accuracy while enabling efficient scaling through LLM assistance (Fig. 2).

Category-Level Expert Templates. For each image category, echocardiography experts define a set of template question-answer pairs that establish the clinical scope and ground truth for that category. These templates encode expert knowledge about quality assessment, expected chamber visibility, LVEF estimation feasibility, and acquisition guidance—ranging from correcting suboptimal views toward a diagnostic A4C (e.g., "Adjust the probe position laterally and angle it toward the patient's left shoulder") to further optimizing already diagnostic images (e.g., "Rotate the probe slightly clockwise to better visualize the septum and right atrium"). These expert-authored templates serve as conditioning context for scalable QA generation.

Fine-Grained QA Generation. Using GPT-4o, we expand each category's expert-defined templates into diverse, image-specific question-answer pairs, conditioning on the category label, the templates, and the image. Categories are assigned by experts, not the model. Critically, while templates provide clinical grounding, the model is prompted to analyze each image directly, as substantial variability exists even within the same category. For instance, images categorized as low-quality A4C may exhibit, among other issues, low contrast obscuring chamber boundaries, partial left ventricular cutoff compromising LVEF estimation while other chambers remain visible, or foreshortening that distorts the true apex position. The model generates questions spanning five types for each ultrasound image: view classification, chamber visibility assessment, image quality evaluation, LVEF estimation feasibility, and acquisition guidance for view optimization. This approach enables efficient scaling—generating thousands of QA pairs per category—while ensuring all generated content remains anchored to both expert-validated clinical knowledge and image-specific observations.

2.3 Expert Validation

Although MLLMs have shown promise in medical imaging, they remain imperfect at specialized medical tasks and may introduce systematic biases in generated content. It is therefore imperative that all generated QA pairs undergo expert review. All 74,819 QA pairs were reviewed by four echocardiography experts; we prioritized full-coverage correctness review over inter-rater agreement on a subset. For each pair, reviewers verified factual correctness, ensured answers were grounded in the specific image rather than reflecting generic category-level knowledge, and corrected or rewrote responses that failed to meet these criteria. They further refined answer phrasing for clinical precision and rejected questions that were ambiguous or inappropriate for a given image. This validation step ensures that despite LLM-assisted generation, the final dataset meets the quality standards expected for clinical use.

3 Parameter-Efficient Medical VQA

Current medical VQA methods operate at a single level of adaptation—modifying either full model weights or low-rank subspaces uniformly. We instead propose a parameter-efficient multi-level prompt framework that injects task-relevant information at three complementary levels: (1) visual adaption prompts [27] that inject visual information into the language model, (2) dataset-level text prompts that encode shared domain knowledge via prompt learning [16]—previously unexplored in multimodal medical settings, and (3) instance-level prompts optimized at inference to capture image-specific context.

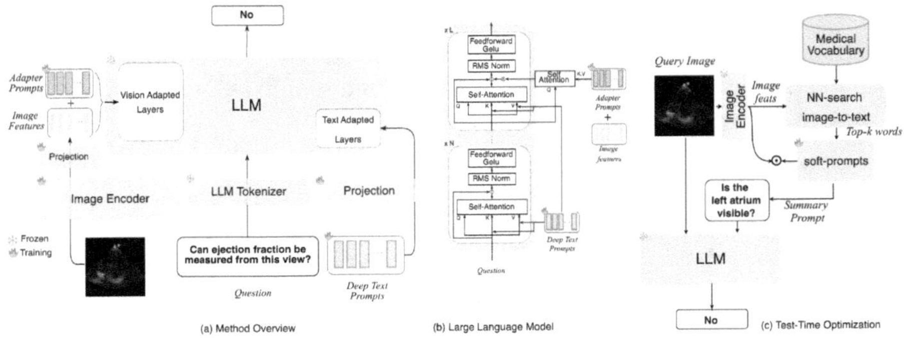

Fig. 3. Proposed method. **(a)** Training pipeline with visual adaption prompts injected into top LLM layers. **(b)** Dataset-level text prompts added to keys and values in the layer's self-attention. **(c)** At inference, instance-level prompts are retrieved from a medical vocabulary, optimized, and injected as a summary prompt.

3.1 Model Architecture

The proposed method and model architecture are illustrated in Fig. 3.

Vision Encoder. The input echocardiography image X^v is encoded by a vision encoder f^v to obtain visual features $Z^v := f^v(X^v) \in \mathbb{R}^{D \times N_v}$, where D is the embedding dimension and N_v is the number of visual tokens. These features are projected to the LLM embedding space via a learnable projection $W_p \in \mathbb{R}^{D_L \times D}$, yielding $\tilde{Z}^v := W_p Z^v \in \mathbb{R}^{D_L \times N_v}$.

Text Embedding. The input question X^q is tokenized into a sequence of S tokens and mapped by the frozen LLM embedding layer f^t to obtain text embeddings $Z^t := f^t(X^q) = (z_1^t, \ldots, z_S^t) \in \mathbb{R}^{D_L \times S}$.

Visual Adaption Prompts. Following [27], we inject visual information into the LLM via learnable adaption prompts $P^a \in \mathbb{R}^{D_L \times N_a}$. The projected visual features are pooled into a global token and added to P^a, then prepended to input sequences at the top L' transformer layers. We use zero-initialized gating with learnable scalars $\alpha^{(l)}$ for stable training.

Dataset-Level Text Prompts. Beyond visual adaption prompts, we introduce learnable text prompts that encode domain-specific knowledge shared across the dataset. These prompts are prepended to the keys and values in the self-attention mechanism. Given an input sequence $H \in \mathbb{R}^{D_L \times K}$, the query, key, and value are computed as: $Q := W^Q H, \quad K := [P^K; W^K H], \quad V := [P^V; W^V H]$, where $W^Q, W^K, W^V \in \mathbb{R}^{D_L \times D_L}$ are the projection matrices, and $P^K, P^V \in \mathbb{R}^{D_L \times N_p}$ are learnable text prompt parameters with N_p prompt tokens. This formulation allows the model to attend to learned domain-specific context without modifying the output sequence length.

3.2 Training

We adopt a two-stage training strategy following prior work in medical multi-modal learning [4,12,26]. In **Stage 1**, we train on medical image-caption datasets to align visual representations with the language model's embedding space. In **Stage 2**, we fine-tune on downstream medical VQA datasets. The model is trained with the standard language modeling objective:

$$\mathcal{L} = -\sum_{i=1}^{|X^a|} \log p(x_i^a | X^v, X^q, x_{<i}^a; \Theta), \tag{1}$$

where Θ denotes all trainable parameters and $x_{<i}^a$ represents answer tokens preceding position i. During both stages, we update the vision encoder f^v, visual projection W_p, adaption prompts P^a, gating scalars $\{\alpha^{(l)}\}$, and text prompt key-value pairs $\mathcal{P}^t$, while keeping the LLM frozen.

3.3 Inference: Instance-Level Prompt Optimization

We introduce instance-level test time prompt optimization to inject image-specific context beyond the dataset-level prompts learned during training.

We construct a **medical vocabulary** $\mathcal{V} = \{v_1, v_2, \ldots, v_M\}$ containing M domain-specific terms from the train set of each downstream dataset. Each term v_m is encoded using BiomedCLIP's text encoder to obtain vocabulary embeddings $E^{\mathcal{V}} \in \mathbb{R}^{D \times M}$. Given a query image X^v, we retrieve the top-k most similar vocabulary terms via cosine similarity: $\mathcal{R} = \text{top-}k\left(\frac{Z^v \cdot E^{\mathcal{V}}}{\|Z^v\| \|E^{\mathcal{V}}\|}\right)$, where $\mathcal{R} = \{r_1, \ldots, r_k\}$ denotes the retrieved terms.

The retrieved terms $\mathcal{R}$ are tokenized and mapped to the LLM embedding space via the frozen embedding layer to obtain $P^{init} := f^t(\mathcal{R}) \in \mathbb{R}^{D_L \times N_r}$, where N_r is the total number of tokens. These embeddings initialize soft prompts P^{soft}, which are **optimized for T iterations** to better align with the visual features: $P^{soft*} = \arg\min_{P^{soft}} \mathcal{L}_{align}(P^{soft}, Z^v)$, where $\mathcal{L}_{align}$ is a cosine similarity loss. After optimization, we select the first retrieved term's embedding as a compact summary prompt capturing the dominant visual semantics. This single summary prompt is concatenated with the question embeddings $\hat{Z}^t = [P^{summary}; Z^t]$, and the final answer is generated by the LLM conditioned on visual features $\tilde{Z}^v$ and the augmented text embeddings $\hat{Z}^t$. The per-query overhead of test-time optimization consists of a single vocabulary retrieval and T forward passes through the frozen LLM to optimize the soft prompt, incurred once per image and independent of answer length.

4 Experiments

Implementation Details. We use BiomedCLIP [28] as our vision encoder and LLaMA-2-7B [24] as the language model. The LLM remains frozen during training while the vision encoder is fine-tuned. We set the number of visual adaption prompts $N_a = 10$ and text prompt tokens $N_p = 10$.

In Stage 1, we train on medical image-caption datasets (ROCO [20], ImageCLEF [7], MEDICAT [22], and MIMIC-CXR [8]). In Stage 2, we fine-tune on each downstream VQA dataset following [4,12,26]. We partition EchoVQA into 70/10/20 train/validation/test splits, stratified by data source and view-quality category to preserve the category and quality distribution. We report all results on the held-out test split. For test-time prompt optimization, we retrieve $k = 10$ terms from the medical vocabulary and optimize for $T = 100$ iterations. All experiments are conducted on 2 NVIDIA H100 GPUs.

Evaluation. We evaluate on three established medical VQA benchmarks: VQA-RAD [10], SLAKE [14], and PathVQA [5]. We additionally evaluate on our proposed EchoVQA benchmark. Following [12,26], we report recall for open-ended questions and exact match accuracy for closed-ended questions.

We compare against several state-of-the-art **baselines**: GPT-4V [18] as a zero-shot baseline; LLaVA [15] as a general-domain MLLM; LoRA finetuning (PeFoMed) [4,6], LLaVA-Med [12] and LLaVA-Tri [26]. Echo-specific models (EchoCLIP [3], EchoPrime [25]) are contrastive/retrieval encoders rather than generative VQA models, hence not comparable under our generative exact-match protocol, while recent generalist medical MLLMs (e.g., MedGemma [21]) fold benchmark splits into multi-stage tuning under looser LLM-as-judge metrics, precluding apples-to-apples comparison.

Table 1. Performance comparison on medical VQA benchmarks. We report recall for open-ended questions and exact match accuracy for closed-ended questions. All methods except GPT-4V have been reproduced by us, using their respective codebases.

Method	Params	VQA-RAD		SLAKE		PathVQA		EchoVQA	
		Open	Closed	Open	Closed	Open	Closed	Open	Closed
GPT-4V	–	39.5	78.9	33.6	43.6	–	–	–	–
LLaVA	7B	63.7	83.4	83.7	84.6	37.1	90.8	56.0	70.6
PeFoMed (LoRA)	33M	58.5	83.4	81.3	82.7	29.5	88.4	54.4	86.9
LLaVA-Med	7B	64.7	84.2	83.0	85.3	**38.5**	91.2	55.5	82.2
LLaVA-Tri	8B	65.1	84.9	82.3	83.2	37.8	**91.8**	55.2	79.7
Ours	104M	**65.8**	**86.8**	**87.3**	**89.7**	37.2	91.3	**60.2**	**90.4**

Results. Table 1 presents results on established medical VQA benchmarks. Our method achieves state-of-the-art performance on VQA-RAD, SLAKE and EchoVQA, outperforming all baselines. On PathVQA we are competitive but not state-of-the-art. The likely cause is a domain gap: our radiology-focused Stage 1 pretraining transfers less well to PathVQA's diverse pathology images, and unlike fully fine-tuned baselines our frozen LLM relies more on the alignment learned during pretraining.

Table 2. Ablation study evaluating the contribution of visual adaption prompts (AP), dataset-level text prompts (TP), and test-time optimization (TTO).

Method	VQA-RAD		SLAKE		PathVQA		EchoVQA	
	Open	Closed	Open	Closed	Open	Closed	Open	Closed
AP	65.6	80.6	87.1	88.1	36.9	90.3	60.0	88.7
AP + TP	65.6	85.7	**87.3**	89.4	**37.2**	91.2	60.0	**90.4**
AP + TP + TTO (Full)	**65.8**	**86.8**	**87.3**	**89.7**	**37.2**	**91.3**	**60.2**	**90.4**

Table 2 isolates the contribution of each component. Visual adaption prompts alone provide a strong baseline. Adding dataset-level text prompts yields consistent gains on closed-ended questions across all benchmarks, with the largest improvement on VQA-RAD (+5.1 closed). Test-time optimization provides further incremental gains.

5 Conclusion and Future Work

We introduced EchoVQA, the first large-scale VQA dataset for echocardiography, spanning diverse image-quality levels and acquisition guidance for POC settings, together with a parameter-efficient method that achieves state-of-the-art performance on EchoVQA and most standard medical VQA benchmarks. Future work includes augmenting the POC cohort with additional subjects spanning broader demographics in order to enable better generalization, a novice-operator user study assessing acquisition-guidance efficacy, inter-rater agreement analysis, full FLOPs/latency profiling of test-time optimization, and extension beyond A4C to additional standard views.

Acknowledgments. This research was funded, in part, by the U.S. Government, under Agreement No. 1AY2AX000062. The views and conclusions contained in this document are those of the authors and should not be interpreted as representing the official policies, either expressed or implied, of the U.S. Government.

Disclosure of Interests. The authors have no competing interests to declare that are relevant to the content of this article.

References

1. Bick, J.S., et al.: Comparison of expert and novice performance of a simulated transesophageal echocardiography examination. Simul. Healthc. J. Soc. Simul. Healthc. **8**(5), 329–334 (2013)
2. Cheitlin, M.D., et al.: ACC/AHA Guidelines for the Clinical Application of Echocardiography. Circulation **95**(6), 1686–1744 (1997)
3. Christensen, M., Vukadinovic, M., Yuan, N., Ouyang, D.: Vision-language foundation model for echocardiogram interpretation. Nat. Med. **30**(5), 1481–1488 (2024)

4. He, J., et al.: Pefomed: Parameter efficient fine-tuning of multimodal large language models for medical imaging. arXiv preprint arXiv:2401.02797 (2024)
5. He, X., et al.: Towards visual question answering on pathology images. In: Zong, C., Xia, F., Li, W., Navigli, R. (eds.) Proceedings of the 59th Annual Meeting of the Association for Computational Linguistics and the 11th International Joint Conference on Natural Language Processing (Volume 2: Short Papers), pp. 708–718. Association for Computational Linguistics, Online (2021)
6. Hu, E.J.: Lora: Low-rank adaptation of large language models. In: International Conference on Learning Representations (ICLR) (2022)
7. Ionescu, B., et al.: Overview of the imageclef 2022: Multimedia retrieval in medical, social media and nature applications. In: Barrón-Cedeño, A., et al. (eds.) Experimental IR Meets Multilinguality, Multimodality, and Interaction. Lecture Notes in Computer Science, vol. 13390, pp. 541–564. Springer, Cham (2022). https://doi.org/10.1007/978-3-031-13643-6_31
8. Johnson, A.E.W., et al.: Mimic-cxr, a de-identified publicly available database of chest radiographs with free-text reports. Sci. Data **6**, 317 (2019)
9. Kornelsen, J., Ho, H., Robinson, V., Frenkel, O.: Rural family physician use of point-of-care ultrasonography: experiences of primary care providers in British Columbia, Canada. BMC Primary Care **24**, 183 (2023)
10. Lau, J.J., Gayen, S., Ben Abacha, A., Demner-Fushman, D.: A dataset of clinically generated visual questions and answers about radiology images. Sci. Data **5**, 180251 (2018)
11. Leclerc, S., et al.: Deep learning for segmentation using an open large-scale dataset in 2d echocardiography. IEEE Trans. Med. Imaging **38**(9), 2198–2210 (2019)
12. Li, C., et al.: Llava-med: Training a large language-and-vision assistant for biomedicine in one day. In: Advances in Neural Information Processing Systems (NeurIPS), Datasets and Benchmarks Track (2023)
13. Lin, Z., et al.: Medical visual question answering: A survey. Artif. Intell. Med. **143**, 102611 (2023)
14. Liu, B., Zhan, L.M., Xu, L., Ma, L., Yang, Y., Wu, X.M.: Slake: A semantically-labeled knowledge-enhanced dataset for medical visual question answering. In: 2021 IEEE 18th International Symposium on Biomedical Imaging (ISBI), pp. 1650–1654. IEEE (2021)
15. Liu, H., Li, C., Wu, Q., Lee, Y.J.: Visual instruction tuning. In: Advances in Neural Information Processing Systems (NeurIPS) (2023)
16. Liu, X., et alJ.: P-tuning: Prompt tuning can be comparable to fine-tuning across scales and tasks. In: Proceedings of the 60th Annual Meeting of the Association for Computational Linguistics (Volume 2: Short Papers), pp. 61–68. Association for Computational Linguistics (2022)
17. Mitchell, C., et al.: Guidelines for Performing a comprehensive transthoracic echocardiographic examination in adults: recommendations from the American society of echocardiography. J. Am. Soc. Echocardiography: Official Publ. Am. Soc. Echocardiograp. **32**(1), 1–64 (2019)
18. OpenAI, et al.: Gpt-4 technical report. arXiv preprint arXiv:2303.08774 (2023)
19. Ouyang, D., et al.: Video-based AI for beat-to-beat assessment of cardiac function. Nature **580**(7802), 252–256 (2020)
20. Pelka, O., Koitka, S., Rückert, J., Nensa, F., Friedrich, C.M.: Radiology objects in context (roco): A multimodal image dataset. In: Multimodal Learning for Clinical Decision Support and Clinical Image-Based Procedures (LABELS / CVII-STENT @ MICCAI). Lecture Notes in Computer Science, vol. 11043, pp. 180–189. Springer, Cham (2018)

21. Sellergren, A., Kazemzadeh, S., Jaroensri, T., Kiraly, A., Traverse, M.: MedGemma technical report (2026)
22. Subramanian, S.: Medicat: A dataset of medical images, captions, and textual references. In: Findings of the Association for Computational Linguistics: EMNLP 2020, pp. 2112–2120. Association for Computational Linguistics, Online (2020)
23. Thapa, R., Let al.: MIMIC-IV-ECHO-Ext-MIMICEchoQA: A Benchmark Dataset for Echocardiogram-Based Visual Question Answering. PhysioNet (2025). version 1.0.0
24. Touvron, H.: Llama 2: Open Foundation and Fine-Tuned Chat Models. ArXiv (2023)
25. Vukadinovic, M., et al.: Comprehensive echocardiogram evaluation with view-primed vision-language AI. Nature **650**(8103), 970–977 (2026)
26. Xie, Y., et al.: Medtrinity-25m: A large-scale multimodal dataset with multigranular annotations for medicine. In: International Conference on Learning Representations (ICLR) (2025)
27. Zhang, R.: LLaMA-Adapter: Efficient Fine-tuning of Language Models with Zero-init Attention (2023). publisher: arXiv Version Number: 3
28. Zhang, S.: Biomedclip: a multimodal biomedical foundation model pretrained from fifteen million scientific image-text pairs. arXiv preprint arXiv:2303.00915 (2023)

FairGE: Gated Expert Routing for Intersectional Fairness in Medical Foundation Models

Yuchen Shao[1], Yutong Xie[2], Jian Chen[1](✉), and Qi Chen[3](✉)

[1] South China University of Technology, Guangzhou, China
ellachen@scut.edu.cn
[2] Mohamed bin Zayed University of Artificial Intelligence, Abu Dhabi, UAE
[3] Australian Institute for Machine Learning, Adelaide University, Adelaide, Australia
qi.chen04@adelaide.edu.au

Abstract. Medical foundation models are increasingly used as generic visual backbones for downstream diagnosis, yet existing fairness studies in medical imaging mostly focus on a single protected attribute (e.g., age or sex) and rarely examine intersectional subgroups defined by multiple demographics. Compared to single-attribute settings, multi-attribute fairness introduces additional challenges: (i) intersectional subgroup data sparsity, where some demographic combinations contain very few samples, making subgroup-specific adaptation unstable and prone to overfitting; and (ii) sensitive attribute leakage (proxy discrimination), where learned representations retain correlated demographic information and allow models to rely on demographic shortcuts rather than clinically relevant visual features. To address this, we propose Fair Gated-Expert adaptation with multi-attribute gating (**FairGE**), a fairness-aware adaptation framework for medical foundation models. FairGE integrates two complementary components: a Multi-attribute Gated distribution-aware Mixture-of-Experts (MA-dMoE) module that employs shared experts with multi-attribute gating to enable intersectional adaptation while avoiding overfitting on small subgroups, and a Multi-attribute Purification (MAP) module that combines a lightweight MLP with multiple adversarial attribute classifiers (one for each sensitive attribute) trained via gradient reversal to prevent the model from encoding sensitive demographic information in the final representation. Experiments on two representative medical benchmarks, HAM10000 and Harvard-FairVLMed, show that FairGE consistently reduces both marginal and intersectional disparities, improves worst-group performance, and preserves overall AUC. Code is available at https://github.com/yuchenshao749/FairGE.

Keywords: Medical foundation models · Fairness · Intersectional fairness · Mixture-of-Experts · Medical multimodal large language models

Y. Shao and Y. Xie—Contributed equally.

G. Yang et al. (Eds.): MICCAI 2026, LNCS 16878, pp. 152–162, 2027.
https://doi.org/10.1007/978-3-032-38059-3_15

1 Introduction

Foundation models have recently transformed medical image analysis, enabling a single pretrained model to support many downstream tasks with modest task-specific data [12,26]. This trend also extends to medical multimodal large language models (MLLMs), where a visual encoder is paired with a language model for clinical reasoning and report generation [1,8,21]. As such models are increasingly used as generic visual feature extractors in diagnostic pipelines, demographic fairness becomes a central concern: unequal performance across patient groups can undermine clinical trust and exacerbate existing health disparities [13,17].

Motivated by these concerns, recent work has begun to evaluate and improve fairness in medical imaging foundation models. FairMedFM [6] benchmarks a broad range of vision foundation models across datasets and sensitive attributes under common usage modes (e.g., zero-shot, linear probing, and parameter-efficient fine-tuning). FairCLIP [10] introduces the Harvard-FairVLMed glaucoma benchmark with fine-grained demographic annotations and studies fairness of CLIP/BLIP-2 style vision-language models [9,15]. At the method level, prior fairness-aware adaptation work includes distribution-aware Mixture-of-Experts (dMoE) [14] for medical image segmentation and MoE-based fairness designs for medical vision-language models [3,4,18,22], highlighting the promise of attribute-aware routing for fairer medical AI systems.

Despite this progress, existing studies are still largely centered on single-attribute fairness, such as considering only one of age, gender, or race. In real-world clinical deployment, however, each patient is characterized by multiple demographics simultaneously, and fairness risks often concentrate in small intersectional groups defined by attribute combinations (e.g., age-sex subgroups) [16,17,25]. This makes multi-attribute fairness a more practical and important target. For this setting, two challenges are particularly critical: (1) intersectional subgroup data sparsity, where some demographic combinations contain very few samples, making subgroup-specific adaptation unstable and prone to overfitting; and (2) sensitive attribute leakage (proxy discrimination), where learned visual representations retain correlated demographic information and allow models to rely on demographic shortcuts rather than clinically relevant visual features.

To address these challenges, we propose Fair Gated-Expert adaptation with multi-attribute gating (**FairGE**), a fairness-aware adaptation framework for frozen medical foundation model visual encoders. FairGE consists of two complementary modules tailored to multi-attribute fairness. First, we introduce a Multi-attribute Gated distribution-aware Mixture-of-Experts (MA-dMoE) module that inserts lightweight dMoE adapters into the last encoder blocks and uses shared experts with multi-attribute gating: for each demographic attribute (e.g., age and sex) we maintain a separate attribute-specific router and, for a given patient, aggregate the routing distributions from all active attributes. This improves intersectional adaptation while avoiding overfitting from training separate adapters for sparse intersectional groups. Second, we introduce

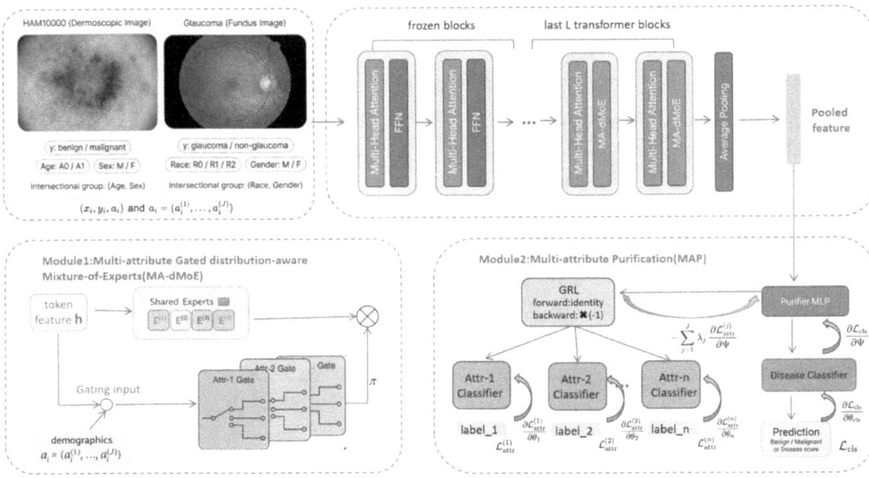

Fig. 1. Overview of **FairGE** with two modules: MA-dMoE inserts multi-attribute gated dMoE adapters into the last few visual-encoder blocks for intersectional adaptation, and MAP applies a Purifier MLP with adversarial attribute classifiers to suppress demographic leakage before disease classification.

a Multi-attribute Purification (MAP) module that applies a compact Purifier MLP to the pooled visual feature and trains multiple adversarial attribute classifiers (one for each sensitive attribute) via gradient reversal to suppress sensitive demographic information in the final representation.We validate FairGE on two representative benchmarks derived from HAM10000 dermoscopy images and the Harvard-FairVLMed glaucoma dataset.

Our contributions are summarized as follows:

- We propose FairGE, a fairness-aware adaptation framework to address subgroup and intersectional fairness under multiple sensitive attributes.
- FairGE integrates MA-dMoE, which combines shared experts with multi-attribute gating to improve subgroup adaptation, and MAP, which uses a lightweight Purifier MLP and multiple adversarial attribute classifiers to suppress sensitive demographic leakage from the final visual representation.
- We conduct comprehensive experiments on two medical imaging benchmarks, showing consistent gains in overall and intersectional subgroup AUC, improved worst-group performance, and a strong fairness–AUC trade-off.

2 Method

Figure 1 illustrates our FairGE. Given a medical image and its demographic attributes (e.g., age and sex), we extract visual features using a pretrained medical foundation model encoder. FairGE then applies two complementary modules:

(1) MA-dMoE, inserted into the last few transformer blocks to improve intersectional adaptation via shared experts and multi-attribute demographic gating; and (2) MAP, which applies a compact Purifier MLP and multiple adversarial attribute classifiers to suppress sensitive demographic information in the final representation. The purified feature is finally fed to a disease classifier.

2.1 Multi-attribute Gated Distribution-Aware Mixture-of-Experts

As shown in Fig. 1, let $\mathcal{D} = \{(x_i, y_i, \mathbf{a}_i)\}_{i=1}^N$ be a labeled dataset, where x_i is a medical image, $y_i \in \{0,1\}$ is the disease label, and $\mathbf{a}_i = (a_i^{(1)}, \dots, a_i^{(J)})$ is a vector of J discrete demographic attributes (e.g., age group and sex), with $a_i^{(j)} \in \mathcal{A}^{(j)}$. We build on a pretrained ViT-style visual encoder and insert dMoE adapters into the last L transformer blocks, while keeping earlier blocks frozen.

For an adapted block ℓ, let $\mathbf{h}_{i,t}^{(\ell)} \in \mathbb{R}^d$ denote the token at position t (after self-attention and before the FFN). We replace the standard FFN with an MA-dMoE layer with E shared experts $\{E_\ell^{(e)}\}_{e=1}^E$ and token-wise sparse routing weights $\pi_{i,t,\cdot}^{(\ell)} \in \mathbb{R}^E$ (Noisy Top-k routing [14]):

$$\tilde{\mathbf{h}}_{i,t}^{(\ell)} = \mathbf{h}_{i,t}^{(\ell)} + \sum_{e=1}^{E} \pi_{i,t,e}^{(\ell)} \, E_\ell^{(e)}\left(\mathbf{h}_{i,t}^{(\ell)}\right). \tag{1}$$

The experts are shared across all demographic groups; only the routing differs. This shared-expert design is parameter-efficient and helps avoid overfitting when some intersectional subgroups are under-represented.

To make routing aware of multiple demographics, for each attribute $j \in \{1, \dots, J\}$ and each group $g \in \mathcal{A}^{(j)}$, we define a group-specific router $G_\ell^{(j,g)}$. For sample x_i, the active router for attribute j is $G_\ell^{(j,a_i^{(j)})}$, which produces a token-wise expert distribution $\pi_{i,t}^{(\ell,j)} \in \mathbb{R}^E$ after Noisy Top-k routing and softmax. We then combine the attribute-specific gates using non-negative weights $\{\alpha_j\}_{j=1}^J$:

$$\tilde{\boldsymbol{\pi}}_{i,t}^{(\ell)} = \sum_{j=1}^{J} \alpha_j \, \pi_{i,t}^{(\ell,j)}, \qquad \pi_{i,t,e}^{(\ell)} = \frac{\tilde{\pi}_{i,t,e}^{(\ell)}}{\sum_{e'=1}^{E} \tilde{\pi}_{i,t,e'}^{(\ell)}}. \tag{2}$$

The fused weights $\pi_{i,t,e}^{(\ell)}$ are used in Eq. (1). This allows FairGE to condition expert usage on multiple attributes jointly (e.g., age and sex) without training a separate adapter for each intersectional subgroup.

Let $\mathbf{H}_i \in \mathbb{R}^{T \times d}$ denote the final token sequence output by the adapted encoder. We obtain the pooled visual representation by

$$\mathbf{z}_i = \mathrm{Pool}(\mathbf{H}_i) \in \mathbb{R}^d. \tag{3}$$

2.2 Multi-attribute Purification

Although multi-attribute gating improves subgroup-specific adaptation, the pooled feature may still encode sensitive demographic information and induce shortcut-based predictions. To mitigate this, we apply a lightweight Purifier MLP $p_\psi : \mathbb{R}^d \to \mathbb{R}^d$ to obtain a purified feature

$$\mathbf{z}'_i = p_\psi(\mathbf{z}_i), \tag{4}$$

and use a disease classifier head $g_{\theta_{cls}} : \mathbb{R}^d \to [0,1]$ for the main task:

$$\hat{y}_i = g_{\theta_{\text{cls}}}(\mathbf{z}'_i), \qquad \mathcal{L}_{\text{cls}} = \frac{1}{|\mathcal{B}|} \sum_{i \in \mathcal{B}} \text{BCE}(\hat{y}_i, y_i). \tag{5}$$

To suppress multi-attribute sensitive attribute leakage, we attach one adversarial classifier for each sensitive attribute. For attribute j, let $h^{(j)}_{\theta_j} : \mathbb{R}^d \to \mathbb{R}^{|\mathcal{A}^{(j)}|}$ be an attribute predictor (e.g., age classifier, sex classifier), and let $\mathcal{R}(\cdot)$ denote a gradient reversal layer (GRL [2]), which acts as the identity in the forward pass and multiplies the backward gradient by -1. The attribute loss is

$$\mathcal{L}^{(j)}_{\text{attr}} = \frac{1}{|\mathcal{B}|} \sum_{i \in \mathcal{B}} \text{CE}\left(h^{(j)}_{\theta_j}(\mathcal{R}(\mathbf{z}'_i)), a^{(j)}_i \right). \tag{6}$$

The overall training objective is

$$\mathcal{L} = \mathcal{L}_{\text{cls}} + \sum_{j=1}^{J} \lambda_j \mathcal{L}^{(j)}_{\text{attr}}, \tag{7}$$

where $\lambda_j \geq 0$ is a weighting coefficient for the j-th adversarial branch. As shown in Fig. 1, the GRL preserves the input in the forward pass and reverses the sign of the gradient in the backward pass. Consequently, each attribute classifier $h^{(j)}_{\theta_j}$ is optimized to predict the corresponding sensitive attribute $a^{(j)}_i$, whereas the gradient propagated to the upstream feature extractor (including the adapted encoder and the Purifier MLP) has the opposite sign, thereby encouraging the final representation $\mathbf{z}'_i$ to discard sensitive-attribute information. The coefficient λ_j controls the relative strength of the j-th adversarial signal in the overall objective.

In all experiments, we initialize the visual encoder from the medical foundation model, train only the last L dMoE-adapted blocks (experts and routers), the Purifier, the disease classifier, and the adversarial attribute classifiers, and keep earlier encoder blocks frozen.

3 Experiments

3.1 Datasets and Experimental Setup

We evaluate on two benchmarks with demographic annotations. HAM10000 [20] contains 9948 dermoscopic images of pigmented skin lesions; we follow prior work and cast it as a binary benign vs. malignant classification task with age and sex as protected attributes. We construct patient-level train/val/test splits, where

Table 1. Sample counts of intersectional demographic subgroups in HAM10000 and Harvard-FairVLMed. For HAM10000, A0/A1 denote age ≤ 60 / > 60.

Dataset	Group	Male	Female	Total
HAM10000	A0	3525	3624	7149
	A1	1875	924	2799
	Total	5400	4548	9948
Harvard-FairVLMed	Asian	393	426	819
	Black	658	833	1491
	White	3318	4372	7690
	Total	4369	5631	10000

age and sex are considered during the splitting process to encourage demographic coverage across splits. Harvard-FairVLMed [10] provides 10,000 fundus images with rich demographic annotations; we construct a binary glaucoma classification task from its glaucoma label and use race and gender as protected attributes.

Table 1 reports the number of samples in each intersectional subgroup used in our experiments. For HAM10000, we summarize age–sex intersections, and for Harvard-FairVLMed, we summarize race–gender intersections.

We first use linear probing with a frozen backbone as a diagnostic baseline to reveal fairness properties of the pretrained visual representation. We then evaluate our proposed FairGE, which performs parameter-efficient adaptation with MA-dMoE and MAP.

We report overall AUC and accuracy on the test sets. For fairness, we evaluate intersectional age-sex subgroups using the max subgroup AUC gap (AUC-gap) and equity-scaled AUC (ES-AUC) [11], and additionally report a difference-in-equalized-odds score (DEOdds) [5] on HAM10000.

3.2 Results on Fairness-Aware Adaptation

We first consider a simple baseline called "Base", which removes any explicit fairness intervention (*i.e.,* without MA-dMoE and MAP for our method). Besides, we compare FairGE against three groups of fairness baselines: (1) Traditional fairness methods: group reweighting (Reweight), and an adversarial classifier (AdvClassifier) that predicts sensitive attributes from intermediate features [7,24]; (2) Recent debiasing methods: FairSDE [19] and a VAE-based debiasing approach [28] applied to foundation model embeddings; and (3) CLIP-based fairness methods: on HAM10000, we additionally compare with CMAC-MMD and DualFairVL [23,27], using their finetuned visual backbones as frozen feature extractors and training a linear classifier on top.

Across both datasets, we observe a consistent pattern. On HAM10000 (Table 2), Base achieves strong overall AUC (90.56) but suffers from a large intersectional gap (AUC-gap 11.98). Traditional debiasing methods offer only modest gains for these groups and leave substantial disparities. In contrast, our FairGE

Table 2. Fairness results on HAM10000 (binary benign vs. malignant). We report overall AUC, the maximum AUC gap across the four age–sex intersectional subgroups (AUC-gap), equity-scaled AUC (ES-AUC), DEOdds, and AUC for each subgroup. Here, A0/A1 denote age ≤ 60 / > 60, and M/F denote male/female.

Method	Overall AUC	Fairness metrics			age-sex intersection			
		AUC-gap	ES-AUC	DEOdds	A0-M	A0-F	A1-M	A1-F
Base	90.56±0.84	11.98±1.36	75.89±1.47	20.63±1.81	94.16	87.79	85.98	82.18
Reweight	90.40±0.92	10.79±1.18	76.29±1.52	17.70±1.64	93.79	87.43	85.67	83.00
AdvClassifier	90.44±0.76	11.22±1.27	76.64±1.43	17.94±1.58	93.70	87.74	86.36	82.48
FairSDE	90.59±0.69	8.71±1.04	77.10±1.29	27.17±1.91	93.77	85.58	86.82	85.06
VAE-debias	88.81±1.12	5.13±0.88	80.07±1.16	15.69±1.38	90.39	85.50	85.26	86.34
DualFairVL	89.93±0.95	11.17±1.45	75.35±1.67	27.83±1.96	93.75	87.64	82.58	84.04
CMAC-MMD	91.09±0.73	5.88±0.94	83.62±1.08	21.50±1.72	92.57	91.56	86.69	88.51
FairGE (ours)	**92.07±0.61**	**3.19±0.57**	**85.81±0.92**	**10.77±1.05**	**92.53**	**90.42**	**89.32**	**89.64**

Table 3. Fairness results on Harvard-FairVLMed glaucoma classification. We report overall AUC, AUC-gap, ES-AUC, and AUC for the six race–gender intersectional subgroups. Here, R0/R1/R2 denote Asian/Black/White, and M/F denote male/female.

Method	Overall AUC	Fairness metrics		Race–gender intersection					
		AUC-gap	ES-AUC	R0-M	R0-F	R1-M	R1-F	R2-M	R2-F
Base	76.03±0.86	14.76±1.42	61.31±1.58	77.36	80.39	78.26	65.63	79.51	73.83
Reweight	75.88±0.91	13.35±1.27	61.49±1.46	73.56	78.91	73.96	67.11	80.46	73.10
AdvClassifier	76.80±0.78	14.90±1.53	62.17±1.39	78.52	82.09	77.44	67.19	80.72	74.45
FairSDE	76.86±0.83	15.80±1.68	61.32±1.62	73.78	81.86	76.23	66.06	80.40	74.57
VAE-debias	76.06±1.04	12.56±1.21	61.59±1.34	82.32	78.91	74.00	69.76	79.93	73.90
FairGE (ours)	**79.08±1.13**	**7.53±0.82**	**68.03±1.06**	**78.09**	**77.21**	**75.35**	**73.93**	**81.46**	**76.96**

adaptation markedly reduces the gap (AUC-gap 3.19), substantially improves ES-AUC (from 75.89 to 85.81), and increases overall AUC to 92.07.

On Harvard-FairVLMed (Table 3), we observe similar long-tail behavior across race–gender intersections: Base attains reasonable overall AUC (76.03) but again exhibits a large subgroup gap (AUC-gap 14.76). FairGE consistently improves fairness, reducing the AUC-gap to 7.53 and raising ES-AUC from 61.31 to 68.03, together with an overall AUC gain to 79.08.

3.3 Ablation Studies

We conduct ablation studies on HAM10000 to examine the contribution of each component in FairGE. Specifically, we evaluate four ablated variants: (1) *w/o MA-dMoE*, which removes the dMoE adapters and keeps only the lightweight Purifier MLP on top of the frozen visual representation; (2) *w/o group-gating*,

Table 4. Component ablation on HAM10000 (binary benign vs. malignant). We report the same metrics as Table 2: overall AUC, AUC-gap, ES-AUC, DEOdds, and AUC for each age-sex intersectional subgroup. All values are reported in percentage form.

Method	AUC	AUC-gap	ES-AUC	DEOdds	A0-M	A0-F	A1-M	A1-F
Base	90.56	11.98	75.89	20.63	94.16	87.79	85.98	82.18
w/o MA-dMoE	90.78	9.42	78.08	18.63	93.16	88.20	86.51	83.74
w/o group-gating	91.05	7.16	80.67	15.84	92.84	88.92	87.47	85.68
w/o MAP	91.34	5.38	82.65	13.68	92.30	89.42	88.13	86.92
FairGE-SA	91.54	4.22	83.94	12.23	92.02	89.76	88.48	87.80
FairGE	**92.07**	**3.19**	**85.81**	**10.77**	**92.53**	**90.42**	**89.32**	**89.64**

Table 5. Effect of the number of adapted blocks L on HAM10000 data.

L	AUC-gap	ES-AUC
2	5.16	81.98
4	4.72	83.06
6	**4.22**	**83.94**
8	5.43	82.52
10	6.78	81.14

Table 6. Effect of age/sex routing weights $(\alpha_{\text{age}}, \alpha_{\text{sex}})$ on HAM10000 data.

$(\alpha_{\text{age}}, \alpha_{\text{sex}})$	AUC-gap	ES-AUC
(1.0, 0.0)	4.22	83.94
(0.9, 0.1)	3.90	84.86
(0.8, 0.2)	**3.19**	**85.81**
(0.7, 0.3)	3.62	84.24
(0.6, 0.4)	4.11	83.25

which keeps the expert FFNs but replaces demographic-specific gates with a single shared gate over the same expert pool; (3) *w/o MAP*, which removes the feature purification module and keeps only the dMoE-based adaptation; and (4) *FairGE-SA*, which disables multi-attribute gating and uses single-attribute (age-based) gating only. For fair comparison, all variants use the same training protocol and evaluation metrics as in Table 2.

As shown in Table 4, all components contribute to both accuracy and AUC-based intersectional fairness. Removing the dMoE adapters (*w/o MA-dMoE*) causes the largest drop in overall and subgroup AUC, confirming expert-based adaptation as the main driver of performance gains. Collapsing demographic-specific gates into a shared gate (*w/o group-gating*) still improves over Base but remains worse than the gated variants, underscoring the value of demographic-aware routing. Dropping the Purifier MLP (*w/o MAP*) leads to a modest yet consistent degradation in fairness metrics, indicating complementary debiasing effects beyond dMoE routing. Among the gated variants, *FairGE* with multi-attribute gating (age and sex) achieves the smallest AUC-gap and highest ES-AUC, outperforming the single-attribute *FairGE-SA*.

3.4 Discussion

To better understand the design choices in our FairGE adapter, we analyze (i) the number L of last transformer blocks adapted with dMoE in the single-attribute variant FairGE-SA, and (ii) the age/sex weights $(\alpha_{\text{age}}, \alpha_{\text{sex}})$ in the multi-attribute gating of FairGE. All experiments in this section are conducted on HAM10000.

As shown in Table 5, when applying FairGE-SA to different numbers of last blocks, fairness metrics improve as we increase the number of adapted blocks from L=2 to L=6, while further expanding adaptation to 8 or 10 blocks yields diminishing or even slightly worse AUC-gap and ES-AUC. We then use L=6 as a default for FairGE, which strikes a good balance between capacity and stability.

Table 6 studies the multi-attribute gating in FairGE. Since age exhibits larger disparities than sex in our main results, we restrict to settings with $\alpha_{\text{age}} \geq 0.6$. A higher weight to age is generally beneficial, and the best trade-off is achieved at $(\alpha_{\text{age}}, \alpha_{\text{sex}}) = (0.8, 0.2)$, which we adopt in our default FairGE configuration.

4 Conclusion

We investigated subgroup and intersectional fairness in medical image classification with frozen medical foundation model encoders on HAM10000 and Harvard-FairVLMed, and proposed **FairGE**, a lightweight fairness-aware adaptation framework built on MA-dMoE and MAP. Across both datasets, FairGE improves the overall fairness–AUC trade-off and yields more balanced performance across intersectional subgroups. Ablations further show that expert-based adaptation, multi-attribute gating, and feature purification provide complementary benefits. Future work will extend FairGE to more medical foundation models, broader protected attributes, and additional clinical tasks to further evaluate its generalizability and fairness in real-world deployment.

Acknowledgments. This research was funded by the National Natural Science Foundation of China (Grant No. 62376099).

Disclosure of Interests. The authors have no competing interests to declare that are relevant to the content of this article.

References

1. Chen, Q., et al.: Act like a radiologist: radiology report generation across anatomical regions. In: Proceedings of the Asian Conference on Computer Vision, pp. 1–17 (2024)
2. Ganin, Y., Lempitsky, V.: Unsupervised domain adaptation by backpropagation. In: International conference on machine learning, pp. 1180–1189. PMLR (2015)
3. Germino, J., Moniz, N., Chawla, N.V.: Fairness-aware mixture of experts with interpretability budgets. In: Bifet, A., Lorena, A.C., Ribeiro, R.P., Gama, J., Abreu, P.H. (eds.) International Conference on Discovery Science, pp. 341–355. Springer, Cham (2023). https://doi.org/10.1007/978-3-031-45275-8_23

4. Germino, J., Moniz, N., Chawla, N.V.: Fairmoe: counterfactually-fair mixture of experts with levels of interpretability. Mach. Learn. **113**(9), 6539–6559 (2024)
5. Hardt, M., Price, E., Srebro, N.: Equality of opportunity in supervised learning. Adv. Neural. Inf. Process. Syst. **29** (2016)
6. Jin, R., et al.: Fairmedfm: fairness benchmarking for medical imaging foundation models. Adv. Neural. Inf. Process. Syst. **37**, 111318–111357 (2024)
7. Kamiran, F., Calders, T.: Data preprocessing techniques for classification without discrimination. Knowl. Inf. Syst. **33**(1), 1–33 (2012)
8. Li, C., et al.: Llava-med: Training a large language-and-vision assistant for biomedicine in one day. Adv. Neural. Inf. Process. Syst. **36**, 28541–28564 (2023)
9. Li, J., Li, D., Savarese, S., Hoi, S.: Blip-2: Bootstrapping language-image pre-training with frozen image encoders and large language models. In: International Conference on Machine Learning, pp. 19730–19742. PMLR (2023)
10. Luo, Y., et al.: Fairclip: Harnessing fairness in vision-language learning. In: Proceedings of the IEEE/CVF Conference on Computer Vision and Pattern Recognition, pp. 12289–12301 (2024)
11. Luo, Y., et al.: Harvard glaucoma fairness: a retinal nerve disease dataset for fairness learning and fair identity normalization. IEEE Trans. Med. Imaging **43**(7), 2623–2633 (2024)
12. Moor, M., et al.: Foundation models for generalist medical artificial intelligence. Nature **616**(7956), 259–265 (2023)
13. Obermeyer, Z., Powers, B., Vogeli, C., Mullainathan, S.: Dissecting racial bias in an algorithm used to manage the health of populations. Science **366**(6464), 447–453 (2019)
14. Oh, Y., et al.: Distribution-aware fairness learning in medical image segmentation from a control-theoretic perspective. arXiv preprint arXiv:2502.00619 (2025)
15. Radford, A., et al.: Learning transferable visual models from natural language supervision. In: International Conference on Machine Learning, pp. 8748–8763. PMLR (2021)
16. Ramachandranpillai, R., Sampath, K., Mohammad, A., Alikhani, M.: Fairness at every intersection: Uncovering and mitigating intersectional biases in multimodal clinical predictions. arXiv preprint arXiv:2412.00606 (2024)
17. Seyyed-Kalantari, L., Zhang, H., McDermott, M.B., Chen, I.Y., Ghassemi, M.: Underdiagnosis bias of artificial intelligence algorithms applied to chest radiographs in under-served patient populations. Nat. Med. **27**(12), 2176–2182 (2021)
18. Sharma, S., Henderson, J., Ghosh, J.: Feamoe: fair, explainable and adaptive mixture of experts. arXiv preprint arXiv:2210.04995 (2022)
19. Tan, X., Wang, Y., Pham, T.H., Zhang, P., Zhang, X.: Achieving fairness without harm via selective demographic experts. In: Proceedings of the AAAI Conference on Artificial Intelligence, vol. 40, pp. 39331–39339 (2026)
20. Tschandl, P., Rosendahl, C., Kittler, H.: The ham10000 dataset, a large collection of multi-source dermatoscopic images of common pigmented skin lesions. Sci. data **5**(1), 180161 (2018)
21. Tu, T., et al.: Towards generalist biomedical ai. Nejm Ai **1**(3), AIoa2300138 (2024)
22. Wang, P., Tong, L., Wu, J., Liu, J., Liu, Z.: Fair-moe: Medical fairness-oriented mixture of experts in vision-language models. In: Gee, J.C., et al. (eds.) International Conference on Medical Image Computing and Computer-Assisted Intervention, pp. 186–196. Springer, Cham (2025). https://doi.org/10.1007/978-3-032-04971-1_18
23. Xia, Y., Ma, B., He, J., Wang, Z., Dou, Q., Xia, Y.: Toward robust medical fairness: Debiased dual-modal alignment via text-guided attribute-disentangled prompt learning for vision-language models. arXiv preprint arXiv:2508.18886 (2025)

24. Xu, Z., Li, J., Yao, Q., Li, H., Zhao, M., Zhou, S.K.: Addressing fairness issues in deep learning-based medical image analysis: a systematic review. npj Digit. Med. **7**(1), 286 (2024)
25. Yang, Y., et al.: Demographic bias of expert-level vision-language foundation models in medical imaging. Sci. Adv. **11**(13), eadq0305 (2025)
26. Zhang, S., Metaxas, D.: On the challenges and perspectives of foundation models for medical image analysis. Med. Image Anal. **91**, 102996 (2024)
27. Zhang, Y., Dunn, A.G., Naseem, U., Kim, J.: Intersectional fairness in vision-language models for medical image disease classification. arXiv preprint arXiv:2512.15249 (2025)
28. Zheng, G., Jacobs, M.A., Braverman, V., Parekh, V.S.: Towards fair medical ai: Adversarial debiasing of 3d ct foundation embeddings. arXiv preprint arXiv:2502.04386 (2025)

FARSIGHT: A Dynamic Multi Modal Medical Image Analysis Framework Powered by a Prior Knowledge Guided Vision Language Model

Jinghan Yang[1,2], Qinghua Li[2,3], Jingxin Liu[1,2], Li Qiu[4], Jie Tian[1,5], Jiaojiao Zhou[4(✉)], and Kun Wang[1(✉)]

[1] CAS Key Laboratory of Molecular Imaging, Institute of Automation, Chinese Academy of Sciences, Beijing, China
kun.wang@ia.ac.cn

[2] School of Artificial Intelligence, University of Chinese Academy of Sciences, Beijing, China

[3] School of Advanced Interdisciplinary Sciences, University of Chinese Academy of Sciences, Beijing, China

[4] Department of Medical Ultrasound, West China Hospital, Sichuan University, Chengdu, China
zhoujiaojiao@wchscu.cn

[5] School of Engineering Medicine, Beihang University, Beijing, China

Abstract. Long-term retrospective medical imaging archives cannot be directly utilized due to the lack of fine-grained annotations, requiring extensive manual curation by expert physicians in most previous medical imaging studies. Furthermore, the absence of standardized archiving protocols and the heavy reliance on subjective judgment for image retention lead to significant variations in the number of images per patient, while the presence of irrelevant frames introduces substantial noise. We propose FARSIGHT, an end-to-end archive-to-prediction framework designed to overcome these universal bottlenecks, evaluated here on renal transplant chronic allograft injury assessment. FARSIGHT integrates the Modality Aware Governance and Optimization System (MAGOS), an offline vision-language-model-guided curation agent, with hierarchical multimodal multi-instance learning (HMMIL) for patient-level prediction. On a retrospective cohort of 4,576 patients comprising 41,203 images across four distinct modalities, MAGOS achieves 0.999 routing accuracy with complete parsing success. Using these curated inputs, HMMIL achieves a test AUC of 0.816 and an accuracy of 0.746 for patient-level prediction, while handling variable-length studies and missing modalities. The code is available at: https://github.com/hanlliumu/Farsight.

Keywords: Automated Data Curation · Vision-Language Models · Multiple Instance Learning · Real-World Data

J. Yang and Q. Li—These authors contributed equally.

G. Yang et al. (Eds.): MICCAI 2026, LNCS 16878, pp. 163–172, 2027.
https://doi.org/10.1007/978-3-032-38059-3_16

1 Introduction

Translating AI advances into large-scale clinical practice remains challenging because real-world medical archives are not model-ready: retrospective data often lack fine-grained annotations, contain unreliable metadata and non-diagnostic vendor overlays, and therefore demand substantial expert curation. Moreover, routine clinical archives are structurally irregular—studies vary widely in image counts, include redundant or noisy frames, and frequently miss or misalign modalities—complicating robust multi-modal learning.

Vision-language models (VLMs) provide a practical bridge from archives to models by recognizing modality-specific cues, interface elements, and layout patterns without task-specific training, enabling archive-level modality routing and ROI-oriented curation [1, 2]. Yet free-form generation is difficult to audit, and institution-specific fine-tuning is often infeasible under strict offline privacy constraints [3, 4]. These constraints motivate an agentic design that combines VLM perception with prior-knowledge rules and structured, verifiable outputs to enable reliable archive-level curation [5, 6]. However, even after curation, clinical studies remain variable in length, partially observed across modalities, and sparsely informative at the frame level. For such settings, multiple-instance learning (MIL) provides a natural weakly supervised formulation in which each patient is represented as a bag of instances with a single patient-level label [7–11]. While MIL has shown strong performance in sparse-evidence tasks such as whole-slide imaging, it typically operates within a single-modality feature space. Clinical archives, in contrast, are multi-modal and heterogeneous, with imbalanced modality distributions and frequent missing modalities. Consequently, effective prediction requires modality-aware aggregation mechanisms that can account for cross-modality variability.

We propose FARSIGHT, an end-to-end archive-to-prediction framework integrating MAGOS with HMMIL. MAGOS performs modality routing and rule-constrained ROI extraction with schema-checked outputs, producing structured inputs without foundation-model fine-tuning [12, 13]. HMMIL then aggregates evidence within and across modalities, accommodating variable-length studies and missing modalities without forced alignment. On renal transplant ultrasound for chronic allograft injury assessment, MAGOS achieves 0.999 routing accuracy on over 40,000 images with complete parsing success, and HMMIL improves test AUC from 0.777 to 0.816 and test accuracy from 0.700 to 0.746 over a strong baseline.

Our main contributions are as follows: We introduce FARSIGHT, a archive-to-prediction framework designed for real-world medical imaging repositories. We present MAGOS, an offline and auditable VLM-plus-rules curation agent that performs modality identification and ROI extraction without foundation-model fine-tuning. We propose HMMIL tailored to variable-length and incompletely paired clinical data, and we demonstrate robust performance under real-world curation noise and missing-modality conditions.

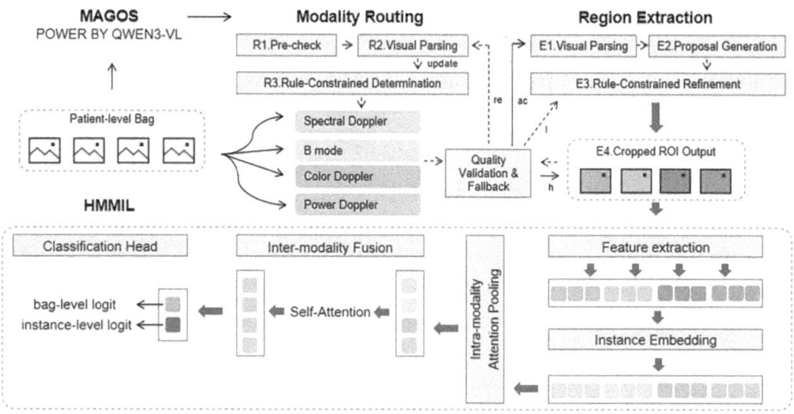

Fig. 1. MAGOS encodes radiologist-derived visual priors as explicit modality-routing rules: R1 validates inputs, R2 detects modality-specific cues, and R3 assigns B, C, P, or S labels. E1–E4 parse images, generate and refine modality-specific ROI proposals, and crop the resulting regions. A quality-validation and fallback module accepts, rejects, or revises unreliable outputs. HMMIL then aggregates the curated instances through intra-modality pooling and inter-modality fusion for patient-level prediction. re, reject; ac, accept; h, high quality; l, low quality.

2 Method

2.1 Dataset and Cohort

This retrospective study included renal transplant recipients evaluated for chronic allograft injury (CAI). The patient-level prediction task was binary classification of CAI versus normal controls, with CAI as the positive class. The primary cohort included 4,576 patients from Hospital A between 2008 and 2025 and was split at the patient level to prevent data leakage into 3,204 training, 686 validation, and 686 internal test patients. These subsets were used exclusively to develop and evaluate HMMIL. The dataset contained 41,203 ultrasound images across four modalities, including 16,456 from patients with CAI and 24,747 from normal controls. Spectral Doppler (S) was the dominant modality with 22,981 images, followed by B-mode (B; 9,669), Color Doppler (C; 8,277), and Power Doppler (P; 276). Because of its limited sample size, Power Doppler was excluded from subsequent predictive analyses. Two additional image-level sets were reserved for independent MAGOS evaluation: Routing Evaluation Set A included 400 images for assessing modality routing, and ROI Evaluation Set B included 200 images for assessing ROI extraction. Neither set was used to optimize MAGOS or to train, validate, or test HMMIL. The cohort retained the irregular structure of real-world clinical examinations. Per-patient frame counts ranged from 1 to 46, with a median of 8. Most patients had three modalities, 133 had all four, and the remainder had one or two, reflecting routine missing-modality patterns in retrospective data. This variability provided a realistic and challenging setting for evaluating the robustness of the proposed multi-instance framework.

2.2 Overview of FARSIGHT System

As illustrated in Fig. 1, FARSIGHT integrates MAGOS with HMMIL to enable robust patient-level diagnosis from raw dataset archives without fine-grained image-level annotations.

MAGOS. To address missing modality labels and ROI annotations in retrospective archives, we develop MAGOS, which decomposes curation into VLM-guided modality routing and deterministic, modality-specific ROI extraction. MAGOS employs Qwen3-VL-8B [14, 15] for privacy-preserving on-prem deployment and reliable structured generation under instruction-following and constrained decoding.

Robust Vision Language Model Guided Modality Routing. The modality router maps each input image X to one of four standardized modality spaces $Y \in \{S, C, P, B\}$. To ensure reproducible and auditable decisions under highly heterogeneous ultrasound interfaces, we encode clinical prior knowledge as an expert-defined visual state space S_{vis} and apply a deterministic rule mapping implemented through prompting. The VLM first predicts key visual states, and the final modality label is obtained by applying the predefined decision rules with greedy decoding at zero temperature. To guarantee valid structured outputs, MAGOS uses JSON-Schema-constrained decoding. Rare residual parsing failures are handled by a lightweight fallback parser consisting of automatic syntax repair, abbreviation normalization, and regex-based recovery from clinical shorthand, enabling stable large-scale deployment within our dataset without crashes.

Spatial Heuristic Adaptive Region Extraction. After modality routing, MAGOS applies a modality-specific computer vision pipeline to extract regions of interest. Importantly, the VLM-based agent does not directly generate or crop bounding boxes. ROI localization is performed exclusively by a deterministic OpenCV-driven module using predefined spatial heuristics. Extracted ROIs are subsequently analyzed by the agent and reviewed by physicians to identify systematic errors or edge cases. Based on this feedback loop, heuristic rules are refined and updated. Thus, the closed-loop optimization updates rule parameters rather than delegating pixel-level localization to the VLM. To enhance cross-device generalization, spatial heuristics are defined using relative image dimensions instead of absolute pixel thresholds. Specifically, Color Doppler and Power Doppler regions are extracted from color overlays while suppressing UI artifacts; Spectral Doppler regions are localized via waveform panel detection; and B-mode regions are obtained by segmenting the principal grayscale scanning field using connected-component and morphology-based criteria [16].

HMMIL. After MAGOS provides modality-standardized, ROI-cropped images, we cast patient-level diagnosis as a weakly supervised multi-modal multi-instance learning problem to handle variable-length exams without image-level annotations. Our key innovations are a two-level hierarchical attention mechanism [17, 18] aligned with clinical modality structure, a masked two-stage Transformer fusion that yields auditable cross-modality interactions under missingness, and an auxiliary Top-K hard-negative instance supervision that improves instance discriminability and enables reliable saliency analysis. For each patient, we build a bag $B = \{(x_{m,i}, m)\}$, where $x_{m,i}$ denotes the i-th ROI image from modality m. The bag is naturally partitioned into modality-specific sub-bags

B_m with size N_m. Missing modalities are represented by a modality mask and excluded from aggregation, enabling inference without fixed modality combinations. Following a lightweight backbone comparison, we adopt ShuffleNet as the instance encoder to produce embeddings $f_{m,i}$ with low memory overhead [19]. To avoid the brittleness of max pooling under noisy exams, we learn modality-aware instance importance within each sub-bag. The network first computes intra modality attention weights $\alpha_{m,i}$ to aggregate instances into a modality specific representation F_m, where w_{intra} and V_{intra} are learnable parameters of the intra modality attention network. This design explicitly matches the clinical acquisition process: instances are first filtered within modality before any cross-modality interaction, improving robustness to modality-specific redundancy and artifacts.

$$\alpha_{m,i} = \frac{exp(w_{intra}^T tanh(V_{intra} f_{m,i}))}{\sum_{j=1}^{N_m} exp(w_{intra}^T tanh(V_{intra} f_{m,j}))}, F_m = \sum_{i=1}^{N_m} \alpha_{m,i} f_{m,i} \quad (1)$$

We introduce a two-stage Transformer fusion pipeline. An intra-modality Transformer first refines the instance sequence within each modality via self-attention, and the resulting modality tokens are then fused by an inter-modality Transformer to model cross-modality interactions. Each modality token is scaled by a learnable scalar weight and aggregated into a patient-level representation for the final classification head. Missing modalities are handled by masking and are excluded from aggregation, enabling robust inference under incomplete modality availability. Standard MIL can assign high attention to non-diagnostic frames, especially in negative patients where "hard negatives" resemble positives. We therefore propose an auxiliary Top-K Hard Negative Instance Loss to sharpen instance-level discriminability under weak supervision. For each bag, we select the Top-K instances by current instance score/attention. In positive bags, these instances serve as pseudo-positives and are optimized with a foreground objective; in negative bags, the Top-K highest-scoring instances are treated as hard negatives and are explicitly penalized. This targeted supervision improves the reliability of learned instance weights, directly benefiting downstream saliency reporting and per-image importance analysis.

3 Results

3.1 Implementation Details

All data preprocessing and model development were conducted on a Linux workstation powered by an Intel Core Ultra 9 processor. The environment was equipped with two NVIDIA GeForce RTX 5090 GPUs, each featuring 32 GB of dedicated memory. The entire experimental framework was implemented utilizing PyTorch version 2.10.0 with comprehensive CUDA acceleration support. Further details on hyperparameters and VLM training available via public links.

3.2 Evaluation of MAGOS Framework

MAGOS Performance. MAGOS was evaluated using the two independent image-level sets defined in Sect. 2.1. An ultrasound physician with five years of clinical experience provided the reference modality labels for Independent Set A and manually delineated the regions of interest for Independent Set B. A second physician independently reviewed a subset of these annotations to assess interrater consistency. All MAGOS components were finalized before evaluation. Modality-routing performance was assessed on Independent Set A, whereas ROI extraction performance was assessed on Independent Set B (Fig. 2).

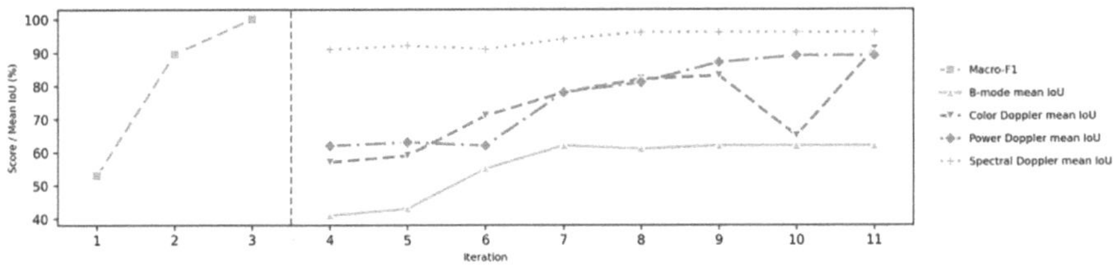

Fig. 2. Performance of the Magos across different iterations, including Macro-F1 score (cyan dashed line with squares), B mode mean IoU (orange solid line with triangles), Color Doppler mean IoU (red dashed line with circles), Power Doppler mean IoU (blue dash-dot line with diamonds), and Spectral Doppler mean IoU (purple dotted line with plus signs). (Color figure removed)

We trained a lightweight supervised baseline utilizing ResNet18 on Test Set A with a training and validation split of seventy to thirty. The network achieved near perfect accuracy on visually distinct modalities (B, C, S), but only 66.7% on P. Conversely, the zero shot vision language model guided routing module achieved superior overall accuracy without requiring a single annotated training sample. Deployed across over 40000 uncurated images, it attained a 99.98 percent overall accuracy derived from the consensus of two physicians who independently analyzed all cases. The system sustained a complete parsing success rate without a single pipeline crash. Among the valid images, comprehensive physician review revealed only three classification errors where Color Doppler cases completely lacking visible color overlays were assigned to B mode.

For quantitative evaluation, we defined algorithmic acceptance as achieving a mean Intersection over Union of at least 0.80 against the delineations of the physician on Test Set B. The finalized spatial heuristic algorithms achieved mean overlap metrics of 91 percent for Color Doppler, 89 percent for Power Doppler, and 96 percent for Spectral Doppler. For B mode images, the measured mean Intersection over Union was 62 percent. The lower mIoU of B is because we prioritized including the entire region to avoid expensive fine annotation, since the kidney occupies a large proportion of the view, and this coarse extraction is sufficient for patient-level prediction tasks.

MIL Aggregation Comparison. As shown in Table 1, HMMIL achieves the best overall performance among all compared methods. It attains the highest AUC on the training, validation, and test sets with values of 0.841, 0.814, and 0.816, respectively, and also delivers the best accuracy with 0.772, 0.756, and 0.746. Compared with the strongest baseline MambaMIL, HMMIL improves test AUC from 0.777 to 0.816 and test ACC from 0.700 to 0.746, demonstrating more robust generalization. In this experiment, we use HMMIL with B-mode and Color Doppler ultrasound as the input modalities and ShuffleNet as the feature extraction backbone; the rationale for these choices is discussed in Sect. 3.4 Ablation Study and the Discussion section.

Table 1. Performance comparison with baselines on the training, validation, and test sets.

Methods	Training		Validation		Test	
	AUC	ACC	AUC	ACC	AUC	ACC
Mean Pooling	0.681	0.643	0.576	0.579	0.555	0.551
Max Pooling	0.648	0.583	0.631	0.620	0.588	0.558
ABMIL [7]	0.791	0.690	0.752	0.678	0.763	0.676
Gated ABMIL [8]	0.777	0.712	0.755	0.713	0.758	0.692
MambaMIL [9]	0.803	0.721	0.774	0.721	0.777	0.700
ACMIL [10]	0.785	0.688	0.752	0.683	0.754	0.677
MHIM-MIL [11]	0.791	0.761	0.761	0.708	0.770	0.692
HMMIL(Ours)	**0.841**	**0.772**	**0.814**	**0.756**	**0.816**	**0.746**

3.3 Ablation Study

We evaluated Magos under local deployment on two additional platforms, a Tesla V100 server and an RTX 5060 Ti workstation. The average latency is 10.42 s/img on V100 and 7.14 s/img on RTX 5060 Ti, compared with 3.04 s/img in our development environment. Despite the slowdown, the achieved throughput remains acceptable for our target deployment scenarios. We adopt HMMIL with B and C as the input modalities and ShuffleNet as the feature extraction backbone. Based on this setting, we conduct ablation experiments to quantify the impact of backbone selection, model components, and modality configuration. The results indicate that the choice of backbone materially affects performance, and the intra-modality module constitutes the primary contributor, as its removal yields the largest degradation. The instance-level loss, the inter-modality module, and hard-negative mining further provide consistent improvements. We analyzed attention weights and hard negative scores in representative patients. In positive cases, high-weighted instances consistently corresponded to B and C images. In contrast, S instances had lower attention weights, consistent with our quantitative modal ablation results. In addition, B is the most informative single modality, S alone is insufficient, and multi-modal fusion enhances discriminative capability. Collectively, these findings support the effectiveness of the proposed HMMIL design and its key components in improving generalization on the validation and test sets. Detailed quantitative results are reported in Table 2.

Table 2. Ablation Study of HMMIL: Backbone, Module, and Modality Analysis (w/o: without; Intra: intra-modality module; Inter: inter-modality module; Linst: instance-level loss; Hard-N: hard-negative mining; B: B-mode; C: Color Doppler; S: Spectral Doppler).

Methods		Training		Validation		Test	
		AUC	ACC	AUC	ACC	AUC	ACC
Backbone analysis	EffientnetbO	0.745	0.684	0.637	0.613	0.595	0.566
	ResNet18	0.626	0.604	0.597	0.584	0.579	0.569
	Mobilenetv2	0.824	0.740	0.797	0.726	0.797	0.714
Module analysis	w/o Intra	0.800	0.733	0.761	0.717	0.730	0.667
	w/o Inter	0.770	0.704	0.721	0.690	0.750	0.709
	w/o Linst	0.796	0.675	0.767	0.682	0.764	0.641
	w/o Hard-N	0.810	0.731	0.760	0.692	0.770	0.699
Modality analysis	B + S	0.806	0.747	0.773	0.720	0.793	0.747
	B + C + S	0.814	0.736	0.770	0.714	0.808	0.750
	Only B	0.808	0.731	0.778	0.717	0.797	0.751
	Only C	0.798	0.734	0.756	0.704	0.781	0.706
	Only S	0.573	0.548	0.580	0.578	0.535	0.527
HMMIL(Ours)		**0.841**	**0.772**	**0.814**	**0.756**	**0.816**	**0.746**

4 Discussion

Our proposed FARSIGHT framework addresses this by providing an end-to-end solution: MAGOS offers zero-fine-tuning data curation, while HMMIL delivers patient-level prediction by handling irregular clinical data structures. This integrated approach bypasses the primary bottleneck of costly human intervention, enabling scalable deployment. From a translational perspective, FARSIGHT represents a preliminary framework for hospital environments, designed to be compatible with existing PACS workflows. It requires no changes to physician workflows, yet dramatically reduces the massive time and cost associated with building multi-modal retrospective cohorts by automating data governance and risk prediction. Regarding diagnostic utility for CAI, our experimental results indicated that C and B images provided significant diagnostic information, whereas Spectral Doppler images offered limited utility when analyzed purely visually without explicit scale information. This observation, that a purely visual approach to S waveforms proved less effective for extracting meaningful diagnostic information, contrasts with clinical practice where physicians routinely assess renal perfusion by analyzing waveform trends and numerical values.

While FARSIGHT demonstrates strong performance on our dataset, effectively addressing real-world challenges, certain limitations define our immediate path forward.

Its current validation necessitates rigorous multi-center external testing and comprehensive evaluation on publicly available datasets to fully confirm Magos's design robustness and generalizability. Furthermore, FARSIGHT's ultimate vision includes leveraging VLM capabilities for image-to-text generation to enhance task interpretability; however, current models like Qwen3-VL struggle with complex medical images due to their general domain training. Moving forward, a primary focus will be validating FARSIGHT's performance on multi-center datasets and benchmark public archives. Concurrently, we will fine-tune small-parameter, domain-specific VLMs or train novel foundation models tailored for medical imaging. This crucial step will enable FARSIGHT to generate insightful, interpretable text explanations, thereby completing its logical loop from data handling to comprehensive, explainable medical image analysis.

Acknowledgments. This work was supported by the National Key Research and Development Program of China (2023YFF1204600), National Natural Science Foundation of China (82441010, 82272029, U25C2048, 92259303, 62027901), and Beijing Natural Science Foundation (Z250003, 20250484964).

Disclosure of Interests. The authors have no competing interests to declare that are relevant to the content of this article.

References

1. Dong, H., Kang, Z., Yin, W., Liang, L., Chao, C., Jiao, R.: Scalable vision language model training via high quality data curation. In: Proceedings of the 63rd Annual Meeting of the Association for Computational Linguistics (ACL), pp. 33272–33293 (2025)
2. Udandarao, V., Burg, M.F., Albanie, S., Bethge, M.: Visual data-type understanding does not emerge from scaling vision-language models. arXiv preprint arXiv:2310.08577 (2023)
3. Zhao, Y., et al.: Med-VLM: enhancing medical image segmentation accuracy through vision-language model. In: Proceedings of the IEEE/CVF International Conference on Computer Vision (ICCV), pp. 7283–7293 (2025)
4. Jiang, S., et al.: Omniv-med: scaling medical vision-language model for universal visual understanding. arXiv preprint arXiv:2504.14692 (2025)
5. Jeong, D.P., Garg, S., Lipton, Z.C., Oberst, M.: Medical adaptation of large language and vision-language models: are we making progress? In: Proceedings of the 2024 Conference on Empirical Methods in Natural Language Processing (EMNLP), pp. 12143–12170 (2024)
6. Silva-Rodríguez, J., Shakeri, F., Bahig, H., Dolz, J., Ben Ayed, I.: Few-shot, now for real: Medical VLMs adaptation without balanced sets or validation. In: International Conference on Medical Image Computing and Computer-Assisted Intervention (MICCAI), LNCS, pp. 237–247. Springer, Heidelberg (2025)
7. Ilse, M., Tomczak, J., Welling, M.: Attention-based deep multiple instance learning. In: Proceedings of the International Conference on Machine Learning (ICML), pp. 2127–2136 (2018)
8. Andersson, A., Koriakina, N., Sladoje, N., Lindblad, J.: End-to-end multiple instance learning with gradient accumulation. In: 2022 IEEE International Conference on Big Data (Big Data), pp. 2742–2746 (2022)
9. Yang, S., Wang, Y., Chen, H.: Mambamil: enhancing long sequence modeling with sequence reordering in computational pathology. In: International Conference on Medical Image Computing and Computer-Assisted Intervention (MICCAI), LNCS, pp. 296–306. Springer, Heidelberg (2024)

10. Zhang, Y., Li, H., Sun, Y., Zheng, S., Zhu, C., Yang, L.: Attention-challenging multiple instance learning for whole slide image classification. In: European Conference on Computer Vision (ECCV), LNCS, pp. 125–143. Springer, Heidelberg (2024)
11. Tang, W., Huang, S., Zhang, X., Zhou, F., Zhang, Y., Liu, B.: Multiple instance learning framework with masked hard instance mining for whole slide image classification. In: Proceedings of the IEEE/CVF International Conference on Computer Vision (ICCV), pp. 4078–4087 (2023)
12. Lu, Y., et al.: Learning to generate structured output with schema reinforcement learning. In: Proceedings of the 63rd Annual Meeting of the Association for Computational Linguistics (ACL), pp. 4905–4918 (2025)
13. Guo, Z., et al.: Glo-vlms: Leveraging vision-language models for fine-grained diseased glomerulus classification. arXiv preprint arXiv:2508.15960 (2025)
14. Bai, S., Cai, Y., Chen, R., et al.: Qwen3-VL technical report. arXiv preprint arXiv:2511.21631 (2025)
15. Liu, L., et al.: SkinFlow: efficient information transmission for open dermatological diagnosis via dynamic visual encoding and staged RL. arXiv preprint arXiv:2601.09136 (2026)
16. Nankivell, B.J., Chapman, J.R., Gruenewald, S.M.: Detection of chronic allograft nephropathy by quantitative Doppler imaging. Transplantation **74**(1), 90–96 (2002)
17. Wang, Q., et al.: Hierarchical attention-based multiple instance learning network for patient-level lung cancer diagnosis. In: 2020 IEEE International Conference on Bioinformatics and Biomedicine (BIBM), pp. 1156–1160 (2020)
18. Li, Z., Jiang, Y., Lu, M., Li, R., Xia, Y.: Survival prediction via hierarchical multimodal co-attention transformer: a computational histology-radiology solution. IEEE Trans. Med. Imaging **42**(9), 2678–2689 (2023)
19. Zhang, X., Zhou, X., Lin, M., Sun, J.: ShuffleNet: an extremely efficient convolutional neural network for mobile devices. In: Proceedings of the IEEE Conference on Computer Vision and Pattern Recognition (CVPR), pp. 6848–6856 (2018)

GLeVE: Graph-Guided Lesion Grounding with Proposal Verification in 3D CT

Shuo Jiang[1], Yuhao Hong[1], Chunbo Jiang[1], Weihong Chen[1], Huangwei Chen[2], Shenghao Zhu[1], Beining Wu[1], Mingxuan Liu[3], Zhu Zhu[4], Feiwei Qin[1(✉)], Min Tan[1], and Yifei Chen[3(✉)]

[1] Zhejiang Key Laboratory of Space Information Sensing and Transmission, Hangzhou Dianzi University, Hangzhou, China
{jiangshuo,qinfeiwei}@hdu.edu.cn
{jiangshuo,qinfeiwei}@hdu.edu.cn
[2] Zhejiang University, Hangzhou, China
[3] Tsinghua University, Beijing, China
justlfc03@gmail.com
justlfc03@gmail.com
[4] Children's Hospital, Zhejiang University School of Medicine, Hangzhou, China

Abstract. Grounding radiology report descriptions to 3D CT volumes is essential for verifiable clinical interpretation, yet remains challenging due to the semantic-spatial gap between free-text narratives and volumetric anatomy. Existing report-assisted and vision-language grounding methods typically rely on phrase-level alignment or dense pixel supervision, resulting in limited lesion-wise correspondence and suboptimal localization accuracy. We propose GLeVE, a graph-guided lesion grounding framework with anatomical prior verification and octree-based autoregressive refinement. GLeVE treats each lesion description as an atomic semantic unit and encodes organ attribution, attributes, and inter-lesion relations through relation-aware graph reasoning to produce discriminative lesion-wise queries. Anatomy-aware proposal generation with region-level verification enforces one-to-one text-lesion alignment, while hierarchical octree refinement progressively improves boundary delineation. Experiments on AbdomenAtlas 3.0 demonstrate consistent gains over classical multimodal foundation models and report-supervised baselines in both segmentation accuracy and lesion-level localization. Our source code is available at https://github.com/JSLiam94/GLeVE.

Keywords: Report Grounding · Lesion Localization · Graph Reasoning · Anatomical Prior · Octree Refinement

1 Introduction

Radiology reports provide structured natural-language descriptions of lesions in 3D Computed Tomography (CT), covering organ attribution, anatomical subregions, and key imaging characteristics such as Hounsfield Units (HU), and

S. Jiang and Y. Hong–Co-first authors.

G. Yang et al. (Eds.): MICCAI 2026, LNCS 16878, pp. 173–183, 2027.
https://doi.org/10.1007/978-3-032-38059-3_17

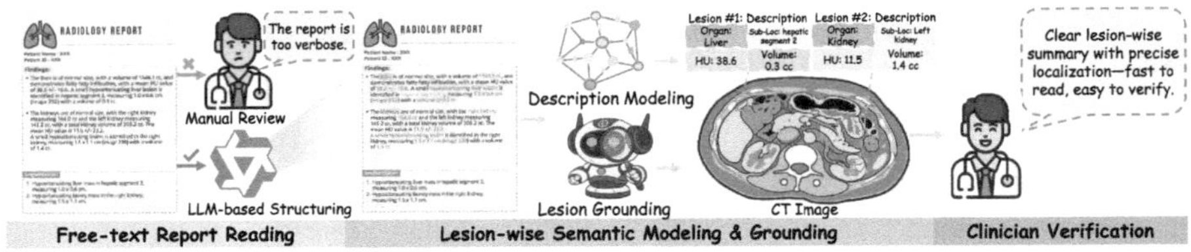

Fig. 1. Overview of lesion-wise report grounding, illustrating structured semantic modeling and precise CT localization for efficient clinical verification.

thus serve as a central interface between imaging evidence and clinical decision-making [4,8,19]. However, in routine clinical workflows, physicians must repeatedly revisit the original CT and search slice by slice around the reported locations to verify critical findings, making lesion localization highly experience-dependent and inefficient [3,11,20]. This burden is further amplified in multi-organ/multi-lesion scenarios: reports grow substantially longer, salient lesion descriptions are diluted by abundant routine statements, and clinicians are hindered in rapidly retrieving key information from lengthy text and stably mapping it to the corresponding image locations, thereby limiting the practical utility of reports for precise decision-making and verifiable interpretation [1,9,26].

Existing general-purpose segmentation models (e.g., SwinUNETR [25] and MedSAM3 [21]) produce global CT masks, hindering interpretable lesion-by-lesion report alignment and requiring dense pixel-level supervision [10]. R-Super [5] uses reports as auxiliary supervision to reduce annotation reliance, but without explicit inference-time matching, it struggles to enforce one-to-one "text-lesion" correspondence. Studies on medical text grounding aim to align report descriptions into the global image coordinate system for explainable localization: Ichinose *et al.* [17] leverage anatomical-structure segmentation and structured report parsing to improve abnormality localization, but their reliance on cross-modal attention alignment still limits spatial precision; Nützel *et al.* [23] employ latent diffusion models for bimodal bias fusion, and Vilouras [27] performs feature selection via cross-attention maps for medical phrase grounding. Nevertheless, these methods mainly perform phrase-level alignment and cannot reliably treat a full lesion description as a coherent semantic unit for precise localization. Furthermore, Multimodal Large Language Model-based methods introduce parameter-efficient fine-tuning and uncertainty-aware multi-hypothesis mechanisms [15,32] to enhance robustness; Yang *et al.* [30] propose a contrastive learning framework with multi-level semantic alignment, but it remains highly sensitive to fine-grained report quality and shows limited cross-scenario generalization. Overall, while prior work has advanced bounding-box-level localization, substantial gaps remain in achieving reliable, pixel-accurate localization of multiple low-contrast small lesions, falling short of core clinical requirements.

To address these issues, we propose **G**raph-Guided **Le**sion Grounding with Proposal **Ve**rification framework (**GLeVE**). **GLeVE** encodes composite semantics in lesion descriptions, including organ attribution, size, and HU, through

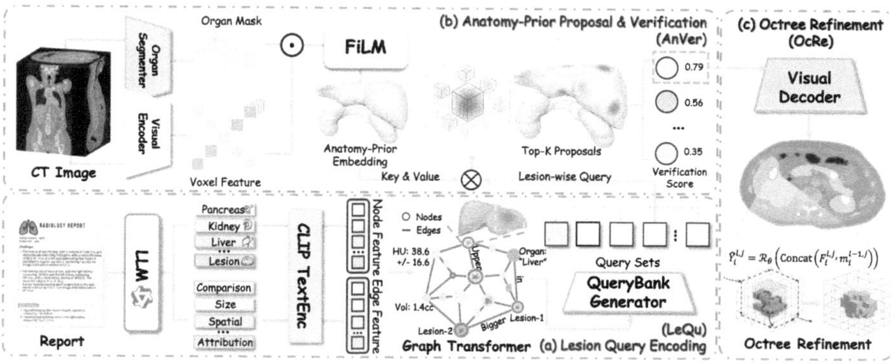

Fig. 2. Architecture of **GLeVE**, including (a) lesion semantic graph modeling, (b) anatomy-aware proposal verification, and (c) octree-based autoregressive refinement.

graph-structured representation learning, and incorporates organ-segmentation priors to construct geometry-aware anatomical embeddings. A multi-candidate verification mechanism is subsequently introduced to suppress boundary noise and spurious responses. This is followed by an octree-based autoregressive refinement strategy that enhances pixel-level localization sensitivity for small lesions. As illustrated in Fig. 1, by optimizing semantic consistency together with spatial precision in complex anatomy, **GLeVE** delivers verifiable pixel-level lesion localization with one-to-one alignment between key report descriptions and imaging evidence, yielding traceable, text-grounded results for reliable clinical interpretation and quantitative evaluation.

2 Proposed Method

2.1 Model Architecture

As illustrated in Fig. 2, **GLeVE** comprises three modules: (i) a graph-structured semantic query generator that converts each lesion description into lesion-wise queries while preserving organ attribution, attributes, and inter-lesion relations; (ii) anatomy-prior feature modulation with voxel-level similarity to generate high-recall candidates, followed by region-level verification for one-to-one text-lesion alignment; and (iii) octree-based autoregressive refinement within selected candidates for high-resolution segmentation, yielding interpretable masks consistent with global spatial coordinates.

2.2 Lesion Semantic Modeling and Querying (LeQu)

For lesion-level semantic completeness and precise cross-modal matching, we use each full lesion description as the atomic reasoning unit, encode its attribute constraints and spatial relations in an explicit graph, and generate corresponding lesion-wise queries for stable, interpretable localization in multi-lesion settings.

Structured Report Parsing and Lesion Unit Construction. Given a report R, we employ a frozen Qwen3-8B [29] to extract a lesion set $\{r_i\}$ with structured fields (organ attribution $o(i)$, sub-location, size, and HU), and incorporate organ-level statistics to explicitly model lesion-background contrast.

Lesion Semantic Graph Construction. The extracted fields are organized into a semantic graph $G_i = (\mathcal{V}_i, \mathcal{E}_i)$ comprising lesion, anatomical, and attribute nodes. Edges encode lesion-organ attribution, lesion-attribute associations, and intra-organ relations, enabling cross-lesion comparison to disambiguate multi-lesion cases and enhance lesion-level localization queries.

Node Initialization and Relation Encoding. For each node $v \in \mathcal{V}_i$, we initialize its embedding $h_v^0 \in \mathbb{R}^d$ with a CLIP text encoder; for each edge $(u \to v) \in \mathcal{E}_i$, we assign a learnable embedding r_{uv} to encode relation types.

Relation-Aware Graph Transformer Reasoning. We adopt a relation-type-aware attention mechanism for message passing, where relation embeddings are injected into the graph reasoning process as attention biases:

$$\begin{aligned} \alpha_{uv}^{\ell} &= \mathrm{softmax}_{u\in\mathcal{N}(v)}\left(\frac{\left(W_q h_v^{\ell-1}\right)^{\top}\left(W_k h_u^{\ell-1} + W_r r_{uv}\right)}{\sqrt{d}}\right), \\ h_v^{\ell} &= \mathrm{FFN}\left(h_v^{\ell-1}\right) + \sum_{u\in\mathcal{N}(v)} \alpha_{uv}^{\ell} W_v h_u^{\ell-1}, \end{aligned} \tag{1}$$

where $h_v^{\ell} \in \mathbb{R}^d$ is the ℓ-th layer embedding of node v, $\mathcal{N}(v)$ denotes its neighborhood, $W_q, W_k, W_v \in \mathbb{R}^{d\times d}$ are the query/key/value projections, and $W_r \in \mathbb{R}^{d\times d_r}$ projects relation embeddings; d is the feature dimension, and $\mathrm{FFN}(\cdot)$ is a residual Feed-Forward Network. This facilitates relation-aware message passing and enhances lesion-level semantic discrimination in multi-lesion settings.

Lesion-Level Query Generation. After graph reasoning, the final representation of the lesion node $h_{v_i^{\mathrm{les}}}^{L}$ is treated as the semantic summary vector z_i for the i-th lesion. A QueryBank Generator (Ψ_{QB}) then maps z_i to a set of lesion queries, $\left\{q_i^{(m)}\right\}_{m=1}^{M} = \Psi_{\mathrm{QB}}(z_i)$, where M denotes the number of queries associated with each lesion.

2.3 Anatomy-Prior Lesion Proposal and Verification (AnVer)

To incorporate anatomical constraints, we combine organ embeddings with feature modulation to render voxel features organ-aware prior to matching, thereby enforcing consistency in candidate generation and region-level verification.

Anatomical Prior Soft Modulation. The visual encoder produces features F. An organ segmenter provides binary masks $\{M_k\}_{k=1}^{K_o}$, where K_o denotes the total number of organ categories, and we Gaussian-smooth them to obtain soft maps $\{\widetilde{M}_k\}_{k=1}^{K_o}$. For each organ, we learn an embedding e_k and form a voxel-wise anatomy token that conditions Feature-wise Linear Modulation (FiLM):

$$\hat{E}(x) = \phi_o\left(\sum_{k=1}^{K_o} \widetilde{M}_k(x)\, e_k\right) \in \mathbb{R}^C,$$
$$(\gamma(x), \beta(x)) = \psi\left(\hat{E}(x)\right), \qquad \widetilde{F}(x) = \gamma(x) \odot F(x) + \beta(x), \tag{2}$$

where C denotes the channel dimension of the visual feature map, ϕ_o is a linear projection from the organ-embedding space to $\mathbb{R}^C$, and ψ is a lightweight MLP that predicts the FiLM parameters $\gamma(x), \beta(x) \in \mathbb{R}^C$. This channel-wise modulation enhances organ-consistent responses while suppressing irrelevant regions, injecting anatomical awareness without altering global spatial structure.

Voxel-Level Similarity Response and Candidate Generation. For the i-th lesion, its query set $\{q_i^{(m)}\}_{m=1}^M$ is matched against the organ-aware feature map $\widetilde{F}$ via voxel-level cosine similarity, yielding a response map S_i:

$$S_i(x) = \max_m \frac{\left\langle q_i^{(m)}, \widetilde{F}(x)\right\rangle}{\left\|q_i^{(m)}\right\| \left\|\widetilde{F}(x)\right\|}. \tag{3}$$

Based on S_i, adaptive thresholding and 3D connected-component analysis are performed to select the Top-K_p high-response components, forming a high-recall candidate set $\mathcal{P}_i = \{P_i^{(k)}\}_{k=1}^{K_p}$. Each candidate $P_i^{(k)}$ retains its spatial indices in the global CT coordinate system and serves as a localization anchor for subsequent verification and segmentation refinement.

Region Consistency Verification. Since voxel-level similarity peaks are prone to noise, we verify candidates at the region level. For lesion description i with candidates $\{P_i^{(k)}\}_{k=1}^{K_p}$, we crop and pool the feature map $\widetilde{F}$ to obtain region features $v_i^{(k)}$, and construct evidence vectors $\xi_i^{(k)}$ encoding anatomical coverage and report-attribute consistency. Specifically, $\xi_i^{(k)}$ includes the overlap ratio with $\widetilde{M}_{o(i)}$ and region statistics from $P_i^{(k)}$, i.e., volume $V_i^{(k)}$ and mean HU $\mu_i^{(k)}$, which are compared with reported volume and HU to derive matching metrics:

$$s_i^{(k)} = \sigma\left(W_s\left[v_i^{(k)}; z_i\right]\right) + \boldsymbol{\eta}^\top \xi_i^{(k)}, \qquad P_i^* = \arg\max_k s_i^{(k)}, \tag{4}$$

where z_i denotes the lesion-level semantic vector, $\sigma(W_s[\cdot])$ produces a scalar semantic matching score, and W_s and $\boldsymbol{\eta}$ are learnable parameters.

To promote a single dominant match per description, we minimize the entropy of the temperature-scaled candidate distribution:

$$\mathcal{L}_{\text{uni}} = -\sum_{i=1}^{N}\sum_{k=1}^{K_p} p_i^{(k)} \log\left(p_i^{(k)} + \epsilon\right), \qquad p_i^{(k)} = \frac{\exp\left(s_i^{(k)}/\tau\right)}{\sum_{j=1}^{K_p} \exp\left(s_i^{(j)}/\tau\right)}, \tag{5}$$

where $p_i^{(k)}$ is the temperature-scaled soft assignment, τ controls distribution sharpness, and ϵ is a small constant for numerical stability.

2.4 Octree Autoregressive Lesion Refinement (OcRe)

We propose an octree-driven autoregressive refinement framework for coarse-to-fine, hierarchical optimization of lesion masks, improving boundary delineation for small lesions with blurred details while remaining computationally efficient.

Octree Construction and Hierarchical Refinement. From the selected candidate P_i^*, we build an octree $\mathcal{T}_i$ with root cube Ω_i enclosing P_i^*; recursive subdivision yields multi-level blocks $\Omega_i^{\ell,j}$ across spatial scales, progressively narrowing the region of interest in a coarse-to-fine, top-down refinement hierarchy.

Autoregressive Mechanism and Conditional Modulation. At each refinement level, we employ a conditional autoregressive strategy, where the prediction from the previous level serves as an explicit structural prior for progressive segmentation refinement. The initial coarse mask $\hat{Y}_i^0$ is initialized from the selected candidate P_i^*. For each octree node $\Omega_i^{\ell,j}$, we crop local organ-aware features, $F_i^{\ell,j} = \text{Crop}(\widetilde{F}, \Omega_i^{\ell,j})$, and use the upsampled prediction from the preceding level as a conditional input, $m_i^{\ell-1,j} = \text{Crop}\left(\text{Up}(\hat{Y}_i^{\ell-1}), \Omega_i^{\ell,j}\right)$. The concatenated representation is then fed into a parameter-shared refinement decoder:

$$\hat{Y}_i^{\ell,j} = R_\theta\left(\text{Concat}(F_i^{\ell,j}, m_i^{\ell-1,j})\right), \tag{6}$$

where R_θ is a lightweight 3D residual refinement head shared across levels and nodes. The node-level predictions $\{\hat{Y}_i^{\ell,j}\}_j$ are merged into $\hat{Y}_i^\ell$ for coarse-to-fine boundary refinement, especially around small and low-contrast lesions.

Loss Under Missing Mask Annotations. Since many clinical samples lack pixel-level masks, we optimize the model on unlabeled data via report-driven weak supervision: for lesion i, the prediction yields V_i and μ_i to match V_i^r and μ_i^r, while $\mathcal{L}_{\text{sep}}$ penalizes inter-lesion overlap; the weak loss is defined as:

$$\mathcal{L}_{\text{weak}} = \lambda_{\text{uni}}\mathcal{L}_{\text{uni}} + \lambda_{\text{con}}\mathcal{L}_{\text{con}} + \lambda_{\text{sep}}\mathcal{L}_{\text{sep}}, \tag{7}$$

where $\mathcal{L}_{\text{con}} = \mathcal{L}_{\text{attr}} + \mathcal{L}_{\text{org}}$, $\mathcal{L}_{\text{attr}} = \sum_{a \in \{V,\mu\}} \sum_{i=1}^{N} \left(\frac{|a_i - a_i^r|}{|a_i^r| + \epsilon}\right)^2$, $\mathcal{L}_{\text{org}}$ penalizes predicted lesion mass outside the binary target-organ mask $M_{o(i)}$ and $\lambda_{\text{uni}}, \lambda_{\text{con}}, \lambda_{\text{sep}}$ weight unimodality, consistency, and separation.

For samples with Ground Truth (GT) masks, a combined Dice and cross-entropy loss $\mathcal{L}_{\text{seg}}$ is employed. Let $\mathcal{D}_m$ denote the subset of samples with pixel-level labels. For each sample x, the total loss is defined as:

$$\mathcal{L}_{\text{total}} = \delta \, \lambda_{\text{seg}} \, \mathcal{L}_{\text{seg}} + \lambda_{\text{weak}} \, \mathcal{L}_{\text{weak}}, \quad \text{s.t.} \quad \delta = \mathbb{I}[x \in \mathcal{D}_m]. \tag{8}$$

3 Experiments

3.1 Dataset and Implementation

Dataset. We conduct experiments on AbdomenAtlas 3.0 [6], comprising 9,262 cases across 138 institutions. Each case includes a report reviewed by 12 board-certified radiologists and voxel-aligned tumor masks for abdominal tumors, including kidney (K, 2,122), liver (L, 1,472), and pancreas (P, 361). We randomly split the cases into training, evaluation, and test sets at an 8:1:1 ratio.

Implementation Details. All experiments are implemented in PyTorch 2.4 and conducted on an NVIDIA H100 GPU. We use MedFormer [12] as the visual encoder-decoder backbone and adopt TotalSegmentator [28] for organ segmentation. We set the batch size to 1 and train for 100 epochs with the random seed of 2026. Model parameters are optimized using AdamW [22] with an initial learning rate of 1×10^{-4} and cosine annealing for learning-rate decay.

Evaluation Metrics. For evaluation, we report Dice and HD95 and additionally adopt Lesion Recall (LR) and Lesion Localization Score (LLS). We extract lesion instances via connected-component analysis on prediction and GT masks, declare a match when Dice ≥ 0.1, and use the Hungarian algorithm for optimal assignment, yielding N_{matched} and N_{gt}. LLS explicitly penalizes centroid displacement, yielding greater sensitivity to lesion-level localization errors.

$$\text{LR} = \frac{N_{\text{matched}}}{N_{\text{gt}}}, \qquad \text{LLS} = \frac{1}{N_{\text{gt}}} \sum_{i=1}^{N_{\text{gt}}} \text{Dice}(i, \pi(i)) \exp\left(-\frac{d_i}{d_0}\right), \tag{9}$$

where d_i is the centroid distance for the i-th match, $d_0 = 20$ mm is the decay scale, and $\text{Dice}(i, \pi(i)) = 0$ for unmatched GT lesions.

3.2 Experiment Results

Comparative Experiments. As shown in Table 1 and Fig. 3, with full mask annotations and report supervision, **GLeVE** consistently surpasses multimodal foundation models (e.g., CT-CLIP [13]), universal segmentation methods (e.g.,

Table 1. Quantitative comparison under varying mask ratios across **GLeVE** and other methods. Red and blue cells denote the best and second-best results, respectively.

Method	$\mathbf{Dice}_{std}$ (%)↑			$\mathbf{HD95}_{std}$ (mm)↓			$\mathbf{LR}_{std}$↑	$\mathbf{LLS}_{std}$ (%)↑		
	K	L	P	K	L	P	Avg	K	L	P
Multimodal Foundation Models										
M3D-LaMed$_{100\%}$ [2]	$35.3_{32.7}$	$13.6_{26.7}$	$19.9_{26.1}$	$95.9_{66.2}$	$103_{45.9}$	$42.1_{32.8}$	$34.6_{41.6}$	$10.4_{14.1}$	$2.62_{5.94}$	$10.6_{15.6}$
Merlin$_{100\%}$ [7]	$47.7_{34.7}$	$27.5_{30.8}$	$18.4_{20.9}$	130_{128}	$121_{87.6}$	$70.0_{36.4}$	$62.9_{41.1}$	$21.7_{22.1}$	$19.6_{18.4}$	$5.77_{9.99}$
SAT$_{100\%}$ [31]	$51.6_{26.4}$	$34.9_{21.3}$	$17.6_{25.5}$	$96.7_{76.2}$	$82.3_{84.7}$	$54.9_{68.1}$	$57.9_{33.9}$	$23.7_{15.7}$	$28.9_{21.3}$	$11.3_{10.5}$
R1Seg-3D$_{100\%}$ [14]	$28.9_{29.7}$	$15.6_{22.6}$	$6.47_{15.0}$	276_{264}	196_{177}	237_{146}	$43.7_{39.6}$	$12.5_{18.1}$	$15.3_{17.8}$	$2.30_{4.83}$
CT-CLIP$_{100\%}$ [13]	$48.9_{28.1}$	$28.6_{24.5}$	$16.4_{19.8}$	$127_{86.8}$	$93.7_{73.1}$	$81.6_{56.4}$	$48.7_{36.8}$	$18.9_{17.3}$	$23.2_{20.4}$	$9.38_{11.5}$
Universal Segmentation Methods										
UNet$_{100\%}$ [24]	$51.3_{30.9}$	$31.7_{24.6}$	$12.7_{16.3}$	180_{152}	234_{164}	105_{151}	$66.2_{38.7}$	$30.2_{25.4}$	$36.3_{22.2}$	$6.83_{10.8}$
nnUNetV2$_{100\%}$ [18]	$55.7_{34.2}$	$34.8_{28.1}$	$21.1_{22.3}$	152_{129}	$72.7_{65.1}$	$109_{85.4}$	$71.4_{40.2}$	$42.2_{29.0}$	$27.2_{26.3}$	$11.6_{12.3}$
STUNet$_{100\%}$ [16]	$53.0_{35.3}$	$30.8_{25.6}$	$22.6_{23.2}$	136_{112}	119_{104}	111_{101}	$65.5_{40.2}$	$33.8_{28.3}$	$33.2_{26.5}$	$12.8_{15.6}$
SwinUNETR$_{100\%}$ [25]	$56.5_{32.2}$	$34.2_{30.2}$	$22.2_{19.5}$	130_{115}	153_{133}	$105_{79.9}$	$71.4_{37.5}$	$36.6_{26.3}$	$31.7_{25.6}$	$13.4_{13.5}$
Medformer$_{100\%}$ [12]	$59.8_{28.0}$	$44.3_{25.2}$	$16.4_{22.3}$	71.0_{149}	90.1_{127}	$36.5_{30.2}$	$73.2_{39.0}$	$39.2_{22.6}$	$35.3_{27.8}$	$15.9_{17.0}$
Med-SAM3$_{100\%}$ [21]	$34.5_{25.4}$	$28.1_{28.9}$	$15.1_{17.9}$	$141_{71.5}$	$120_{67.8}$	$127_{79.3}$	$68.5_{40.1}$	$23.3_{19.6}$	$25.9_{23.6}$	$11.8_{13.3}$
Advanced Methods										
R-Super$_{10\%}$ [5]	$53.3_{32.4}$	$37.5_{27.8}$	$15.1_{19.6}$	162_{193}	128_{139}	$99.6_{85.3}$	$73.9_{36.9}$	$32.9_{25.5}$	$33.5_{19.8}$	$6.71_{12.5}$
R-Super$_{25\%}$ [5]	$54.3_{32.0}$	$39.6_{31.3}$	$20.5_{25.3}$	101_{128}	$74.4_{73.5}$	$85.1_{73.2}$	$74.4_{36.4}$	$37.2_{26.8}$	$37.1_{23.6}$	$15.3_{14.8}$
R-Super$_{100\%}$ [5]	$63.9_{28.5}$	$49.2_{26.9}$	$22.6_{25.5}$	81.1_{118}	$66.6_{53.2}$	$33.4_{26.9}$	$74.6_{38.3}$	$42.9_{23.7}$	$40.9_{28.7}$	$15.5_{17.4}$
GLeVE (Ours)$_{10\%}$	$53.9_{32.1}$	$38.2_{29.4}$	$18.4_{23.0}$	113_{148}	96.1_{131}	$62.7_{92.4}$	$74.1_{40.1}$	$35.4_{25.1}$	$33.9_{23.7}$	$12.8_{18.4}$
GLeVE (Ours)$_{25\%}$	$56.6_{29.8}$	$42.9_{27.1}$	$19.3_{21.4}$	98.4_{132}	83.3_{118}	$54.6_{85.7}$	$74.9_{37.8}$	$38.1_{24.5}$	$37.7_{21.4}$	$15.5_{18.6}$
GLeVE (Ours)$_{100\%}$	$65.8_{26.2}$	$48.7_{23.9}$	$27.8_{19.6}$	69.5_{109}	$60.2_{96.4}$	$29.5_{64.8}$	$76.2_{35.3}$	$44.3_{21.2}$	$38.9_{20.4}$	$17.8_{14.8}$

MedFormer [12]), and an advanced report-assisted method R-Super [5]. Averaged across organs, it increases Dice by 7.26% over the best universal segmentation method and reduces HD95 by 7.3 mm relative to the advanced method, indicating improved global geometric consistency and fewer extreme boundary errors. Lesion-level performance is likewise strong, reaching 76.2% LR and 33.7% LLS. Under weak supervision with only 10% or 25% masks, **GLeVE** remains competitive and often outperforms most alternatives, highlighting robustness to annotation scarcity and improved data efficiency.

Ablation Study. As reported in Table 2 and Fig. 4, replacing LeQu with a text encoder applied to structured descriptions reduces LR by 2.9%, indicating that explicit relational modeling preserves lesion semantics and enhances representation discriminability. Disabling AnVer reduces the organ-averaged LLS by 7.3%, indicating that candidate filtering and verification are critical for semantic alignment. Removing OcRe increases the organ-averaged HD95 by 4.2 mm, highlighting the contribution of hierarchical recursive refinement to accurate boundary delineation. Overall, the three modules are complementary in semantic modeling, verification, and refinement, collectively enabling robust localization.

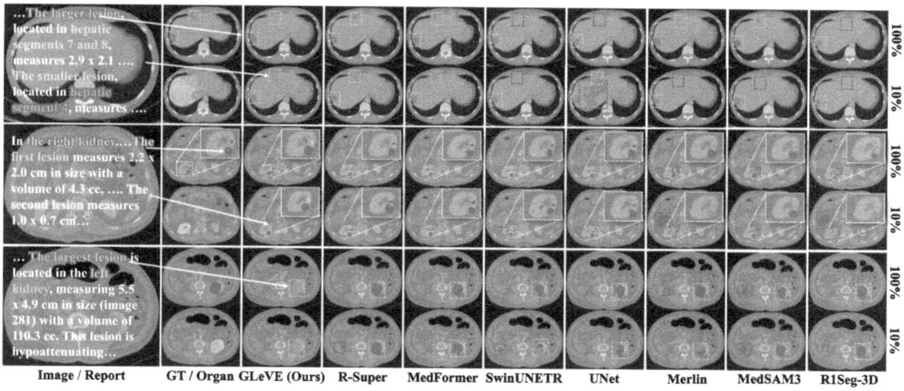

Fig. 3. Qualitative comparison of grounding results under 10% and 100% mask supervision, highlighting localization accuracy and multi-lesion disambiguation.

Table 2. Ablation analysis of key components under 100% mask supervision.

Method	Dice_{std} (%)↑			HD95_{std} (mm)↓			LR_{std}↑	LLS_{std} (%)↑		
	K	L	P	K	L	P	Avg	K	L	P
w/o LeQu	$59.5_{28.4}$	$42.5_{25.7}$	$23.5_{20.9}$	84.6_{124}	72.0_{108}	$39.3_{76.5}$	$73.3_{37.2}$	$40.9_{22.1}$	$32.8_{20.7}$	$16.4_{16.2}$
w/o AnVer	$55.9_{31.6}$	$39.7_{28.9}$	$20.3_{23.1}$	89.8_{134}	75.1_{121}	$47.6_{84.9}$	$69.3_{40.5}$	$34.6_{24.6}$	$30.5_{22.9}$	$13.9_{17.8}$
w/o OcRe	$63.7_{26.9}$	$44.3_{24.1}$	$25.4_{19.8}$	74.8_{112}	$64.0_{99.6}$	$33.0_{69.2}$	$75.1_{36.6}$	$42.2_{21.7}$	$33.9_{19.8}$	$17.6_{15.1}$
GLeVE (Ours)	$65.8_{26.2}$	$48.7_{23.9}$	$27.8_{19.6}$	69.5_{109}	$60.2_{96.4}$	$29.5_{64.8}$	$76.2_{35.3}$	$44.3_{21.2}$	$38.9_{20.4}$	$17.8_{14.8}$

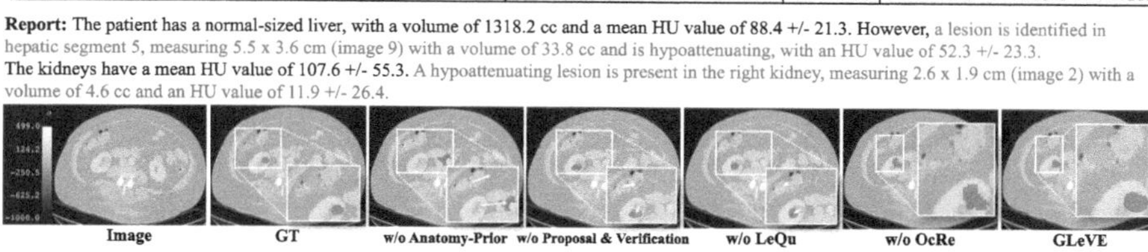

Fig. 4. Qualitative ablation results of GLeVE, showing the effect of removing Anatomical-Prior, Proposal & Verification, LeQu, and OcRe on lesion grounding and boundary refinement. Arrows indicate localization center offsets.

4 Conclusion

In this study, we propose **GLeVE** for verifiable, pixel-accurate lesion grounding from radiology reports in 3D CT. By modeling each lesion description as an atomic unit, **GLeVE** learns relation-aware lesion queries, enforces anatomy-consistent one-to-one text-lesion matching via soft modulation and region-level verification, and refines masks with octree-based autoregressive coarse-to-fine decoding; a report-driven objective further enables training under missing masks. Experiments on AbdomenAtlas 3.0 demonstrate consistent improvements over advanced methods, supporting robust grounding in challenging multi-lesion settings. Future work will explore extending the framework to broader anatomical regions and more diverse disease categories.

Acknowledgements. This work was supported by the National Natural Science Foundation of China (No. 62472133) and the Fundamental Research Funds for the Provincial Universities of Zhejiang (No. GK259909299001-006).

Disclosure of Interests. The authors declare no competing interests.

References

1. Alabed, S., Anderson, A., Maiter, A.: Large language models for simplifying radiology reports: a systematic review and meta-analysis of patient, public, and clinician evaluations. Lancet Digital Health (2026)
2. BAI, F., Du, Y., Huang, T., q.-h. Meng, M., Zhao, B.: M3D: advancing 3D medical image analysis with multi-modal large language models. In: International Conference on Learning Representations (2024)
3. Bai, X., Liu, M., Chen, Y., Yang, H., Tian, Q.: Chest-OMDL: organ-specific multidisease detection and localization in chest computed tomography using weakly supervised deep learning from free-text radiology report. In: Medical Imaging with Deep Learning (2025)
4. Bai, X., Liu, M., Song, T.: EXACT: an explainable anomaly-aware vision foundation model for analysis of 3D chest CT. arXiv preprint arXiv:2604.24146 (2026)
5. Bassi, P.R., Li, W., Chen, J.: Learning segmentation from radiology reports. In: Proceedings of Medical Image Computing and Computer Assisted Intervention, pp. 305–315 (2025)
6. Bassi, P.R., et al.: RadGPT: constructing 3D image-text tumor datasets. In: Proceedings of the IEEE/CVF International Conference on Computer Vision, pp. 23720–23730 (2025)
7. Blankemeier, L., Cohen, J.P., Kumar, A., et al.: Merlin: a vision language foundation model for 3D computed tomography. Res. Square pp. rs–3 (2024)
8. Busch, F., Hoffmann, L., Dos Santos, D.P., et al.: Large language models for structured reporting in radiology: past, present, and future. Eur. Radiol. **35**(5), 2589–2602 (2025)
9. Chen, H., et al.: MMLNB: multi-modal learning for neuroblastoma subtyping classification assisted with textual description generation. arXiv preprint arXiv:2503.12927 (2025)
10. Chen, Y., Zou, B., Guo, Z., et al.: SCUNet++: swin-UNet and CNN bottleneck hybrid architecture with multi-fusion dense skip connection for pulmonary embolism CT image segmentation. In: Proceedings of the IEEE/CVF Winter Conference on Applications of Computer Vision, pp. 7759–7767 (2024)
11. Dong, F., Nie, S., Chen, M., et al.: Keyword-based AI assistance in the generation of radiology reports: a pilot study. NPJ Dig. Med. **8**(1), 490 (2025)
12. Gao, Y., Zhou, M., Liu, D., Yan, Z., Zhang, S., Metaxas, D.N.: A data-scalable Transformer for medical image segmentation: architecture, model efficiency, and benchmark. arXiv preprint arXiv:2203.00131 (2023)
13. Hamamci, I.E., Er, S., Wang, C., Almas, F.: Generalist foundation models from a multimodal dataset for 3D computed tomography. Nat. Biomed. Eng. 1–19 (2026)
14. Hao, Q., Yu, L., Tian, S., Ye, X., Zhang, L.: R1Seg-3D: rethinking reasoning segmentation for medical 3D CTs. In: Proceedings of Medical Image Computing and Computer Assisted Intervention, pp. 415–425 (2025)

15. He, J., Li, P., Liu, G., Zhong, S.: Parameter-efficient fine-tuning medical multimodal large language models for medical visual grounding. In: International Symposium on Biomedical Imaging, pp. 1–5. IEEE (2025)
16. Huang, Z., Wang, H., Deng, Z., Ye, J., Su, Y., Sun, H., et al.: STU-net: scalable and transferable medical image segmentation models empowered by large-scale supervised pre-training. arXiv preprint arXiv:2304.06716 (2023)
17. Ichinose, A., Hatsutani, T., Nakamura, K.: Visual grounding of whole radiology reports for 3D CT images. In: Proceedings of Medical Image Computing and Computer Assisted Intervention, pp. 611–621 (2023)
18. Isensee, F., Jaeger, P.F., Kohl, S.A., Petersen, J., Maier-Hein, K.H.: nnU-Net: a self-configuring method for deep learning-based biomedical image segmentation. Nat. Methods **18**(2), 203–211 (2021)
19. Kulsoom, U., Glavin, F.G., Bendechache, M.: Natural language processing and machine learning for analysis of radiology reports-a systematic review. IEEE Access **13**, 112215–112254 (2025)
20. Li, Y., Kong, C., Zhao, G., Zhao, Z.: Automatic radiology report generation with deep learning: a comprehensive review of methods and advances. Artif. Intell. Rev. **58**(11), 344 (2025)
21. Liu, A., Xue, R., Cao, X.R.: MedSAM3: delving into segment anything with medical concepts. arXiv preprint arXiv:2511.19046 (2025)
22. Loshchilov, I., Hutter, F.: Decoupled weight decay regularization. In: International Conference on Learning Representations (2017)
23. Nützel, F., Dombrowski, M., Kainz, B.: Generate to ground: multimodal text conditioning boosts phrase grounding in medical vision-language models. In: Medical Imaging with Deep Learning (2025)
24. Ronneberger, O., Fischer, P., Brox, T.: U-net: convolutional networks for biomedical image segmentation. In: Navab, N., Hornegger, J., Wells, W.M., Frangi, A.F. (eds.) MICCAI 2015. LNCS, vol. 9351, pp. 234–241. Springer, Cham (2015). https://doi.org/10.1007/978-3-319-24574-4_28
25. Tang, Y., etr al.: Self-supervised pre-training of swin transformers for 3D medical image analysis. In: Proceedings of the IEEE/CVF Conference on Computer Vision and Pattern Recognition, pp. 20730–20740 (2022)
26. Tejani, A.S., Klontzas, M.E., Gatti, A.A., et al.: Checklist for artificial intelligence in medical imaging (CLAIM): 2024 update. Radiol. Artif. Intell. **6**(4), e240300 (2024)
27. Vilouras, K., Sanchez, P., O'Neil, A.Q., Tsaftaris, S.A.: Zero-shot medical phrase grounding with off-the-shelf diffusion models. IEEE J. Biomed. Health Inf. **29**(12), 9051–9059 (2025)
28. Wasserthal, J., Breit, H.C., Meyer, M.T., et al.: TotalSegmentator: robust segmentation of 104 anatomic structures in CT images. Radiol. Artif. Intell. **5**(5), e230024 (2023)
29. Yang, A., Li, A., Yang, B., et al.: Qwen3 technical report. arXiv preprint arXiv:2505.09388 (2025)
30. Yang, H., Zhou, H.Y., Liu, J.: A multimodal vision-language model for generalizable annotation-free pathology localization. Nat. Biomed. Eng. 1–15 (2026)
31. Zhao, Z., et al.: Large-vocabulary segmentation for medical images with text prompts. NPJ Dig. Med. **8**(1), 566 (2025)
32. Zou, K., Bai, Y., Liu, B., et al.: Uncertainty-aware medical diagnostic phrase identification and grounding. IEEE Trans. Pattern Anal. Mach. Intell. **47**(12), 11315–11329 (2025)

Hallucination Detection and Correction in Medical VLMs via Counter-Evidence Verification

Nan Zhou[1], Ke Zou[2], Meng Liu[1], Linchao He[1], Jiaqi Zhu[4], Yi Zhang[3], Hu Chen[1(✉)], and Huazhu Fu[5(✉)]

[1] The College of Computer Science, Sichuan University, Chengdu, China
huchen@scu.edu.cn

[2] Yong Loo Lin School of Medicine, National University of Singapore, Singapore, Singapore

[3] The School of Cyber Science and Engineering and the Key Laboratory of Data Protection and Intelligent Management, Ministry of Education, Sichuan University, Chengdu, China

[4] National Key Laboratory of Autonomous Intelligent Unmanned Systems, Beijing Institute of Technology, Beijing, China

[5] The Institute of High Performance Computing (IHPC), Agency for Science, Technology and Research (A*STAR), Singapore, Singapore
hzfu@ieee.org

Abstract. Vision-Language models (VLMs) reliability in medical diagnosis is challenged by trust-undermining hallucinations. Existing hallucination detection approaches mainly focus on identifying factual inconsistencies between generated text and reference data. While some studies analyze where models attend in images, they seldom verify whether such attention truly reflects the visual evidence supporting the generated text. To address this gap, we propose **Counter-Evidence Verification (CoEV)**, a training-free plug-and-play framework that detects and corrects hallucinations through evidence-based factual consistency verification. CoEV performs bidirectional verification between textual assertions and visual evidence, testing whether each statement is supported by its corresponding evidence region, and assigns each statement into a four-quadrant diagnostic map capturing combinations of text factuality and visual grounding. CoEV detects hallucinated content and serves as a post hoc refinement tool, correcting hallucinations without retraining. Extensive experiments on four medical datasets show that CoEV combats hallucinations in VLMs.For hallucination detection, CoEV consistently outperforms existing methods, improving average PR-AUC and ROC-AUC by 3.0% and 3.9% absolute points respectively, with notable gains of up to 18.5% in specific VQA scenarios. For hallucination correction, it improves Micro-F1 by up to 12.5%, reduces hallucination rates by over 11.9% on medical report generation, and also boosts medical VQA accuracy. These results show that CoEV enables reliable detection and correction of hallucinations, providing clinicians with dependable, evidence-based cues for diagnosis. Code is avaliable at https://github.com/Zhounan1222/CoEV.

G. Yang et al. (Eds.): MICCAI 2026, LNCS 16878, pp. 184–194, 2027.
https://doi.org/10.1007/978-3-032-38059-3_18

Keywords: Hallucination Detection · Counterfactual Intervention · Medical Vision-Language Models

1 Introduction

Medical Vision-Language Models (VLMs) [12,21,22,26] have demonstrated transformative potential in clinical workflows, enabling automated interpretation across diverse tasks such as Medical Visual Question Answering (Med-VQA), Radiology Report Generation (MRG), and precise phrase grounding [4,15,28,29]. By bridging high-dimensional visual features with complex clinical vocabulary, these models facilitate rapid diagnostic assistance. However, the deployment of VLMs in high-stakes clinical environments remains constrained by persistent hallucinations, where models generate clinically inaccurate, contradictory, or fabricated descriptions that lack support from underlying visual evidence [16]. In the medical domain, such ungrounded outputs are not merely linguistic errors; they represent significant safety risks, causing erroneous treatment recommendations and undermining clinician trust in AI-assisted decision-making [16].

Existing methods for hallucination detection primarily leverage uncertainty estimation, cross-model consistency checks [7,13,17,25], or traditional linguistic similarity metrics like BLEU [19] and ROUGE [14]. While these approaches provide a coarse-grained measure of output quality, they are often insensitive to the factual grounding required for medical accuracy. Specifically, textual similarity metrics fail to capture nuanced hallucinations arising from flawed visual reasoning, where a report may be fluently written but medically incorrect [7,17]. On the other hand, interpretability tools such as attention visualization or gradient-based saliency maps provide spatial intuition [1,20]; however, they merely highlight statistical correlations and do not verify the rigorous factual alignment between specific textual assertions and localized image regions. Consequently, there remains a need for a diagnostic framework that transcends simple detection to provide traceable evidence, identifying not only the presence of hallucinations but also localizing the underlying visual regions responsible for them.

To address this, we propose Counter-Evidence Verification (CoEV), a plug-and-play and training-free framework for detecting and correcting hallucinations in medical VLMs. Unlike methods requiring external models or fine-tuning [9,23], CoEV performs diagnosis-informed interventions by selectively masking visual regions as counterfactual probes. This tests whether assertions depend on visual evidence or linguistic priors. By analyzing behavioral shifts, CoEV introduces a Four-Quadrant Diagnostic Map to systematically categorize claims into states of factual consistency. As a post-hoc mechanism, CoEV enables effective mitigation during inference without retraining. Evaluations across four datasets and multiple VLMs demonstrate that CoEV significantly improves diagnostic accuracy and consistency, providing an interpretable pathway for clinical AI.

2 Method

2.1 Preliminaries

Hallucination Definition. Building on [16], we define hallucination $y \in \{0, 1\}$ as a binary phenomenon conditioned on both textual and visual factors. Given a generated text T_{gen} and its corresponding image I, CoEV models the hallucination detection process as:

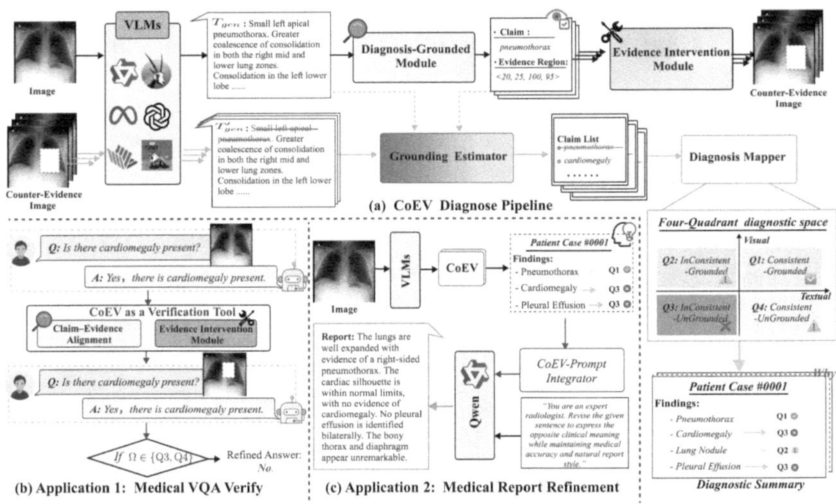

Fig. 1. (a) **CoEV Overview.** The CoEV diagnostic process performs counter-evidence verification and four-quadrant mapping to identify hallucinated claims. (b) Med-VQA Verification: Using diagnostic signals to validate answers and detect hallucinations. (c) Report Refinement: Rewriting hallucinated sentences using CoEV guidance.

$$y = \mathbf{1}\big[(f(T_{\text{gen}}) = 0) \vee (z(T_{\text{gen}}, I) = 0)\big], \tag{1}$$

where $f(\cdot)$ measures factual correctness and $z(\cdot)$ evaluates grounding consistency with visual evidence. Specifically, $y = 1$ indicates a hallucination, which defined as a claim that is either factually incorrect ($f = 0$) or insensitive to the masking of referenced visual regions ($z = 0$). Conversely, a grounded claim ($y = 0$) must be both factually accurate and visually dependent. Figure 1 illustrates how the CoEV integrate verification, reasoning, and visualization into an actionable diagnostic workflow for factual and visual correctness.

Diagnosis-Grounded Module (DGM). To operationalize the "What & Where" linkage, the DGM establishes grounding between claims and visual evidence. Let $\mathcal{P}(\cdot)$ be a semantic parser that extracts positive factual claims C_{pos} from generated text T_{gen}:

$$C_{\text{pos}} = \{c_i \mid \mathcal{P}(T_{\text{gen}})[i] = 1\}, \tag{2}$$

where label 1 indicates a positive clinical finding. For each $c_i \in C_{\text{pos}}$, a radiology expert $\mathcal{G}$ localizes a visual region $B_i \subset I$ with confidence $s_i \in [0, 1]$:

$$(B_i, s_i) = \mathcal{G}(I, c_i). \tag{3}$$

We define a grounding indicator $\delta_i \in \{0, 1\}$ as the basis for subsequent probing: $\delta_i = 1$ if $s_i > 0$, and $\delta_i = 0$ otherwise. This structured output from DGM serves as the prerequisite for our proposed counterfactual verification.

2.2 Counter-Evidence Consistency Evaluation (CECE).

The CECE module evaluates whether a generated claim is grounded in visual evidence or primarily driven by linguistic priors. CECE integrates two sub-processes: **Evidence Intervention Module (EIM) and Grounding Estimator (GE),** which jointly assess a claim's generation depends on visual support or linguistic bias. For each identified claim ($\delta_i = 1$), a counter-evidence image is constructed by masking its corresponding visual region B_i:

$$I_i' = \mathcal{M}(I, B_i), \tag{4}$$

where $\mathcal{M}$ is an intervention operator that masks B_i with black pixels while preserving the remaining image context. The VLM is then re-evaluated on the modified input to obtain $T_i' = g_\theta(I_i')$.

Let $\phi(c_i, T_i') \in \{0, 1\}$ indicate if claim c_i persists in the modified output T_i'. Specifically, T_i' is fed into the semantic parser ($\mathcal{P}(\cdot)$), and $\phi(c_i, T_i') = 1$ if c_i is still identified as a positive finding in T_i'; otherwise, $\phi(c_i, T_i') = 0$. The grounding attribution $z_i \in \{0, 1\}$ is defined as:

$$z_i = \begin{cases} 0, & \text{if } \phi(c_i, T_i') = 1, \\ 1, & \text{if } \phi(c_i, T_i') = 0, \end{cases} \tag{5}$$

Specifically, $z_i = 1$ confirms visual grounding via evidence reliance, while $z_i = 0$ reveals a prior-driven hallucination where the claim persists despite intervention. Unlike correctness-only methods, CECE explicitly verifies the *causal* dependence of claims on their supporting visual context.

2.3 Diagnosis Mapper (DM)

DM integrates factual correctness $f_i \in \{0, 1\}$ and grounding attribution $z_i \in \{0, 1\}$ into a unified diagnostic mapping $\mathcal{D}(f_i, z_i) \in \{Q_1, Q_2, Q_3, Q_4\}$:

$$\mathcal{D}(f_i, z_i) = \begin{cases} \text{Q1}, & f_i = 1, z_i = 1 \\ \text{Q2}, & f_i = 0, z_i = 1 \\ \text{Q3}, & f_i = 0, z_i = 0 \\ \text{Q4}, & f_i = 1, z_i = 0 \end{cases} \tag{6}$$

A claim is grounded ($y_i = 0$) if $(f_i, z_i) \in Q_1$, and hallucinated ($y_i = 1$) otherwise. These quadrants categorize assertions into factual alignment (Q_1), language-driven bias (Q_2), visual misinterpretation (Q_3), and total failure (Q_4). This taxonomy distinguishes medical inaccuracies from lack of visual dependence.

2.4 CoEV-Guided Correction Applications

CoEV enables correction across two downstream tasks: (1) verification for close-ended Med-VQA and (2) refinement for open-ended medical reports.

Med-VQA Verification. For a predicted answer $A = g_\theta(I, Q)$, CoEV evaluates its textual and visual integrity via $\Omega = \mathcal{F}(I, Q, A)$, ensuring the response is both accurate and visually grounded. where $\mathcal{F}$ denote the CoEV, and $\Omega \in \{Q_1, Q_2, Q_3, Q_4\}$. We define a verification operator that produces a corrected or validated answer:

$$\hat{A} = A \oplus \mathbb{I}[\Omega \in Q_2, Q_3] \tag{7}$$

This mechanism enables evidence-grounded verification of close-ended predictions and refines visually unsupported or hallucinated answers.

Report-Level Refinement. For MRG, CoEV enables sentence-level correction of hallucinated content. Let the baseline report be $R_{\text{gen}} = \{s_1, s_2, \dots, s_n\}$, and the diagnostic outcomes $\{(c_i, \Omega_i)\}$ be obtained by applying CoEV to each claim c_i in the report. Each sentence flagged as hallucinated ($\Omega_i \in \{Q_2, Q_3, Q_4\}$) is revised via a refinement operator $s_i' = \mathcal{R}(s_i, \mathrm{p})$, which prompts the LLM to rewrite the statement based on validated visual evidence, ensuring the final report is both factually accurate and grounded.

The refined report is reconstructed by replacing hallucinated sentences while leaving others unchanged:

$$\hat{R} = \begin{cases} s_i', & \text{if } \Omega_i \in \{Q_2, Q_3\}, \\ s_i, & \text{otherwise.} \end{cases} \tag{8}$$

This process leverages CoEV diagnostics to guide automatic correction, ensuring the final report aligns with factual and visual grounding consistency.

3 Experiments

3.1 Experimental Setup

Tasks and Datasets. We evaluate CoEV on two medical vision-language tasks: **(1) Med-VQA**: Conducted on VQA-RAD [11],and MIMIC-VQA [2], focusing on visually grounded clinical assertions.**(2) MRG:** Evaluated on MIMIC-CXR [10] and IU-Xray [6] datasets using official test splits. We utilized the GREEN model [18] to generate hallucination ground-truth labels by referencing gold-standard answers.

Table 1. Hallucination detection performance (%) across MRG and Med-VQA tasks. **PR**: PR-AUC; **ROC**: ROC-AUC. Best results are **bolded**, second-best are underline.

Method	MIMIC-CXR		IU-Xray		VQA-RAD		MIMIC-VQA	
	PR ↑	ROC ↑	PR ↑	ROC ↑	PR ↑	ROC ↑	PR ↑	ROC ↑
General-purpose Model: InternVL3-2B [27]								
RadFlag [25]	70.61	54.02	77.02	57.97	65.32	54.09	70.54	62.71
SE [7]	76.96	52.54	78.62	59.37	66.51	52.58	60.93	64.10
VASE [13]	81.66	54.43	80.00	61.76	68.48	58.11	62.79	68.06
CoEV (Ours)	**82.35**	**56.75**	**80.21**	**61.89**	**72.35**	**67.45**	**81.31**	**76.95**
General-purpose Model: Qwen2-3B [3]								
RadFlag [25]	73.41	56.68	87.56	62.82	82.99	**74.37**	79.91	61.08
SE [7]	78.74	58.96	89.20	62.63	86.16	70.15	61.66	60.18
VASE [13]	75.18	52.19	**89.92**	**68.09**	89.71	61.81	71.01	64.34
CoEV (Ours)	**79.20**	**60.04**	89.18	64.61	**92.16**	69.25	**90.89**	**72.65**
Medical-tuned Model: Lingshu-7B [24]								
RadFlag [25]	83.29	55.31	72.67	40.21	87.71	52.39	84.10	61.37
SE [7]	84.73	59.48	77.53	44.68	81.85	52.47	66.69	61.79
VASE [13]	**87.59**	61.87	77.89	49.66	87.45	58.78	65.29	65.51
CoEV (Ours)	87.10	63.67	**79.90**	**49.69**	**89.73**	**65.63**	**85.10**	**70.60**
Medical-tuned Model: LLaVA-Med [12]								
RadFlag [25]	91.53	53.13	70.61	54.02	71.86	61.12	79.54	54.11
SE [7]	92.76	54.94	66.70	52.54	63.00	64.34	62.40	62.00
VASE [13]	**94.26**	59.34	75.94	56.50	63.02	67.01	62.57	71.08
CoEV (Ours)	93.94	**61.97**	**76.17**	**59.95**	**86.57**	**68.94**	**72.06**	**76.12**

Evaluation Metrics. For hallucination detection, we employ PR-AUC and ROC-AUC. Report quality is assessed via diagnostic accuracy, following [4] (CheXpert-style F_1 scores: Mi-$F_{1\text{-}5}$ and Mi-$F_{1\text{-}14}$) and hallucination metrics: CHAIR-I/S [8] for instance/sentence-level errors, and MediHall [5] for clinically weighted penalties. VQA performance is measured by model accuracy. To assess generalizability, we benchmark CoEV across diverse VLMs, spanning general-purpose and medical-tuned architectures.

3.2 Results and Analyses

Hallucination Detection Performance. As shown in Table 1, CoEV consistently outperforms baselines across diverse model scales. Compared to the second-best method, CoEV achieves an average absolute improvement of 3.0% in PR-AUC and 3.1% in ROC-AUC. While SE exhibits instability in larger models and VASE yields moderate results, CoEV provides robust detection. Notably, on InternVL3-2B for MIMIC-VQA, CoEV improves PR-AUC and ROC-AUC

Table 2. Hallucination detection performance (%). **F5/F14**: diagnostic F_1 scores; **C-I/C-S**: CHAIR-I/S ↓; **MH**: MediHall ↑. ✓ denotes our CoEV framework.

Model	CoEV	MIMIC-CXR					IU-Xray				
		Acc ↑		Hallu			Acc ↑		Hallu		
		F_5	F_{14}	C-I↓	C-S↓	MH↑	F_5	F_{14}	C-I↓	C-S↓	MH↑
Qwen2-3B	✗	16.95	16.46	69.68	16.25	42.80	12.50	17.41	90.80	18.82	70.73
	✓	19.69	18.64	65.92	14.66	43.80	32.78	57.63	86.93	14.32	72.74
InternVL-8B	✗	49.80	40.30	49.10	21.33	70.78	24.84	56.64	61.33	7.71	86.24
	✓	64.96	47.89	38.67	14.63	72.24	30.65	57.86	49.47	6.09	87.71
LLava-Med	✗	18.76	21.12	74.06	27.08	34.01	22.83	43.41	86.98	23.77	76.50
	✓	23.80	28.29	72.92	23.43	36.35	12.50	57.50	86.35	20.35	77.91
Lingshu-7B	✗	30.35	24.93	47.35	9.53	85.38	25.97	57.97	56.45	1.90	90.88
	✓	34.15	26.66	37.08	6.47	87.09	30.77	57.55	41.30	10.79	91.70

Table 3. Performance comparison on MIMIC-VQA and VQA-RAD. "Ours" indicates applying our refinement. Improvement (↑) is shown in absolute %.

Model	MIMIC-VQA			VQA-RAD		
	w/o Ours	w/ Ours	↑	w/o Ours	w/ Ours	↑
Qwen2-3B	51.83	78.99	**27.16**	71.31	78.09	**6.78**
InternVL-8B	60.38	74.13	**13.75**	76.89	82.47	**5.58**
LLaVA-Med	68.56	78.99	**10.43**	71.35	85.66	**14.31**
Lingshu-7B	64.42	78.22	**13.80**	78.09	91.24	**13.15**

by 10.77% and 8.89% respectively. This highlights that our framework effectively isolates ungrounded clinical claims where purely statistical metrics fail. The performance gain across both open-set reporting and closed-set VQA tasks underscores CoEV's generalizability in safety-critical medical environments.

CoEV as a Post-hoc Refinement Tool. Beyond detection, CoEV serves as an effective pipeline for downstream clinical tasks. **(1) Medical Report Refinement:** As shown in Table 2, CoEV significantly enhances diagnostic accuracy. On average, $F_{1\text{-}5}$ improves by 7.01% on MIMIC-CXR and 8.16% on IU-Xray. Notably, for InternVL-8B, $F_{1\text{-}5}$ increases from 49.80% to 64.96% (+15.16%) on MIMIC-CXR. The reduction in hallucinations is evidenced by decreased CHAIR scores (e.g., C-I drops by 11.86% for IU-Xray on Qwen2-3B) and improved MediHall indices, confirming that CoEV successfully prunes unsupported clinical assertions while preserving factual findings. **(2) Med-VQA Verification:** According to Table 3, CoEV yields a consistent mean accuracy gain of 13.04% across evaluated datasets. Smaller models exhibit the most pronounced gains, such as a +27.16% absolute increase for Qwen2-3B on MIMIC-VQA.

Table 4. Ablation of the contribution of CoEV's two diagnostic axes on MIMIC-CXR.

Textual Axis	Visual Axis	Precision	Recall	F1-Score
✗	✗	0.5233	0.4095	0.5064
✗	✓	0.6307	0.4532	0.5460
✓	✗	0.7474	0.5042	0.5986
✓	✓	0.7750	0.5046	0.6112

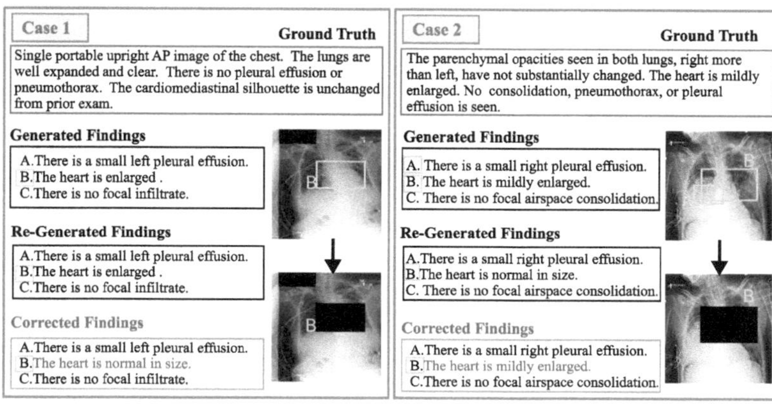

Fig. 2. Representative cases. **Case 1** (Q_2): description persists after masking, indicating ungrounded bias. **Case 2** (Q_1): description disappears post-masking, reflecting evidence-based reasoning. CoEV refines reports by aligning claims with visual evidence.

Even for specialized models like Lingshu-7B, CoEV provides steady improvements (avg. +13.47%), underscoring its architecture-agnostic robustness. These results demonstrate that CoEV effectively mitigates hallucinations to improve the reliability of medical visual reasoning.

3.3 Ablation Study

We ablate CoEV's two core diagnostic dimensions: (1) the *textual axis*, which evaluates textual consistency, and (2) the *visual axis*, which evaluates whether a claim depends on the visual evidence.

As shown in Table 4, removing both axes yields the lowest performance with an F1-Score of 0.5064. Using only the visual axis increases the F1-Score to 0.5460, indicating that visual dependence carries strong discriminative value for identifying ungrounded assertions. Introducing only the textual axis improves precision by more than 20%, yet it remains limited because visually unsupported but plausible statements cannot be distinguished. Combining both axes achieves the highest overall F1-Score of 0.6112. This joint model consistently improves precision and recall simultaneously, demonstrating that factual correctness and visual dependence provide complementary signals. This confirms that CoEV's two-axis decomposition offers a meaningful and more discriminative structure for characterizing hallucination types. Figure 2 illustrates CoEV's ability to distinguish and refine unsupported claims. In multi-disease scenarios, masking region A impacts only claim A, confirming that CoEV captures localized causal effects instead of global OOD noise, thus ensuring reliable refinement.

4 Conclusion

We present **CoEV**, a unified **Counter-Evidence Verification** framework for detecting and correcting hallucinations in medical VLMs. Unlike prior methods that constrain outputs during generation, CoEV enables a plug-and-play post-hoc evaluation that integrates seamlessly into existing clinical workflows. By intervening on visual evidence and mapping claims onto a four-quadrant diagnostic space, CoEV provides fine-grained, interpretable explanations that help clinicians trace the origin of model errors. Extensive experiments on MRG and Med-VQA benchmarks demonstrate that CoEV consistently outperforms baselines, significantly enhancing factual accuracy while reducing hallucination rates across various architectures. Ultimately, CoEV establishes a robust foundation for evidence-based reliability in safety-critical AI. Looking forward, we aim to extend this framework to address multi-modal inconsistencies and explore its deployment in real-time clinical decision support systems.

Acknowledgements. This work was supported in part by the Sichuan Natural Science Foundation under Grant 2026NSFSC0426, the National Natural Science Foundation of China under Grant U25A20439, the National Natural Science Foundation of China under Grant 624B2027, and the Singapore National Medical Research Council under Grant MOH-001745-00.

Disclosure of Interests. The authors have no competing interests to declare that are relevant to the content of this article.

References

1. An, J., Joe, I.: Attention map-guided visual explanations for deep neural networks. Appl. Sci. **12**(8), 3846 (2022)
2. Bae, S., et al.: Mimic-ext-mimic-cxr-vqa: a complex, diverse, and large-scale visual question answering dataset for chest x-ray images (2024)
3. Bai, J., et al.: Qwen-vl: a frontier large vision-language model with versatile abilities, **1**(2), 3 (2023). arXiv preprint arXiv:2308.12966
4. Bannur, S., et al.: Maira-2: grounded radiology report generation. arXiv preprint arXiv:2406.04449 (2024)
5. Chen, J., et al.: Detecting and evaluating medical hallucinations in large vision language models. arXiv preprint arXiv:2406.10185 (2024)

6. Demner-Fushman, D., et al.: Preparing a collection of radiology examinations for distribution and retrieval. J. Am. Med. Inf. Assoc. **23**(2), 304–310 (2015)
7. Farquhar, S., Kossen, J., Kuhn, L., Gal, Y.: Detecting hallucinations in large language models using semantic entropy. Nature **630**(8017), 625–630 (2024)
8. Gu, Z., Chen, J., Liu, F., Yin, C., Zhang, P.: Medvh: toward systematic evaluation of hallucination for large vision language models in the medical context. Adv. Intell. Syst. 2500255 (2025)
9. Gunjal, A., Yin, J., Bas, E.: Detecting and preventing hallucinations in large vision language models. In: Proceedings of the AAAI Conference on Artificial Intelligence, vol. 38, pp. 18135–18143 (2024)
10. Johnson, A.E., et al.: Mimic-cxr, a de-identified publicly available database of chest radiographs with free-text reports. Sci. Data **6**(1), 317 (2019)
11. Lau, J.J., Gayen, S., Ben Abacha, A., Demner-Fushman, D.: A dataset of clinically generated visual questions and answers about radiology images. Sci. Data **5**(1), 1–10 (2018)
12. Li, C., et al.: Llava-med: training a large language-and-vision assistant for biomedicine in one day. Adv. Neural. Inf. Process. Syst. **36**, 28541–28564 (2023)
13. Liao, Z., Hu, S., Zou, K., Fu, H., Zhen, L., Xia, Y.: Vision-amplified semantic entropy for hallucination detection in medical visual question answering. In: International Conference on Medical Image Computing and Computer-Assisted Intervention, pp. 669–679. Springer, Heidelberg (2025). DOI: https://doi.org/10.1007/978-3-032-04971-1_63
14. Lin, C.Y.: Rouge: a package for automatic evaluation of summaries. In: Text Summarization Branches Out, pp. 74–81 (2004)
15. Liu, B., et al.: Gemex: a large-scale, groundable, and explainable medical vqa benchmark for chest x-ray diagnosis. In: Proceedings of the IEEE/CVF International Conference on Computer Vision, pp. 21310–21320 (2025)
16. Liu, H., et al.: A survey on hallucination in large vision-language models. arXiv preprint arXiv:2402.00253 (2024)
17. Moslonka, C., Randrianarivo, H., Garnier, A., Malherbe, E.: Learned hallucination detection in black-box llms using token-level entropy production rate. arXiv preprint arXiv:2509.04492 (2025)
18. Ostmeier, S., et al.: Green: generative radiology report evaluation and error notation. In: Findings of the Association for Computational Linguistics: EMNLP 2024, pp. 374–390 (2024)
19. Papineni, K., Roukos, S., Ward, T., Zhu, W.J.: Bleu: a method for automatic evaluation of machine translation. In: Proceedings of the 40th Annual Meeting of the Association for Computational Linguistics, pp. 311–318 (2002)
20. Raghavan, K., B, S., v, K.: Attention guided grad-cam: an improved explainable artificial intelligence model for infrared breast cancer detection. Multimedia Tools Appl. **83**(19), 57551–57578 (2024)
21. Sellergren, A., et al.: Medgemma technical report. arXiv preprint arXiv:2507.05201 (2025)
22. Thawkar, O., et al.: Xraygpt: chest radiographs summarization using medical vision-language models. arXiv preprint arXiv:2306.07971 (2023)
23. Xiao, W., et al.: Detecting and mitigating hallucination in large vision language models via fine-grained ai feedback. In: Proceedings of the AAAI Conference on Artificial Intelligence, vol. 39, pp. 25543–25551 (2025)
24. Xu, W., et al.: Lingshu: a generalist foundation model for unified multimodal medical understanding and reasoning. arXiv preprint arXiv:2506.07044 (2025)

25. Zhang, S., et al.: Radflag: a black-box hallucination detection method for medical vision language models. arXiv preprint arXiv:2411.00299 (2024)
26. Zhang, S., et al.: Biomedclip: a multimodal biomedical foundation model pretrained from fifteen million scientific image-text pairs. arXiv:2303.00915 (2023)
27. Zhu, J., et al.: Internvl3: exploring advanced training and test-time recipes for open-source multimodal models. arXiv preprint arXiv:2504.10479 (2025)
28. Zou, K., et al.: Medrg: medical report grounding with multi-modal large language model. arXiv e-prints pp. arXiv–2404 (2024)
29. Zou, K., et al.: Uncertainty-aware medical diagnostic phrase identification and grounding. IEEE Trans. Pattern Anal. Mach. Intell. (2025)

Hepato-LLaVA: An Expert MLLM with Sparse Topo-Pack Attention for Hepatocellular Pathology Analysis on Whole Slide Images

Yuxuan Yang[1,3], Zhonghao Yan[1,3], Yi Zhang[2], Bo Yun[1,3], Muxi Diao[1,3], Guowei Zhao[2], Kongming Liang[1,3(✉)], Wenbin Li[2(✉)], and Zhanyu Ma[1,3]

[1] Beijing University of Posts and Telecommunications, Beijing, China
{wssf51379,zhonghao.yan,liangkongming}@bupt.edu.cn
[2] National Cancer Center/National Clinical Research Center for Cancer/Cancer Hospital, Chinese Academy of Medical Sciences and Peking Union Medical College, Beijing, China
{zhangyi,liwenbin}@cicams.ac.cn
[3] Beijing Key Laboratory of Multimodal Data Intelligent Perception and Governance, Beijing, China

Abstract. Hepatocellular Carcinoma diagnosis relies heavily on the interpretation of gigapixel Whole Slide Images. However, current computational approaches are constrained by fixed-resolution processing mechanisms and inefficient feature aggregation, which inevitably lead to either severe information loss or high feature redundancy. To address these challenges, we propose Hepato-LLaVA, a specialized Multi-modal Large Language Model designed for fine-grained hepatocellular pathology analysis. We introduce a novel Sparse Topo-Pack Attention mechanism that explicitly models 2D tissue topology. This mechanism effectively aggregates local diagnostic evidence into semantic summary tokens while preserving global context. Furthermore, to overcome the lack of multi-scale data, we present HepatoPathoVQA, a clinically grounded dataset comprising 33K hierarchically structured question-answer pairs validated by expert pathologists. Our experiments demonstrate that Hepato-LLaVA achieves state-of-the-art performance on HCC diagnosis and captioning tasks, significantly outperforming existing methods. Our code and implementation details are available at https://github.com/PRIS-CV/Hepto-LLaVA.

Keywords: Hepatocellular Carcinoma · Multi-modal Large Language Models · Whole Slide Images · Digital Pathology

---

Y. Yang, Z. Yan and Y. Zhang—Equal Contribution.
Z. Yan—Project Lead.

G. Yang et al. (Eds.): MICCAI 2026, LNCS 16878, pp. 195–205, 2027.
https://doi.org/10.1007/978-3-032-38059-3_19

1 Introduction

Hepatocellular Carcinoma (HCC), a leading cause of global cancer mortality, relies on histopathological Whole Slide Images (WSIs) examination as the gold standard. However, HCC's high heterogeneity and complex microenvironment render gigapixel WSIs interpretation labor-intensive. Manual analysis is prone to inter-observer variability, especially for subtle early-stage lesions. Current Multiple Instance Learning (MIL) approaches [3,10,12,18] are often limited to classification, lacking clinical diagnostic capacity. This has catalyzed WSI-based Multi-modal Large Language Models (MLLMs) [2,7,9,15,17], which align visual features with language to enable Visual Question Answering (VQA).

A key design choice in these MLLMs is how to represent gigapixel WSIs for MLLMs. Existing methods typically fall into two categories: thumbnail-based approaches [9,17] that resize the gigapixel image into a megapixel image, and slide-encoder-based approaches [2,7] that aggregate thousands of patches into global tokens. However, thumbnail-based approaches inevitably lose patch-level details, while both methods ingest only WSIs inputs, consequently less equipped with patch-level capabilities essential for diagnosis. We investigate two primary research questions regarding the slide encoder, which acts as a bottleneck:

- **RQ1:** How can the slide encoder compress HCC WSIs into representations that preserve critical diagnostic details while minimizing redundancy?
- **RQ2:** How can the slide encoder accommodate variable-resolution inputs to generate features for multi-scale diagnosis in complex liver tissues?

To address **RQ1**, we design a **Sparse Topo-Pack Attention** mechanism that simulates the local aggregation and global juxtaposition of pathological diagnosis. By modeling 2D tissue topology, the slide encoder extracts topology-informed representations while significantly reducing spatial redundancy. A light-weight connector then condenses these features into fixed-length query tokens, effectively bridging the modality gap without losing critical diagnostic details. To address **RQ2**, we leverage hierarchical features to develop **HepatoPathoVQA**, a comprehensive multi-scale VQA dataset specifically for HCC. This dataset contains over 33K expert-validated QA pairs covering the full clinical workflow across three distinct scales: WSIs, Region of Interest (ROI, 5× mag.), and Patch (10× and 20× mag.). Finally, through Low-Rank Adaptation (LoRA) fine-tuning on this unified framework, we present **Hepato-LLaVA**, a specialized MLLM optimized for fine-grained hepatocellular pathology analysis.

In summary, our three main contributions are:

1. We construct **HepatoPathoVQA**, the first multi-scale WSI dataset for HCC, comprising over 33K QA pairs across three scales, which improves multi-scale modeling and bridges data with real-world clinical practice.
2. We introduce a **Sparse Topo-Pack Attention** mechanism that models 2D tissue topology, extracting global-context-aware representations for WSIs while mitigating information redundancy.

3. We present **Hepato-LLaVA**, a specialized MLLM optimized via a three-stage pipeline, achieving an improvement of 20% in average diagnostic accuracy in HCC over existing open-source pathology MLLMs.

2 Methods

2.1 Preliminary

Patch Encoding. Let $\mathbf{I}_{raw} \in \mathbb{R}^{H_{raw} \times W_{raw} \times 3}$ denote raw WSIs. We tessellate the tissue region into non-overlapping patches of size $P \times P$. This forms a grid layout of dimensions $H \times W$, where $H = \lfloor H_{raw}/P \rfloor$ and $W = \lfloor W_{raw}/P \rfloor$. Let $x_{i,j}$ represent the patch at the spatial coordinate (i, j) for $1 \leq i \leq H, 1 \leq j \leq W$. A frozen feature encoder f_{enc} then maps each patch into a D-dimensional embedding space, yielding a feature grid:

$$\mathbf{H}_{grid} = \left\{ \mathbf{h}_{i,j} \in \mathbb{R}^D \mid \mathbf{h}_{i,j} = f_{enc}(x_{i,j}) \right\}_{1 \leq i \leq H,\, 1 \leq j \leq W} \tag{1}$$

Slide Encoding. Since the raw feature $\mathbf{H}_{grid}$ contains thousands of patches, directly utilizing it for slide-level tasks introduces excessive dimensionality and redundant information. To address this, a slide encoder $\mathcal{T}(\cdot)$ is employed to extract semantic features from the flattened patch sequence $\mathbf{S}_{patch} = \text{Flatten}(\mathbf{H}_{grid})$:

$$\mathbf{H}_{slide} = \mathcal{T}(\mathbf{S}_{patch}) \tag{2}$$

In conventional MIL, $\mathcal{T}$ acts as a global aggregation function, where the output $\mathbf{H}_{slide} \in \mathbb{R}^{1 \times D}$ is a single compact token. In contrast, for Transformer-based approaches, $\mathcal{T}$ typically functions as a self-attention network where the output $\mathbf{H}_{slide} \in \mathbb{R}^{N \times D}$, often requiring a subsequent pooling to obtain a global token.

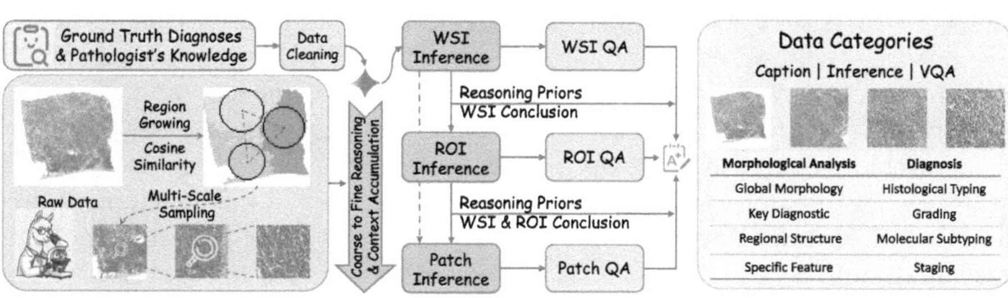

Fig. 1. Overview of the **HepatoPathoVQA** construction pipeline: (1) Extracts ROIs and Patches from WSIs using MST-based clustering and triangular seed-point selection. (2) Employs Gemini-3-flash for hierarchical inference by integrating macroscopic descriptions as context for subsequent microscopic analysis. (3) Generates multi-scale QA pairs and captions for instruction tuning and alignment.

2.2 HepatoPathoVQA Dataset

Dataset Construction. We collected 200 WSIs containing HCC, with all patient identifiers removed to ensure privacy. From these WSIs, we constructed **HepatoPathoVQA**, a multi-scale dataset featuring 33K QA pairs centered on morphological analysis and diagnosis. As shown in Fig. 1, following the standard diagnostic workflow of pathologists, we developed a data generation pipeline using Gemini-3-flash [4]. This pipeline simulates the clinical reasoning process, transitioning from macroscopic observation to microscopic details. Each step in the pipeline is designed as an atomic operation. The data construction pipeline consists of three stages. **(1) Hierarchical Multi-Scale Sampling.** WSIs serve as the raw input. We identify ROIs using a triangular seed-point algorithm and a Minimum Spanning Tree (MST). The MST aggregates adjacent patches based on cosine similarity to form three candidate regions. These ROIs represent a 5$\times$ magnification based on medical priors of cell adjacency. We manually remove uninformative regions. Finally, 10$\times$ and 20$\times$ local patches are randomly sampled around the ROI centers. **(2) Context-Injected Reasoning Generation.** Since Gemini cannot process gigapixel images, all regions are resized to 2048$\times$2048 pixels. We design scale-specific prompts based on clinical focus and diagnostic status. Gemini generates descriptions in a hierarchical manner. Descriptions from higher scales serve as context for the next scale to ensure logical consistency from macroscopic to microscopic levels. **(3) Scale-aware Data Synthesis.** Leveraging the generated hierarchical descriptions, we generate diverse multi-scale QA pairs for instruction tuning. These pairs cover all three resolutions, enabling the model to perform both granular morphological analysis and holistic clinical diagnosis. Furthermore, these descriptions are utilized to produce high-quality image-text captions, which serve as the foundation for connector pre-training to achieve robust vision-language alignment.

Dataset Statistics. HepatoPathoVQA comprises 33,332 QA pairs. At our three proposed spatial scales, we follow the WSI-LLaVA [7] setup to prompt the model to generate reasoning chains along two axes—morphology and diagnosis. Each dimension is further subdivided into four subcategories, thereby supporting hierarchical diagnosis from overall patterns to cellular details. Beyond the VQA pairs, we provide **HepatoPathoCaption** (3,288 pairs) for alignment pre-training and **HepatoPathoBench** (3,056 pairs) split from HepatoPathoVQA for evaluation. The training and validation sets were strictly separated at the patient level. The entire dataset construction process was validated by three expert pathologists under a blind evaluation protocol, yielding high inter-rater consistency (Pearson $r = 0.96$) and a low rejection rate ($< 2\%$).

2.3 Sparse Topo-Pack Attention

Topological Mismatch in Existing Modeling. Existing pathology-specific MLLM predominantly rely on Longnet to extract features, treating the WSIs as a flattened 1D sequence. This assumption neglects the intrinsic 2D topological properties of pathological tissues. Biologically, semantic dependencies vary significantly across scales: local regions share a common semantic meaning(e.g., tumor margins crossing multiple patches), whereas distant regions are distinct with lower semantic coupling. To address this, we propose a hierarchical sparse attention mechanism that restores topological priors (Fig. 2, upper panel).

Hierarchical Sequence Construction. Unlike standard flattening, we construct a structured input sequence that explicitly encodes the grid hierarchy. We organize this grid into M Summary Packs in row-major order. Each pack $\mathbf{P}_m$ represents a local $k \times k$ window. We generate the pack summary token $\mathbf{s}_m$ by resizing the local image region of pack m and passing it through the patch encoder f_{enc} for semantic anchors. Similarly, the global token $\mathbf{g}_{global}$ is derived by resizing entire WSIs and encoding it. The total sequence $\mathbf{S}_{in}$ is formed as:

$$\mathbf{S}_{in} = [\ \mathbf{g}_{global},\ \underbrace{\mathbf{h}^1_{1,1}, \dots, \mathbf{h}^1_{k,k}, \mathbf{s}_1}_{\text{Pack } \mathbf{P}_1},\ \dots,\ \underbrace{\mathbf{h}^M_{1,1}, \dots, \mathbf{h}^M_{k,k}, \mathbf{s}_M}_{\text{Pack } \mathbf{P}_M}\ ] \tag{3}$$

where $\mathbf{h}^m_{i,j}$ represents the fine-grained patch tokens within the m-th pack. In our implementation, we set $k = 3$, resulting in 10 tokens per pack. Through the hierarchical attention mechanism, the summary token functions as a dynamic query that aggregates diagnostic evidence from patch tokens $\mathbf{h}^m_{i,j}$.

Hierarchical Sparse Mask. We define a hierarchical mask $\mathcal{M}$ to enforce the specific interaction rules. Let Ω_P and Ω_S denote the sets of all patch tokens and all summary tokens, respectively. Let $\mathcal{P}(t)$ denote the pack index of a token t. For any query token t_i and key token t_j in $\mathbf{S}_{in}$, the attention mask $\mathcal{M}_{i,j}$ is:

$$\mathcal{M}_{i,j} = \begin{cases} 0 & \text{if } t_j = \mathbf{g}_{global} \quad \text{(Global Sink)} \\ 0 & \text{if } t_i, t_j \in \Omega_P \wedge \mathcal{P}(t_i) = \mathcal{P}(t_j) \quad \text{(Intra-Pack Dense)} \\ 0 & \text{if } t_i \in \Omega_S \wedge t_j \in \Omega_P \wedge \mathcal{P}(t_i) = \mathcal{P}(t_j) \quad \text{(Aggregation)} \\ 0 & \text{if } t_i, t_j \in \Omega_S \quad \text{(Summary-Level Interaction)} \\ -\infty & \text{otherwise} \end{cases} \tag{4}$$

This hierarchical architecture effectively models pathological topology by harmonizing dual-scale interactions. The global token provides a macro-level reference, enabling local windows to focus on aggregating dense features into summary tokens. Summary tokens then facilitate long-range modeling, preserving the structural integrity of tissue regions across entire slide. In our implementation ($k = 3$), this sparsity reduces the theoretical attention overhead to $\approx$ **1%** of the dense counterpart. The end-to-end feature extraction takes 4.3 min per slide on average, compared to the 19.3 mins/slide required by Prov-GigaPath [13].

2.4 Hepato-LLaVA

MAE Pretrain. As shown in the lower panel of Fig. 2, we applied MAE pre-training on a composite dataset consisting of TCGA [11] and our internal data to optimize feature extraction. The dataset was augmented via spatial chunking and geometric flipping. The core innovation lies in a two-stage curriculum masking strategy: **Phase 1** implements patch-wise masking to reconstruct missing patches within intact packs, focusing on capturing tissue textures; **Phase 2** advances to pack-wise masking, where entire packs are removed to enforce long-range dependency modeling, thereby capturing high-level structural patterns.

MoCo Pretrain. Following MAE pre-training, this stage utilizes the Momentum Contrast (MoCo) [5] architecture. We utilize the un-augmented dataset to prevent the inclusion of augmented views as negative samples. This stage aims to align visual representations at the **Summary Token level**, focusing on tissue semantics rather than holistic descriptors. Addressing the high I/O costs of WSIs processing, we adapt the MoCo framework to operate on feature-level, generating positive pairs by injecting noise directly into token features. Coupled with a momentum-updated encoder and a queue containing cross-case negative samples, the model is optimized via InfoNCE loss to discriminate specific regions.

Multi-scale Instruction Tuning. To effectively bridge the modality gap between the pre-trained, frozen slide encoder and the MLLM, we introduce a Q-Former [6] Connector. To handle variable-length pack summary token sequences, this lightweight module employs a set of learnable query vectors to interact with inputs. The instruction tuning process is further divided into two sub-phases: (1) Visual-Language Alignment: We first train the connector on the **HepatoPath-Caption** dataset while keeping both the slide encoder and MLLM frozen. This

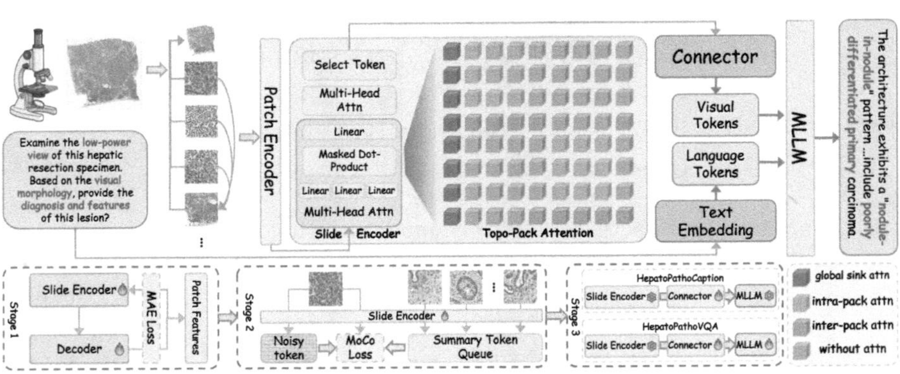

Fig. 2. Overview of the **Hepato-LLaVA** framework: (Upper) Incorporates Sparse Topo-Pack Attention into the model architecture. (Lower) Implements a three-stage training pipeline: MAE pre-training, MoCo pre-training, and instruction tuning (via LoRA). The sparse attention mask defines three topological interactions: (1) Global Sink for macro-context broadcasting, (2) Intra-Pack for local dense interactions, and (3) Inter-Pack for summary-level connections across packs.

stage aligns visual representations with linguistic semantics, enabling the connector to translate pathological features. (2) Diagnostic Instruction Tuning: We fine-tune both the connector and the MLLM on the **HepatoPath-VQA** dataset. This stage optimizes the model for diagnostic inference, enabling it to interpret multi-scale visual evidence and generate clinically precise responses.

3 Experiments

3.1 Experimental Settings

Datasets. The slide encoder is pre-trained on 10K WSIs aggregated from TCGA [11], HCMI [8], and internal dataset. For instruction tuning, we utilize HepatoPathoCaption (3K pairs) and HepatoPathoVQA (33K pairs). Evaluation is performed on HepatoPathoBench (3K pairs), strictly disjoint from the training.

Models. We conduct a comparison across three types of 7B models. For general medical models, we evaluate HuatuoGPT [16] and Lingshu [14]. For thumbnail-based pathology MLLMs, we include Quilt-LLaVA [9] and Patho-R1 [17](inputs resized to their native square resolutions). For WSI-based pathology MLLMs with sparse attention, we compare against SlideChat [2] and WSI-LLaVA [7].

Metrics. To evaluate the clinical accuracy of open-ended responses, we employ METEOR [1] and WSI-P [7], which leverage LLM to evaluate predictions, providing a more reliable proxy for pathologist expertise. For single-choice questions, we adopt standard Accuracy based on a strict match. For multiple-choice questions, predictions that are correct but incomplete are assigned a score of 0.5.

Table 1. Evaluation on HepatoPathoBench. Baselines include general medical MLLMs and pathology-specific MLLMs. Single/Multi refer to single/multiple choice tasks. WSI, ROI, and Patch denote the spatial scales. The Avg score is calculated across all tasks. Best results are **bold**; second-best are <u>underlined</u>

Model	Input	Morphological Analysis				Diagnosis				Multi-scale			Avg
		Open		Close		Open		Close					
		WSI-P	METEOR	Single	Multi	WSI-P	METEOR	Single	Multi	WSI	ROI	Patch	
Lingshu	Thumbnail	0.53	0.17	0.38	0.44	<u>0.73</u>	0.18	0.39	0.38	0.52	0.52	0.49	0.50
Huatuo-GPT	Thumbnail	<u>0.74</u>	<u>0.24</u>	0.81	0.45	0.70	<u>0.23</u>	0.59	0.32	0.60	0.65	0.65	0.65
Quilt-LLaVA	Thumbnail	0.64	0.22	0.47	0.32	0.56	0.15	0.57	0.37	0.57	0.60	0.55	0.57
Patho-R1	Thumbnail	0.66	0.19	<u>0.87</u>	<u>0.50</u>	0.20	0.05	0.59	<u>0.45</u>	0.55	0.55	0.54	0.55
SlideChat	WSI	0.70	0.17	<u>0.87</u>	0.47	0.72	0.14	0.63	0.39	<u>0.66</u>	<u>0.68</u>	<u>0.66</u>	<u>0.66</u>
WSI-LLaVA	WSI	0.69	0.20	0.84	0.46	0.67	0.16	<u>0.65</u>	0.36	0.65	0.67	0.64	0.65
Hepato-LLaVA	WSI	**0.79**	**0.33**	**0.97**	**0.88**	**0.75**	**0.33**	**0.87**	**0.68**	**0.82**	**0.83**	**0.83**	**0.83**

3.2 Main Results

Hepato-LLaVA is initialized from WSI-LLaVA, which is fine-tuned using LoRA (r=128), while the connector is trained from scratch. We evaluated Hepato-LLaVA on HepatoPathoBench against six baselines. As shown in Table 1, while Thumbnail-based models yielded suboptimal results (Avg 0.50–0.57) due to severe information loss, Hepato-LLaVA achieved state-of-the-art performance (Avg **0.83**). This establishes a substantial lead of 0.17 over the best WSI-based runner-up SlideChat (0.66). In open-ended tasks, Hepato-LLaVA achieved a WSI-P of 0.79 in morphology and 0.75 in diagnosis, surpassing the SlideChat (0.70, 0.72). In close-ended tasks, our model attained 0.97 accuracy in morphological single-choice and 0.88 in multi-choice, significantly outperforming Patho-R1 (0.87, 0.50), a strong baseline enhanced by reinforcement learning. Regarding multi-scale consistency, Hepato-LLaVA demonstrated robust performance across WSI (0.82), ROI (0.83), and Patch (0.83) levels. This represents an improvement over our backbone model WSI-LLaVA (0.65, 0.67, 0.64), validating that the proposed sparse attention and connector effectively overcomes the scale variance limitations inherent in the base architecture. Qualitative evidence in Fig. 3 confirms its ability to interpret critical morphological evidence for precise diagnosis.

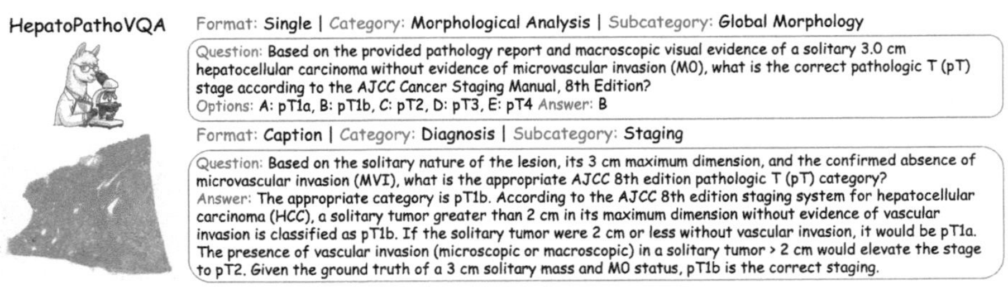

Fig. 3. A representative diagnostic case from HepatoPathoVQA, illustrating Hepato-LLaVA's capability to provide precise by interpreting critical morphological evidence.

Table 2. Ablation of connector types and two-stage training strategy. For two-stage setting, we pre-train the connector on HepatoPathCaption with MLLM frozen.

Connector	Two-Stage	WSI	ROI	Patch
wsi-llava	✗	81.61	88.05	86.83
mlp	✗	74.43	82.83	79.97
mlp	✓	83.33	87.24	83.66
qformer	✗	**88.51**	87.82	85.34
qformer	✓	**88.51**	**88.75**	**88.10**

Table 3. Ablation of connector types and token numbers, reporting accuracy at different spatial scales (WSI/ROI/Patch).

Connector	Token Num	WSI	ROI	Patch
wsi-llava	all	81.61	88.05	86.83
mlp	all	83.33	87.24	83.66
mlp	32	85.34	**89.10**	86.13
mlp	1	85.34	87.47	87.25
qformer	500	**89.37**	88.40	85.25
qformer	32	88.51	88.75	**88.10**
qformer	1	87.64	88.52	86.65

3.3 Ablation Study

To validate the effectiveness of our architectural design, we conducted ablation studies on the HepatoPathoBench with identical hyperparameters.

Effectiveness of Connector and Training Strategy. We compared our Q-Former-based connector against the MLP used in baseline methods like WSI-LLaVA [7]. As shown in Table 2, the Q-Former architecture significantly outperforms the MLP baseline across all scales (+5.18% on WSI). It also exhibits superior stability ($std = 0.26$) compared to the MLP ($std = 1.77$), validating its effectiveness in resampling variable-length visual sequences into fixed query embeddings. Furthermore, the results indicate that the Two-Stage setting yields performance gains (+2.76% on ROI) compared to direct fine-tuning. We also report the performance of the MLP connector initialized from WSI-LLaVA weights, which benefits from broader pre-training and serves as a strong baseline.

Analysis of Feature Redundancy and Token Efficiency. We hypothesize (RQ2) that standard slide encoders generate excessively redundant representations from gigapixel WSIs. To validate this, we evaluated the impact of token quantity on model performance. As detailed in Table 3, increasing the token count introduces noise and degrades performance. The Q-Former with 32 learnable queries achieves peak accuracy (88.75% on ROI, 88.10% on Patch), outperforming the 500-token configuration. A similar trend is observed in the MLP baseline, where even single-token(via max pooling) surpasses the use of all pack tokens. These results corroborate the sparsity of diagnostic signals in WSIs and underscore the necessity of condensing visual features into pack summaries. While the 500-query Q-Former excels in handling the extended sequences typical of WSIs, we select the 32-query setting to strike an optimal balance between inference efficiency and robust performance across varying spatial scales.

4 Conclusion

We present Hepato-LLaVA, a specialized MLLM for hepatocellular pathology analysis. To mitigate information redundancy in WSIs, we propose a Sparse Topo-Pack Attention mechanism that explicitly models 2D tissue topology. We also establish HepatoPathoVQA, a comprehensive multi-scale VQA dataset specifically for HCC WSIs. Extensive experiments demonstrate that our model achieves state-of-the-art performance, validating that topology-aware representations outperform high-dimensional redundant features. We believe this work demonstrates the efficacy of embedding pathological priors into deep learning frameworks, advancing the frontier of efficient AI for precision pathology.

Acknowledgments. This work was supported by the National Natural Science Foundation of China (Grant Nos. U23B2052, 62476029, 62225601, and 82541023), the National Key Research and Development Program of China (Grant No. 2026ZD0557402), the Beijing Natural Science Foundation (Grant No. JQ24045), the

Fundamental Research Funds for the Beijing University of Posts and Telecommunications (Grant No. 2025TSQY08), and the Beijing Nova Program and the Beijing Key Laboratory of Multimodal Data Intelligent Perception and Governance.

Disclosure of Interests. The authors have no competing interests to declare that are relevant to the content of this article.

References

1. Banerjee, S., Lavie, A.: Meteor: An automatic metric for mt evaluation with improved correlation with human judgments. In: Proceedings of the ACL Workshop on Intrinsic and Extrinsic Evaluation Measures for Machine Translation and/or Summarization, pp. 65–72 (2005)
2. Chen, Y., et al.: Slidechat: A large vision-language assistant for whole-slide pathology image understanding. In: Proceedings of the Computer Vision and Pattern Recognition Conference, pp. 5134–5143 (2025)
3. Fourkioti, O., De Vries, M., Jin, C., Alexander, D.C., Bakal, C.: Camil: Context-aware multiple instance learning for cancer detection and subtyping in whole slide images. arXiv preprint arXiv:2305.05314 (2023)
4. Google: Gemini 3 flash (free tier). https://gemini.google.com/ (2026). Accessed 02 Feb 2026
5. He, K., Fan, H., Wu, Y., Xie, S., Girshick, R.: Momentum contrast for unsupervised visual representation learning. In: Proceedings of the IEEE/CVF Conference on Computer Vision and Pattern Recognition, pp. 9729–9738 (2020)
6. Li, J., Li, D., Savarese, S., Hoi, S.: Blip-2: Bootstrapping language-image pre-training with frozen image encoders and large language models. In: International Conference on Machine Learning, pp. 19730–19742. PMLR (2023)
7. Liang, Y., et al.: Wsi-llava: A multimodal large language model for whole slide image. In: Proceedings of the IEEE/CVF International Conference on Computer Vision, pp. 22718–22727 (2025)
8. National Cancer Institute, National Human Genome Research Institute, European Bioinformatics Institute: Human cancer models initiative (hcmi). www.cancer.gov/ccg/research/functional-genomics/hcmi (2017). Accessed 03 Feb 2026
9. Seyfioglu, M.S., Ikezogwo, W.O., Ghezloo, F., Krishna, R., Shapiro, L.: Quilt-llava: Visual instruction tuning by extracting localized narratives from open-source histopathology videos. In: Proceedings of the IEEE/CVF Conference on Computer Vision and Pattern Recognition, pp. 13183–13192 (2024)
10. Song, A.H., Chen, R.J., Ding, T., Williamson, D.F., Jaume, G., Mahmood, F.: Morphological prototyping for unsupervised slide representation learning in computational pathology. In: Proceedings of the IEEE/CVF Conference on Computer Vision and Pattern Recognition, pp. 11566–11578 (2024)
11. Weinstein, J.N., et al.: The cancer genome atlas pan-cancer analysis project. Nat. Genet. **45**(10), 1113–1120 (2013)
12. Wu, J., et al.: Learning heterogeneous tissues with mixture of experts for gigapixel whole slide images. In: Proceedings of the Computer Vision and Pattern Recognition Conference, pp. 5144–5153 (2025)
13. Xu, H., et al.: A whole-slide foundation model for digital pathology from real-world data. Nature **630**(8015), 181–188 (2024)

14. Xu, W., et al.: Lingshu: A generalist foundation model for unified multimodal medical understanding and reasoning. arXiv preprint arXiv:2506.07044 (2025)
15. Yan, Z., et al.: Medreasoner: Reinforcement learning drives reasoning grounding from clinical thought to pixel-level precision. arXiv preprint arXiv:2508.08177 (2025)
16. Zhang, H., et al.: Huatuogpt, towards taming language model to be a doctor. In: Findings of the Association for Computational Linguistics: EMNLP 2023. pp. 10859–10885 (2023)
17. Zhang, W., et al.: Patho-r1: A multimodal reinforcement learning-based pathology expert reasoner. arXiv preprint arXiv:2505.11404 (2025)
18. Zhuang, Z., Zhou, F., Wang, L.: Libra-mil: Multimodal prototypes stereoscopic infused with task-specific language priors for few-shot whole slide image classification . arXiv preprint arXiv:2511.07941 (2025)

Hierarchical Memory for Radiology Report Generation with Rich Context

Qingyue Jiao[1], Jiahao Zheng[1], Jun Zhuang[2], and Yiyu Shi[1](✉)

[1] University of Notre Dame, Notre Dame, IN 46556, USA
{qjiao,jzheng7,yshi4}@nd.edu
[2] Boise State University, Boise, ID 83725, USA
junzhuang@boisestate.edu

Abstract. Radiology report generation (RRG) increasingly relies on rich evidence, most commonly retrieved similar reports from other patients and longitudinal history from the same patient. Prior approaches often concatenate evidence into a long prompt, which can degrade performance. Such long contexts are typically noisy and only partially relevant, making it challenging for the backbone decoder to reliably select the most relevant evidence for the current query. We propose **HM-RRG**, a hierarchical-memory framework that organizes and selects evidence instead of treating all context as a flat sequence. HM-RRG builds an image-conditioned retrieval memory bank by compressing each retrieved report into a compact memory vector grounded on the current image. It then processes the patient's prior reports as a chronological stream, maintains segment-level longitudinal memories, and performs hierarchical retrieval over the union of cross-patient and within-patient memories to form an evidence prompt for each segment during decoding. This design enables selective recall of relevant information while suppressing irrelevant or conflicting details under long contexts. Experiments on MIMIC-CXR with Longitudinal-MIMIC show that HM-RRG improves both text quality and clinical accuracy. In particular, HM-RRG achieves the best factuality metrics among state-of-the-art RRG models, demonstrating that structured memory is effective for radiology report generation. Our code is available at https://github.com/QingyueJ-nd/HM-RRG.

Keywords: Radiology Report Generation · Longitudinal Data · LLM

1 Introduction

Automated radiology report generation with large language models (LLMs) has been extensively studied [16], and recent systems increasingly rely on external evidence to improve factuality and clinical completeness. In practice, two evidence sources are used most often: (i) retrieved similar reports from other patients and (ii) longitudinal history from the same patient. Retrieval-augmented approaches mainly improve the retriever and filtering stage, then condition the

G. Yang et al. (Eds.): MICCAI 2026, LNCS 16878, pp. 206–216, 2027.
https://doi.org/10.1007/978-3-032-38059-3_20

decoder on a small set of retrieved evidence. For example, Prefilling [23] integrates multimodal history with cross-attention and a memory-driven transformer to reduce hallucinations, while Diff-RRG [20] uses disease-wise patch differences to guide change-aware generation. MLRG [9] further models longitudinal disease progression by fusing current multi-view chest X-rays with prior-visit information in a multi-view longitudinal fusion network and pretraining via multi-view longitudinal contrastive learning aligned to the corresponding report.

However, despite these advancements, a key gap remains: how to use both evidence sources together during decoding. Cross-patient retrieval can provide strong priors for rare findings, but it may miss patient-specific trends without longitudinal context. Longitudinal history is often most relevant to the current patient, as temporal information in radiology reports captures disease progression across examinations and supports clinical decision-making [18], yet it can bias the model toward copying prior statements and overlooking new changes. To the best of our knowledge, there is still no unified report-generation pipeline that effectively leverages both retrieved similar reports and longitudinal history within a single, evidence-driven decoding framework.

Motivated by real clinical workflows, we aim to combine both evidence sources in a way that remains robust in long and noisy contexts. Radiologists do not treat prior notes and similar cases as a flat sequence of text; instead, they organize, retain, and recall information selectively: they consult a patient's history to identify changes, and they draw on similar cases to resolve uncertainty and refine the diagnosis. In contrast, most LLM-based report-generation systems use external resources as a single flat, concatenated prompt, forcing the backbone to perform evidence selection implicitly. This "flat-memory" setup is brittle: as context grows, models are more likely to overlook critical evidence, over-weight irrelevant snippets, or fail to resolve conflicts.

To address this, we propose Hierarchical Memory Radiology Report Generation (HM-RRG), a rich-context framework that explicitly structures retrieved evidence and longitudinal history into a hierarchical memory. Our method is inspired by the Hierarchical Memory Transformer (HMT) [2], which improves long-context processing by maintaining segment-level memory recurrence and enabling selective recall of previously seen information. Concretely, we segment input evidence into manageable units, construct memory-augmented representations that persist across segments, and condition memory formation and retrieval on the current image so that the model can prioritize clinically relevant content while suppressing noise and contradictions. This design enables HM-RRG to leverage large evidence sets without relying on the backbone to solve long-context selection purely through prompt concatenation. Our contributions are:

1. We propose **HM-RRG**, a rich-context radiology report generation framework that mirrors real clinical reasoning by integrating both cross-patient retrieved reports and longitudinal patient history to generate more fact-grounded reports (Fig. 1).

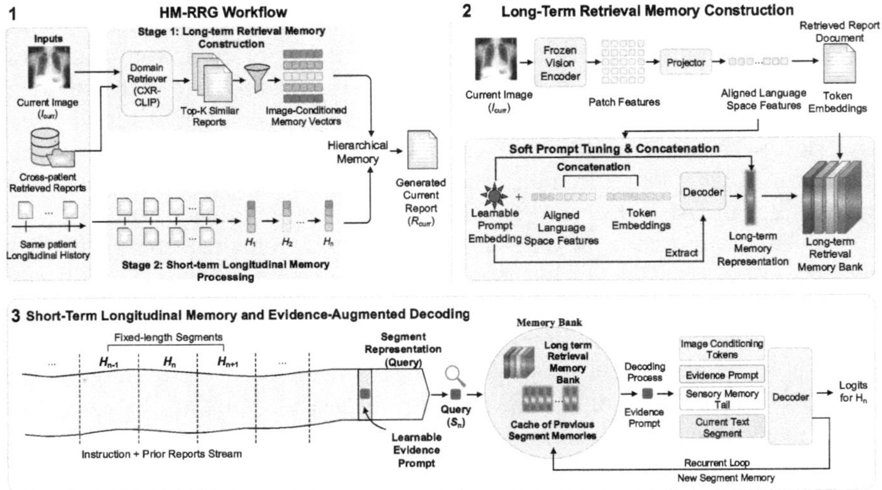

Fig. 1. Overall architecture of HM-RRG. The upper-left panel provides an overview of the HM-RRG workflow. The upper-right panel illustrates long-term memory construction, and the bottom panel depicts short-term longitudinal memory and evidence-augmented decoding.

2. We introduce current-image-conditioned hierarchical memory tokens over retrieved evidence and longitudinal context, enabling robust evidence selection under long contexts.
3. We evaluate HM-RRG on the Longitudinal-MIMIC dataset, demonstrating consistent improvements in report quality over state-of-the-art methods when leveraging rich context.

2 Method

In this section, we present the overall workflow of HM-RRG, a hierarchical-memory framework for radiology report generation that integrates retrieved evidence from other patients and longitudinal history from the same patient. Our design is motivated by the observation that these two information sources exhibit fundamentally different structures: retrieved reports form an unordered and often noisy set, whereas longitudinal priors constitute an ordered temporal stream. Treating both sources with a single uniform context mechanism can waste context budget on irrelevant retrieved details and dilute the clinically meaningful temporal signal.

HM-RRG decomposes report generation into two stages. The first stage is the construction of long-term retrieval memory. Given the current image I_{curr}, we use a domain retriever to obtain the top-K similar reports $\{R_i^{\text{retr}}\}_{i=1}^{K}$. We then transform each retrieved report into an image-conditioned memory representation, forming a retrieval memory bank that distills "what is similar to the current case" while filtering irrelevant content. The second stage is short-term longitudinal memory processing. We process the patient's prior reports $\left\{R_t^{\text{prior}}\right\}_{t=1}^{T-1}$ in chronological order as a sequential stream, and at each step construct compact segment representations that are adaptively augmented by both the retrieval memory bank and recent longitudinal context. The resulting hierarchical memory state conditions the decoder to generate the current report R_{curr} for I_{curr}, with an explicit mechanism to focus on clinically salient evidence and temporal changes.

Long-Term Retrieval Memory. Given the current study image I_{curr}, we first construct an image-conditioned long-term retrieval memory bank that summarizes evidence from similar cases. This design explicitly addresses two challenges of retrieval-augmented report generation: retrieved reports form an unordered set and often contain irrelevant details, and injecting the full retrieved text into the generation context is costly and can dilute clinically salient signals. Instead, we compress each retrieved report into a compact memory vector in the decoder's hidden embedding space, conditioned on I_{curr}, yielding a lightweight evidence block that can be queried repeatedly during longitudinal decoding.

We employ a domain-specific chest X-ray retriever CXR-CLIP [19] to retrieve the top-K most similar reports based on the current image I_{curr}. We denote the retrieved reports as $\{R_i^{\text{retr}}\}_{i=1}^{K}$. We then encode the input image I_{curr} with a pre-trained frozen vision encoder to obtain patch features $\tilde{V}(I_{\text{curr}}) = \text{VisionEnc}(I_{\text{curr}})$. We then project them to the language space via $V(I_{\text{curr}}) = \text{Proj}(\tilde{V}(I_{\text{curr}}))$. We adopt the CXR-CLIP vision encoder and train a lightweight projector to match the decoder's hidden size d.

For each retrieved report R_i^{retr}, we obtain its token embeddings via the language embedding layer $E(R_i^{\text{retr}})$. We then summarize the retrieved report conditioned on the current image by soft prompt tuning [10]. We augment the retrieved report embedding with t_{mem}, a summarization prompt embedding. Specifically, it is a learnable parameter embedding to prompt the backbone decoder to summarize the current embedding. We input the report embedding with t_{mem} into the backbone decoder, denoted as LM: $H_i = \text{LM}\Big(t_{\text{mem}} \parallel V(I_{\text{curr}}) \parallel E(R_i^{\text{retr}}) \parallel t_{\text{mem}}\Big)$ The long-term memory representation for the i-th retrieved report is taken as the hidden state at the memory token position: $M_i^{\text{retr}} = H_i[-1]$. Finally, we stack all retrieved memories to form the long-term retrieval memory bank $M^{\text{retr}} = \left[M_1^{\text{retr}}; \ldots; M_K^{\text{retr}}\right]$, which provides a compact, image-grounded representation of retrieved evidence that can be efficiently queried during longitudinal decoding.

Short-Term Longitudinal Memory. Next, HM-RRG processes a patient's longitudinal history as a short-term, temporally ordered memory stream and integrates it with the long-term retrieval memory bank. The core adaptation from HMT [2] for radiology report generation is to preserve chronological structure in prior reports and enable selective access to retrieved evidence without concatenating full retrieved text into the decoder context.

Stream Construction and Segmentation. Given the current image I_{curr} and the set of prior reports $\{R_t^{\text{prior}}\}$ ordered by time, we form a single textual stream $\mathcal{X} = [\text{Instr} \,\|\, R_1^{\text{prior}} \,\|\, \cdots \,\|\, R_{T-1}^{\text{prior}}]$, where Instr denotes the instruction prompt. We tokenize $\mathcal{X}$ and map token IDs to embeddings using the language embedding layer, then partition the resulting embedding sequence into N fixed-length segments of L tokens $H_1, \ldots, H_N$. To avoid splitting visual tokens across segments, we treat image embeddings as segment-level conditioning (prefixed during decoding) rather than inserting them into the segmented text stream.

Segment Representation for Evidence Querying. For each segment H_n, we compute a compact segment representation S_n that serves as a query for selecting relevant evidence. Concretely, we attach a small learnable summary token t_{sum} to the segment prefix and use the decoder to produce a summary vector, similar to the long-term memory construction, with S_n extracted from the summary-token position. This design provides a stable, fixed-size descriptor of the current local context, even when segments cut across report boundaries.

Hierarchical Memory Retrieval. In addition to the long-term retrieval memory bank M^{retr}, we maintain a cache of segment-level longitudinal memories produced from earlier segments of the same patient, $\{M_1^{\text{seg}}, \ldots, M_{n-1}^{\text{seg}}\}$. At segment n, we form a unified hierarchical memory set by concatenating cross-patient evidence and within-patient summaries: $M_n^{\text{all}} = [\,M^{\text{retr}}\,;\, M_{1:n-1}^{\text{seg}}\,]$. We then compute an evidence prompt P_{mem} by attention over M_n^{all} using the segment query S_n: $Q_n = S_n W_q$, $K_n = M_n^{\text{all}} W_k$, $\alpha_n = \text{softmax}\Big(\frac{Q_n K_n^\top}{\sqrt{d_h}}\Big)$, and $P_{\text{mem}} = \alpha_n M_n^{\text{all}}$. This construction yields a single, compact prompt vector that adaptively mixes cross-patient evidence distilled from retrieved reports and salient cues accumulated from earlier longitudinal segments.

Local Sensory Memory. Longitudinal narratives often span several segments. To preserve local information, we carry a short-range contextual tail from the previous segment. We denote the last k token embeddings of segment $n-1$ as $C_n = H_{n-1}[L-k:L]$ with C_1 initialized to zeros. In all experiments we set $k{=}32$, which provides local continuity while keeping overhead small [2].

Evidence-augmented Decoding and Memory Update. For segment n, the decoder receives an augmented input that combines the evidence prompt P_{mem}, the sensory memory C_n, the current text segment H_n, and the current-image conditioning tokens $V(I_{\text{curr}})$: $\mathcal{X}_n = [\,V(I_{\text{curr}}) \,\|\, P_{\text{mem}} \,\|\, C_n \,\|\, H_n \,\|\, P_{\text{mem}}\,]$. Compared with flat concatenation, this formulation avoids feeding all retrieved and longitudinal reports into the decoder as one long sequence. Instead, the decoder processes the current segment together with compact memory prompts, while evidence selection is performed over the compressed memory set, reducing the self-attention cost of long-context evidence. The decoder produces hidden states Y_n, from which we compute logits for the current segment and extract a compact segment memory M_n^{seg}. We cache $\{M_n^{\text{seg}}\}$ as short-term longitudinal memory for subsequent segments, enabling the model to keep clinically relevant information as generation progresses. Unlike the original HMT formulation, which is designed for a long (over 30k tokens) and potentially infinite input stream and thus employs a fixed-size sliding window over memory states, we have a bounded clinical setting with a capped number of longitudinal reports and retrieved cases. Therefore, we retain all segment memories within each example by default to avoid removing clinically important temporal changes and to simplify training and evaluation. Importantly, this is an implementation choice rather than a limitation: the same retrieval formulation naturally supports an optional sliding-window cache when scaling to substantially longer histories.

3 Experiments

Datasets and Implementation. We conduct experiments on MIMIC-CXR-JPG [5] and Longitudinal-MIMIC [23]. The MIMIC-CXR dataset contains 377,110 chest X-ray images corresponding to 227,835 radiographic reports from 65,379 patients. Longitudinal-MIMIC is derived from MIMIC-CXR and includes 26,625 patients and 94,169 samples with at least two visits. We use MIMIC-CXR as the retrieval source and Longitudinal-MIMIC to construct longitudinal report generation samples with prior visits. We follow the official train, validation, and test splits as Longitudinal-MIMIC and ensure patient-level separation across splits to prevent leakage. For retrieval augmentation, we additionally restrict the retrieval source to the training split only. Each Longitudinal-MIMIC example provides a current study and its most recent prior. To better reflect realistic clinical history, we aggregate all available prior studies for the same patient within the split, order them chronologically, and form a single training instance per current study. The model input includes the current image, an instruction prompt, and longitudinal prior reports. The target is the current report.

We use CXR-CLIP [19] as the domain retriever to retrieve top-K similar reports from other patients, with K=5 in all experiments. We follow Diff-RRG [20] by using the same metrics and the BioMistral-7B decoder backbone [6], paired with the CXR-CLIP vision encoder [19]. We train a lightweight projector that maps CXR-CLIP vision features into the decoder embedding space, and fine-tune the decoder using LoRA [3] with rank 16 and α=16. We train using an AdamW optimizer [11] with peak learning rate 2×10^{-4}, linear decay, warmup ratio 0.03, and gradient clipping with max grad norm 1.0. Unless otherwise stated, we use gradient accumulation to reach an effective batch size of 16 per GPU. To train the hierarchical memory, we adopt a two-stage training. We set the segment length $L = 128$, the segment-summary prefix length $j = L/2 = 64$, and the sensory memory length $k = 32$. In Stage 1, we train the long-term memory construction for 2,000 steps with a learning rate of 5×10^{-6}. In Stage 2, we initialize from the Stage-1 checkpoint and continue training with hierarchical

Table 1. Evaluation results on the MIMIC-CXR dataset. Baseline evaluations are cited from the original publications and MLRG [9]. Bold and underlined values indicate the highest and second-highest scores, respectively. The last row reports the percentage difference between HM-RRG and the best prior method.

Input	Method	NLG Metrics↑			CE Metrics↑			
		BLEU	MTR	R-L	RG	P	R	F1
SI	R2GenGPT [17]	0.191	0.125	0.240	–	0.331	0.253	0.287
SI	B-LLM [8]	0.243	0.175	0.291	–	0.465	<u>0.482</u>	0.473
SI	LLaVA-Rad [21]	0.168	–	<u>0.292</u>	0.208	–	–	0.475
Long	Prefilling [23]	0.198	0.137	0.274	–	0.506	0.364	0.423
Long	HERGen [15]	0.234	0.156	0.285	–	0.421	0.289	0.343
Long	Diff-RRG [20]	0.236	0.164	0.276	–	0.528	0.430	0.474
MVL	MLRG [9]	<u>0.262</u>	<u>0.176</u>	**0.320**	<u>0.291</u>	<u>0.549</u>	0.468	<u>0.505</u>
R&L	HM-RRG	**0.263**	**0.177**	<u>0.292</u>	**0.299**	**0.561**	**0.600**	**0.580**
–	Δ (%)↑	+0.4	+0.6	-8.8	+2.7	+2.2	+24.4	+14.9

retrieval enabled. The decoder LoRA weights, image projector, learnable memory and summary tokens, and memory-retrieval projections are trainable, while the base BioMistral-7B decoder and CXR-CLIP vision encoder remain frozen. We train Stage 2 with backpropagation through time (BPTT) [12] to a depth of 2, use a micro-batch size of 1, and optimize the hierarchical-memory parameters for 8,000 steps with a learning rate of 5×10^{-6}. All hierarchical-memory-related hyperparameters follow the sensitivity analysis in [2], unless otherwise noted. All experiments are run on a single NVIDIA A10 GPU.

We report natural language generation (NLG) similarity and clinical efficacy (CE) metrics. For the main comparison, we report average BLEU [13], ROUGE-L [7], and METEOR [1] between generated and reference reports. For clinical accuracy, we compute the micro-average of Precision (P), Recall (R), and F1-score (F1) of CheXbert labelling [14], F1-score of RadGraph [4], which is the overlap in radiological entities and clinical relations.

Results. We compare HM-RRG with recent MIMIC-CXR report generation methods across three input types: single-image (SI), longitudinal history (Long), and multi-view longitudinal (MVL). Our HM-RRG models use both retrieved reports and longitudinal history (R&L). As shown in Table 1, HM-RRG outperforms prior methods across these input settings on clinical accuracy metrics, demonstrating its ability to generate clinically meaningful reports by effectively leveraging both retrieved and longitudinal context via hierarchical memory. Figure 2 shows the HM-RRG generated report with one prior report and one retrieved report used as evidence. This example demonstrates that HM-RRG captures disease progression from the prior report and selectively extracts relevant information while filtering noise from the retrieved report.

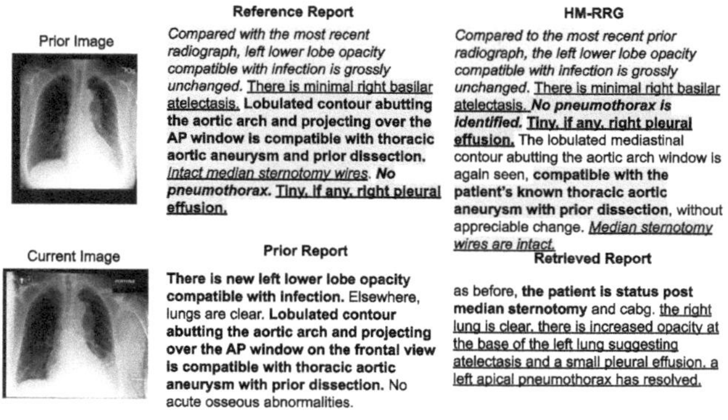

Fig. 2. HM-RRG generated report. Matching findings between the reference and generated reports are indicated with the same colors and text styles. Relevant and irrelevant evidence are marked in bold and underlined, respectively.

Table 2. Ablation results on the MIMIC-CXR dataset.

Model	BLEU	ROUGE-L	BERTScore	F1-RadGraph	F1-CheXbert
Baseline	0.175	0.215	0.399	0.214	0.463
+ Prior Reports	0.227	0.267	0.434	0.225	0.507
+ top-1 Retr.	0.171	0.221	0.413	0.217	0.539
+ top-5 Retr.	0.176	0.224	0.415	0.242	0.500
+ Prior & top-5 Retr.	0.177	0.220	0.416	0.211	0.483
w/ HM-RRG	**0.263**	**0.292**	**0.521**	**0.299**	**0.580**

Ablation Results. To test whether hierarchical memory is necessary for using rich context, we conduct ablation studies and report the results in Table 2. For the ablation study, we additionally report BERTScore [22] to assess semantic similarity between generated and reference reports. We compare a baseline that uses only the current image with an instruction prompt, adding longitudinal context, adding top-1 or top-5 retrieved reports, a flat combination of priors and retrieval, and the full HM-RRG. Adding prior reports consistently improves NLG and clinical accuracy, demonstrating that longitudinal history provides useful cues for disease progression. In contrast, retrieval-only inputs may introduce irrelevant clinical context, leading to mixed results across metrics. A key finding is that simply concatenating longitudinal and retrieved context does not reliably improve performance. This suggests that when the context becomes long and noisy, the model struggles to focus on the right evidence, and important signals can be diluted. HM-RRG resolves this issue by converting unordered retrieved reports into current-image-conditioned memory vectors and allowing each longitudinal segment to query a compact unified memory, making evidence selection explicit rather than relying on the decoder to resolve a long, noisy sequence. Overall, these results confirm the need for hierarchical memory: when both longitudinal and retrieved contexts are available, hierarchical memory can selectively leverage useful evidence while suppressing irrelevant content.

4 Conclusion

In this paper, we propose HM-RRG, a hierarchical memory framework for radiology report generation that stores, organizes, and selects key evidence from both cross-patient retrieval and within-patient longitudinal history. Unlike prior approaches that rely on flat context concatenation, HM-RRG leverages structured memory and adaptive evidence selection to better leverage rich clinical context and generate clinically meaningful reports. The hierarchical memory structure and decoding mechanism are model-independent and can be readily applied to other LLM and VLM decoders with lightweight training overhead. Future work can extend HM-RRG to richer retrieval settings, longer longitudinal histories, external clinical datasets, and additional decoder backbones, further expanding its applicability to diverse medical settings.

Acknowledgments. The authors thank Professor Jason Cong at the University of California, Los Angeles, for his valuable advice.

Disclosure of Interests. The authors have no competing interests to declare that are relevant to the content of this article.

References

1. Banerjee, S., Lavie, A.: METEOR: an automatic metric for MT evaluation with improved correlation with human judgments. In: Goldstein, J., Lavie, A., Lin, C.Y., Voss, C. (eds.) Proceedings of the ACL Workshop on Intrinsic and Extrinsic Evaluation Measures for Machine Translation and/or Summarization, pp. 65–72. Association for Computational Linguistics, Ann Arbor (2005). https://aclanthology.org/W05-0909/
2. He, Z., Cao, Y., Qin, Z., Prakriya, N., Sun, Y., Cong, J.: HMT: hierarchical memory transformer for efficient long context language processing. In: Chiruzzo, L., Ritter, A., Wang, L. (eds.) Proceedings of the 2025 Conference of the Nations of the Americas Chapter of the Association for Computational Linguistics: Human Language Technologies, pp. 8068–8089. Association for Computational Linguistics (2025)
3. Hu, E.J., et al.: LoRA: low-rank adaptation of large language models. In: International Conference on Learning Representations (ICLR) (2022)
4. Jain, S., et al.: RadGraph: extracting clinical entities and relations from radiology reports. PhysioNet (2021)
5. Johnson, A., et al.: MIMIC-CXR-JPG - chest radiographs with structured labels. PhysioNet (2024)
6. Labrak, Y., Bazoge, A., Morin, E., Gourraud, P.A., Rouvier, M., Dufour, R.: BioMistral: a collection of open-source pretrained large language models for medical domains. In: Ku, L.W., Martins, A., Srikumar, V. (eds.) Findings of the Association for Computational Linguistics: ACL 2024, pp. 5848–5864. Association for Computational Linguistics, Bangkok, Thailand (2024)

7. Lin, C.Y.: ROUGE: a package for automatic evaluation of summaries. In: Text Summarization Branches Out, pp. 74–81. Association for Computational Linguistics (2004)
8. Liu, C., Tian, Y., Chen, W., Song, Y., Zhang, Y.: Bootstrapping large language models for radiology report generation. In: Proceedings of the Thirty-Eighth AAAI Conference on Artificial Intelligence and Thirty-Sixth Conference on Innovative Applications of Artificial Intelligence and Fourteenth Symposium on Educational Advances in Artificial Intelligence, AAAI 2024/IAAI 2024/EAAI 2024. AAAI Press (2024). https://doi.org/10.1609/aaai.v38i17.29826
9. Liu, K., Ma, Z., Kang, X., Li, Y., Xie, K., Jiao, Z., Miao, Q.: Enhanced contrastive learning with multi-view longitudinal data for chest X-ray report generation. In: 2025 IEEE/CVF Conference on Computer Vision and Pattern Recognition (CVPR), pp. 10348–10359 (2025)
10. Liu, P., Yuan, W., Fu, J., Jiang, Z., Hayashi, H., Neubig, G.: Pre-train, prompt, and predict: a systematic survey of prompting methods in natural language processing. ACM Comput. Surv. (2023)
11. Loshchilov, I., Hutter, F.: Decoupled weight decay regularization. In: International Conference on Learning Representations (2019)
12. Mozer, M.C.: A focused backpropagation algorithm for temporal pattern recognition, pp. 137–169. L. Erlbaum Associates Inc., USA (1995)
13. Papineni, K., Roukos, S., Ward, T., Zhu, W.J.: BLEU: a method for automatic evaluation of machine translation. In: Proceedings of the 40th Annual Meeting on Association for Computational Linguistics, pp. 311–318. Association for Computational Linguistics, USA (2002)
14. Smit, A., Jain, S., Rajpurkar, P., Pareek, A., Ng, A., Lungren, M.: CheXbert: combining automatic labelers and expert annotations for accurate radiology report labeling using BERT. In: Proceedings of the 2020 Conference on Empirical Methods in Natural Language Processing (EMNLP), pp. 1500–1519. Association for Computational Linguistics, Online (2020)
15. Wang, F., Du,: S., Yu, L.: HerGen: elevating radiology report generation with longitudinal data. In: Computer Vision – ECCV 2024: 18th European Conference. Milan, Italy, 29 September–4 October 2024, Proceedings, Part LV, pp. 183–200. Springer, Heidelberg (2024)
16. Wang, X., Figueredo, G., Li, R., Zhang, W.E., Chen, W., Chen, X.: A survey of deep-learning-based radiology report generation using multimodal inputs. Med. Image Anal. 103627 (2025)
17. Wang, Z., Liu, L., Wang, L., Zhou, L.: R2GenGPT: radiology report generation with frozen LLMs. Meta-Radiol. 100033 (2023)
18. Xu, G., et al.: A comprehensive survey of ai agents in healthcare. J. Biomed. Inform. **179**, 105045 (2026). https://doi.org/10.1016/j.jbi.2026.105045, https://www.sciencedirect.com/science/article/pii/S1532046426000699
19. You, K., et al.: CXR-CLIP: toward large scale chest X-ray language-image pre-training. In: proceedings of Medical Image Computing and Computer Assisted Intervention, pp. 101–111. Springer (2023)
20. Yun, H., Maeng, J., Kang, E., Suk, H.I.: Diff-RRG: longitudinal disease-wise patch difference as guidance for LLM-based radiology report generation. In: proceedings of Medical Image Computing and Computer Assisted Intervention (2025)
21. Zambrano Chaves, J.M., Huang, S.C.: A clinically accessible small multimodal radiology model and evaluation metric for chest X-ray findings. Nat. Commun. 3108 (2025)

22. Zhang, T., Kishore, V., Wu, F., Weinberger, K.Q., Artzi, Y.: BertScore: evaluating text generation with BERT. In: International Conference on Learning Representations (2020)
23. Zhu, Q., Mathai, T.S., Mukherjee, P., Peng, Y., Summers, R.M., Lu, Z.: Utilizing longitudinal chest X-rays and reports to pre-fill radiology reports. In: Proceedings of Medical Image Computing and Computer Assisted Intervention, pp. 189–198 (2023)

HiPath: Hierarchical Vision-Language Alignment for Structured Pathology Report Prediction

Ruicheng Yuan[1,2], Zhenxuan Zhang[2], Anbang Wang[2], Liwei Hu[2], Xiangqian Hua[3], Yaya Peng[4], Jiawei Luo[1(✉)], and Guang Yang[2(✉)]

[1] College of Computer Science and Electronic Engineering, Hunan University, Changsha, Hunan, China
{raytion,luojiawei}@hnu.edu.cn

[2] Department of Bioengineering and Imperial-X, Imperial College London, London, UK
g.yang@imperial.ac.uk

[3] Department of Pathology, Xiangtan Maternal and Child Health Hospital, Xiangtan, Hunan, China

[4] Department of Pathology, The First People's Hospital of Xiangtan City, Xiangtan, Hunan, China

Abstract. Pathology reports are structured, multi-granular documents encoding diagnostic conclusions, histological grades, and ancillary test results across one or more anatomical sites; yet existing pathology vision-language models (VLMs) reduce this output to a flat label or free-form text. We present HiPath, a lightweight VLM framework built on frozen UNI2 and Qwen3 backbones that treats structured report prediction as its primary training objective. Three trainable modules totalling 15 M parameters address complementary aspects of the problem: a Hierarchical Patch Aggregator (HiPA) for multi-image visual encoding, Hierarchical Contrastive Learning (HiCL) for cross-modal alignment via optimal transport, and Slot-based Masked Diagnosis Prediction (Slot-MDP) for structured diagnosis generation. Trained on 749 K real-world Chinese pathology cases from three hospitals, HiPath achieves 68.9% strict and 74.7% clinically acceptable accuracy with a 97.3% safety rate, outperforming all baselines under the same frozen backbone. Cross-hospital evaluation confirms generalisation with only a 3.4 pp drop in strict accuracy while maintaining 97.1% safety. Code and evaluation protocol: https://github.com/raytions/hipath.

Keywords: Computational Pathology · Vision-Language Model · Structured Diagnosis · Contrastive Learning

1 Introduction

Recent progress in computational pathology has been driven by large-scale self-supervised foundation models [3,10,18] and vision-language models (VLMs) [7,8]

G. Yang et al. (Eds.): MICCAI 2026, LNCS 16878, pp. 217–227, 2027.
https://doi.org/10.1007/978-3-032-38059-3_21

that connect histopathology images with clinical text. These models have become strong generic feature extractors and support a wide range of downstream tasks, including tissue classification, biomarker prediction, and zero-shot recognition [7, 12].

However, the clinical output of pathology is rarely a single label. A routine report is a structured record that jointly specifies (i) a primary diagnosis, (ii) a histological grade, and (iii) immunohistochemistry (IHC) results, often repeated across multiple anatomical sites for the same patient [13]. This structure is central to decision-making: different fields have different clinical consequences, so an evaluation that collapses all errors into flat accuracy can obscure safety-critical failures. At the same time, open-ended text generation offers no guarantee that the output is slot-consistent or reliably parseable.

This leaves an open question for pathology VLMs: can we train and evaluate a model whose primary objective is structured report prediction, rather than flat classification or unconstrained generation? To our knowledge, existing pathology VLMs are not trained or benchmarked for structured, slot-level diagnosis prediction in this setting. A related gap concerns language and reporting conventions. All major pathology VLMs to date are trained exclusively on English corpora [7,8,12], despite substantial clinical demand in non-English contexts. China accounts for approximately 24% of global cancer cases [1,6] and faces an estimated shortage of 90,000 pathologists [20]. Chinese reports also follow distinct conventions, including site-prefixed segment identifiers, standardised grading terminology, and structured IHC formatting. While Chinese groups have introduced notable MIL methods [16], pathology foundation models [19], and VLMs [17], none target structured Chinese diagnosis prediction at scale.

In this work, we propose HiPath (Fig. 1), a lightweight framework that treats structured report prediction as the central learning objective. HiPath is built on frozen UNI2 [3] and Qwen3 [21] backbones and adds 15 M trainable parameters. The model (1) aggregates multi-image evidence selected by pathologists via hierarchical cross-attention (HiPA), (2) uses segment and case-level image-text alignment via optimal transport as an auxiliary training signal (HiCL) [5], and (3) predicts typed diagnostic slots by matching against a closed vocabulary in the frozen LLM embedding space (Slot-MDP). Free-text reports are used only during training; at inference, only frozen vocabulary embeddings and visual features are required. Trained on 749 K cases from three hospitals, HiPath achieves 68.9% strict accuracy, 74.7% clinically acceptable accuracy, and 97.3% safety. Vision-only classifiers on the same frozen visual features plateau at 30–32%, and a non-hierarchical variant with identical text access trails by 12.9 pp. Cross-hospital evaluation (training on two hospitals, testing on the third) shows a 3.4 pp drop while safety remains above 97%.

Our contributions are as follows. (1) We formalise structured slot-level report prediction as a primary objective for pathology VLMs and show that it substantially outperforms flat classification. (2) We introduce a lightweight architecture on frozen backbones that combines hierarchical multi-image aggregation, OT-based local alignment, and vocabulary-grounded slot prediction for structured

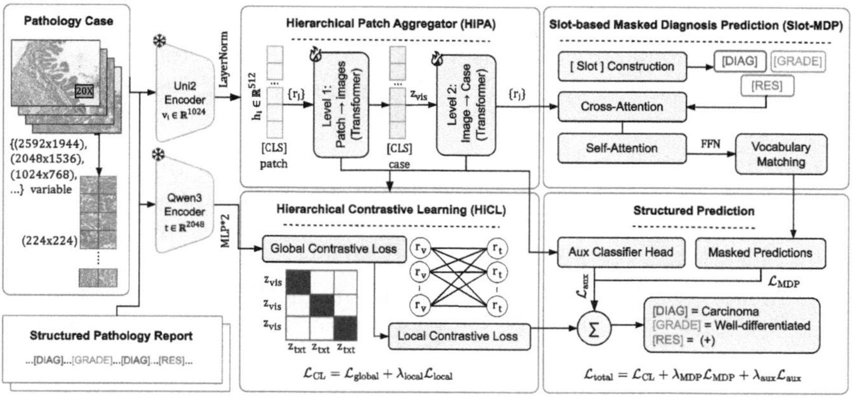

Fig. 1. Overview of HiPath. Frozen UNI2 patch features and Qwen3 text embeddings are processed by three trainable modules: HiPA aggregates patches hierarchically into case representations; HiCL aligns modalities at case and segment level via optimal transport; Slot-MDP fills typed diagnostic slots from visual features alone. Free-text reports are used only during training.

diagnosis. (3) We propose a four-level clinical evaluation protocol with safety-aware metrics and validate it on 749 K cases from three hospitals. The evaluation toolkit, including the 304-term vocabulary, synonym mappings, coarse taxonomy, and safety boundary definitions, is publicly released to facilitate reproducible research in structured pathology diagnosis.

2 Methods

Problem Formulation. Each case c consists of J physician-selected ROI images and a structured report parsed into anatomical-site segments. In routine Chinese pathology practice, reports follow a conventional structure organised by anatomical site, with each site recording a primary diagnosis, an optional histological grade, and any ancillary findings such as IHC results. We exploit this regularity by parsing historical reports into segments and replacing diagnostic terms with typed masks `[DIAG]`, `[GRADE]`, `[RES]`, yielding K slots $\{(m_k, \tau_k)\}_{k=1}^{K}$, where $m_k \in \mathcal{V}_{\tau_k}$ is the ground-truth term and τ_k its type.

At inference, the model requires a report template specifying the number of segments, slot positions, and slot types. This structure is derivable from information available before pathologist examination: the clinical requisition records the specimen source and number of tissue blocks, while the IHC panel is ordered at accessioning. For example, a breast biopsy with ER/PR testing yields a predictable template of `[DIAG]`+`[GRADE]`+`[RES(ER)]`+`[RES(PR)]`, but this structure reveals only which fields need to be filled, not whether the tissue is benign or malignant, nor whether any marker is positive or negative. Slot queries attend exclusively to visual features, so all diagnostic predictions are grounded in morphological evidence. In our experiments, templates are derived from the ground-truth report structure; in deployment, they would be constructed from clinical requisition and accessioning metadata.

Training Data. We collected 749 K pathology cases from three tertiary hospitals in China spanning 2011–2025. Hospitals A and C are general hospitals; Hospital B is a maternal-child hospital. The top five organs are uterus (9.3%), cervix (8.1%), stomach (6.1%), colon (5.0%), and breast (3.1%), covering 42 distinct organ sites. Each case pairs one or more pathologist-selected ROI images (median 1, P99 = 8) with a structured Chinese report. The vocabulary was curated to 304 standardised terms (184 DIAG + 93 GRADE + 27 RES) after manual synonym consolidation by two senior pathologists and frequency filtering ($\geq$10 occurrences); the top 10% of terms cover 90.8% of predictions. Cases were split 80/10/10 at the patient level independently per hospital, yielding approximately 600 K/75 K/75 K cases for training, validation, and testing. The 304-term vocabulary and safety boundaries were defined based on clinical standards prior to splitting. This study was conducted under IRB approval; all data were de-identified.

Feature Extraction. Visual features: frozen UNI2 (ViT-Giant/14) tiles each image into 224×224 patches and encodes them to native 1536-d vectors, which we project to 1024-d. Text features: frozen Qwen3-30B-A3B [21] (MoE, 30B total/3B activated) encodes each report segment independently, yielding 2048-d embeddings $\{\mathbf{s}_l\}_{l=1}^{L}$ via mean pooling. Feature pre-extraction cost ~120 GPU-h (H100).

2.1 Hierarchical Patch Aggregator (HiPA)

A pathology case often contains multiple ROI images capturing different tissue regions. Unlike whole-slide methods that apply self-attention within a single gigapixel image [2], HiPA uses cross-attention with learnable queries to aggregate across multiple independent images selected by pathologists. Patch embeddings are projected: $\mathbf{h}_i = \mathrm{LN}(\mathbf{W}_{\mathrm{proj}}\mathbf{v}_i) \in \mathbb{R}^d$ $(d = 512)$. Two cross-attention levels aggregate them:

Level 1 (Patch→Image). A learnable `[CLS]` token queries patches within image j:

$$\mathbf{r}_j = \mathrm{TransBlock}([\mathbf{q}_{\mathrm{patch}}; \mathbf{h}_1, \ldots, \mathbf{h}_{N_j}])\big|_{\mathtt{CLS}} \tag{1}$$

Level 2 (Image→Case). A second token aggregates per-image representations:

$$\mathbf{z}_{\mathrm{vis}} = \mathrm{TransBlock}([\mathbf{q}_{\mathrm{case}}; \mathbf{r}_1, \ldots, \mathbf{r}_J])\big|_{\mathtt{CLS}} \tag{2}$$

yielding case embedding $\mathbf{z}_{\mathrm{vis}} \in \mathbb{R}^d$ and per-image representations $\{\mathbf{r}_j\}$ for downstream modules.

2.2 Hierarchical Contrastive Learning (HiCL)

HiCL learns a shared vision-language embedding space via contrastive alignment. It serves as an auxiliary training signal that regularises HiPA's visual representations and enables image-text retrieval. Qwen3 segment embeddings are projected via a two-layer MLP to case-level $\mathbf{z}_{\text{txt}}$ and segment-level $\{\mathbf{s}_l\}_{l=1}^{L}$.

Global Loss. Symmetric InfoNCE on case-level pairs:

$$\mathcal{L}_{\text{global}} = \tfrac{1}{2}\left[\text{CE}(\mathbf{S} \cdot e^{\alpha}, \mathbf{y}) + \text{CE}(\mathbf{S}^{\top} \cdot e^{\alpha}, \mathbf{y})\right] \tag{3}$$

where $S_{ij} = \bar{\mathbf{z}}_{\text{vis},i}^{\top} \bar{\mathbf{z}}_{\text{txt},j}$ (L2-normalised), $\mathbf{y}$ is the identity target, and α is a learnable log-temperature clamped to $[1, 2.7]$.

Local Loss. Per-image $\{\mathbf{r}_j\}_{j=1}^{J_c}$ and per-segment $\{\mathbf{s}_l\}_{l=1}^{L_c}$ embeddings form an unordered many-to-many correspondence with $J_c \neq L_c$ in general. We adopt optimal transport with entropic regularisation [5] rather than hard matching (*e.g.*, Hungarian algorithm, which requires $J_c = L_c$) or independent dot-product attention [15] (which treats each pair independently and does not enforce a global assignment). Sinkhorn OT produces a differentiable, dense transport plan that distributes probability mass across all image-segment pairs while respecting marginal constraints. The cost matrix $C_{jl} = 1 - \bar{\mathbf{r}}_j^{\top} \bar{\mathbf{s}}_l$ is solved with Sinkhorn-Knopp (ϵ=0.05, 3 iterations, uniform marginals). Let $\mathcal{B}_{\text{multi}}$ denote mini-batch cases with J_c>1 and L_c>1:

$$\mathcal{L}_{\text{local}} = \frac{1}{|\mathcal{B}_{\text{multi}}|} \sum_{c \in \mathcal{B}_{\text{multi}}} \frac{\langle \mathbf{T}_c^*, \mathbf{C}_c \rangle}{\min(J_c, L_c)} \tag{4}$$

where $\mathbf{T}_c^* \in \mathbb{R}^{J_c \times L_c}$ is the optimal transport plan. The combined loss is $\mathcal{L}_{\text{HiCL}} = \mathcal{L}_{\text{global}} + 0.5\,\mathcal{L}_{\text{local}}$.

2.3 Slot-Based Masked Diagnosis Prediction (Slot-MDP)

Slot-MDP is the primary prediction module. Unlike masked language modelling over open vocabularies, it operates over a closed, type-specific vocabulary: each slot predicts from $\mathcal{V}_{\tau_k}$ via cosine matching against fixed Qwen3 embeddings $\mathbf{e}_w \in \mathbb{R}^{2048}$.

Each masked slot receives a learnable query $\mathbf{q}_k = \mathbf{E}_{\text{pos}}(k) + \mathbf{E}_{\text{type}}(\tau_k)$. Slots attend to per-image representations via cross-attention, then communicate via self-attention to encourage diagnostic consistency:

$$\hat{\mathbf{q}}_k = \text{CrossAttn}(\mathbf{q}_k, \{\mathbf{r}_j\}) + \mathbf{q}_k \tag{5}$$

$$\tilde{\mathbf{q}}_k = \text{SelfAttn}(\hat{\mathbf{q}}_k, \{\hat{\mathbf{q}}_{k'}\}) + \hat{\mathbf{q}}_k \tag{6}$$

Each slot is projected to $\mathbf{p}_k = \text{FFN}(\tilde{\mathbf{q}}_k)$ and scored:

$$p(w \mid k) = \frac{\exp(\beta \cdot \bar{\mathbf{p}}_k^{\top} \bar{\mathbf{e}}_w)}{\sum_{w' \in \mathcal{V}_{\tau_k}} \exp(\beta \cdot \bar{\mathbf{p}}_k^{\top} \bar{\mathbf{e}}_{w'})} \tag{7}$$

Table 1. Main results (%). Top-1 = strict exact match. Accept. = clinically acceptable accuracy (Eq. 9). TP = trainable parameters. All UNI2-based methods share frozen features, data, and split. † Uses Qwen3 text during training only; all methods use only visual features and frozen vocabulary embeddings at inference. HiPath Cross = trained on Hospitals A+C, tested on Hospital B.

Method	Encoder	TP	Top-1	Top-5	Accept.	Safety
Text-only (no images)	Qwen3	0	21.52	54.27	–	–
Linear probe	PLIP	156 K	10.64	45.13	33.91	78.15
	CONCH	156 K	14.63	55.71	42.68	85.62
	UNI2	312 K	30.64	71.81	58.54	87.50
ABMIL [9]	UNI2	603 K	32.37	73.54	60.29	87.21
Deep MLP	UNI2	15.9 M	32.49	73.86	61.76	87.46
Flat CrossAttn†	UNI2	15.7 M	55.98	83.83	68.92	92.18
HiPath†	UNI2	15.0 M	**68.89**	**88.68**	**74.74**	**97.32**
HiPath Cross†	UNI2	15.0 M	65.47	80.99	68.52	97.06

where β is learnable (initialised to 10). Training uses a focal loss [11] over the ground-truth term m_k of each slot:

$$\mathcal{L}_{\mathrm{MDP}} = \frac{1}{K}\sum_{k=1}^{K}\big(1 - p(m_k \mid k)\big)^{\gamma}\,\mathrm{CE}_{\epsilon}\big(p(\cdot \mid k),\, m_k\big), \tag{8}$$

with focusing parameter γ=2 and label-smoothed cross-entropy CE_ϵ (ϵ=0.1).

Training Objective. The total loss is $\mathcal{L} = \mathcal{L}_{\mathrm{HiCL}} + 0.5\,\mathcal{L}_{\mathrm{MDP}} + 0.1\,\mathcal{L}_{\mathrm{aux}}$, where $\mathcal{L}_{\mathrm{aux}}$ is BCE for coarse category prediction (12 classes). Trainable parameters total 15.0 M (HiPA 5.2 M, HiCL 4.1 M, Slot-MDP 5.7 M). We trained for 50 epochs with AdamW (lr 10^{-4}, cosine schedule, batch 32) on 4×H100 (∼5.6 h). All experiments are repeated over three random seeds; we report mean values (std ≤0.3 pp for all metrics).

Evaluation Protocol. We define four per-slot indicators in a hierarchical cascade: (1) strict: exact match; (2) semantic: match after synonym mapping; (3) coarse: same category (12 classes); (4) safe: no crossing of clinically dangerous boundaries (benign↔malignant, non-adjacent grade jump, IHC polarity reversal). These satisfy strict ⇒ semantic ⇒ coarse and strict ⇒ safe, but coarse and safe are independent. Clinically acceptable accuracy is defined as:

$$\mathrm{Acc}_{\mathrm{accept}} = \frac{1}{N}\Big|\{k : \mathrm{strict}(k) \vee \mathrm{semantic}(k) \vee (\mathrm{coarse}(k) \wedge \mathrm{safe}(k))\}\Big| \tag{9}$$

An exact or synonym match is always acceptable; a coarse match is acceptable only if safe. This reflects clinical practice [13,14]: identifying the broad diagnostic category suffices for the correct clinical pathway, whereas crossing the benign-malignant boundary causes direct harm. Safety boundaries were defined by consensus of two senior pathologists (>15 years each). All metrics are computed per slot (N = 109,523 in the test set).

3 Results and Discussion

Table 2. Per-slot-type performance breakdown for HiPath (%). N = number of test slots of each type. Vocab = vocabulary size.

Slot type	Vocab	N	Strict	Semantic	Accept.	Safety
DIAG	184	95,372	69.68	74.25	76.14	96.93
GRADE	93	4,600	71.30	76.11	76.11	100.00
RES	27	9,551	59.88	59.98	59.98	100.00
All	304	109,523	68.89	73.08	74.74	97.32

Table 3. Ablation study with component configuration. ✓/× = present/removed. Flat CrossAttn from Table 1 included for reference. Removing HiPA also disables local OT (no per-image representations to align).

	HiPA		HiCL		Slot-MDP	Top-1	Accept.	i2t R@1	4+img Top-1
	L1	L2	Glob.	Local					
Full HiPath	✓	✓	✓	✓	✓	**68.9**	**74.7**	**21.2**	**55.1**
w/o HiCL	✓	✓	×	×	✓	67.6 (−1.3)	64.3 (−10.4)	0.0	53.2 (−1.9)
w/o HiPA	×	×	✓	×	✓	67.5 (−1.4)	63.3 (−11.4)	22.6	51.4 (−3.6)
w/o Slot-MDP	✓	✓	✓	✓	×	–	64.6 (−10.1)	21.1	–
Flat CrossAttn†	×	×	✓	×	✓	56.0	68.9	14.3	40.7

Baseline Design. The baselines in Table 1 are designed to progressively isolate the factors contributing to HiPath's performance. (i) Linear probes (PLIP, CONCH, UNI2) and ABMIL [9] are vision-only flat classifiers over the full 304-term vocabulary, establishing the ceiling achievable without text supervision or structured prediction. Deep MLP, the highest-capacity vision-only baseline, applies a 4-block 2048-d LN/GELU/Dropout MLP with a 304-way focal cross-entropy head to mean-pooled UNI2 features; the Text-only row instead classifies from frozen Qwen3 report embeddings without any image input, probing the language prior available in report text alone. English-pretrained encoders (PLIP, CONCH) reach ≤14.6% on Chinese data, reflecting the combined effect of domain shift and label-taxonomy mismatch. UNI2 classifiers plateau at 30–32% regardless of capacity (312 K → 15.9 M parameters yields +1.9 pp), indicating that frozen features alone are insufficient for structured prediction. (ii) Flat

CrossAttn† adds text-supervised training and slot-based prediction (Slot-MDP + global contrastive loss) while replacing the two-level HiPA with a single [CLS] over all patches and removing local OT. It reaches 56.0%, isolating the combined contribution of text alignment and the slot formulation as ~24 pp over the vision-only baselines. (iii) The w/o HiCL variant in Table 3 retains HiPA and Slot-MDP but removes all contrastive text alignment, serving as a purely vision-based structured predictor. It achieves 67.6%, confirming that the slot formulation with hierarchical aggregation is the primary driver of performance. Both HiPath and Flat CrossAttn use Qwen3 text only during training (marked †); at inference, only frozen vocabulary embeddings and visual features are used. The remaining 12.9 pp gap between Flat CrossAttn and the full model is decomposed by the ablation: hierarchical aggregation (HiPA: +1.4 pp overall, +3.6 pp on 4+ image cases) and contrastive alignment (HiCL: +1.3 pp strict, +10.4 pp acceptable).

Per-Slot-Type Analysis. Table 2 reports performance broken down by slot type. DIAG slots, which draw from the largest vocabulary (184 terms) and require the most nuanced visual interpretation, show the lowest strict accuracy. GRADE and RES slots both maintain 100% safety: all grading errors are between adjacent grades, and non-strict RES errors are predominantly semi-quantitative confusions (*e.g.*, (1+) vs. (2+)) that do not alter clinical decisions. Among the 3.1% of DIAG slots classified as unsafe, the top confusions are: chronic inflammation↔dysplasia (847 slots), adenoma↔adenocarcinoma (312 slots), and leiomyoma↔leiomyosarcoma (201 slots). These are morphologically challenging distinctions with documented high inter-observer variability among expert pathologists [14].

Cross-Hospital Generalisation. HiPath Cross is trained on Hospitals A+C and tested on Hospital B, with no overlapping patients, staining protocols, or scanners. It achieves 65.5% strict accuracy and 97.1% safety (Table 1), only 3.4 pp below the in-distribution result. Hospital B is a maternal-child hospital with a narrower organ distribution (12 sites vs. 38–41 for A and C), so this experiment primarily tests generalisation to a different clinical population and staining protocol rather than to entirely unseen organ types.

Long-Tail Analysis. Head terms (top 10%, covering 90.8% of predictions) achieve 70.2% strict accuracy; mid-frequency terms (109 terms) reach 44.4%; tail terms (136 terms, 0.2% of predictions) drop to 2.6%. Three rebalancing strategies (square-root resampling, Focal loss adjustment, class-balanced loss [4]) all degraded head accuracy by 2–5 pp without meaningful tail improvement (+0.8 pp). Inspection of the top-10 tail-term confusions confirms that they are predominantly confused with morphologically similar head terms (*e.g.*, rare carcinoma subtypes misclassified as the dominant "adenocarcinoma" category), consistent with the hypothesis that tail errors reflect visual ambiguity rather than sampling imbalance.

Ablation. Table 3 confirms that each module targets a distinct capability. Removing HiCL collapses retrieval (R@1: 21.2 → 0) and reduces acceptable accuracy by 10.4 pp, while strict accuracy drops more modestly (−1.3 pp) as Slot-MDP compensates; HiCL's value is therefore primarily in cross-modal alignment and the auxiliary regularisation it provides to visual representations. Removing HiPA costs 3.6 pp on multi-image cases (4+ images) but only 0.8 pp on single-image cases, confirming that hierarchical aggregation specifically benefits multi-ROI interpretation. The gap between Flat CrossAttn (56.0%) and w/o HiPA (67.5%) shows that HiPA and local OT interact: without per-image representations, OT has nothing to operate on, creating a compounding deficit.

We note two limitations of the current experimental design. First, we do not compare against a generative baseline (*e.g.*, a visual-adapter-augmented Qwen3 with constrained decoding over the same vocabulary. Such a comparison would require fine-tuning the LLM backbone, which substantially exceeds the 15 M parameter budget of our framework; we consider this an important direction for future work. Second, while we motivate Sinkhorn OT by its ability to handle unequal-sized sets differentiably (Sect. 2.2), a systematic comparison against alternative soft-assignment strategies (*e.g.*, softmax attention) is not included.

4 Conclusion

We presented HiPath, a lightweight vision-language framework that treats structured pathology report prediction as its primary learning objective. With only 15 M trainable parameters on frozen backbones, HiPath achieves 68.9% strict accuracy, 74.7% clinically acceptable accuracy, and 97.3% safety on 749 K cases, with robust cross-hospital generalisation. Several limitations point to future work: the current evaluation covers three hospitals in one region (broader multi-centre validation is needed); the model operates on pathologist-selected ROIs rather than whole-slide images; and cases outside the 304-term vocabulary are assigned the nearest term rather than rejected. We release the evaluation protocol, including the vocabulary, synonym mappings, coarse taxonomy, and safety boundary definitions, to support reproducibility and further research.

Acknowledgments. R. Yuan was supported by the China Scholarship Council (CSC). G. Yang was supported in part by the ERC IMI (101005122), the H2020 (952172), the MRC (MC/PC/21013), the Royal Society (IEC/NSFC/211235), the NVIDIA Academic Hardware Grant Program, the SABER project supported by Boehringer Ingelheim Ltd, the NIHR Imperial Biomedical Research Centre (RDA01), the Wellcome Leap Dynamic Resilience program (co-funded by Temasek Trust), UKRI guarantee funding for Horizon Europe MSCA Postdoctoral Fellowships (EP/Z002206/1), a UKRI MRC Research Grant, TFS Research Grants (MR/U506710/1), the Swiss National Science Foundation (Grant No. 220785), and the UKRI Future Leaders Fellowship (MR/V023799/1, UKRI2738). J. Luo was supported in part by the National Natural Science Foundation of China (No. 62372165).

Disclosure of Interests. The authors have no competing interests to declare that are relevant to the content of this article.

References

1. Bray, F., et al.: Global cancer statistics 2022: GLOBOCAN estimates of incidence and mortality worldwide for 36 cancers in 185 countries. CA: Cancer J. Clin. **74**(3), 229–263 (2024). https://doi.org/10.3322/caac.21834
2. Chen, R.J., et al.: Scaling vision transformers to gigapixel images via hierarchical self-supervised learning. In: IEEE/CVF Conference on Computer Vision and Pattern Recognition (CVPR), pp. 16144–16155 (2022)
3. Chen, R.J., Ding, T., Lu, M.Y., Williamson, D.F., et al.: Towards a general-purpose foundation model for computational pathology. Nat. Med. **30**(3), 850–862 (2024). https://doi.org/10.1038/s41591-024-02857-3
4. Cui, Y., Jia, M., Lin, T.Y., Song, Y., Belongie, S.: Class-balanced loss based on effective number of samples. In: IEEE/CVF Conference on Computer Vision and Pattern Recognition (CVPR), pp. 9268–9277 (2019)
5. Cuturi, M.: Sinkhorn distances: lightspeed computation of optimal transport. In: Advances in Neural Information Processing Systems (NeurIPS), pp. 2292–2300 (2013)
6. Han, B., et al.: Cancer incidence and mortality in China, 2022. J. Natl. Cancer Cent. **4**(1), 47–53 (2024). https://doi.org/10.1016/j.jncc.2024.01.006
7. Huang, Z., et al.: A visual-language foundation model for pathology image analysis using medical Twitter. Nat. Med. **29**(9), 2307–2316 (2023). https://doi.org/10.1038/s41591-023-02504-3
8. Ikezogwo, W.O.: Quilt-1M: one million image-text pairs for histopathology. In: Advances in Neural Information Processing Systems (NeurIPS) (2023)
9. Ilse, M., Tomczak, J.M., Welling, M.: Attention-based deep multiple instance learning. In: International Conference on Machine Learning (ICML), pp. 2127–2136 (2018)
10. Karasikov, M., et al.: Training state-of-the-art pathology foundation models with orders of magnitude less data. In: Medical Image Computing and Computer Assisted Intervention – MICCAI 2025. Lecture Notes in Computer Science, vol. 15967, pp. 573–583. Springer (2025). https://doi.org/10.1007/978-3-032-04984-1_55
11. Lin, T.Y., Goyal, P., Girshick, R., He, K., Dollár, P.: Focal loss for dense object detection. In: IEEE International Conference on Computer Vision (ICCV) (2017)
12. Lu, M.Y., Chen, B., et al.: A visual-language foundation model for computational pathology. Nat. Med. **30**(3), 863–874 (2024). https://doi.org/10.1038/s41591-024-02856-4
13. Nagtegaal, I.D., et al.: The 2019 WHO classification of tumours of the digestive system. Histopathology **76**(2), 182–188 (2019). https://doi.org/10.1111/his.13975
14. Raab, S.S., Grzybicki, D.M., Janosky, J.E., Zarbo, R.J., Meier, F.A., Jensen, C., Geyer, S.J.: Clinical impact and frequency of anatomic pathology errors in cancer diagnoses. Cancer **104**(10), 2205–2213 (2005). https://doi.org/10.1002/cncr.21431
15. Radford, A., et al.: Learning transferable visual models from natural language supervision. In: International Conference on Machine Learning (ICML) (2021)
16. Shao, Z., et al.: TransMIL: transformer based correlated multiple instance learning for whole slide image classification. In: Advances in Neural Information Processing Systems (NeurIPS), vol. 34, pp. 2136–2147 (2021)
17. Sun, Y., et al.: PathAsst: a generative foundation AI assistant towards artificial general intelligence of pathology. In: Proceedings of the AAAI Conference on Artificial Intelligence, vol. 38, pp. 5034–5042 (2024). https://doi.org/10.1609/aaai.v38i5.28308

18. Vorontsov, E., et al.: Virchow: a million-slide digital pathology foundation model (2023). https://doi.org/10.48550/arXiv.2309.07778
19. Wang, X., Zhao, J., et al.: A pathology foundation model for cancer diagnosis and prognosis prediction. Nature **634**(8035), 970–978 (2024). https://doi.org/10.1038/s41586-024-07894-z
20. Xu, C., Li, Y., Chen, P., Pan, M., Bu, X.: A survey on the attitudes of Chinese medical students towards current pathology education. BMC Med. Educ. **20**(1), 259 (2020). https://doi.org/10.1186/s12909-020-02167-5
21. Yang, A., et al.: Qwen3 technical report (2025). https://doi.org/10.48550/arXiv.2505.09388

IRIS: An Intelligent Vision-Language System for Ocular Surface Diseases via Topic Tree and Scene-Driven VQA Generation

Hao Wei[1], Wenjin Qi[2], Dasen Dai[1], Minqing Zhang[1], and Wu Yuan[1](✉)

[1] The Chinese University of Hong Kong, Shatin, Hong Kong, China
wyuan@cuhk.edu.hk

[2] Westlake University, Xihu District, Hangzhou, Zhejiang, China

Abstract. While Large Vision-Language Models (VLMs) demonstrate remarkable generic capabilities, their clinical reasoning in specialized domains like ocular surface diseases (OSDs) is severely hindered by a paucity of high-fidelity, multimodal instruction-tuning data. To dismantle this data bottleneck, we introduce IRIS, an Intelligent Recognition and Interaction System tailored for fine-grained OSD understanding via external eye photography. First, we curate IRIS-120K, the largest and most comprehensive OSD visual question-answering (VQA) dataset to date. Crucially, to overcome the semantic shallowness of conventional image-caption pairs, we propose a synergistic data generation paradigm to explicitly inject clinical priors. Our data engine operates via a dual-branch framework: 1) a Topic Finding Tree (TFT) that hierarchically anchors visual features to precise anatomical and pathological concepts, enforcing rigorous medical deduction logic; and 2) a Scene-driven strategy that synthesizes role-adaptive clinical dialogues to ensure pragmatic generalization. By explicitly aligning a compact 4B-parameter VLM on this structurally enriched corpus, IRIS achieves highly competitive performance and outperforms both generalist and specialized medical VLMs with up to 34B parameters within this dataset. Our findings underscore that structured knowledge injection effectively bridges the domain gap, unlocking the potential for resource-efficient, expert-level AI deployment on mobile edge devices for scalable OSD screening. Code, datasets, and model weights will be publicly released by this repo.

Keywords: Vision-Language Models · Ocular Surface Diseases · Clinical Reasoning · Interpretability · Mobile Edge AI

1 Introduction

Ocular surface diseases (OSDs), including dry eye disease (DED) and infectious keratitis, represent a formidable global health challenge and the fifth leading

H. Wei and W. Qi—Equal contribution.

G. Yang et al. (Eds.): MICCAI 2026, LNCS 16878, pp. 228–238, 2027.
https://doi.org/10.1007/978-3-032-38059-3_22

cause of blindness [15,26]. DED afflicts up to 34% of adults, while infectious keratitis is a leading cause of blindness, which is largely preventable if diagnosed early [6,15]. Despite this prevalence, expert-level diagnosis traditionally mandates specialized stationary equipment (e.g., slit-lamp biomicroscopes) [13,26], exacerbating healthcare disparities in resource-limited settings. Consequently, the ubiquity of high-resolution smartphone cameras has catalyzed mobile eye health (MeHealth) into a transformative frontier, offering an unprecedented opportunity for accessible, large-scale OSD screening via external eye photography [23].

Ophthalmic AI has rapidly evolved from task-specific classifiers (e.g., CorneAI [14,15]) towards multimodal foundation models like RETFound [28], VisionFM [17] and EyeCLIP [19]. While these architectures demonstrate robust generalization for retinal conditions, a critical modality disparity persists: current models exhibit a profound bias towards the posterior segment. For instance, within the 2.77-million-image EyeCLIP dataset, external eye photographs account for a negligible 0.5%. Similarly, EyecareGPT [9] predominantly targets fundus and OCT imagery. This leaves a severe data and reasoning vacuum in the fine-grained understanding of external eye pathologies.

While generic Large Vision-Language Models (VLMs) show potential for interactive diagnostics [21,25], recent benchmarking exposes their stark clinical limitations in ophthalmology, often producing hallucinated interpretations lacking precise anatomical grounding [10,25]. Even domain-specific medical VLMs (e.g., HuatuoGPT-Vision [4], Lingshu [24]) are inherently bottlenecked by the flat, unstructured nature of PubMed image-caption pairs. Consequently, they consistently fail to spatially anchor morphological features to specific anatomical structures (e.g., distinguishing a corneal infiltrate from an adjacent iris nodule)—a rigorous spatial deduction step critical for accurate OSD diagnosis [8].

To bridge this diagnostic gap, we introduce **IRIS** (**I**ntelligent **R**ecognition & **I**nteraction **S**ystem for OSDs). We curate **IRIS-120K**, the largest comprehensive OSD visual question-answering (VQA) dataset to date. To explicitly overcome the semantic misalignment of generic VLMs [8,25], we engineer a Clinically-driven Data Engine composed of two novel paradigms: 1) **Topic Finding Tree (TFT):** A structured framework that hierarchically anchors visual features to 10 predefined ocular regions and pathological findings, forcing the model to internalize rigorous, step-by-step clinical deduction. 2) **Scene-Driven Generation:** A role-adaptive paradigm simulating clinical interactions across three primary users (Doctor, Patient, Student) and 12 scenarios, ensuring pragmatic utility in real-world MeHealth deployments.

By fine-tuning a highly compact 4B-parameter VLM on this enriched corpus, IRIS demonstrates strong performance on our curated dataset, overwhelmingly surpassing generalist and medical models possessing up to 34B parameters. Our findings underscore a vital paradigm shift: meticulous structural knowledge injection and explicit anatomical grounding—rather than mere parameter scaling—are effective strategies to enabling highly efficient, expert-level AI deployment on mobile edge devices for scalable OSD screening [8,23].

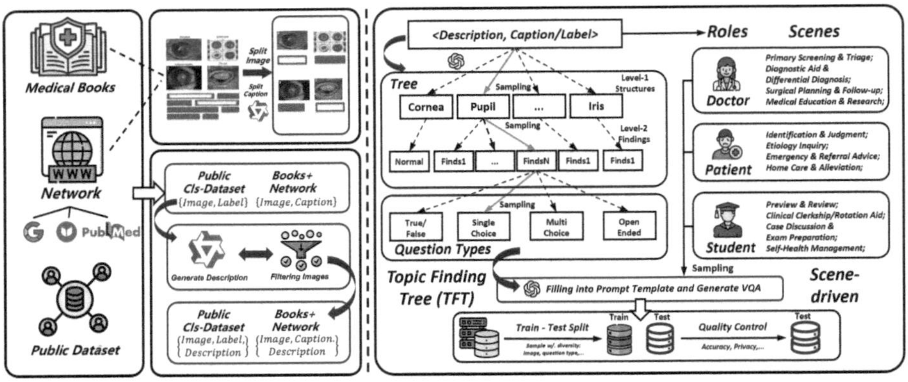

Fig. 1. Overview of our Clinically-driven Data Engine. The pipeline aggregates heterogeneous ocular images, and utilizes a dual-branch VQA generation paradigm: Topic Finding Tree (TFT) for rigorous clinical reasoning and Scene-driven generation for role-adaptive interactions.

The primary contributions of this work are summarized as follows:

- **A Dedicated OSD Interactive Corpus:** We curate and release IRIS-120K, successfully filling a major void in high-fidelity, multimodal instruction-tuning data specifically tailored for external eye photography.
- **Knowledge-Injected Data Engines:** We pioneer the TFT and Scene-Driven strategies to explicitly model spatial ocular anatomy and clinical personas, fundamentally resolving the interpretability and pragmatism deficits inherent in existing medical VLMs.
- **Highly Efficient Foundation Model:** We open-source IRIS-4B, a lightweight clinical VLM optimized for resource-constrained environments. It demonstrates unprecedented parameter efficiency and fine-grained visual-textual grounding compared to massively scaled baselines.

2 Methodology

To bridge the gap between the poor zero-shot performance of existing VLMs and the clinical demand for OSD screening, we propose a novel Clinically-driven Data Engine (Fig. 1). Rather than simply scaling up model parameters, our framework curates **IRIS-120K**, the largest external eye VQA dataset to date. By innovating Topic Finding Tree (TFT) and Scene-driven generation paradigms, we inject rigorous medical reasoning and role-adaptive empathy into lightweight models, enabling highly efficient edge deployment on patient smartphones.

2.1 Large-Scale Data Curation and Preprocessing

To construct our Data Engine, we aggregate external eye images into *image-caption* or *image-label* pairs from three heterogeneous sources: **Medical Books**

(3.1K cases), **Network Data** (93.5K samples from WeChat, PubMed, and Google Lens), and **Public Datasets** (35.4K). Note that these counts are the results before any filtering. To resolve domain noise and format inconsistencies, we design source-specific preprocessing pipelines.

Books and PubMed Literature. Medical literature contains high-value but complex multi-panel figures that severely degrade VLM image-text alignment. Sourcing from textbooks and BIOMEDICA [12] (a biomedical image-caption archive), we deploy a rigorous three-stage pipeline [1]. First, Qwen2.5-VL-7B [3] filters non-external eye images and categorizes data into Paper-Single (standalone) and Paper-Multi (compound). For *Paper-Multi*, a YOLOv7-based detector [22] segments sub-figures, and Qwen2.5-VL-72B [3] uses these bounding boxes to semantically map the overarching monolithic caption to specific sub-figure crops. A secondary VLM filter eliminates irrelevant content (e.g., fundus) exposed after splitting. Captions are further enriched using main-text references. Book data undergoes an identical fine-grained alignment.

Web Resources. To capture real-world long-tail diversity, we use curated book images as visual seeds for Google Lens reverse searches, extracting alt-texts as captions. Concurrently, we scrape ophthalmic WeChat articles, which natively provide structured *(image, disease name, symptom description)* triplets. Both sources undergo strict VLM-based filtering to guarantee external-eye relevance.

Public Datasets Adaptation. We integrate high-quality classification and segmentation datasets. To adapt segmentation data, we introduce an area-based heuristic: if over 50% of the images contain lesion bounding boxes occupying $\geq$50% of the total image area (meaning localized pathology visually dominates), the entire dataset is repurposed for classification. All data is standardized as *(image, label)* pairs.

Quality-Aware Stratification. Given varying raw source fidelity, we conduct sampling-based human evaluations on resolution, image-text alignment, and semantic richness to derive a strict quality hierarchy: **WeChat** > **Books** > **Paper-Single** > **Public Cls** > **Paper-Multi** > **Web (Google Lens)**. This empirically established prior fundamentally directs our "Quality-Aware Dynamic Sampling" strategy (Sect. 2.2), ensuring high-fidelity data dominates the training distribution of our lightweight models.

2.2 Knowledge-Driven VQA Generation & Quality-Aware Sampling

Traditional "image-caption" pairs generate flat representations, failing to impart the "anatomy-to-sign" deduction logic required in clinical practice. To overcome this, we transform the curated images into interpretable VQA pairs using a dual-branch strategy. To enrich the semantic context and mitigate LLM hallucinations, we first employ Qwen3-VL-32B [2] to generate a detailed visual description T_{desc} by the image and its original source information, including captions, disease labels, symptom descriptions, or textbook/public-dataset annotations.

Combined with the original caption/label T_{cap}, this forms a rich textual surrogate $x = (T_{cap}, T_{desc})$, serving as the foundational semantic anchor for downstream generation.

Topic Finding Tree (TFT) Generation. To force the model to "think step-by-step like an ophthalmologist", we introduce TFT, a structured anatomical decomposition framework. The TFT hierarchically parses x into a depth-2 logical tree: the first level comprises 10 predefined ocular anatomical regions $a_i \in \mathcal{A}$ (*Pupil, Cornea, Iris, Sclera-Conjunctiva Region, Eyelashes, Upper Eyelid, Lower Eyelid, Superior Palpebral Conjunctiva, Inferior Palpebral Conjunctiva, and Others*), and the second level extracts region-specific clinical findings $f \in F_{a_i}$. This explicitly establishes anatomical-pathological paths $\mathcal{T} = \{(a_i, f) \mid a_i \in \mathcal{A}, f \in F_{a_i}\}$.

For each instantiated path, an LLM generates diverse question formats (e.g., True/False, Single-choice, Multiple-choice or Open-ended). Crucially, we prompt the LLM to construct a structured **4-stage reasoning chain** ($Think$): *(1) Visual Observation* (identifying morphological features); *(2) Clinical Correlation*; *(3) Logical Deduction* (excluding differential diagnoses); and *(4) Conclusion*. This explicit mapping ensures the generated VQA tuples $(Q, A, Think)$ are strictly grounded in localized clinical evidence.

Scene-Driven Adaptation. To accommodate the diverse interaction needs of real-world edge deployments, we simulate conversational contexts across three primary user roles: $\mathcal{R} = \{\text{Patient, Doctor, Student}\}$. For each role, we define tailored clinical scenes $\mathcal{S}_r$ (e.g., triage advice for patients, differential diagnosis for doctors, as shown in the right panel of the Fig. 1). Using scene-specific templates, the LLM generates interactions incorporating a sequential clinical chain-of-thought–*Query Analysis, Evidence Extraction, Clinical Reasoning, Response Formulation*–ensuring the response tone and jargon density closely align the target user's medical literacy.

Quality-Aware Dynamic Sampling and Dataset Finalization. Exhaustive generation across all TFT nodes and clinical scenes would yield over 2 million highly redundant VQA pairs. This contradicts our fundamental objective of training a highly efficient, small-parameter model (2B/4B) capable of offline edge deployment. Therefore, we implement a quality-aware dynamic sampling protocol. Instead of uniform generation, we probabilistically assign combination strategies based on the predefined data quality hierarchy. High-quality sources trigger denser generation (e.g., sampling multiple TFT regions and scenarios simultaneously), while lower-quality web data is sparsely sampled to maximize information yield without redundancy. For each image, 1–3 prompts are generated using either the TFT or Scene-Driven method.

This balanced protocol generates approximately 130K initial VQA pairs. We apply pHash [27] for strict image deduplication to prevent data leakage. An initial test pool of 9K VQA pairs undergoes a rigorous final quality control pass using GPT-5-mini to filter out factual hallucinations and privacy anomalies. To further validate data quality, a random subset was reviewed by an ophthalmologist,

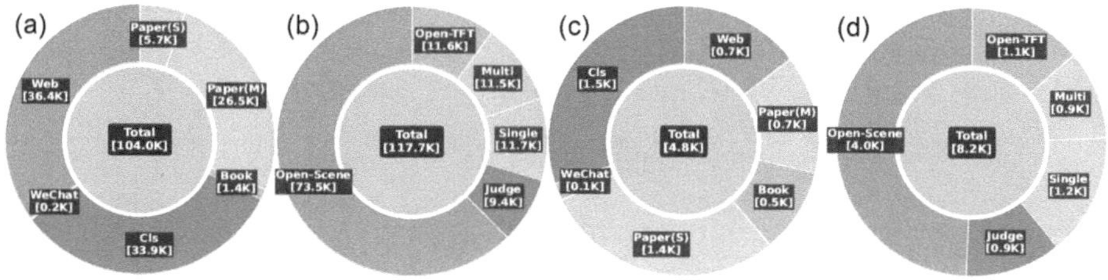

Fig. 2. Dataset statistics of IRIS-120K. (a)/(c) Distribution of image sources across Train/Test sets; (b)/(d) Distribution of generated VQA types (TFT vs. Scene-driven) in train and test, respectively.

achieving an 87/100 factual accuracy for image descriptions and a 4.4/5.0 rating for the generated VQA pairs. This ultimately yields the **IRIS-120K**, comprising **117.7K** training pairs and **8.2K** test set (detailed statistics in Fig. 2).

3 Experiments

Experimental Setup. First, we fine-tune Qwen3-VL via Low-Rank Adaptation on four A6000 GPUs to get IRIS series models, more details can be found our repo. To comprehensively evaluate multi-modal capabilities, we benchmark IRIS against 16 contemporary VLMs, encompassing General-domain (GPT-5-mini [20], InternVL-3.5 series [5], Qwen3-VL series [2]) and Medical-domain models (HuatuoGPT-Vision [4], HuluMed [7], Lingshu [24], Med-Gemma [18]), which are evaluated in a zero-shot manner on the current test set. We employ Accuracy for closed-ended objective tasks (Judge, Single/Multi-choice) and standard generative metrics (BLEU-1 [16], ROUGE-1/L-F [11]) for open-ended interactions (Tree/Scene-VQA).

Main Results. Quantitative evaluations are summarized in Table 1. The IRIS demonstrates highly competitive performance across all question types. Our optimal model, **IRIS-4B**, achieves near-perfect objective reasoning accuracies (97.25 on Judge, 98.52 on Single-choice) and dominates complex generative tasks (e.g., 42.36 on Tree-VQA B-1). Specifically, across the five primary metrics (Accuracies and B-1 scores), IRIS-4B attains an outstanding overall average of **74.26**.

Crucially, IRIS exhibits profound parameter efficiency. IRIS-4B significantly outperforms baselines up to 8× larger, outperforming the strongest medical VLM (Lingshu-32B, overall avg. 55.00) by 19.26 absolute points, and the best generic VLM (Qwen3-VL-32B, overall avg. 50.61) by 23.65 points. Furthermore, even our minimalist IRIS-2B (avg. 69.57) consistently beats all 32B/34B competitors. Notably, the performance gap between IRIS-4B and its direct pretrained counterpart, Qwen3-VL-4B, effectively isolates the impact of domain-specific supervision within the same backbone family.

Ablation Studies. 1) Model Scaling: Evaluating the architecture across 2B, 4B, and 8B scales reveals that the performance peaks at 4B. Scaling further to

Table 1. Performance comparison of General (Gen), Medical (Med), and Our models across all five question types. *Gen. Avg.* denotes the average score across the six open-ended generative metrics (**B**LEU and **R**OUGE). Best results are in **bold**.

Group	Model	Judge	Single	Multi	Tree-VQA			Scene-VQA			Gen.Avg.
					B-1	R-1-F	R-L-F	B-1	R-1-F	R-L-F	
Gen	GPT-5-mini	69.52	72.08	67.48	16.48	11.31	21.81	13.73	22.02	11.81	16.19
	Internvl3.5-14B	77.35	75.62	54.59	11.82	19.80	11.59	18.50	20.22	11.34	15.55
	Internvl3.5-8B	72.59	75.37	49.41	10.52	17.94	10.64	15.42	18.68	10.17	13.90
	Qwen3-vl-32B	77.57	79.98	62.64	13.65	20.79	13.06	19.22	20.36	11.84	16.49
	Qwen3-vl-8B	69.52	73.06	57.41	7.77	16.30	8.88	13.33	17.09	9.18	12.09
	Qwen3-vl-4B	62.43	70.68	51.98	6.19	14.24	7.83	10.86	16.39	8.63	10.69
	Qwen3-vl-2B	69.31	62.36	44.15	4.82	15.88	9.06	7.15	16.59	8.96	10.41
Med	HGPT-V-7B	32.70	25.62	8.45	21.34	22.65	15.02	19.58	18.53	11.59	18.12
	HGPT-V-34B	66.56	68.29	50.36	28.60	27.18	18.79	24.40	22.02	13.89	22.48
	HuluMed-7B	83.07	49.09	26.61	31.91	30.17	20.58	26.12	25.49	15.84	25.02
	HuluMed-14B	83.92	59.59	25.63	32.48	30.68	20.80	25.98	25.70	15.70	25.22
	HuluMed-32B	83.70	60.05	30.39	33.02	31.06	21.11	26.02	25.92	15.90	25.51
	Lingshu-7B	84.97	64.25	37.70	21.63	25.88	19.33	13.59	20.64	13.46	19.09
	Lingshu-32B	84.76	84.43	59.71	28.04	29.85	21.92	18.08	23.66	15.13	22.78
	MG-4B	66.88	60.38	36.22	16.13	22.68	16.52	17.76	19.94	12.27	17.55
	MG1.5-4B	71.01	58.24	31.14	11.52	17.19	9.98	11.72	16.36	8.26	12.51
Ours	IRIS-8B	97.04	98.19	88.55	42.27	**39.68**	**29.64**	44.19	**36.32**	**24.74**	**36.14**
	IRIS-4B	**97.25**	**98.52**	**88.96**	**42.36**	39.29	29.03	**44.23**	36.08	24.45	35.91
	IRIS-4B-TFT	96.51	98.19	88.26	41.86	39.18	28.95	40.46	28.86	18.17	32.91
	IRIS-4B-Scene	93.12	83.20	48.42	37.02	34.31	24.48	44.13	36.01	24.41	33.39
	IRIS-2B	96.19	97.03	83.98	40.40	38.06	27.77	30.28	26.70	16.86	30.01

8B introduces a slight performance plateau, indicating that 4B is the optimal capacity sweet-spot for learning our specialized representations without overfitting, ideal for edge deployment. **2) Data Composition:** We ablated the training components of IRIS-4B. Training exclusively on structured TFT data (*IRIS-4B-TFT*) preserves strong logical deduction but degrades Scene-VQA generation. Conversely, relying solely on role-playing Scene data (*IRIS-4B-Scene*) precipitates a substantial degradation in objective precision (e.g., Multi-choice accuracy plummets from 88.96 to 48.42). This confirms the essential synergy of our framework: TFT builds the rigorous medical logic, while Scene-adaptation provides the clinical interaction interface.

Qualitative Analysis and Interpretability. Fig.ure 3 qualitatively validates the clinical reliability and transparency of IRIS. In complex closed-ended tasks (Fig. 3, top), IRIS-4B accurately diagnoses *Granular corneal dystrophy* and verifies stromal involvement, overcoming the diagnostic inconsistencies and hallucinations that plague much larger medical baselines (e.g., Lingshu-32B).

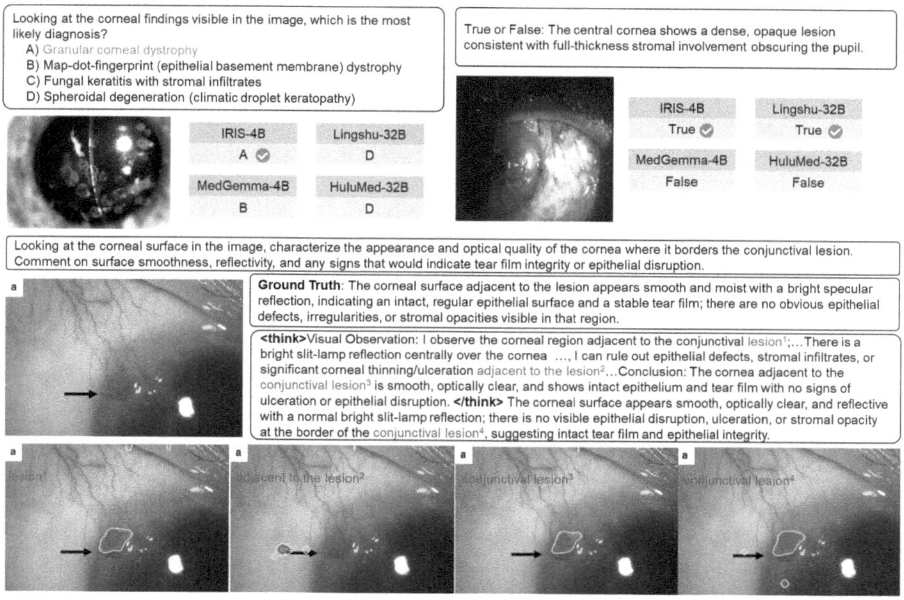

Fig. 3. Qualitative evaluation of IRIS-4B. **Top:** Accurate predictions on complex objective tasks, outperforming substantially larger medical baselines. **Bottom:** Demonstration of transparent reasoning and visual grounding. IRIS-4B articulates its logic via a <think> block and visually anchors generated clinical terms (red text) to localized pathological regions (attention contours: > 25% yellow, > 50% red). (Color figure online)

Crucially, for open-ended interactions (Fig. 3, bottom), IRIS-4B transcends "black-box" answering. Driven by our structured TFT training paradigm, it generates an explicit <think> block to articulate a rigorous clinical chain-of-thought—detailing visual observations and logical deductions—prior to formulating its final conclusion. Furthermore, IRIS-4B exhibits robust *visual grounding*. Generated clinical terms (e.g., "lesion", "adjacent to the lesion") are spatially anchored to the input image via attention maps (delineated by > 25% yellow and > 50% red importance contours). While functioning conceptually akin to Class Activation Mapping (CAM) rather than exact pixel-level segmentation, this precise spatial-textual correlation confirms that IRIS-4B explicitly anchors its reasoning to authentic morphological cues rather than language priors, offering vital interpretability for reliable clinical triage.

4 Conclusion

In this paper, we introduced IRIS, a highly efficient and interpretable Vision-Language Model tailored for OSD screening. To overcome the severe domain gap in existing VLMs, we curated IRIS-120K, the largest external eye VQA dataset to date, and proposed a Clinically-driven Data Engine that combines a Topic Finding Tree with role-adaptive, scene-driven interactions. This design encourages the model to learn structured clinico-pathological reasoning rather than relying on superficial language priors. Extensive evaluations show that our optimal model, IRIS-4B, consistently outperforms general and medical baselines up to eight times its size across diverse multimodal tasks, supporting the key insight that high-quality, structurally curated domain data can outweigh sheer parameter scaling. Moreover, IRIS integrates explicit step-by-step reasoning chains (<think>) with attention-based visual grounding, improving interpretability, and aligning with clinical diagnostic workflows. Importantly, its efficiency enables low-cost, fully offline edge deployment on patient devices, offering a practical path toward large-scale, privacy-preserving screening.

Despite these promising results, our study has several limitations. First, our evaluation is conducted in a closed-loop setting where the model is fine-tuned and evaluated on data splits generated by the same VLM/LLM-based data engine. Consequently, the reported results reflect benchmark-specific performance and distributional alignment rather than universal clinical superiority over all baselines. Second, while open-ended metrics (e.g., BLEU and ROUGE) are widely adopted to measure text alignment with references, they serve as weak proxies and are not complete proofs of diagnostic correctness or clinical safety. Moving forward, rigorous external clinical validation with expert multi-reader adjudication remains a necessary future work to confirm the real-world diagnostic reliability and practical utility of IRIS.

Acknowledgments. This work was supported in part by the Research Grants Council (RGC) of Hong Kong SAR (GRF14213125, GRF14201824, GRF14216222), the Innovation and Technology Fund (ITF) of Hong Kong SAR (ITS/252/23).

Disclosure of Interests. The authors have no competing interests to declare that are relevant to the content of this article.

References

1. Baghbanzadeh, N., et al.: Advancing medical representation learning through high-quality data. In: International Conference on Medical Image Computing and Computer-Assisted Intervention, pp. 24–33. Springer (2025)
2. Bai, S., Cai, Y., Chen, R., Chen, K., et.al, X.C.: Qwen3-VL technical report. arXiv preprint arXiv:2511.21631 (2025)
3. Bai, S., et al.: Qwen2.5-VL technical report. arXiv preprint arXiv:2502.13923 (2025)
4. Chen, J., et al.: HuatuoGPT-vision, towards injecting medical visual knowledge into multimodal LLMs at scale. arXiv preprint arXiv:2406.19280 (2024)
5. Chen, Z., et al.: InternVL: scaling up vision foundation models and aligning for generic visual-linguistic tasks. In: Proceedings of the IEEE/CVF Conference on Computer Vision and Pattern Recognition, pp. 24185–24198 (2024)
6. Graham, A.D., et al.: A machine learning approach to predicting dry eye-related signs, symptoms and diagnoses from meibography images. Heliyon **10**(17) (2024)

7. Jiang, S., et al.: Hulu-med: a transparent generalist model towards holistic medical vision-language understanding. arXiv preprint arXiv:2510.08668 (2025)
8. Jin, K., et al.: A systematic review of vision and vision-language foundation models in ophthalmology. Adv. Ophthalmol. Pract. Res. (2025)
9. Li, S., et al.: EyecareGPT: boosting comprehensive ophthalmology understanding with tailored dataset, benchmark and model. In: Proceedings of the 33rd ACM International Conference on Multimedia, pp. 3893–3902 (2025)
10. Li, Z., et al.: Large language model-based multimodal system for detecting and grading ocular surface diseases from smartphone images. Front. Cell Dev. Biol. **13**, 1600202 (2025)
11. Lin, C.Y.: ROUGE: a package for automatic evaluation of summaries. In: Text Summarization Branches Out, pp. 74–81 (2004)
12. Lozano, A., et al.: Biomedica: an open biomedical image-caption archive, dataset, and vision-language models derived from scientific literature. In: Proceedings of the Computer Vision and Pattern Recognition Conference, pp. 19724–19735 (2025)
13. Lu, T.C., Huang, C.H., Lin, I.C.: Artificial intelligence application in cornea and external diseases. Diagnostics **15**(24), 3199 (2025)
14. Maehara, H., et al.: Artificial intelligence support improves diagnosis accuracy in anterior segment eye diseases. Sci. Rep. **15**(1), 5117 (2025)
15. Maehara, H., et al.: The importance of clinical experience in ai-assisted corneal diagnosis: verification using intentional ai misleading. Sci. Rep. **15**(1), 1462 (2025)
16. Papineni, K., Roukos, S., Ward, T., Zhu, W.J.: BLEU: a method for automatic evaluation of machine translation. In: Proceedings of the 40th Annual Meeting of the Association for Computational Linguistics, pp. 311–318 (2002)
17. Qiu, J., et al.: VisionFM: a multi-modal multi-task vision foundation model for generalist ophthalmic artificial intelligence. arXiv preprint arXiv:2310.04992 (2023)
18. Sellergren, A., et al.: MedGemma technical report. arXiv preprint (2025). arXiv:2507.05201 (2025)
19. Shi, D., et al.: A multimodal visual-language foundation model for computational ophthalmology. NPJ Digit. Med. **8**(1), 381 (2025)
20. Singh, A., Fry, A., Perelman, A., et al.: OpenAI GPT-5 system card (2025). https://arxiv.org/abs/2601.03267. Accessed 21 Feb 2026
21. Srinivasan, S., et al.: can off-the-shelf visual large language models detect and diagnose ocular diseases from retinal photographs? BMJ Open Ophthalmol. **10**(1) (2025)
22. Sun, Y., et al.: PathAsst: a generative foundation ai assistant towards artificial general intelligence of pathology. In: Proceedings of the AAAI Conference on Artificial Intelligence, vol. 38, pp. 5034–5042 (2024)
23. Wang, M.H., et al.: Leveraging artificial intelligence and clinical laboratory evidence to advance mobile health applications in ophthalmology: taking the ocular surface disease as a case study. iLABMED **3**(1), 64–85 (2025)
24. Xu, W., et al.: Lingshu: a generalist foundation model for unified multimodal medical understanding and reasoning. arXiv preprint arXiv:2506.07044 (2025)
25. Yeh, C.H., et al.: Insight: a multi-modal diagnostic pipeline using LLMs for ocular surface disease diagnosis. In: International Conference on Medical Image Computing and Computer-Assisted Intervention, pp. 711–721. Springer (2024)
26. Yonehara, M., et al.: Use of explainable ai on slit-lamp images of anterior surface of eyes to diagnose allergic conjunctival diseases. Allergol. Int. **74**(1), 86–96 (2025)
27. Zauner, C.: Implementation and benchmarking of perceptual image hash functions. Master's thesis, University of Applied Sciences Hagenberg, Hagenberg, Aus-

tria (2010). https://www.phash.org/docs/pubs/thesis_zauner.pdf. pHash library implementation
28. Zhou, Y., et al.: A foundation model for generalizable disease detection from retinal images. Nature **622**(7981), 156–163 (2023)

KMP-MIL: Knowledge Memory Pool Multiple Instance Learning with Foundation Model for Continual Whole Slide Image Classification

Lingling Yuan[1,3], Zhaoxia Yin[3], Yan Han[1], Jinghua Zhang[4], Marcin Grzegorzek[5], and Chen Li[1,2](✉)

[1] Group of Microscopic Image and Medical Image Analysis, College of Medicine and Biological Information Engineering, Northeastern University, Shenyang, China
lichen@bmie.neu.edu.cn
[2] Software College, Northeastern University, Shenyang, China
[3] Shanghai Key Laboratory of Multidimensional Information Processing, School of Information and Electronic Engineering, East China Normal University, Shanghai, China
[4] College of Computer Science and Software Engineering, Hohai University, Nanjing, China
[5] Institute of Medical Informatics, University of Lübeck, Lübeck, Germany

Abstract. Continual learning (CL) for pathological whole slide image (WSI) classification is challenging because gigapixel images, sparse diagnostic cues, and cross-organ domain shifts exacerbate catastrophic forgetting. In addition, existing methods often rely on rehearsal buffers, which are difficult to deploy in clinical practice under privacy and governance constraints. We propose knowledge memory pool multiple instance learning (KMP-MIL), a rehearsal-free continual MIL framework that couples a frozen pathology foundation model (FM) with parameterized memory-based adaptation. KMP-MIL introduces: 1) a growing KMP that incrementally stores compact task-specific memory units instead of replayed samples, and 2) query-based memory retrieval with prototype-conditioned feature calibration (PFC) and textual calibration (TC) in a shared vision-language space to improve adaptation. The framework supports task-CL with task identity (ID) and class-CL with task-agnostic inference through a universal memory pool that retrieves task-relevant knowledge without task ID. Experiments on sequential cross-organ WSI classification across breast, lung, kidney, and esophagus datasets show strong and stable performance. KMP-MIL achieves 89.08% average accuracy in task-CL and competitive class-CL results with reduced forgetting. It is also stable under reverse-order evaluation and storage-efficient, with persistent memory growing from 0.80 MB to 3.21 MB over four tasks, about 3.14% of a replay buffer storing 10 WSIs. Our code is available at https://github.com/Lingling-Yuan/KMP-MIL.

Keywords: Whole slide image · Continual learning · Foundation model

G. Yang et al. (Eds.): MICCAI 2026, LNCS 16878, pp. 239–249, 2027.
https://doi.org/10.1007/978-3-032-38059-3_23

1 Introduction

Whole slide images (WSIs) provide cellular-resolution evidence that is critical for tumor diagnosis and subtyping. However, a single slide can reach gigapixel-level resolution, while diagnostic cues are sparse and unevenly distributed across the tissue [19]. These properties make dense annotation over WSIs impractical, due to high cost and engineering burden [3,25]. As a result, clinical-grade WSI classification commonly uses weakly supervised multiple instance learning (MIL). In MIL, each WSI is split into patches and treated as a bag, and the model learns from slide-level labels. Aggregation modules, such as attention mechanisms or transformers, combine patch features to produce a slide-level prediction without pixel-level annotations [9,16,22]. In real deployments, data collection rarely stays fixed. Systems often grow incrementally, expanding from pilot sites to more hospitals, and from a single organ to multiple organs [7,23]. With changing clinical needs and collaborations, the model must continuously incorporate data from new sources, forming a sequential task stream as shown in Fig. 1A. Therefore, models need continual learning (CL), rather than one-time offline training, to keep performance on earlier tasks while learning new ones [6,18].

When organ-specific tasks arrive sequentially, CL faces greater catastrophic forgetting risk. Large differences in tissue structure, staining appearance, and scanning protocols across organs and cancer types can cause the model to overwrite useful cues from earlier tasks when learning later tasks, worsening forgetting in cross-organ settings [24]. Existing pathology CL methods often rely on rehearsal buffers that store representative slides or sample-level embeddings, and combine replay with regularization or distillation to stabilize MIL aggregation across tasks [8,11–13,17]. However, compliance and data governance can restrict centralized storage and replay of historical WSIs, limiting the applicability of buffer-based methods. This motivates rehearsal-free CL methods that retain knowledge without storing or replaying past slides or sample-level features.

Motivated by these constraints, we study rehearsal-free continual WSI classification with sequential organ-specific tasks, while considering that task identity (ID) may be unavailable at inference. We propose *knowledge memory pool multiple instance learning* (KMP-MIL), as shown in Fig. 1. KMP-MIL couples a frozen pathology foundation model (FM) with parameterized memory-based adaptation in a vision-language space. Instead of replaying historical WSIs, it maintains a growing *knowledge memory pool* (KMP) of compact task-specific memory units. For each slide, a WSI-level query retrieves Top-K units to provide knowledge-conditioned context. The retrieved units are used for *prototype-conditioned feature calibration* (PFC) before MIL aggregation, while *textual calibration* (TC) uses class text prototypes to calibrate bag representations and improve semantic alignment. KMP-MIL supports both task-CL and class-CL. In task-CL, retrieval is restricted to the task-specific KMP segment when task ID is available; in class-CL, the model queries a universal pool built from all seen tasks for task-agnostic prediction. Our main contributions are as follows:

- We propose KMP-MIL, which uses a growing KMP to store compact task-specific memory units, thereby avoiding the replay of historical WSIs.

- We design a query-based mechanism for WSI-specific Top-K memory selection, and combine it with PFC and TC in a shared vision-language space to improve adaptation and cross-organ semantic alignment.
- We evaluate KMP-MIL on sequential cross-organ WSI classification across breast, lung, kidney, and esophagus datasets under both task-CL and class-CL settings, including reverse-order evaluation. KMP-MIL achieves strong performance with reduced forgetting and good storage efficiency.

2 Proposed Method

We study rehearsal-free continual WSI classification with sequentially introduced organ-specific tasks. The learning stream contains T tasks indexed by $t \in \{1, \ldots, T\}$, and each task provides training data from a new organ domain. We propose KMP-MIL, which maintains a growing KMP for rehearsal-free CL and applies PFC and TC to improve cross-domain robustness.

2.1 Training Stage

Knowledge Memory Pool (KMP). For each task t, we maintain a task-specific $\text{KMP}_t = \{(k_m^t, p_m^t)\}_{m=1}^M$, as shown in Fig. 1B. Each memory unit contains a learnable key $k_m^t \in \mathbb{R}^C$ in the aligned visual-text space and prompt tokens $p_m^t \in \mathbb{R}^{L \times D}$ for WSI aggregation, where C denotes the visual-text embedding dimension, and L and D denote the prompt length and the hidden dimension of the frozen text tower E_{txt}, respectively. As tasks arrive, the memory pool grows as $\text{KMP}_{1:t} = \bigcup_{t'=1}^{t} \text{KMP}_{t'}$, enabling rehearsal-free CL by storing compact parameterized memory units instead of historical WSIs. For a WSI bag $\mathcal{X}_i$, we extract patch embeddings h_{ij} by the frozen pathology FM image encoder E_{img} and form $H_i = [h_{i1}, \ldots, h_{iN_i}]^\top \in \mathbb{R}^{N_i \times C}$. We obtain a WSI-level query $q_i \in \mathbb{R}^C$ by max pooling over patch embeddings, using it only as a lightweight retrieval router to preserve sparse diagnostic cues without adding trainable query modules. To inject task knowledge into retrieval, we use E_{txt} to encode a task-level descriptor d^t that captures organ-specific histological patterns and diagnostic morphological cues. The resulting embedding initializes the newly added keys in KMP_t, providing semantic priors that improve early routing stability by reducing ambiguity between current-task keys and historical keys. In parallel, class text embeddings are used as class prototypes for TC. We ℓ_2-normalize q_i and all keys, so cosine similarity is equivalent to an inner product. The retrieval score for the m-th key in task t is:

$$s_{i,m}^t = \cos(q_i, k_m^t) = q_i^\top k_m^t, \quad m \in \{1, \ldots, M\}, \tag{1}$$

where higher $s_{i,m}^t$ indicates the query is more consistent with the memory unit. We select Top-K indices $\mathcal{I}_i^t = \text{TopK}(\{s_{i,m}^t\}_{m=1}^M)$ from current-task keys and use $\mathcal{L}_{\text{match}} = 1 - \frac{1}{BK} \sum_{i=1}^{B} \sum_{m \in \mathcal{I}_i^t} s_{i,m}^t$ to align queries with retrieved keys, where

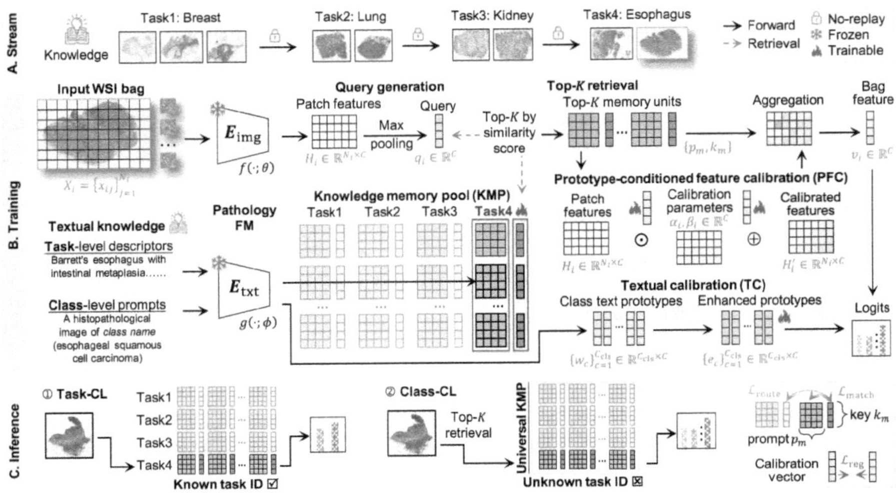

Fig. 1. Overview of KMP-MIL for rehearsal-free continual WSI classification. A: Data stream. Sequential stream of four organ-specific WSI tasks. **B: Training.** Frozen E_{img} extracts patch features; query q retrieves Top-K KMP units for PFC and prototype-guided aggregation, and TC-enhanced prototypes from E_{txt} are used for classification. **C: Inference.** Task-CL retrieves within the task-specific KMP using task ID, while class-CL retrieves from a universal KMP without task ID.

K is the retrieval size and B is the batch size. For calibration and aggregation, we convert the Top-K retrieval similarities into normalized weights $w_{i,m}$ using softmax, so that more relevant memory units receive larger weights while non-selected units are assigned zero weight. To stabilize retrieval, we encourage each query to prefer current-task keys over past-task keys using a routing loss:

$$\mathcal{L}_{\text{route}} = \frac{1}{B} \sum_{i=1}^{B} \text{softplus}\left(\frac{m_i^{\text{past}} - m_i^{\text{cur}} + \mu}{\tau} \right), \tag{2}$$

where $m_i^{\text{cur}} = \max_m q_i^\top k_m^t$ and $m_i^{\text{past}} = \max_{t'<t,\, m} q_i^\top k_m^{t'}$, and μ and τ are the margin and temperature, respectively. For the first task, we set $\mathcal{L}_{\text{route}} = 0$. The KMP loss is $\mathcal{L}_{\text{KMP}} = \mathcal{L}_{\text{match}} + \lambda_{\text{route}} \mathcal{L}_{\text{route}}$. The retrieved memory units provide knowledge-conditioned context for downstream MIL aggregation, enabling the model to focus on task-relevant patterns without accessing past training data. During continual training, only the newly added memory units in KMP_t are updated, while memory units from previous tasks are frozen and reused.

Prototype-Conditioned Feature Calibration (PFC). To mitigate domain shift across sequential organ tasks, we calibrate patch features using the Top-K memory units retrieved from KMP. Given a query q_i, we reuse the KMP retrieval outputs, including the Top-K index set $\mathcal{I}_i^t$ and the normalized retrieval weights $w_{i,m}$. Inspired by FiLM [20], we propose the PFC module, which attaches learnable calibration parameters to each KMP memory unit, as shown in Fig. 1B.

Specifically, for task t, we learn two channel-wise parameter pools $\{\alpha_m^t\}_{m=1}^M$ and $\{\beta_m^t\}_{m=1}^M$ with $\alpha_m^t, \beta_m^t \in \mathbb{R}^C$, where each pair (α_m^t, β_m^t) serves as an affine modulator in the shared embedding space. Because these parameters are indexed by retrieved KMP units, the modulation is compact and retrieval-aware, which is suitable for rehearsal-free CL. For each bag, we fuse the retrieved calibration parameters using the retrieval weights $\alpha_i = \sum_{m \in \mathcal{I}_i^t} w_{i,m}\alpha_m^t, \beta_i = \sum_{m \in \mathcal{I}_i^t} w_{i,m}\beta_m^t$, and calibrate the patch embedding matrix $H_i \in \mathbb{R}^{N_i \times C}$ as:

$$H_i' = H_i \odot \alpha_i + \mathbf{1}_{N_i}\beta_i^\top, \tag{3}$$

where $\alpha_i, \beta_i \in \mathbb{R}^C$ are broadcast to all N_i patches, $\mathbf{1}_{N_i} \in \mathbb{R}^{N_i}$ is an all-ones vector, and $\odot$ denotes channel-wise multiplication. To avoid overly aggressive modulation, we regularize the fused scaling and bias by encouraging α_i to stay close to the identity transform and β_i to stay close to zero:

$$\mathcal{L}_{\text{reg}} = \frac{1}{BC}\sum_{i=1}^{B}\sum_{c=1}^{C}(\alpha_{i,c} - 1)^2 + \frac{1}{BC}\sum_{i=1}^{B}\sum_{c=1}^{C}(\beta_{i,c})^2. \tag{4}$$

The calibrated patch embeddings H_i' are then aggregated by a prototype-guided MIL module using multi-prototype cosine-weighted pooling, where cosine similarities between patch features and retrieved KMP prototypes are normalized across patches to compute the bag feature v_i.

Textual Calibration (TC). To align prediction with the shared vision-language embedding space, we build class text prototypes using a textual template ensemble. For each class c, we construct multiple textual descriptions by combining template sentences with class-name variants, and encode them with E_{txt} to obtain a base class text prototype w_c. We then introduce a residual TC module for CL. Specifically, as shown in Fig. 1B, a learnable calibration vector $u_c \in \mathbb{R}^C$ is added to the base prototype to form the enhanced prototype $e_c = w_c + \gamma u_c$, where γ is a residual scaling factor. This residual design preserves the semantic prior of w_c while allowing task-adaptive refinement under CL. Given the bag representation v_i, we ℓ_2-normalize v_i and e_c and compute class logits by cosine similarity $\hat{y}_{i,c}$. We optimize the classification branch with $\mathcal{L}_{\text{cls}}$ and a class similarity regularizer to reduce prototype collapse $\mathcal{L}_{\text{sim}} = \frac{1}{C_{\text{cls}}(C_{\text{cls}}-1)}\sum_{a \neq b}\left(e_a^\top e_b + 1\right)$, where C_{cls} is the number of classes, and a and b index different class prototypes.

Training Objective. We optimize only the new KMP_t, PFC calibration parameters α_m^t and β_m^t, and TC prototype residual u_c. The overall objective is $\mathcal{L} = \mathcal{L}_{\text{cls}} + \mathcal{L}_{\text{KMP}} + \lambda_{\text{reg}}\mathcal{L}_{\text{reg}} + \lambda_{\text{sim}}\mathcal{L}_{\text{sim}}$, where $\mathcal{L}_{\text{cls}}$ is the cross-entropy loss.

2.2 Inference Stage

As shown in Fig. 1C, KMP-MIL uses the same forward pipeline in both evaluation settings, but differs in retrieval scope. For a test WSI, we compute a

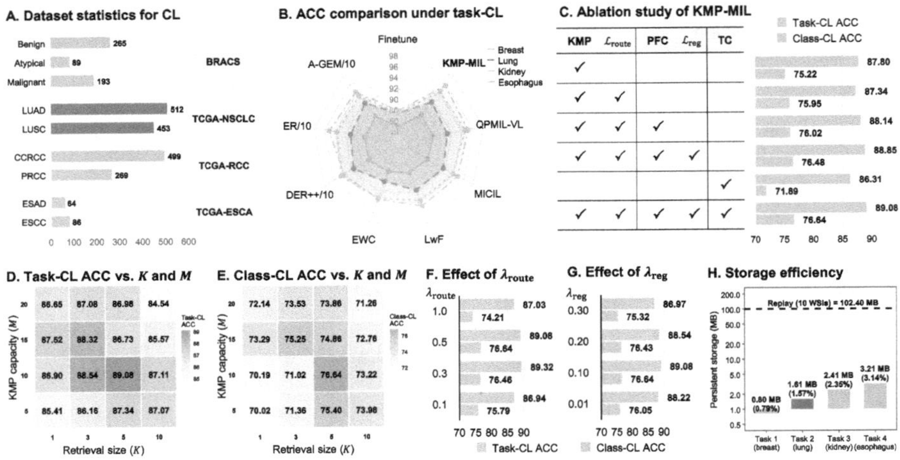

Fig. 2. Dataset statistics and related experimental results for continual WSI learning. **A:** Dataset statistics of the four sequential organ-specific WSI tasks used for CL. **B:** ACC comparison across different CL methods under task-CL. **C:** Ablation study of KMP-MIL under task-CL and class-CL. **D-E:** Sensitivity analysis of K and M under task-CL and class-CL. **F-G:** Sensitivity analysis of λ_{route} and λ_{reg} under task-CL and class-CL. **H:** Storage efficiency comparison between KMP-MIL and replay-based methods across sequential tasks.

query q_i and retrieve Top-K memory units from KMP to guide feature calibration and prototype-guided aggregation, producing the bag representation v_i and scaled cosine logits with TC-enhanced class prototypes. By reusing stored memory units from previous tasks, the model leverages knowledge at test time without storing or replaying historical WSIs, alleviating forgetting. In task-CL, retrieval is restricted to the task-specific pool KMP_t to match the known organ domain. In class-CL, task ID is unavailable, so we retrieve from a universal pool $\text{KMP}_{1:t}$ formed by all seen tasks, enabling task-agnostic inference while reusing learned memory units and calibration parameters.

3 Experiments

3.1 Experimental Setup

Datasets. We used four organ-specific WSI datasets covering breast, lung, kidney, and esophagus. Unless otherwise specified (i.e., reverse-order evaluation), the sequential task order is breast → lung → kidney → esophagus. The breast dataset is from BRACS [1] with three categories, and the other three datasets are from TCGA[1], including non-small cell lung carcinoma (NSCLC), renal cell carcinoma (RCC), and esophageal carcinoma (ESCA), each with two categories. Detailed statistics are shown in Fig. 2A. We cropped non-overlapping 256 × 256 patches at 10× magnification using CLAM [16].

[1] https://portal.gdc.cancer.gov.

Table 1. Performance comparison under class-CL (task ID unavailable). Results are reported as mean ± std, with the best performance highlighted in **bold**.

Method	Performance (↑)			Forgetting (↓)		
	ACC	AUC	PR-AUC	$\mathcal{F}_{\text{ACC}}$	$\mathcal{F}_{\text{AUC}}$	$\mathcal{F}_{\text{PR-AUC}}$
Finetune	15.83±1.14	68.83±1.80	67.53±1.65	41.71±2.68	22.63±0.94	21.65±0.76
A-GEM/10 [4]	31.49±12.55	78.05±8.96	74.95±8.24	12.14±5.27	12.76±8.07	11.94±7.21
ER/10 [21]	73.88±3.86	91.20±2.19	88.29±2.26	14.91±2.50	5.47±2.10	5.61±2.08
DER++/10 [2]	75.79±3.13	**93.74**±1.56	90.48±1.97	12.81±2.26	3.00±1.32	3.50±1.63
EWC [10]	13.07±6.17	84.08±4.28	81.73±3.94	11.25±2.92	3.03±0.92	3.05±0.64
LwF [14]	51.10±4.17	93.61±0.96	90.15±1.17	21.10±3.39	3.80±0.49	2.48±0.72
MICIL [17]	20.74±7.25	90.77±1.68	87.74±1.31	15.53±5.70	4.06±0.88	4.54±1.05
QPMIL-VL [5]	75.55±3.38	92.49±1.22	90.05±1.37	8.18±1.74	2.91±0.60	2.75±0.50
Ours (KMP-MIL)	**76.64**±2.27	93.56±0.98	**90.64**±0.93	**7.15**±1.72	**2.13**±0.62	**2.08**±0.54

Implementation Details. All experiments were conducted on an NVIDIA RTX 4090 GPU. We trained the model sequentially for 20 epochs per task and reported the mean and standard deviation (std) over 10-fold cross-validation with patient-level splits. Due to variable-sized MIL bags, we used batch size 1 with gradient accumulation over 16 iterations per optimizer step. For a fair comparison, all methods were evaluated using the pre-trained CONCH encoder E_{img} [15] to extract patch features. Trainable parameters were optimized with Adam (lr $=$ 5e−2, weight decay $=$ 5e−4). Unless otherwise stated, we set $M = 10$, $K = 5$, $\mu = 0.05$, $\tau = 0.07$, $\lambda_{\text{route}} = 0.5$, $\lambda_{\text{reg}} = 0.1$, $\lambda_{\text{sim}} = 0.5$, and $\gamma = 0.5$.

Evaluation Metrics. We evaluated KMP-MIL under task-CL and class-CL using classification metrics: accuracy (ACC), area under the ROC curve (AUC), and area under the precision-recall curve (PR-AUC). To measure old-task degradation, we used the corresponding forgetting rates, $\mathcal{F}_{\text{ACC}}$, $\mathcal{F}_{\text{AUC}}$, and $\mathcal{F}_{\text{PR-AUC}}$, when the base metrics were reported. These rates are especially informative in class-CL, where unavailable task IDs make cross-task confusion central.

3.2 Results

Comparison with SOTA. We compared KMP-MIL with representative CL baselines under both task-CL and class-CL settings, including rehearsal-based methods (A-GEM [4], ER [21], and DER++ [2]), regularization methods (EWC [10] and LwF [14]), and pathology-oriented CL methods (MICIL [17] and QPMIL-VL [5]). Under task-CL, KMP-MIL achieves the best ACC on three of four organs, namely breast, lung, and esophagus, while remaining competitive on kidney, as shown in Fig. 2B. It reaches 73.91% on breast, 91.06% on lung, and 96.71% on esophagus, and attains the highest average ACC across organs at 89.08%, surpassing QPMIL-VL at 88.59%. Under class-CL, where task ID is unavailable at inference, KMP-MIL achieves the best overall balance between performance and forgetting, as shown in Table 1. It obtains the highest ACC and PR-AUC, while maintaining a highly competitive AUC that is close to the best reported result. In addition, KMP-MIL yields the lowest forgetting metrics, including $\mathcal{F}_{\text{ACC}}$, $\mathcal{F}_{\text{AUC}}$, and $\mathcal{F}_{\text{PR-AUC}}$.

Table 2. Performance comparison under reverse-order CL (sequence: esophagus → kidney → lung → breast) for both task-CL and class-CL settings. Results are reported as mean ± std, with the best performance highlighted in **bold**.

Method	Task-CL setting		Class-CL setting			
	ACC (↑)	AUC (↑)	ACC (↑)	AUC (↑)	$\mathcal{F}_{\text{ACC}}$ (↓)	$\mathcal{F}_{\text{AUC}}$ (↓)
Finetune	64.58±3.04	88.29±2.21	12.19±3.20	66.37±2.46	28.97±3.26	5.51±2.44
A-GEM/10 [4]	81.27±2.39	89.28±1.20	36.32±9.39	75.57±5.43	21.94±4.99	4.82±2.00
ER/10 [21]	84.83±2.07	92.78±1.33	70.32±3.36	93.88±2.04	19.38±3.47	2.15±1.17
DER++/10 [2]	85.18±1.94	93.13±1.02	70.14±3.56	93.60±1.63	18.90±3.42	2.36±0.51
EWC [10]	78.72±1.88	90.25±2.01	13.97±5.94	82.58±3.93	26.79±4.20	4.81±1.23
LwF [14]	88.35±2.87	**95.90±1.14**	49.03±4.74	93.90±1.22	25.37±4.83	5.32±1.43
MICIL [17]	87.73±1.03	93.20±1.29	23.44±8.32	92.54±1.28	22.66±4.03	3.21±1.84
QPMIL-VL [5]	86.27±2.93	92.92±0.82	69.48±2.33	92.21±1.82	19.62±3.20	2.15±0.91
Ours (KMP-MIL)	**88.46±2.44**	95.79±0.91	**71.90±3.77**	**94.11±1.18**	**16.16±2.93**	**1.40±0.78**

Ablation Study. The ablation results in Fig. 2C show that KMP is the core contributor in both task-CL and class-CL, as it preserves task-specific knowledge through compact memory units. Starting from the KMP-only variant, adding $\mathcal{L}_{\text{route}}$ improves class-CL by 0.73% while slightly reducing task-CL by 0.46%, indicating reduced routing ambiguity under task-agnostic inference. Adding PFC further improves task-CL and class-CL by 0.80% and 0.07%, respectively, and $\mathcal{L}_{\text{reg}}$ brings additional gains of 0.71% and 0.46%, suggesting that retrieval-aware calibration mitigates cross-organ feature shift while avoiding over-modulation. TC alone performs worse than KMP-only by 1.49% in task-CL and 3.33% in class-CL, showing that textual prototypes alone cannot replace memory retrieval. When combined with KMP-MIL, TC provides complementary semantic alignment, yielding the best overall results.

Hyperparameter Sensitivity Analysis. We analyzed the key hyperparameters of KMP-MIL, including KMP capacity M, retrieval size K, and the weights of $\mathcal{L}_{\text{route}}$ and $\mathcal{L}_{\text{reg}}$ in Fig. 2D-G. M and K balance memory cost and retrieval coverage, where too small values may miss task knowledge and overly large values may introduce redundant memories. KMP-MIL remains stable across different settings, with strong balanced performance around $M = 10$, $K = 5$, $\lambda_{\text{route}} = 0.5$, and $\lambda_{\text{reg}} = 0.1$. λ_{route} improves task-agnostic routing in class-CL, while λ_{reg} stabilizes PFC by preventing overly aggressive feature modulation.

Reverse-Order Evaluation. We further evaluated KMP-MIL under a reverse-order CL setting, as shown in Table 2. KMP-MIL achieves the best ACC in both task-CL and class-CL, reaching 88.46% and 71.90%, respectively. Under class-CL, KMP-MIL also achieves the best AUC at 94.11% and the lowest forgetting metrics, including $\mathcal{F}_{\text{ACC}}$ of 16.16% and $\mathcal{F}_{\text{AUC}}$ of 1.40%. These results show that KMP-MIL is less sensitive to this reversed task order and can effectively reduce forgetting in this more challenging CL scenario.

Rehearsal-Free Storage Efficiency. To evaluate the rehearsal-free advantage of our method in continual WSI learning, we further compared the persistent storage cost of KMP-MIL with replay-based methods. As shown in Fig. 2H, as tasks are added sequentially, the cumulative storage of KMP-MIL increases only from 0.80 MB to 3.21 MB. In contrast, a replay-based method requires about 102.40 MB when storing 10 WSIs in the buffer. This means that after four tasks, KMP-MIL uses only about 3.14% of the replay storage budget, which greatly reduces storage cost in CL. These results highlight the rehearsal-free nature of KMP-MIL: even without replaying historical samples, it can still retain and transfer knowledge effectively with very small persistent storage overhead.

4 Conclusion

In this paper, we proposed KMP-MIL as a rehearsal-free continual MIL framework for WSI classification. It couples a frozen pathology FM with a growing KMP. By retrieving compact task-specific memory units and combining PFC with TC, KMP-MIL reuses knowledge without storing or replaying historical WSIs, thereby reducing catastrophic forgetting in cross-organ CL. KMP-MIL supports both task-CL and class-CL settings, including task-agnostic inference when task ID is unavailable. Experiments show stable performance, reduced forgetting, robustness under reverse-order evaluation, and favorable storage efficiency. One limitation is that the current KMP uses a fixed memory budget per task and grows linearly with the number of tasks. Adaptive memory allocation, memory merging, pruning, compression, Top-K task-origin tracing, and visualization of retrieved memory units, feature shifts, and prototype shifts are future directions for longer task streams and better interpretability.

Acknowledgements. This work was supported by the National Natural Science Foundation of China (Grant No. 82220108007).

Disclosure of Interests. The authors have no competing interests to declare that are relevant to the content of this article.

References

1. Brancati, N., Anniciello, A.M., Pati, P., Riccio, D., Scognamiglio, G., et al.: BRACS: a dataset for breast carcinoma subtyping in H&E histology images. Database **2022**, baac093 (2022)
2. Buzzega, P., Boschini, M., Porrello, A., Abati, D., Calderara, S.: Dark experience for general continual learning: a strong, simple baseline. In: Advances in Neural Information Processing Systems, vol. 33, pp. 15920–15930 (2020)
3. Campanella, G., Hanna, M.G., Geneslaw, L., Miraflor, A., Werneck Krauss Silva, V.: Clinical-grade computational pathology using weakly supervised deep learning on whole slide images. Nat. Med. **25**(8), 1301–1309 (2019)
4. Chaudhry, A., Ranzato, M., Rohrbach, M., Elhoseiny, M.: Efficient lifelong learning with A-GEM. In: Proceedings of the International Conference on Learning Representations (2019)

5. Gou, J., Ji, L., Liu, P., Ye, M.: Queryable prototype multiple instance learning with vision-language models for incremental whole slide image classification. In: Proceedings of the AAAI Conference on Artificial Intelligence, vol. 39, pp. 3158–3166 (2025)
6. Hao, X., Ni, W., Jiang, X., Tan, W., Yan, B.: Addressing imbalance for class incremental learning in medical image classification. In: Proceedings of the ACM International Conference on Multimedia, pp. 2467–2476 (2024)
7. Hartman, D.J., Pantanowitz, L., McHugh, J.S., Piccoli, A.L., O'Leary, M.J., et al.: Enterprise implementation of digital pathology: feasibility, challenges, and opportunities. J. Digit. Imaging **30**(5), 555–560 (2017)
8. Huang, Y., Zhao, W., Wang, S., Fu, Y., Jiang, Y.: ConSlide: asynchronous hierarchical interaction transformer with breakup-reorganize rehearsal for continual whole slide image analysis. In: Proceedings of the IEEE/CVF International Conference on Computer Vision, pp. 21349–21360 (2023)
9. Ilse, M., Tomczak, J.M., Welling, M.: Attention-based deep multiple instance learning. In: Proceedings of the International Conference on Machine Learning, vol. 80, pp. 2127–2136 (2018)
10. Kirkpatrick, J., Pascanu, R., Rabinowitz, N., Veness, J., Desjardins, G., et al.: Overcoming catastrophic forgetting in neural networks. Proc. Natl. Acad. Sci. **114**(13), 3521–3526 (2017)
11. Kumari, P., Reisenbüchler, D., Bozorgpour, A., Schaadt, N.S., Feuerhake, F.: Attention-based generative latent replay: a continual learning approach for WSI analysis. arXiv preprint (2025)
12. Kumari, P., Reisenbüchler, D., Luttner, L., Schaadt, N.S., Feuerhake, F.: Continual domain incremental learning for privacy-aware digital pathology. In: Medical Image Computing and Computer Assisted Intervention - MICCAI 2024, vol. 15012, pp. 34–44. Springer (2024)
13. Li, X., Cui, Y., Li, J., Chan, A.B.: Advancing multiple instance learning with continual learning for whole slide imaging. In: Proceedings of the IEEE/CVF Conference on Computer Vision and Pattern Recognition, pp. 20800–20809 (2025)
14. Li, Z., Hoiem, D.: Learning without forgetting. IEEE Trans. Pattern Anal. Mach. Intell. **40**(12), 2935–2947 (2018)
15. Lu, M.Y., Chen, B., Williamson, D.F.K., Chen, R.J., Liang, I., et al.: A visual-language foundation model for computational pathology. Nat. Med. **30**(3), 863–874 (2024)
16. Lu, M.Y., Williamson, D.F.K., Chen, T.Y., Chen, R.J., Barbieri, M., et al.: Data-efficient and weakly supervised computational pathology on whole-slide images. Nat. Biomed. Eng. **5**(6), 555–570 (2021)
17. Meseguer, P., del Amor, R., Naranjo, V.: MICIL: multiple-instance class-incremental learning for skin cancer whole slide images. Artif. Intell. Med. **152**, 102870 (2024)
18. Nguyen, A.T., Byeon, K., Kim, K., Song, B., Chae, S.W., et al.: CAMP: continuous and adaptive learning model in pathology. NPJ Artif. Intell. **2**(1), 2 (2026)
19. Niazi, M.K.K., Parwani, A.V., Gurcan, M.N.: Digital pathology and artificial intelligence. Lancet Oncol. **20**(5), e253–e261 (2019)
20. Perez, E., Strub, F., de Vries, H., Dumoulin, V., Courville, A.: FiLM: visual reasoning with a general conditioning layer. In: Proceedings of the AAAI Conference on Artificial Intelligence, vol. 32 (2018)
21. Rolnick, D., Ahuja, A., Schwarz, J., Lillicrap, T.P., Wayne, G.: Experience replay for continual learning. In: Advances in Neural Information Processing Systems, vol. 32, pp. 348–358 (2019)

22. Shao, Z., Bian, H., Chen, Y., Wang, Y., Zhang, J., et al.: TransMIL: transformer-based correlated multiple instance learning for whole slide image classification. Adv. Neural. Inf. Process. Syst. **34**, 2136–2147 (2021)
23. Treanor, D., Williams, B.: The Leeds guide to digital pathology. The Leeds Teaching Hospitals NHS Trust and University of Leeds, Leeds (2019)
24. Wu, X., Xu, Z., Tong, R.K.Y.: Continual learning in medical image analysis: a survey. Comput. Biol. Med. **182**, 109206 (2024)
25. Yuan, L., Gu, Y., Han, Y., Du, T., Grzegorzek, M., et al.: Exploring transparency in pathological image analysis: a comprehensive review of explainable artificial intelligence (XAI) techniques. Comput. Sci. Rev. **60**, 100863 (2026)

Learning to Read Where to Look: Disease-Aware Vision–Language Pretraining for 3D CT

Simon Ging[1,2(✉)], Philipp Arnold[3], Sebastian Walter[4], Hani Alnahas[1], Hannah Bast[4], Elmar Kotter[3], Jiancheng Yang[5,6], Behzad Bozorgtabar[2], and Thomas Brox[1]

[1] Computer Vision Group, University of Freiburg, Freiburg im Breisgau, Germany
gings@cs.uni-freiburg.de
[2] Adaptive & Agentic AI (A3) Lab, Aarhus University, Aarhus, Denmark
[3] Department of Radiology, Medical Center – University of Freiburg, Freiburg im Breisgau, Germany
[4] Chair of Algorithms and Data Structures, University of Freiburg, Freiburg im Breisgau, Germany
[5] ELLIS Institute Finland, Espoo, Finland
[6] School of Electrical Engineering, Aalto University, Espoo, Finland

Abstract. Recent 3D CT vision–language models align volumes with reports via contrastive pretraining, but typically rely on limited public data and provide only coarse global supervision. We train a 3D CT vision–language model on 98k report–volume pairs (50k patients) collected at a single hospital, combined with public datasets, using SigLIP-style contrastive pretraining together with prompt-based disease supervision in the shared vision–text embedding space. On CT-RATE, our model achieves state-of-the-art text-to-image retrieval (R@10 31.5 vs. 22.2) and competitive disease classification (AUC 83.8 vs. 83.8), with consistent results on Rad-ChestCT (AUC 77.0 vs. 77.3). We further observe that radiologists routinely reference specific images within their reports (e.g., "series X, image Y"), linking textual descriptions to precise axial locations. We automatically mine 262k such snippet–slice pairs and introduce the task of intra-scan snippet localization—predicting the axial depth referred to by a text snippet—reducing mean absolute error to 36.3 mm at 12 mm feature resolution, compared with 67.0 mm for the best baseline. Adding this localization objective leaves retrieval and classification broadly unchanged within confidence bounds, yielding a single unified model for retrieval, classification, and intra-scan grounding. Code and models are available at https://github.com/lmb-freiburg/radfinder.

Keywords: 3D CT · Vision–language pretraining · Contrastive learning

G. Yang et al. (Eds.): MICCAI 2026, LNCS 16878, pp. 250–259, 2027.
https://doi.org/10.1007/978-3-032-38059-3_24

1 Introduction

The volume of radiological imaging continues to grow, with CT examinations increasing steadily, while the number of radiologists has not kept pace [14]. Each CT study requires careful analysis: a radiologist reviews hundreds of axial slices, identifies findings, and dictates a free-text report. Foundation models that learn general-purpose representations from large-scale imaging data offer a path toward assisting radiologists across diverse tasks—from automated retrieval and triage to report generation—without task-specific annotation.

Recent 3D CT vision–language models (VLMs) learn such representations by aligning CT volumes with radiology reports via contrastive pretraining. CT-CLIP [6] pioneered this direction on the public CT-RATE dataset of 50k chest CT volumes. SPECTRE [3] scaled self-supervised and cross-modal pretraining to 230k volumes and achieved strong retrieval results, while other approaches leverage structured entity extraction [10] or hierarchical vision architectures [1]. These methods align entire volumes with entire reports, providing only coarse, global supervision. Other approaches improve supervision granularity by aligning at the organ level: MedVista3D [11] and fVLM [16] both use an external segmentation model to identify anatomical regions and contrast each region's visual features against LLM-generated region descriptions. While effective, this requires a pretrained segmentation model and curated region-level text that may not capture the specific findings a radiologist highlights.

In this work, we train *RadFinder*, a 3D CT VLM on 98k report–volume pairs (50k patients) from a single hospital, combined with public datasets. The model uses SigLIP [20] contrastive pretraining on full reports with pretrained medical vision and text encoders, and integrates structured disease labels as text prompts within the contrastive objective (Sect. 3.2). RadFinder achieves state-of-the-art text-to-image retrieval and competitive disease classification on CT-RATE and Rad-ChestCT [4], demonstrating strong cross-dataset transfer.

We additionally observe that radiology reports contain a largely untapped form of local supervision: radiologists frequently reference specific slices within a scan—for example, "*hepatic lesion, see series 4, image 38*". We mine 262k such snippet–slice pairs and propose the task of *intra-scan snippet localization*: given a text snippet, predict the axial depth it refers to within a CT volume. While slice-level matching from reports has been explored at small scale [8], we formulate this in 3D feature space and achieve results substantially better than simple baselines. Training localization jointly with global objectives does not degrade retrieval or classification, yielding a unified model for all three tasks.

Contributions

- We build a large-scale 3D report–volume dataset (98k pairs, 50k patients) from clinical routine at a single hospital and automatically mine 262k snippet–slice pairs that provide weak local supervision without additional annotation.
- Using this local supervision signal, we propose the task of intra-scan snippet localization, establish baselines, and show that localization can be trained jointly with global objectives without degrading retrieval or classification.

- We train a 3D CT VLM by combining contrastive pretraining on reports with prompt-based disease label training and snippet localization. Our model achieves state-of-the-art retrieval and competitive disease classification on external benchmarks, demonstrating strong cross-dataset transfer. Code and pretrained models will be made publicly available.

2 Dataset

We collect 97,760 report–volume pairs from 50,474 patients at a single hospital, spanning 13 years of clinical practice. The dataset covers chest (46%), abdomen (22%), and combined chest–abdomen (33%) studies. The radiologist report for the study contains detailed findings and a summary impression, and we select the single largest axial series from each study. In-plane resolution and slice thickness medians are 0.71 mm and 3.0 mm, respectively.

Snippet Mining. Radiologists frequently reference specific slices when dictating reports, where *series* identifies the volume and *image* the axial slice number (e.g., "hepatic lesion, see series 4, image 38" or "pulmonary nodule in the right lower lobe (3/72)"). We extract these references via pattern matching heuristics. Each snippet associates a short textual finding with a precise axial position in the scan, providing weak local supervision without additional annotation. To verify our pipeline, we compare against manually extracted references from 100 reports and find that our heuristics achieve 99.4% precision and 90.2% recall (F1 94.6%). To ensure correct spatial alignment, we keep only slices where slicing the 3D volume at the referenced position produces identical content to the original 2D image file stored by the scanner. After filtering, this yields 261,800 snippet–slice pairs across all scans, 2.7 per scan on average.

Text Processing. We anonymize all text by removing patient and physician identifiers via rule-based matching and converting absolute dates into relative references (e.g., "Spine injury 5 years ago"). Reports are translated from German to English using Gemma 3 27B [5]; all training is conducted on the English translations. We apply the RATE protocol from Pillar-0 [1], using LLM-based question answering to classify each report into 93 chest and 226 abdomen binary findings across 30 organ categories; these structured labels serve as input for prompt-based training (Sect. 3.2). Data is split 80/10/10 by randomized patient IDs, with no patient overlap with public evaluation datasets.

3 Method

3.1 Global Vision–Language Pretraining

We initialize from the pretrained SPECTRE [3] model, which uses a two-stage vision encoder: a ViT-Large local backbone processes each volume in separate $128 \times 128 \times 32$-voxel windows at $0.75 \times 0.75 \times 3.0\,\text{mm}^3$ spacing, and a 4-layer global feature combiner aggregates the window-level representations. The text encoder is Qwen3-Embedding [12] (0.6B) with LoRA adapters. SigLIP [20] projection heads map both modalities to a 512-dim shared space. We freeze the local vision backbone and fine-tune all other modules with SigLIP contrastive loss (Fig. 1). Unlike SPECTRE, which crops to a fixed grid, we process full volumes with variable input shapes. We train on up to four datasets totaling 159k report–volume pairs: *RefCT* (internal, 78k in the training split), CT-RATE [6] (47k), Merlin [2] (15k), and INSPECT [7] (19k). During training, we apply three text augmentations, each with probability 0.2: replacing the full report with concatenated organ-level descriptions from the RATE pipeline [1], removing historic comparisons ("compared to prior exam...") via LLM parsing following CT-RATE [6], or dropping the *findings* section.

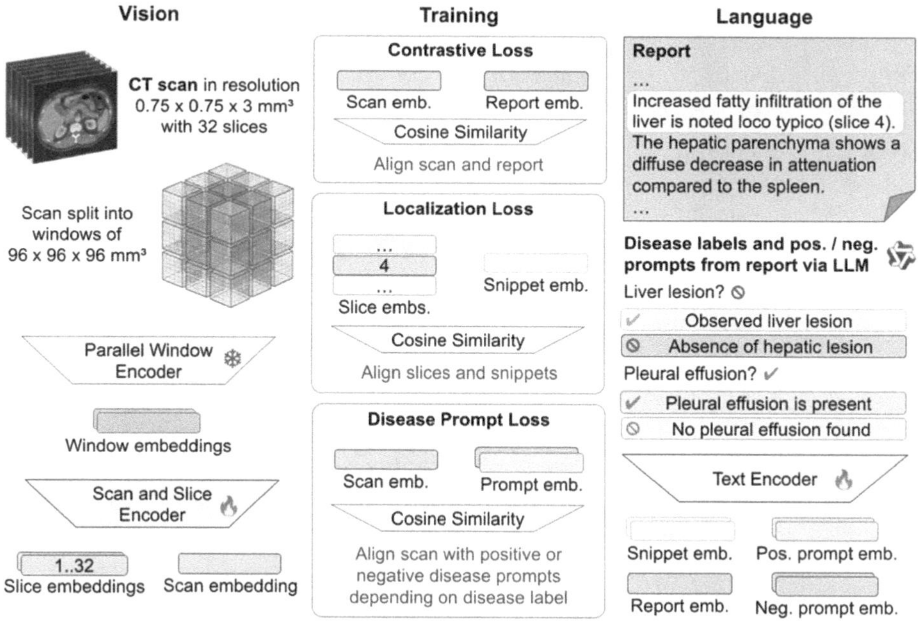

Fig. 1. Overview of the RadFinder architecture and training pipeline. **Left (Vision):** A 3D CT scan is processed by a frozen parallel window encoder, followed by a trainable encoder to extract both a global volume embedding and local slice-level embeddings. **Right (Language):** A trainable text encoder processes full radiology reports, localized text snippets (e.g., the sentence that references the fourth slice out of the 32 slices in this example scan), and LLM-extracted positive/negative disease prompts. **Center (Training):** The model is optimized via three contrastive objectives in a shared embedding space: a global Contrastive Loss aligning the full scan and report, an intra-scan Localization Loss aligning text snippets with their specific slices, and a Disease Prompt Loss aligning the global scan with corresponding disease prompt descriptions

3.2 Prompt-Based Disease Label Training

To improve text-based disease classification, we augment the contrastive loss with disease label prompts. During training, labels are represented as text prompts and supervised via a BCE loss on the cosine similarity difference between the volume embedding and positive/negative prompt embeddings. We do not train a separate classification head; supervision is applied entirely in the shared vision–text embedding space.

Labels. Using the RATE protocol [1], we extract binary findings from each report via LLM-based question answering with Qwen3-30B-A3B [12] (93 chest and 226 abdomen findings). We additionally derive binary labels for the 18 CT-RATE disease classes by mapping the 93 RATE chest findings.

Prompts. For each finding q, we construct three positive and three negative text prompt variants (e.g., "Pleural effusion is present.", "No pleural effusion is identified."). During training, one variant is sampled randomly; at inference, the three embeddings are averaged.

Loss. Let $\mathbf{z}$ denote the L2-normalized image embedding and $\mathbf{p}_q^+, \mathbf{p}_q^-$ the positive and negative prompt embeddings for finding q. We classify each finding via the logit difference scaled by the shared SigLIP temperature τ:

$$\mathcal{L}_{\text{prompt}} = \frac{1}{|\mathcal{M}|} \sum_{q \in \mathcal{M}} w_q \Big(- \alpha_q \, y_q \, \log \sigma(x_q) - (1 - y_q) \log\big(1 - \sigma(x_q)\big) \Big) \tag{1}$$

$$x_q = (\mathbf{z}^\top \mathbf{p}_q^+ - \mathbf{z}^\top \mathbf{p}_q^-)/\tau, \qquad \alpha_q = \min(n_q^+/n_q^-, 20) \tag{2}$$

where $y_q \in \{0, 1\}$ is the binary finding label, σ is the logistic sigmoid, $\mathcal{M}$ is the set of valid image–question pairs, w_q upweights the 18 CT-RATE classes to balance them against the larger RATE label set, and α_q handles per-question label imbalance given n_q^+ positive and n_q^- negative training examples. The prompt loss is weighted by λ relative to the SigLIP loss.

The idea of using label text as prompts in contrastive training originates in CXR-CLIP [19] and UniCL [18]. The closest prior work in 3D CT either extracts free-text disease descriptions from reports with GPT-4 for contrastive training (BrgSA [9]) or trains an explicit classification head on LLM-extracted labels before contrastive alignment (MPS-CT [10]). Our approach uses structured binary labels directly as text prompts within the contrastive objective.

3.3 Intra-Scan Localization Loss

To encourage depth-aware representations, we introduce an intra-scan localization objective. Given a text snippet describing a radiological finding and the corresponding CT volume, the model classifies which depth position along the axial axis the snippet refers to. Crucially, negatives come from other depth positions *within the same scan*, avoiding cross-sample false negatives that arise when different patients share similar pathology.

Depth Features. Let $\mathbf{z}_d \in \mathbb{R}^E$ denote the L2-normalized image embedding at depth position $d \in \{1, \ldots, D\}$, with one position covering 12mm in the axial plane, obtained by processing backbone patch features through the global feature combiner, averaging over the coronal and sagittal dimensions, and projecting through the SigLIP head. Let $\mathbf{t} \in \mathbb{R}^E$ denote the L2-normalized projected text embedding of the snippet.

Gaussian Soft Target. Each snippet is associated with a specific axial slice, whose physical position determines a ground-truth depth index $d^* \in \{1, \ldots, D\}$ in the feature grid. Since annotations are inherently imprecise—a finding typically extends to neighboring slices—we define a one-hot indicator $\mathbf{m} \in \{0, 1\}^D$ with $m_{d^*} = 1$ and convolve it with a normalized 1-D Gaussian kernel g of standard deviation $\sigma = 2$:

$$\tilde{\mathbf{m}} = \frac{(\mathbf{m} * g)}{\|\mathbf{m} * g\|_1}, \qquad g_k = \frac{1}{Z} \exp\Big(-\frac{k^2}{2\sigma^2}\Big), \quad |k| \leq \lceil 3\sigma \rceil. \tag{3}$$

Loss. We compute cosine-similarity logits between the snippet embedding and each depth feature, scaled by a temperature τ=0.1, and minimize the cross-entropy against the soft target $\tilde{\mathbf{m}}$:

$$\ell_d = \mathbf{z}_d^\top \mathbf{t} / \tau, \qquad \mathcal{L}_{\text{loc}} = -\sum_{d=1}^{D} \tilde{m}_d \log \frac{\exp(\ell_d)}{\sum_{d'} \exp(\ell_{d'})}. \tag{4}$$

The total training objective combines the global report–volume contrastive loss $\mathcal{L}_{\text{global}}$, the prompt loss $\lambda \mathcal{L}_{\text{prompt}}$ (Eq. 1), and the localization loss $\beta \mathcal{L}_{\text{loc}}$.

Table 1. Text-to-image retrieval on CT-RATE, Rad-ChestCT (RC) and Merlin validation sets. Training datasets: A: RefCT (ours, 78k), B: CT-RATE [6] (47k), C: Merlin [2] (15k), D: INSPECT [7] (19k). *fVLM additionally trains on MedVL-CT69 [16] (272k). We highlight **values within the confidence interval of the highest score** separately for the comparison of our main model with baselines (top) and our model with our ablations (bottom).

Method	Data				CT-RATE			RC	Merlin R@1	
	A	B	C	D	R@10	MAP@5	AUC	AUC	Findings	Impr.
Random Chance					0.3	0.1	50.0	50.0	0.8	0.8
CT-CLIP [6]		✓			5.0	68.3	73.1	62.9	—	—
MedVista3D-ViT [11]		✓			10.7	—	77.8	—	—	—
MedVista3D-UniMISS		✓			8.7	—	78.2	—	—	—
Merlin [2]			✓		2.7	62.6	72.8	64.4	59.4	19.4
fVLM* [16]		✓			—	—	77.8	68.0	—	—
BrgSA [9]		✓			22.2	70.4	82.9	74.2	—	—
MPS-CT [10]		✓			16.3	71.3	**83.8**	**77.3**	—	—
SPECTRE [3]		✓	✓	✓	18.2 1.4	**76.8** 1.7	54.3 1.1	55.7 0.6	44.5 0.7	29.2 0.6
RadFinder (Ours)	✓	✓	✓	✓	**31.5** 1.6	**78.0** 1.7	**83.8** 0.7	**77.0** 0.5	**69.0** 0.9	**40.3** 0.7
No localization loss	✓	✓	✓	✓	**31.5** 1.6	**78.2** 1.7	83.6 0.7	76.9 0.5	69.3 0.9	40.2 0.6
Global loss only	✓	✓	✓	✓	29.4 1.6	**76.6** 1.7	56.9 1.1	62.8 0.6	**70.7** 0.7	**42.2** 0.7
Prompt loss only	✓	✓	✓	✓	5.6 0.8	**77.9** 1.7	**84.8** 0.7	**78.6** 0.5	13.1 0.6	10.9 0.4
RefCT dataset	✓				26.3 1.6	**76.7** 1.7	80.4 0.8	77.7 0.6	60.5 0.9	35.8 0.7
Public datasets, no loc.		✓	✓	✓	26.9 1.6	**77.1** 1.7	83.2 0.7	73.6 0.6	67.5 0.9	40.2 0.8

Table 2. Results for snippet localization on the RefCT test set. *SigLIP2 finetuned on RefCT snippet–slice pairs via contrastive loss.

Method	MAE (mm)	<6mm	<18mm	<30mm
Random slice	126.9 5.7	4.7 1.1	11.8 1.6	17.8 2.0
Middle slice	95.8 3.6	4.6 1.2	13.0 1.8	20.8 2.2
CLIP ViT-B/32 [13]	124.6 5.5	3.2 0.8	9.4 1.4	14.8 1.7
BiomedCLIP [21]	86.6 5.2	8.3 1.3	20.2 1.9	30.5 2.2
MedSigLIP-448 [15]	75.6 5.0	9.7 1.4	26.2 2.1	40.1 2.3
SigLIP2 [17]	82.4 4.7	7.8 1.3	19.3 1.8	31.5 2.3
SigLIP2 finetuned*	67.0 4.9	17.4 1.8	35.8 2.4	48.7 2.3
RadFinder (Ours)	**36.3** 2.9	**20.3** 2.0	**45.3** 2.5	**61.8** 2.4
Loc. loss only	**36.0** 2.9	**19.9** 2.0	**45.2** 2.5	**62.7** 2.4

4 Results

Implementation Details. We fine-tune for 10 epochs with AdamW (lr 2×10^{-4}, 1 epoch linear warmup, cosine decay to 1×10^{-6}) and a batch size of 4096 on a single NVIDIA H100 (96 GB, 32 h). We use loss weights $\lambda = 8$ and $\beta = 1$. We report 95% bootstrap confidence intervals (B = 10,000) over test set samples.

Datasets and Protocols. We evaluate on three external benchmarks. CT-RATE [6] provides 1564 reports paired with 3039 volumes (multiple reconstructions per study). We report text-to-image retrieval (R@10) over the full 3039-volume pool, volume-to-volume retrieval (MAP@5) using disease-label IoU as relevance, and disease classification (AUC) over 18 pathology classes. For the disease class prediction, we use the 7 prompt templates (e.g., "{a} is present.") proposed by CT-RATE and classify by softmax over cosine similarities between the volume embedding and the averaged positive/negative prompt embeddings. RadChestCT [4] provides 3630 chest CT scans with disease labels (no reports); we evaluate AUC with the same prompt protocol. Merlin [2] provides 5125 report–volume pairs; we report R@1 separately for findings and impressions sections (R@1_{find}, R@1_{impr}) with the protocol by the original authors, which averages the metric over 100 trials with pools of 128 randomly sampled volumes. RefCT (internal) is used to evaluate snippet localization at 12 mm axial resolution. The test set contains 1,564 volumes with 1 snippet–slice pair each.

Retrieval (Table 1). Training on RefCT alone already surpasses all published baselines on CT-RATE retrieval (R@10 26.3 vs. 22.2 for BrgSA)—demonstrating strong cross-dataset transfer. Adding public datasets further improves retrieval to R@10 31.5. On the Merlin dataset, our model achieves R@1_{find} 69.0, surpassing the original Merlin model (59.4).

Disease Classification (Table 1). Without prompt-based training, our model achieves only AUC ∼57 on CT-RATE—comparable to SPECTRE and well below methods that incorporate disease knowledge. Adding prompts yields AUC 83.8 on CT-RATE and 77.0 on Rad-ChestCT, matching MPS-CT [10] (83.8/77.3) within confidence bounds. Training on RefCT alone already reaches AUC 80.4/77.7, showing that disease knowledge transfers from our internal data. Conversely, training with prompt loss only yields the highest classification AUC (84.8) but sacrifices retrieval (R@10 5.6), as expected when the model sees only binary disease labels without free-text reports. The combined model achieves both: strong retrieval and competitive disease classification.

Public Datasets (Table 1, Last Row). Training our method only on public datasets reaches R@10 26.9/AUC 83.2 on CT-RATE and AUC 73.6 on Rad-ChestCT, showing that both retrieval and disease classification benefit significantly from our training setup compared to the original SPECTRE method. By adding our internal dataset, we gain R@10 4.6 points on CT-RATE and AUC 3.4 points on Rad-ChestCT; the contributions of our method and our internal dataset are complementary.

Localization (Table 2). Naive baselines (middle slice, random slice) achieve MAE 96 and 127 mm. Pretrained 2D VLMs reduce this to 87 mm (Biomed-CLIP [21]) and 76 mm (MedSigLIP [15]); finetuning SigLIP2 [17] on our snippet–slice pairs reaches 67 mm. RadFinder achieves MAE 36.3 mm (20.3% within 6 mm), nearly halving the error of the best baseline. Training with localization loss alone yields comparable results (MAE 36.0 mm), confirming that the global objectives do not interfere with local grounding. Conversely, adding localization to the full model leaves retrieval and classification unchanged within confidence bounds (Table 1, RadFinder vs. No loc.), yielding a single unified model for all three tasks.

5 Conclusion

We presented RadFinder, a 3D CT vision–language model that combines contrastive pretraining on reports with prompt-based disease label supervision, achieving state-of-the-art retrieval and competitive disease classification on external benchmarks. We additionally showed that snippet–slice references mined from radiology reports provide effective local supervision for intra-scan localization, and that this task can be trained jointly with global objectives in a single unified model without degrading retrieval or classification.

Limitations. The localization resolution is limited to 12 mm along the slice axis, which may be insufficient for precisely localizing small findings. Furthermore, our snippet mining relies on explicit slice references in reports, which limits coverage to institutions where such references are part of the reporting practice.

Future Work. Promising directions include investigating whether local supervision can also improve global representations, collecting data from additional hospitals to study multi-institutional training, evaluating bilingual inference on the original German reports, and improving localization resolution through finer-grained feature maps. Coupling the pretrained encoder with a language model for grounded report generation is another natural extension.

Acknowledgments. This research was funded by the Deutsche Forschungsgemeinschaft (DFG, German Research Foundation) 417962828, 539134284, and 499552394 – SFB 1597 – Small Data, as well as through EFRE (FEIH_2698644) and the state of Baden-Württemberg.

Baden-Württemberg Co-funded by the European Union

Disclosure of Interests. The authors have no competing interests to declare that are relevant to the content of this article.

References

1. Agrawal, K.K., et al.: Pillar-0: a new frontier for radiology foundation models. arXiv preprint arXiv:2511.17803 (2025)
2. Blankemeier, L., Kumar, A., Cohen, J.P., Liu, J., Liu, L., et al.: Merlin: a computed tomography vision-language foundation model and dataset. Nature **652**(8112), 1318–1328 (2026). https://doi.org/10.1038/s41586-026-10181-8
3. Claessens, C., Viviers, C., D'Amicantonio, G., Bondarev, E., van der Sommen, F.: Scaling self-supervised and cross-modal pretraining for volumetric CT transformers. arXiv preprint arXiv:2511.17209 (2025)
4. Draelos, R.L., Dov, D., Mazurowski, M.A., Lo, J.Y., Henao, R., et al.: Machine-learning-based multiple abnormality prediction with large-scale chest computed tomography volumes. Med. Image Anal. **67**, 101857 (2021). https://doi.org/10.1016/j.media.2020.101857
5. Gemma Team: Gemma 3 technical report. arXiv preprint arXiv:2503.19786 (2025)
6. Hamamci, I.E., Er, S., Wang, C., Almas, F., Simsek, A.G., et al.: Generalist foundation models from a multimodal dataset for 3D computed tomography. Nat. Biomed. Eng. (2026). https://doi.org/10.1038/s41551-025-01599-y
7. Huang, S.C., Huo, Z., Steinberg, E., Chiang, C.C., Langlotz, C.: INSPECT: a multimodal dataset for patient outcome prediction of pulmonary embolisms. In: Advances in Neural Information Processing Systems, Datasets and Benchmarks Track, vol. 36 (2023)
8. Huemann, Z., Church, S., Warner, J.D., Tran, D., Tie, X., et al.: ConTEXTual Net 3D: vision-language modeling in PET/CT for visual grounding of positive findings. J. Imaging Inform. Med. (2026). https://doi.org/10.1007/s10278-026-01879-2
9. Lai, H., Jiang, Z., Yao, Q., Wang, R., He, Z., et al.: Bridged semantic alignment for zero-shot 3D medical image diagnosis. IEEE J. Biomed. Health Inform. 1–14 (2025). https://doi.org/10.1109/JBHI.2025.3629096
10. Li, Y., Lai, H., Zhou, X., Ming, S., Ma, W., et al.: More performant and scalable: Rethinking contrastive vision-language pre-training of radiology in the LLM era. In: Gee, J.C., Alexander, D.C., Hong, J., Iglesias, J.E., Sudre, C.H., et al. (eds.) Medical Image Computing and Computer Assisted Intervention - MICCAI 2025.

LNCS, vol. 15966, pp. 348–357. Springer, Cham (2025). https://doi.org/10.1007/978-3-032-04981-0_33
11. Li, Y., Chen, Y., Lai, Y., Zhong, J., Wildman, V., Yang, X.: MedVista3D: vision-language modeling for reducing diagnostic errors in 3D CT disease detection, understanding and reporting. arXiv preprint arXiv:2509.03800 (2025)
12. Qwen Team: Qwen3 technical report. arXiv preprint arXiv:2505.09388 (2025)
13. Radford, A., Kim, J.W., Hallacy, C., Ramesh, A., Goh, G.: Learning transferable visual models from natural language supervision. In: International Conference on Machine Learning, pp. 8748–8763. PMLR (2021)
14. Rimmer, A.: Radiologist shortage leaves patient care at risk, warns royal college. BMJ **359**, j4683 (2017). https://doi.org/10.1136/bmj.j4683
15. Sellergren, A., Kazemzadeh, S., Jaroensri, T., Kiraly, A., Traverse, M.: MedGemma technical report. arXiv preprint arXiv:2507.05201 (2025)
16. Shui, Z., Zhang, J., Cao, W., Wang, S., Guo, R.: Large-scale and fine-grained vision-language pre-training for enhanced CT image understanding. In: The Thirteenth International Conference on Learning Representations (2025)
17. Tschannen, M., Gritsenko, A., Wang, X., Naeem, M.F., Alabdulmohsin, I.: SigLIP 2: multilingual vision-language encoders with improved semantic understanding, localization, and dense features. arXiv preprint arXiv:2502.14786 (2025)
18. Yang, J., Li, C., Zhang, P., Xiao, B., Liu, C., et al.: Unified contrastive learning in image-text-label space. In: Proceedings of the IEEE/CVF Conference on Computer Vision and Pattern Recognition, pp. 19141–19151. IEEE (2022). https://doi.org/10.1109/CVPR52688.2022.01857
19. You, K., Gu, J., Ham, J., Park, B., Kim, J., et al.: CXR-CLIP: toward large scale chest X-ray language-image pre-training. In: Greenspan, H., Madabhushi, A., Mousavi, P., Salcudean, S.E., Duncan, J., et al. (eds.) Medical Image Computing and Computer Assisted Intervention – MICCAI 2023. LNCS, vol. 14221, pp. 101–111. Springer, Cham (2023). https://doi.org/10.1007/978-3-031-43895-0_10
20. Zhai, X., Mustafa, B., Kolesnikov, A., Beyer, L.: Sigmoid loss for language image pre-training. In: Proceedings of the IEEE/CVF International Conference on Computer Vision, pp. 11941–11952. IEEE (2023). https://doi.org/10.1109/ICCV51070.2023.01100
21. Zhang, S., Xu, Y., Usuyama, N., Xu, H., Bagga, J., et al.: A multimodal biomedical foundation model trained from fifteen million image–text pairs. NEJM AI **2**(1), AIoa2400640 (2025). https://doi.org/10.1056/AIoa2400640

Learning to Trim: End-to-End Causal Graph Pruning with Dynamic Anatomical Feature Banks for Medical VQA

Zibo Xu[1], Qiang Li[1], Weizhi Nie[2(✉)], and Yuting Su[2]

[1] School of Microelectronics, Tianjin University, Tianjin 300072, China
[2] School of Electrical and Information Engineering, Tianjin University, Tianjin 300072, China
weizhinie@tju.edu.cn

Abstract. Medical Visual Question Answering (MedVQA) models often exhibit limited generalization due to reliance on dataset-specific correlations, such as recurring anatomical patterns or question-type regularities, rather than genuine diagnostic evidence. Existing causal approaches are typically implemented as static adjustments or post-hoc corrections. To address this issue, we propose a Learnable Causal Trimming (LCT) framework that integrates causal pruning into end-to-end optimization. We introduce a Dynamic Anatomical Feature Bank (DAFB), updated via a momentum mechanism, to capture global prototypes of frequent anatomical and linguistic patterns, serving as an approximation of dataset-level regularities. We further design a differentiable trimming module that estimates the dependency between instance-level representations and the global feature bank. Features highly correlated with global prototypes are softly suppressed, while instance-specific evidence is emphasized. This learnable mechanism encourages the model to prioritize causal signals over spurious correlations adaptively. Experiments on VQA-RAD, SLAKE, SLAKE-CP and PathVQA demonstrate that LCT consistently improves robustness and generalization over existing debiasing strategies.

Keywords: Medical VQA · Causal Inference · Causal Trimming · Dynamic Feature Bank

1 Introduction

Medical Visual Question Answering (MedVQA) connects medical imaging with natural language to support clinical decision-making [1,12,16,25]. Despite recent progress, existing models often suffer significant performance drops under Out-of-Distribution (OOD) settings. This issue largely arises from reliance on spurious correlations (e.g., associating question types with background textures) instead of true causal mechanisms [23].

G. Yang et al. (Eds.): MICCAI 2026, LNCS 16878, pp. 260–270, 2027.
https://doi.org/10.1007/978-3-032-38059-3_25

Recent approaches incorporate causal inference by modeling spurious correlations as confounders and applying backdoor adjustment with static dictionaries [10,12,18]. However, spurious correlations arising from variable anatomical features are inherently dynamic, rendering static dictionaries inadequate. Meanwhile, Test-Time Adaptation (TTA) methods attempt to prune biased features during inference [15], but they incur additional computational cost and treat the trained model as a black box, without improving intrinsic robustness. To address these limitations, we propose Learnable Causal Trimming (LCT), an end-to-end framework that integrates causal intervention directly into training. Instead of static confounder sets, we construct a Dynamic Anatomical Feature Bank (DAFB) updated via momentum to capture evolving anatomical priors and dataset biases. A Causal Trimming (CT) module operates as a soft gate, estimating the dependency between instance features and DAFB prototypes, and adaptively reweighting confounder-correlated components while preserving informative evidence. This learnable intervention encourages the model to disentangle causal signals from confounding noise during optimization.

To our knowledge, we are the first to embed dynamic causal trimming as a trainable module within MedVQA. Our contributions are threefold:

- We propose the LCT framework, transforming causal adjustment in MedVQA from static or test-time strategies into an end-to-end trainable paradigm.
- We design a Dynamic Anatomical Feature Bank with a Causal Trimming module to track and mitigate confounders during training without external annotations.
- Experiments on four datasets demonstrate that LCT achieves state-of-the-art performance and improves OOD generalization.

2 Methodology

2.1 Causal Formulation: The Confounding Problem

We analyze bias in MedVQA using a Structural Causal Model (SCM) [20], as shown in Fig. 1(c). I, Q, and A denote the medical image, question, and answer. Ideally, the model should estimate $P(A|I,Q)$ through the causal paths $I \rightarrow A$ and $Q \rightarrow A$ [24]. However, medical datasets contain strong priors. We introduce a latent confounder C to represent dataset-level regularities. The backdoor paths $I \leftarrow C \rightarrow A$ and $Q \leftarrow C \rightarrow A$ induce spurious correlations. For example, if C encodes lung-related priors, the model may predict "Pneumonia" from lung appearance alone, ignoring lesion-specific evidence in I. Standard supervised learning on the observational distribution may therefore capture shortcuts induced by C. To mitigate this issue, we simulate a causal intervention to reduce the confounding effect of C on answer prediction. As C is unobserved, we approximate it with a *Dynamic Anatomical Feature Bank* and apply *Causal Trimming* to reweight features associated with C during representation learning.

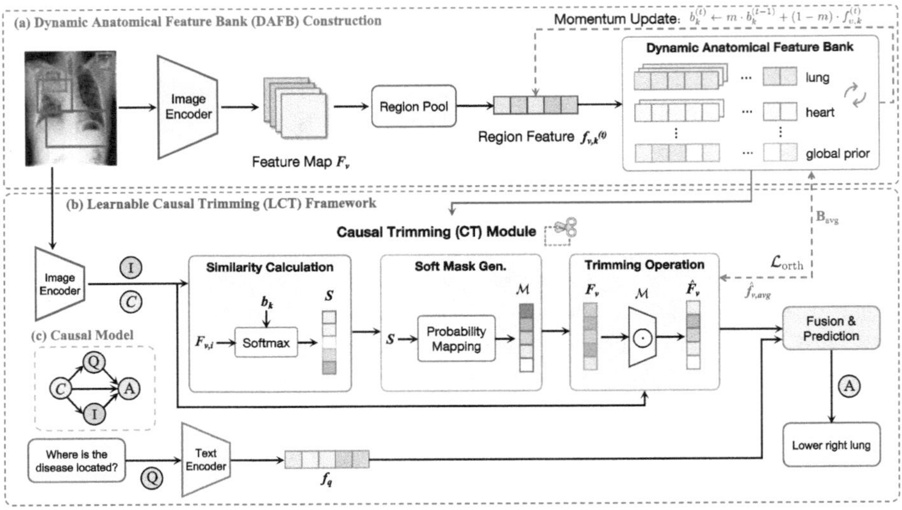

Fig. 1. The proposed Learnable Causal Trimming (LCT) Framework with Dynamic Anatomical Feature Bank.

2.2 Learnable Causal Trimming (LCT) Framework

Based on Sect. 2.1, the key to achieving robust MedVQA is executing the causal intervention by eliminating the confounding effect of C. To this end, we propose the Learnable Causal Trimming (LCT) framework. Unlike previous Test-Time Adaptation methods [15] that perform hard pruning on frozen models during inference, LCT internalizes the causal intervention mechanism into an end-to-end training pipeline. The overall architecture is depicted in Fig. 1.

The framework processes the inputs through three main stages:

Dual-Stream Encoding: The input medical image I and question Q are processed by an image encoder and a text encoder to extract the regional visual feature map $F_v \in \mathbb{R}^{N \times D}$ and the question embedding f_q, respectively.

Confounder Proxy Construction: A confounder bank is maintained via momentum updates to aggregate global, high-frequency anatomical priors across the dataset, serving as a proxy of the unobserved confounder C.

Causal Intervention via Trimming: The Causal Trimming (CT) Module acts as a soft gate. It evaluates the similarity between the instance-specific visual features F_v and the confounder prototypes in the DAFB, dynamically fading out spurious correlations to output the purified visual representation $\hat{F}_v$.

Finally, the trimmed visual features are fused with f_q for answer prediction, supervised by a joint loss function that enforces causal disentanglement.

2.3 Dynamic Anatomical Feature Bank (DAFB)

As defined in our SCM, the confounder C represents dataset-specific priors, such as the frequent co-occurrence of specific anatomical backgrounds and diagnostic answers. Since C is latent and constantly evolving during training, we propose to explicitly model it using the **DAFB**.

Let $\mathbf{B} \in \mathbb{R}^{K \times D}$ denote the feature bank, where K is the bank size and D is the feature dimension. The bank $\mathbf{B} = \{b_1, b_2, \ldots, b_K\}$ stores global prototypes representing common visual patterns (i.e., potential confounders). Instead of using a static dictionary, we employ a **Momentum Update** mechanism to maintain the stability of $\mathbf{B}$, ensuring it captures the long-term distribution of the dataset rather than the noisy fluctuations of a single mini-batch. Specifically, at training iteration t, we update the specific prototypes in the bank using the region-level features from the current batch. Let $f_{v,k}^{(t)}$ denote the aggregated visual feature of the current batch associated with the k-th prototype. We update the individual prototype $b_k \in \mathbf{B}$ via a momentum mechanism:

$$b_k^{(t)} \leftarrow m \cdot b_k^{(t-1)} + (1 - m) \cdot f_{v,k}^{(t)}, \tag{1}$$

where $m \in [0, 1)$ is a momentum coefficient. A large m (e.g., 0.99) ensures that the bank evolves smoothly, effectively filtering out instance-specific details and retaining only the robust, high-frequency anatomical priors (e.g., generic lung shapes or background artifacts) that constitute the confounder C.

2.4 Causal Trimming (CT) Module

With the confounder proxy $\mathbf{B}$ constructed, the next critical step is to reduce the confounding effect of C on the final prediction. To achieve this, we design the **Causal Trimming (CT) Module**, which adaptively reweights regional features according to their correlations with the dataset priors $\mathbf{B}$. As shown in Fig. 1, this module operates in three sequential steps:

1. Similarity Calculation: Given the input visual feature map $F_v \in \mathbb{R}^{N \times D}$ (where N corresponds to the number of patch-level feature tokens extracted by the visual encoder), we first compute the Confounder Relevance Score matrix $\mathbf{S} \in \mathbb{R}^{N \times K}$ by measuring the similarity between each image region and the bank prototypes:

$$\mathbf{S}_{i,k} = \frac{\exp(F_{v,i} \cdot b_k^T / \tau)}{\sum_{j=1}^{K} \exp(F_{v,i} \cdot b_j^T / \tau)}, \tag{2}$$

where τ is a temperature hyperparameter. A high score $\mathbf{S}_{i,k}$ indicates that the i-th region strongly resembles a common prototype in the bank, implying it is likely associated with dataset-level priors.

2. Soft Mask Generation: To make the trimming process differentiable, we map the relevance scores into a Soft Trimming Mask $\mathcal{M} \in \mathbb{R}^{N \times 1}$. We aggregate the scores across all prototypes to determine the total confounding degree of each region:

$$\mathcal{M}_i = 1 - \sigma\left(\psi\left(\sum_{k=1}^{K} \mathbf{S}_{i,k} \cdot w_k\right)\right), \tag{3}$$

where $\sigma(\cdot)$ is the Sigmoid function, ψ is a learnable projection layer, and w_k are learnable weights for the prototypes. The resulting mask $\mathcal{M}_i$ controls the residual gain of each region: it approaches 0 for regions highly correlated with confounder prototypes, reducing their additional amplification, and approaches 1 for unique, instance-specific diagnostic regions, enhancing their contribution.

3. Trimming Operation: Finally, the reweighted Trimmed Feature $\hat{F}_v$ is obtained via element-wise modulation combined with a residual connection:

$$\hat{F}_v = F_v \odot \mathcal{M} + F_v. \tag{4}$$

The residual connection is crucial here; it preserves gradient flow and retains the fundamental anatomical context, while the mask selectively emphasizes diagnostic evidence over confounder-correlated patterns.

2.5 Optimization Objectives

To ensure accurate answers and causally independent representations, LCT optimizes a joint objective comprising a BCE classification loss ($\mathcal{L}_{\text{vqa}}$) and an orthogonality loss ($\mathcal{L}_{\text{orth}}$):

$$\mathcal{L}_{\text{vqa}} = -\sum_{j=1}^{|\mathcal{A}|} y_j \log(P(a_j|\hat{F}_v, f_q)), \quad \mathcal{L}_{\text{orth}} = \left\| \frac{\hat{f}_{v,\text{avg}} \cdot \mathbf{B}_{\text{avg}}^T}{||\hat{f}_{v,\text{avg}}||\,||\mathbf{B}_{\text{avg}}||} \right\|^2, \tag{5}$$

where y is the ground truth. To explicitly enforce causal intervention, $\mathcal{L}_{\text{orth}}$ minimizes the cosine similarity between the mean vectors of the trimmed features ($\hat{f}_{v,\text{avg}}$) and the confounder bank ($\mathbf{B}_{\text{avg}}$), penalizing any residual correlation. The total objective is $\mathcal{L}_{\text{total}} = \mathcal{L}_{\text{vqa}} + \lambda\mathcal{L}_{\text{orth}}$, where λ balances accuracy and causal disentanglement.

3 Experiments

3.1 Datasets and Implementation Details

Datasets. VQA-RAD [6] contains 3,148 QA pairs on 315 radiology images. **SLAKE** [14] is a semantically-labeled bilingual dataset with 7,033 QA pairs on 642 radiology images. **PathVQA** [5] comprises 32,799 QA pairs on 4,998 pathology images, evaluating the model's generalizability across drastically different medical domains. Furthermore, to explicitly evaluate out-of-distribution (OOD) robustness, we utilize **SLAKE-CP** [25].

Implementation Details. Our framework is implemented in PyTorch. We leverage the pre-trained PMC-CLIP [11] for dual-stream encoding to extract initial text embeddings and patch-level visual features. For the causal intervention, the DAFB is initialized with $K = 256$ prototypes and updated with a momentum coefficient of m=0.99 to capture stable dataset priors. For the trimming module, the temperature scaling factor τ is set to 0.07. The entire network is trained end-to-end using the AdamW optimizer with a learning rate of $1e-4$ for 50 epochs. The balancing weight for the orthogonality loss, λ, is empirically set to 0.1.

Table 1. Comparison of accuracy (%) on VQA-RAD and SLAKE datasets. Best results are highlighted in **bold**.

Methods	VQA-RAD			SLAKE		
	Open	Closed	Overall	Open	Closed	Overall
MEVF+SAN [17]	49.2	73.9	64.1	75.3	78.4	76.5
MEVF+BAN [17]	49.2	77.2	66.1	77.8	79.8	78.6
CPRD+BAN [13]	52.5	77.9	67.8	79.5	83.4	81.1
M2I2 [9]	61.8	81.6	73.7	74.7	**91.1**	81.2
MISS [2]	**71.8**	80.4	76.1	81.5	82.9	82.0
M3AE [3]	67.2	83.5	77.0	80.3	87.8	83.3
CCIS-MVQA [1]	68.8	79.2	75.1	80.1	86.7	84.1
PMC-CLIP [11]	67.0	84.0	77.6	81.9	88.0	84.3
CLIPQCR [4]	58.0± 1.4	79.6± 1.1	71.1± 1.2	78.2± 1.3	82.6± 1.5	80.1± 1.3
DeBCF [25]	58.6± 1.1	80.9± 0.8	71.6± 1.0	80.8± 0.9	84.9± 0.7	82.6± 0.9
DeCoCT [23]	67.1± 0.5	85.7± 0.4	78.3± 0.5	82.5± 0.3	87.0± 0.6	84.9± 0.5
CIMB-MVQA [12]	69.3± 0.2	86.2± 0.2	79.4± 0.2	82.1± 0.1	89.4± 0.1	85.1± 0.2
LCT (Ours)	70.9± 0.6	**87.9**± 0.4	**81.2**± 0.4	**83.4**± 0.3	89.9± 0.5	**86.0**± 0.4

3.2 Comparison with State-of-the-Arts

Table 1 compares LCT with SOTA methods on VQA-RAD and SLAKE. LCT achieves the best Overall accuracy on both datasets, surpassing CIMB-MVQA by 1.8% and 0.9%, respectively. For **Closed-ended questions**, LCT obtains the highest accuracy on VQA-RAD (87.9%). Although M2I2 achieves 91.1% on SLAKE Closed, it drops notably on Open questions (74.7%). In contrast, LCT maintains balanced performance, achieving 89.9% on Closed and a leading 83.4% on Open questions. For **Open questions**, LCT reaches 70.9% on VQA-RAD and achieves SOTA on SLAKE Open. While slightly below MISS on VQA-RAD Open, LCT exceeds MISS by 5.1% in Overall accuracy, indicating stronger holistic reasoning performance.

To further assess cross-modality generalization, we evaluate on PathVQA (Table 2). LCT achieves the highest Overall accuracy of 65.6%, outperforming MUMC (65.1%). Notably, LCT establishes a new SOTA on Open-ended questions (41.1%), while achieving comparable Closed accuracy (89.9%). These results suggest that explicitly modeling dynamic confounders via causal trimming improves robustness, particularly for fine-grained pathological reasoning.

3.3 Out-of-Distribution Generalization on SLAKE-CP

Table 3 presents results on SLAKE-CP, a challenging OOD benchmark for evaluating diagnostic robustness. LCT achieves the best performance across all metrics, reaching 51.0% Overall, 28.9% on Open questions, and 71.6% on Closed questions. These results indicate that the proposed method improves the model's ability to mitigate dataset-specific biases under distribution shifts.

Table 2. Comparison with SOTA methods on PathVQA.

Methods	PathVQA		
	Open	Closed	Overall
MEVF-BAN [17]	8.1	81.4	44.8
MedGemma [21]	20.4	65.1	47.7
AMAM [19]	18.2	84.4	50.4
LLaVA-Med [7]	32.7	89.8	61.3
M2I2 [9]	36.3	88.0	62.2
CLIP-ViT [22]	40.0	87.0	63.6
MUMC [8]	39.0	**90.4**	65.1
LCT (Ours)	**41.1**	89.9	**65.6**

Table 3. Comparison with SOTA methods on SLAKE-CP.

Methods	SLAKE-CP		
	Open	Closed	Overall
MEVF+BAN [17]	$13.0_{\pm 1.4}$	$29.8_{\pm 1.2}$	$29.1_{\pm 1.3}$
CLIPQCR [4]	$13.4_{\pm 1.2}$	$30.5_{\pm 1.1}$	$30.0_{\pm 1.2}$
CPRD+BAN [13]	$13.9_{\pm 1.3}$	$31.2_{\pm 1.5}$	$30.4_{\pm 1.5}$
DeBCF [25]	$18.6_{\pm 1.1}$	$35.4_{\pm 1.0}$	$34.2_{\pm 1.2}$
MISS [2]	$17.3_{\pm 1.1}$	$49.2_{\pm 1.0}$	$33.8_{\pm 1.1}$
M2I2 [9]	$17.3_{\pm 1.1}$	$51.9_{\pm 0.9}$	$35.2_{\pm 1.0}$
M3AE [3]	$24.4_{\pm 1.0}$	$65.1_{\pm 0.9}$	$45.4_{\pm 1.0}$
DeCoCT [23]	$27.3_{\pm 0.5}$	$69.6_{\pm 0.5}$	$49.2_{\pm 0.6}$
LCT (Ours)	$\mathbf{28.9}_{\pm 0.6}$	$\mathbf{71.6}_{\pm 0.4}$	$\mathbf{51.0}_{\pm 0.5}$

Under the OOD setting, conventional representation learning models exhibit substantial performance degradation. For example, MISS and M2I2 drop to 33.8% and 35.2%, respectively. In comparison, LCT outperforms M2I2 by 15.8% in Overall accuracy, demonstrating stronger robustness to distribution shifts. Compared with recent debiasing and causal approaches, LCT also shows consistent improvements. Relative to DeCoCT, LCT improves Overall accuracy by 1.8%, Open accuracy by 1.6%, and Closed accuracy by 2.0%. In addition, LCT maintains a low standard deviation (±0.5% Overall), suggesting stable performance across runs. These findings support that the proposed learnable trimming strategy enhances both robustness and training stability in OOD scenarios.

3.4 Ablation Study on Key Components

We conduct ablation studies on VQA-RAD, SLAKE, and PathVQA to evaluate the individual contributions of each component. Results are in Table 4.

Baseline + DAFB (w/o CT): We remove the causal trimming mechanism and directly concatenate DAFB features with the joint vision-language embeddings, followed by an MLP classifier. Using DAFB alone brings moderate improvements (+1.5% on VQA-RAD and +1.3% on PathVQA), suggesting that enriched anatomical priors are beneficial but insufficient to fully mitigate dataset biases when applied through standard feature fusion.

Baseline + CT (w/o DAFB): We remove the dynamic feature bank and compute trimming masks using global visual pooled features. Applying CT alone leads to larger gains(+3.5% on VQA-RAD and +2.2% on PathVQA). This indicates that explicitly suppressing spurious correlations contributes significantly to robustness, even without precise anatomical prototypes.

Full LCT: Combining DAFB and CT achieves the best performance across all datasets. The consistent improvements over individual variants demonstrate that

Table 4. Ablation study on the key components of the proposed LCT framework. **CT**: Causal Trimming module. **DAFB**: Dynamic Anatomical Feature Bank.

Baseline	DAFB	CT	VQA-RAD			SLAKE			PathVQA
			Open	Closed	Overall	Open	Closed	Overall	Overall
✓			66.5	83.1	76.5	81.6	87.7	84.0	61.4
✓	✓		68.2	84.6	78.0	82.2	88.2	84.5	62.7
✓		✓	69.3	87.1	80.0	82.2	88.9	84.8	63.6
✓	✓	✓	**70.9**	**87.9**	**81.2**	**83.4**	**89.9**	**86.0**	**65.6**

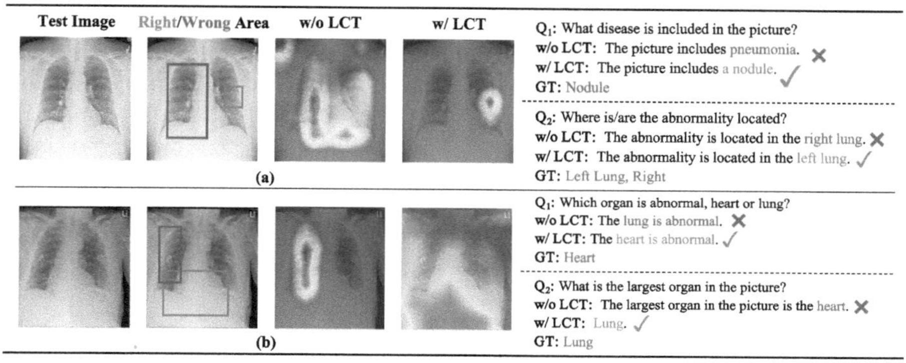

Fig. 2. Qualitative comparison of Grad-CAM visualizations and QA results. (Color figure online)

the two modules are complementary: DAFB enhances anatomical representation, while CT performs causal trimming. Their integration yields stronger and more stable diagnostic reasoning performance.

3.5 Qualitative Analysis and Visualization

To illustrate how LCT mitigates spurious correlations, we present Grad-CAM visualizations and QA outputs in Fig. 2. Without LCT, the baseline often attends to irrelevant or biased regions. In (a), attention is diffusely distributed over the right lung (red box), failing to localize the actual lesion. In Example (b), the baseline focuses on an upper lung region instead of the target organs.

With LCT, the attention maps become more anatomically focused. In the "w/ LCT" column, activations concentrate on the relevant structures, such as the small nodule in the left lung (Example a) and the enlarged cardiac silhouette (Example b), indicating improved visual grounding.

The refined attention is reflected in the predicted answers. In Example (a), the baseline predicts "pneumonia" and incorrectly identifies the "right lung", whereas LCT correctly recognizes the "nodule" and localizes it in the "left lung". In Example (b), LCT accurately identifies the heart as the abnormal organ and the lung as the largest organ, while the baseline produces inconsistent responses. These results suggest that causal trimming improves anatomical grounding and reduces reliance on dataset-specific shortcuts.

4 Conclusion

In this paper, we introduced the Learnable Causal Trimming (LCT) framework to overcome the spurious correlations and dataset biases in MedVQA. By explicitly purifying visual representations at the feature level, LCT integrates a Dynamic Anatomical Feature Bank (DAFB) to continuously capture domain-specific medical prototypes and a Causal Trimming (CT) module to dynamically filter out confounding visual noise while preserving authentic pathological signals. Extensive experiments demonstrate that our approach not only establishes new state-of-the-art performance across multiple in-distribution benchmarks—including VQA-RAD, SLAKE, and PathVQA—but also exhibits exceptional robustness on the out-of-distribution SLAKE-CP dataset. Ultimately, LCT forces the network to ground its predictions in genuine anatomical evidence rather than exploiting statistical shortcuts, representing a significant step toward developing highly reliable multi-modal AI systems for clinical applications.

Acknowledgments. This work was supported in part by the National Natural Science Foundation of China under Grant 62272337.

Disclosure of Interests. The authors have no competing interests to declare that are relevant to the content of this article.

References

1. Cai, L., Fang, H., Xu, N., Ren, B.: Counterfactual causal-effect intervention for interpretable medical visual question answering. IEEE Trans. Med. Imaging **43**(12), 4430–4441 (2024)
2. Chen, J., Yang, D., Jiang, Y., Lei, Y., Zhang, L.: Miss: A generative pre-training and fine-tuning approach for med-vqa. In: International Conference on Artificial Neural Networks, pp. 299–313 (2024)
3. Chen, Z., et al.: Multi-modal masked autoencoders for medical vision-and-language pre-training. In: International Conference on Medical Image Computing and Computer-Assisted Intervention (2022)
4. Eslami, S., Meinel, C., De Melo, G.: Pubmedclip: How much does clip benefit visual question answering in the medical domain? In: Findings of the Association for Computational Linguistics: EACL 2023, pp. 1151–1163 (2023)
5. He, X., Zhang, Y., Mou, L., Xing, E., Xie, P.: Pathvqa: 30000+ questions for medical visual question answering. arXiv preprint arXiv:2003.10286 (2020)
6. Lau, J.J., Gayen, S., Ben Abacha, A., Demner-Fushman, D.: A dataset of clinically generated visual questions and answers about radiology images. Sci. Data **5**(1), 180251 (2018)
7. Li, C., et al.: Llava-med: Training a large language-and-vision assistant for biomedicine in one day. Adv. Neural. Inf. Process. Syst. **36**, 28541–28564 (2023)

8. Li, P., Liu, G., He, J., Zhao, Z., Zhong, S.: Masked vision and language pre-training with unimodal and multimodal contrastive losses for medical visual question answering. In: International Conference on Medical Image Computing and Computer-Assisted Intervention, pp. 374–383 (2023)
9. Li, P., Liu, G., Tan, L., Liao, J., Zhong, S.: Self-supervised vision-language pre-training for medical visual question answering. In: 2023 IEEE 20th International Symposium on Biomedical Imaging (ISBI), pp. 1–5 (2023)
10. Li, Q., Liu, M., Chang, R., Nie, W., Bai, S., Liu, A.: Multilabel chest x-ray image classification via category disentangled causal learning. IEEE Trans. Artif. Intell. **7**(2), 1048–1061 (2026)
11. Lin, W., et al.: Pmc-clip: Contrastive language-image pre-training using biomedical documents. In: International Conference on Medical Image Computing and Computer-Assisted Intervention, pp. 525–536 (2023)
12. Liu, B., Liu, L., Ding, J., Yang, X., Peng, W., Liu, L.: Cimb-mvqa: Causal intervention on modality-specific biases for medical visual question answering. Med. Image Anal. 103850 (2025)
13. Liu, B., Zhan, L.M., Wu, X.M.: Contrastive pre-training and representation distillation for medical visual question answering based on radiology images. In: International conference on medical image computing and computer-assisted intervention, pp. 210–220 (2021)
14. Liu, B., Zhan, L.M., Xu, L., Ma, L., Yang, Y., Wu, X.M.: Slake: A semantically-labeled knowledge-enhanced dataset for medical visual question answering. In: 2021 IEEE 18th International Symposium on Biomedical Imaging (ISBI), pp. 1650–1654 (2021)
15. Liu, Y., Qiao, R., Lee, M.L., Hsu, W.: Test-time adaptation by causal trimming. In: The Thirty-Ninth Annual Conference on Neural Information Processing Systems (2025)
16. Lu, Z., Zeng, Q., Lu, M., Chen, G., Xia, Y.: Bridging the semantic gap in medical visual question answering with prompt learning. IEEE Trans. Med. Imaging **44**(11), 4605–4616 (2025)
17. Nguyen, B.D., Do, T.T., Nguyen, B.X., Do, T., Tjiputra, E., Tran, Q.D.: Overcoming data limitation in medical visual question answering. In: International Conference on Medical Image Computing and Computer-Assisted Intervention, pp. 522–530 (2019)
18. Nie, W., Zhang, C., Song, D., Bai, Y., Xie, K., Liu, A.A.: Chest x-ray image classification: A causal perspective. In: Medical Image Computing and Computer Assisted Intervention - MICCAI 2023, pp. 25–35 (2023)
19. Pan, H., He, S., Zhang, K., Qu, B., Chen, C., Shi, K.: Amam: an attention-based multimodal alignment model for medical visual question answering. Knowl.-Based Syst. **255**, 109763 (2022)
20. Pearl, J.: Causal inference in statistics: An overview (2009)
21. Sellergren, A., et al.: Medgemma technical report. arXiv preprint arXiv:2507.05201 (2025)
22. Van Sonsbeek, T., Derakhshani, M.M., Najdenkoska, I., Snoek, C.G., Worring, M.: Open-ended medical visual question answering through prefix tuning of language models. In: International Conference on Medical Image Computing and Computer-Assisted Intervention, pp. 726–736 (2023)
23. Wan, X., Teng, Q., Chen, J., Lu, Y., Yuan, D., Liu, Z.: Eliminating language bias for medical visual question answering with counterfactual contrastive training. In: International Conference on Medical Image Computing and Computer-Assisted Intervention, pp. 194–204 (2025)

24. Xu, Z., Li, Q., Nie, W., Wang, W., Liu, A.: Structure causal models and llms integration in medical visual question answering. IEEE Trans. Med. Imaging **44**(8), 3476–3489 (2025)
25. Zhan, C., et al.: Debiasing medical visual question answering via counterfactual training. In: International Conference on Medical Image Computing and Computer-Assisted Intervention, pp. 382–393 (2023)

LoFi: Location-Aware Fine-Grained Representation Learning for Chest X-Ray

Myeongkyun Kang[1,2], Yanting Yang[1,2], and Xiaoxiao Li[1,2](✉)

[1] The University of British Columbia, Vancouver, BC V6T 1Z4, Canada
[2] Vector Institute, Toronto, ON M5G 0C6, Canada
myeongkyun.kang@ubc.ca, xiaoxiao.li@ece.ubc.ca

Abstract. Fine-grained representation learning is crucial for retrieval and phrase grounding in chest X-rays, where clinically relevant findings are often spatially confined. However, existing methods remain limited on these tasks, as contrastive models lack explicit region-level supervision and large vision language models often fail to generalize fine-grained representations under external validation. To address these limitations, we propose Location-aware Fine-grained representation learning (LoFi), which jointly optimizes sigmoid, captioning, and location-aware captioning losses using a lightweight large language model. The location-aware captioning loss enables region-level supervision through grounding and dense captioning objectives, thereby facilitating fine-grained representation learning. Building upon these representations, we integrate a fine-grained encoder into retrieval-based in-context learning to enhance chest X-ray grounding across diverse settings. Extensive experiments demonstrate that our method achieves superior retrieval and phrase grounding performance on MIMIC-CXR and PadChest-GR.

Keywords: Fine-Grained Representation Learning · Retrieval · Phrase Grounding · In-Context Learning · Chest X-ray

1 Introduction

Clinically relevant findings in medical imaging are often spatially confined rather than semantically distinct, necessitating fine-grained representation learning [7]. Retrieval tasks identify relevant chest X-rays among similar candidates, requiring fine-grained discrimination [28], and phrase grounding tasks localize objects based on textual descriptions, demanding fine-grained cross-modal alignment [2]. However, medical contrastive models [12,28] and large vision language models (LVLMs) [1,5] remain limited in their ability to capture subtle and fine-grained representations, leading to suboptimal performance on these tasks.

Contrastive chest X-ray models have been proposed to effectively leverage paired radiology reports for fine-grained, clinically relevant representation learning [9,12]. However, these models rely on implicit attention-based local image-text alignment without explicit region-level supervision (e.g., bounding boxes);

Code available at https://github.com/myeongkyunkang/lofi.

G. Yang et al. (Eds.): MICCAI 2026, LNCS 16878, pp. 271–280, 2027.
https://doi.org/10.1007/978-3-032-38059-3_26

incorporating such supervision can alleviate this limitation. Nevertheless, compared with prior web-scale pretraining approaches [23], training autoregressive models from scratch on image-text-box triples remains challenging in the medical domain due to the scarcity of region-level annotations. Therefore, leveraging a lightweight large language model (LLM) [21] and a region-level pretrained encoder [22] is desirable to address the challenges of medical image-text-box alignment.

A further challenge is that existing models, including LVLMs, remain limited in generalizing fine-grained representations under external validation [13]. To address this challenge, retrieval-based in-context learning (ICL) has been proposed to enable adaptation to new tasks by leveraging retrieved demonstrations at inference time [25]. However, its performance largely depends on the quality of the retrieved demonstrations, underscoring the need for fine-grained representations and learning objectives that facilitate them.

We propose **Lo**cation-aware **Fi**ne-grained representation learning (LoFi), which jointly optimizes sigmoid, captioning, and location-aware captioning losses with a lightweight LLM [21] and a region-level pretrained encoder [22]. The captioning loss facilitates learning from long-form text, while the location-aware captioning loss enables region-level supervision through grounding and dense captioning objectives, thereby enhancing fine-grained representation learning. Building upon these representations, we integrate a fine-grained encoder into retrieval-based ICL to enhance chest X-ray grounding under external validation. We train our model on 208,949 radiology reports and 394,835 text-box pairs from MIMIC-CXR, and demonstrate the superiority of our fine-grained representations through retrieval tasks on MIMIC-CXR, as well as internal and external phrase grounding tasks on PadChest-GR.

In summary, the contributions are as follows: *(a)* We propose LoFi, which leverages a lightweight LLM to jointly optimize sigmoid, captioning, and location-aware captioning losses for fine-grained chest X-ray representation learning. *(b)* We integrate a fine-grained encoder into retrieval-based ICL, providing a practical solution for real-world chest X-ray grounding. *(c)* We demonstrate the effectiveness of our approach on retrieval and grounding tasks through comprehensive internal and external validation.

2 Method

Location-Aware Fine-Grained Representation Learning (LoFi). The proposed method is shown in Fig. 1(a). Given a dataset $\mathcal{D}$ of image-text-box triplets (I, T, B), where box annotations may be optional, we aim to train an image encoder E_I and a text encoder E_T, leveraging a lightweight LLM LLM. We employ 64 attention pooling modules and a projection layer to align the E_I outputs with the LLM input dimension (notation omitted for simplicity). We implement three objectives to facilitate learning from long-form text and region-level supervision: (a) a sigmoid loss, (b) a captioning loss, and (c) a location-aware captioning loss consisting of a grounding loss and a dense captioning loss. We adopt a sigmoid loss [22] as our contrastive objective, encouraging the model

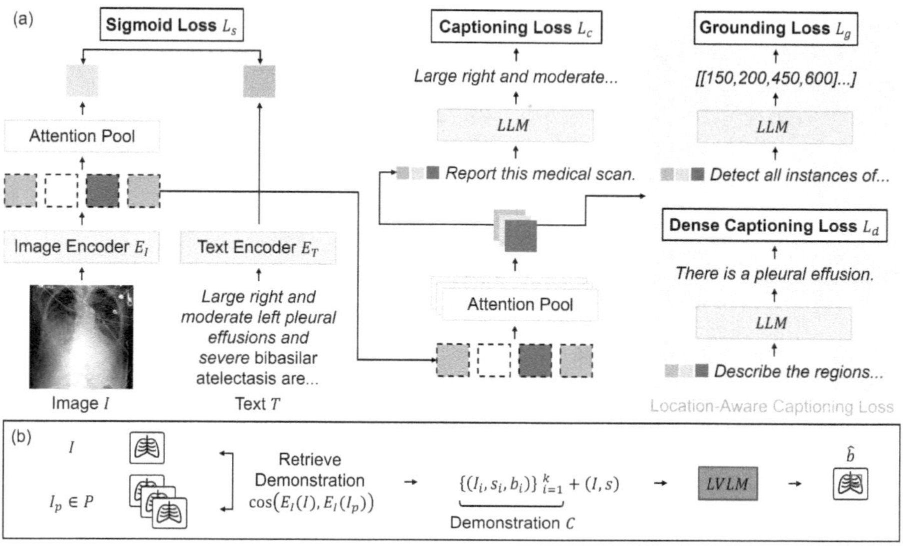

Fig. 1. Illustration of (a) location-aware fine-grained representation learning frameworks and (b) a retrieval-based in-context learning pipeline.

to assign higher similarity scores to matched pairs and thus learn representations that are well suited for retrieval. Formally, the sigmoid loss is defined as

$$\mathcal{L}_s = -\mathbb{E}_{(I,T)\sim\mathcal{D}}\left[\log\sigma\big(\delta(I,T)\,E_I(I)^\top E_T(T)\big)\right], \tag{1}$$

where $\sigma(\cdot)$ denotes the sigmoid, and $\delta(I,T) \in \{+1,-1\}$ is a sign indicator that equals $+1$ for matched positive pairs and -1 for in-batch negative pairs. However, the sigmoid loss alone is limited in leveraging long-form text and region-level supervision of fine-grained object locations during training. To address this issue, we adapt prior works [22,23] to our learning objective using a lightweight LLM. We denote the conditional distribution induced by the LLM as p_{LLM}, and define the autoregressive captioning loss as

$$\mathcal{L}_c = -\mathbb{E}_{(I,T)\sim\mathcal{D}}\left[\log p_{LLM}(T \mid E_I(I), \mathcal{P}_c)\right], \tag{2}$$

where $\mathcal{P}_c$ denotes a fixed prompt (e.g., *"Report this medical scan."*). Next, to incorporate location information of fine-grained objects during training, we adapt the grounding objective and the dense captioning objective to our autoregressive loss. Note that a text (e.g., a radiology report) T consists of sentences, and we denote a sentence as $s \in T$. We denote the set of bounding boxes associated with s by $b \subseteq B$, where each box is represented by the coordinates of its top-left and bottom-right corners $(x_{\min}, y_{\min}, x_{\max}, y_{\max})$. Formally, we define the grounding loss as

$$\mathcal{L}_g = -\mathbb{E}_{(I,T,B)\sim\mathcal{D}_B,\, s\sim T}\left[\log p_{LLM}(b \mid E_I(I), \mathcal{P}_g, s)\right], \tag{3}$$

where $\mathcal{D}_B \subseteq \mathcal{D}$ denotes samples with bounding boxes, and $\mathcal{P}_g$ denotes a fixed prompt (e.g., *"Detect all instances of..."*). For the dense captioning loss, we swap the prediction target from b to s. Formally, we define the dense captioning loss as $\mathcal{L}_d = -\mathbb{E}_{(I,T,B)\sim\mathcal{D}_B,\, s\sim T}\left[\log p_{LLM}(s \mid E_I(I), \mathcal{P}_d, b)\right]$, where $\mathcal{P}_d$ denotes a fixed prompt (e.g., *"Describe the regions..."*). In summary, the total loss is defined as $\mathcal{L}_t = \mathcal{L}_s + \lambda \cdot \mathcal{L}_\tau, \tau \in \{c, g, d\}$, where τ indicates the loss type (captioning, grounding, or dense captioning), and λ is a hyperparameter for weighting the autoregressive loss (set to 5). When (I, T, B) are available, $\mathcal{L}_c$ and the location-aware captioning loss $\mathcal{L}_{g,d}$ contribute equally to the total loss.

Retrieval-Based In-Context Learning for Grounding. The retrieval-based in-context learning (ICL) is illustrated in Fig. 1(b). Given an image query I and a sentence (or phrase) query s, the instruction-tuned model $LVLM$ predicts a set of bounding boxes $\hat{b}$ on I corresponding to s, i.e., $\hat{b} = LVLM(I, s)$. We use ICL to adapt $LVLM$ to new tasks without fine-tuning by leveraging a small set of demonstrations at inference time. The $LVLM$ predicts $\hat{b}$ conditioned on the set of demonstrations C, i.e., $\hat{b} = LVLM(I, s \mid C)$, where $C = \{(I_i, s_i, b_i)\}_{i=1}^{k}$ denotes a set of k image-sentence-box triples. Following retrieval-based approaches [25], we construct C by selecting the top-k most relevant triplets from the candidate pool. Given a candidate pool P and an image encoder E_I, we retrieve relevant demonstrations by computing $\cos\left(E_I(I), E_I(I_p)\right)$ for all $I_p \in P$, where $\cos(\cdot,\cdot)$ denotes cosine similarity. Since ICL performance relies on the quality of retrieved demonstrations, the fine-grained representations captured by E_I play a crucial role in effective grounding, underscoring the value of fine-grained representation learning in our framework.

Discussion. In phrase grounding, inter-observer variability leads to inconsistent bounding box annotations, posing challenges for model training and evaluation [2]. Retrieval-based ICL is particularly effective in this setting, as it mitigates annotation inconsistency at inference time without additional fine-tuning. In contrast, when training is feasible, models can be fine-tuned with $\mathcal{L}_g$, thereby further enhancing performance on downstream tasks. Across both settings, strong fine-grained representations are essential for achieving high performance, which aligns with our training objective.

Implementation Details. We adopted a pretrained SigLIP2-400M [22] with an input size of 512 as E_I and E_T, and used Gemma-3-270M [21] as the LLM. Following prior work [12, 24], we randomly sampled 4–8 sentences from radiology reports for contrastive learning due to the 64 token limit of the text encoder. We normalized bounding box coordinates to [0, 1000] and converted them into strings to compute autoregressive $\mathcal{L}_g$ and $\mathcal{L}_d$. We applied LoRA [8] with rank 16 to the query, key, value, and output layers of E_I and E_T, and with rank 4 to the query and value layers of the LLM. We trained the model for 10 epochs with a batch size of 16 using cosine annealing with an initial learning rate of 3e-4.

Table 1. Retrieval performance on the MIMIC-CXR dataset. R@K denotes Recall at K, where K is the cutoff rank.

Model	Image-to-Text					Text-to-Image				
	R@1	R@5	R@10	R@20	R@40	R@1	R@5	R@10	R@20	R@40
SigLIP2 [22]	0.08	0.95	1.56	2.59	4.32	0.27	0.98	1.60	2.95	4.69
BiomedCLIP [27]	0.82	2.10	3.78	7.07	12.66	0.58	1.77	3.17	6.00	11.02
BMC-CLIP [15]	0.12	0.53	1.15	2.55	4.44	0.05	0.80	1.65	2.63	4.78
MedSigLIP [19]	4.61	14.27	21.92	32.69	44.82	2.38	10.33	17.47	26.43	38.48
CARZero [12]	11.64	29.52	40.17	51.19	63.98	9.67	26.02	36.24	48.40	62.37
RadIR [28]	7.36	22.62	32.48	44.70	57.44	7.11	22.70	32.61	45.44	59.91
Ours	13.90	35.57	47.29	60.20	72.53	11.91	31.30	42.36	54.46	66.71

For retrieval-based ICL, we used the instruction-tuned MedGemma-1.5-4B [19] with an input size of 896, motivated by our empirical observation that general-purpose LVLMs (e.g., Qwen [10]) have limited instruction-following capabilities for chest X-rays. Other LVLMs [1,4,5,13,14] had limited context lengths (e.g., 4096 tokens) and therefore could not support ICL.

3 Experiments and Results

Datasets. MIMIC-CXR is a dataset consisting of 377,110 chest X-rays and 227,835 radiology reports [11]. We utilized the findings and impression sections annotated by an LLM in [26] and excluded non-frontal and low-quality chest X-rays for our experiments using [6]. For grounding annotations, a total of 394,835 A^{++}-rated text-box pairs from 153,602 studies in MIMIC-Ext [17] were used. As the dataset provides machine-generated silver-standard ground-truth annotations, we applied a weighted box fusion strategy [20] to merge duplicate boxes. Notably, unlike typical phrase grounding datasets for abnormal findings (e.g., PadChest-GR [2]), MIMIC-Ext consists of anatomical-level bounding box annotations, providing comparatively larger bounding boxes. The dataset was used to train our model on the official training split, and the model's fine-grained representations were evaluated on 2,432 patient studies from the official test split via retrieval tasks. **PadChest-GR** is a dataset consisting of 4,555 chest X-rays paired with grounding reports containing text-box annotations [2]. We exclusively used annotations with non-empty bounding boxes in our phrase grounding experiments. We used 1,238 text-box pairs from 604 chest X-rays in the official test split for evaluation and 4,341 text-box pairs from 2,096 chest X-rays for training (or as the candidate pool).

3.1 Retrieval Experiments

We evaluated the fine-grained representations of our model on the MIMIC-CXR dataset using image-to-text and text-to-image retrieval tasks. We split each radiology report into five-sentence sub-reports and evaluated report-level retrieval

Table 2. Phrase grounding performance on the PadChest-GR dataset for external validation using retrieval-based ICL with MedGemma and k demonstrations.

Model	k=20					k=40				
	Ro/L	Ro/S	F@0.5	P@0.5	R@0.5	Ro/L	Ro/S	F@0.5	P@0.5	R@0.5
SigLIP2 [22]	52.50	42.14	17.75	18.61	16.96	56.87	44.70	21.47	22.53	20.51
BiomedCLIP [27]	56.19	42.39	18.43	19.57	17.42	60.14	45.40	22.06	22.94	21.24
BMC-CLIP [15]	53.24	42.20	17.15	18.16	16.24	56.71	44.66	20.92	21.95	19.99
MedSigLIP [19]	58.79	44.69	21.77	22.62	20.97	61.63	46.89	24.51	25.12	23.93
RadIR [28]	56.03	42.13	17.50	18.54	16.57	56.34	41.72	17.04	17.77	16.37
Ours	59.10	44.49	22.48	23.40	21.63	63.55	48.00	25.34	25.98	24.72

performance using Recall at K (R@K), where K is the cutoff rank [24]. We compared our method against several contrastive-based representation learning models, including SigLIP2 [22], BiomedCLIP [27], and BMC-CLIP [15], as well as models trained on MIMIC-CXR, such as MedSigLIP [19], CARZero [12], and RadIR [28]. We loaded the publicly released checkpoints of the comparison methods and evaluated them on the test set.

Table 1 shows the retrieval performance on the MIMIC-CXR dataset. General medical contrastive models achieve lower retrieval performance than models tailored to chest X-ray data, highlighting the advantage of modality-specific approaches. MedSigLIP exhibits lower accuracy than CARZero and RadIR, which can be attributed to its training on medical data alongside a large proportion of non-medical images. Our method outperforms state-of-the-art contrastive models trained on MIMIC-CXR in both retrieval directions, demonstrating its effectiveness in learning fine-grained representations.

3.2 Phrase Grounding Experiments

We conducted phrase grounding experiments on the PadChest-GR dataset to evaluate fine-grained representations in external validation using retrieval-based ICL and downstream performance in internal validation using pretrained parameters. In external validation, PadChest-GR was used exclusively for evaluation, whereas in internal validation, it was used for both model training and evaluation. We reported the localization (Ro/L) and shape-matching (Ro/S) scores of robust detection outcome [16], as well as precision (P@0.5), recall (R@0.5), and F1 score (F@0.5) at an intersection over union threshold of 0.5 (rather than average precision, which relies on confidence scores) [10]. All models were evaluated under a single-class setting to facilitate comparison with each other.

External Validation. We evaluated the phrase grounding performance of contrastive models [15,19,22,27,28] using retrieval-based ICL with MedGemma [19]. We assessed performance across different settings by varying the number of

Table 3. Phrase grounding performance on the PadChest-GR dataset for external validation.

Model	Ro/L	Ro/S	F@0.5	P@0.5	R@0.5
MedRPG [3]	31.44	24.00	2.10	2.34	1.91
ChEX [18]	39.73	30.98	3.09	2.97	3.22
GEMeX [14]	27.93	16.10	9.36	15.15	6.77
MedGemma	38.66	27.01	2.03	2.27	1.84
K2Sight [13]	49.81	25.61	8.68	9.46	8.02
Ours *w/* ICL	**63.55**	**48.00**	**25.34**	**25.98**	**24.72**

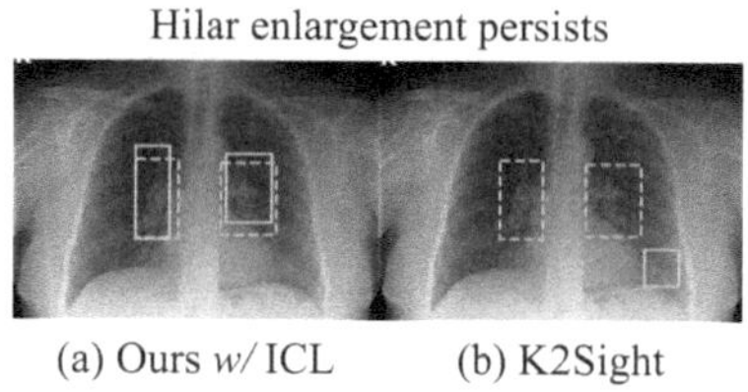

Fig. 2. Qualitative results with ground truth (dashed) and predictions (solid).

demonstrations (k=20 and k=40). CARZero [12] was excluded as it only supports cross-modal similarity. Additionally, we compared our method (*w/* ICL) against existing chest X-ray grounding models, including MedRPG [3], ChEX [18], GEMeX [14], K2Sight [13], and MedGemma [19].

Table 2 shows the phrase grounding performance on the PadChest-GR dataset for external validation using retrieval-based ICL. Compared to MedGemma's performance without ICL in Table 3, incorporating retrieval-based ICL with contrastive models leads to a significant performance improvement, and additional gains are observed as the number of demonstrations increases. Among these models, our method achieves the best performance, demonstrating its effectiveness in capturing fine-grained representations for real-world settings.

Table 3 shows the phrase grounding performance of chest X-ray grounding models. Although MedRPG and ChEX are designed for grounding abnormal findings, they exhibit limited performance due to inter-observer variability. Similarly, models trained on anatomical-level MIMIC-CXR annotations, including GEMeX and MedGemma (*w/o* ICL), show degraded performance. K2Sight, designed to improve reliability on PadChest-GR through interpretable visual attribute decomposition, similarly shows limited performance (see Fig. 2(b)). Overall, our method with retrieval-based ICL (k=40) outperforms all baselines, highlighting its potential for real-world chest X-ray grounding.

Table 4. Phrase grounding performance on the PadChest-GR dataset for internal validation.

Model	Ro/L	Ro/S	F@0.5	P@0.5	R@0.5
CheXagent [4]	27.19	14.93	11.30	20.81	7.76
MAIRA-2 [1]	53.57	35.87	32.93	**40.09**	27.94
RadVLM [5]*	70.21	44.85	29.27	27.70	31.03
FT *w/o* Ours	65.54	46.60	28.39	27.81	28.99
FT ***w/ Ours***	**70.42**	**47.97**	**33.44**	33.76	**33.14**

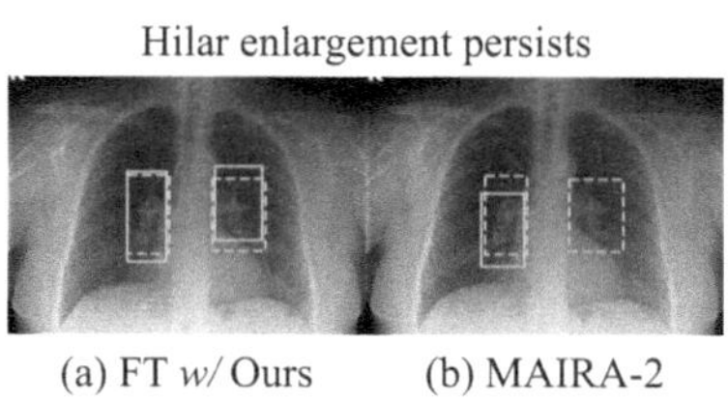

Fig. 3. Qualitative results with ground truth (dashed) and predictions (solid).

Internal Validation. We compared our method (FT *w/* Ours) against chest X-ray grounding LVLMs trained on PadChest-GR, including CheXagent [4], MAIRA-2 [1], and RadVLM [5]. The performance of RadVLM may be inflated, as the model was trained on the full PadChest-GR dataset. Additionally, we evaluated a variant of our model without MIMIC-CXR pre-training (FT *w/o* Ours) to quantify its impact.

Table 4 shows the phrase grounding performance on the PadChest-GR dataset for internal validation. MAIRA-2 shows relatively higher P@0.5 but lower R@0.5 and Ro/L performance, likely due to its cautious bounding box predictions (see Fig. 3(b)). FT *w/o* Ours outperforms MAIRA-2 on the Ro/L metric but achieves lower performance on F@0.5. By leveraging pretrained parameters, FT *w/* Ours improves F@0.5 by 5% compared to FT *w/o* Ours, surpassing MAIRA-2 on both Ro/L and F@0.5 metrics. Although the performance gains in F@0.5 are marginal compared to MAIRA-2, our model is more parameter-efficient, using 270M parameters, whereas MAIRA-2 uses 7B. In summary, the proposed objective facilitates the learning of pretrained parameters that significantly improve downstream fine-grained grounding performance.

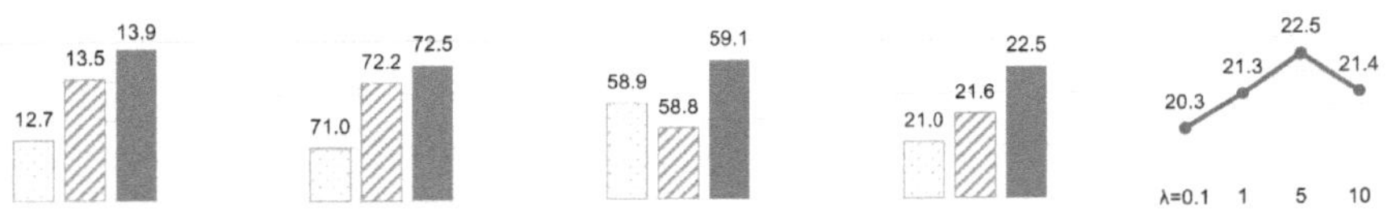

(a) R@1 (b) R@40 (c) Ro/L (d) F@0.5 (e) F@0.5 (λ)

Fig. 4. (a)–(d) Ablation results of our method using all losses (filled), without $\mathcal{L}_{g,d}$ (hatched), and without both $\mathcal{L}_c$ and $\mathcal{L}_{g,d}$ (dotted). (e) Ablation results for different λ values (line graph).

Ablation Studies. We performed ablation studies on the retrieval task in MIMIC-CXR and the grounding task in PadChest-GR. We evaluated variants of our method without $\mathcal{L}_{g,d}$ and without both $\mathcal{L}_c$ and $\mathcal{L}_{g,d}$. Additionally, we evaluated our method using different values of λ. Since the dataset contains samples without bounding boxes, we did not consider the variant without $\mathcal{L}_c$. We reported (a) R@1, (b) R@40, (c) Ro/L (k=20), and (d)–(e) F@0.5 (k=20).

Figure 4(a)–(d) show the ablation results on both tasks. These results suggest that incorporating the location-aware loss is beneficial for fine-grained representation learning. Figure 4(e) shows performance with varying λ. These results indicate that λ=5 performs best and that assigning a moderately higher weight to the proposed components benefits training. In summary, each loss plays an important role in fine-grained representation learning.

4 Conclusion

We propose LoFi to address the limitations of existing medical contrastive models and LVLMs in chest X-ray retrieval and phrase grounding. By jointly optimizing sigmoid, captioning, and location-aware captioning losses with a lightweight LLM, our model facilitates learning from long-form text and region-level supervision, enhancing fine-grained representations. Extensive experiments demonstrate that our method consistently achieves superior performance in retrieval on MIMIC-CXR and in phrase grounding on PadChest-GR under both internal and external validation settings. The computational overhead introduced by ICL may limit the scalability of the proposed framework, highlighting the need for future research to alleviate this bottleneck and enable more efficient deployment.

Acknowledgments. The research is supported, in part, by the NSERC Discovery Grant RGPIN-2022-05316, NSERC Alliance Grant ALLRP 602633-24, Tri-Agency Canada; Canada CIFAR AI Chair Awards and Canada Research Chair Fellowship; Genome British Columbia (ISI003), the Terry Fox Research Institute, Health Research BC; UBC AI and Health Network Fellowship; Google Gemini Research Awards; IITP grant, the Ministry of Science and ICT (RS-2024-00445087, RS-2025-25464461), funded by the Korea government (MSIT); the National Research Foundation of Korea (NRF) grant funded by the Korea government (MSIT) (RS-2025-00515536).

Disclosure of Interests. The authors have no competing interests to declare that are relevant to the content of this article.

References

1. Bannur, S., et al.: Maira-2: Grounded radiology report generation. arXiv preprint arXiv:2406.04449 (2024)
2. de Castro, D.C., et al.: Padchest-gr: A bilingual chest x-ray dataset for grounded radiology report generation. NEJM AI **2**(7), AIdbp2401120 (2025)
3. Chen, Z., et al.: Medical phrase grounding with region-phrase context contrastive alignment. In: International Conference on Medical Image Computing and Computer-Assisted Intervention, pp. 371–381. Springer, Cham (2023)
4. Chen, Z., et al.: Chexagent: towards a foundation model for chest x-ray interpretation. In: AAAI 2024 Spring Symposium on Clinical Foundation Models (2024)
5. Deperrois, N., et al.: Radvlm: A multitask conversational vision-language model for radiology. arXiv preprint arXiv:2502.03333 (2025)
6. Gaggion, N., et al.: Chexmask: a large-scale dataset of anatomical segmentation masks for multi-center chest x-ray images. Sci. Data **11**(1), 511 (2024)
7. Haghighi, F., Taher, M.R.H., Gotway, M.B., Liang, J.: Self-supervised learning for medical image analysis: discriminative, restorative, or adversarial? Med. Image Anal. **94**, 103086 (2024)
8. Hu, E.J., Shen, Y., Wallis, P., Allen-Zhu, Z., Li, Y., Wang, S., Wang, L., Chen, W., et al.: Lora: low-rank adaptation of large language models. ICLR **1**(2), 3 (2022)
9. Huang, S.C., Shen, L., Lungren, M.P., Yeung, S.: Gloria: a multimodal global-local representation learning framework for label-efficient medical image recognition. In: Proceedings of the IEEE/CVF International Conference on Computer Vision, pp. 3942–3951 (2021)

10. Jiang, Q., et al.: Detect anything via next point prediction. In: Proceedings of the IEEE/CVF Conference on Computer Vision and Pattern Recognition, pp. 25472–25483 (2026)
11. Johnson, A.E., et al.: Mimic-cxr, a de-identified publicly available database of chest radiographs with free-text reports. Sci. Data, **6**(1), 317 (2019)
12. Lai, H., et al.: Carzero: cross-attention alignment for radiology zero-shot classification. In: Proceedings of the IEEE/CVF Conference on Computer Vision and Pattern Recognition, pp. 11137–11146 (2024)
13. Li, J., et al.: Knowledge to sight: reasoning over visual attributes via knowledge decomposition for abnormality grounding. In: Proceedings of the IEEE/CVF Winter Conference on Applications of Computer Vision, pp. 2359–2369 (2026)
14. Liu, B., et al.: Gemex: a large-scale, groundable, and explainable medical vqa benchmark for chest x-ray diagnosis. In: Proceedings of the IEEE/CVF International Conference on Computer Vision, pp. 21310–21320 (2025)
15. Lozano, A., et al.: Biomedica: an open biomedical image-caption archive, dataset, and vision-language models derived from scientific literature. In: Proceedings of the Computer Vision and Pattern Recognition Conference, pp. 19724–19735 (2025)
16. Meissen, F., Müller, P., Kaissis, G., Rueckert, D.: Robust detection outcome: a metric for pathology detection in medical images. In: Medical Imaging with Deep Learning, pp. 568–585. PMLR (2024)
17. Müller, P., Jungmann, F., Kaissis, G., Rueckert, D.: A structured, tagged, and localized visual question answering dataset with full sentence answers and scene graphs for chest x-ray images. In: ICLR (2026)
18. Müller, P., Kaissis, G., Rueckert, D.: Chex: interactive localization and region description in chest x-rays. In: European Conference on Computer Vision, pp. 92–111. Springer, Cham (2024)
19. Sellergren, A., et al.: Medgemma technical report. arXiv preprint arXiv:2507.05201 (2025)
20. Solovyev, R., Wang, W., Gabruseva, T.: Weighted boxes fusion: ensembling boxes from different object detection models. Image Vis. Comput. **107**, 104117 (2021)
21. Team, G., et al.: Gemma 3 technical report. arXiv preprint arXiv:2503.19786 (2025)
22. Tschannen, M., et al.: Siglip 2: multilingual vision-language encoders with improved semantic understanding, localization, and dense features. arXiv preprint arXiv:2502.14786 (2025)
23. Wan, B., et al.: Locca: visual pretraining with location-aware captioners. Adv. Neural. Inf. Process. Syst. **37**, 116355–116387 (2024)
24. Xiao, R., Kim, S., Georgescu, M.I., Akata, Z., Alaniz, S.: Flair: vlm with fine-grained language-informed image representations. In: Proceedings of the Computer Vision and Pattern Recognition Conference, pp. 24884–24894 (2025)
25. Yang, Z., et al.: An empirical study of gpt-3 for few-shot knowledge-based VQA. In: Proceedings of the AAAI Conference on Artificial Intelligence, vol. 36, pp. 3081–3089 (2022)
26. Zambrano Chaves, J.M., et al.: A clinically accessible small multimodal radiology model and evaluation metric for chest x-ray findings. Nat. Commun. **16**(1), 3108 (2025)
27. Zhang, S., et al.: A multimodal biomedical foundation model trained from fifteen million image–text pairs. NEJM AI **2**(1), AIoa2400640 (2025)
28. Zhang, T., et al.: Radir: a scalable framework for multi-grained medical image retrieval via radiology report mining. In: International Conference on Medical Image Computing and Computer-Assisted Intervention, pp. 508–518. Springer, Cham (2025)

Measuring What VLMs Don't Say: Validation Metrics Hide Clinical Terminology Erasure in Radiology Report Generation

Aditya Parikh(✉), Aasa Feragen, Sneha Das, and Stella Frank

DTU Compute, Technical University of Denmark, Kongens Lyngby, Denmark
adipa@dtu.dk

Abstract. Reliable deployment of Vision-Language Models (VLMs) in radiology requires validation metrics that go beyond surface-level text similarity to ensure clinical fidelity and demographic fairness. This paper investigates a critical blind spot in current model evaluation: the use of decoding strategies that lead to high aggregate token-overlap scores despite succumbing to **template collapse**, in which models generate only repetitive, safe generic text and omit clinical terminology. Unaddressed, this blind spot can lead to **metric gaming**, where models that perform well on benchmarks prove clinically uninformative. Instead, we advocate for lexical diversity measures to check model generations for clinical specificity. We introduce Clinical Association Displacement (CAD), a vocabulary-level framework that quantifies shifts in demographic-based word associations in generated reports. Weighted Association Erasure (WAE) aggregates these shifts to measure the clinical signal loss across demographic groups. We show that deterministic decoding produces high levels of semantic erasure, while stochastic sampling generates diverse outputs but risks introducing new bias, motivating a fundamental rethink of how "optimal" reporting is defined. Code available at [19,20].

Keywords: Radiology Report Generation · Validation · Metrics · Fairness · Bias · Semantic Erasure · VLMs · ReXGradient

1 Introduction and Related Work

Radiology report generation (RRG) aims to automate the conversion of medical images into clinically actionable text, reduce documentation burden, and support diagnostic decision-making [5,23]. While recent VLMs have made RRG more viable in clinical workflows [16], in this paper we demonstrate that there is a major gap between scoring well on standard text similarity metrics and providing better clinical reporting or more equitable outcomes across patient populations.

Metrics That Enable Gaming. Contemporary RRG evaluation relies on word-overlap metrics like BLEU [18], ROUGE [13], BERTScore [28], measuring

G. Yang et al. (Eds.): MICCAI 2026, LNCS 16878, pp. 281–291, 2027.
https://doi.org/10.1007/978-3-032-38059-3_27

aggregate agreement between generated and reference (ground-truth) reports. Models evaluated on such metrics are vulnerable to **template collapse**: generating high-frequency "normal finding" templates that maximise token overlap despite omitting patient-specific details [26]. As normal radiographs dominate datasets, such generic outputs can achieve deceptively high scores with low diagnostic value [7]. Standard metrics **cannot detect terminology redistribution**, in which group-specific clinical terms are suppressed, over-generated, or reassigned across demographic groups. The risk increases with VLMs' documented tendency to under-utilize visual evidence in favor of language priors [10], which can lead to more generic outputs reproducing demographic biases. Current evaluation protocols focus on aggregate accuracy, leaving such biases undetected.

Decoding as a Confounding Variable. The inference strategy (deterministic/stochastic decoding) directly shapes the diversity and specificity of generated reports [22]. Deterministic decoding optimises for mode selection, converging to a small set of high-probability templates; stochastic decoding introduces controlled randomness to explore lower-probability tokens. Prior work [12] has motivated stochastic decoding based on marginal gains in BLEU and ROUGE scores. However, selecting decoding strategies solely on average-metric improvements obscures critical fairness trade-offs: while stochasticity may reduce template collapse, it can simultaneously introduce or amplify demographic associations that were absent in reference reports [22]. Existing evaluation frameworks provide no mechanism to audit these lexical shifts.

The Need for Complementary Metrics. Recent work has called for evaluation methods that go beyond surface similarity to measure clinical utility, demographic equity, and semantic fidelity [6]. Clinician-aligned scoring (e.g., RaTEScore [30]) offers one avenue by incorporating expert judgment, but it does not directly quantify how models redistribute clinical terminology across demographic groups. Fairness metrics in NLP often rely on group-wise F1 gaps or counterfactual perturbation tests. What is missing is a **vocabulary-level diagnostic** that explicitly tracks whether the demographic association structure of clinical language is preserved, erased, or distorted during generation.

Our Contributions. (i) We propose **Clinical Association Displacement**, a vocabulary-level framework quantifying shifts in group-associated word usage between human-authored and model-generated reports, identifying erased terminology as well as emergent and preserved bias; (ii) We introduce **Weighted Association Erasure**, a summary statistic that measures the global magnitude of association displacement, diagnosing whether a model systematically avoids clinically informative language; (iii) Using **our radiology VLM ReX-QWEN** as a testbed, we conduct an inference sensitivity analysis, showing that deterministic decoding inflates scores via *template collapse* (high erasure) while stochastic decoding recovers vocabulary but may increase bias; (iv) our baseline comparisons reveal fairness failures invisible to conventional metrics.

Our Work Establishes that the choice of decoding strategy and metric jointly confounds clinical evaluation. By auditing both what *models say* and systematically *don't say*, we enable transparent evaluation of whether RRG progress reflects genuine clinical utility or evaluation artifacts.

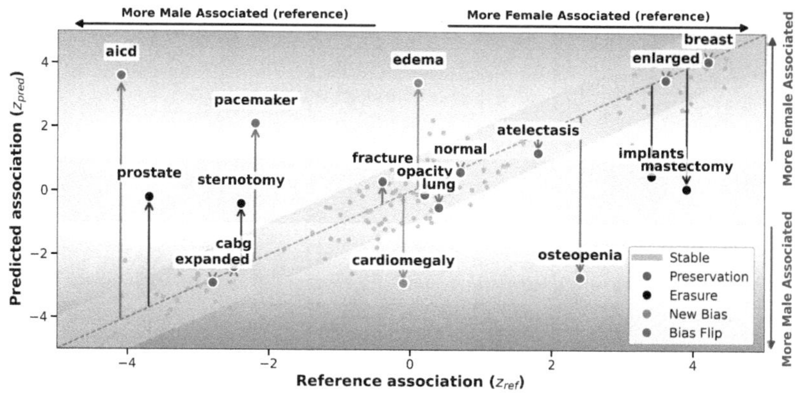

Fig. 1. Clinical interpretation of CAD. Each point is a vocabulary term, mapped from its demographic association in the reference corpus (z_{ref}, x-axis) to its association in predictions (z_{pred}, y-axis). Points near the diagonal (gray band) indicate stable associations, while off-diagonal points highlight terms whose demographic association has shifted, reflecting erasure, new bias, or bias flips.

2 Methods

We introduce a corpus-level diagnostic of global lexical shift, quantifying how vocabulary is preserved, erased, or newly introduced during generation. Such shifts may involve clinically meaningful terms that are differentially associated with demographic groups and therefore relevant for diagnostic interpretation. We therefore also introduce diagnostics for demographic-specific shifts, exposing shifts in lexical usage that are not captured by aggregate metrics. We formalize this through: Clinical Association Displacement (CAD) (Sect. 2.1) and Weighted Association Erasure (WAE) (Sect. 2.2).

2.1 Clinical Association Displacement (CAD)

CAD is a vocabulary-level diagnostic framework that quantifies how the demographic association, both in polarity and magnitude, of individual clinical terms shifts between a human-reference corpus and generated outputs (see Fig. 1).

Problem Formulation and Log Ratio: Let $\mathcal{D} \in \{\text{ref}, \text{pred}\}$ denote the reference and prediction corpora, where each report is associated with a binary group, here sex: $g \in \{F, M\}$. For each corpus $\mathcal{D}$, we compute, for each word w in the vocabulary $\mathcal{V}$, token counts $c_g^{(\mathcal{D})}(w)$ counting the occurrences of w in reports with sex g.

The total token count per group is $N_g^{(\mathcal{D})} = \sum_{w \in \mathcal{V}} c_g^{(\mathcal{D})}(w)$.

To measure how strongly a word w is associated with female vs. male reports in a corpus $\mathcal{D}$, we use a Dirichlet-smoothed log ratio of unigram probabilities [15]. Let $\alpha > 0$ (default: $\alpha = 0.1$) be a smoothing parameter (where $\tilde{}$ denotes α-smoothed estimates) and $V = |\mathcal{V}|$ the vocabulary size. We define:

$$\forall g \in \{F, M\}: \quad p_g^{(\mathcal{D})}(w) = \frac{c_g^{(\mathcal{D})}(w)+\alpha}{N_g^{(\mathcal{D})}+\alpha V} = \frac{\tilde{c}_g^{(\mathcal{D})}(w)}{\sum_{w' \in \mathcal{V}} \tilde{c}_g^{(\mathcal{D})}(w')}, \quad s^{(\mathcal{D})}(w) = \log \frac{p_F^{(\mathcal{D})}(w)}{p_M^{(\mathcal{D})}(w)}.$$

$s^{(\mathcal{D})}(w) > 0$ indicates a *female* association for w, while $s^{(\mathcal{D})}(w) < 0$ indicates that w is more associated to *male*. We apply delta-method approximation under a multinomial model to estimate the variance of $s^{(\mathcal{D})}(w)$:

$$\tilde{\mathrm{Var}}[s^{(\mathcal{D})}(w)] \approx \frac{1}{\tilde{c}_F^{(\mathcal{D})}(w)} + \frac{1}{\tilde{c}_M^{(\mathcal{D})}(w)}, \quad z^{(\mathcal{D})}(w) = \frac{s^{(\mathcal{D})}(w)}{\sqrt{\tilde{\mathrm{Var}}[s^{(\mathcal{D})}(w)]}},$$

yielding a standardized association $z^{(\mathcal{D})}(w)$ adjusting for frequency variance. We denote $s^{(\mathrm{ref})}(w)$, $s^{(\mathrm{pred})}(w)$ and $z^{(\mathrm{ref})}(w)$, $z^{(\mathrm{pred})}(w)$ for reference and predictions.

Quantifying Displacement: We define **Association Displacement** $z_{\mathrm{disp}}(w)$ as the standardized difference between prediction and reference associations, and a critical threshold z_{crit} for significance testing:

$$z_{\mathrm{disp}}(w) = \frac{s^{(\mathrm{pred})}(w) - s^{(\mathrm{ref})}(w)}{\sqrt{\tilde{\mathrm{Var}}[s^{(\mathrm{pred})}(w)] + \tilde{\mathrm{Var}}[s^{(\mathrm{ref})}(w)]}}, \quad z_{\mathrm{crit}} = \Phi^{-1}\left(1 - \frac{p^\star}{2}\right).$$

Assuming $\mathcal{D} \in \{\mathrm{ref}, \mathrm{pred}\}$ are independently identically distributed (IID), the null hypothesis is $z_{\mathrm{disp}}(w) \sim \mathcal{N}(0,1)$. We convert $z_{\mathrm{disp}}(w)$ to two-sided p-values, apply Benjamini–Hochberg FDR correction [2], and flag words with $|z_{\mathrm{disp}}(w)| > z_{\mathrm{crit}}$ at significance level $p^\star$ (default: $p^\star = 0.05$), where Φ^{-1} is the standard normal quantile function.

Clinical Categories of Shift: For each vocabulary term w we compute $z^{(\mathrm{ref})}(w)$, $z^{(\mathrm{pred})}(w)$ and $z_{\mathrm{disp}}(w)$. Words with $|z_{\mathrm{disp}}| < |z_{\mathrm{crit}}|$ are deemed **stable** (no-displacement) and excluded from further audit. Displaced terms fall into three categories: **(i) Erasure:** A word is strongly associated in the reference ($|z^{(\mathrm{ref})}| > z_{\mathrm{strong}}$) but becomes weak, neutral or is not generated ($|z^{(\mathrm{pred})}| < z_{\mathrm{neutral}}$), indicating loss of clinical signal. **(ii) Emergent Bias:** The model introduces group-associations absent in the reference (**New Bias**: $|z^{(\mathrm{ref})}| < z_{\mathrm{neutral}}$, $|z^{(\mathrm{pred})}| \geq z_{\mathrm{neutral}}$) or reverses polarity (**Bias Flip**: $\mathrm{sign}(z^{(\mathrm{ref})}) \neq \mathrm{sign}(z^{(\mathrm{pred})})$, both non-neutral), indicating hallucinated demographic correlations. **(iii) Preservation:** Polarity is conserved ($\mathrm{sign}(z^{(\mathrm{ref})}) = \mathrm{sign}(z^{(\mathrm{pred})})$) without loss of demographic association.

We set $z_{\mathrm{neutral}} = 1$ & $z_{\mathrm{strong}} = 2$ (corresponding to two-tailed normal cutoffs of $p \approx 0.32$ & $p \approx 0.05$), so only statistically reliable deviations from neutrality are treated as clinically meaningful associations.

2.2 Weighted Association Erasure (WAE)

While CAD tracks displacement for individual words, WAE aggregates these term-level shifts to provide a single, global measure of clinical signal loss across the entire vocabulary, weighted by clinical relevance.

$$\text{WAE} = \frac{\sum_{w \in \mathcal{V}} \omega(w) \cdot z_{\text{disp}}(w)^2}{\sum_{w \in \mathcal{V}} \omega(w)}$$

where the numerator represents the *aggregate displacement energy* scaled by a term importance factor $\omega(w)$ (approximated by term frequency). The denominator is normalization factor to adjust for the total number of predicted tokens, enabling WAE to be compared across output corpora of different lengths.

We define the weight $\omega(w) = c^{(\text{pred})}(w)$ (prediction-based) or $\omega(w) = c^{(\text{ref})}(w)$ (reference-based). Prediction-based weighting quantifies shifts within text the model *actually produces*, avoiding penalties for omitted vocabulary. Reference-based weighting reveals which clinically important terms (frequent in human reports) are systematically avoided. Aggregating over terms associated with group g in the reference yields a group-specific score WAE_g, and disparity is defined as ΔWAE (e.g., $\Delta\text{WAE} = \Delta\text{WAE}_F - \Delta\text{WAE}_M$).

Uncertainty is estimated via nonparametric bootstrap; statistical calibration is confirmed via label-permutation tests, where observed WAE consistently exceeds the null distribution under random demographic assignment.

3 Experiments

Dataset: We use **ReXGradient-160K**, the largest publicly available multi-site chest X-ray dataset [29], which includes clinician-authored findings and impressions. We filter frontal views using a foundational model [14] and create a stratified test-subset of $N = 2,000$ with balanced sex representation.

Models: ReX-QWEN: We fine-tune Qwen3-VL-8B-Instruct [25] on ReXGradient using LoRA [9](rank $r = 16$, $\alpha = 32$) targeting attention and MLP layers. Input images are resized to 512×512 RGB. Our primary model, **ReX-BB** (Balanced Baseline) is trained on 60K sex-balanced examples (50% male, 50% female) as our primary model. Two sex-skewed variants: ReX-MB (67% male) and ReX-FB (67% female) validate CAD/WAE sensitivity to demographic shifts. We compare ReX-QWEN against three models under identical evaluation protocols: **CXR-LLaVA** [11], **LLaVA-Rad** [27] (medical VLMs for radiology reporting); and **OpenAI GPT-4.1** [17], prompted zero-shot for report generation.

Inference Strategy Study: We hypothesise that semantic erasure stems from inference strategies that systematically suppress clinical terminology to minimise generation risk. We partition inference strategies into **deterministic** (greedy, beam search), which select the highest-probability sequences and may lead to template collapse, and **stochastic** (temperature T, top-k, nucleus sampling), which sample from truncated distributions to increase lexical diversity.

Specifically, we evaluate: (i) Greedy (argmax), (ii) Beam Search ($k = 4$), (iii) Conservative (Low) Sampling $T = 0.3$, top-$p = 0.95$), (iv) Rich Sampling ($T = 0.7$, top-$p = 0.95$), (v) Top-k ($k = 50$, $T = 1.0$), and (vi) Nucleus Sampling (top-$p = 0.9$, $T = 1.0$). All use max tokens=512 with caching enabled.

Evaluation Metrics: We evaluate models across three dimensions: **Clinical Accuracy:** We report BERTScore (F1, Precision, Recall) using CheXbert as the backbone model [24], ROUGE-1/L, and RateScore to measure semantic and surface-level similarity against reference reports. **Textual Diversity:** Following [8], we compute: (i) Template Diversity (fraction of unique reports), (ii) Type-Token Ratio (TTR), (iii) 1-Self-BLEU, and (iv) Semantic Diversity. **Our Metrics:** We report: (i) our proposed CAD-flagged clinical categories (Erasure, New Bias, Bias Flip) quantifying vocabulary-level association shifts, (ii) our proposed WAE summarizing global displacement magnitude, (iii) the existing WEAT [1,3] and its effect size (ES) as a comparative word-association baseline, and (iv) Disparity Impact Ratio (ΔDIR) [21], the change in male-to-female token ratio between reference and prediction. Stop words are removed prior to all association-based metric computation to ensure focus on clinical terms.

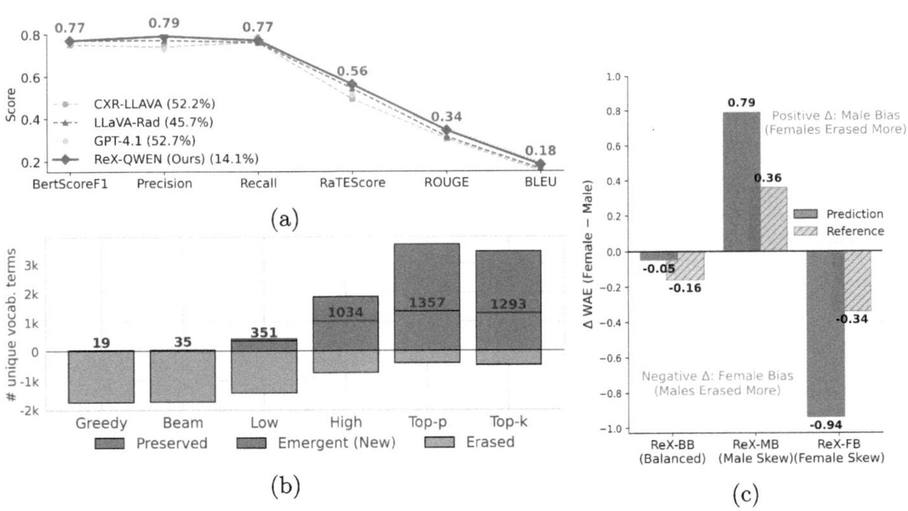

Fig. 2. (a) Clinical performance comparison (w/ Template Diversity %). (b) Global lexical shift across inference strategies. (c) ΔWAE captures semantic shifts caused by sex-skewed training.

4 Results and Discussion

We establish that while ReX-QWEN achieves competitive baseline performance (see Fig. 2a), severe template collapse remains evident (Template Diversity 14.1–52.7%). This collapse motivates our subsequent lexical analysis, specifically investigating how decoding choices drive a fundamental trade-off between semantic erasure and emergent bias. To evaluate these dynamics, we compare ReX-BB (sex-balanced) across six distinct inference strategies (Table 1).

Global Lexical Collapse. Deterministic strategies exhibit severe lexical attrition, generating vocabularies of just 22–38 unique terms, 5% of the original reference vocabulary (see Fig. 2b), collapsing instead to "normal-finding" templates. Stochastic approaches preserve vocabulary size but introduce emergent terms absent in references, raising validity concerns: are these recovered clinical signals or hallucinated noise? To determine if these shifts disproportionately affect specific patient groups, we turn to CAD to audit for semantic erasure and emergent bias in sex-associated terms.

Deterministic Decoding Inflates Similarity via Collapse. Greedy and beam search produce 0.3–0.5% unique reports, presenting a case of *template collapse*. Despite BERTScore F1=0.78, CAD reveals severe sex-association erasure: 32–35 erased terms, corresponding to nearly all the highly-biased terms in the references. For example, *AICD*, *prostate* (male-associated), and *vasculature*, *pregnancy*, *mastectomy* (strongly female-associated) shift their predicted association to neutral. This metric-gaming artifact confirms that high aggregate scores can mask systematic lexical erasure of terminology present in reference reports.

Stochastic Decoding Restores Vocabulary But Introduces bias. Temperature scaling increases diversity (36.3–100%) and reduces erasure but introduces 25–65 new bias terms and elevates WAE_{pred} (2.50 → 33.81) due to the emergence of newly biased terms. ΔDIR widens, signaling verbosity imbalances across sex groups. For instance, the pathology *pneumothorax* is neutral in human-reference ($z_{ref} = 0.15$); however low temperature generation actively hallucinated a strong association with male patients ($z_{pred} = -5.69$); more examples of induced biases: *calcification* (+1.12 → −0.24), *pectoralis* (−0.17 → 2.88), *asthma* (−2.96 → 2.54), *contours* (+7.47 → −3.79), *hyperinflation* (−0.22 → −3.77), *fractures* (−8.04 → 1.93). WEAT evaluates static latent embeddings and therefore remains entirely blind to decoding-induced biases.

Implications. CAD identifies two opposing failure modes: **(i) semantic erasure** under deterministic decoding: systematic vocabulary suppression invisible to BERTScore/ROUGE; **(ii) emergent bias** under high-temperature sampling: new demographic associations absent in references. No single strategy simultaneously optimises accuracy, diversity, and fairness, motivating fairness-aware decoding selection tuned jointly on clinical accuracy *and* CAD/WAE diagnostics rather than defaulting to greedy inference. **We recommend** augmenting standard evaluation with CAD categories, WAE, and lexical diversity metrics to prevent interpreting template-collapse artifacts as genuine clinical improvement.

Table 1. Inference strategy sensitivity. (**Bold**: best; <u>Underline</u>: worst).

Metric Category	Ref.	Deterministic		Stochastic (Increasing Entropy)			
		Greedy	Beam	$T = 0.3$	$T = 0.7$	Top-k	Top-p
Surface Level Similarity							
BERTScore F1 ↑	–	**0.78**	**0.78**	0.76	0.73	0.71	<u>0.69</u>
BERTScore P/R ↑	–	0.80/0.75	**0.81/0.76**	0.78/0.75	0.73/0.74	<u>0.68/0.74</u>	0.70/0.73
ROUGE (1/L) ↑	–	0.34/0.29	**0.35/0.30**	0.32/0.26	0.26/0.19	<u>0.20/0.14</u>	0.21/0.15
RateScore ↑	–	**0.55**	**0.55**	0.54	0.48	<u>0.44</u>	<u>0.44</u>
Textual Diversity & Collapse							
Template Div. (%) ↑	82.45	<u>0.3</u>	0.5	36.3	97.05	**100.0**	**100.0**
TTR (%) ↑	2.915	<u>0.006</u>	0.131	0.654	2.128	2.540	**2.781**
1 - Self-BLEU (%) ↑	35.70	1.05	<u>0.20</u>	10.78	33.78	46.73	**46.84**
Semantic Div. (%) ↑	85.23	12.16	<u>5.41</u>	55.51	83.75	85.80	**86.19**
Clinical Categories (Ours)							
Erasure (#) ↓	–	32	<u>35</u>	17	4	5	**3**
New Bias (#) ↓	–	**0**	3	41	45	25	<u>65</u>
Bias Flip (#) ↓	–	**0**	1	16	33	44	<u>69</u>
Fairness							
WAE (pred) (Ours) ↓	–	**2.49**	2.50	4.95	33.81	12.90	<u>13.27</u>
WAE (ref) (Ours) ↓	–	**0.19**	0.60	3.43	9.06	4.60	<u>14.70</u>
WEAT (S/ES) ↓	–	-1.46/-0.64	1.10/0.03	-1.35/-0.01	4.30/-0.07	-2.13/-0.02	0.27/0.00
ΔDIR ↓	–	-0.03	**0.01**	-0.08	-0.01	-0.20	<u>-0.32</u>

Generalisation and Robustness: LLaVA-Rad on ReXGradient and CheXpert [4] confirms the pattern (Table 2): deterministic decoding reduces template diversity (45.7% vs. 91.1%) with higher erasure (19 terms), while stochastic sampling introduces bias flips (5→12). WAE drops modestly (1.17→1.03 on ReXGradient), confirming the trade-off persists across architectures and datasets. Sex-skewed training (ReX-MB, ReX-FB) leaves conventional metrics stable (F1 (BertScore)$\in$ [0.76, 0.78], RateScore$\in$ [0.54, 0.57], WEAT-ES$\approx$ 0). However, $\Delta\text{WAE}_{\text{pred}}$ goes from near-zero (ReX-BB: -0.05) to +0.79 for ReX-MB and −0.94 for ReX-FB, confirming WAE captures semantic shifts invisible to existing metrics (refer Fig. 2c). **For Ablation**, we sweep $p^{\star}_{\text{disp}}$ across its full range strictly preserving the relative ranking of all decoding strategies. This ablation confirms that the erasure-versus-bias trade-off is a fundamental generation mechanic, rather than an artifact of metric thresholding.

Table 2. External validation confirms the persistence of the erasure-versus-bias trade-off using LLaVA-Rad with different decoding strategies Text similarity metrics (BERTScore, ROUGE etc.) are comparable ($\Delta \leq 2\%$, unshown), while Diversity metrics show clear differences in template usage (Div. Temp.) and erasure rates. (NB: New Bias; BF: Bias Flip)

Metrics	ReXGradient		CheXpert	
	Greedy	Rich	Greedy	Rich
Div. (Temp./Sem.%) ↑	45.70/70.32	91.05/79.49	94.05/80.60	100/83.13
Erasure/NB/BF ↓	19/6/5	11/5/11	6/1/0	1/1/4
WAE (pred/ref) ↓	1.17/0.84	1.03/0.93	0.76/0.60	0.84/0.69
WEAT (S/ES) ↓	1.13/0.01	-0.36/-0.00	-0.78/0.11	-0.14/0.17

5 Conclusion and Future Work

We demonstrate that decoding strategy and metric choice jointly confound evaluation in radiology report generation. By introducing CAD and WAE, we provide a framework to audit semantic erasure and emergent bias that remain completely invisible to conventional metrics. Our findings are diagnostic at the corpus and lexical level. Together, these tools enable transparent evaluation, allowing the field to distinguish genuine clinical utility from the illusion of performance caused by *template collapse*. **Future Work:** Our analysis focuses on binary sex associations; future work should extend CAD to age, ethnicity, and intersectional groups. Clinician validation is needed to confirm whether CAD-flagged erasures correspond to clinically meaningful omissions.

Acknowledgment. This work was funded by the Novo Nordisk Foundation (project number 0087102).

Disclosure of Interests. The authors have no competing interests to declare.

References

1. Badilla, P., Bravo-Marquez, F., Zambrano, M.J., Pérez, J.: Wefe: a python library for measuring and mitigating bias in word embeddings. J. Mach. Learn. Res. **26**(156), 1–6 (2025). http://jmlr.org/papers/v26/22-1133.html
2. Benjamini, Y., Hochberg, Y.: Controlling the false discovery rate: a practical and powerful approach to multiple testing. J. Roy. Stat. Soc.: Ser. B (Methodol.) **57**(1), 289–300 (1995)
3. Caliskan, A., Bryson, J.J., Narayanan, A.: Semantics derived automatically from language corpora contain human-like biases. Science **356**(6334), 183–186 (2017)
4. Chambon, P., et al.: Chexpert plus: augmenting a large chest x-ray dataset with text radiology reports, patient demographics and additional image formats. arXiv preprint arXiv:2405.19538 (2024)

5. Demner-Fushman, D., Chapman, W.W., McDonald, C.J.: What can natural language processing do for clinical decision support? J. Biomed. Inform. **42**(5), 760–772 (2009)
6. Du, X., et al.: Testing and evaluation of generative large language models in electronic health record applications: a systematic review. J. Am. Med. Inform. Assoc., ocaf233 (2026)
7. Gao, L., Zhang, L., Liu, C., Wu, S.: Handling imbalanced medical image data: a deep-learning-based one-class classification approach. Artif. Intell. Med. **108**, 101935 (2020)
8. Guo, Y., Shang, G., Vazirgiannis, M., Clavel, C.: The curious decline of linguistic diversity: Training language models on synthetic text. In: Findings of the Association for Computational Linguistics: NAACL 2024, pp. 3589–3604 (2024)
9. Hu, E.J., et al.: Lora: low-rank adaptation of large language models. ICLR **1**(2), 3 (2022)
10. Lee, K.i., et al.: Vlind-bench: measuring language priors in large vision-language models. In: Findings of the Association for Computational Linguistics: NAACL 2025, pp. 4129–4144 (2025)
11. Lee, S., Youn, J., Kim, H., Kim, M., Yoon, S.H.: CXR-LLAVA: a multimodal large language model for interpreting chest x-ray images. Eur. Radiol. **35**(7), 4374–4386 (2025)
12. Li, Q.: Harnessing the power of pre-trained vision-language models for efficient medical report generation. In: Proceedings of the 32nd ACM International Conference on Information and Knowledge Management. CIKM 2023, pp. 1308–1317. Association for Computing Machinery, New York, NY, USA (2023). https://doi.org/10.1145/3583780.3614961
13. Lin, C.Y.: Rouge: a package for automatic evaluation of summaries. In: Text Summarization Branches Out, pp. 74–81 (2004)
14. Ma, D., Pang, J., Gotway, M.B., Liang, J.: A fully open ai foundation model applied to chest radiography. Nature **643**(8071), 488–498 (2025)
15. Monroe, B.L., Colaresi, M.P., Quinn, K.M.: Fightin'words: lexical feature selection and evaluation for identifying the content of political conflict. Polit. Anal. **16**(4), 372–403 (2008)
16. Nam, Y., et al.: Multimodal large language models in medical imaging: current state and future directions. Korean J. Radiol. **26**(10), 900 (2025)
17. OpenAI: Introducing gpt-4.1 in the API (2025). https://openai.com/index/gpt-4-1/. Accessed 24 Feb 2026
18. Papineni, K., Roukos, S., Ward, T., Zhu, W.J.: Bleu: a method for automatic evaluation of machine translation. In: Proceedings of the 40th Annual Meeting of the Association for Computational Linguistics, pp. 311–318 (2002)
19. Parikh, A.: Clinical-cad: clinical association displacement (2026). https://pypi.org/project/clinical-cad/
20. Parikh, A.: ReX-QWEN: radiology VLM for clinical association displacement framework (2026). https://github.com/ADE-17/ReX-QWEN---Radiology-VLM-for-Clinical-Association-Displacement
21. Parikh, A., Das, S., Feragen, A.: Investigating label bias and representational sources of age-related disparities in medical segmentation. arXiv preprint arXiv:2511.00477 (2025)
22. Presacan, O., Nik, A., Thambawita, V., Ionescu, B., Riegler, M.: A comparative study of decoding strategies in medical text generation. In: International Conference on Multimedia Modeling, pp. 1–15. Springer (2026)

23. Rocha, D.M., Brasil, L.M., Lamas, J.M., Luz, G.V., Bacelar, S.S.: Evidence of the benefits, advantages and potentialities of the structured radiological report: an integrative review. Artif. Intell. Med. **102**, 101770 (2020)
24. Smit, A., Jain, S., Rajpurkar, P., Pareek, A., Ng, A.Y., Lungren, M.P.: Chexbert: combining automatic labelers and expert annotations for accurate radiology report labeling using Bert (2020)
25. Team, Q.: Qwen3 technical report (2025). https://arxiv.org/abs/2505.09388
26. Yun, L., An, C., Wang, Z., Peng, L., Shang, J.: The price of format: diversity collapse in LLMs. arXiv preprint arXiv:2505.18949 (2025)
27. Zambrano Chaves, J.M., et al.: A clinically accessible small multimodal radiology model and evaluation metric for chest x-ray findings. Nat. Commun. **16**(1), 3108 (2025)
28. Zhang, T., Kishore, V., Wu, F., Weinberger, K.Q., Artzi, Y.: Bertscore: evaluating text generation with Bert. arXiv preprint arXiv:1904.09675 (2019)
29. Zhang, X., Acosta, J.N., Miller, J., Huang, O., Rajpurkar, P.: Rexgradient-160k: a large-scale publicly available dataset of chest radiographs with free-text reports. arXiv preprint arXiv:2505.00228 (2025)
30. Zhao, W., Wu, C., Zhang, X., Zhang, Y., Wang, Y., Xie, W.: Ratescore: a metric for radiology report generation. In: Proceedings of the 2024 Conference on Empirical Methods in Natural Language Processing, pp. 15004–15019 (2024)

MEC: A Multi-expert Consultation Framework for Synergizing Frozen Foundation Models in Whole Slide Image Analysis

Zongyi Chen[1], Yu Liang[2], Weiping Lin[3], Rui Chen[4], Baptiste Magnier[5,6], Xi Wang[7], and Liansheng Wang[1,3,8](✉)

[1] National Institute for Data Science in Health and Medicine, Xiamen University, Xiamen, China
zongyichen@stu.xmu.edu.cn
[2] Institute of Artificial Intelligence, Xiamen University, Xiamen, China
[3] Department of Computer Science, School of Informatics, Xiamen University, Xiamen, China
wplin@stu.xmu.edu.cn, lswang@xmu.edu.cn
[4] Renji Hospital, Shanghai Jiao Tong University School of Medicine, Shanghai, China
drchenrui@foxmail.com
[5] EuroMov Digital Health in Motion, Univ Montpellier, IMT Mines Ales, Ales, France
baptiste.magnier@mines-ales.fr
[6] Service de Médecine Nucléaire, Centre Hospitalier Universitaire de Nîmes, University of Montpellier, Nîmes, France
[7] Department of Computer Science and Engineering, Hong Kong University of Science and Technology, Hong Kong, China
vancywangxi@ust.hk
[8] Chongqing Jinfeng Laboratory, Chongqing, China

Abstract. The integration of Pathology Foundation Models (PFMs) with Multiple Instance Learning (MIL) has emerged as a prevailing paradigm for Whole Slide Image (WSI) analysis. However, due to diverse pre-training architectures and data, PFMs exhibit intrinsic heterogeneity and inconsistent performance across tasks. This limitation mirrors the challenge of clinical inter-observer variability, where relying on a single pathologist introduces the risk of subjective bias. To address this, we propose MEC, a Multi-Expert Consultation framework compatible with arbitrary weight-based MIL aggregators, designed to simulate the clinical consultation workflow. Specifically, MEC leverages diverse frozen PFMs as digital experts to independently localize task-specific discriminative regions within heterogeneous feature spaces. To broaden the overall scope of collective diagnostic evidence, we introduce a Stochastic Evidence Masking strategy that suppresses top-ranking features, thereby surfacing a wider variety of underlying cues. Finally, MEC synthesizes a robust, multi-view diagnostic decision by aggregating the collective consultation consensus with individual expert opinions. Extensive experiments demonstrate that MEC outperforms individual PFM baselines and

G. Yang et al. (Eds.): MICCAI 2026, LNCS 16878, pp. 292–302, 2027.
https://doi.org/10.1007/978-3-032-38059-3_28

alternative ensemble strategies, offering a practical framework to harness collective diagnostic potential without requiring additional large-scale pre-training. Code is available at https://github.com/JoeyFe/MEC.

Keywords: Whole slide image · Pathology Foundation Models · Multiple Instance Learning

1 Introduction

Computational pathology has revolutionized diagnostic workflows by facilitating automated histological sub-typing and prognostic analysis on gigapixel Whole Slide Images (WSIs) [6,21]. To mitigate the computational burden imposed by the massive resolution of WSIs, a prevailing paradigm has emerged that employs frozen Pathology Foundation Models (PFMs) for feature extraction, followed by Multiple Instance Learning (MIL) aggregation [25]. This strategy leverages the robust representation capabilities of PFMs to achieve superior performance in downstream tasks [2,15,17].

However, due to diverse pre-training data and architectures, PFMs exhibit intrinsic heterogeneity and distinct inductive biases [2]. Recent comprehensive benchmarks show that no single model is consistently superior [2,15,17], mirroring clinical inter-observer variability [5,18] where single-pathologist reliance risks subjective bias. Leveraging collaborative consensus among heterogeneous experts should yield superior performance. Existing solutions generally fall into three categories (Fig. 1). First, model selection [8,26] heuristically searches for the optimal PFM-MIL combination; however, this process is computationally exhaustive and discards complementary expertise inherent in sub-optimal models. Second, feature alignment [9,13,14,27] constructs a unified feature space from multiple experts (e.g., knowledge distillation); however, these task-agnostic methods often overlook task-specific nuances and require extensive post-training resources. Third, naive ensembles [2,15] combine predictions via voting or averaging, overlooking deep feature interactions and lacking robustness as noise from weaker experts frequently dilutes reliable insights.

Fundamentally, existing ensemble approaches function by directly fusing latent features or final predictions in a black-box manner. Such a process frequently overlooks that the localization of key diagnostic evidence represents a crucial component of both clinical workflows [10,16] and MIL paradigms [7,12]. To systematically capture the diverse pathological cues essential for multifaceted clinical diagnosis, we propose MEC, a Multi-Expert Consultation framework that emulates the workflow of evidence-based joint diagnosis. In contrast to conventional ensembles, MEC employs PFMs as digital experts, enabling the independent localization of task-specific regions to capture multi-view evidence. Furthermore, we introduce a Stochastic Evidence Masking (SEM) strategy that encourages the discovery of a broader range of complementary patterns. Finally, MEC dynamically aggregates the opinions of individual experts to produce a robust diagnostic consensus.

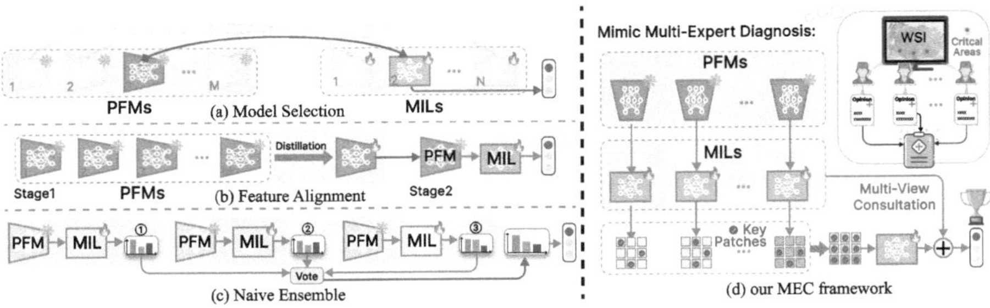

Fig. 1. Comparison of (a) Model Selection, (b) Feature Alignment, (c) Naive Ensembles, and (d) our MEC framework. MEC mimics clinical joint diagnosis, dynamically consulting diverse experts for an evidence-driven consensus.

Our contributions are summarized as follows: (1) We propose MEC, a generic WSI analysis framework compatible with arbitrary weight-based MIL aggregators. By leveraging frozen foundation models, it efficiently adapts to downstream tasks without requiring additional large-scale training. (2) We reveal the inherent complementarity of distinct PFMs in Region of Interest (ROI) mining and accordingly introduce a multi-view key patch mining strategy that fuses diverse evidence to construct robust task-specific representations. (3) Extensive benchmarking across multiple datasets demonstrates the consistent superiority of MEC over individual PFM baselines and ensemble strategies, validating it as a practical, general-purpose framework for harnessing collective diagnostic potential.

2 Method

The proposed Multi-Expert Consultation framework structurally simulates the clinical consensus process by synergizing heterogeneous PFMs to achieve robust and multi-view WSI analysis. The overall architecture is illustrated in Fig. 2.

2.1 Feature Extraction with Heterogeneous PFMs

Following standard WSI preprocessing protocols [12], the tissue area of a WSI is tessellated into N non-overlapping patches $\{p_i\}_{i=1}^{N}$. To exploit complementary representations, we employ a set of M heterogeneous and frozen PFMs, denoted by $\{f_m\}_{m=1}^{M}$. Each expert model f_m projects the patches into a distinct feature space $\mathbb{R}^{D_m}$. Specifically, $\mathbf{H}^{(m)} = [\mathbf{h}_1^{(m)}, \ldots, \mathbf{h}_N^{(m)}]^\top \in \mathbb{R}^{N \times D_m}$ represents the bag of features derived from the m-th expert, where $\mathbf{h}_i^{(m)} \in \mathbb{R}^{D_m}$ denotes the embedding of the i-th patch.

2.2 Task-Specific Evidence Localization

We employ a multi-branch architecture, where each frozen PFM is paired with a learnable MIL backbone. For the m-th branch, the feature bag $\mathbf{H}^{(m)} \in \mathbb{R}^{N \times D_m}$

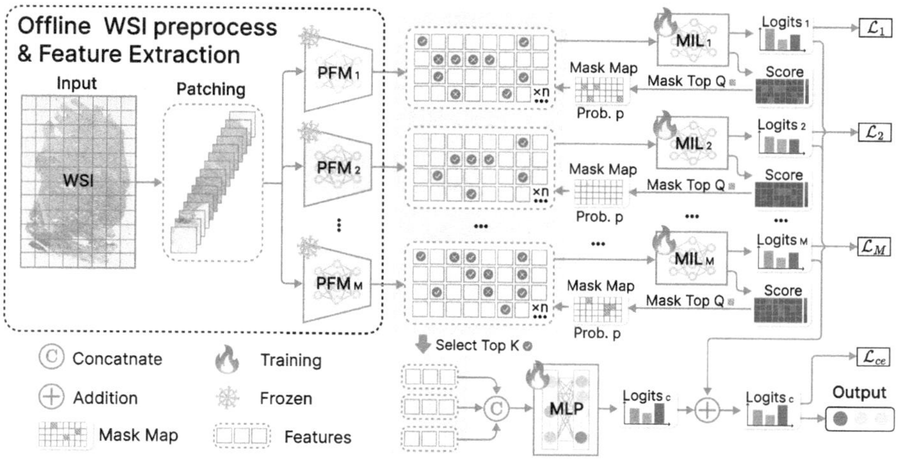

Fig. 2. Overview of the MEC framework utilizing frozen foundation model features as input for task-specific adaptation.

is processed by a dedicated MIL network. Our framework accommodates diverse MIL architectures that produce instance-level importance weights. Let $\Phi_m(\cdot;\theta_m)$ denote the instance scoring network (e.g., the widely adopted attention network) of the m-th expert. This network maps input embeddings to importance scores $\mathbf{A}^{(m)} \in \mathbb{R}^{1\times N}$, quantifying the diagnostic contribution of each patch:

$$\mathbf{A}^{(m)} = \Phi_m(\mathbf{H}^{(m)};\theta_m). \tag{1}$$

To drive the learning of task-aware representations, we impose independent supervision on each expert branch. Specifically, within the m-th branch, weighted patch-level features are aggregated into a slide-level representation $\mathbf{z}_{slide}^{(m)}$ via a global function $\mathcal{G}_m(\cdot)$, which accommodates diverse MIL architectures:

$$\mathbf{z}_{slide}^{(m)} = \mathcal{G}_m(\mathbf{H}^{(m)}, \mathbf{A}^{(m)};\theta_m). \tag{2}$$

Subsequently, the embedding $\mathbf{z}_{slide}^{(m)}$ is fed into a classifier $g_m(\cdot)$ to generate logits $\hat{\mathbf{y}}^{(m)}$. We enforce discriminative learning by applying a Cross-Entropy loss $\mathcal{L}_{CE}$ to each expert:

$$\mathcal{L}_{experts} = \frac{1}{M}\sum_{m=1}^{M} \mathcal{L}_{CE}(\hat{\mathbf{y}}^{(m)}, y), \tag{3}$$

where y denotes the ground-truth label. This explicit supervision aligns feature aggregation with the ground truth, yielding more reliable importance scores $\mathbf{A}^{(m)}$ for subsequent mining.

2.3 Stochastic Evidence Masking

We introduce a stochastic masking strategy inspired by regularization techniques [28,29]. While prior works employ this mechanism to prevent overfitting in single-view MIL, we repurpose it to expand the collective pool of diagnostic evidence. Specifically, in the training phase, we generate a binary mask $\mathbf{M}^{(m)} \in \{0,1\}^N$ to suppress the instances with the top-Q scores with a probability p. The masked importance scores $\tilde{\mathbf{A}}^{(m)}$ are then obtained via element-wise multiplication:

$$\tilde{\mathbf{A}}^{(m)} = \mathbf{A}^{(m)} \odot \mathbf{M}^{(m)}. \tag{4}$$

By preventing the model from collapsing onto a narrow set of dominant features, this perturbation ensures a richer basis for the subsequent multi-view integration.

2.4 Synthesizing Multi-view Predictions

Following the evidence mining phase, we aggregate multi-view diagnostic cues to establish a holistic consensus. Importance score distributions in MIL are inherently sparse, with a small subset of highly discriminative patches dominating the slide-level prediction [22,28]. Motivated by this observation, we focus on capturing these critical instances. Specifically, for the m-th expert branch, we utilize the masked importance scores $\tilde{\mathbf{A}}^{(m)}$ (derived in Eq. 4) to retrieve the top-K most critical instances. Let $\mathcal{I}^{(m)}$ denote the indices corresponding to the K highest values in $\tilde{\mathbf{A}}^{(m)}$, sorted in descending order. To fully leverage the rich semantic representations inherent in the foundation models, we extract the raw feature embeddings from the original feature bag $\mathbf{H}^{(m)}$ at these indices to construct a sequence of critical evidence $\mathbf{S}^{(m)} = [\mathbf{h}_i^{(m)}]_{i \in \mathcal{I}^{(m)}} \in \mathbb{R}^{K \times D_m}$. These selected instance features $\mathbf{S}^{(m)}$ are subsequently flattened into a vector $\mathbf{v}^{(m)} \in \mathbb{R}^{K \cdot D_m}$. To synthesize a unified representation, we concatenate the flattened vectors from all M experts:

$$\mathbf{V}_{fused} = \text{Concat}\left([\mathbf{v}^{(1)}, \mathbf{v}^{(2)}, \ldots, \mathbf{v}^{(M)}]\right). \tag{5}$$

The fused vector $\mathbf{V}_{fused}$, which encapsulates complementary evidence derived from diverse inductive biases, is processed by a Multi-Layer Perceptron (MLP) to obtain a high-level consensus representation. We formulate this feature transformation and the subsequent classification as:

$$\mathbf{z}_{fused} = \mathcal{F}_{mlp}(\mathbf{V}_{fused}; \Theta_{mlp}), \quad \hat{\mathbf{y}}_{consensus} = \mathcal{G}_{cls}(\mathbf{z}_{fused}), \tag{6}$$

where $\mathcal{F}_{mlp}(\cdot)$ denotes the non-linear mapping parameterized by Θ_{mlp}, and $\mathcal{G}_{cls}(\cdot)$ represents the final classification head projecting the refined features $\mathbf{z}_{fused}$ to the target class logits. Inspired by the bias-variance decomposition principles underlying ensemble learning [4], we employ a residual-style aggregation strategy in logit space to derive the final slide-level prediction $\hat{\mathbf{y}}_{final}$. This design fuses the high-capacity collaborative consensus $\hat{\mathbf{y}}_{consensus}$ with the stable ensemble of individual expert predictions, yielding a more robust final decision.

$$\hat{\mathbf{y}}_{final} = \hat{\mathbf{y}}_{consensus} + \frac{1}{M} \sum_{m=1}^{M} \hat{\mathbf{y}}^{(m)}. \tag{7}$$

Table 1. Performance evaluation of different feature extractors combined with various MIL models across diverse pathological tasks. Best results are highlighted in **bold**, and the best single-model results are underlined.

	Dataset	Camelyon17		TCGA-GBMLGG		TCGA-NSCLC		STS	
	Method	Acc	F1-score	Acc	F1-score	Acc	F1-score	Acc	F1-score
ABMIL	UNI	79.17±3.88	50.45±3.30	83.25±2.29	80.80±2.61	91.95±1.60	91.91±1.63	69.12±7.73	58.14±5.11
ABMIL	CONCHv1.5	83.53±1.53	57.74±4.07	83.69±2.60	81.21±2.85	92.71±1.92	92.68±1.95	72.57±7.02	61.78±3.68
ABMIL	Virchow2	83.79±2.29	59.55±4.63	83.31±2.94	81.01±4.10	92.97±1.30	92.95±1.31	73.30±7.11	64.95±3.38
CLAM	UNI	83.37±2.96	56.43±1.90	82.66±1.60	80.02±1.74	92.21±1.51	92.19±1.52	69.68±8.62	56.08±6.01
CLAM	CONCHv1.5	83.36±1.15	57.51±2.12	81.37±1.38	78.64±1.64	92.04±1.73	92.01±1.76	69.00±10.95	56.07±7.59
CLAM	Virchow2	84.79±2.30	59.57±1.67	83.37±1.86	81.06±2.09	93.36±1.45	93.33±1.47	73.39±7.38	63.52±6.25
TransMIL	UNI	69.50±3.44	38.96±5.22	80.86±2.76	78.18±3.05	89.97±3.21	89.92±3.26	70.52±6.86	60.87±3.70
TransMIL	CONCHv1.5	82.99±3.34	60.67±3.96	85.12±2.69	82.51±2.14	90.42±1.44	90.39±1.45	70.01±7.63	59.91±6.32
TransMIL	Virchow2	71.13±5.52	44.51±7.10	83.21±4.27	81.06±3.93	91.38±2.28	91.36±2.31	71.07±7.85	61.72±5.78
RRTMIL	UNI	78.13±5.95	51.68±7.38	83.90±3.12	81.86±3.43	92.03±1.47	92.02±1.49	71.28±7.12	60.90±4.99
RRTMIL	CONCHv1.5	81.95±4.92	60.69±3.18	85.05±2.81	82.78±2.08	92.04±2.34	92.02±2.37	70.73±9.37	60.54±5.76
RRTMIL	Virchow2	82.22±6.17	58.16±7.11	85.20±4.35	83.33±4.01	92.41±1.66	92.39±1.68	70.84±8.59	60.21±4.95
TDAMIL	UNI	73.84±5.91	47.82±12.24	85.21±2.01	83.33±1.57	90.89±1.65	90.85±1.68	70.84±7.71	60.29±4.62
TDAMIL	CONCHv1.5	83.79±5.19	59.40±4.47	84.12±2.67	81.91±2.01	91.57±2.33	91.53±2.38	71.90±6.15	61.84±7.58
TDAMIL	Virchow2	80.21±3.37	55.38±4.19	85.49±3.52	83.87±2.87	92.32±1.61	92.28±1.64	71.81±7.50	63.11±4.19
	MEC$_{ABMIL}$	**86.96±3.10**	**68.44±2.64**	87.79±1.92	86.05±2.20	93.94±1.72	93.91±1.75	77.16±5.80	68.72±4.23
	MEC$_{ABMIL-G}$	85.20±2.47	64.26±1.96	87.61±1.69	86.08±1.96	93.75±1.61	93.72±1.62	76.56±4.26	68.87±5.41
	MEC$_{CLAM}$	86.31±3.83	66.90±4.07	**87.80±2.55**	**86.20±2.69**	**94.04±2.04**	**94.02±2.05**	**77.16±5.52**	**70.20±5.06**

The total objective function $\mathcal{L}_{total}$ is defined as:

$$\mathcal{L}_{total} = \mathcal{L}_{CE}(\hat{\mathbf{y}}_{final}, y) + \mathcal{L}_{experts}. \tag{8}$$

3 Experiments

Datasets. We evaluate the proposed framework on four real-world datasets: Camelyon17 (500 WSIs, four-class lymph node metastasis detection) [1], TCGA-NSCLC (1,053 WSIs, binary lung cancer subtyping) [12], TCGA-GBMLGG (1,525 WSIs, three-class glioma molecular subtyping) [24], and STS (813 WSIs, fine-grained classification of five soft tissue sarcoma subtypes). We use a standard 5-fold cross-validation protocol, ensuring no patient overlap across training, validation, and testing sets.

Table 2. Performance comparison of different ensemble methods across diverse pathological tasks. Best in **bold**, second best is underlined.

Dataset	Camelyon17		TCGA-GBMLGG		TCGA-NSCLC		STS	
Method	Acc	F1-score	Acc	F1-score	Acc	F1-score	Acc	F1-score
Mean	85.00±2.44	63.91±7.05	86.66±2.36	84.75±2.81	92.98±1.38	92.94±1.39	75.32±5.35	65.62±3.62
Concat	83.19±3.64	60.06±4.18	86.44±2.20	84.76±2.26	92.98±1.55	92.94±1.58	75.18±5.49	65.70±3.33
Logit	85.78±3.15	64.59±2.51	84.29±2.72	81.84±3.40	93.17±1.55	93.14±1.57	74.37±6.50	64.99±5.41
Vote	85.42±2.53	61.77±2.85	85.05±3.08	82.79±1.26	93.26±1.43	93.24±1.14	74.45±6.59	64.84±2.61
COBRA	75.91±3.35	43.16±2.83	85.05±1.82	83.31±1.21	92.33±1.78	92.30±1.81	72.93±8.22	65.61±7.49
GPFM	82.78±2.32	60.36±3.52	86.26±2.91	84.29±3.62	92.97±2.03	92.93±2.03	74.21±6.40	65.37±6.40
MEC	**86.96±3.10**	**68.44±2.64**	**87.79±1.92**	**86.05±2.20**	**93.94±1.72**	**93.91±1.75**	**77.16±5.80**	**68.72±4.23**

* We use official pre-trained weights to extract features for COBRA and GPFM. COBRA follows the recommended linear probing method, whereas ABMIL is used as the backbone for all other methods.

Implementation Details. Following [15], we use PrePATH for WSI preprocessing, tissue regions are segmented into non-overlapping 512 × 512 patches at 20× magnification. For feature extraction, we employ three frozen pathology foundation models: UNI [3], Virchow2 [30], and CONCHv1.5 [11]. For the consensus branch, the concatenated evidence vector $\mathbf{V}_{fused}$ is fed into an MLP classifier with two hidden layers of dimensions 512 and 128, where each hidden layer is followed by a ReLU activation and dropout with a rate of 0.3. To ensure fair comparisons, all methods are optimized using the Adam optimizer coupled with a cosine annealing learning rate schedule.

Comparison with Single-PFM Baselines. We benchmark 15 combinations, pairing three PFMs (UNI, CONCHv1.5, Virchow2) with five MIL architectures (ABMIL [7], CLAM [12], TransMIL [20], RRTMIL [23], TDAMIL [19]). As presented in Table 1, our MEC framework consistently outperforms all single-PFM baselines. These results support that integrating complementary inductive biases can alleviate the limitations of any individual PFM. Besides, consistent with prior observations [3,14,26], performance varies only marginally with the choice of MIL architecture when expressive PFMs are used. This observation suggests that rich features enable standard attention mechanisms to localize evidence effectively, thereby justifying our use of simple attention scores for patch mining. Finally, the consistent improvements observed with ABMIL-GATE and CLAM demonstrate that our framework remains MIL-agnostic, effectively supporting diverse aggregation networks without altering the consensus mechanism.

Comparison with Ensemble Methods. We compare MEC with feature-level fusion (Mean, Concat), prediction-level fusion (Logit, Vote), and feature alignment models (GPFM [14], COBRA [9]). As shown in Table 2, MEC consistently outperforms all baselines. Its superiority over static ensembles validates our dynamic consensus mechanism in effectively filtering noise and amplifying discriminative cues across diverse expert representations, whereas surpassing COBRA/GPFM proves the efficacy of task-driven probing without heavy pre-training. **In terms of efficiency**, multi-PFM feature extraction increases offline preprocessing time (49.8 s/WSI on a single GPU vs. 28.0 s for GPFM), but this

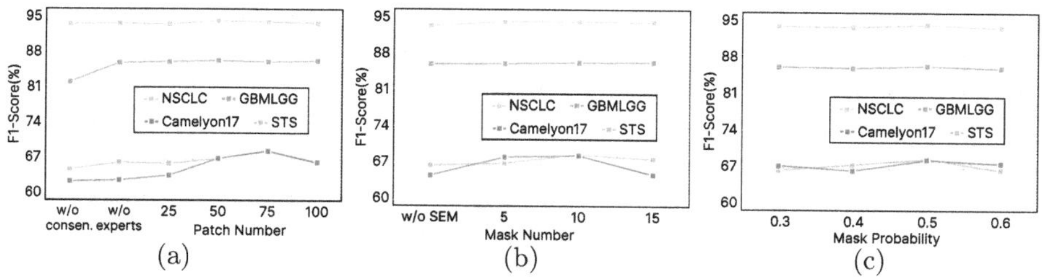

Fig. 3. Ablation study on key components. (a) Impact of instance selection number K. (b) Impact of mask number. (c) Impact of mask probability.

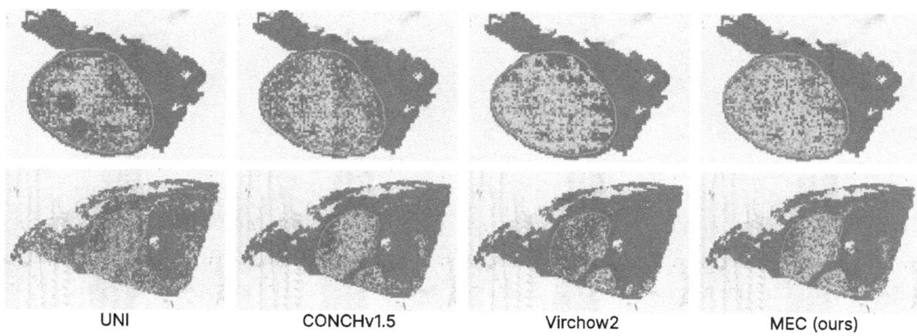

Fig. 4. Visualization of WSI sample heatmaps on the STS dataset using an ABMIL backbone with single foundation models versus MEC. pathologist-annotated Regions of Interest (ROIs) are circled in green. (Color figure online)

cost is highly parallelizable (30.3 s/WSI with multi-GPU). Crucially, MEC's MIL inference requires merely 2.02 ms. This renders the accuracy-compute trade-off highly justified for offline clinical WSI workflows, where maximizing diagnostic precision on complex tasks strictly outweighs the need for absolute latency.

Ablation Study. To validate the architectural design of MEC, we compare the full framework against three variants: (1) *w/o consen.*, which removes the consensus branch and relies solely on aggregated expert logits; (2) *w/o experts*, which utilizes only the consensus branch prediction; and (3) *w/o SEM*, which removes the stochastic evidence masking mechanism. As presented in Figs. 3a and 3b, the full MEC framework consistently outperforms all ablated variants across all datasets. This confirms that the synergy between expert guidance and consensus recalibration is critical for robust diagnosis.

We further perform ablation studies on instance selection (K). As presented in Fig. 3a, TCGA-NSCLC and TCGA-GBMLGG achieve optimal performance at $K = 50$, whereas Camelyon17 and STS favor a larger context ($K = 75$). Overall, performance is relatively stable across a reasonable range of K, suggesting that MEC is not overly sensitive to the exact threshold as long as sufficient contextual evidence is retained. Regarding masking intensity in SEM (Figs. 3b and 3c), results consistently favor moderate regularization. In practice, setting the mask number to 10 and the mask probability to 50% is generally sufficient for achieving near-optimal performance.

Analysis of Attention Regions. To qualitatively assess our collaborative mechanism, we visualize attention heatmaps produced by individual PFMs and by MEC. As illustrated in Fig. 4, different PFMs attend to markedly different regions, reflecting model-specific preferences. In contrast, MEC yields a more comprehensive and coherent attention distribution that covers complementary diagnostically relevant areas. This suggests that the consultation-based consensus can synthesize evidence across experts, reducing reliance on any single model's bias and leading to more robust slide-level decisions.

4 Conclusions

We propose MEC, a generic framework simulating clinical multi-expert consultation to address PFM heterogeneity. Treating frozen PFMs as digital experts with distinct inductive biases, MEC employs a multi-branch architecture to parallelly localize task-specific regions. Extensive experiments demonstrate that MEC consistently outperforms individual PFMs and standard ensembles. Results validate MEC as a practical, computation-efficient solution to harness the collective potential of foundation models without extra large-scale pre-training.

Acknowledgments. This work was supported by the Ministry of Science and Technology of the People's Republic of China (STI2030-Major Projects 2021ZD0201900), the National Natural Science Foundation of China (Grant No. 62371409), and the Fujian Provincial Natural Science Foundation of China (Grant No. 2023J01005).

Disclosure of Interests. The authors have no competing interests to declare that are relevant to the content of this article.

References

1. Bandi, P., et al.: From detection of individual metastases to classification of lymph node status at the patient level: the camelyon17 challenge. IEEE Trans. Med. Imaging **38**(2), 550–560 (2019)
2. Bareja, R., et al.: Evaluating vision and pathology foundation models for computational pathology: a comprehensive benchmark study. medRxiv (2025)
3. Chen, R.J., et al.: Towards a general-purpose foundation model for computational pathology. Nat. Med. (2024)
4. Dietterich, T.G.: Ensemble methods in machine learning. In: Multiple Classifier Systems, pp. 1–15. Springer Berlin Heidelberg (2000)
5. Elmore, J.G., et al.: Diagnostic concordance among pathologists interpreting breast biopsy specimens. JAMA **313**(11), 1122–1132 (2015)
6. Hosseini, M.S., et al.: Computational pathology: a survey review and the way forward. J. Pathol. Inf. **15**, 100357 (2024)

7. Ilse, M., Tomczak, J., Welling, M.: Attention-based deep multiple instance learning. In: ICML. vol. 80, pp. 2127–2136. PMLR (2018)
8. Lee, J., Lim, J., Byeon, K., Kwak, J.T.: Benchmarking pathology foundation models: adaptation strategies and scenarios. Comput. Biol. Med. **190**, 110031 (2025)
9. Lenz, T., Neidlinger, P., Ligero, M., Wölflein, G., van Treeck, M., Kather, J.N.: Unsupervised foundation model-agnostic slide-level representation learning. In: CVPR, pp. 30807–30817 (2025)
10. Louis, D.N., et al.: The 2016 world health organization classification of tumors of the central nervous system: a summary. Acta Neuropathol. **131**(6), 803–820 (2016)
11. Lu, M.Y., et al.: A visual-language foundation model for computational pathology. Nat. Med. **30**, 863–874 (2024)
12. Lu, M.Y., Williamson, D.F.K., Chen, T.Y., Chen, R.J., Barbieri, M., Mahmood, F.: Data-efficient and weakly supervised computational pathology on whole-slide images. Nat. Biomed. Eng. **5**(6), 555–570 (2021)
13. Luo, X., Wang, X., Eweje, F.: Ensemble learning of foundation models for precision oncology. arXiv preprint 2508.16085 (2025)
14. Ma, J., Guo, Z., Zhou, F.: A generalizable pathology foundation model using a unified knowledge distillation pretraining framework. Nat. Biomed. Eng. (2025)
15. Ma, J., Xu, Y., Zhou, F.: Pathbench: a comprehensive comparison benchmark for pathology foundation models towards precision oncology. arXiv preprint arXiv:2505.20202 (2025)
16. Mercan, E., Aksoy, S., Shapiro, L.G., Weaver, D.L., Brunyé, T.T., Elmore, J.G.: Localization of diagnostically relevant regions of interest in whole slide images: a comparative study. J. Digit. Imaging **29**(4), 496–506 (2016)
17. Neidlinger, P., et al.: Benchmarking foundation models as feature extractors for weakly supervised computational pathology. Nat. Biomed. Eng. (2025)
18. Niazi, M.K.K., Parwani, A.V., Gurcan, M.N.: Digital pathology and artificial intelligence. Lancet Oncol. **20**(5), e253–e261 (2019)
19. Reisenbüchler, D., Deng, R., Matek, C., Feuerhake, F., Merhof, D.: Top-down attention-based multiple instance learning for whole slide image analysis . In: MICCAI. vol. LNCS 15960, pp. 651–660 (2025)
20. Shao, Z., et al.: TransMIL: transformer based correlated multiple instance learning for whole slide image classification. In: Advances in Neural Information Processing Systems. vol. 34, pp. 2136–2147 (2021)
21. Song, A.H., et al.: Artificial intelligence for digital and computational pathology. Nat. Rev. Bioeng. **1**(12), 930–949 (2023)
22. Tang, W., et al.: Revisiting end-to-end learning with slide-level supervision in computational pathology. arXiv preprint arXiv:2506.02408 (2025)
23. Tang, W., Zhou, F., Huang, S., Zhu, X., Zhang, Y., Liu, B.: Feature re-embedding: towards foundation model-level performance in computational pathology. In: CVPR, pp. 11343–11352 (2024)
24. Wang, W., Zhao, Y., Teng, L., et al.: Neuropathologist-level integrated classification of adult-type diffuse gliomas using deep learning from whole-slide pathological images. Nat. Commun. **14**(1), 6359 (2023)
25. Xiong, C., Chen, H., Sung, J.J.Y.: A survey of pathology foundation model: progress and future directions. arXiv:2504.04045 arXiv preprint (2025)
26. Xu, H., et al.: When multiple instance learning meets foundation models: advancing histological whole slide image analysis. Med. Image Anal. **101**, 103456 (2025)
27. Yu, Z., et al.: FM2: fusing multiple foundation models for pathology image analysis via disentangled consensus-divergence representation. Inf. Fusion **127**, 103840 (2026)

28. Zhang, Y., Li, H., Sun, Y., Zheng, S., Zhu, C., Yang, L.: Attention-challenging multiple instance learning for whole slide image classification. In: ECCV, pp. 125–143 (2024)
29. Zhu, W., et al.: How effective can dropout be in multiple instance learningâĂŕ? In: ICML. vol. 267, pp. 80090–80106. PMLR (2025)
30. Zimmermann, E., et al.: Virchow2: scaling self-supervised mixed magnification models in pathology. arXiv preprint arXiv:2408.00738 (2024)

MEC: Medical Evidence Capsules for Retrieval-Augmented Generation in Medical Multimodal Question Answering

Zhenghua Xu[1], Xinwei Dai[1], Bo Wang[2(✉)], and Tian Tian[2]

[1] School of Health Sciences and Biomedical Engineering, Hebei University of Technology, Tianjin, China
[2] School of Artificial Intelligence, Hebei University of Technology, Tianjin, China
wangbo@hebut.edu.cn

Abstract. In medical multimodal question answering (QA), retrieval-augmented generation (RAG) serves as a core paradigm for improving model accuracy and mitigating hallucinations, yet existing approaches typically rely on a single source of evidence and thus struggle to support reliable and consistent reasoning. In practice, medical questions often require the integration of multiple evidence carriers, including narrative medical texts such as textbooks and clinical guidelines, structured medical knowledge systems, experiential priors implicitly encoded in expert models, and multimodal clinical examinations such as medical images and their associated reports. These evidence sources differ substantially in representation and granularity, making it difficult to accurately bind conclusions to traceable evidence snippets during retrieval, thereby increasing the risk of hallucinated answers. We propose Medical Evidence Capsules (MEC), an evidence organization mechanism for medical multimodal QA. MEC reformulates evidence from heterogeneous sources into compact capsule structures, enabling unified utilization of multiple evidence sources. We further construct a medical multimodal retrieval corpus that integrates both textual and imaging evidence, and evaluate MEC on five medical QA benchmarks. Experimental results demonstrate that MEC consistently yields performance improvements across both text-only QA and visual question answering (VQA) benchmarks. The resources of our work are available at https://github.com/fruitbitlab/MEC.

Keywords: Medical Question Answering · Retrieval-Augmented Generation · Large Language Models · Vision-Language Models

1 Introduction

Large language models (LLMs) and Large vision models (LVMs) have achieved significant progress in medical question answering (QA) [19,22], and retrieval-augmented generation (RAG) has been widely adopted to improve accuracy

G. Yang et al. (Eds.): MICCAI 2026, LNCS 16878, pp. 303–312, 2027.
https://doi.org/10.1007/978-3-032-38059-3_29

and mitigate hallucinations in high-risk scenarios [26]. However, in medical QA, the core challenge lies in whether retrieved evidence can be reliably used. If conclusions cannot form stable correspondences with explicit evidence, even linguistically plausible outputs may introduce hallucinations due to mismatched evidence granularity or insufficient alignment [25].

In medical multimodal QA, RAG introduces external medical knowledge to constrain the generation process, yet existing methods handle multi-source evidence in a relatively coarse manner. Most approaches retrieve textual, structured, and imaging information separately and directly concatenate them as input [23], assuming that the model can perform knowledge selection and alignment on its own. However, this often leads to mismatched evidence granularity [21,24] and incomparable similarity signals, making it difficult to stably bind generated conclusions to citable evidence snippets, thereby weakening answer consistency and traceability [18].

The key bottleneck of generative models in the medical domain is not the lack of information, but how to establish controllable usage relationships among multi-source evidence [15]. Medical questions often involve textual resources, structured knowledge, parameterized experience from models, and imaging materials, which differ substantially in granularity and representation [9]. Existing approaches typically concatenate these sources as independent contexts [17] without explicitly constraining how evidence should be used [21], forcing the model to rely on implicit alignment during generation and thereby increasing the risk of hallucinations.

Based on these issues, we refocus the problem from "how much evidence to retrieve" to "how retrieved evidence is used," and propose Medical Evidence Capsules (MEC), an evidence organization mechanism for medical multimodal QA. MEC restructures information from heterogeneous sources into compact evidence capsules, each explicitly comprising a medical assertion, a retrieval cue, and a citable evidence snippet. With this structure, textual resources, parameterized experience from models, and imaging evidence can be retrieved and invoked collaboratively under a unified interface, so that retrieval returns structured evidence components that can directly support reasoning rather than loosely concatenated contexts, thereby improving answer consistency and traceability (Fig. 1).

Building upon this framework, we construct a large-scale medical multimodal retrieval corpus that integrates medical texts, structured knowledge, parameterized experience from models, as well as imaging data and corresponding reports. We evaluate MEC on text-only QA and visual QA benchmarks, where it consistently demonstrates stable improvements across different settings. In summary, our contributions are as follows: (1) we propose MEC, which explicitly constrains evidence usage through a compact capsule structure, providing a unified retrievable evidence representation for medical multimodal QA; (2) we build a retrieval corpus containing over 550,000 medical assertions, offering reusable resources and evaluation foundations for medical multimodal QA.

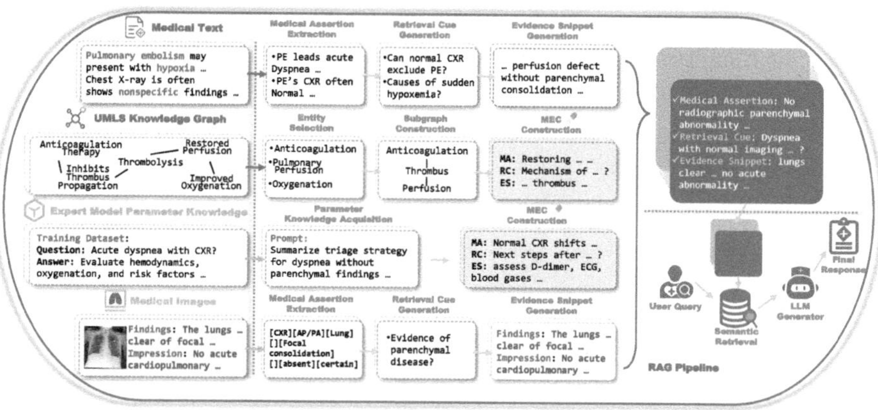

Fig. 1. Overview of Medical Evidence Capsules. Heterogeneous evidence sources are normalized into capsules to support unified retrieval and constrained generation.

2 Methods

2.1 Overview of Medical Evidence Capsule

In this section, we introduce the overall design of Medical Evidence Capsules (MEC). MEC is an evidence organization and invocation structure for medical multimodal QA, designed to support reliable utilization of multi-source evidence in RAG. With MEC, the retrieval stage returns structured evidence capsules rather than loosely concatenated contexts, providing clearer evidence constraints for the generation stage.

2.2 Evidence from Medical Texts

Medical textbooks, clinical guidelines, and review articles provide important knowledge sources for medical QA, yet their long-form narratives are difficult to directly leverage in RAG. To improve the usability of textual evidence, we build assertion-centric textual evidence capsules, enabling textual knowledge to be reliably retrieved and directly used for reasoning.

We first extract semantically coherent text segments from the textbook corpus, forming the collection $\mathcal{T} = \{t_i\}$. For each segment, we employ a LLM to distill its core medical assertion [12]:

$$a_i = \mathcal{F}_{\text{assert}}(t_i). \tag{1}$$

Medical assertions compactly capture the key facts or judgments expressed in the text, reducing semantic ambiguity during generation. Based on each assertion, we further construct a paired evidence snippet and retrieval cue, which provide necessary context and characterize typical invocation patterns:

$$(e_i, c_i) = \mathcal{F}_{\text{support}}(t_i, a_i). \tag{2}$$

Finally, each textual evidence item is organized as a capsule consisting of the assertion a_i, the retrieval cue c_i, and the evidence snippet e_i. This design ensures that retrieval returns structured evidence components rather than lengthy raw contexts, providing clearer evidence constraints for generation.

2.3 Evidence from Structured Resources

Structured medical knowledge provides explicit entity definitions and relational constraints, serving as an important evidence source beyond narrative texts. We build graph-based evidence capsules from the UMLS [2], transforming medical entities and their relations into retrievable and citable structured evidence.

We first select entities with unique concept identifiers and well-defined semantic types from UMLS to form an entity set $\mathcal{V}$. To avoid isolated nodes and overly redundant hub entities, we choose evidence anchors by jointly considering structural importance and semantic specificity. For an entity $u \in \mathcal{V}$, its structural importance is defined as:

$$c(u) = \alpha \log(1 + \deg(u)) + (1 - \alpha)\,\mathrm{PR}(u). \tag{3}$$

where $\deg(u)$ denotes node degree, $\mathrm{PR}(u)$ is the PageRank score, and α balances local connectivity and global centrality. Meanwhile, we quantify medical specificity using the semantic type $s(u)$ via the information content:

$$i(u) = -\log \frac{|\{v \in \mathcal{V} : s(v) = s(u)\}|}{|\mathcal{V}|}. \tag{4}$$

which prioritizes more discriminative entities in the medical knowledge system. Around each selected anchor, we construct a scale-controlled local subgraph $\mathcal{G}_u$ and employ a LLM to convert the structured relations into an medical evidence capsule:

$$(a, e, c) = \mathcal{F}_{\text{graph}}(\mathcal{G}_u). \tag{5}$$

where a is the medical assertion, e is the evidence snippet describing the relations, and c is the retrieval cue. This process organizes graph knowledge into MEC, enabling structured evidence to participate in multi-source reasoning within a unified framework.

2.4 Evidence from Model Parameters

In addition to explicit texts and structured resources, expert-level language models also encode substantial medical experiential priors in their parameters [11]. To avoid the uncontrolled inference that may arise from directly invoking such implicit knowledge, we convert parametric knowledge into retrievable and constrained MEC evidence capsules.

Specifically, given QA pairs (q, a) from the training split of datasets, we generate explanatory evidence under explicit answer-alignment constraints:

$$e = \mathcal{F}_{\text{param}}(q, a). \tag{6}$$

Table 1. Comparison of different RAG methods, retrieval return fields, and LLM scales on Text QA. Bold values indicate the best performance across models and datasets.

Retrieval Evidence	Method	MecQA			Med-FlashCards			HfMedQA		
		Qw-3B	Qw-7B	Lm-8B	Qw-3B	Qw-7B	Lm-8B	Qw-3B	Qw-7B	Lm-8B
—	Wo-RAG	0.2550	0.2653	0.2384	0.1470	0.1473	0.1854	0.1285	0.1274	0.1770
Original Text	SR-RAG	0.3682	0.3899	0.3212	0.3310	0.3499	0.3381	0.2834	0.3015	0.2654
	FLARE	0.3416	0.3508	0.3356	0.2839	0.3007	0.3224	0.2370	0.2418	0.2344
	DRAGIN	0.3578	0.3588	0.3122	0.3144	0.3135	0.3122	0.3092	0.3014	0.2709
Evidence Snippet	SR-RAG	0.3905	0.3989	0.3330	0.3348	0.3435	0.3312	0.2845	0.3067	0.2676
	FLARE	0.3492	0.3526	0.3372	0.2827	0.2982	0.3255	0.2371	0.2424	0.2340
	DRAGIN	0.3598	0.3607	0.3374	0.3154	0.3118	0.3149	0.3125	0.3003	0.2720
MEC (Ours)	SR-RAG	**0.3982**	**0.4489**	0.3580	**0.4034**	**0.4100**	**0.3952**	0.3113	0.3493	0.2800
	FLARE	0.3517	0.3653	0.3531	0.3164	0.3252	0.3568	0.2516	0.2756	0.2570
	DRAGIN	0.3684	0.3895	**0.3593**	0.3434	0.3522	0.3380	**0.3305**	**0.3709**	**0.2962**

This process enforces consistency with the known answer and avoids unconstrained expansions beyond the question context, thereby reducing the risk of hallucinations. Building on the generated evidence, we further distill a corresponding medical assertion and construct a matching retrieval cue, so that parametric knowledge can be incorporated into the MEC framework. Through this conversion, implicit experience encoded in model parameters is made explicit as citable evidence, enabling it to be jointly used with textual, graph-based, and imaging evidence under a unified structure.

2.5 Evidence from Medical Images

Medical images and their reports provide the most direct objective evidence for medical QA. However, their descriptions are highly structured and dense with negation and uncertainty, making them difficult to use on the same semantic scale as textual or graph-based knowledge. To address this, we reorganize imaging information into MEC-compliant imaging evidence capsules, enabling unified retrieval and constrained generation-time reasoning.

We use approximately 7,000 image–report pairs from MIMIC-CXR [8]. From each report, we extract medical assertions [6] and use the *findings* section as the citable evidence snippet, preserving the observational context without altering the original semantics. To improve retrievability and consistency, we normalize each imaging assertion into a fixed eight-tuple:

$$a_{\text{img}} = (m, p, o, l, f, s, e, u), \tag{7}$$

where m denotes the imaging modality, p the acquisition position, o the organ or structure, l the spatial region or laterality, f the core finding, s the severity or morphological descriptor, e the existence polarity, and u the uncertainty level. This representation covers both positive and negative findings and explicitly distinguishes confirmed observations from speculative conclusions, thereby reducing implicit inference during generation.

Based on these normalized imaging assertions, we generate corresponding retrieval cues and organize the assertion, retrieval cue, and the report's *findings* and *impression* into an imaging evidence capsule.

Table 2. Comparison of different retrieval return fields, and VLM scales on Visual QA. Bold values indicate the best performance across models and datasets.

Datasets	Retrieval Evidence	Models			
		Qw-VL-3B	Qw-VL-7B	MedG-4B	Lshu-7B
SLAKE	Wo-RAG	0.5541	0.5811	0.4838	0.7865
	Findings	0.5608	0.5864	0.4905	0.7838
	Image+Findings	0.5716	0.5932	0.5014	0.7919
	MEC (Ours)	**0.5838**	**0.6027**	**0.5108**	**0.7986**
VQA-RAD	Wo-RAG	0.4971	0.5714	0.5886	0.6000
	Findings	0.5029	0.5771	0.5886	0.5943
	Image+Findings	0.5086	0.5829	0.5909	0.6057
	MEC (Ours)	**0.5257**	**0.5952**	**0.6000**	**0.6114**

Table 3. Tasks, models, datasets, and evaluation metrics used in experiments.

Task Type	Model	Datasets	Evaluation Metric
Text QA	Qwen2.5-3B-Instruct [28]	MecQA (In-domain)	ROUGE-L [13]
	Qwen2.5-7B-Instruct [28]	Med-FlashCards [4]	
	Llama-3.1-8B-Instruct [3]	HfMedQA	
Visual QA	Qwen2.5-VL-3B-Instruct [1]	VQA-RAD (CXR) [10]	Accuracy
	medgemma-1.5-4b-it [16]		
	Qwen2.5-VL-7B-Instruct [1]	SLAKE (CXR) [14]	
	Lingshu-7B [27]		

3 Experiments and Results

3.1 Experimental Setups

Datasets and Settings. In our experiments, we construct a MEC retrieval corpus for medical multimodal QA, supporting both text-only QA and visual QA. The textual and structured evidence is built from medical textbook corpora aligned with MedQA [7] and UMLS [2], comprising approximately 500,000 medical assertions. The imaging evidence is derived from MIMIC-CXR [8], where about 50,000 imaging-related assertions are extracted from roughly 7,000 image–report pairs. Text-only QA uses the in-domain MecQA (sampled from our corpus) and cross-domain Med-FlashCards [4] and HfMedQA[1]; VQA uses VQA-RAD [10] and SLAKE [14]. Models and metrics are in Table 3. GPT-4o [5] is used to construct all evidence capsules.

Baselines. We adopt several representative text-based RAG methods [11, 20] as baselines and integrate MEC into these frameworks under identical generation models and parameter settings to evaluate its evidence representation capability. For visual QA, we compare against no-retrieval settings and analyze the contribution of imaging evidence capsules through different MEC component combinations. All experiments are conducted with consistent model configurations and generation settings to ensure fair comparison.

3.2 Main Results Analysis

Text QA results are shown in Table 1. MEC consistently improves performance across different retrieval strategies and model scales, achieving the best results on all benchmarks. By organizing evidence into assertion-driven structured forms, MEC aligns retrieved content with QA semantics and reduces the implicit alignment burden during generation. This is particularly evident on medium-scale models, suggesting that structured evidence enhances multi-source knowledge efficiency. Visual QA results are in Table 2. When compared to no-retrieval or direct report text usage, MEC's organization of imaging assertions yields more stable accuracy improvements, with consistent trends across different VLMs. By structuring imaging data with explicit polarity and uncertainty markers, MEC enables seamless collaboration between imaging evidence and evidence from other sources, ensuring robust benefits across models.

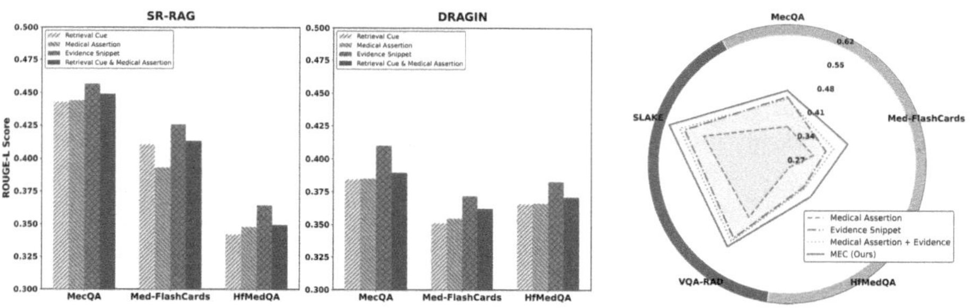

Fig. 2. Ablations on matching fields (bar) and return contents (radar). MEC performs best across text QA and medical VQA, with a better efficiency–accuracy trade-off.

[1] https://huggingface.co/datasets/asmitha26/medical-data.

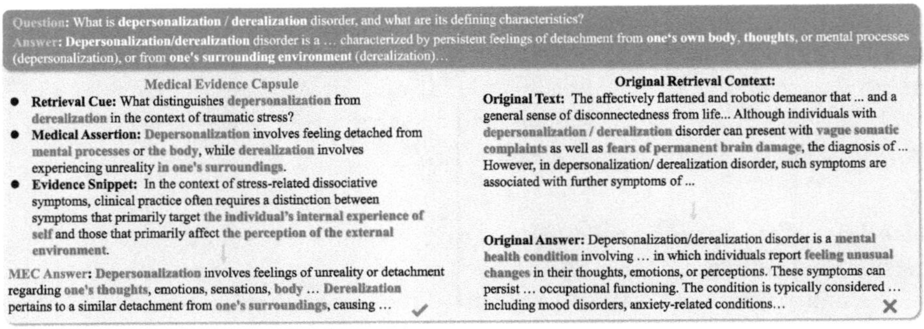

Fig. 3. Case study comparing Original retrieval with MEC retrieval. MEC provides a clearer reasoning scaffold, yielding more focused and traceable answers.

3.3 Ablations and Case Discussion

Effect of Retrieval and Return Fields. As shown in Fig. 2, we analyze the effect of MEC from the matching fields (bar chart) and the returned contents (radar chart). Using evidence snippets as the matching signal is typically the best, while matching only with medical assertions and retrieval cues yields only a slight performance drop. Moreover, since assertions and cues are more compact, this strategy substantially reduces the computational cost of vectorization and matching. Under a fixed matching strategy, the full MEC achieves the best results on all five benchmarks, outperforming returning only assertions or only evidence snippets. These results indicate that by jointly providing structured assertions, retrieval cues, and citable evidence snippets, MEC imposes more complete semantic and evidential constraints during generation, leading to a more robust trade-off between efficiency and effectiveness.

Case Study. We further illustrate the practical effect of MEC through a case study, as shown in Fig. 3. Compared with retrieval based on raw text, the evidence returned under the MEC framework provides a clearer semantic scaffold during generation. The medical assertion directly captures the core discriminative information required by the question, the retrieval cue strengthens semantic alignment between the query and the evidence, and the evidence snippet supplies necessary contextual support and conditional constraints, thereby reducing the need for implicit filtering and alignment within lengthy contexts. In contrast, raw text retrieval often returns content that is descriptively rich but less task-relevant, increasing the reasoning burden and weakening the focus of the generated answer.

4 Conclusion

This paper proposes Medical Evidence Capsules (MEC), an evidence organization mechanism for medical multimodal QA, which explicitly constrain evidence retrieval and invocation through a compact structure, mitigating evidence granularity mismatch and improving the traceability of conclusions in medical multimodal QA. We further construct a large-scale retrieval corpus integrating multi-source medical knowledge and evaluate MEC on medical text QA and visual QA benchmarks. Experimental results demonstrate that MEC consistently improves performance across different models and retrieval settings, validating its effectiveness in medical multimodal RAG.

Acknowledgments. This work was supported by the National Natural Science Foundation of China under Grant 62276089, by the Natural Science Foundation of Tianjin Municipality, China, under Grant 24JCJQJC00200, by the Natural Science Foundation of Hebei Province, China, under Grant F2024202064, and by the Ministry of Human Resources and Social Security, China, under Grant RSTH-2023-135-1.

Disclosure of Interests. The authors have no competing interests to declare that are relevant to the content of this article.

References

1. Bai, S., et al.: Qwen2.5-vl technical report. arXiv preprint arXiv:2502.13923 (2025)
2. Bodenreider, O.: The unified medical language system (UMLs): integrating biomedical terminology. Nucleic Acids Res. **32**(suppl_1), D267–D270 (2004)
3. Grattafiori, A., et al.: The llama 3 herd of models. arXiv preprint arXiv:2407.21783 (2024)
4. Han, T., et al.: Medalpaca-an open-source collection of medical conversational AI models and training data. arXiv preprint arXiv:2304.08247 (2023)
5. Hurst, A., et al.: Gpt-4o system card. arXiv preprint arXiv:2410.21276 (2024)
6. Jain, S., et al.: Radgraph: extracting clinical entities and relations from radiology reports. arXiv preprint arXiv:2106.14463 (2021)
7. Jin, D., Pan, E., Oufattole, N., Weng, W.H., Fang, H., Szolovits, P.: What disease does this patient have? a large-scale open domain question answering dataset from medical exams. arXiv preprint arXiv:2009.13081 (2020)
8. Johnson, A.E., et al.: Mimic-CXR, a de-identified publicly available database of chest radiographs with free-text reports. Sci. Data **6**(1), 317 (2019)
9. Karim, A.R., Uzuner, O.: Masonnlp at mediqa-wv 2025: multimodal retrieval-augmented generation with large language models for medical VQA. In: Proceedings of the 7th Clinical Natural Language Processing Workshop, pp. 84–94 (2025)
10. Lau, J.J., Gayen, S., Ben Abacha, A., Demner-Fushman, D.: A dataset of clinically generated visual questions and answers about radiology images. Sci. Data **5**(1), 180251 (2018)
11. Lewis, P., et al.: Retrieval-augmented generation for knowledge-intensive nlp tasks. Adv. Neural. Inf. Process. Syst. **33**, 9459–9474 (2020)
12. Li, L., Zhou, X., Zhang, Y., Wu, X.: From retrieval to generation: unifying external and parametric knowledge for medical question answering. arXiv preprint arXiv:2510.18297 (2025)
13. Lin, C.Y.: Rouge: a package for automatic evaluation of summaries. In: Text Summarization Branches Out, pp. 74–81 (2004)
14. Liu, B., Zhan, L.M., Xu, L., Ma, L., Yang, Y., Wu, X.M.: Slake: a semantically-labeled knowledge-enhanced dataset for medical visual question answering. In: 2021 IEEE 18th International Symposium on Biomedical Imaging (ISBI), pp. 1650–1654. IEEE (2021)

15. Ning, Y., Sun, Y., Luo, L., Wang, Y., Pan, Y., Lin, H.: Medtrust-rag: evidence verification and trust alignment for biomedical question answering. arXiv preprint arXiv:2510.14400 (2025)
16. Sellergren, A., et al.: Medgemma technical report. arXiv preprint arXiv:2507.05201 (2025)
17. Shaaban, M.A., Zarei, M.R., Khan, A., Akkasi, A., Yaqub, M., Komeili, M.: Towards multimodal retrieval-augmented generation for medical visual question answering. Res. Square (2025). https://doi.org/10.21203/rs.3.rs-7752202/v1. preprint
18. Shang, Y., et al.: Medusa: cross-modal transferable adversarial attacks on multimodal medical retrieval-augmented generation. arXiv preprint arXiv:2511.19257 (2025)
19. Sohn, J., et al.: Rationale-guided retrieval augmented generation for medical question answering. In: Proceedings of the 2025 Conference of the Nations of the Americas Chapter of the Association for Computational Linguistics: Human Language Technologies (Volume 1: Long Papers), pp. 12739–12753 (2025)
20. Su, W., Tang, Y., Ai, Q., Wu, Z., Liu, Y.: Dragin: dynamic retrieval augmented generation based on the real-time information needs of large language models. In: Proceedings of the 62nd Annual Meeting of the Association for Computational Linguistics (Volume 1: Long Papers), pp. 12991–13013 (2024)
21. Sun, L., Zhao, J.J., Han, W., Xiong, C.: Fact-aware multimodal retrieval augmentation for accurate medical radiology report generation. In: Proceedings of the 2025 Conference of the Nations of the Americas Chapter of the Association for Computational Linguistics: Human Language Technologies (Volume 1: Long Papers), pp. 643–655 (2025)
22. Wang, C., Chen, Y.: Evaluating large language models for evidence-based clinical question answering. arXiv preprint arXiv:2509.10843 (2025)
23. Wu, J., et al.: Medical graph rag: evidence-based medical large language model via graph retrieval-augmented generation. In: Proceedings of the 63rd Annual Meeting of the Association for Computational Linguistics (Volume 1: Long Papers), pp. 28443–28467 (2025)
24. Xia, P., et al.: Rule: reliable multimodal rag for factuality in medical vision language models. In: Proceedings of the 2024 Conference on Empirical Methods in Natural Language Processing, pp. 1081–1093 (2024)
25. Xiong, G., Jin, Q., Lu, Z., Zhang, A.: Benchmarking retrieval-augmented generation for medicine. In: Findings of the Association for Computational Linguistics: ACL 2024, pp. 6233–6251 (2024)
26. Xu, S., et al.: Unsupervised information refinement training of large language models for retrieval-augmented generation. arXiv preprint arXiv:2402.18150 (2024)
27. Xu, W., et al.: Lingshu: a generalist foundation model for unified multimodal medical understanding and reasoning. arXiv preprint arXiv:2506.07044 (2025)
28. Yang, A., et al.: Qwen2.5 technical report. arXiv preprint arXiv:2412.15115 (2024)

Med-CAP: Counterfactual Evidence and Adaptive Prior Suppression for Robust Medical Visual Question Answering

Zaiqiang Huang[1,2], Zhihong Zhu[2], Zixuan Huang[2], Yunyan Zhang[2], Hui Zhang[1], and Xian Wu[2(✉)]

[1] Tsinghua University, Beijing, China
[2] Tencent Jarvis Lab, Beijing, China
kevinxwu@tencent.com

Abstract. Medical Visual Question Answering (Med-VQA) aims to provide clinical decision support by jointly reasoning over medical images and text. While existing research has made significant strides in addressing data scarcity through advanced representation learning and ensemble-based debiasing, current models still suffer from two critical issues: imbalance bias arising from skewed training distributions and modality shortcut bias where models rely on superficial linguistic patterns. In this paper, we propose Med-CAP, a counterfactual-and-calibration framework for robust Med-VQA. Specifically, to mitigate modality shortcuts, we introduce Counterfactual Evidence Modeling (CEM), which extracts prediction-relevant visual features via gradient attribution and enforces decision-making to be visually grounded through sufficiency-and-necessity constraints. To alleviate imbalance bias, we propose Adaptive Prior Suppression (APS), which calibrates fused logits by dynamically subtracting sample-adaptive language priors based on evidence strength. Experiments on SLAKE, VQA-RAD, SLAKE-CP and VQA-RAD-CP benchmarks show that Med-CAP achieves state-of-the-art robustness and accuracy across both standard and out-of-domain (OOD) settings. The code is available at https://github.com/ZaiqiangHuang/Med-CAP.

1 Introduction

Medical Visual Question Answering (Med-VQA) [9,13] aims to answer clinical questions by jointly reasoning over medical images and text, which has gained increasing attention [1,2,8,19]. Unlike general-domain models that benefit from massive data, Med-VQA models are prone to overfitting on limited samples. Therefore, models often generate predictions based on superficial associations rather than deep pathological reasoning, severely limiting their deployment in high-stakes clinical environments and motivating more robust approaches.

G. Yang et al. (Eds.): MICCAI 2026, LNCS 16878, pp. 313–323, 2027.
https://doi.org/10.1007/978-3-032-38059-3_30

To address the data limitations in Med-VQA, existing research has achieved significant milestones in representation learning and bias mitigation [15]. For instance, methods like MEVF [20] have successfully introduced hybrid visual features to enhance image understanding when annotated data is scarce. Simultaneously, ensemble-based approaches, such as RUBi [5], have demonstrated the effectiveness of using a question-only branch to explicitly identify and subtract inherent linguistic biases from the joint predictions. Furthermore, the introduction of data re-weighting strategies has proven beneficial in balancing the model's attention across various answer categories [4,7,11,22,23].

Although these methods have achieved promising results, we still discover two critical issues that undermine the robustness of current Med-VQA systems. First, from the data perspective, the persistent imbalance bias acts as a root cause of language priors; models tend to over-rely on frequent answer distributions in the training set, which leads to catastrophic performance drops on rare but clinically vital cases [9,13,22,23]. Second, from the reasoning perspective, models often succumb to modality shortcut bias, an effect where they bypass deep pathological visual evidence and instead rely on superficial linguistic patterns from standardized question templates [4,5,7,11]. As a result, even state-of-the-art models often fail to generalize to out-of-distribution (OOD) clinical scenarios, where the visual evidence contradicts the historical language patterns [22,23].

To address these challenges, we propose Med-CAP, a counterfactual-and-calibration framework for robust Med-VQA. Targeting the modality shortcut bias, we introduce Counterfactual Evidence Modeling (CEM). CEM extracts prediction-relevant visual evidence via feature-gradient attribution and constructs evidence-keeping and evidence-dropping counterfactual inputs. By enforcing a margin-based sufficiency-and-necessity constraint, CEM encourages the model to make decisions grounded in causally relevant visual evidence rather than superficial linguistic patterns. To alleviate the data imbalance bias, we propose Adaptive Prior Suppression (APS). APS calibrates the fused logits by subtracting a sample-adaptive question-only prior, where the suppression strength is dynamically determined by language prior confidence and counterfactual evidence strength. This dual-mechanism approach ensures that the model not only looks at the right visual regions but also remains resilient to skewed training distributions.

Our main contributions are summarized as follows: (1) We propose Med-CAP, a new Med-VQA framework that addresses medical language biases from both imbalance and shortcut perspectives. (2) We introduce CEM to ensure decisions are visually grounded via counterfactual constraints, and APS to calibrate predictions to alleviate dataset imbalance. (3) Experiments on standard (SLAKE, VQA-RAD) and bias-sensitive (SLAKE-CP, VQA-RAD-CP) [22] benchmarks demonstrate that Med-CAP achieves state-of-the-art robustness in OOD scenarios while maintaining competitive accuracy.

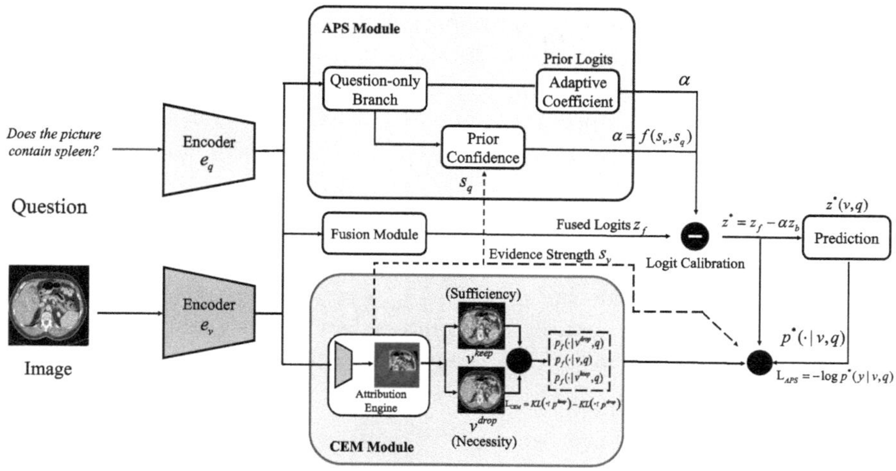

Fig. 1. Overview of the proposed Med-CAP framework for mitigating medical language biases. The framework consists of two complementary modules: Counterfactual Evidence Modeling (CEM) and Adaptive Prior Suppression (APS). The CEM module generates counterfactual feature representations to quantify the sufficiency and necessity of visual evidence, guiding the model to focus on clinically relevant cues rather than spurious correlations. The APS module dynamically adjusts prediction logits to suppress over-confident question-only priors, effectively mitigating dataset-specific biases.

2 Methods

2.1 Preliminaries

We consider a medical VQA dataset $\mathcal{D} = \{(v^{(n)}, q^{(n)}, a^{(n)})\}_{n=1}^{N}$, where each sample consists of object-level visual features $v \in \mathbb{R}^{K \times d}$, a question represented as a token sequence $q \in \mathbb{N}^{T}$, and a one-hot encoded answer $a \in \{0, 1\}^{|\mathcal{A}|}$. The goal is to learn a predictive model $f : \mathcal{V} \times \mathcal{Q} \rightarrow \mathbb{R}^{|\mathcal{A}|}$ that outputs logits corresponding to all candidate answers. However, existing Med-VQA models often suffer from medical language bias, where predictions can be dominated by question-only shortcuts rather than grounded visual evidence. To mitigate this issue, we introduce two plug-in modules, **Counterfactual Evidence Modeling (CEM)** and **Adaptive Prior Suppression (APS)**, which jointly promote evidence-grounded reasoning by enforcing counterfactual evidence consistency and suppressing over-confident question priors.

2.2 Counterfactual Evidence Modeling (CEM)

Motivated by counterfactual reasoning [16], we propose CEM to encourage causally grounded visual decision-making. The key idea is that if a prediction truly depends on visual evidence, then retaining such evidence should preserve the confidence of the correct answer, while suppressing it should reduce confidence.

Given a training sample with question q and ground-truth answer class $y \in \mathcal{A}$, the image is represented by K object-level features $v = \{v_k\}_{k=1}^{K}$, where $v_k \in \mathbb{R}^d$. Let z_y denote the fused logit corresponding to class y produced by the backbone. We estimate object-level importance via gradient-based attribution [18], for each object k, the attribution score is computed as the gradient magnitude of the target logit w.r.t. the object feature, i.e., $s_k = \left\|\frac{\partial z_y}{\partial v_k}\right\|_2$, where $\|\cdot\|_2$ is the ℓ_2 norm. We normalize $\{s_k\}_{k=1}^{K}$ within each sample and select the top-ρ objects to obtain a binary evidence mask $M \in \{0,1\}^K$, where $M_k = 1$ indicates that object k is treated as evidence and $\rho \in (0,1)$ controls sparsity. Using M, we construct counterfactual features by suppressing visual signals with a scaling factor $\beta \in (0,1)$, i.e., $\tilde{v} = \beta v$. We then form evidence-keeping and evidence-dropping inputs:

$$v^{keep} = M \odot v + (1-M) \odot \tilde{v}, \quad v^{drop} = (1-M) \odot v + M \odot \tilde{v}, \tag{1}$$

where $\odot$ denotes element-wise multiplication (broadcast along the feature dimension). Intuitively, v^{keep} preserves evidence objects while attenuating non-evidence objects, whereas v^{drop} attenuates evidence objects and preserves the remaining context. Let $p^{keep}(\cdot) = p(\cdot|v^{keep}, q)$ and $p^{drop}(\cdot) = p(\cdot|v^{drop}, q)$ denote the predictive distributions after softmax normalization of the corresponding logits. We enforce a margin between their confidence on the correct class y:

$$\mathcal{L}_{CEM} = \max\Big(0,\ m - \big[\log p^{keep}(y) - \log p^{drop}(y)\big]\Big), \tag{2}$$

where $m > 0$ is a margin hyperparameter. Minimizing $\mathcal{L}_{CEM}$ promotes sufficiency via increasing $p^{keep}(y)$ and necessity via decreasing $p^{drop}(y)$.

2.3 Adaptive Prior Suppression (APS)

Although CEM encourages evidence-based reasoning, Med-VQA models may still exploit question-conditioned priors. Following the question-only prior paradigm in debiased VQA [5,6], we introduce APS to suppress such priors in an instance-adaptive manner. Specifically, we add an auxiliary question-only branch that outputs logits $z_b(q) \in \mathbb{R}^{|\mathcal{A}|}$, from which we obtain the prior distribution $p_b(\cdot|q) = \text{softmax}(z_b(q))$. We compute a prior-strength score s_q from $p_b(\cdot|q)$, where larger s_q indicates a more confident (lower-entropy and peakier) question-only prediction. To reflect the reliability of visual evidence, we compute an evidence-strength score using CEM counterfactual predictions as $s_v = p^{keep}(y) - p^{drop}(y)$, where larger s_v indicates that keeping evidence increases confidence for the correct answer. The sample-adaptive suppression coefficient is defined as:

$$\alpha(v,q) = \text{clip}\big(\alpha_0 + \alpha_1 s_q - \alpha_2 s_v,\ \alpha_{\min}, \alpha_{\max}\big), \tag{3}$$

where α_0 is a base level, α_1 and α_2 control the sensitivity to s_q and s_v, and $\text{clip}(\cdot, \alpha_{\min}, \alpha_{\max})$ constrains α to a fixed interval. Let $z_f(v,q) \in \mathbb{R}^{|\mathcal{A}|}$ denote the fused logits from the backbone. We calibrate them by subtracting the scaled

Table 1. Performance comparison on SLAKE-CP and VQA-RAD-CP datasets.

Methods	SLAKE-CP			VQA-RAD-CP		
	All	Open	Closed	All	Open	Closed
SAN [19]	26.02	48.30	6.42	16.29	59.73	6.05
MFB [20]	30.56	55.70	8.44	22.53	72.12	10.84
BAN [8]	17.50	30.90	5.72	17.30	67.70	5.42
UpDn [1]	31.45	59.90	6.42	26.67	74.78	15.33
MEVF+SAN [15]	18.62	32.60	6.33	22.11	68.14	11.26
MEVF+BAN [15]	19.33	35.00	5.54	19.07	62.39	8.86
RUBi [5]	33.88	60.30	10.64	81.27	60.62	86.13
LPF [11]	40.34	43.70	37.38	41.52	65.04	35.97
GGE-iter [7]	35.05	61.30	11.96	21.60	51.33	14.60
RMLVQA [4]	76.42	60.50	**90.41**	89.45	69.03	94.26
CEDO [22]	79.27	66.70	90.33	92.07	73.45	**96.45**
Med-BiasX [23]	78.33	64.80	90.24	91.48	76.11	95.10
Qwen2.5-VL-7B-Instruct [3]	70.85	61.60	78.98	65.63	53.50	75.30
InternVL3-8B [24]	70.43	62.50	77.40	65.19	50.50	76.89
Med-Flamingo [14]	30.65	43.15	16.42	45.68	36.00	53.39
HuatuoGPT-Vision-7B [21]	68.65	62.30	74.23	67.18	58.00	74.50
MedGemma-1.5-4B-IT [17]	65.56	65.40	65.70	62.08	45.50	75.30
LLaVA-Med-v1.5-Mistral-7B [10]	50.02	57.12	41.94	45.68	35.00	54.18
HealthGPT [12]	25.18	23.30	26.82	58.54	47.00	67.73
Med-CAP	**79.97**	**68.30**	90.24	**92.24**	**76.99**	95.83
Improvement ↑	+0.70%	+1.60%	-0.17%	+0.17%	+0.88%	-0.62%

prior logits, $z^*(v,q) = z_f(v,q) - \alpha(v,q)\, z_b(q)$ and $p^*(\cdot|v,q) = \text{softmax}(z^*(v,q))$, then APS is optimized with cross-entropy on the calibrated distribution:

$$\mathcal{L}_{APS} = -\log p^*(y|v,q), \tag{4}$$

where $p^*(y|v,q)$ denotes the calibrated probability assigned to the golden class.

2.4 Overall Objective

For each training sample, the final loss is defined as

$$\mathcal{L} = \mathcal{L}_{base} + \mathcal{L}_{CEM} + \mathcal{L}_{APS}, \tag{5}$$

where $\mathcal{L}_{base}$ follows the training objective of RMLVQA [4], and the underlying Med-VQA backbone is instantiated with the UpDn architecture [1]. The overall framework is illustrated in Fig. 1, and we name the proposed method **Med-CAP**.

3 Experiments

3.1 Datasets and Implementation Details

We evaluate Med-CAP on two widely-used Med-VQA benchmarks, SLAKE [13] and VQA-RAD [9]. To assess robustness against language priors, we additionally adopt two bias-sensitive benchmarks, SLAKE-CP and VQA-RAD-CP [22]. Following previous works, we report accuracy (%) on *Open*, *Closed*, and *All* question types. We implement all models in PyTorch and train on a single NVIDIA A100 GPU. We adopt UpDn [1] as the backbone architecture and use AdamW with weight decay 1×10^{-3}. The batch size is set to $B = 60$ and the learning rate is 2×10^{-3}. All results are obtained by averaging the scores over three runs with different random seeds. Following the standard VQA evaluation protocol adopted by prior Med-VQA debiasing works [4,22,23], accuracy for each question type is computed over the entire test set; samples whose ground-truth answers are absent from the training answer vocabulary contribute zero to their type's accuracy.

3.2 Comparison with State-of-the-Art Methods

We compare with: (i) classical VQA models (SAN, MFB, BAN, UpDn), (ii) medical debiasing baselines (MEVF-based), (iii) natural-image debiasing methods adapted to Med-VQA (RUBi, LPF, GGE-iter, RMLVQA, CEDO, Med-BiasX), and (iv) recent vision-language models (VLM/medical VLM) including Qwen2.5-VL, InternVL3, Med-Flamingo, HuatuoGPT-Vision, MedGemma, LLaVA-Med, and HealthGPT. All reported numbers are from the original papers or reproduced under the same protocol for fair and consistent comparison.

Table 1 summarizes results on SLAKE-CP and VQA-RAD-CP. Our method achieves the best overall performance on both datasets. Notably, on SLAKE-CP, our model improves the overall accuracy from 79.27 to 79.97 compared with the strongest competitor, and also yields consistent gains on Open questions. On VQA-RAD-CP, our method achieves 92.24 overall accuracy, demonstrating improved robustness to language priors. Table 2 reports results on the original SLAKE and VQA-RAD benchmarks. Our approach achieves the best overall performance, indicating that the proposed debiasing strategy improves robustness to bias without compromising standard predictive performance, thereby striking a favorable balance between bias mitigation and in-distribution accuracy.

Table 2. Performance comparison on SLAKE and VQA-RAD datasets.

Methods	SLAKE			VQA-RAD		
	All	Open	Closed	All	Open	Closed
SAN [19]	76.00	74.00	79.10	52.89	31.64	65.50
MFB [20]	73.89	71.63	77.40	54.10	41.90	62.13
BAN [8]	76.25	75.97	76.68	55.43	48.60	59.93
UpDn [1]	81.34	79.84	83.65	66.74	51.40	76.47
MEVF+SAN [15]	75.97	74.72	77.88	60.71	40.65	74.05
MEVF+BAN [15]	77.76	75.97	80.53	62.34	43.09	75.14
RUBi [5]	78.42	76.43	81.49	51.22	36.87	60.66
LPF [11]	75.59	73.33	79.09	56.32	49.72	60.66
GGE-iter [7]	79.83	79.22	80.77	65.19	49.16	75.74
RMLVQA [4]	81.43	80.47	82.93	65.41	49.16	76.10
CEDO [22]	83.41	81.09	87.02	67.41	**58.66**	73.16
Med-BiasX [23]	82.47	80.62	86.34	67.42	50.64	**78.46**
Qwen2.5-VL-7B-Instruct [3]	68.43	66.76	76.62	65.85	52.50	76.49
InternVL3-8B [24]	73.02	71.02	83.10	65.41	52.00	76.10
Med-Flamingo [14]	23.78	22.31	30.99	37.47	32.00	41.83
HuatuoGPT-Vision-7B [21]	66.92	63.88	71.63	67.63	58.00	75.30
MedGemma-1.5-4B-IT [17]	69.87	68.83	75.21	60.31	47.00	70.92
LLaVA-Med-v1.5-Mistral-7B [10]	51.43	50.72	54.93	44.12	32.50	53.39
HealthGPT [12]	26.12	24.72	32.96	32.37	17.50	44.22
Med-CAP	**84.26**	**81.65**	**88.43**	**69.18**	55.31	78.31
Improvement ↑	+0.85%	+0.56%	+1.41%	+1.76%	-3.35%	-0.15%

3.3 Ablation Study

We conduct ablations to isolate the contribution of each component. Table 3 reports results on the bias-sensitive benchmarks (SLAKE-CP and VQA-RAD-CP), while Table 4 reports results on the standard benchmarks (SLAKE and VQA-RAD). Adding either CEM or APS consistently improves the baseline, and combining both yields the best performance across datasets. These results validate that CEM provides effective counterfactual supervision, while APS further stabilizes predictions by suppressing question-only shortcuts when visual evidence is weak. These ablation results confirm that the proposed components highlight the effectiveness and generalizability of the full model design.

Table 3. Ablation study of CEM and APS on SLAKE-CP and VQA-RAD-CP.

Methods	CEM	APS	SLAKE-CP			VQA-RAD-CP		
			All	Open	Closed	All	Open	Closed
Baseline	–	–	76.42	64.80	**90.41**	89.54	73.89	93.22
w/ CEM	✓	–	78.15	65.00	89.71	91.39	72.12	**95.93**
w/ APS	–	✓	78.10	64.20	90.32	91.05	73.45	95.20
Med-CAP	✓	✓	**79.97**	**68.30**	90.24	**92.24**	**76.99**	95.83

Table 4. Ablation study of CEM and APS on SLAKE and VQA-RAD.

Methods	CEM	APS	SLAKE			VQA-RAD		
			All	Open	Closed	All	Open	Closed
Baseline	–	–	81.43	80.47	82.93	65.41	49.16	76.10
w/ CEM	✓	–	83.13	80.87	86.74	68.29	54.75	77.21
w/ APS	–	✓	82.28	79.15	87.10	67.85	54.19	76.84
Med-CAP	✓	✓	**84.26**	**81.65**	**88.43**	**69.18**	**55.31**	**78.31**

3.4 Parameter Analysis

We conduct a parameter analysis of Med-CAP on SLAKE-CP to study the influence of the key hyperparameters in CEM and APS. As shown in Fig. 2, Med-CAP achieves the best performance with $\beta = 0.2$ and $(\alpha_0, \alpha_1, \alpha_2) = (0.5, 0.3, 0.2)$. This suggests that a moderate counterfactual feature suppression in CEM, together with an appropriate adaptive prior suppression schedule in APS, is crucial for improving robustness and overall accuracy.

3.5 Case Analysis

To qualitatively assess Med-CAP, we compare it with the baseline on SLAKE-CP (Fig. 3). The baseline tends to predict the majority answers from the training distribution, indicating a reliance on language priors rather than visual evidence. For instance, it incorrectly follows dominant training biases (e.g., predicting frequent answers such as "2" or "Yes") despite opposite ground truths and shifted test distributions, and its attention regions are less aligned with clinically relevant cues. Med-CAP suppresses question-only priors and produces predictions grounded in relevant visual evidence. Consequently, it focuses on more appropriate regions and correctly answers both prior-opposed cases, demonstrating improved robustness to language bias under distribution shift. These qualitative results further corroborate the quantitative findings, highlighting that Med-CAP improves interpretability and robustness to language bias, particularly under distribution shift scenarios commonly encountered in medical VQA tasks.

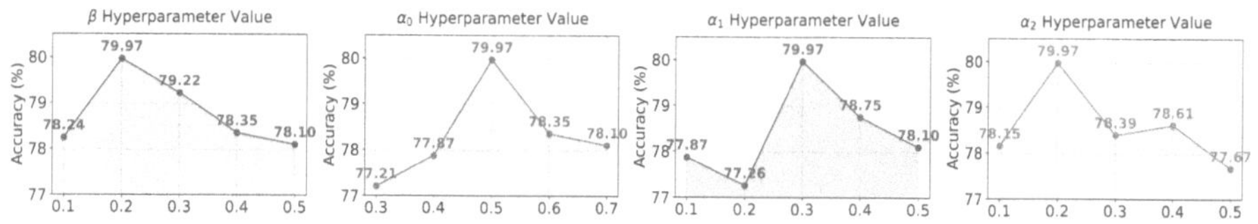

Fig. 2. Accuracy comparison on SLAKE-CP under different hyperparameter settings.

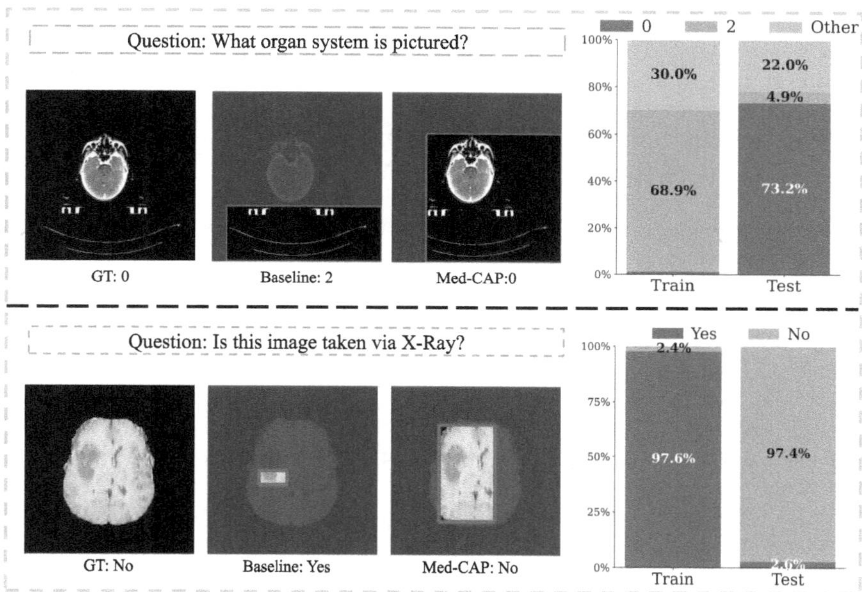

Fig. 3. Case analysis of the proposed Med-CAP. Two representative examples are shown for the "Yes/No" (up) and "How many" (down) question types.

4 Conclusion

In this paper, we propose Med-CAP, a counterfactual-and-calibration framework for robust medical visual question answering. By integrating Counterfactual Evidence Modeling (CEM) and Adaptive Prior Suppression (APS), our method promotes evidence-grounded reasoning while mitigating medical language bias. CEM enforces the sufficiency and necessity of visual evidence through counterfactual supervision, whereas APS adaptively suppresses over-confident question-only priors based on prior strength and evidence strength. Extensive experiments on both standard benchmarks (SLAKE/VQA-RAD) and bias-sensitive protocols (SLAKE-CP/VQA-RAD-CP) demonstrate that Med-CAP improves robustness to language bias while maintaining competitive in-distribution accuracy. Future work will explore extending the framework to larger-scale medical vision-language models and more complex clinical scenarios.

Disclosure of Interests. The authors have no competing interests to declare that are relevant to the content of this article.

References

1. Anderson, P., He, X., Buehler, C., Teney, D., Johnson, M., Gould, S., Zhang, L.: Bottom-up and top-down attention for image captioning and visual question answering. In: Proceedings of the IEEE Conference on Computer Vision and Pattern Recognition, pp. 6077–6086 (2018)
2. Antol, S., Agrawal, A., Lu, J., Mitchell, M., Batra, D., Zitnick, C.L., Parikh, D.: VQA: visual question answering. In: Proceedings of the IEEE International Conference on Computer Vision, pp. 2425–2433 (2015)
3. Bai, S., et al.: Qwen2.5-vl technical report (2025). https://arxiv.org/abs/2502.13923
4. Basu, A., Addepalli, S., Babu, R.V.: Rmlvqa: a margin loss approach for visual question answering with language biases. In: Proceedings of the IEEE/CVF Conference on Computer Vision and Pattern Recognition, pp. 11671–11680 (2023)
5. Cadene, R., Dancette, C., Cord, M., Parikh, D.: Rubi: reducing unimodal biases for visual question answering. Adv. Neural. Inf. Process. Syst. **32** (2019)
6. Clark, C., Yatskar, M., Zettlemoyer, L.: Don't take the easy way out: ensemble based methods for avoiding known dataset biases. In: Proceedings of the 2019 Conference on Empirical Methods in Natural Language Processing and the 9th International Joint Conference on Natural Language Processing (EMNLP-IJCNLP), pp. 4069–4082 (2019)
7. Han, X., Wang, S., Su, C., Huang, Q., Tian, Q.: Greedy gradient ensemble for robust visual question answering. In: Proceedings of the IEEE/CVF International Conference on Computer Vision, pp. 1584–1593 (2021)
8. Kim, J.H., Jun, J., Zhang, B.T.: Bilinear attention networks. In: Advances in Neural Information Processing Systems, p. 31 (2018)
9. Lau, J.J., Gayen, S., Ben Abacha, A., Demner-Fushman, D.: A dataset of clinically generated visual questions and answers about radiology images. Sci. data **5**(1), 180251 (2018)
10. Li, C., et al.: Llava-med: training a large language-and-vision assistant for biomedicine in one day. Adv. Neural. Inf. Process. Syst. **36**, 28541–28564 (2023)
11. Liang, Z., Hu, H., Zhu, J.: LPF: a language-prior feedback objective function for debiased visual question answering. In: Proceedings of the 44th International ACM SIGIR Conference on Research and Development in Information Retrieval, pp. 1955–1959 (2021)
12. Lin, T., et al.: HealthGPT: a medical large vision-language model for unifying comprehension and generation via heterogeneous knowledge adaptation. In: Forty-second International Conference on Machine Learning (2025) openReview: WbP2OwMULq
13. Liu, B., Zhan, L.M., Xu, L., Ma, L., Yang, Y., Wu, X.M.: Slake: a semantically-labeled knowledge-enhanced dataset for medical visual question answering. In: 2021 IEEE 18th International Symposium on Biomedical Imaging (ISBI), pp. 1650–1654. IEEE (2021)
14. Moor, M., et al.: Med-flamingo: a multimodal medical few-shot learner. In: Machine Learning for Health (ML4H), pp. 353–367. PMLR (2023)

15. Nguyen, B.D., Do, T.T., Nguyen, B.X., Do, T., Tjiputra, E., Tran, Q.D.: Overcoming data limitation in medical visual question answering. In: International Conference on Medical Image Computing and Computer-assisted Intervention, pp. 522–530. Springer (2019)
16. Niu, Y., Tang, K., Zhang, H., Lu, Z., Hua, X.S., Wen, J.R.: Counterfactual VQA: a cause-effect look at language bias. In: Proceedings of the IEEE/CVF Conference on Computer Vision and Pattern Recognition, pp. 12700–12710. (2021)
17. Sellergren, A., et al.: Medgemma technical report (2025). arXiv:2507.05201 arXiv preprint
18. Selvaraju, R.R., Cogswell, M., Das, A., Vedantam, R., Parikh, D., Batra, D.: Grad-cam: Visual explanations from deep networks via gradient-based localization. In: Proceedings of the IEEE International Conference on Computer Vision, pp. 618–626 (2017)
19. Yang, Z., He, X., Gao, J., Deng, L., Smola, A.: Stacked attention networks for image question answering. In: Proceedings of the IEEE Conference on Computer Vision and Pattern Recognition, pp. 21–29 (2016)
20. Yu, Z., Yu, J., Fan, J., Tao, D.: Multi-modal factorized bilinear pooling with co-attention learning for visual question answering. In: Proceedings of the IEEE International Conference on Computer Vision, pp. 1821–1830 (2017)
21. Zhang, H., et al.: Huatuogpt, towards taming language model to be a doctor. In: Findings of the association for computational linguistics: EMNLP 2023, pp. 10859–10885 (2023)
22. Zhu, H., Liu, Y., Fang, X., Lu, G., Chen, B.: Cause-effect driven optimization for robust medical visual question answering with language biases (2025). arXiv:2506.17903 arXiv preprint
23. Zhu, H., Liu, Y., Zhou, C., Lu, G., Chen, B.: Med-biasx: robust medical visual question answering with language biases. In: International Conference on Medical Image Computing and Computer-Assisted Intervention, pp. 369–378. Springer (2025)
24. Zhu, J., et al.: Internvl3: exploring advanced training and test-time recipes for open-source multimodal models (2025). arXiv:2504.10479 arXiv preprint

MedFuse-Seg: Multi-level Visual and Semantic Context Fusion for Segmentation-Based Medical Reasoning

Keetawan Limaroon[1], Monrada Chiewhawan[2,3], Watcharapong Timklaypachara[2,3], Peerapon Vateekul[4(✉)], and Titipat Achakulvisut[2(✉)]

[1] Department of Computer Engineering, Faculty of Engineering, King Mongkut's University of Technology Thonburi, Bangkok, Thailand

[2] Department of Biomedical Engineering, Faculty of Engineering, Mahidol University, Nakhon Pathom, Thailand
titipat.ach@mahidol.ac.th

[3] Faculty of Medicine Ramathibodi Hospital, Mahidol University, Bangkok, Thailand

[4] Department of Computer Engineering, Faculty of Engineering, Chulalongkorn University, Bangkok, Thailand
peerapon.v@chula.ac.th

Abstract. Language-driven medical image segmentation is critical for clinical workflows, yet existing models struggle with linguistic variability and complex diagnostic queries. These architectures often process semantic and spatial information in isolation, making it difficult to merge abstract clinical reasoning with precise anatomical boundaries. To address these limitations, we introduce **MedFuse-Seg**, an architecture that combines complex reasoning via Q&A generation with multi-level visual feature injection for prompt-based medical image segmentation. First, we construct **Med-ReasonSeg**, a large-scale reasoning segmentation dataset consisting of 539,383 image-mask-Q&A triplets and encompassing 9 modalities from 16 datasets. We generate referring and semantic Q&A pairs that are post-processed with a hierarchical LLM-based pipeline to ensure logical fidelity and reduce hallucinations. Second, we propose a **Multi-Level Context Fusion** mechanism that injects hierarchical features from MedSigLIP (L6-L24) directly into the MedSAM encoder, while leveraging MedGemma reasoning to guide mask decoding. MedFuse-Seg outperforms zero-shot BiomedParse and the fine-tuned LISA-7B by 13.49% and 4.89% in Dice Similarity Coefficient (DSC), and reduces 95th percentile Hausdorff Distance (HD95) by 54.04 and 15.29 pixels, respectively. These results demonstrate that combining reasoning-augmented training data with multi-level feature injection can address semantic and spatial gaps in medical segmentation. Our code and dataset are publicly available at https://github.com/biodatlab/medfuse-seg.

G. Yang et al. (Eds.): MICCAI 2026, LNCS 16878, pp. 324–334, 2027.
https://doi.org/10.1007/978-3-032-38059-3_31

Keywords: Medical Image Segmentation · Multimodal Large Language Model · Reasoning Segmentation · Large-scale Dataset Construction

1 Introduction

Medical image segmentation is essential for clinical workflows, from diagnosis to treatment planning [12]. While traditional models like U-Net [10,23] established a strong baseline, they are limited to a fixed set of categories. To address this, foundation models like MedSAM [3,11,17] introduced zero-shot segmentation using spatial prompts (such as points or boxes). However, this approach remains inefficient, as it requires clinicians to provide precise location markers manually and often fails to capture irregular lesions due to fixed geometric shapes [5]. Consequently, the field is shifting toward language-driven frameworks that ground segmentation directly in diagnostic descriptions.

Despite their promise, current frameworks face distinct challenges. Generalist foundation models ranging from detectors like Grounding DINO [14] to text-promptable segmenters like SAM 3 [2] excel in natural images but struggle with medical anatomy [17]. Conversely, specialist models like BiomedParse [28] are hindered by limited data diversity. These methods rely on rigid templates, using large language models (LLMs) only for synonym generation in the training set. Coupled with the shallow reasoning of standard text encoders [25], these models are highly sensitive to how prompts are phrased. This results in poor performance when faced with the varied language used in real-world clinical queries [22].

To bridge the reasoning gap in segmentation, recent works, like LISA [13,27], adapt multimodal large language models (MLLMs) to compress visual data into a single segmentation token. However, these architectures often process understanding and localizing in isolation, losing fine-grained boundary details during compression [16,20]. We hypothesize that complex segmentation queries require models that not only reason effectively but also perceive with expert-level depth. While standard pipelines rely on final-layer outputs that prioritize abstract meaning over spatial precision, we leverage multi-level visual features to preserve the fine-grained anatomical details often lost during alignment [1].

We introduce **MedFuse-Seg**, an architecture integrating complex reasoning with multi-level visual features for medical segmentation. We construct **MedReasonSeg**, a large-scale reasoning dataset of 539,383 image-mask-Q&A triplets from 16 datasets, refined via LLM verification to ensure logical accuracy and minimize hallucinations. To connect semantic and spatial context, we introduce a **Multi-Level Context Fusion** mechanism that injects detailed features from MedSigLIP into the MedSAM encoder. Our model improves prompt-based segmentation by 13.49% and 4.89% in DSC compared to zero-shot BiomedParse and fine-tuned LISA-7B baselines, respectively. These contributions establish a more practical and resilient framework for language-driven medical image analysis.

2 Med-ReasonSeg Dataset

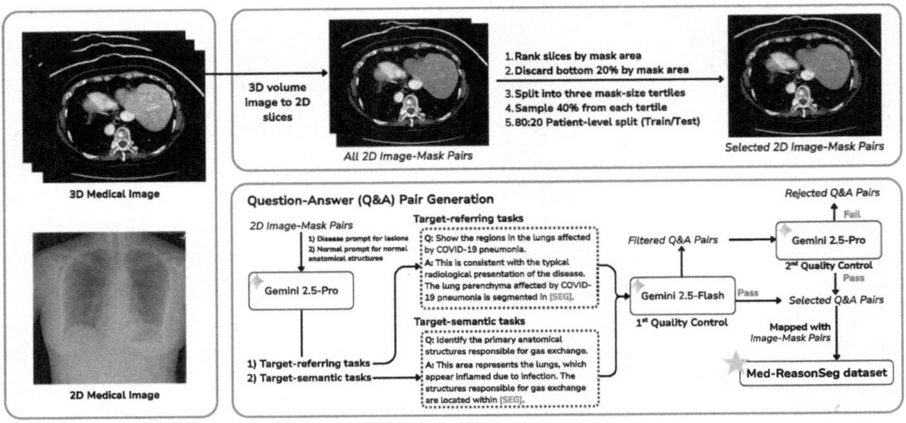

Fig. 1. Overview of the Med-ReasonSeg dataset construction pipeline. The process involves: (i) Data Curation, employing slice sampling from 3D volumes based on mask size; (ii) Instruction Generation, synthesizing target-referring and target-semantic prompts via Gemini 2.5-Pro; and (iii) LLM-Based Q&A Verification, a two-stage text quality control using Gemini 2.5-Flash and Pro.

Robust language-driven segmentation models require contextual reasoning beyond simple entity referral. While BiomedParse established a unified multi-modality parsing standard, its prompts rely on rule-based templates with GPT-4 used only for lexical rephrasing [28]. To address this, we constructed **Med-ReasonSeg**, aggregating 16 open-access datasets primarily curated from the BiomedParse collection [28] across 9 imaging modalities. The final dataset comprises 539,383 image–mask–Q&A triplets derived from 90,021 distinct scans. MRI (48.6%) and CT (27.3%) constitute the majority, followed by X-ray (9.9%), with the remaining 14.2% distributed across dermoscopy, fundus, endoscopy, OCT, mammography, and ultrasound. The construction process involves three key stages: (i) Data Curation, (ii) Question-and-Answer (Q&A) Pair Generation, and (iii) LLM-Based Q&A Verification (Fig. 1).

2.1 Data Curation

We adopted the data distribution protocol from BiomedParse [28] to align with established benchmarks. For 3D volumes, we processed them as 2D slices, discarding the bottom 20% based on mask area rankings to exclude uninformative samples with negligible anatomy. We then stratified the remaining slices into three mask-size tertiles and sampled 40% from each to mitigate inter-slice redundancy while balancing training efficiency. To strictly prevent data leakage, all data were split into training and test sets (80:20) by patient ID, ensuring that all slices from the same patient remained in the same split.

2.2 Question-and-Answer (Q&A) Pair Generation

We leveraged Multimodal Large Language Models (MLLMs) to generate Q&A pairs for model training and evaluation. We designed two task types adapted for biomedical contexts: **(i) Target-referring tasks** directly name the segmentation target with linguistic variations (e.g., Q: "Show the regions affected by COVID-19", A: "This pattern is consistent with viral pneumonia. The affected lung parenchyma is segmented in `[SEG]`."). This task evaluates robustness to variation and ensures coverage of synonyms, abbreviations, and multi-word expressions for anatomical structures or diseases. **(ii) Target-semantic tasks** describe targets through their functional, spatial, or visual properties without explicit naming (e.g., Q: "Identify the structures for gas exchange", A: "This area represents the lungs, which appear inflamed. These structures are located within `[SEG]`."). This task type allows the model to learn implicit reasoning. By jointly training on both tasks, we aim to align explicit and implicit textual prompts with segmentation targets across diverse clinical inputs.

To generate these pairs, we employed Gemini 2.5-Pro, selected for its superior multimodal reasoning [4], using both images and their corresponding masks as inputs. This visual grounding allowed us to synthesize six distinct instructions per unique mask (three referring, three semantic) to enhance generalization, scaling the dataset to 541,752 pairs (429,798 training and 111,954 testing).

2.3 LLM-Based Q&A Verification

To ensure dataset integrity, we verified 541,752 pairs via a two-stage pipeline. Following a rule-based check to exclude corrupted images, null masks, or malformed responses missing the `[SEG]` token, we narrowed the pool to 541,662 candidates and employed a hierarchical filtering strategy. First, Gemini 2.5-Flash served as an efficient screener to process the extensive dataset, flagging 5,186 pairs for potential logical inconsistencies or hallucinations. Subsequently, leveraging its advanced reasoning capabilities, Gemini 2.5-Pro acted as a specialist to re-evaluate these flagged instances. This expert review recovered 2,907 over-penalized samples, while discarding 2,279 noisy pairs. This rigorous process yielded a final high-quality dataset of 539,383 pairs (427,861 training and 111,522 testing), each comprising the question, answer, modality, and target organ.

3 MedFuse-Seg Architecture

MedFuse-Seg architecture integrates three foundation models including MedSigLIP [24], MedGemma-4B [24] and MedSAM [17] through two core mechanisms: (i) Multi-Level Context Fusion and (ii) Reasoning-Guided Mask Decoding (Fig. 2). First, for reasoning-enhanced text encoding, we utilize MedSigLIP's visual features inherently coupled with MedGemma's language backbone (Based on Gemma 3 [6]) and fine-tune this architecture to generate reasoning tokens. This enables the model to interpret clinical language by grounding text prompts

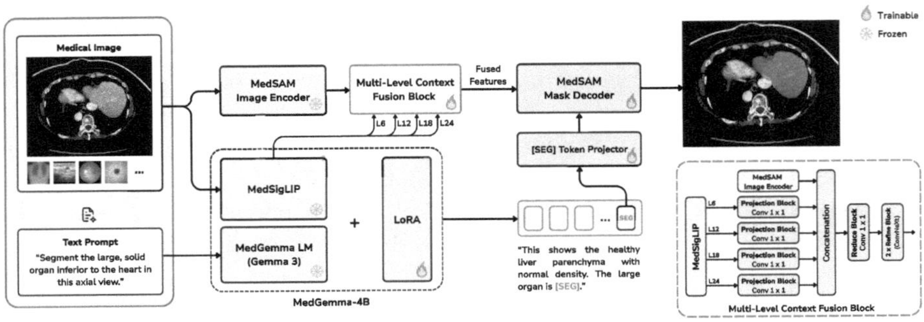

Fig. 2. Overview of the MedFuse-Seg architecture. The proposed framework integrates three foundation models through two core mechanisms: (i) Multi-Level Context Fusion for integrating MedSigLIP's medical knowledge with MedSAM's spatial features, and (ii) Reasoning-Guided Mask Decoding via MedGemma-4B for reasoning-enhanced segmentation.

in image content. Second, for spatial segmentation, we introduce a Multi-Level Context Fusion Block that injects MedSigLIP's hierarchical features directly into MedSAM's encoder at multiple scales. Our design maximizes the utility of specialized medical representations while maintaining parameter efficiency through shared visual encoding.

3.1 Multi-level Context Fusion

Recent work shows that intermediate vision encoder layers retain spatial details often lost in final outputs [10], aligning with hierarchical medical segmentation architectures like UNETR [8] and Swin UNETR [7]. We translate this into a hierarchical fusion strategy (Fig. 2) that integrates MedSigLIP semantics with MedSAM spatial priors. We extract multi-level features from MedSigLIP layers $L6$, $L12$, $L18$, and $L24$, where early layers capture low-level details (edges, shapes) and deeper layers encode high-level medical semantics. Each feature is projected via 1×1 *convolution* to align dimensions, then concatenated with MedSAM embeddings to create a unified representation. A Reduce Block (1×1 *convolution*) synthesizes these concatenated features through cross-channel interaction, combining semantic and spatial cues while maintaining compatibility with the MedSAM encoder. Finally, a Refine Block utilizing two cascaded large-kernel ConvNeXt blocks [26] captures long-range dependencies to refine feature representations before mask decoding. This strategy mitigates semantic ambiguity by injecting MedSigLIP's medical priors into MedSAM's boundary delineation, establishing a robust foundation for reasoning-guided decoding.

3.2 Reasoning-Guided Mask Decoding

While the fusion block provides rich visual features, the segmentation target is defined by the LLM. Following LISA [13], we add a special `[SEG]` token to the vocabulary. When the model identifies a target region, it generates this token, whose hidden state is projected as the prompt embedding. The MedSAM mask decoder then receives the refined image features and reasoning prompt to generate the final mask. By conditioning on the LLM's output, MedFuse-Seg aligns clinical reasoning to segmentation.

4 Experiments and Results

Implementation Details. We implemented our framework using PyTorch [19] and DeepSpeed [21] on 4 NVIDIA A100 GPUs. We adopted a parameter-efficient fine-tuning strategy via Low-Rank Adaptation (LoRA) [9], applying adapters ($r = 64, \alpha = 128$) to all linear layers in the MedSigLIP encoder and MedGemma model. The pre-trained backbones were frozen, while we optimized the LoRA parameters, projectors, and segmentation decoder. The model was trained for 5 epochs with a global batch size of 32 using the AdamW optimizer [15] configured with an initial learning rate of 10^{-4} and a cosine annealing scheduler with a 3% linear warmup. To maximize computational throughput, we utilized mixed precision training (bfloat16) and gradient checkpointing. All fine-tuned models used identical training settings.

Data Augmentation. We applied augmentations to enhance generalization while preserving consistency between images and Q&A pairs, such as directional terms (e.g., "left", "top"). We used soft geometric transformations ($p = 0.5$) with limited ranges: rotation ($\pm 15°$), shift (5%), and scale (20%). We added Gaussian blur and noise ($p = 0.15$), and brightness/contrast adjustments ($p = 0.5$).

Loss Function and Evaluation Metrics. We minimized a joint objective for text generation and mask segmentation: $\mathcal{L} = \mathcal{L}_{text} + \mathcal{L}_{mask}$. For text generation, we employed standard auto-regressive cross-entropy ($\mathcal{L}_{text}$) to ensure semantic consistency. For segmentation, we combined Focal ($\gamma = 2, \alpha = 0.25$) and Dice losses to address class imbalance while maximizing structural precision: $\mathcal{L}_{mask} = 4\mathcal{L}_{Focal} + 2\mathcal{L}_{Dice}$. For evaluation on the Med-ReasonSeg test set, we assessed performance using the Dice Similarity Coefficient (DSC) to measure region overlap and the 95th percentile Hausdorff Distance (HD95) to capture boundary adherence and shape discrepancies [18]. All metrics are computed at the model's standard resolution (1024^2 pixels) and reported as macro-averages.

4.1 Main Results

We benchmark MedFuse-Seg against state-of-the-art approaches in Table 1. The generalist foundation model (SAM 3) and the hybrid approach (G-DINO +

Table 1. Main Benchmarking Results. Performance comparison on Target-referring and Target-semantic tasks across various architectures. Metrics: DSC (↑) and HD95 (↓, px). Experimental settings are tailored to each model's design: *Oracle* (MedSAM w/ GT boxes), *Zero-shot* for foundation models (SAM 3, G-DINO + MedSAM, Biomed-Parse) to evaluate generalization, and *Fine-tuning* for LISA-7B to ensure a competitive reasoning baseline. Best results are in **bold**, second best are <u>underlined</u>.

Method	Type	Referring		Semantic		Average	
		DSC ↑	HD95 ↓	DSC ↑	HD95 ↓	DSC ↑	HD95 ↓
Traditional / Oracle							
MedSAM (w/ GT Box) [17]	Specialist	-	-	-	-	0.6329	<u>58.41</u>
Foundation / Zero-shot							
SAM 3 [2]	Foundation	0.1425	373.52	0.1167	370.50	0.1296	372.01
BiomedParse [28]	Specialist	0.6703	105.23	0.6344	115.97	0.6524	110.60
Detection-Guided Models							
G-DINO [14] + MedSAM [17]	Hybrid	0.1424	494.06	0.1461	488.56	0.1442	491.31
Reasoning Segmentation							
LISA-7B (ft) [13]	Generalist	<u>0.7398</u>	<u>71.55</u>	<u>0.7370</u>	<u>72.13</u>	<u>0.7384</u>	71.84
MedFuse-Seg (Ours)	**Specialist**	**0.7879**	**56.46**	**0.7867**	**56.65**	**0.7873**	**56.55**

MedSAM) struggle significantly in the zero-shot setting (DSC < 0.15), likely due to the substantial domain gap. While the medical specialist BiomedParse shows improved generalization, it lags behind our method, particularly in semantic tasks. Overall, MedFuse-Seg outperforms zero-shot BiomedParse by 13.49% in DSC and 54.04 pixels in HD95. Notably, even when the LISA-7B baseline is fully fine-tuned on our dataset, our model maintains superior margins of 4.89% in DSC and 15.29 pixels in HD95. These gains show our Multi-Level Context Fusion effectively integrates visual and textual features, outperforming bottleneck-only fusion in existing reasoning architectures.

Qualitative analysis on the test set shows strong model performance in both segmentation and text output reasoning. The model can handle both descriptive (Fig. 3A and 3B) and direct segmentation requests (Fig. 3C). Moreover, the model can perform complex segmentation and incorporate clinical disease knowledge into its responses (Fig. 3C and 3D). Preliminary expert review of these outputs suggests overall clinical soundness. However, performance remains limited for some underrepresented irregular lesions. Incorporating more diverse, disease-focused training data may further improve performance on these challenging cases.

4.2 Ablation Studies

Impact of Visual Feature Fusion Strategy. We observe that low-level features ($L6$, $L12$) contribute more to DSC from 0.7373 to 0.7865, whereas high-level features ($L18$, $L24$) contribute to HD95 from 69.74 to 56.55. Incorporating both

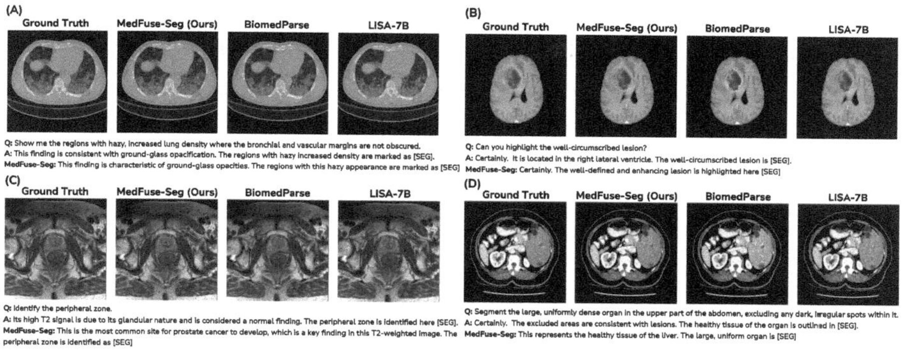

Fig. 3. Sample segmentation results of MedFuse-Seg, BiomedParse, and LISA-7B against ground truth, with queries, key answers, and MedFuse-Seg's responses.

Table 2. Comprehensive MedFuse-Seg Ablation Study. The default configuration uses MedSAM (Frozen), Multi-Level Fusion, Verified Data, and End-token positioning. The best results are in **bold**, second best are <u>underlined</u>.

Component	Variant / Setting	DSC ↑	HD95 ↓
Feature Fusion	Baseline (LISA-Style)	0.7373	69.74
	Spatial Focus ($L6, L12$)	<u>0.7865</u>	58.96
	Semantic Focus ($L18, L24$)	0.7786	<u>58.81</u>
	Multi-Level ($L6, L12, L18, L24$) (Ours)	**0.7873**	**56.55**
Visual Backbone	SAM (Unfrozen)	0.7090	98.27
	SAM (Frozen)	0.7740	70.24
	MedSAM (Unfrozen)	<u>0.7844</u>	<u>57.22</u>
	MedSAM (Frozen) (Ours)	**0.7873**	**56.55**
Data Verification	Raw / Unverified	0.7853	59.07
	LLM-Verified (Ours)	**0.7873**	**56.55**
[SEG] Position	Start	0.7851	57.63
	End (Ours)	**0.7873**	**56.55**

yields the highest performance, achieving DSC and HD95 of 0.7873 and 56.55, respectively (Table 2). This demonstrates that the fusion mechanism successfully captures intricate boundary precision in complex anatomical structures, rather than relying solely on text reasoning.

Impact of Visual Backbone and Training Strategy. MedSAM outperforms the generalist SAM, likely due to pretraining on large-scale medical images. Keeping the MedSAM frozen achieves higher performance (DSC = 0.7873) than full fine-tuning (DSC = 0.7844), indicating that preserving pretrained domain-specific representations is crucial for instruction-tuning stability.

Impact of LLM-Based Verification. Models trained on our LLM-verified dataset achieve superior segmentation stability over raw synthetic data (Table 2). LLM-based verification significantly improves DSC from 0.7853 to 0.7873 ($t = 6.91$, $p < 0.001$) and HD95 from 59.07 to 56.55 ($t = 8.99$, $p < 0.001$), confirming data quality is critical for minimizing noise and enhancing robustness.

Impact of [SEG] Token Positioning. Our experiment shows that appending `[SEG]` at the end (e.g., `description... [SEG]`) reduces HD95 from 57.63 to 56.55 ($t = 9.89, p < 0.001$) compared to placing it at the start (e.g., `[SEG] description...`). This demonstrates that leveraging the autoregressive nature of MLLMs in a "think-before-act" style may have an effect on the outputs that should be further explored.

5 Conclusion

MedFuse-Seg bridges clinical reasoning and precise segmentation by injecting hierarchical visual features into MLLMs, outperforming state-of-the-art models on both DSC and HD95. With our large-scale Med-ReasonSeg dataset, the model establishes a robust, medically-aware foundation for language-driven medical image analysis. Future work should explore volumetric and video segmentation with inter-slice consistency, comparisons with a broader range of medical reasoning models, and policy optimization for refining clinical reasoning.

Acknowledgments. This research has received funding support from the National Science, Research and Innovation Fund (NSRF) via the Program Management Unit for Human Resources & Institutional Development, Research and Innovation under Grant B13F680099.

Disclosure of Interests. The authors have no competing interests to declare that are relevant to the content of this article.

References

1. Bolya, D., et al.: Perception encoder: the best visual embeddings are not at the output of the network. In: The Thirty-ninth Annual Conference on Neural Information Processing Systems (2026)
2. Carion, N., et al.: SAM 3: segment anything with concepts. In: The Fourteenth International Conference on Learning Representations (2026)
3. Cheng, J., et al.: SAM-Med2D. arXiv preprint arXiv:2308.16184 (2023)
4. Comanici, G., et al.: Gemini 2.5: pushing the frontier with advanced reasoning, multimodality, long context, and next generation agentic capabilities (2025). arXiv:2507.06261 arXiv preprint
5. Du, Y., Bai, F., Huang, T., Zhao, B.: SegVol: universal and interactive volumetric medical image segmentation. Adv. Neural. Inf. Process. Syst. **37**, 110746–110783 (2024)
6. Gemma team: gemma 3 technical report (2025). arXiv:2503.19786 arXiv preprint

7. Hatamizadeh, A., Nath, V., Tang, Y., Yang, D., Roth, H.R., Xu, D.: Swin UNETR: swin transformers for semantic segmentation of brain tumors in MRI images. In: International MICCAI brainlesion workshop, pp. 272–284. Springer (2021)
8. Hatamizadeh, A., et al.: UNETR: transformers for 3d medical image segmentation. In: Proceedings of the IEEE/CVF Winter Conference on Applications of Computer Vision, pp. 574–584. (2022)
9. Hu, E.J., et al.: LoRA: low-rank adaptation of large language models. ICLR **1**(2), 3 (2022)
10. Isensee, F., Jaeger, P.F., Kohl, S.A., Petersen, J., Maier-Hein, K.H.: NNU-Net: a self-configuring method for deep learning-based biomedical image segmentation. Nat. Methods **18**(2), 203–211 (2021)
11. Kirillov, A., et al.: Proceedings of the IEEE/CVF International Conference on Computer Vision, pp. 4015–4026 (2023). Segment anything
12. Kumar, S.: Advancements in medical image segmentation: a review of transformer models. Comput. Electr. Eng. **123**, 110099 (2025)
13. Lai, X., Tian, Z., Chen, Y., Li, Y., Yuan, Y., Liu, S., Jia, J.: LISA: reasoning segmentation via large language model. In: Proceedings of the IEEE/CVF Conference on Computer Vision and Pattern Recognition, pp. 9579–9589 (2024)
14. Liu, S., et al.: Grounding DINO: marrying dino with grounded pre-training for open-set object detection. In: European Conference on Computer Vision, pp. 38–55. Springer (2024)
15. Loshchilov, I., Hutter, F.: Decoupled weight decay regularization. In: International Conference on Learning Representations (2019)
16. Lu, Y., et al.: RSVP: reasoning segmentation via visual prompting and multi-modal chain-of-thought. In: Che, W., Nabende, J., Shutova, E., Pilehvar, M.T. (eds.) Proceedings of the 63rd Annual Meeting of the Association for Computational Linguistics (Volume 1: Long Papers), pp. 14699–14716. Association for Computational Linguistics, Vienna, Austria (2025)
17. Ma, J., et al.: Segment anything in medical images. Nat. Commun. **15**(1), 654 (2024)
18. Maier-Hein, L., et al.: Metrics reloaded: recommendations for image analysis validation. Nat. Methods **21**(2), 195–212 (2024)
19. Paszke, A., et al.: Pytorch: an imperative style, high-performance deep learning library. Adv. Neural. Inf. Process. Syst. **32** (2019)
20. Qian, R., Yin, X., Dou, D.: Reasoning to attend: try to understand how <seg> token works. In: Proceedings of the Computer Vision and Pattern Recognition Conference, pp. 24722–24731 (2025)
21. Rasley, J., Rajbhandari, S., Ruwase, O., He, Y.: DeepSpeed: system optimizations enable training deep learning models with over 100 billion parameters. In: Proceedings of the 26th ACM SIGKDD international conference on knowledge discovery & data mining, pp. 3505–3506 (2020)
22. Rokuss, M., et al.: VoxTell: free-text promptable universal 3d medical image segmentation. In: Proceedings of the IEEE/CVF Conference on Computer Vision and Pattern Recognition (CVPR), pp. 37538–37557 (2026)
23. Ronneberger, O., Fischer, P., Brox, T.: U-Net: convolutional networks for biomedical image segmentation. In: International Conference on Medical image computing and computer-assisted intervention, pp. 234–241. Springer (2015)
24. Sellergren, A., et al.: Medgemma technical report (2025). arXiv:2507.05201 arXiv preprint
25. Wei, J., et al.: Chain-of-thought prompting elicits reasoning in large language models. Adv. Neural. Inf. Process. Syst. **35**, 24824–24837 (2022)

26. Woo, S., Debnath, S., Hu, R., Chen, X., Liu, Z., Kweon, I.S., Xie, S.: ConvNeXt V2: co-designing and scaling convnets with masked autoencoders. In: Proceedings of the IEEE/CVF Conference on Computer Vision and Pattern Recognition (CVPR), pp. 16133–16142 (2023)
27. Yang, S., Qu, T., Lai, X., Tian, Z., Peng, B., Liu, S., Jia, J.: LISA++: an improved baseline for reasoning segmentation with large language model. arXiv preprint arXiv:2312.17240 (2023)
28. Zhao, T., et al.: A foundation model for joint segmentation, detection and recognition of biomedical objects across nine modalities. Nat. Methods **22**(1), 166–176 (2025)

Medical Image Spatial Grounding with Semantic Sampling

Andrew Seohwan Yu[1,2], Mohsen Hariri[1], Kunio Nakamura[2], Mingrui Yang[2], Xiaojuan Li[1,2], and Vipin Chaudhary[1(✉)]

[1] Case Western Reserve University, Cleveland, OH 44106, USA
vipin@case.edu
[2] Cleveland Clinic, Cleveland, OH 44106, USA

Abstract. Vision language models (VLMs) have shown significant promise in visual grounding for images as well as videos. In medical imaging research, VLMs represent a bridge between object detection and segmentation, and report understanding and generation. However, spatial grounding of anatomical structures in the three-dimensional space of medical images poses many unique challenges. In this study, we examine image modalities, slice directions, and coordinate systems as differentiating factors for vision components of VLMs, and the use of anatomical, directional, and relational terminology as factors for the language components. We then demonstrate that visual and textual prompting systems such as labels, bounding boxes, and mask overlays have varying effects on the spatial grounding ability of VLMs. To enable measurement and reproducibility, we introduce **MIS-Ground**, a benchmark that comprehensively tests a VLM for vulnerabilities against specific modes of **M**edical **I**mage **S**patial **Ground**ing. We release MIS-Ground to the public at github.com/asy51/mis-ground. In addition, we present **MIS-SemSam**, a low-cost, inference-time, and model-agnostic optimization of VLMs that improves their spatial grounding ability with the use of **Sem**antic **Sam**pling. We find that MIS-SemSam improves the accuracy of Qwen3-VL-32B on MIS-Ground by 13.06%.

Keywords: multi-modal large language models · vision language models · 3D medical imaging · visual grounding · spatial grounding

1 Introduction

Merging the embedding spaces of vision and language models through contrastive learning [18] was a revelation that gave rise to the modern vision language model (VLM). Video VLMs were the logical next step, extending these ideas to spatiotemporal data where models must align sequences of frames with natural language. Despite the sheer scale of information in videos relative to images,

A. S. Yu and M. Hariri—Equal contribution.

G. Yang et al. (Eds.): MICCAI 2026, LNCS 16878, pp. 335–344, 2027.
https://doi.org/10.1007/978-3-032-38059-3_32

substantial progress has been made in both video understanding and multimodal reasoning [14].

Medical imaging VLMs have been a different matter. As narrow vision models continue to mature in the medical field, medical VLMs are in their early stages of development. Because the available data is limited, models, benchmarks, and training objectives tend to revolve around report generation [8,12] and visual question answering [11,13,15,20], especially disease identification. Though the reported results can be impressive, it is difficult to gauge the practicality of these VLMs in a setting as high-stakes as the medical field. Sambara et al. [19] showed that despite using the correct terminology and making references to anatomical components and lesions, VLMs failed to display reasoning capabilities that were spatially grounded in objects that were actually present in the image in question.

There has been recent work to introduce rigor in spatial grounding to medical VLM training and benchmarking, demonstrating that general-purpose 2D VLMs fail at even simple visual grounding of medical images, due in large part to language-only pretraining that contains ground truths about human anatomy, which may differ from the radiological image in question [23]. Methods for linking anatomical components to language using contrastive alignment in 2D images have also been introduced [10].

To push the current understanding of 3D VLM spatial grounding capabilities, this study proposes **M**edical **I**mage **S**patial **Ground**ing (**MIS-Ground**), a benchmark designed to identify the extent to which video VLMs are able to identify, localize, and differentiate anatomical structures in 3D medical images. MIS-Ground covers a wide variety of language and vision inputs that are likely to occur in medical imaging scenarios. These include different text and visual prompts to align the object of interest to text, multiple views of the same 3D volume, as well as various combinations of question classes and target answers (Fig. 1). Because a 3D volume can be serialized as an ordered stack of slices, video-native VLMs offer a practical interface for it; MIS-Ground is nonetheless not video-specific, and we also evaluate 3D-enabled medical VLMs (Sect. 2.3). To our knowledge, MIS-Ground is the first benchmark to probe 3D anatomical spatial grounding under a controlled, factorial design that jointly varies modality, slice direction, coordinate convention, prompt type, and terminology, whereas prior medical-grounding studies report aggregate failures [19,23] without isolating these factors. This study systematically evaluated five key dimensions of spatial grounding, organized as three research questions (RQ) and two ablations (AB): (RQ1) their capacity for 3D understanding across slice directions; (RQ2) their preference for anatomical over colloquial direction terms; (RQ3) the conditional impact of visual prompts based on orientation and terminology; (AB1) their reliance on pretrained anatomical knowledge priors; and (AB2) their ability to perform abstract spatial reasoning in the absence of medical images.

This study also proposes **M**edical **I**mage **S**patial **Sem**antic **Sam**pling (**MIS-SemSam**), which enhances the reasoning capabilities of the models by employing a sampling strategy based on groups of semantically similar tokens instead of individual ones. This study tested open-weights models Qwen [2–4,17,21],

MedGemma [20], Molmo [5], and paid model Gemini [6,7] on MIS-Ground; M3D [1] and Med3DVLM [24], two leading 3D medical VLMs, were also tested. With an overall accuracy of 66.5%, Qwen3-VL-32B with MIS-SemSam demonstrated a strong ability in 3D medical image spatial grounding, despite being trained on videos of natural scenes.

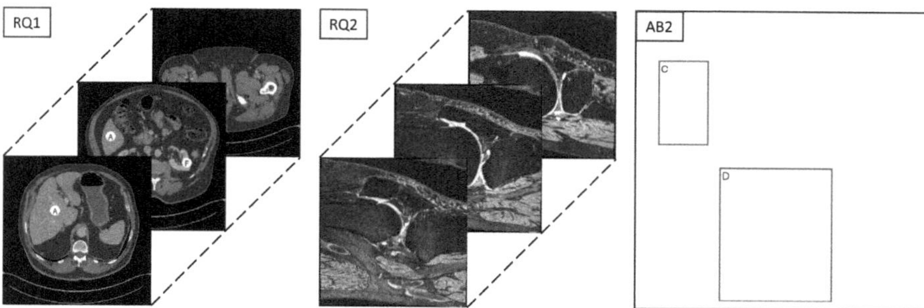

Q: Does the Liver (A) appear before the left kidney? (F) A: True	Q: What is the anatomical spatial relationship between patella bone and the femur bone? A: Anterior	Q: What is the spatial relationship between C and D? A: Above and to the left
Q: What is the slice direction of this image? A: Axial	Q: Is coronal the slice direction of this image? A: False	Q: Is C above D? A: True
Q: What is the name of F? A: Left kidney	Q: Is the patella bone above the femur bone? A: True	Q: Is D to the right of C? A: True
Q: Is A the label of the left kidney? A: False	Q: Is the femur bone posterior to the patella bone? A: True	Q: What is the spatial relationship between D and C? A: Below and to the right

Fig. 1. Sample question-answer pairs from MIS-Ground. For each 2D or 3D vision input (top), multiple questions are generated (bottom). Video VLMs display the ability to discern anatomical structure relationships across slices (RQ1); they prefer anatomical direction terms like anterior (RQ2); and they are capable of abstract spatial reasoning (AB2). Note: the CT image is shown in axial RAS standard viewing orientation, whereas the MRI is shown in sagittal RAS storage mode.

2 Methods

MIS-Ground was designed to systematically test the spatial grounding capabilities of VLMs against a variety of scenarios in medical imaging. This requires a careful curation of the dataset and data pipeline to query the VLMs effectively and efficiently. These integrations are included as part of MIS-Ground. VLMs with and without MIS-SemSam are treated as separate models and thus can be tested on MIS-Ground individually.

2.1 Dataset Preparation

Knee MRI from the Osteoarthritis Initiative (OAI) [16] and torso CT from TotalSegmentator (CT dataset) [22] were used in this study. With access to the MRI dataset under the appropriate data use agreement, 209 double echo steady state (DESS) MR images of the knee joint were collected. For each of these scans, 3D bounding boxes for 10 anatomical components in the knee were gathered from 3DReasonKnee-Bench [19], an open license (CC BY 4.0) benchmark dataset. These components included bone, cartilage, ligaments, and other regions in the knee joint that are associated with osteoarthritis. For the CT dataset, 1228 full-dose axial scans of the torso were collected partly under an open license (CC BY 4.0) and a non-commercial license from TotalSegmentator [22]. Under the same licenses, up to 117 matching segmentation masks of anatomical components were collected for each scan. Prioritizing visibility and ease of prompting, 67 of the larger and more commonly named components were selected for this study, and scans with fewer than 200 or greater than 450 slices were omitted. For preprocessing, MRI and CT datasets underwent percentile-based and soft-tissue windowing, respectively. Multi-planar reconstruction with trilinear interpolation was used to render the scans in alternative slice directions. Finally, the scans were placed in one of two orientations: RAS (right-anterior-superior) coordinate system storage (where $[0, 0, 0]$ is the RAS-most point) and standard viewing orientations.

2.2 Benchmark Curation

MIS-Ground consists primarily of questions evaluating the relative positions of anatomical structures. Questions were defined by four parameters: (1) Visual prompts were derived from ground-truth annotations, including masks, bounding boxes, and centroids (points) for the CT dataset, and bounding boxes for the MRI dataset. Prompts were uniquely colored and/or labeled with a letter A-F. (2) Text prompts included names of anatomical structures and/or the colors or letters of the visual labels. (3) Target type determined the target answers of the generated queries: structure names, labels, or spatial relationships, classified as either anatomical (e.g. superior) or colloquial (above). (4) Question type formatted the queries as open-ended, closed (True), or closed-inverted (False). Ablations included omitting visual inputs entirely (text-only) or omitting the medical image itself (visual prompts on a white background). To ensure broad coverage, pairs of anatomical structures were randomly selected for each parameter combination, yielding an average of 683 and 2,757 questions per MRI and CT scan, respectively.

2.3 VLM Evaluation

A variety of video-native open-weights VLMs were benchmarked on MIS-Ground, including Alibaba's Qwen2.5-VL-Instruct (3B, 7B, 32B, 72B) [4] and Qwen3-VL-Instruct (2B, 4B, 8B, 32B) [2]; and AllenAI's Molmo 2 (4B, 8B) [5]. In addition, 3D-enabled medical VLMs were tested, including Google's MedGemma (4B, 27B) and MedGemma 1.5 (4B) (based on Gemma 3) [20]; M3D (based on Llama 2 7B) [1]; and Med3DVLM (based on Qwen2.5 7B) [24]. For individual model performances, see Fig. 2. Additionally, closed-source models, Google's

Gemini models (2.5 Flash, 2.5 Flash Lite, 3 Flash) were benchmarked as well. To ensure fairness in benchmark comparisons and to promote reproducibility, inference configurations were tuned for the same sampling configuration: reasoning with max new tokens = 8,912 and temperature = 0.5. To report performance robustly under limited trials, we report the Bayesian credible interval [9]. Due to model design, inputs to M3D and Med3DVLM were resized to $[32, 256, 256]$ and $[128, 256, 256]$ respectively with trilinear interpolation, and may have resulted in reduced performance.

2.4 Semantic Sampling

As discussed in Sect. 1, many failures in medical image spatial grounding are not due to a complete lack of visual recognition, but rather to **language-side brittleness**: small decoding perturbations can flip directional terms (e.g., *superior* vs. *inferior*), select awkward morphological variants, or drift into semantically adjacent but incorrect anatomical phrasing. This effect becomes more pronounced under long-form reasoning (chain of thought), where a single locally plausible token can derail the subsequent trajectory. To mitigate this issue at inference time without additional training, we introduce **MIS-SemSam**, an adaptation of Semantic Sampling to medical VLM spatial grounding.

In Semantic Sampling, instead of selecting the next token purely by its own probability, Semantic Sampling prefers tokens that are supported by probability mass in their local semantic neighborhood in token-embedding space. Intuitively, if the model is unsure among several semantically similar realizations of the same concept (common in directional and anatomical terminology), aggregating neighborhood mass stabilizes decoding toward semantically coherent regions of the vocabulary.

Offline Semantic Neighborhoods. Let the language component of a VLM have an input embedding matrix $E \in \mathbb{R}^{V_{\text{emb}} \times d}$ and a tokenizer vocabulary V_{tok}. Following Semantic Sampling decoding, we partition token ids into *content tokens* C (ordinary text) and *non-content tokens* U (special, added or control tokens and any reserved embedding rows not produced by the tokenizer). This distinction is particularly important for VLMs because modality and chat-template markers (e.g., image delimiters, role tokens) behave as control symbols and can form embedding space "hubs" that contaminate nearest-neighbor structure.

We therefore construct neighborhoods *only over content tokens*. Let $E_C = E[C]$ denote the content-token embeddings. We row-normalize

$$\hat{e}_i = \frac{E_C[i]}{\|E_C[i]\|_2 + \epsilon}, \quad i \in C, \tag{1}$$

so that cosine similarity is $\cos(i, j) = \hat{e}_i^\top \hat{e}_j$. For each $i \in C$, we precompute its K nearest neighbors in C under cosine similarity (including i itself) and store two lookup tables:

$$S_{\text{tid}} \in \mathbb{N}^{|C| \times K}, \quad S_{\text{val}} \in \mathbb{R}^{|C| \times K}, \tag{2}$$

where $S_{\text{tid}}[i,k]$ is the k-th neighbor id and $S_{\text{val}}[i,k]$ its cosine similarity. These tables are computed once per embedding space and reused across all MIS-Ground queries; moreover, when multiple checkpoints share the same tokenizer indexing, the same neighbor *ids* can be reused as in Semantic Sampling.

Inference-Time Semantic Rescoring. At decoding step t, the VLM outputs logits $\ell_t \in \mathbb{R}^{V_{\text{emb}}}$ over the next text token, which induce (temperature-scaled) probabilities

$$p_t(v) = \frac{\exp(\ell_t(v)/T)}{\sum_{v'} \exp(\ell_t(v')/T)}. \tag{3}$$

We first apply any standard truncation filter $\textsc{Filter}(\cdot)$ (e.g., top-M, top-p) to obtain a candidate set $I_t \subseteq V_{\text{emb}}$. Because semantic neighborhoods are defined only for content tokens, if $I_t \cap U \neq \emptyset$, we skip MIS-SemSam for this step and defer to the model's default decoding rule (greedy for T=0, or the configured sampler for T>0). Otherwise, $I_t \subseteq C$ and we compute a semantic score for each candidate $c \in I_t$ by aggregating probability mass over its neighborhood:

$$\text{Score}_t(c) \;=\; \sum_{k\in\mathcal{K}(c)} w_{c,k}\, p_t\big(S_{\text{tid}}[c,k]\big), \qquad w_{c,k} = \max\big(0,\, S_{\text{val}}[c,k]\big), \tag{4}$$

where $\mathcal{K}(c)$ denotes the kept neighbor indices (either the top-K' neighbors, $K' \leq K$, or a similarity-threshold prefix), and negative cosine values are clamped to zero to prevent cancellation.

Finally, the next token is chosen from I_t using either: (i) deterministic selection $x_{t+1} = \arg\max_{c\in I_t} \text{Score}_t(c)$ (used to maximize reproducibility), or (ii) stochastic selection by sampling from a softmax over Score_t (used in chain of thought settings to preserve controlled exploration). MIS-SemSam requires no additional forward passes and adds only $O(|I_t| \cdot |\mathcal{K}(c)|)$ table lookups per token.

MIS-SemSam is applied as a drop-in replacement for the final "pick next token" step in generation, leaving the vision encoder, prompting strategy, and truncation filter unchanged. In our experiments, we treat a model with MIS-SemSam enabled as a separate inference configuration (Sect. 2.3), enabling apples-to-apples comparisons between default decoding and semantic-neighborhood-supported decoding on identical MIS-Ground questions. The reported improvement is therefore a paired comparison by construction: the model, prompts, images, questions, and inference settings are held fixed, and only the final token-selection rule changes. We use *model-agnostic* in the algorithmic sense—MIS-SemSam applies to any VLM that exposes next-token logits and input embeddings, and requires neither training nor additional forward passes—rather than as a claim of broad empirical validation; our strongest empirical evidence is on the Qwen3-VL family, while MIS-Ground itself spans many model families.

3 Results

A total of 1160 3D volumes, 2320 2D slices, and 33864 questions were generated across both CT and MRI subsets, which contained 67 and 10 labeled anatomical components per sample respectively. In an effort to balance the dataset for each anatomical component, the CT subset was sampled more and accounted for 84% of the entire dataset. Consequently, the aggregate score is CT-weighted, and results should be read at the benchmark level rather than as uniform cross-modality or clinical-readiness claims. Sample question-answer pairs are shown in Fig. 1. Model response was instructed to be placed at the end in special tags, after all the reasoning tokens. If the tags were missing, the question was omitted from analysis; this omission reflects instruction-following rather than spatial correctness, and predominantly affected the smallest models and the 3D medical VLMs M3D and Med3DVLM, whose outputs were frequently malformed.

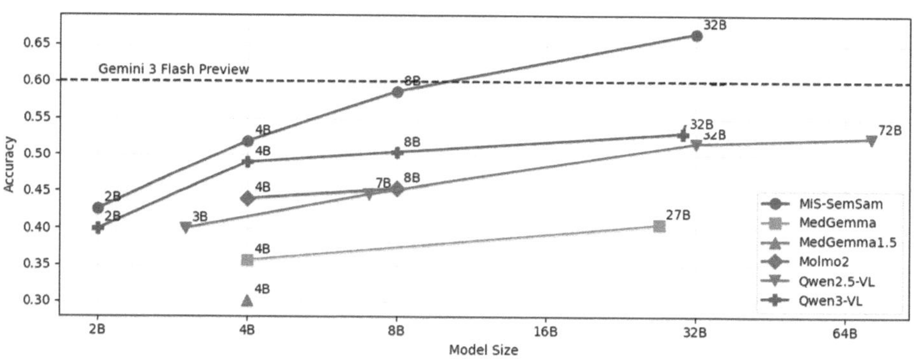

Fig. 2. Overall accuracy on the MIS-Ground benchmark, by model family and size. Note that Med3DVLM (0.0%) and M3D (15.9%) were tested but omitted from this graph. The MIS-SemSam family is derived from the Qwen3-VL family.

Model version and size was strongly correlated to overall performance, with Qwen3-VL-32B outperforming all other open-weights models by a comfortable margin. The overview of these results are found in Fig. 2. M3D and Med3DVLM both failed to give valid responses, often giving long-winded explanations for unrelated concepts, often in non-English characters. These failures likely reflect aggressive input resizing and rigid response formatting (Sect. 2.3) rather than an absence of spatial capability, and should not be read as a general verdict on 3D medical VLMs. By contrast, the closed-source Gemini models served as a strong reference point, underscoring that MIS-Ground is useful beyond our own method. Medgemma did manage to follow the prompt for the most part, though it still gave poor responses. On the other hand, Qwen and Molmo models had much more success, though the very small models (2B, 3B and 4B) still struggled to give valid responses. Large Qwen2.5 models, especially 72B, underperformed relative to its size, which confirmed Qwen's iterative improvements between versions. Otherwise, no strongly discernible patterns emerged regarding model class and specific performances. As the best performing open-weights model, MIS-SemSam (Qwen3-VL-32B with semantic sampling) was used to evaluate specific spatial grounding capabilities across the following research questions and ablations.

(RQ1) 3D Understanding: MIS-SemSam effectively processed 3D medical images, accurately predicting slice direction in 77.6% of cases. The model successfully determined spatial relationships across slices (68.0% accuracy) nearly as well as in-plane relationships (71.7%). Overall performance on 3D images (56.9%) remained competitive with 2D images (58.8%).

(RQ2) Direction Terms: The model demonstrated a strong preference for anatomical directional terms over colloquial terms. In standard viewing orientations, anatomical terms outperformed colloquial terms (69.4% vs. 57.8%). However, the model shifted its preference to colloquial terms (59.85% vs. 50.2%) in RAS storage mode, where standard medical priors conflict with the visual presentation.

(RQ3) Visual Prompts: The impact of visual prompts varied depending on orientation and terminology. In standard RAS viewing orientation, adding points, bounding boxes, or masks yielded only modest improvements (+2.97%, +2.33%, and +2.52% respectively) over the 57.6% baseline. For RAS storage mode utilizing anatomical terms, visual prompts severely degraded a strong 75.3% baseline performance (-13.33% for points, −9.22% for boxes) due to conflicting information. Conversely, when forced to use colloquial terms, visual prompts significantly aided the model, improving a weak 45.1% baseline by up to +8.23%.

(AB1) Anatomy Prior: The model relied heavily on pretrained anatomical knowledge. In text-only ablations (no image), it achieved 69.3% accuracy on anatomical relationships (compared to 74.0% with the image). This prior did not extend to colloquial terms, where text-only accuracy dropped to 40.4%.

(AB2) Abstract Reasoning: The model demonstrated capable abstract spatial reasoning. When tested on visual prompts placed against a blank background without medical scans, the model still accurately determined spatial relationships with 64.2% (points) and 60.4% (boxes) accuracy.

4 Conclusions

With MIS-Ground, VLMs demonstrated measurable capabilities in spatial grounding of anatomical structures in medical imaging. This is in contrast to previous studies, which reported apparent random guessing and failure to follow instructions [19,23]. This is due to careful considerations made in the construction of MIS-Ground, which prioritized the balance and diversity of image modality, visual prompts, text prompts, target answer types, question types, image orientation, slice direction, 2D and 3D, terminology use, and ablations. MIS-SemSam also represents an exploration into the optimization of the VLMs in medical image spatial grounding: an inference-time method with negligible costs and significant performance improvements.

More improvements and applications of VLMs lie ahead. For one, fine-tuning is a crucial task for a domain as highly specialized as medical imaging. Unfortunately, many current such paradigms seem to overfit the task of existing medical visual question answering benchmarks and fail to translate to general capabilities like spatial grounding. New developments in such benchmarks and training datasets are necessary for this field to continue to grow. Also, spatial grounding of anatomical structures is only a partial coverage of what is needed in clinical practice, such as report generation. There is still a lack of alignment between the model capabilities and practical use. To help the community build toward these goals, we release our benchmark MIS-Ground to the public at github.com/asy51/mis-ground.

Acknowledgments. This research was supported in part by NSF awards 2117439 and 2320952.

Disclosure of Interests. The authors declare they have no competing interests.

References

1. Bai, F., Du, Y., Huang, T., Meng, M.Q.H., Zhao, B.: M3d: advancing 3d medical image analysis with multi-modal large language models. arXiv preprint arXiv:2404.00578 (2024)
2. Bai, S., et al.: Qwen3-vl technical report. arXiv preprint arXiv:2511.21631 (2025)
3. Bai, S., et al.: Qwen-vl: A versatile vision-language model for understanding, localization, text reading, and beyond. arXiv preprint arXiv:2308.12966 (2023)
4. Bai, S., et al.: Qwen2.5-vl technical report. arXiv preprint arXiv:2502.13923 (2025)
5. Clark, C., et al.: Molmo2: Open weights and data for vision-language models with video understanding and grounding. arXiv preprint arXiv:2601.10611 (2026)
6. Gemini Team: Gemini 2.5: Pushing the frontier with advanced reasoning. arXiv preprint arXiv:2507.06261 (2025)
7. Google DeepMind: Gemini 3 flash (model documentation) (2026). https://deepmind.google/models/gemini/
8. Hamamci, I.E., et al.: Developing generalist foundation models from a multimodal dataset for 3d computed tomography. arXiv preprint arXiv:2403.17834 (2024)
9. Hariri, M., Samandar, A., Hinczewski, M., Chaudhary, V.: Don't pass@k: a bayesian framework for large language model evaluation. In: The Fourteenth International Conference on Learning Representations (2026). https://openreview.net/forum?id=PTXi3Ef4sT
10. Hashmi, A.U.R., Saeed, N., Lippert, C.: Anatomix, an anatomy-aware grounded multimodal large language model for chest x-ray interpretation. arXiv preprint arXiv:2601.03191 (2026)
11. He, X., Zhang, Y., Mou, L., Xing, E., Xie, P.: Pathvqa: 30000+ questions for medical visual question answering. arXiv preprint arXiv:2003.10286 (2020)
12. Johnson, A.E., et al.: Mimic-cxr, a de-identified publicly available database of chest radiographs with free-text reports. Sci. Data, **6**(1), 317 (2019)
13. Lau, J.J., Gayen, S., Ben Abacha, A., Demner-Fushman, D.: A dataset of clinically generated visual questions and answers about radiology images. Sci. Data **5**(1), 180251 (2018)

14. Lin, B., et al.: Video-llava: learning united visual representation by alignment before projection. In: Proceedings of the 2024 Conference on Empirical Methods in Natural Language Processing, pp. 5971–5984 (2024)
15. Liu, B., Zhan, L.M., Xu, L., Ma, L., Yang, Y., Wu, X.M.: Slake: a semantically-labeled knowledge-enhanced dataset for medical visual question answering. In: 2021 IEEE 18th International Symposium on Biomedical Imaging (ISBI), pp. 1650–1654. IEEE (2021)
16. Peterfy, C.G., Schneider, E., Nevitt, M.: The osteoarthritis initiative: report on the design rationale for the magnetic resonance imaging protocol for the knee. Osteoarthritis Cartilage **16**(12), 1433–1441 (2008). https://doi.org/10.1016/j.joca.2008.06.016
17. Qwen Team: Qwen3 technical report. arXiv preprint arXiv:2505.09388 (2025)
18. Radford, A., et al.: Learning transferable visual models from natural language supervision. In: International Conference on Machine Learning, pp. 8748–8763. PMLR (2021)
19. Sambara, S., et al.: 3dreasonknee: advancing grounded reasoning in medical vision language models. In: Biocomputing 2026: Proceedings of the Pacific Symposium, pp. 99–113. World Scientific (2025)
20. Sellergren, A., et al.: Medgemma technical report. arXiv preprint arXiv:2507.05201 (2025)
21. Wang, P.: Qwen2-vl technical report. arXiv preprint arXiv:2409.12191 (2024)
22. Wasserthal, J., et al.: Totalsegmentator: robust segmentation of 104 anatomic structures in CT images. Radiology: Artif. Intell. **5**(5), e230024 (2023)
23. Wolf, D., Hillenhagen, H., Taskin, B., Bäuerle, A., Beer, M., Götz, M., Ropinski, T.: Your other left! vision-language models fail to identify relative positions in medical images. In: International Conference on Medical Image Computing and Computer-Assisted Intervention, pp. 691–701. Springer, Cham (2025)
24. Xin, Y., Ates, G.C., Gong, K., Shao, W.: Med3dvlm: an efficient vision-language model for 3d medical image analysis. IEEE J. Biomed. Health Inform. (2025)

MedObvious: Exposing the Medical Moravec's Paradox in VLMs via Clinical Triage

Ufaq Khan[1(✉)], Umair Nawaz[1], Lekkala Sai Teja[2], Numan Saeed[1], Muhammad Bilal[3], Yutong Xie[1], Mohammad Yaqub[1], and Muhammad Haris Khan[1]

[1] Mohamed bin Zayed University of Artificial Intelligence, Abu Dhabi, UAE
ufaq.khan@mbzuai.ac.ae
[2] National Institute of Technology, Silchar, India
[3] Birmingham City University, Birmingham, UK

Abstract. Vision Language Models (VLMs) are increasingly used for tasks like medical report generation and visual question answering. However, fluent diagnostic text does not guarantee safe visual understanding. In clinical practice, interpretation begins with *pre-diagnostic sanity checks*: verifying that the input is valid to read (correct modality and anatomy, plausible viewpoint and orientation, and no obvious integrity violations). Existing benchmarks largely assume this step is solved, and therefore miss a critical failure mode: a model can produce plausible narratives even when the input is inconsistent or invalid. We introduce **MedObvious**, a 1,880-task benchmark that isolates input validation as a set-level consistency capability over small multi-panel image sets: the model must identify whether any panel violates expected coherence. MedObvious spans five progressive tiers, from basic orientation/modality mismatches to clinically motivated anatomy/viewpoint verification and triage-style cues, and includes five evaluation formats to test robustness across interfaces. Evaluating 17 different VLMs, we find that sanity checking remains unreliable: several models hallucinate anomalies on normal (negative-control) inputs, performance degrades when scaling to larger image sets, and measured accuracy varies substantially between multiple-choice and open-ended settings. These results show that pre-diagnostic verification remains unsolved for medical VLMs and should be treated as a distinct, safety-critical capability before deployment.

Keywords: Medical Vision-Language Models · Pre-diagnostic Sanity Checking · Clinical Triage · Visual Grounding · AI Safety

1 Introduction

Vision-Language Models (VLMs) are increasingly being used to interpret medical images. Recent systems can generate radiology-style descriptions, answer clini-

Project Page: MedObvious-Website, Github: MedObvious.

G. Yang et al. (Eds.): MICCAI 2026, LNCS 16878, pp. 345–355, 2027.
https://doi.org/10.1007/978-3-032-38059-3_33

cal questions, and perform multi-step reasoning over images and text, driven by both general-purpose models such as GPT-4o [1], Flamingo [4] and LLaVA [20] and medical adaptations such as LLaVA-Med [15], RadFM [26], and others [13,21,22]. In parallel, these models are being explored as the core perception for *visual AI agents* [10,16] that can also interact with imaging software (e.g., navigating viewers, selecting series, adjusting visualization, and triggering downstream tools). This progress has motivated their potential use as assistants for clinical reporting and decision support. However, fluent language generation does not guarantee reliable visual perception. VLMs may produce coherent diagnostic narratives while failing basic sanity checks, such as detecting incorrect orientation, mismatched anatomy, unexpected modality, or physically implausible artifacts. We refer to this mismatch as the **Medical Moravec's Paradox**, extending Moravec's observation [2] that perception and spatial reasoning, trivial for humans, can be disproportionately difficult for machines even when higher-level outputs appear plausible. In medical imaging, this gap is consequential because failures occur before diagnosis: when the input is invalid or inconsistent, downstream reports become clinically uninterpretable.

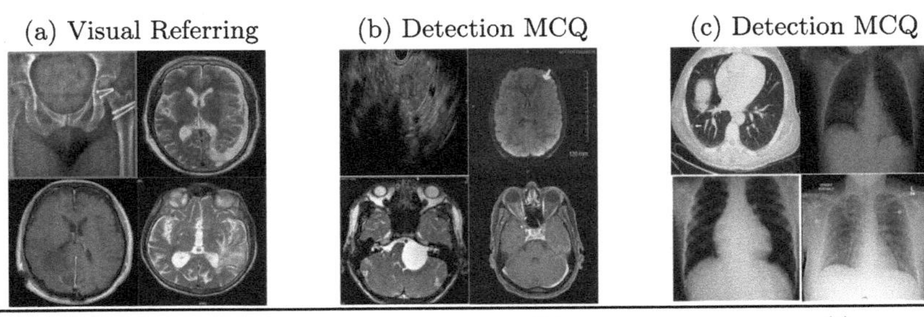

	(a)	(b)	(c)
Question	Is this highlighted scan the clinical outlier; i.e., does it differ in modality, anatomy, or pathology from the others? Answer yes/no	You are reviewing a 2 × 2 grid of medical scans. One scan is the clinical outlier as it differs in modality, anatomy, or pathology. Which position contains the outlier? *A)* top-left *B)* top-right *C)* bottom-left *D)* bottom-right *E)* None	In this 2 × 2 medical image grid, one scan contains a clinical anomaly (chest CT). Which position is it? *A)* top-left *B)* top-right *C)* bottom-left *D)* bottom-right *E)* None
Ground truth	Yes	A (Top-left)	A (Top-left)
LLaVA-1.5	No (✗)	B (✗)	B (✗)
Qwen3-VL	Yes (✓)	D (✗)	D (✗)
Lingshu	Yes (✓)	B (✗)	A (✓)

Fig. 1. Qualitative MedObvious examples for pre-diagnostic visual triage. Each column corresponds to a grid in (a)–(c). We report the task question, ground truth, and predictions from representative VLMs.

Clinical interpretation begins with pre-diagnostic triage: clinicians first verify body part, view, modality, laterality, orientation, and basic image integrity, and they do not proceed to diagnosis if these checks fail. This requirement is amplified in *multi-image* and *AI-agentic* settings, where decisions depend on consistency across a set of inputs, such as multiple fetal ultrasound views or long CT/MRI slice sequences and series. A single mis-acquired view, corrupted slice, mismatched series, or orientation error can compromise study-level reasoning, especially for models that aggregate evidence across images. Moreover, common tools such as 3D Slicer [11] and ITK-SNAP [28] support multi-panel layouts (e.g., axial, sagittal, coronal, and 3D) similar to a 2×2 display. VLM-based agents operating in these viewers must detect obvious panel-level inconsistencies to avoid acting on the wrong series, anatomy, or invalid inputs. Despite their importance, pre-diagnostic competencies are rarely evaluated explicitly. Standard medical VLM benchmarks such as VQA-RAD [14], PathVQA [12], PMC-VQA [29], VQA-Med [6], and SLAKE [17] primarily assess the correctness of final answers, while hallucination-focused benchmarks such as Med-HallMark [7] emphasize factual consistency of the generated text. These settings largely assume the input has been correctly perceived and can therefore miss failures on visually obvious sanity checks, allowing models to score well while remaining brittle and potentially unsafe in real multi-image or agentic workflows.

We introduce **MedObvious**, a benchmark for pre-diagnostic visual sanity checking in medical images. MedObvious asks a question that precedes diagnosis: *Is the input coherent and appropriate to interpret?* For this, we present small grids (2×2 or 3×3) and require models to either identify an outlier panel or state that no outlier exists, as shown in Fig. 1. Although it is not a clinical reading interface, this provides a controlled probe of set-level consistency, mirroring requirements in multi-view ultrasound, multi-slice CT/MRI, and multi-panel viewer agentic workflows. MedObvious evaluates two axes. The *Clinical Safety* axis targets real failure modes that should be caught before any diagnostic verdict, including wrong body parts, flipped images, viewpoint/anatomy mismatches, and grossly apparent major pathology. The *Visual Grounding* axis uses synthetic inconsistencies (e.g., inserting modality-incompatible textures into an image) to test whether models check the visual input itself rather than relying on language priors for downstream tasks. MedObvious also includes explicit **negative controls** where all panels are consistent, so the correct response is that no outlier exists, directly measuring false alarms. Furthermore, it is organized into five progressive tiers (T1–T5) that increase in difficulty and clinical specificity as depicted in Fig. 2, ranging from basic orientation/modality mismatches to anatomy/viewpoint inconsistencies and high-saliency triage failures. In our zero-shot evaluation across *7 general*, *4 medical*, and *6 proprietary* VLMs, performance remains uneven as the best mean accuracy reaches 63.2%, yet negative-control accuracy spans a wide range, indicating that false alarms on normal inputs remain common. We also observe strong format sensitivity, with large gaps between multiple-choice and open-ended variants of the same underlying capability. Our main contributions are:

- We formalize the *Medical Moravec's Paradox* for medical VLMs, highlighting a gap between fluent diagnostic language and reliable pre-diagnostic visual sanity-checking, especially in multi-image and agentic-viewer settings.

- We present first **MedObvious**, a 1,880 task benchmark spanning 5 progressive tiers, multiple grid configurations, five evaluation modes, and systematic negative controls, designed to evaluate pre-diagnostic visual triage independently of diagnosis.
- We benchmark 7 representative open-source, 6 closed-source, and 4 medically specialized VLMs under zero-shot inference.

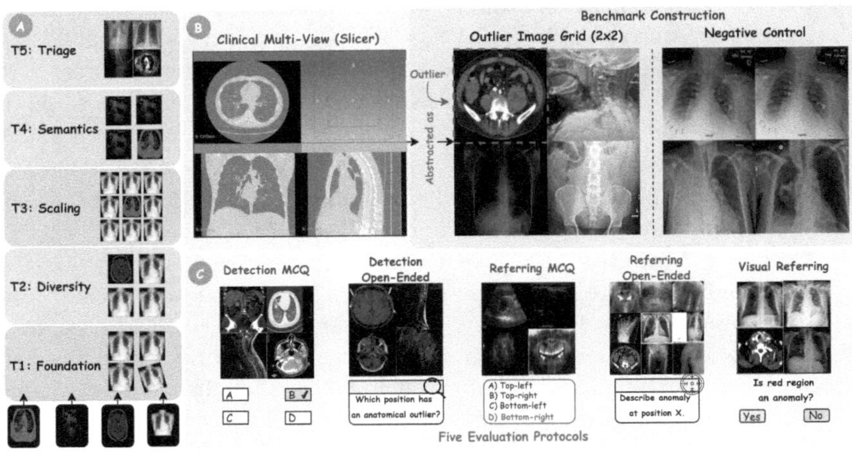

Fig. 2. MedObvious overview. (**A**) Five progressive tiers (T1–T5). (**B**) Construction: multi-view studies are abstracted into small grids with either a single outlier or a negative-control (no outlier). (**C**) Five evaluation protocols: Detection (MCQ/Open), Referring (MCQ/Open), and Visual Referring (Yes/No).

2 MedObvious Construction

MedObvious is a benchmark designed to test whether medical VLMs can recognize *obvious* input-level inconsistencies before attempting any diagnosis, as inspired by [9]. It acts as a prerequisite for safe medical VLM deployment by *pre-diagnostic visual sanity checking.* Before interpreting pathology, clinicians first verify that an image is valid to read, including modality, anatomical region, viewpoint/orientation, and basic integrity. If these checks fail, the subsequent diagnosis is unreliable regardless of how fluent the generated text appears. MedObvious isolates this input-validation step and explicitly evaluates it.

Motivation. The clinical need is inherently *set-based.* Many studies are interpreted across multiple views, slices, or series, where a single inconsistent element can compromise the conclusions of the study. In fetal ultrasound, assessment of the fetal heart relies on multiple standard views. One mis-acquired plane or corrupted view may change the interpretation. In CT/MRI, clinicians reason over long slice sequences and multi-series studies. In longitudinal assessment, such as reviewing the progression of Multiple Sclerosis from two or more brain MRI scans, an incorrectly positioned slice could lead to a completely different clinical decision. A flipped series, corrupted slice, or modality/anatomy mismatch is a coherence break that should be detected prior to diagnosis. This set-level checking is also reflected in common imaging software (e.g., 3D Slicer and ITK-SNAP), which presents multi-planar views in multi-panel layouts resembling 2×2 grid. Our grid-based tasks are therefore a controlled abstraction of this real requirement by detecting whether any element in a small visual set violates expected consistency before proceeding to downstream reasoning or agentic actions.

Problem Formulation. Each MedObvious instance presents a grid of n images, $G = \{I_1, \dots, I_n\}$ with $n \in \{4, 9\}$. The model must either identify the index of the inconsistent image or state that no outlier exists ($y \in \{1, \dots, n\} \cup \{\varnothing\}$) where $y = k$ denotes I_k as the outlier and $y = \varnothing$ denotes a clean and consistent grid. This setup evaluates input validity and set-level coherence, i.e., whether images that should belong to the same study context (across views, slices/series, or timepoints) are mutually consistent, rather than evaluating diagnosis.

Table 1. MedObvious composition. Tiers increase in clinical specificity.

Tier	Name	Grid	#Mod.	Tasks	Pos.	Neg.	Clinical analog
T1	FOUNDATION	2×2	2	440	275	165	Basic QC (orientation, modality)
T2	DIVERSITY	2×2	2	480	300	180	Robustness across sources
T3	SCALING	3×3	4	360	225	135	Multi-view/-slice comparison
T4	SEMANTICS	2×2	3	320	200	120	Anatomy/viewpoint verification
T5	TRIAGE	Mixed	4	280	175	105	"Stop and verify" safety cues
Total			**4**	**1,880**	**1,175**	**705**	

Datasets. We primarily use curated subsets of ROCO [23] to construct anatomically and modality-defined tasks (e.g., chest radiographs, CT, MRI, ultrasound) using metadata filtering. To reduce modality shortcuts and enforce modality awareness, we additionally include non-radiological images, including endoscopy from Kvasir [24], to increase visual diversity and discourage reliance on generic grayscale radiology priors.

Template-Based Generation. Each task is generated from a shared template: we sample $n - 1$ *inlier* panels from a reference category (defined by modality, and when available, anatomy/viewpoint), and then either (i) insert one *outlier* from a different category or via a controlled integrity violation (e.g., orientation change or physically inconsistent composite), or (ii) create a *negative control* by sampling all n panels from same reference category so the correct answer is $\varnothing$.

Progressive Tiers. MedObvious comprises five tiers (1,880 tasks) with increasing clinical specificity (Table 1): T1 FOUNDATION (2×2; orientation/modality mismatches), T2 DIVERSITY (2×2; finer distinctions across sources), T3 SCALING (3×3; many distractors and modality diversity), T4 SEMANTICS (2×2; anatomy/viewpoint mismatches and integrity violations), and T5 TRIAGE; high-saliency failures (like gross abnormalities), and cross-modality coherence breaks.

Evaluation Protocols. Each grid is evaluated in 5 formats to separate visual capability from response-format effects: Detection MCQ (*pick the outlier*), Detection Open (*state the outlier position*), Referring MCQ (*choose an outlier description given its position*), Referring Open (*describe the outlier given its position*), and Visual Referring (*Yes/No for highlighted region*). Using both multiple-choice and open-ended settings exposes selection biases and over-generation.

Negative Controls. To measure false alarms, MedObvious includes explicit negative controls where all panels are consistent and no outlier exists (37.5% of tasks; 705/1,880). The correct label is $y = \varnothing$ (or "No" for binary verification).

Table 2. MedObvious: Per-task accuracy (%). Task A–E correspond to Detection MCQ, Detection Open, Referring MCQ, Referring Open, and Visual Referring, respectively. "Positive (Pos)"/"Negative (Neg)" indicate accuracy on anomalous/non-anomalous samples. Avg is the macro-average of Task A–E, Pos(+), and Neg(-). Avg confidence intervals are 95% case-level bootstrap CIs. Best results per task are **bolded**.

Model	Task-A	Task-B	Task-C	Task-D	Task-E	Pos(+)	Neg(−)	Avg [95% CI]	Var. (10^{-4})
General open-source VLMs									
LLaVA-1.5-7B	40.4	22.3	22.1	35.7	50.0	37.5	31.9	34.3 [32.4, 36.2]	0.94
Qwen2-VL-7B	32.3	49.5	72.7	25.1	53.1	56.3	28.7	45.4 [43.3, 47.6]	1.23
Qwen2.5-VL-7B	**58.7**	**82.1**	75.3	29.7	**65.1**	60.9	**70.7**	**63.2 [60.5, 66.0]**	**1.98**
Qwen3-VL-8B	31.2	32.7	**80.8**	**38.7**	52.9	**68.9**	2.9	44.0 [41.9, 46.3]	1.30
InternVL2.5-8B	56.3	56.3	69.7	26.8	51.7	59.7	42.5	51.9 [49.3, 54.4]	1.72
InternVL3-8B	38.5	43.6	80.4	27.6	50.2	64.6	16.6	45.9 [43.5, 48.4]	1.60
Pixtral-12B	31.0	22.5	76.6	26.3	51.9	59.5	5.3	39.0 [37.0, 41.1]	1.12
Medical open-source VLMs									
LLaVA-Med-7B	10.0	36.8	21.2	23.4	50.0	37.1	17.5	28.0 [26.1, 30.0]	0.97
Fleming-8B	26.8	23.8	78.3	23.8	50.0	57.4	5.3	37.9 [36.1, 39.8]	0.86
Medgemma1.5-4B-IT	23.6	**86.1**	43.4	19.5	**66.6**	44.6	**64.1**	49.7 [47.7, 51.8]	1.07
Lingshu-7B	**39.3**	78.5	**79.5**	**26.8**	61.9	**66.8**	43.8	**56.6 [54.6, 58.8]**	**1.15**
Proprietary VLMs									
Gemini-2.0-Flash	54.2	42.7	**85.9**	**35.7**	**69.3**	**75.4**	25.6	**55.5 [53.3, 57.9]**	**1.39**
Gemini-2.5-Flash	**67.2**	45.5	80.4	31.9	55.9	74.1	**26.3**	54.4 [52.2, 56.8]	1.35
GPT-4o	47.2	**50.4**	62.8	26.3	61.7	68.0	22.7	48.4 [46.1, 50.9]	1.54
GPT-4.1-nano	25.3	16.8	26.3	18.3	55.1	34.9	21.4	28.3 [26.1, 30.6]	1.29
GPT-4.1-mini	41.9	32.9	53.1	29.3	64.2	64.0	13.6	42.7 [40.4, 45.1]	1.40
GPT-5-nano	43.4	41.7	82.5	28.9	63.4	73.1	14.8	49.6 [47.5, 51.7]	1.14
Human expert	82.1	85.7	82.1	90.9	92.9	89.4	95.7	88.4	–

3 Experiments and Results

MedObvious targets a pre-diagnostic requirement, i.e., before interpretation, the model must verify that the input is coherent and safe to reason over. We therefore evaluate VLMs as potential *pre-diagnostic gatekeepers* using the following clinically grounded research questions:

RQ1 (Gatekeeping and False Alarms). Can models detect gross input violations (wrong modality or anatomy) and decide whether to proceed or abstain? Also, can models correctly determine that no anomaly is present when given normal, internally consistent inputs?

RQ2 (Set-Level Consistency). How does performance change as the candidate set grows, requiring systematic comparison across images?

Table 3. Paired comparisons as statistical significance. Avg is the mean of Task A–E, Pos(+), and Neg(-). Δ denotes Avg(Model A) − Avg(Model B), with paired case-level bootstrap 95% confidence intervals. Exact McNemar Holm-adjusted p-values are reported for binary item-level correctness comparisons.

Model A	Model B	Avg A	Avg B	Δ Avg [95% CI]	Paired units	Holm p
Within-category leader vs. runner-up						
Qwen2.5-VL-7B	InternVL2.5-8B	63.25	51.88	11.37 [9.12, 13.59]	235	$< 10^{-8}$
Lingshu-7B	Medgemma1.5-4B-IT	56.70	49.74	6.96 [4.62, 9.29]	235	2.65×10^{-4}
Gemini-2.0-Flash	Gemini-2.5-Flash	55.61	54.51	1.10 [-0.81, 3.02]	235	1.00
Cross-category leader comparisons						
Qwen2.5-VL-7B	Lingshu-7B	63.25	56.70	6.56 [4.52, 8.53]	235	4.60×10^{-7}
Qwen2.5-VL-7B	Gemini-2.0-Flash	63.25	55.61	7.65 [5.11, 10.18]	235	1.92×10^{-6}
Lingshu-7B	Gemini-2.0-Flash	56.70	55.61	1.09 [-1.22, 3.41]	235	1.00
Representative proprietary-model contrasts						
Gemini-2.0-Flash	GPT-5-nano	55.61	49.63	5.98 [3.94, 8.04]	235	5.10×10^{-7}
GPT-5-nano	GPT-4o	49.63	48.49	1.14 [-0.93, 3.25]	235	1.00

RQ3 (Clinical Semantics). Do models reliably detect clinically meaningful mismatches (e.g., anatomy/viewpoint) rather than relying on superficial cues, and do they confuse such mismatches with pathology?

RQ4 (Grounding). Under physically implausible or cross-modality inconsistencies, do models reject the input or rationalize it with plausible narratives?

RQ5 (Interface Robustness). Are sanity-check decisions consistent across binary, localization, and free-text interfaces, or strongly format-dependent?

Evaluation Pipeline. All models are evaluated on the full MedObvious benchmark in a **zero-shot** setting, without fine-tuning, retrieval augmentation, or few-shot exemplars. We evaluate three model groups: **general open-source VLMs** (LLaVA-1.5-7B [19], Qwen2-VL-7B [25] Qwen2.5-VL-7B, Qwen3-VL-8B [5], Intern VL2.5-8B [8], InternVL3-8B [30], Pixtral-12B [3]), **medical open-source VLMs** (LLaVA-Med-7B [15], MedGemma1.5-4B-IT, Lingshu-7B [27]), Fleming-8B [18], and **proprietary VLMs** (Gemini-2.0-Flash, Gemini-2.5-Flash, GPT-4o, GPT-4.1-mini/nano, GPT-5-nano). Each instance is evaluated in five formats, each with format-specific prompts. Outputs are constrained and parsed into a closed label space (option letter, grid position, or Yes/No) to ensure consistent scoring across models. Moreover, open-source models are evaluated with a unified inference pipeline on NVIDIA A100 (40 GB) GPU. Proprietary models are queried via public APIs using the same prompts and output normalization.

Evaluation Metrics. We report accuracy for each format, as well as **Positive** accuracy (outlier present), **Negative** accuracy (no outlier), and **Overall** accuracy. Reporting Positive and Negative separately is clinically important as a model can appear strong on anomaly-present cases, yet remain unsafe due to false alarms on normal inputs.

Results. We summarize results by research question. Table 2 reports performance on the 5 evaluation modes, Table 4 reports performance on 5 tiers, and Table 3 reports the pairwise statistical signifcance of the benchmarked VLMs. It shows consistent paired rankings: general open-source models often perform better, while closed models are not statistically separable.

Table 4. Per-tiers overall accuracy (%) for different VLMs.

Model	T1	T2	T3	T4	T5	All
General open-source VLMs						
LLaVA-1.5-7B	35.2	40.4	21.6	42.5	36.7	35.2
Qwen2-VL-7B	50.9	47.2	37.2	52.8	39.6	45.5
Qwen2.5-VL-7B	**68.8**	**67.2**	**50.5**	**84.0**	49.2	**63.9**
Qwen3-VL-8B	47.2	48.9	34.7	53.4	32.8	43.4
InternVL2.5-8B	58.8	56.0	33.0	67.2	**49.3**	52.8
InternVL3-8B	50.4	50.4	37.2	57.1	33.9	45.8
Pixtral-12B	45.0	41.4	26.3	47.8	33.2	38.7

Model	T1	T2	T3	T4	T5	All
Medical open-source VLMs						
LLaVA-Med-7B	32.7	32.5	28.8	26.8	25.0	29.1
Fleming-8B	41.8	43.9	29.1	39.0	31.4	37.0
Medgemma1.5-4B	59.7	53.7	37.2	53.7	**53.5**	51.5
Lingshu-8B	**62.9**	**64.7**	**48.8**	**63.1**	46.0	**57.1**

Model	T1	T2	T3	T4	T5	All
Proprietary VLMs						
Gemini-2.0-Flash	**59.3**	**61.0**	**56.3**	**66.8**	34.6	55.6
Gemini-2.5-Flash	59.0	62.7	55.2	62.1	35.0	54.8
GPT-4o	54.3	56.0	45.2	59.6	34.6	49.9
GPT-4.1-nano	29.3	31.8	20.2	31.8	**37.5**	30.1
GPT-4.1-mini	47.7	52.2	36.1	50.9	33.5	44.1
GPT-5-nano	52.5	56.6	46.6	61.2	33.2	50.0

RQ1 (Gatekeeping and False Alarms). Table 2 shows that overall accuracy is still far from a reliable pre-diagnostic gate, with large variance between models. The most safety-relevant signal is the **Pos(+)** and **Neg(−)** split. Several models achieve high Pos(+) accuracy while collapsing on Neg(−), indicating a "*always-find-something*" bias that would be unacceptable for a gatekeeper. In contrast, a smaller subset achieves substantially higher Neg(−), demonstrating that abstention is learnable, but not consistently present across model families. Importantly, the proprietary scale does not automatically fix this failure mode, as the negative accuracy remains modest for many such models, suggesting that normal-case calibration is a distinct problem from diagnostic fluency.

RQ2 (Set-Level Consistency). The tier analysis in Table 4 reveals that the scaling from 2×2 to 3×3 is not a minor increase in difficulty but a **qualitative failure point**. The systematic drop in T3 (SCALING) suggests that many models do not reliably perform an exhaustive comparison across candidates. Instead, they appear to rely on a small number of salient cues. This matters clinically because multi-view and multi-slice studies require consistent reasoning across many related images (CT slices or multi-view fetal ultrasound), not just the detection of a single obvious frame.

RQ3 (Clinical Semantics). Models often rebound on T4 (SEMANTICS), which emphasizes anatomy/viewpoint mismatches. This indicates that clinically significant distribution shifts (e.g., chest vs. abdomen, frontal vs. lateral) can be easier to detect than distractor-heavy scaling, likely because they produce greater global changes in appearance. However, this "semantic strength" does not imply safety: a model can detect an anatomy mismatch yet still hallucinate anomalies.

RQ4 (Grounding). We test whether models reject invalid inputs rather than rationalize them. High false-alarm rates on negative controls, together with strong interface sensitivity, suggest that models do not behave as conservative verifiers and often predict an outlier even when the correct response is "none".

RQ5 (Interface Robustness). Performance is strongly format-dependent. Many models are high on Referring MCQ but low on Referring Open, suggesting option selection is easier than producing grounded descriptions. Similarly, the reverse pattern also appears for some models. Overall, sanity checking is not interface-invariant, so models should be evaluated for consistency across binary, localization, and explanation-style outputs rather than a single prompt format.

Discussion. MedObvious shows that fluent report-style generation does not imply reliable pre-diagnostic verification. Across models, the main failures are **false alarms** on normal grids, **degradation under scaling**, and strong **format sensitivity**. These failures directly limit the use of VLMs as sanity-check assistants. The implications are stronger for *agentic* medical AI, where models act on multi-panel, multi-image inputs and verify set-level consistency of modality, anatomy, and integrity before safe action. A model that hallucinates anomalies or fails under larger candidate sets can propagate errors while being confidently fluent. Overall, MedObvious isolates this prerequisite layer and highlights the need to reduce false alarms and improve set-level comparison under scaling.

Limitations. MedObvious is a controlled, template-based benchmark and therefore does not capture the full complexity of real clinical workflows. It also does not directly test downstream diagnostic performance or cover the full space of VLM-specific failure modes, which remain important directions for future work.

4 Conclusion

We present **MedObvious**, a benchmark for pre-diagnostic visual sanity checking in medical VLMs. Across progressive tiers, formats, and negative controls, we find that current models remain unreliable gatekeepers, with frequent false alarms, scaling degradation, and strong format sensitivity, motivating pre-diagnostic triage as a prerequisite for safe clinical and agentic deployment. Future work can extend to full multi-series volumes and interactive viewer-based evaluation.

Disclosure of Interests. The authors have no competing interests to declare that are relevant to the content of this article.

References

1. Achiam, J., et al.: GPT-4 technical report. arXiv preprint arXiv:2303.08774 (2023)
2. Agrawal, K.: To study the phenomenon of the Moravec's paradox. arXiv preprint arXiv:1012.3148 (2010)
3. Agrawal, P., et al.: Pixtral 12b. arXiv preprint arXiv:2410.07073 (2024)
4. Alayrac, J.B.: Flamingo: a visual language model for few-shot learning. In: Advances in Neural Information Processing Systems, vol. 35, pp. 23716–23736 (2022)
5. Bai, S., et al.: Qwen3-VL technical report. arXiv preprint arXiv:2511.21631 (2025)
6. Ben Abacha, A., Hasan, S.A., Datla, V.V., Demner-Fushman, D., Müller, H.: Vqa-med: Overview of the medical visual question answering task at imageclef 2019. In: Proceedings of CLEF (Conference and Labs of the Evaluation Forum) 2019 Working Notes. 9-12 September 2019 (2019)
7. Chen, J., et al.: Detecting and evaluating medical hallucinations in large vision language models. arXiv preprint arXiv:2406.10185 (2024)
8. Chen, Z., et al.: Expanding performance boundaries of open-source multimodal models with model, data, and test-time scaling. arXiv preprint arXiv:2412.05271 (2024)
9. Dahou, Y., Huynh, N.D., Le-Khac, P.H., Para, W.R., Singh, A., Narayan, S.: Vision-language models can't see the obvious. arXiv preprint arXiv:2507.04741 (2025)
10. Fallahpour, A., Ma, J., Munim, A., Lyu, H., Wang, B.: Medrax: medical reasoning agent for chest x-ray. arXiv preprint arXiv:2502.02673 (2025)
11. Fedorov, A., et al.: 3d slicer as an image computing platform for the quantitative imaging network. Magn. Reson. Imaging **30**(9), 1323–1341 (2012)
12. He, X., Zhang, Y., Mou, L., Xing, E., Xie, P.: Pathvqa: 30000+ questions for medical visual question answering. arXiv preprint arXiv:2003.10286 (2020)
13. Jiang, S., et al.: Omniv-med: scaling medical vision-language model for universal visual understanding. arXiv preprint arXiv:2504.14692 (2025)
14. Lau, J.J., Gayen, S., Ben Abacha, A., Demner-Fushman, D.: A dataset of clinically generated visual questions and answers about radiology images. Sci. data **5**(1), 180251 (2018)
15. Li, C., et al.: Llava-med: training a large language-and-vision assistant for biomedicine in one day. In: Advances in Neural Information Processing Systems, vol. 36, pp. 28541–28564 (2023)
16. Lin, K.Q., et al.: Showui: one vision-language-action model for GUI visual agent. In: Proceedings of the Computer Vision and Pattern Recognition Conference, pp. 19498–19508 (2025)

17. Liu, B., Zhan, L.M., Xu, L., Ma, L., Yang, Y., Wu, X.M.: Slake: a semantically-labeled knowledge-enhanced dataset for medical visual question answering. In: 2021 IEEE 18th international symposium on biomedical imaging (ISBI), pp. 1650–1654. IEEE (2021)
18. Liu, C., Li, D., Shu, Y., Chen, R., Duan, D., Fang, T., Dai, B.: Fleming-r1: toward expert-level medical reasoning via reinforcement learning. arXiv preprint arXiv:2509.15279 (2025)
19. Liu, H., Li, C., Li, Y., Lee, Y.J.: Improved baselines with visual instruction tuning. In: Proceedings of the IEEE/CVF Conference on Computer Vision and Pattern RecognitionD pp. 26296–26306 (2024)
20. Liu, H., Li, C., Wu, Q., Lee, Y.J.: Visual instruction tuning. In: Advances in Neural Information Processing Systems, vol. 36, pp. 34892–34916 (2023)
21. Nath, V., et al.: Vila-m3: Enhancing vision-language models with medical expert knowledge. In: Proceedings of the Computer Vision and Pattern Recognition Conference, pp. 14788–14798 (2025)
22. Pan, J., et al.: Medvlm-r1: incentivizing medical reasoning capability of vision-language models (VLMs) via reinforcement learning. In: International Conference on Medical Image Computing and Computer-Assisted Intervention, pp. 337–347. Springer (2025)
23. Pelka, O., Koitka, S., Rückert, J., Nensa, F., Friedrich, C.M.: Radiology objects in context (ROCO): a multimodal image dataset. In: Intravascular Imaging and Computer Assisted Stenting and Large-Scale Annotation of Biomedical Data and Expert Label Synthesis, pp. 180–189. Springer, Cham (2018)
24. Pogorelov, K., et al.: Kvasir: a multi-class image dataset for computer aided gastrointestinal disease detection. In: Proceedings of the 8th ACM on Multimedia Systems Conference, pp. 164–169 (2017)
25. Wang, P., et al.: Qwen2-VL: Enhancing vision-language model's perception of the world at any resolution. arXiv preprint arXiv:2409.12191 (2024)
26. Wu, C., et al.: Towards generalist foundation model for radiology by leveraging web-scale 2D&3D medical data. Nat. Commun. **16**(1), 7866 (2025)
27. Xu, W., et al.: Lingshu: a generalist foundation model for unified multimodal medical understanding and reasoning. arXiv preprint arXiv:2506.07044 (2025)
28. Yushkevich, P.A., et al.: User-guided 3d active contour segmentation of anatomical structures: significantly improved efficiency and reliability. Neuroimage **31**(3), 1116–1128 (2006)
29. Zhang, X., et al.: PMC-VQA: visual instruction tuning for medical visual question answering. arXiv preprint arXiv:2305.10415 (2023)
30. Zhu, J., et al.: Internvl3: exploring advanced training and test-time recipes for open-source multimodal models. arXiv preprint arXiv:2504.10479 (2025)

MedPruner: Training-Free Hierarchical Token Pruning for Efficient 3D Medical Image Understanding in Vision-Language Models

Shengyuan Liu[1], Zanting Ye[2,3], Yunrui Lin[3], Chen Hu[2,4], Wanting Geng[5], Xu Han[6], Bulat Ibragimov[7], Yefeng Zheng[2(✉)], and Yixuan Yuan[1(✉)]

[1] Chinese University of Hong Kong, Hong Kong, China
yxyuan@ee.cuhk.edu.hk
[2] Westlake University, Hangzhou, China
zhengyefeng@westlake.edu.cn
[3] Southern Medical University, Guangzhou, China
[4] Jiangnan University, Wuxi, China
[5] Dalian University of Technology, Dalian, China
[6] Shanghai Jiao Tong University, Shanghai, China
[7] University of Copenhagen, Copenhagen, Denmark

Abstract. While specialized Medical Vision-Language Models (VLMs) have achieved remarkable success in interpreting 2D and 3D medical modalities, their deployment for 3D volumetric data remains constrained by significant computational inefficiencies. Current architectures typically suffer from massive anatomical redundancy due to the direct concatenation of consecutive 2D slices and lack the flexibility to handle heterogeneous information densities across different slices using fixed pruning ratios. To address these challenges, we propose MedPruner, a training-free and model-agnostic hierarchical token pruning framework specifically designed for efficient 3D medical image understanding. MedPruner introduces a two-stage mechanism: an Inter-slice Anchor-based Filtering module to eliminate slice-level temporal redundancy, followed by a Dynamic Information Nucleus Selection strategy that achieves adaptive token-level compression by quantifying cumulative attention weights. Extensive experiments on three 3D medical benchmarks and across three diverse medical VLMs reveal massive token redundancy in existing architectures. Notably, MedPruner enables models such as MedGemma-1.5 to maintain or even exceed their original performance while retaining fewer than 5% of visual tokens, thereby reducing visual-token overhead and validating the necessity of dynamic token selection for practical clinical deployment. Our code is available at here.

Keywords: Medical Vision-Language Models · 3D Medical Imaging · Token Pruning

S. Liu, Z. Ye, and Y. Lin—Equal contributions.

G. Yang et al. (Eds.): MICCAI 2026, LNCS 16878, pp. 356–365, 2027.
https://doi.org/10.1007/978-3-032-38059-3_34

1 Introduction

Vision-Language Models (VLMs), ranging from proprietary systems like GPT-4o [9], Gemini [6] to high-performance open-source models such as the LLaVA [17] and Qwen-VL [3] series, have demonstrated extraordinary universal perceptual and reasoning capabilities. Drawing upon these advancements, specialized medical VLMs [5,8,15,23,24,28,29,31] such as Med-PaLM M [25] and LLaVA-Med [15], have achieved exceptional proficiency in 2D medical image interpretation, providing critical support for accurate diagnosis and clinical decision-making [19,22,26]. Beyond 2D modalities, specialized architectures [1,2,13,14,20] have been developed to navigate 3D volumetric data (e.g., CT and MRI), enabling the analysis of volumetric anatomical structures and temporal dynamics that are critical for complex clinical scenarios. Recently, advanced medical VLMs, such as Hulu [11], MedGemma-1.5 [23], and RadFM [27], have further extended their capabilities to simultaneously process both 2D and 3D medical inputs, aiming to consolidate multimodal medical image understanding within a unified framework. Despite these advancements, extending these models from 2D images to 3D clinical scenarios remains challenging, as high-resolution volumetric medical images can induce a substantial increase in token count. This soaring computational demand necessitates an intelligent token pruning mechanism to maintain reasoning integrity while ensuring clinical-grade inference speeds.

However, current processing pipelines and general-purpose pruning methods exhibit critical limitations when applied to 3D medical inputs. First, their architectures [3,11,23,29] typically rely on feeding 2D slices along a single axis into the model, where the generated tokens are directly concatenated. Since consecutive slices in 3D volumes share extreme spatial similarity, this direct concatenation introduces massive redundancy that exhausts the LLM's context window and hinders the processing of auxiliary clinical information. Second, existing token pruning methods [11,12,18,30] typically employ a static, predefined pruning ratio, which fails to account for the inherent heterogeneity of information density. While certain slices capture the intricate boundaries of a tumor, others may only contain uniform tissue with minimal diagnostic value; a fixed ratio either risks losing fine-grained pathological details or wastes tokens on irrelevant backgrounds. Crucially, these static approaches ignore the variance in attention distributions across different vision backbones. Different models exhibit distinct perceptual biases toward medical features, rendering model-agnostic pruning sub-optimal.

To address these challenges, we propose MedPruner, a training-free and model-agnostic hierarchical token pruning approach tailored for 3D medical image understanding in VLMs. Specifically, we first incorporate an inter-slice anchor-based filtering module to effectively manage the high temporal redundancy inherent in 3D medical volumes. Additionally, we develop a dynamic information nucleus selection strategy to achieve adaptive compression across slices with varying information densities by quantifying the cumulative attention weights contributed by visual tokens. To validate the efficacy of our proposed method, we conduct extensive experiments on three 3D medical benchmarks and three VLMs. Our results reveal substantial token redundancy in current medical vision-language models. Notably, MedPruner allows MedGemma-1.5 to maintain or even exceed its original performance while using less than 5% of the original visual tokens. This extreme compression highlights a highly skewed attention distribution in medical VLMs, demonstrating that dynamic token selection is essential for effectively filtering background noise and capturing critical diagnostic signals. Our contributions can be summarized as follows:

- We introduce MedPruner, which, to the best of our knowledge, is the first training-free and model-agnostic hierarchical token pruning framework tailored to 3D medical VLMs.
- We employ a training-free, two-stage mechanism to dynamically prune redundant information at both the slice and token levels.
- We conduct comprehensive experiments across 3 datasets and 3 VLMs, consistently demonstrating the effectiveness and robustness of our approach.

2 Methods

2.1 3D Vision-Language Models

Existing VLM architectures generally consist of three components: a visual encoder, a modality projector, and an LLM backbone. In 3D medical imaging, a common approach is to slice a 3D volume $V \in \mathbb{R}^{D\times H\times W}$ along the axial axis, resulting in a sequence of 2D slices $\mathcal{I} = \{I_1, I_2, \ldots, I_D\}$. Each slice I_i is independently processed by a visual encoder $\mathcal{E}_v$ and a modality projector $\mathcal{P}$. The final visual representation H_v is the concatenation of the projected tokens from all slices:

$$H_v = \bigoplus_{i=1}^{D} \mathcal{P}(\mathcal{E}_v(I_i)), \tag{1}$$

where $\bigoplus$ denotes sequence concatenation. If each slice yields M tokens, the total visual token count $D \times M$ leads to a severe sequence length explosion. To mitigate this, we propose MedPruner, a hierarchical framework consisting of two components: Inter-slice Anchor-based Filtering (IAF), which reduces inter-slice redundancy by tracking content evolution, and Dynamic Information Nucleus Selection (DINS), which adaptively prunes uninformative tokens based on attention distribution. The overview of MedPruner is shown in Fig. 1.

2.2 Inter-slice Anchor-Based Filtering

To effectively manage the high temporal redundancy inherent in 3D medical volumes, we introduce Inter-slice Anchor-based Filtering (IAF). Rather than employing a static or fixed-interval sampling rate, IAF utilizes a dynamic, content-aware strategy to adaptively identify slices that provide significant anatomical information. The process operates sequentially by maintaining a dynamic anchor slice, denoted as I_{anc}. We initialize this filtering process by

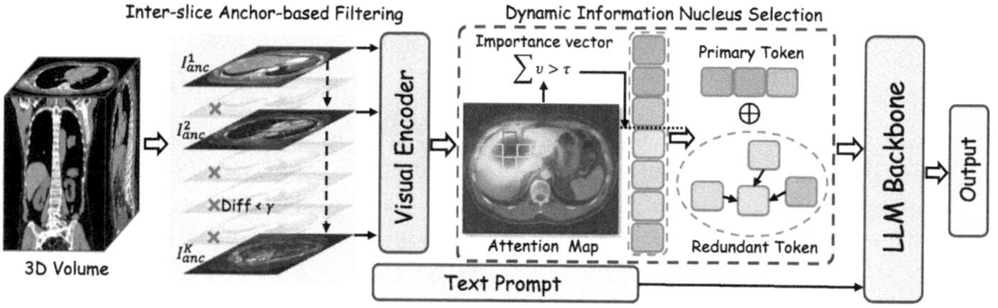

Fig. 1. The overview of our MedPruner.

setting the first slice of the volume as the starting anchor, such that $I_{anc} = I_1$. As we traverse the remaining sequence, we evaluate the informational divergence of each incoming slice I_i $(i > 1)$ relative to the current active anchor. This divergence is quantified using the pixel-wise mean L_1 distance:

$$\Delta(I_i, I_{anc}) = \frac{1}{N} \sum_{j=1}^{N} |I_{i,j} - I_{anc,j}|, \tag{2}$$

where N denotes the total number of pixels in a slice. This distance $\Delta(I_i, I_{anc})$ serves as a proxy for morphological change. A small distance indicates that the structural information in I_i is already well-represented by I_{anc}, rendering the current slice redundant. The core of the IAF mechanism lies in its threshold-driven update logic. We continuously compare the distance $\Delta(I_i, I_{anc})$ against a predefined sensitivity threshold γ. If the distance exceeds γ, the slice I_i is deemed to contain significant novel anatomical features. Consequently, I_i is preserved and immediately takes over as the new active anchor ($I_{anc} \leftarrow I_i$) for evaluating the remaining sequence. Conversely, if the distance falls below the threshold, the slice lacks sufficient new information and is entirely filtered out. Through this continuous traversal and updating process, the original dense volume of length D is distilled down to a sparse, informative subsequence comprising only the dynamically preserved anchor frames $\mathcal{I}_{filtered} = \{I^1_{anc}, I^2_{anc}, \dots, I^k_{anc}\}$, k is the final number of retained slices ($k < D$). By discarding the non-anchor intermediate frames, IAF compresses the sequence length. This ensures that the model concentrates its computational budget solely on regions with high structural variance, including the boundaries of organs or the appearance of lesions. Consequently, it prepares a highly condensed and representative sequence for the subsequent token-level optimization.

2.3 Dynamic Information Nucleus Selection

Following the inter-slice filtering, we further optimize the token density within each preserved slice by deriving token importance directly from the self-attention layers of the vision encoder. First, we calculate the attention score for each head as follows:

$$S_h = \text{Softmax}\left(\frac{Q_h K_h^\top}{\sqrt{D_h}}\right), \tag{3}$$

where D_h is the head dimension, and Q_h and K_h represent the query and key matrices, respectively. By averaging these scores across all heads, we obtain an aggregated attention matrix S_{avg}. To evaluate the raw significance of each visual token, we compute the average of S_{avg} along the sequence dimension, yielding an initial importance vector $\hat{v} \in \mathbb{R}^M$, where M is the number of tokens in the slice. To transform these raw scores into a comparable probability distribution and allow for adjustable selection sensitivity, we apply a temperature-scaled softmax normalization such that:

$$v_i = \frac{\exp(\hat{v}_i/T)}{\sum_{j=1}^{M} \exp(\hat{v}_j/T)}, \tag{4}$$

where the temperature coefficient T serves as a smoothing factor; a lower T sharpens the distribution to emphasize high-scoring tokens, while a higher T retains broader contextual information.

The inherent heterogeneity of information density across medical slices suggests that a fixed pruning ratio is suboptimal, as it fails to distinguish between salient anatomical features and uninformative backgrounds. To address this, we utilize a strategy inspired by nucleus filtering to adaptively capture the essential core of each slice by sorting the normalized weights in v in descending order to obtain v_{sorted}. We then dynamically select the minimal set of top-ranked tokens, designated as *primary tokens* $\mathcal{K}$, whose cumulative attention mass reaches a predefined information threshold τ:

$$\mathcal{K} = \{\text{Top-}k \text{ tokens} \mid \min k \quad \text{s.t.} \quad \sum_{j=1}^{k} v_{\text{sorted},j} \geq \tau\}. \tag{5}$$

By anchoring the selection boundary to the cumulative probability mass, this mechanism ensures that slices with concentrated attention are significantly compressed, while those with dispersed, critical details retain a larger token set to maintain diagnostic integrity. Token dimensions remain unchanged.

Finally, the remaining unselected tokens are treated as *redundant tokens*. To retain the global structural context without increasing the sequence length, we apply a bipartite matching and clustering operation following [30]. These clustered redundant tokens are subsequently concatenated with the primary tokens and passed to the modality projector for the final VLM inference.

3 Experiments

3.1 Experiment Setting

Datasets. In our experiments, we evaluate our approach on three 3D medical benchmarks: M3D [2], 3D-RAD [7], and AMOS-MM [10]. Both M3D and 3D-RAD are comprehensive 3D Medical Visual Question Answering (VQA) datasets

Table 1. Quantitative resutls on the 3DRad [7] and M3D [2] VQA datasets. **R-Rate** denotes the token retention rate. **Bold** and underlined values represent the best and second-best performance, respectively.

Method	M3D						3DRad					
	Acc	Rouge-1	Rouge-L	BLEU-1	BLEU-4	R-Rate	Acc	Rouge-1	Rouge-L	BLEU-1	BLEU-4	R-Rate
Hulu-Med-7B [11]												
Original	75.317	34.188	34.047	31.820	9.541	100.00%	78.767	23.280	22.917	18.760	5.695	100.00%
Hulu-L1 [11]	**77.790**	47.334	47.167	45.234	12.491	66.03%	78.082	23.765	23.487	20.389	6.280	45.06%
VisionZip [30]	77.623	**47.596**	**47.418**	**45.451**	12.549	69.96%	79.232	25.618	25.336	22.391	6.949	69.74%
HiPrune [18]	77.616	47.072	46.199	45.417	12.574	**22.30%**	79.061	25.721	25.559	21.339	6.759	**22.30%**
MedPruner	77.452	47.434	47.262	45.304	**12.580**	52.10%	**79.280**	**26.247**	**25.982**	**22.865**	**7.123**	51.88%
MedGemma1.5-4B [23]												
Original	32.717	6.797	5.888	3.741	0.835	100.00%	59.155	4.865	4.094	2.629	0.516	100.00%
Hulu-L1 [11]	44.638	7.810	4.473	2.779	0.651	65.98%	57.868	5.009	1.596	1.034	0.217	44.89%
VisionZip [30]	**45.428**	7.925	4.364	2.707	0.639	69.98%	56.537	5.091	1.155	0.754	0.165	69.82%
HiPrune [18]	44.420	8.201	4.447	2.792	0.664	22.30%	55.959	**6.109**	2.045	**1.365**	**0.304**	22.30%
MedPruner	43.718	**8.983**	**5.845**	**3.923**	**1.005**	**4.87%**	**60.843**	5.168	**2.057**	1.317	0.277	**4.62%**

that support both open-ended and closed-ended evaluations. Specifically, M3D queries detailed anatomical and pathological attributes, such as imaging planes, contrast phases, specific organs, and abnormalities. 3D-RAD focuses specifically on radiology CT scans and introduces complex reasoning challenges across six diverse VQA tasks, including anomaly detection, medical computation, and multi-stage temporal diagnosis. Additionally, we utilize the AMOS-MM benchmark, which comprises CT and MRI of abdominal organs, primarily to assess the model's performance in 3D medical report generation.

Implementation Details. We evaluate MedPruner across three VLMs which support 3D medical imaging input, including a general-purpose model, Qwen3-VL-8B [3], and two medical domain-specific models, Hulu-Med-7B [11] and MedGemma-1.5-4B [23]. For the evaluation metrics, we utilize Accuracy for closed-set VQA tasks to rigorously measure classification performance. For open-set reasoning and 3D medical report generation, we employ standard Natural Language Generation metrics including BLEU [21] (BLEU-1, BLEU-4), Rouge [16] (Rouge-1, Rouge-L), and METEOR [4]. All experiments are conducted on 8 $\times$ NVIDIA H20 GPUs. Hyperparameters are set as $\gamma = 0.05$ and $\tau = 0.9$.

3.2 Comparison Results

In this section, we conduct a comprehensive evaluation of MedPruner against three existing training-free token reduction methods: the L1-compression method proposed in Hulu-Med [11] (Hulu-L1), VisionZip [30], and HiPrune [18].

Table 1 presents quantitative results on the 3DRad and M3D VQA datasets using Hulu-Med and MedGemma-1.5. In cases where datasets involve massive slice counts, such as M3D with an average of 87 and a maximum of over 600 slices, MedPruner plays a critical role in mitigating information overload. Notably, on

Table 2. Quantitative results on the AMOS-MM dataset [10]. **R-Rate** denotes the token retention rate. **Average** represents the mean percentage of performance across all metrics relative to the original model. **Speed** indicates the average processing time per sample in seconds (s). **Bold** and underlined values represent the best and second-best performance, respectively.

Method	Rouge-1	Rouge-L	BLEU-1	BLEU-4	METEOR	Average↑	R-Rate↓	Speed↓
Hulu-Med-7B [11]								
Original	39.518	28.763	37.317	13.375	31.729	100.00%	100.00%	9.212
Hulu-L1 [11]	33.063	24.826	30.754	9.799	26.212	81.65%	**16.35%**	8.039
VisionZip [30]	38.519	28.216	36.876	12.705	30.738	97.25%	49.69%	8.435
HiPrune [18]	39.093	27.956	35.764	12.513	31.086	96.70%	22.30%	9.111
MedPruner	**39.255**	**28.860**	**37.339**	**13.120**	**31.136**	**99.19%**	54.20%	**7.931**
MedGemma1.5-4B [23]								
Original	13.791	9.906	5.845	0.253	9.721	100.00%	100.00%	38.001
Hulu-L1 [11]	**13.999**	9.787	5.843	0.241	9.782	99.25%	16.07%	36.193
VisionZip [30]	13.830	10.043	**6.050**	**0.260**	10.129	**102.42%**	49.61%	36.682
HiPrune [18]	13.369	9.782	5.828	0.236	9.939	98.21%	21.85%	37.619
MedPruner	13.706	**10.234**	5.985	0.235	**10.252**	100.65%	**2.46%**	**35.889**
Qwen3-VL-8B [3]								
Original	18.717	18.317	26.994	1.077	18.821	100.00%	100.00%	11.179
Hulu-L1 [11]	18.201	17.808	25.839	0.870	17.788	93.10%	**16.50%**	9.569
VisionZip [30]	18.590	**18.047**	25.604	0.925	18.116	94.98%	49.61%	10.654
HiPrune [18]	18.217	17.255	25.886	0.919	**18.600**	94.33%	22.30%	10.052
MedPruner	**19.469**	17.908	**26.119**	**0.937**	17.648	**95.86%**	43.28%	**9.044**

the M3D dataset, MedPruner frequently outperforms the uncompressed baseline. This occurs because directly concatenating hundreds of slices introduces significant background noise and redundant structures that can overwhelm the LLM's context window. By filtering these uninformative tokens, MedPruner achieves the highest BLEU-4 scores on M3D (12.580) and 3DRad (7.123) with Hulu-Med while maintaining competitive accuracy and reducing the token retention rate (R-Rate) to approximately 52%.

Table 2 reports the performance on the AMOS-MM dataset. Across all tested architectures, MedPruner achieves the optimal balance between accuracy and efficiency, delivering the fastest inference speeds while maintaining exceptional performance. In several instances, it even surpasses the original baselines, such as reaching a 100.65% average score on MedGemma-1.5. These results highlight the model-agnostic robustness of MedPruner in optimizing diverse VLMs under real-world computational constraints.

A notable observation is the extreme token compression achieved by MedPruner on the MedGemma-1.5 model. As shown in the tables, MedPruner maintains high performance while requiring fewer than 5% of the visual tokens across all three datasets, reaching an incredibly low R-Rate of 2.46% on AMOS-MM. Upon analyzing the model's behavior, we discovered that MedGemma-1.5's attention weights are highly concentrated on a single or a very small subset of tokens. By dynamically anchoring the selection threshold to the cumulative attention mass, our module naturally adapts to this highly skewed distribution. This phenomenon strongly validates the necessity of our dynamic selection strategy. Unlike fixed-ratio methods that arbitrarily discard useful tokens or retain unnecessary ones (e.g., HiPrune fixed at 22.30%), MedPruner intelligently scales the token retention rate based on the intrinsic attention distribution, ensuring optimal efficiency tailored to each specific slice and model architecture.

3.3 Ablation Study

Component Analysis. We first evaluate the contribution of each MedPruner component on the AMOS-MM dataset, with results summarized in Table 3. As shown in Table 3, IAF lowers average inference time from 9.2 s to 7.7 s. While this slice-level filtering initially leads to a performance drop, the introduction of Primary token selection and Redundant token clustering effectively restores diagnostic accuracy. Specifically, incorporating primary tokens and redundant clustering recovers the Average score to over 100% of the baseline. The full MedPruner configuration achieves an optimal balance, maintaining 99.19% of the original performance while reaching the strongest compression with a 54.20% R-Rate.

Sensitivity of Information Threshold τ. Figure 2 illustrates the impact of the information threshold τ on the Hulu-Med-7B model. As τ increases, the token retention rate rises, leading to improved Rouge-L and METEOR scores. However, these performance gains gradually plateau, indicating that the primary tokens identified by our nucleus selection strategy have already captured the most critical diagnostic features. This confirms that further increasing the

Table 3. Ablation study on the AMOS-MM dataset [10] using the Hulu-Med-7B model [11].

IAF	Primary	Redundant	Average↑	R-Rate↓	Speed↓
✗	✗	✗	100.00%	100.00%	9.212
✓	✗	✗	92.13%	<u>60.33%</u>	**7.751**
✗	✓	✗	98.73%	83.11%	8.894
✗	✓	✓	**100.07%**	88.14%	9.586
✓	✓	✓	<u>99.19%</u>	**54.20%**	<u>7.931</u>

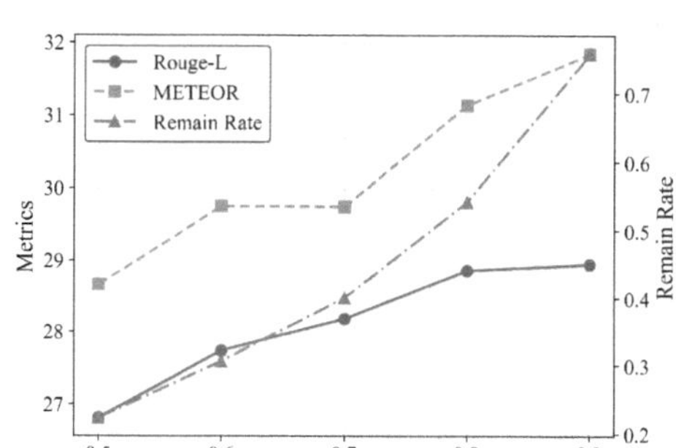

Fig. 2. Ablation study of τ.

token count yields diminishing returns, validating the efficiency of our adaptive selection mechanism.

4 Conclusion

In this paper, we propose MedPruner, a training-free hierarchical pruning framework that bridges the gap between high-performance 3D medical VLMs and the constraints of real-time clinical deployment. By adaptively managing the non-uniform information density across volumetric data, our approach ensures that critical diagnostic details are prioritized over redundant anatomical backgrounds. Extensive evaluations show that MedPruner improves the efficiency-performance trade-off without compromising diagnostic integrity, offering a scalable and model-agnostic solution for the practical integration of VLMs into complex medical workflows.

Acknowledgments.. This work was supported by National Natural Science Foundation of China (No. 62522124), Hong Kong Research Grants Council General Research Fund 14206725 and Hong Kong Innovation and Technology Commission Innovation and Technology Fund PRP/082/24FX.

Disclosure of Interests. The authors have no competing interests to declare that are relevant to the content of this article.

References

1. Agrawal, K.K., et al.: Pillar-0: a new frontier for radiology foundation models. arXiv preprint arXiv:2511.17803 (2025)
2. Bai, F., Du, Y., Huang, T., Meng, M.Q.H., Zhao, B.: M3D: advancing 3D medical image analysis with multi-modal large language models. arXiv preprint arXiv:2404.00578 (2024)
3. Bai, S., et al.: Qwen2.5-VL technical report. arXiv preprint arXiv:2502.13923 (2025)
4. Banerjee, S., Lavie, A.: METEOR: An automatic metric for MT evaluation with improved correlation with human judgments. In: Proceedings of ACL Workshop, pp. 65–72 (2005)
5. Chen, J., et al.: Towards injecting medical visual knowledge into multimodal LLMs at scale. In: Proceedings of the 2024 Conference on Empirical Methods in Natural Language Processing, pp. 7346–7370 (2024)
6. Comanici, G., et al.: Gemini 2.5: Pushing the frontier with advanced reasoning, multimodality, long context, and next generation agentic capabilities. arXiv preprint arXiv:2507.06261 (2025)
7. Gai, X., Liu, J., Li, Y., Meng, Z., Wu, J., Liu, Z.: 3D-RAD: A comprehensive 3D radiology Med-VQA dataset with multi-temporal analysis and diverse diagnostic tasks. arXiv preprint arXiv:2506.11147 (2025)
8. He, S., Nie, Y., Chen, Z., Cai, Z., Wang, H., Yang, S., Chen, H.: MedDR: diagnosis-guided bootstrapping for large-scale medical vision-language learning. CoRR (2024)

9. Hurst, A., et al.: GPT-4o system card. arXiv preprint arXiv:2410.21276 (2024)
10. Ji, Y., et al.: AMOS: a large-scale abdominal multi-organ benchmark for versatile medical image segmentation. Adv. NeurIPS **35**, 36722–36732 (2022)
11. Jiang, S., et al.: Hulu-Med: a transparent generalist model towards holistic medical vision-language understanding. arXiv preprint arXiv:2510.08668 (2025)
12. Khaki, S., et al.: SparseVILA: decoupling visual sparsity for efficient VLM inference. In: Proceedings of the ICCV, pp. 23784–23794 (2025)
13. Lai, H., et al.: Med3D-R1: incentivizing clinical reasoning in 3D medical vision-language models for abnormality diagnosis. arXiv preprint arXiv:2602.01200 (2026)
14. Li, C.Y., et al.: Towards a holistic framework for multimodal LLM in 3D brain CT radiology report generation. Nat. Commun. **16**(1), 2258 (2025)
15. Li, C., et al.: LLaVA-Med: training a large language-and-vision assistant for biomedicine in one day. arXiv preprint arXiv:2306.00890 (2023)
16. Lin, C.Y.: ROUGE: A package for automatic evaluation of summaries. In: Text summarization branches out, pp. 74–81 (2004)
17. Liu, H., Li, C., Wu, Q., Lee, Y.J.: Visual instruction tuning. Adv. NeurIPS **36**, 34892–34916 (2023)
18. Liu, J., Du, F., Zhu, G., Lian, N., Li, J., Chen, B.: HiPrune: training-free visual token pruning via hierarchical attention in vision-language models. arXiv preprint arXiv:2508.00553 (2025)
19. Liu, S., Zheng, B., Chen, W., Peng, Z., Yin, Z., Shao, J., Hu, J., Yuan, Y.: EndoBench: A comprehensive evaluation of multi-modal large language models for endoscopy analysis (2025)
20. Vepa, A.M., et al.: A multimodal LLM approach for visual question answering on multiparametric 3D brain MRI. arXiv e-prints, pp. arXiv–2509 (2025)
21. Papineni, K., Roukos, S., Ward, T., Zhu, W.J.: BLEU: a method for automatic evaluation of machine translation. In: Proceedings of the ACL, pp. 311–318 (2002)
22. Peng, Z., et al.: Omnibrainbench: a comprehensive multimodal benchmark for brain imaging analysis across multi-stage clinical tasks. In: Proceedings of the CVPR, pp. 42732–42743 (2026)
23. Sellergren, A., et al.: MedGemma technical report. arXiv preprint arXiv:2507.05201 (2025)
24. Tanno, R., et al.: Collaboration between clinicians and vision-language models in radiology report generation. Nat. Med. **31**(2), 599–608 (2025)
25. Tu, T., et al.: Towards generalist biomedical AI. Nejm AI **1**(3), AIoa2300138 (2024)
26. Wang, W., et al.: Medical reasoning in the era of LLMs: a systematic review of enhancement techniques and applications. arXiv preprint arXiv:2508.00669 (2025)
27. Wu, C., et al.: Towards generalist foundation model for radiology by leveraging web-scale 2D&3D medical data. Nat. Commun. **16**(1), 7866 (2025)
28. Xu, H., et al.: A whole-slide foundation model for digital pathology from real-world data. Nature **630**(8015), 181–188 (2024)
29. Xu, W., et al.: Lingshu: a generalist foundation model for unified multimodal medical understanding and reasoning. arXiv preprint arXiv:2506.07044 (2025)
30. Yang, S., et al.: VisionZip: longer is better but not necessary in vision language models. In: Proceedings of the CVPR, pp. 19792–19802. (2025)
31. Chaves, J.M.Z., et al.: A clinically accessible small multimodal radiology model and evaluation metric for chest X-ray findings. Nat. Commun. **16**(1), 3108 (2025)

MedTri: Structured Medical Report Normalization for Enhanced Vision–Language Pretraining

Yuetan Chu, Xinghua Ma, Xinran Jin, Gongning Luo, and Xin Gao(✉)

King Abdullah University of Science and Technology (KAUST), Thuwal, Saudi Arabia
xin.gao@kaust.edu.sa

Abstract. Medical vision–language pretraining increasingly relies on medical reports as large-scale supervisory signals; however, raw reports often exhibit substantial stylistic heterogeneity, variable length, and a considerable amount of image-irrelevant content. Although text normalization is frequently adopted as a preprocessing step in prior work, its design principles and empirical impact on vision-language pretraining remain insufficiently and systematically examined. In this study, we present **MedTri**, a deployable normalization framework for medical vision-language pretraining that converts free-text reports into a unified *[Anatomical Entity: Radiologic Description + Diagnosis Category]* triplet. This structured, anatomy-grounded normalization preserves essential morphological and spatial information while removing stylistic noise and image-irrelevant content, providing consistent and image-grounded textual supervision at scale. Across multiple datasets spanning both X-ray and computed tomography (CT) modalities, we demonstrate that *structured, anatomy-grounded text normalization is an important factor in medical vision–language pretraining quality*, yielding consistent improvements over raw reports and existing normalization baselines. In addition, we illustrate how this normalization can easily support targeted text-level augmentation strategies, including knowledge enrichment and anatomy-grounded counterfactual supervision, which provide complementary gains in robustness and generalization without altering the core normalization process. Together, our results position structured text normalization as a critical and generalizable preprocessing component for medical vision–language learning, and establish MedTri as a practical implementation of this approach. The source code is available at https://github.com/Arturia-Pendragon-Iris/MedTri.

Keywords: Medical vision-language pretraining · Large-language model · Contrastive learning

1 Introduction

vision-language pretraining (VLP) that leverages naturally paired medical images and medical reports has emerged as a powerful paradigm in medical

G. Yang et al. (Eds.): MICCAI 2026, LNCS 16878, pp. 366–376, 2027.
https://doi.org/10.1007/978-3-032-38059-3_35

Raw Report Irrelevant Information Mixed structures

FINAL REPORT
AP CHEST, 5 A.M., ___
HISTORY: ___-year-old man after cardiac arrest. Evaluate for possible pneumonia.
IMPRESSION: AP chest compared to chest radiographs since ___ through ___, 4:41 a.m., read in conjunction with chest CTA on ___: Pulmonary edema resolved on ___, 4:41 a.m. and has returned, moderate in severity, right lung greater than left. Though pneumonia could be present concurrently, there is no reason to invoke that diagnosis to explain the findings. Severe cardiomegaly is stable. Mediastinum is widened by an extremely tortuous and only mildly dilated aorta. The pleural effusion is small, if any. No pneumothorax. Right subclavian line ends low in the SVC. Right pleural tube still in place in the lower hemi thorax.

RadGraph

edema returned moderate [pulmonary][definitely present];
greater [right lung left] [definitely present];
pneumonia [uncertain];
severe cardiomegaly stable [definitely present];
widened [mediastinum] [definitely present];
extremely tortuous [aorta] [definitely present];
mildly dilated [aorta] [definitely present];
effusion small [pleural] [uncertain];
pneumothorax [definitely absent];
line low [right subclavian, svc] [definitely present];
tube in place [right pleural, lower hemithorax] [definitely present]

MedTri

lung: Moderate pulmonary edema, right greater than left; possible pneumonia.
heart: Severe cardiomegaly; stable.
aorta: Extremely tortuous and mildly dilated; no significant findings.
pleura: Small pleural effusion, if any; no pneumothorax.
mediastinum: Widened due to tortuous aort; no acute findings.
subclavian line: Ends low in the SVC; no complications noted.
pleural tube: Right pleural pigtail catheter in lower hemithorax; in place.

Fig. 1. Illustration of normalization differences across raw reports, RadGraph, and MedTri. The raw clinical report contains irrelevant content and mixed structures. RadGraph extracts only diagnostic entities with limited imaging descriptions. MedTri produces anatomically anchored, image-grounded triplets that preserve morphological and spatial detail while removing stylistic and clinically irrelevant text.

image analysis [14,22]. Such pairs provide large-scale semantic supervision without additional annotation effort [25,28], as reports contain expert interpretations grounded in patient-specific visual findings. By jointly modeling visual content and textual descriptions, medical VLP strategies have demonstrated strong capabilities in capturing disease-relevant semantics, improving representation quality, and enhancing generalization across diverse downstream tasks [19]. This synergy positions vision-language alignment as a foundational component for modern medical image pretraining.

Despite the success of VLP, the textual supervision provided by raw clinical reports introduces several obstacles for effective multimodal alignment. Medical reports vary widely in style, verbosity, and structure, often mixing image-grounded observations with unrelated clinical history or management recommendations [17]. This heterogeneity reduces image-relevant signals, inflates sequence length, and weakens the fine-grained correspondence between visual findings and textual descriptions. As a result, recent studies increasingly adopt text normalization to standardize report content and enhance the stability of vision-language learning [19,20]. However, text normalization is often introduced as a preprocessing component and combined with other architectural designs, while its standalone design principles and empirical impact on vision–language pretraining remain insufficiently examined. Moreover, existing normalization methods, such as schema-based or NER-driven systems (e.g., RadGraph [13]), primarily focus on entity extraction, whereas cloud-based LLM methods rely on large-scale generative rewriting at the expense of increased computational overhead and potential privacy concerns. Consequently, the field still lacks a lightweight, anatomically expressive, and locally deployable normalization solution.

In this study, we present **MedTri**, a structured normalization approach for medical vision–language pretraining. MedTri converts free-text radiology reports into a unified *[Anatomical Entity: Radiologic Description + Diagnosis Category]* triplet that preserves essential morphological and spatial information while removing stylistic noise and image-irrelevant content [8,9]. This design enables consistent, efficient, and privacy-preserving report normalization suitable for large-scale pretraining. Beyond the normalization itself, MedTri further provides a modular interface that allows additional text-level augmentation on top of the normalized triplet. Using this interface, we instantiate two optional examples: knowledge enrichment and anatomy-grounded counterfactual augmentation. Across multiple datasets covering both X-ray and CT modalities, we systematically demonstrate that *structured, anatomy-grounded text normalization is an important factor in medical vision–language pretraining quality*, with MedTri consistently outperforming raw reports and existing normalization approaches. The optional augmentation modules provide further, complementary gains in performance and generalization. Together, these results position MedTri as a practical and anatomically expressive normalization approach for medical vision–language learning (Fig. 1).

Table 1. Overview of the report dataset used to develop and evaluate the MedTri normalization platform.

Dataset	# Reports	Image Modality	Body Region
MIMIC-CXR [14]	65,000	X-ray	Chest
CT-RATE [11]	20,000	CT	Chest
CT-INSPECTION [12]	10,000	CT	Chest
AbdomenAtlas [18]	5,000	CT	Abdomen
Parrot [17]	2,658	CT, MRI, X-ray, etc	Head, abdomen, chest, etc

2 Method

2.1 Structured Triplet

To support stable normalization for medical vision-language pretraining, we adopt a structured triplet that decomposes each report into a set of clinically grounded triplets:

[*Anatomical Entity: Radiologic Description* + *Diagnosis Category*]

The schema captures the minimal semantic unit of radiologic reasoning, consisting of an anatomical anchor, objective imaging attributes, and an associated diagnostic interpretation when present. By explicitly disentangling these components, MedTri converts heterogeneous free-text reports into anatomy-level alignment units, reducing lexical and stylistic variability while preserving semantically discriminative, image-grounded information for vision-language pretraining.

2.2 Local Model Development

Clinical reports from different institutions exhibit substantial variability in style, grammar, and diagnostic phrasing. To ensure that MedTri remains robust to this heterogeneity, we established a dataset with more than 100,000 reports selected from multiple publicly available datasets covering diverse imaging modalities and anatomical regions (Table 1). Using this dataset, we first generated structured reference summaries through a standardized ChatGPT-5.1 prompt specifically designed for our triplet schema. The prompt was iteratively refined on approximately 1,000 cases and then applied to the full dataset, yielding paired samples (x_i, y_i) in which x_i is the original report and y_i is its normalized triplet (Table 2).

To enable scalable and privacy-preserving deployment, we train a lightweight biomedical encoder–decoder model (BioBart [27]) to approximate the cloud LLM-distilled supervision. The model is fine-tuned on the paired dataset using standard cross-entropy loss, following a traditional Seq2Seq formulation [15,26]. This locally deployable text-transfer model forms the core of the MedTri approach, allowing rapid and consistent normalization without relying on large cloud-based language models. A total of 500 reports were randomly sampled from the established dataset for model testing, whereas the remaining reports were used for model training and validation.

Table 2. Example of the standardized ChatGPT-5.1 prompt used to generate structured triplets from free-text reports.

Input:	IMPRESSION: 1. No evidence of pulmonary embolism. No acute intrathoracic pathology. 2. Mild bronchial thickening, nonspecific, but may reflect infectious or inflammatory process. 3. Main pulmonary artery is dilated measuring 39 mm, possibly reflecting background pulmonary hypertension.
Prompt:	You will receive one paragraph containing radiology findings (image-based description) and a final diagnosis. Your ONLY task is to convert this information into a structured radiology summary using the specified format. Each line must follow this exact format: [Anatomical Term]: [Radiologic Description + Diagnosis Category]. List only the anatomical regions mentioned in the findings or diagnosis. Maintain the anatomical hierarchy (e.g., lung before airway, heart before aorta) if multiple structures are present. For each anatomical term, include radiologic findings (location, size, attenuation, density, texture, etc.) and diagnosis category if explicitly stated. Use objective imaging language only. Use direct descriptive phrasing. Keep each anatomical line concise and single-sentence, using semicolons to separate multiple findings.
Output:	Lung: No evidence of pulmonary embolism; no acute intrathoracic pathology. Bronchi: Mild bronchial thickening; nonspecific; may reflect infectious or inflammatory process. Main pulmonary artery: Dilated; measures 39 mm; possibly reflecting background pulmonary hypertension.

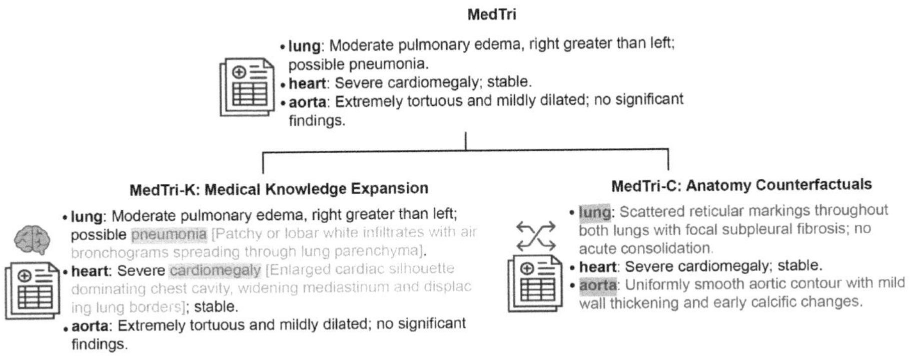

Fig. 2. Illustration of MedTri's optional augmentation modules. MedTri-K enriches each triplet with clinically validated radiologic signatures to improve visual interpretability. MedTri-C generates anatomically inconsistent counterfactuals through controlled perturbations to strengthen fine-grained visual–semantic discrimination.

2.3 Optional Text-Level Augmentation on MedTri

Medical Knowledge Expansion (MedTri-K). Prior work in vision–language learning has shown that explicitly grounding diagnostic terms in visual attributes can improve semantic alignment [19]. However, such strategies can be difficult to apply reliably to raw radiology reports due to stylistic variability and entangled narrative structure. Leveraging the structured triplet produced by MedTri, we can integrate a lightweight knowledge expansion mechanism (Fig. 2, left). For each normalized triplet, the diagnosis is augmented with a concise description of its characteristic radiological appearance, retrieved from our created dictionary of standard medical definitions covering over 100 common radiological findings and diagnoses. For example, pneumonia is associated with parenchymal consolidation or high-attenuation opacity within the affected lobe. These dictionary entries are created by board-certified radiologists and normalized for terminological consistency, allowing them to be seamlessly integrated into MedTri without altering its underlying structure. Importantly, this module operates entirely on normalized text and does not modify the core normalization process.

Anatomy-Grounded Counterfactuals (MedTri-C). Counterfactual supervision and hard negative sampling have been widely explored to encourage fine-grained discrimination in vision–language models [4,6,16]. The anatomically grounded triplet structure produced by MedTri enables a controlled and fine-grained instantiation of counterfactual text augmentation through localized perturbations. Specifically, we generate counterfactual reports by modifying the descriptions of several anatomical entities (n = 2 in our study) at a time, replacing them with semantically incompatible counterparts randomly sampled from other normalized triplets (Fig. 2, right). Replacement is constrained within the same anatomical hierarchy level to ensure structural consistency while altering local semantic alignment. These substitutions preserve the overall syntactic format and global report semantics, while deliberately disrupting the local factual alignment between anatomy and imaging findings.

The resulting counterfactual texts are paired with the original images and treated as hard negatives during contrastive training. By introducing fine-grained, anatomy-level inconsistencies rather than global semantic shifts, this strategy forces the model to attend to localized visual evidence and fine-grained anatomical features, instead of relying on coarse diagnostic signals. As with MedTri-K, this module is optional and applied only during training, without affecting the normalization backbone.

3 Experiments and Results

3.1 Evaluation of MedTri for Normalization

To assess the quality and deployability of the proposed approach, we compare MedTri against two representative baselines: (1) ChatGPT-5.1-based structured rewriting, which serves as a high-quality but non-deployable reference for distilled supervision, and (2) Qwen2.5-14B [1], a compact open-source model commonly used in local summarization workflows. Quantitative results are summarized in Table 3.

To evaluate clinical validity beyond surface-level textual similarity and to mitigate potential reference bias, we conduct a physician expert evaluation as the primary assessment. Twenty board-certified physicians independently rated 50 randomly selected cases in a double-blind manner using a five-point Likert scale, focusing on anatomical correctness and image groundedness. As shown in Table 3, MedTri achieves expert scores comparable to the cloud-based LLM reference and substantially higher than the open-source baseline, indicating that it preserves clinically meaningful and image-grounded information while remaining deployable in typical research and clinical settings. The evaluation also showed high inter-rater reliability, with ICC values of 0.983 and 0.990 for anatomical correctness and image groundedness, respectively, and Fleiss' kappa values of 0.656 and 0.634, supporting both the reliability of averaged scores and agreement among physicians.

We also report automatic text similarity metrics, including BERT score, BLEU, and ROUGE [7,10,21], computed against ChatGPT-5.1-generated references. These metrics primarily measure the degree to which MedTri approximates the distilled reference normalization and are therefore used as proxy indicators of consistency rather than as direct measures of clinical correctness. Quantitative results for the open-source baseline are included for reference.

3.2 MedTri for Improving Downstream Tasks

Experiment Settings and Datasets. We adopt Swin Transformer and Vision Transformer (ViT) [3,5] as the image encoders, and BiomedVLP-CXR [2] as the text encoder. All reports were truncated to a fixed maximum length of 512

Table 3. Comparison of computational efficiency, expert evaluation, and normalization accuracy across different systems.

Metric	ChatGPT-5.1	Qwen2.5	MedTri
Anatomical correctness (1-5)	4.888	3.363	4.838
Image groundedness (1-5)	4.775	2.938	4.763
VRAM usage (GB)	–	15.04	1.50
Time/report (s)	3.25±2.14	1.21±0.67	0.45±0.18
BERT score	–	0.734±0.073	0.906±0.072
BLEU	–	0.097	0.541
ROUGE-1	–	0.566±0.127	0.824±0.146
ROUGE-2	–	0.307±0.122	0.720±0.190

tokens, which sufficiently covers the majority of radiology reports [11] in our datasets and avoids disproportionately truncating raw reports, ensuring a fair comparison. Model training is conducted using the InfoNCE objective for contrastive learning. We apply data augmentation, including random horizontal flipping and Gaussian noise injection (sigma=0.05). All experiments are executed on an Ubuntu workstation equipped with a single NVIDIA A6000 GPU.

Pretraining is conducted separately for two imaging modalities. For 2D radiographs, we use the MIMIC-CXR dataset [14] (n = 370k), while for 3D volumetric imaging, we adopt the CT-RATE dataset [11] (n= 42k) for CT pretraining.

For X-ray downstream evaluation, we conduct multi-label classification on three datasets: MIMIC-CXR, NIH ChestX-ray14 [24] (n = 112k), and RSNA-Pneumonia [23] (n = 30 k). From each dataset, we randomly sample 1,024 studies for evaluation.

CT downstream tasks are evaluated on the CT-RATE dataset, with the training and testing splits following the official protocol described in [11]. Due to the severe class imbalance commonly observed in medical imaging datasets, we determine the decision threshold that maximizes the F1 score and report the corresponding accuracy. Both F1 score and accuracy are computed on a per-label basis and then macro-averaged across all labels [19].

The quantitative results are presented in Tables 4 and 5, respectively. Across all X-ray and CT benchmarks, MedTri and its variants consistently outperform raw reports and RadGraph across different data scales and visual backbones. The improvements are observed for both accuracy and F1 score, with particularly pronounced gains in low-data regimes (1% and 10%), indicating that structured normalization substantially improves sample efficiency for vision-language pretraining. The consistent performance gains across datasets and architectures

Table 4. Downstream classification performance of Swin Transformer (SwinT) and ViT on different X-ray datasets under different text preprocessing strategies. Best performance is marked in bold.

SwinT	MIMIC						NIH						Pneumonia					
	1%		10%		100%		1%		10%		100%		1%		10%		100%	
	ACC	F1	ACC	F1	ACC	F1	ACC	F1	ACC	F1	ACC	F1	ACC	F1	ACC	F1	ACC	F1
Raw report	0.612	0.218	0.701	0.263	0.781	0.302	0.655	0.221	0.722	0.266	0.762	0.293	0.812	0.673	0.833	0.699	0.848	0.722
RadGraph	0.584	0.205	0.673	0.251	0.748	0.284	0.628	0.207	0.703	0.249	0.747	0.265	0.804	0.665	0.822	0.690	0.840	0.715
MedTri	0.628	0.229	0.716	0.279	0.792	0.314	0.672	0.233	0.739	0.281	0.779	0.304	0.817	0.676	0.836	0.701	0.850	0.723
MedTri-K	**0.645**	**0.241**	**0.732**	0.286	0.784	0.305	**0.689**	**0.242**	0.746	0.284	0.771	0.292	0.823	**0.681**	**0.840**	**0.703**	0.851	0.722
MedTri-C	0.629	0.226	0.731	**0.291**	**0.803**	**0.318**	0.674	0.227	**0.751**	**0.290**	**0.801**	**0.321**	**0.824**	0.679	0.838	0.701	**0.855**	**0.728**

ViT	MIMIC						NIH						Pneumonia					
	1%		10%		100%		1%		10%		100%		1%		10%		100%	
	ACC	F1	ACC	F1	ACC	F1	ACC	F1	ACC	F1	ACC	F1	ACC	F1	ACC	F1	ACC	F1
Raw report	0.609	0.215	0.703	0.265	0.774	0.282	0.655	0.221	0.720	0.262	0.787	0.284	0.794	0.644	0.827	0.683	0.847	0.718
RadGraph	0.587	0.207	0.674	0.250	0.741	0.262	0.631	0.209	0.702	0.248	0.738	0.266	0.785	0.649	0.813	0.678	0.834	0.707
MedTri	0.629	0.229	0.717	0.280	0.787	0.299	0.673	0.235	0.738	0.281	0.792	0.306	0.801	0.660	0.832	0.693	0.846	0.721
MedTri-K	**0.647**	**0.241**	**0.734**	0.288	0.782	0.289	**0.690**	**0.244**	**0.746**	0.285	0.792	0.306	**0.806**	**0.671**	0.841	**0.706**	0.842	0.723
MedTri-C	0.625	0.226	**0.734**	**0.291**	**0.801**	**0.307**	0.677	0.238	0.744	**0.289**	**0.808**	**0.312**	0.803	0.669	**0.846**	0.704	**0.855**	**0.728**

demonstrate that the benefits of MedTri are robust and stem primarily from improved textual supervision. Wilcoxon signed-rank tests further showed significant improvements of MedTri over raw reports and RadGraph on X-ray ($p = 0.000231$ and $p = 0.000194$, respectively) and on CT ($p = 0.03125$).

Table 5. Downstream classification performance of Swin Transformer (SwinT) and ViT on CT-RATE datasets under different text preprocessing strategies. Best performance is marked in bold.

	SwinT						ViT					
	1%		10%		100%		1%		10%		100%	
	ACC	F1	ACC	F1	ACC	F1	ACC	F1	ACC	F1	ACC	F1
Raw report	0.719	0.493	0.741	0.518	0.775	0.525	0.696	0.473	0.733	0.511	0.745	0.518
RadGraph	0.714	0.480	0.735	0.513	0.744	0.523	0.694	0.471	0.722	0.508	0.716	0.509
MedTri	0.723	0.499	0.753	0.520	0.785	0.538	0.704	0.488	0.745	0.518	0.775	0.537
MedTri-K	**0.732**	**0.510**	0.757	0.519	0.789	0.536	**0.718**	**0.498**	0.743	**0.519**	0.775	0.530
MedTri-C	0.726	0.501	**0.767**	**0.523**	**0.794**	**0.557**	0.704	0.485	**0.751**	**0.519**	**0.782**	**0.543**

The two optional augmentation modules exhibit complementary behaviors. MedTri-K (knowledge expansion) tends to achieve the best or near-best performance under limited data settings (1% and 10%), suggesting that explicitly linking diagnostic terms to characteristic imaging appearances provides additional semantic grounding when training data is scarce. However, its gains diminish at full-data scale (100%), where the model can already learn such associations implicitly from abundant image–text pairs. In contrast, MedTri-C (counterfactual construction) shows limited improvement in the 1% setting, likely because extremely limited data prevents the model from effectively exploiting fine-grained counterfactual distinctions. As data availability increases (10% and 100%), counterfactual supervision becomes more effective, leading to stronger gains in medium- and full-data regimes by encouraging finer-grained visual-semantic discrimination. The improvements of MedTri-K and MedTri-C over MedTri were also statistically significant ($p = 0.00733$ and $p = 0.000245$).

4 Discussion and Conclusion

In this work, we present MedTri, a lightweight and deployable approach that normalizes free-text medical reports into structured, anatomically grounded triplets for vision-language pretraining. Experiments on X-ray and CT datasets show that structured normalization consistently improves downstream performance across data scales, datasets, and visual backbones, offering a practical alternative to cloud LLM-dependent pipelines for scalable and privacy-preserving clinical and research use.

This study has several limitations. Our evaluation focuses on CLIP-style contrastive vision-language pretraining and does not cover generative, instruction-tuned, or task-specific multimodal frameworks. MedTri is also evaluated only on radiology reports and imaging modalities, leaving its applicability to other medical domains or non-radiological clinical narratives unexplored. Finally, the proposed triplet schema is one principled design choice, and alternative schemas or decomposition strategies warrant further investigation.

Acknowledgments. This work was supported by KAUST Office of ORA under Awards RGC/3/6655-01-01, RGC/3/6208-01-01, RGC/3/5679-01-01, REI/1/5289-01-01, REI/1/5992-01-01, URF/1/6713-01-01, FCC/1/5932-12-11, URF/1/6599-01-02, RGC/3/6295-01-01, URF/1/6993-01-01, KCSH Award 5932, and Center of Excellence on Generative AI Award 5940.

Disclosure of Interests. The authors declare no competing interests.

References

1. Bai, S., et al.: Qwen2. 5-vl technical report. arXiv preprint arXiv:2502.13923 (2025)
2. Boecking, B., et al.: Making the most of text semantics to improve biomedical vision-language processing (2022)
3. Cardoso, M.J., et al.: Monai: an open-source framework for deep learning in healthcare. arXiv preprint arXiv:2211.02701 (2022)
4. Chu, Y., et al.: Ct-based AI system for quantitative and integrated management of acute respiratory distress syndrome in critical care. npj Digit. Med. (2026)
5. Chu, Y., et al.: Improving representation of high-frequency components for medical visual foundation models. IEEE Trans. Med. Imaging (2025)
6. Feng, C., Patras, I.: Maskcon: masked contrastive learning for coarse-labelled dataset. In: Proceedings of the IEEE/CVF Conference on Computer Vision and Pattern Recognition (CVPR) (2023)

7. Feng, C., Shen, M., Balashankar, A., Gerner-Beuerle, C., Rodrigues, M.R.D.: Noisy but valid: robust statistical evaluation of LLMs with imperfect judges. In: International Conference on Learning Representations (ICLR) (2026)
8. Feng, C., Tzimiropoulos, G., Patras, I.: Clipcleaner: cleaning noisy labels with clip. In: Proceedings of the 32nd ACM International Conference on Multimedia (ACM MM) (2024)
9. Feng, C., Tzimiropoulos, G., Patras, I.: Noisebox: towards more efficient and effective learning with noisy labels. IEEE Trans. Circuits Syst. Video Technol. (2024)
10. Gao, Y., Lu, Y., Zhang, Z., Nie, J., Yu, S., Xuan, Q.: DSPC: dual-stage progressive compression framework for efficient long-context reasoning. arXiv preprint arXiv:2509.13723 (2025)
11. Hamamci, I.E., et al.: A foundation model utilizing chest CT volumes and radiology reports for supervised-level zero-shot detection of abnormalities. CoRR (2024)
12. Huang, S.C., et al.: Inspect: a multimodal dataset for pulmonary embolism diagnosis and prognosis. In: Thirty-seventh Conference on Neural Information Processing Systems Datasets and Benchmarks Track
13. Jain, S., et al.: Radgraph: extracting clinical entities and relations from radiology reports. In: Thirty-fifth Conference on Neural Information Processing Systems Datasets and Benchmarks Track (Round 1)
14. Johnson, A.E.W., et al.: Mimic-CXR, a de-identified publicly available database of chest radiographs with free-text reports. Sci. Data **6**(1), 317 (2019)
15. Ke, Z., Cao, Y., Chen, Z., Yin, Y., He, S., Cheng, Y.: Early warning of cryptocurrency reversal risks via multi-source data. Finance Res. Lett., 107890 (2025)
16. Lai, C., Song, S., Yan, S., Hu, G.: Improving vision and language concepts understanding with multimodal counterfactual samples. In: European Conference on Computer Vision, pp. 174–191. Springer (2024)
17. Le Guellec, B., et al.: Parrot, an open multilingual radiology reports dataset. European J. Radiol. Artifi. Intell., 100066 (2025)
18. Li, W., et al.: Abdomenatlas: a large-scale, detailed-annotated, & multi-center dataset for efficient transfer learning and open algorithmic benchmarking. Med. Image Anal. **97**, 103285 (2024)
19. Liang, X., et al.: Medfilip: medical fine-grained language-image pre-training. IEEE J. Biomed. Health Inform. **29**(5), 3587–3597 (2025)
20. Lin, J., et al.: Ct-glip: 3d grounded language-image pretraining with CT scans and radiology reports for full-body scenarios. arXiv preprint arXiv:2404.15272 (2024)
21. Lu, Y., et al.: Reassessing layer pruning in LLMs: new insights and methods. arXiv preprint arXiv:2411.15558 (2024)
22. Radford, A., et al.: Learning transferable visual models from natural language supervision. In: International Conference on Machine Learning, pp. 8748–8763. PMLR (2021)
23. Shih, G., et al.: Augmenting the national institutes of health chest radiograph dataset with expert annotations of possible pneumonia. Radiol. Artif. Intell. **1**(1), e180041 (2019)
24. Summers, R.: Nih Chest x-ray Dataset of 14 Common Thorax Disease Categories. NIH Clinical Center: Bethesda, MD, USA (2019)
25. Xiao, C., Xu, T., Ma, S., Jiang, Y., Gao, H., Wu, Y.: Reversible primitive-composition alignment for continual vision-language learning. In: The Fourteenth International Conference on Learning Representations (2026)
26. Yang, Y., et al.: Abot-m0: VLA foundation model for robotic manipulation with action manifold learning. arXiv preprint arXiv:2602.11236 (2026)

27. Yuan, H., Yuan, Z., Gan, R., Zhang, J., Xie, Y., Yu, S.: Biobart: pretraining and evaluation of a biomedical generative language model. In: Proceedings of the 21st Workshop on Biomedical Language Processing, pp. 97–109 (2022)
28. Zhou, H., et al.: Comem: compositional concept-graph memory for vision-language adaptation. In: The Fourteenth International Conference on Learning Representations (2026)

MedTriage-LM: Anatomically Grounded Visual Phenotype Synthesis for Interpretable ED Triage

Zhixiang Lu[1,4], Xiwei Liu[2], Jinfeng Wang[3], Mian Zhou[1], Anh Nguyen[4], Jionglong Su[1(✉)], Imran Razzak[2(✉)], and Sifan Song[1(✉)]

[1] Xi'an Jiaotong-Liverpool University, Suzhou, China
zhixiang@liverpool.ac.uk
[2] Mohamed bin Zayed University of Artificial Intelligence, Abu Dhabi, UAE
[3] Kunming University of Science and Technology, Kunming, China
[4] University of Liverpool, Liverpool, UK

Abstract. Accurate emergency severity assessment fundamentally dictates patient survival and resource allocation. However, existing automated triage pipelines rely exclusively on tabular data and clinical text, neglecting the visual semantics of patient distress. A paramount barrier to anatomically informed reasoning is the absolute absence of visual modalities in public cohorts like the MIMIC-IV-ED dataset. To overcome this limitation, this study introduces MedTriage-LM, an anatomically grounded Multimodal Large Language Model (MLLM) that algorithmically injects visual priors without requiring real patient images. The architecture maps latent clinical representations onto a canonical human template via a weakly supervised Gaussian field generator, yielding continuous, interpretable Visual Phenotype Maps (VPMs). Synergizing these spatial representations with clinical embeddings via cross-attention shifts the paradigm from numerical risk scoring toward actionable three-class triage instruction prediction: Life-saving, High-Risk Assessment, and Resource Estimation. MedTriage-LM attains state-of-the-art performance among fine-tuned open-source models and approaches the capabilities of proprietary MLLMs, while additionally providing explicit, anatomically grounded interpretability. Crucially, explicit spatial grounding empowers an efficient 8B-parameter model to approach large proprietary multimodal foundation models, generating spatially grounded textual rationales that improve clinical trustworthiness.

Keywords: Multimodal Learning · Synthetic Anatomical Prior Generation · Multimodal Clinical Reasoning · Emergency Department Triage

Z. Lu and X. Liu—Equal contribution.
Z. Lu is the project leader.

G. Yang et al. (Eds.): MICCAI 2026, LNCS 16878, pp. 377–387, 2027.
https://doi.org/10.1007/978-3-032-38059-3_36

1 Introduction

Emergency department (ED) triage is a time-critical and high-stakes decision process that directly affects patient outcomes and resource allocation [7]. The Emergency Severity Index is a widely adopted triage scale that stratifies patients by acuity and anticipated resource needs [9]. However, despite standardized protocols, real-world triage relies heavily on the subjective judgment of medical staff under extreme time pressure, often leading to inconsistencies and suboptimal resource allocation [24].

Large-scale electronic health record (EHR) datasets have catalyzed data-driven clinical decision support. The MIMIC-IV database provides comprehensive records spanning emergency medicine [16], and its dedicated module MIMIC-IV-ED [15] offers vital signs and triage information. While prior machine learning approaches utilizing these databases have demonstrated the value of automated triage support [11,21], they are predominantly unimodal and restricted by the fact that MIMIC-IV-ED completely lacks visual records. To overcome this fundamental limitation, our study utilizes the MIMIC-IV-Ext Triage Instruction Corpus, known as the MIETIC dataset [23]. MIETIC provides structured triage-time variables, chief complaints, instruction-answer pairs, and expert-validated triage records. Since it does not provide native bedside visual data, we synthesize anatomically grounded visual priors from clinical variables.

The inclusion of visual cues is essential, as clinicians routinely rely on visual distress signals at the bedside. Furthermore, effective clinical decision-making must be highly actionable. Therefore, rather than predicting standard numerical grades, our primary classification objective aligns with the three-class Instruction Types defined in the MIETIC dataset: Life-saving, High-Risk Assessment, and Resource Estimation. This three-class target provides a more directed and practical triage directive. To enable anatomically grounded multimodal representation learning and address potential scenarios with missing visual information, we introduce MedTriage-LM, a hybrid learned-and-deterministic pipeline that generates an interpretable Visual Phenotype Map (VPM). The VPM projects clinical evidence onto a canonical human silhouette using a Gaussian field generator, providing an explicit spatial inductive bias. MedTriage-LM then integrates structured triage variables, clinical texts, and these visual maps within a unified Transformer-based architecture [17,26]. By employing strong clinical text encoders [1,10] and a vision encoder [6], our model deeply aligns cross-modal features to perform three-class triage instruction prediction. The main contributions of this work are as follows:

- We propose a hybrid learned-and-deterministic Visual Phenotype Map algorithm for synthetic anatomical visual prior generation, effectively translating sparse clinical evidence into anatomically grounded visual priors to bridge the visual gap in public emergency datasets [23,29].
- We introduce a cross-representation clinical reasoning architecture enhanced by attention mechanisms, which dynamically aligns clinical state embeddings with visual features to achieve competitive performance in the three-class triage instruction classification task [23,27].

- We design a rationale generation framework that leverages synthesized visual features to generate clinically actionable text, providing transparent and trustworthy intermediate reasoning for triage decisions [18].

2 Method

We propose MedTriage-LM, a hybrid learned-and-deterministic, interpretable framework designed to bridge the gap between sparse clinical records and visual anatomical intuition. As illustrated in Fig. 1, our architecture comprises four tightly integrated modules: (1) Clinical State Encoder, (2) Anatomical Region Mapper, (3) Anatomical Field Generator and Heatmap Rendering, and (4) Multimodal Integration and Prediction.

2.1 Multimodal Clinical State Encoder

Given a patient encounter during triage, we denote the structured time-series and tabular data, such as vital signs and demographics, as $\mathbf{x}_s \in \mathbb{R}^{D_s}$. The unstructured clinical text, including chief complaints, is denoted as $\mathbf{x}_t$. To capture the complex physiological interactions within the structured variables, we employ an FT-Transformer [10], yielding a structured latent representation $\mathbf{h}_s = f_{\text{FT}}(\mathbf{x}_s) \in \mathbb{R}^{d_s}$. Concurrently, the unstructured clinical text is tokenized and encoded via a pre-trained Language Model (ClinicalBERT) [1,13]. We extract the `[CLS]` token as the holistic textual embedding $\mathbf{h}_t = f_{\text{BERT}}(\mathbf{x}_t)_{[\text{CLS}]} \in \mathbb{R}^{d_t}$.

A dedicated Fusion Layer subsequently integrates these disparate modalities into a unified latent clinical state embedding $\mathbf{z} \in \mathbb{R}^d$:

$$\mathbf{z} = \sigma\left(\mathbf{W}_f[\mathbf{h}_s \parallel \mathbf{h}_t] + \mathbf{b}_f\right), \tag{1}$$

where $\parallel$ denotes concatenation, and $\sigma(\cdot)$ represents a non-linear activation. This comprehensive latent vector $\mathbf{z}$ encapsulates the patient's entire clinical profile, serving as the foundation for the subsequent anatomical reasoning.

2.2 Anatomical Region Mapper

To introduce an explicit anatomical inductive bias without relying on real photographs, we predefine K clinical anatomical regions. An Adaptive Region Mapping Layer, parameterized by a multi-layer perceptron (MLP), projects the clinical state $\mathbf{z}$ to a region intensity vector $\boldsymbol{\mu} \in [0,1]^K$:

$$\boldsymbol{\mu} = \text{Sigmoid}(\mathbf{W}_r\mathbf{z} + \mathbf{b}_r), \tag{2}$$

where each scalar μ_k quantifies the predicted physiological distress level in region k. Given the inherent absence of ground-truth regional distress labels in public cohorts, we extract rule-based weak labels $\mathbf{y}^{\text{weak}} \in \{0,1\}^K$ through Chief Complaint Mapping, effectively linking explicit complaints like chest pain to the thoracic region. We enforce spatial alignment by minimizing a Weak Supervision Loss using Binary Cross-Entropy (BCE):

$$\mathcal{L}_{\text{weak}} = \sum_{k=1}^{K} \text{BCE}(\mu_k, y_k^{\text{weak}}). \tag{3}$$

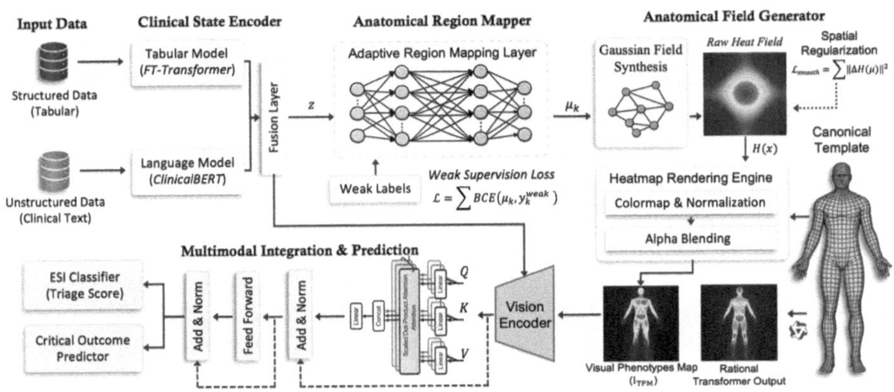

Fig. 1. Overview of the proposed MedTriage-LM framework.

2.3 Anatomical Field Generator and Heatmap Rendering

We algorithmically transform the discrete regional intensities $\boldsymbol{\mu}$ into a continuous two-dimensional visual manifold via Gaussian Field Synthesis. Assigning each region k a fixed spatial coordinate $\mathbf{c}_k$ and variance σ_k^2 on a canonical two-dimensional template Ω, we synthesize a raw heat field $H(\mathbf{u})$ for any pixel coordinate $\mathbf{u} \in \Omega$:

$$H(\mathbf{u}) = \sum_{k=1}^{K} \mu_k \exp\left(-\frac{\|\mathbf{u} - \mathbf{c}_k\|_2^2}{2\sigma_k^2}\right). \tag{4}$$

To ensure physiological coherence and strictly suppress high-frequency topological noise, we impose a spatial regularization term $\mathcal{L}_{\text{smooth}}$ over the gradient field:

$$\mathcal{L}_{\text{smooth}} = \sum_{\mathbf{u} \in \Omega} \|\nabla H(\mathbf{u})\|^2. \tag{5}$$

Finally, the Heatmap Rendering Engine applies Colormap and Normalization $\mathcal{C}(\cdot)$ to the raw field, followed by Alpha Blending with a canonical human silhouette $\mathbf{I}_{\text{canonical}}$. This produces the final Visual Phenotype Map I_{VPM}:

$$I_{\text{VPM}} = \alpha \cdot \mathcal{C}(H) + (1 - \alpha) \cdot \mathbf{I}_{\text{canonical}}. \tag{6}$$

2.4 Transformer Multimodal Architecture and Prediction

This pivotal module aligns the synthesized visual evidence with the clinical reasoning context. The rendered I_{VPM} is processed by a Vision Encoder [6] to extract a sequence of spatial visual tokens $\mathbf{V} \in \mathbb{R}^{N_v \times d_v}$.

Cross-Modal Vision-Language Alignment. To optimally fuse the synthesized visual perception with the clinical text and tabular context, we utilize a cross-attention mechanism within the Transformer architecture [17]. The clinical embedding $\mathbf{z}$ is linearly projected to form the Query ($\mathbf{Q}$), while the visual tokens $\mathbf{V}$ are projected to form Keys ($\mathbf{K}$) and Values ($\mathbf{V}_{val}$):

$$\text{Attention}(\mathbf{Q}, \mathbf{K}, \mathbf{V}_{val}) = \text{Softmax}\left(\frac{\mathbf{Q}\mathbf{K}^\top}{\sqrt{d_k}}\right)\mathbf{V}_{val}. \tag{7}$$

This Scaled Dot-Product Alignment allows the clinical state to dynamically localize relevant regions within the visual phenotype, effectively bridging the semantic gap between modalities. Following standard Transformer topology [26], the attended output traverses an Add and Norm layer, a Feed-Forward Network, and a subsequent Add and Norm layer, yielding an enriched cross-modal representation $\mathbf{h}_{\text{multi}}$.

Rationale Generation via MLLM Backbone. To generate clinically actionable text rationales, we deploy a multimodal large language model backbone (Qwen3-VL-8B [2]). The cross-modal representation $\mathbf{h}_{\text{multi}}$ serves as a conditioning soft-prompt. Driven by the instruction-answer pairs from the MIETIC dataset [23], the backbone autoregressively decodes the rationale sequence $\mathbf{R} = \{w_1, w_2, \ldots, w_T\}$. The generative probability is formulated as:

$$p(\mathbf{R} \mid \mathbf{h}_{\text{multi}}) = \prod_{t=1}^{T} p(w_t \mid w_{<t}, \mathbf{h}_{\text{multi}}), \tag{8}$$

where w_t represents the generated token at time step t. This process is optimized using a standard causal language modeling loss $\mathcal{L}_{\text{gen}}$.

Prediction Heads and Optimization. Diverging from conventional basic severity scores, our primary classifier explicitly targets the three-class Triage Instructions defined by the MIETIC dataset: Life-saving, High-Risk Assessment, and Resource Estimation. We formulate the primary instruction probability distribution $\hat{\mathbf{y}}_{\text{inst}}$ and the secondary critical outcome probability $\hat{y}_{\text{crit}}$ as independent analytical blocks to ensure clarity:

$$\begin{aligned} \hat{\mathbf{y}}_{\text{inst}} &= \text{Softmax}(\text{MLP}_{\text{inst}}(\mathbf{h}_{\text{multi}})), \\ \hat{y}_{\text{crit}} &= \text{Sigmoid}(\text{MLP}_{\text{crit}}(\mathbf{h}_{\text{multi}})). \end{aligned} \tag{9}$$

The end-to-end framework is jointly optimized by minimizing the classification and generation losses alongside the anatomical regularization terms:

$$\mathcal{L}_{\text{total}} = \mathcal{L}_{\text{CE}}(\hat{\mathbf{y}}_{\text{inst}}, \mathbf{y}_{\text{inst}}) + \lambda_1 \mathcal{L}_{\text{BCE}}(\hat{y}_{\text{crit}}, y_{\text{crit}}) + \lambda_2 \mathcal{L}_{\text{weak}} + \lambda_3 \mathcal{L}_{\text{smooth}} + \lambda_4 \mathcal{L}_{\text{gen}}. \tag{10}$$

Table 1. Quantitative comparison on the MIETIC dataset. Results are reported as Mean ± 95% Confidence Interval over five runs. Models are grouped by architectural paradigm and sorted chronologically. Best results are **bolded**, and second best results are underlined. Vis. indicates synthesized VPMs, not native patient images.

Model	Modalities			Metrics (Mean ± 95% CI)				
	Tab.	Txt.	Vis.	Acc. ↑	Macro F1 ↑	AUROC ↑	AUPRC ↑	BERTScore ↑
Traditional and Deep Tabular Models								
MLP (1986) [22]	√	×	×	0.502 ± 0.035	0.498 ± 0.042	0.821 ± 0.028	0.791 ± 0.033	N/A
Random Forest (2001) [3]	√	×	×	0.463 ± 0.024	0.458 ± 0.031	0.817 ± 0.018	0.785 ± 0.022	N/A
AutoML (2015) [8]	√	×	×	0.694 ± 0.011	0.553 ± 0.014	0.838 ± 0.012	0.811 ± 0.015	N/A
XGBoost (2016) [4]	√	×	×	0.687 ± 0.015	0.546 ± 0.019	0.824 ± 0.014	0.793 ± 0.017	N/A
TabTransformer (2020) [14]	√	×	×	0.713 ± 0.009	0.579 ± 0.012	0.849 ± 0.010	0.816 ± 0.011	N/A
Specialized Triage Pipelines								
Med-UniC (2023) [27]	×	√	√	0.728 ± 0.014	0.667 ± 0.016	0.852 ± 0.012	0.824 ± 0.015	N/A
ED-Copilot (2024) [25]	√	√	√	0.746 ± 0.011	0.683 ± 0.013	0.864 ± 0.010	0.837 ± 0.012	N/A
SDTA (2025) [23]	×	√	√	0.761 ± 0.008	0.712 ± 0.009	0.879 ± 0.007	0.846 ± 0.008	N/A
Zero-Shot Multimodal Large Language Models (with CoT [28])								
GPT-4o (2024) [19]	√	√	√	0.879 ± 0.006	0.882 ± 0.007	0.912 ± 0.005	0.887 ± 0.006	0.894 ± 0.021
GPT-5.2 (2025) [20]	√	√	√	**0.912 ± 0.003**	**0.910 ± 0.004**	**0.961 ± 0.003**	**0.942 ± 0.004**	**0.925 ± 0.013**
GLM-5 (2026) [31]	√	√	√	0.824 ± 0.008	0.831 ± 0.009	0.886 ± 0.006	0.862 ± 0.007	0.873 ± 0.025
Fine-Tuned Foundation Models								
LLaMA3.1-8B (2024) [12]	√	√	×	0.836 ± 0.007	0.843 ± 0.008	0.876 ± 0.006	0.852 ± 0.007	0.864 ± 0.036
Gemma3-4B (2025) [5]	√	√	×	0.821 ± 0.009	0.829 ± 0.011	0.864 ± 0.008	0.841 ± 0.010	0.859 ± 0.037
Qwen3-VL-8B (2025) [2]	√	√	√	0.867 ± 0.005	0.871 ± 0.006	0.906 ± 0.005	0.883 ± 0.006	0.886 ± 0.034
MedTriage-LM (Ours)	√	√	√	0.898 ± 0.002	0.895 ± 0.001	0.948 ± 0.001	0.925 ± 0.003	0.908 ± 0.009

This multi-task objective encourages MedTriage-LM to learn anatomically aligned visual priors that support reliable clinical triage instructions and transparent textual rationales.

3 Experiments

3.1 Datasets and Evaluation Metrics

Datasets. MedTriage-LM is evaluated in the MIMIC-IV-Ext Triage Instruction Corpus, referred to as the MIETIC dataset [23]. Inputs are strictly constrained to triage time variables, comprising structured hemodynamics and unstructured clinical complaints. To align with practical clinical workflows, our primary task is the three-class triage instruction classification: Life Saving, High Risk Assessment, and Resource Estimation.

Evaluation Metrics. Classification performance is evaluated via Accuracy and Macro F1. For critical risk stratification, specifically predicting intensive care unit admission, the Area Under the Receiver Operating Characteristic (AUROC) and Precision Recall Curve (AUPRC) are utilized, with AUPRC being essential for imbalanced clinical datasets. Rationale semantic quality is objectively measured via BERTScore [30]. All quantitative results are reported as the mean and 95% Confidence Interval (95% CI) across five independent runs.

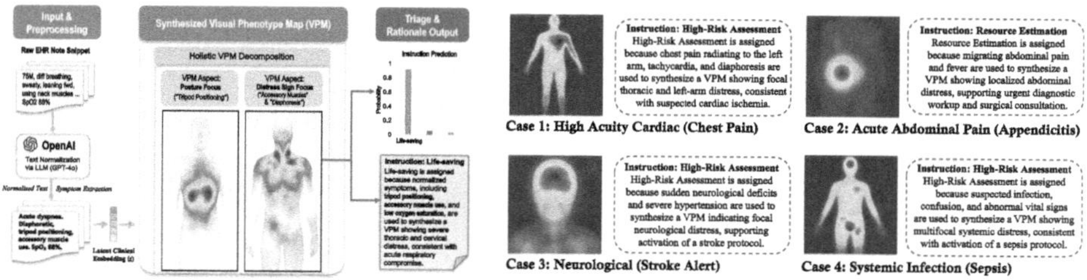

(a) Inference pipeline (b) Diverse clinical scenarios

Fig. 2. MedTriage-LM architecture and clinical generalization. **(a)** The end-to-end inference pipeline, illustrated with an acute respiratory distress case. **(b)** Synthesized Visual Phenotype Maps (VPMs) accurately reflecting distinct anatomical activations across diverse high-acuity scenarios.

3.2 Implementation Details

The MedTriage-LM architecture is implemented in PyTorch. The vision encoder is initialized with pretrained ViT weights, while the multimodal rationale generation backbone is driven by Qwen3-VL-8B. End-to-end training is executed on a single NVIDIA A100 GPU employing mixed precision. Optimization is performed via the AdamW optimizer, configured with a base learning rate of 2×10^{-5} and regulated by a cosine annealing scheduler. The loss weighting hyperparameters are empirically established through a comprehensive grid search on the validation set, yielding optimal values of $\lambda_1 = 0.5$, $\lambda_2 = 1.0$, $\lambda_3 = 0.5$, and $\lambda_4 = 1.0$.

3.3 Main Quantitative Results

The proposed framework is benchmarked against a comprehensive suite of models. Baselines span traditional machine learning, advanced tabular architectures, specialized triage pipelines, state-of-the-art large language models, and fine-tuned foundation models, including LLaMA3.1-8B [12], Gemma3-4B [5], and Qwen3-VL-8B [2]. To ensure fair evaluation, all zero-shot models employ a Zero-Shot Chain of Thought prompting strategy [28], with complete prompt templates detailed in the supplementary material.

As demonstrated in Table 1, pure tabular methods achieve moderate AUROC but inherently lack the capacity for semantic rationale generation. Conversely, advanced zero-shot multimodal models like GPT-5.2 [20] exhibit strong reasoning capabilities with Macro F1 scores exceeding 0.9. However, relying solely on massive parameter scaling often obscures explicit clinical spatial physiology and incurs prohibitive computational costs. MedTriage-LM addresses this efficiency and interpretability gap. By synthesizing the intermediate VPM to anchor the fine-tuned Qwen3-VL-8B backbone, the specialized 8B framework achieves performance highly competitive with the large proprietary model, GPT-5.2.

Table 2. Comprehensive ablation study integrating modality progression and hyper-parameter sensitivity on the MIETIC dataset. The upper section demonstrates the sequential addition of modalities and optimization constraints (λ_1–λ_4). The lower section isolates the sensitivity of the spatial smoothness regularization (λ_3) to demonstrate its qualitative and quantitative impact. All metrics are reported as Mean ± 95% CI over five independent runs. **Bold** denotes the optimal full configuration.

Ablation Variant	Modalities			Loss Weights				Metrics (Mean ± 95% CI)		
	Tab.	Txt.	Vis.	λ_1	λ_2	λ_3	λ_4	Macro F1	AUROC	BERTScore
Progressive Architecture and Joint Optimization										
Tabular Features Only	√	×	×	0	0	0	0	0.753 ± 0.009	0.874 ± 0.008	0.791 ± 0.024
+ Clinical Text Features	√	√	×	0	0	0	0	0.820 ± 0.006	0.892 ± 0.005	0.866 ± 0.028
+ Visual Map (Unconstrained)	√	√	√	0	0	0	0	0.851 ± 0.005	0.910 ± 0.004	0.880 ± 0.019
+ Auxiliary Outcome Prediction Task (λ_1)	√	√	√	0.5	0	0	0	0.868 ± 0.004	0.922 ± 0.004	0.883 ± 0.013
+ Weak Anatomical Supervision (λ_2)	√	√	√	0.5	1.0	0	0	0.877 ± 0.004	0.931 ± 0.003	0.892 ± 0.021
+ Spatial Smoothness Regularization (λ_3)	√	√	√	0.5	1.0	0.5	0	0.883 ± 0.003	0.941 ± 0.003	0.899 ± 0.016
Full Joint Opt. (+ Gen Loss λ_4)	√	√	√	0.5	1.0	0.5	1.0	**0.895 ± 0.001**	**0.948 ± 0.001**	**0.908 ± 0.009**
Spatial Regularization (λ_3) Sensitivity Analysis										
Insufficient Smoothness ($\lambda_3 = 0.1$)	√	√	√	0.5	1.0	0.1	1.0	0.862 ± 0.004	0.915 ± 0.004	0.882 ± 0.017
Minor Blurring ($\lambda_3 = 1.0$)	√	√	√	0.5	1.0	1.0	1.0	0.872 ± 0.003	0.926 ± 0.003	0.890 ± 0.015
Over Smoothness ($\lambda_3 = 5.0$)	√	√	√	0.5	1.0	5.0	1.0	0.811 ± 0.008	0.863 ± 0.007	0.844 ± 0.022

Notably, although advanced baselines such as GPT-4o [19] and the Qwen3-VL-8B also receive synthesized VPM inputs, MedTriage-LM outperforms them. Ultimately, MedTriage-LM closely approaches the strong GPT-5.2 baseline across all predictive metrics (trailing by merely 0.014 in Accuracy and 0.017 in AUPRC) while generating interpretable rationales (BERTScore 0.908) with stable performance. This highlights that directly providing synthetic visual inputs to general-purpose multimodal models is clinically insufficient. Instead, explicit anatomical grounding acts as a powerful inductive bias, enabling an efficient architecture to approach the performance of large proprietary models while retaining explicit anatomical interpretability.

3.4 Ablation Study

Progressive Architecture and Loss Ablation. Reconstructing MedTriage-LM (Table 2, upper) demonstrates that while text integration improves rationale quality, VPM synthesis provides a notable discriminative gain. Sequentially, the auxiliary outcome loss (λ_1) enhances high-risk feature learning, weak supervision (λ_2) anchors anatomical activations, and spatial smoothness (λ_3) suppresses topological noise. Joint optimization with generation loss (λ_4) achieves the best performance within the ablation suite while preserving rationale integrity.

Spatial Regularization Sensitivity. Unlike standard cross entropy objectives ($\lambda_1, \lambda_2, \lambda_4$) which remain resilient to scalar shifts, the spatial smoothness weight (λ_3) strictly dictates visual manifold topology (Table 2, lower). Insufficient weighting ($\lambda_3 = 0.1$) collapses the field into high-frequency artifacts, degrading Macro F1 to 0.860 and AUROC to 0.915. Conversely, excessive penalization ($\lambda_3 = 5.0$) forces distributions toward a uniform mean, washing out focal clinical signals and degrading rationale quality. The optimum ($\lambda_3 = 0.5$) suggests that explicit anatomical grounding is important for accurate prediction.

Clinical Interpretability. Beyond quantitative metrics, the inference pipeline (Fig. 2a) provides exceptional transparency. The synthesized VPMs qualitatively reflect distinct anatomical activation patterns across diverse high-acuity scenarios (Fig. 2b). For example, respiratory distress consistently triggers thoracic activation, whereas systemic infections exhibit biologically plausible multifocal heat distributions. This explicit spatial alignment improves the transparency of the triage decision support prior to critical triage interventions.

4 Conclusion

This study presented MedTriage-LM[1], a hybrid learned-and-deterministic, anatomically interpretable multimodal framework for emergency severity assessment. The introduction of the Visual Phenotype Map algorithm successfully bridges the critical visual modality gap prevalent in public electronic health record cohorts. By synergizing synthesized anatomical priors with tabular and textual clinical states via a cross-attention architecture, the framework achieves strong performance among efficient fine-tuned models and narrows the gap to proprietary MLLMs for three-class triage instruction prediction on the MIETIC dataset. Crucially, explicit anatomical grounding empowers a highly efficient 8B parameter model to achieve predictive accuracy approaching large proprietary models, while generating transparent textual rationales grounded in the synthesized anatomical representation. Future work will focus on validating the generalized robustness of these visual maps across diverse multi-institutional clinical environments and integrating dynamic temporal sequences for continuous patient monitoring.

Disclosure of Interests. The authors have no competing interests to declare that are relevant to the content of this article.

References

1. Alsentzer, E., et al.: Publicly available clinical Bert embeddings. In: Proceedings of the 2nd Clinical Natural Language Processing Workshop, pp. 72–78 (2019). https://doi.org/10.18653/v1/W19-1909
2. Bai, S.: Qwen3-vl technical report. arXiv preprint arXiv:2511.21631 (2025)
3. Breiman, L.: Random forests. Mach. Learn. **45**(1), 5–32 (2001)
4. Chen, T., Guestrin, C.: Xgboost: a scalable tree boosting system, pp. 785–794 (2016). https://doi.org/10.1145/2939672.2939785
5. DeepMind: Gemma 3 technical report (2025). https://arxiv.org/abs/2503.19786

[1] The code is available at https://github.com/Leo1998-Lu/MedTriage-LM.

6. Dosovitskiy, A., et al.: An image is worth 16x16 words: transformers for image recognition at scale. In: International Conference on Learning Representations (2021). https://openreview.net/forum?id=YicbFdNTTy
7. Eitel, D.R., et al.: The emergency severity index triage algorithm version 2 is reliable and valid. Acad. Emerg. Med. **10**(10), 1070–1080 (2003). https://doi.org/10.1111/j.1553-2712.2003.tb00577.x
8. Feurer, M.: Efficient and robust automated machine learning. In: Advances in Neural Information Processing Systems (2015)
9. Gilboy, N., Tanabe, P., Travers, D., Rosenau, A.: Emergency severity index (ESI): a triage tool for emergency department care, version 4: Implementation handbook. Agency for Healthcare Research and Quality (AHRQ) (2011). aHRQ Publication No. 12-0014
10. Gorishniy, Y., et al.: Revisiting deep learning models for tabular data. In: Advances in Neural Information Processing Systems (2021)
11. Goto, T., Camargo, C.A., Jr., Faridi, M.K., Freishtat, R.J., Hasegawa, K.: Machine learning-based prediction of clinical outcomes for children during emergency department triage. JAMA Netw. Open **2**(1), e186937 (2019). https://doi.org/10.1001/jamanetworkopen.2018.6937
12. Grattafiori, A., et al.: The llama 3 herd of models (2024). https://arxiv.org/abs/2407.21783
13. Huang, K., Altosaar, J., Ranganath, R.: Clinicalbert: modeling clinical notes and predicting hospital readmission. arXiv preprint arXiv:1904.05342 (2019)
14. Huang, X., Khetan, A., Cvitkovic, M., Karnin, Z.: Tabtransformer: tabular data modeling using contextual embeddings (2020). https://arxiv.org/abs/2012.06678
15. Johnson, A., Pollard, T., Mark, R.: Mimic-iv-ed (emergency department) database. PhysioNet (2022). https://physionet.org/content/mimic-iv-ed/
16. Johnson, A.E.W., et al.: Mimic-iv, a freely accessible electronic health record dataset. Sci. Data **10**, 1–15 (2023). https://doi.org/10.1038/s41597-022-01899-x
17. Lu, Z., et al.: Deepgb-tb: a risk-balanced cross-attention gradient-boosted convolutional network for rapid, interpretable tuberculosis screening. In: Proceedings of the AAAI Conference on Artificial Intelligence vol. 40, no. 46, pp. 38989–38997 (2026). https://doi.org/10.1609/aaai.v40i46.41245
18. Lundberg, S.M., Lee, S.I.: A unified approach to interpreting model predictions. In: Advances in Neural Information Processing Systems (2017)
19. OpenAI: Gpt-4o system card (2024) . https://openai.com/index/gpt-4o-system-card/
20. OpenAI: Openai gpt-5 system card (2025). https://arxiv.org/abs/2601.03267
21. Raita, Y., Goto, T., Faridi, M.K., Brown, D.F.M., Camargo, C.A., Jr., Hasegawa, K.: Emergency department triage prediction of clinical outcomes using machine learning models. Crit. Care **23**(1), 64 (2019). https://doi.org/10.1186/s13054-019-2351-7
22. Rumelhart, D.E., Hinton, G.E., Williams, R.J.: Learning representations by back-propagating errors. Nature **323**, 533–536 (1986)
23. Shen, Q., Zhang, X., Ren, H., Guo, Q., Yi, Z.: Knowledge-embedded large language models for emergency triage. Knowl.-Based Syst. **318**, 113431 (2025). https://doi.org/10.1016/j.knosys.2025.113431
24. Souza-Pereira, L., Ouhbi, S., Pombo, N.: A process model for quality in use evaluation of clinical decision support systems. J. Biomed. Inf. **123**(C) (2021). https://doi.org/10.1016/j.jbi.2021.103917

25. Sun, L., et al.: Ed-copilot: reduce emergency department wait time with language model diagnostic assistance. In: International Conference on Machine Learning (2024)
26. Vaswani, A., et al.: Attention is all you need. In: International Conference on Neural Information Processing Systems, pp. 6000–6010 (2017)
27. Wan, Z., et al.: Med-UNIC: unifying cross-lingual medical vision-language pre-training by diminishing bias. In: International Conference on Neural Information Processing Systems (2023)
28. Wei, J., et al.: Chain-of-thought prompting elicits reasoning in large language models. In: Advances in Neural Information Processing Systems, vol. 35, pp. 24824–24837 (2022)
29. Xie, F., et al.: Benchmarking emergency department prediction models with machine learning and public electronic health records. Sci. Data **9**, 1–11 (2022). https://doi.org/10.1038/s41597-022-01782-9
30. Zhang, T., Kishore, V., Wu, F., Weinberger, K.Q., Artzi, Y.: Bertscore: evaluating text generation with Bert. In: International Conference on Learning Representations (2020). https://openreview.net/forum?id=SkeHuCVFDr
31. ZhipuAI: GLM-5: from vibe coding to agentic engineering (2026). https://arxiv.org/abs/2602.15763

MedXEdit-GRPO: Reasoning-Aware Online RL for Counterfactual Medical Image Generation

Yaning Pan[1], Renwei Xiao[1], Qingqiu Li[1], Xiaoyi Li[1], Xiaoling Li[2], Rui Feng[1,2,3(✉)], and Xiaobo Zhang[2(✉)]

[1] College of Computer Science and Artificial Intelligence, Shanghai Key Laboratory of Intelligent Information Processing, Fudan University, Shanghai, China
fengrui@fudan.edu.cn

[2] Children's Hospital of Fudan University, National Children's Medical Center, Shanghai, China
zhangxiaobo0307@163.com

[3] College of Intelligent Robotics and Advanced Manufacturing, Fudan University, Shanghai, China

Abstract. Counterfactual medical image generation enables "what-if" clinical exploration through instruction-conditioned editing. However, existing methods often lack the multimodal reasoning required to align abstract directives with precise visual modifications, failing to balance instruction compliance with anatomical preservation. Reinforcement learning offers a promising solution, but it remains largely unexplored in this field. In this paper, we propose **MedXEdit-GRPO**, a reinforcement learning framework for counterfactual medical image editing. To provide stable and comprehensive feedback, we curate MedXEdit-Reward, a large-scale dataset of 30,000 samples with rationales, and develop MedXEdit-Score, a reward model enforcing a "reasoning-before-scoring" paradigm for semantically grounded optimization. We optimize our generative policy via Group Relative Policy Optimization (GRPO), which effectively harmonizes diverse clinical objectives into a unified, stable training process. Quantitative and qualitative evaluations demonstrate that MedXEdit-GRPO significantly outperforms state-of-the-art methods in both instruction compliance and anatomical consistency, highlighting its potential for robust and interpretable medical analysis.

Keywords: Counterfactual Generation · Disease Progression · Image Editing · Reinforcement Learning

1 Introduction

Counterfactual medical image generation synthesizes hypothetical "what-if" scenarios (e.g., disease progression) through the instruction-conditioned editing of pathological features [4,13]. This methodology provides a critical instrument

G. Yang et al. (Eds.): MICCAI 2026, LNCS 16878, pp. 388–397, 2027.
https://doi.org/10.1007/978-3-032-38059-3_37

for evaluating model robustness and enhancing classifier performance via targeted synthetic data augmentation [12,21]. Furthermore, it facilitates systematic anomaly detection [7] and the visual exploration of diverse clinical hypotheses [1,26], providing a controlled and interpretable mechanism for simulating complex pathological variations across distinct patient profiles.

Recent works in counterfactual medical image generation have demonstrated impressive results by leveraging autoregressive or diffusion-based architectures to synthesize plausible pathological edits [10,14,17,18]. However, current methods are predominantly constrained by the Supervised Fine-Tuning (SFT) paradigm, which primarily optimizes for pixel-level distribution matching rather than explicit semantic-pathological alignment. To bridge this gap, Reinforcement Learning (RL) offers a promising framework for direct, goal-oriented optimization. However, its application in medical image editing remains challenging by the scarcity of effective reward mechanisms. RL necessitates a reliable reward model capable of providing stable and multidimensional feedback [20,23], yet existing imaging-based reward functions, typically restricted to hand-crafted pixel heuristics (e.g., MSE) [8] or task-specific proxy classifiers, face two challenges. First, optimizing for multiple independent rewards often introduces optimization instability, where conflicting gradients can trigger "reward hacking", leading the model to exploit individual metrics at the expense of overall synthesis fidelity. Second, these isolated rewards provide a reductionist assessment that fails to capture the holistic and nuanced nature of expert clinical judgment. Consequently, a unified and reasoning-aware reward model is essential to provide stable, multidimensional feedback for RL-based medical editing.

To address these challenges, we first curate **MedXEdit-Reward**, a large-scale dataset specifically designed for reward modeling in medical CXR image editing, using a Gemini-based distillation pipeline (Fig. 1). MedXEdit-Reward contains 30,000 quintuplets, each of which includes a source image, an edited image, an edit instruction, and corresponding expert-level rationales paired with comprehensive scores. Second, leveraging this dataset, we develop **MedXEdit-Score**, a specialized reward model that enforces a "reasoning-before-scoring" paradigm to provide semantically grounded feedback for generative edits. Building upon this evaluator, we propose **MedXEdit-GRPO**, a reinforcement learning framework to optimize medical image editing through a unified optimization process (Fig. 2). By employing Group Relative Policy Optimization (GRPO), our framework harmonizes diverse clinical objectives into a single, stable training signal. This approach effectively overcomes the optimization instability and reductionist assessment inherent in fragmented reward systems, ensuring superior alignment with clinical directives. Our contributions can be summarized as follows:

1. We introduce MedXEdit-Reward, a large-scale dataset specifically designed for reward modeling in medical image editing, containing 30,000 quintuplets that pair clinical instructions and images with expert-level rationales.
2. We develop MedXEdit-Score, a logic-driven reward model that enforces a "reasoning-before-scoring" paradigm, providing interpretable and semanti-

cally grounded feedback to bridge the gap between abstract clinical directives and visual modifications.
3. We propose MedXEdit-GRPO, a reinforcement learning framework that optimizes medical image editing via GRPO, effectively harmonizing diverse clinical objectives into a unified training signal.
4. Quantitative and qualitative evaluations demonstrate that MedXEdit-GRPO significantly outperforms state-of-the-art methods in both instruction compliance and anatomical consistency, highlighting its potential for robust and reliable counterfactual medical analysis.

2 Related Work

General Image Editing. Driven by the rapid advancement of diffusion models [15,22] and unified multimodal architectures [28], general image editing has achieved remarkable progress [2,30]. Recent works [6,28] have demonstrated the capability to precisely understand instructions and efficiently accomplish image editing tasks. Building upon this, to further enhance model performance on complex editing instructions, researchers have begun incorporating reinforcement learning [16,24] into this field to achieve more precise alignment with user editing intent. Nevertheless, this highly promising technical paradigm remains in its early exploratory stages for applications in high-stakes specialized scenarios such as medical image editing.

Counterfactual Medical Image Generation. Early works such as BiomedJourney [10] and PIE [14] established the foundation for modeling longitudinal clinical trajectories through image-editing paradigms. To refine spatial precision, InstructX2X [18] introduces region-specific editing to prevent unintended modifications to non-target anatomical or demographic attributes. ProgEmu [17] further addresses the "silent prediction" problem by jointly synthesizing counterfactual images and their corresponding textual interpretations. More recently, UniMedVL [19] has emerged as a unified medical foundation model, where counterfactual generation is treated as one of its diverse multimodal tasks. Despite these architectural advancements, existing methodologies predominantly rely on the SFT paradigm, which often fails to fully elicit the model's multimodal reasoning capabilities for instruction-conditioned editing. Our MedXEdit-GRPO bridges this gap by introducing RL for reasoning-aware policy alignment, ensuring superior generative fidelity and instruction compliance.

3 Methodology

In this section, we present the MedXEdit-GRPO framework, which integrates clinical logic into generative optimization through two main stages: (1) the construction of a reasoning-aware reward model, MedXEdit-Score, to provide semantically grounded feedback, and (2) the iterative alignment of a multimodal generative policy using GRPO.

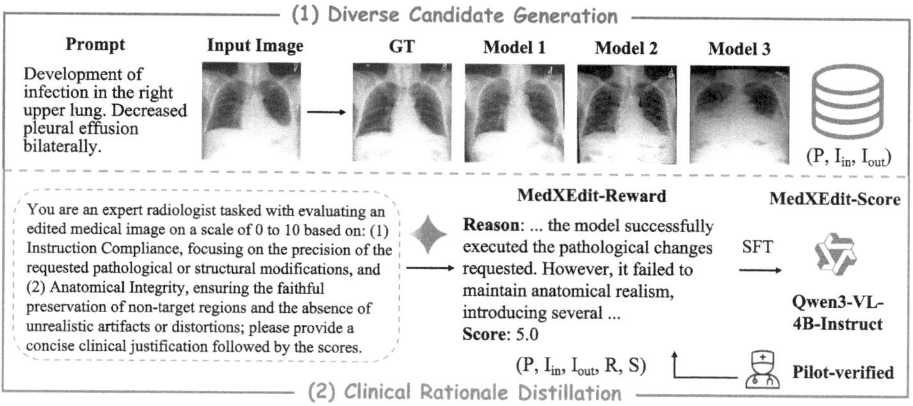

Fig. 1. The MedXEdit-Score construction pipeline. (1) Diverse Candidate Generation: Multiple generative models synthesize varied edited outputs to curate training triplets. (2) Clinical Rationale Distillation: Clinical rationales and scores are distilled from Gemini-3-Flash to form the MedXEdit-Reward dataset. Finally, the MedXEdit-Score model is developed by fine-tuning the Qwen3-VL-4B backbone on this pilot-verified data.

3.1 Reward Modeling

Problem Formulation. Inspired by [16,27], we formulate reward modeling as a conditional textual generation task. We fine-tune the Qwen3-VL-4B backbone using a standard autoregressive objective to serve as a specialized medical evaluator. Given the input triplet (P, I_{in}, I_{out}), where P is the editing instruction, I_{in} is the source medical image, and I_{out} is the edited output image, our MedXEdit-Score is trained to output a sequence consisting of a clinical rationale R followed by a scalar score S. This "reasoning-before-scoring" mechanism not only provides transparent clinical justifications but also leverages the model's internal logic to significantly enhance the sensitivity and accuracy of the final reward.

MedXEdit-Reward Curation Pipeline. MedXEdit-Reward is derived from the ICG-CXR [17] dataset, which comprises 10,679 training and 760 test samples. Each original sample in ICG-CXR is a triplet (P, I_{in}, I_{target}), where I_{target} denotes the ground-truth edited image. To broaden the variety of editing quality for reward learning, the diverse edited outputs I_{out} are generated by three medical generative models: ProgEmu [17], InstructX2X [18], and UniMedVL [19]. For each model, we systematically adjust the inference hyperparameters to produce a wide spectrum of image qualities. As shown in Fig. 1, for each candidate output, we employ Gemini-3-Flash, which has demonstrated robust clinical reasoning capabilities [3], to annotate numerical scores ranging from 0 to 10 and provide rationales considering instruction following and anatomical consistency. Next, we apply filtering based on group-wise maximum scores to exclude unachievable tasks and group standard deviation to remove cases with low discriminability, resulting in a final 30,000 training samples.

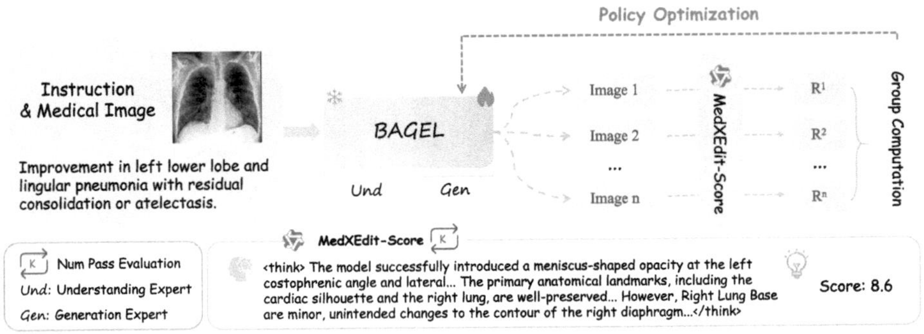

Fig. 2. Overview of the MedXEdit-GRPO. BAGEL, with its understanding expert frozen and generation expert trained, receives instructions and medical images as input to generate a group of edited images. The MedXEdit-Score model then evaluates these images against the instructions using an N-Pass strategy, assigning scores accompanied by clinical rationales. Consequently, these rewards drive the MedXEdit-GRPO policy optimization to refine the BAGEL iteratively.

Reward Model Evaluation. To evaluate the model's discriminative ability, we follow the evaluation protocol of [16] to transform the 760 test samples into a set of 2,000 pairwise preference tuples. Specifically, for a given input image and instruction, we form pairs of candidate edits (A, B), where A is preferred over B based on the oracle's assessment. To ensure the reliability of this gold standard, 200 pairs were randomly selected for verification by board-certified radiologists, achieving an inter-rater agreement of 92.41%. The evaluation is quantified by the preference prediction accuracy, calculated as the proportion of pairs where the model assigns a higher score to the preferred output, i.e., $S(A) > S(B)$.

3.2 MedXEdit-GRPO

As shown in Fig. 2, we validate the effectiveness of MedXEdit-Reward based on Bagel [6], a multimodal foundation architecture with understanding and generation capabilities. Given the tuple (P, I_{in}), the source image I_{in} is first projected into a sequence of visual tokens $V \in \mathbb{R}^{k \times d}$ using a vision encoder. These visual tokens, along with the text, simultaneously condition the generation expert to generate the edited image.

Reinforcement learning (RL) aims to improve the model's performance by maximizing expected cumulative reward. Group Relative Policy Optimization (GRPO) eliminates the value function and estimates the advantage in a group-relative manner compared to PPO [23]. For each input (P, I_{in}), we sample a group of G independent candidate outputs $\{z_0^1, \dots, z_0^G\}$ from the current policy π_θ. The Advantage A_i is computed by normalizing the MedXEdit-Score reward r_i across the group, as follows:

$$A_i = \frac{r_i - \text{mean}(r_1, \dots, r_G)}{\text{std}(r_1, \dots, r_G)} \tag{1}$$

Table 1. Performance comparison of different methods for disease progression image editing. The best results are highlighted in **bold** and the second-best results are underlined.

Method	Instruct Alignment			Identity		FID ↓	Holistic Score↑
	AUC↑	CLIP-T↑	CLIP-I↑	Race↑	Age↑		
Real images (GT)	-	38.08	100.00	98.95	91.06	-	-
PIE [14]	77.35	34.03	-	-	-	-	-
BiomedJourney [10]	83.85	34.59	-	-	-	-	-
CXR-IRGen [25]	52.36	32.51	75.18	-	-	-	-
ProgEmu [17]	78.42	36.19	89.79	93.56	81.86	0.64	5.58
InstructX2X [18]	85.25	37.08	89.95	96.51	84.66	1.19	4.69
UniMedVL [19]	82.46	**38.48**	90.02	94.56	81.90	0.49	6.88
MedXEdit-GRPO	**86.40**	37.97	**91.60**	**97.93**	**86.74**	**0.40**	**7.84**

The GRPO objective function is formulated as:

$$\mathcal{L}_{GRPO}(\theta) = -\frac{1}{G}\sum_{i=1}^{G}\left[\min\left(\rho_i(\theta)A_i, \text{clip}(\rho_i(\theta), 1-\epsilon, 1+\epsilon)A_i\right) - \beta D_{KL}(\pi_\theta \| \pi_{ref})\right], \quad (2)$$

where $\rho_i(\theta) = \frac{\pi_\theta(z^i|X)}{\pi_{\theta_{old}}(z^i|X)}$ is the importance sampling ratio. KL divergence penalty D_{KL} is incorporated to constrain the policy π_θ from deviating excessively from the reference model π_{ref}.

In RL training, we enhance reward accuracy through two strategies: (1) We require the model to first think across two dimensions before outputting a score: instruction following and consistency. (2) We perform an *N-Pass* evaluation, scoring each sample multiple times and averaging the results to derive the final reward, which helps mitigate the volatility of individual assessments.

4 Experiments

4.1 Experimental Setup

Implementation Details. For the MedXEdit-Score model, we conduct full-parameter SFT utilizing the LLaMA-Factory framework. Training was performed for 5 epochs with a batch size of 32, employing the AdamW optimizer at a learning rate of 10^{-5}. For the MedXEdit-GRPO model, we initialize its parameters with pre-trained weights from the medical foundation model UniMedVL [19] and perform RL training via LoRA with a rank of $r = 64$. All medical images are processed at a resolution of 512×512. We train our model on 24 GPUs with a group size of 8 and a learning rate of 10^{-4} using bf16 mixed precision. For the diffusion process, we employ 25 timesteps during training and 50 steps

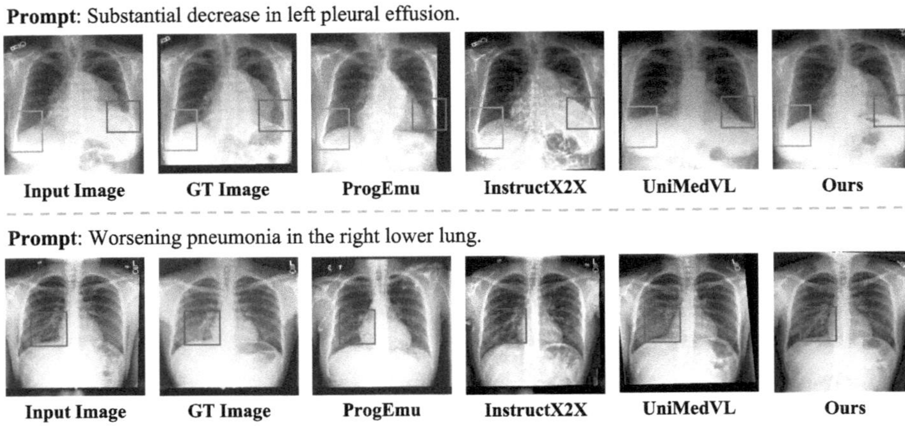

Fig. 3. Qualitative results of counterfactual images, where red and green bounding boxes highlight the targeted modifications and the preserved non-target regions, respectively. (Color figure online)

for evaluation. During inference, the classifier-free guidance scales for text and image are set to 4.0 and 1.0, respectively, to balance editing fidelity and visual quality.

Evaluation Metrics. We evaluate our model across four key dimensions: edit-instruction alignment, demographic identity preservation, generative quality, and holistic assessment. Specifically, edit-instruction alignment is measured by AUC from a pathology classifier [4] for clinical accuracy, and CLIP-T/CLIP-I scores via BiomedCLIP [29] for semantic and image consistency. To ensure demographic identity preservation, we use the state-of-the-art race classifier [9] and age classifier [11] to evaluate identity drift. Generative synthesis quality is assessed using Fréchet Inception Distance (FID) with XRV [5] to evaluate visual realism. Finally, we report a Holistic Score from MedXEdit-Score as an integrative measure of overall editing performance and clinical plausibility.

4.2 Experimental Results

Quantitative Comparison with State-of-the-Art. We compare MedXEdit-GRPO to six state-of-the-art methods for counterfactual generation. As shown in Table 1, MedXEdit-GRPO achieves superior results across nearly all metrics. It leads in Instruct Alignment with an AUC of 86.40 and a CLIP-I of 91.60, while remaining highly competitive in CLIP-T. Notably, our model excels in Identity Preservation, reaching the highest Race (97.93) and Age (86.74) scores, effectively suppressing unintended demographic modifications.

Furthermore, it attains the best Visual Fidelity with a remarkably low FID (0.40), demonstrating that our RL-based approach successfully harmonizes precise editing with robust anatomical consistency.

Table 2. Reward model performance comparison across different medical LVLMs.

Method	Lingshu-I	HuatuoGPT-V	Hulu-Med-4B	Qwen3-VL-4B	Ours (w/o reas.)	Ours (Full)
Score ↑	31.97	36.96	37.72	37.85	71.48	**74.04**

Qualitative Comparison of Counterfactual Images. Figure 3 demonstrates MedXEdit-GRPO's capability to balance precise pathological editing with anatomical preservation. Compared to baselines, our model achieves a more realistic reduction of pleural effusion (red boxes) while maintaining strict consistency in unmentioned regions (green boxes). Similarly, for "worsening pneumonia", our method generates consolidations that best align with the clinical directives with superior visual fidelity.

Analysis of Reward Model Performance. As shown in Table 2, MedXEdit-Score achieves a peak of 74.04, nearly doubling its backbone (37.85) and significantly surpassing specialized models like Hulu-Med-4B. This margin suggests that general LVLMs lack the sensitivity required for complex editing evaluation. Furthermore, our ablation confirms that the "reasoning-before-scoring" paradigm further elevates performance (74.04 vs. 71.48), enhancing the model's discriminatory power and providing a more accurate reward signal for optimization.

Table 3. Ablation study for reward model.

Reward Model	AUC↑	CLIP-T↑	CLIP-I↑	Race↑	Age↑	FID↓
Hybrid Reward	83.87	37.46	89.74	92.69	79.27	0.50
MedXEdit-Score	**86.40**	**37.97**	**91.60**	**97.93**	**86.74**	**0.40**

Ablation Study. We evaluate the necessity of a unified reward model by replacing MedXEdit-Score with a heuristic baseline, which uses a weighted average of the pathological classifier's AUC, MSE, and SSIM as the reward signal. As shown in Table 3, the significant performance gap proves that conventional, disjoint metrics are insufficient for capturing the complex clinical logic and multifaceted constraints inherent in medical image editing, underscoring the necessity of a unified evaluator for stable policy optimization.

5 Conclusion

We introduce MedXEdit-GRPO, a reinforcement learning framework for counterfactual medical image editing. To enable reasoning-aware optimization, we curate a large-scale MedXEdit-Reward dataset and develop MedXEdit-Score, a specialized reward model enforcing a "reasoning-before-scoring" paradigm. By optimizing the generative policy through GRPO, our framework effectively harmonizes diverse clinical objectives into a unified training signal. Experimental results demonstrate that MedXEdit-GRPO significantly outperforms existing methods in both instruction compliance and anatomical consistency, highlighting its potential for robust medical analysis.

Acknowledgments. This work was supported by National Natural Science Foundation of China (No. 62576107), the decision support system for Shanghai Municipal Health Commission, early warning and auxiliary diagnosis and treatment of pediatric severe pneumonia (No. 2025ZHYL040), Shanghai Municipal Commission of Economy and Informatization, Corpus Construction for Large Language Models in Pediatric Respiratory Diseases (No. 2024-GZL-RGZN-01013), and 2025 National Major Science and Technology Project—Noncommunicable Chronic Diseases-National Science and Technology Major Project, Research on the Pathogenesis of Pancreatic Cancer and Novel Strategies for Precision Medicine (No. 2025ZD0552303). Rui Feng and Xiaobo Zhang are corresponding authors.

Disclosure of Interests. The authors have no competing interests to declare that are relevant to the content of this article.

References

1. Bedel, H.A., Çukur, T.: Dreamr: diffusion-driven counterfactual explanation for functional MRI. IEEE Trans. Med. Imaging **44**(9), 3654–3669 (2024)
2. Brooks, T., Holynski, A., Efros, A.A.: Instructpix2pix: Learning to follow image editing instructions. In: Proceedings of the IEEE/CVF Conference on Computer Vision and Pattern Recognition, pp. 18392–18402 (2023)
3. Choi, B., Bae, S., Kweon, S., Choi, E.: Kormedmcqa-v: A multimodal benchmark for evaluating vision-language models on the Korean medical licensing examination. arXiv preprint arXiv:2602.13650 (2026)
4. Cohen, J.P., et al.: Gifsplanation via latent shift: a simple autoencoder approach to counterfactual generation for chest x-rays. In: Medical imaging with deep learning, pp. 74–104. PMLR (2021)
5. Cohen, J.P., et al.: Torchxrayvision: a library of chest x-ray datasets and models. In: International Conference on Medical Imaging with Deep Learning, pp. 231–249. PMLR (2022)
6. Deng, C., et al.: Emerging properties in unified multimodal pretraining. arXiv preprint arXiv:2505.14683 (2025)
7. Fontanella, A., Mair, G., Wardlaw, J., Trucco, E., Storkey, A.: Diffusion models for counterfactual generation and anomaly detection in brain images. IEEE Trans. Med. Imaging (2024)
8. Furuta, R., Inoue, N., Yamasaki, T.: Fully convolutional network with multi-step reinforcement learning for image processing. In: Proceedings of the AAAI conference on artificial intelligence, vol. 33, pp. 3598–3605 (2019)
9. Gichoya, J.W., et al.: Ai recognition of patient race in medical imaging: a modelling study. Lancet Digit. Health **4**(6), e406–e414 (2022)
10. Gu, Y., et al: Biomedjourney: counterfactual biomedical image generation by instruction-learning from multimodal patient journeys. arXiv:2310.10765 (2023)
11. Ieki, H., et al.: Deep learning-based age estimation from chest x-rays indicates cardiovascular prognosis. Commun. Med. **2**(1), 159 (2022)

12. Ktena, I., et al.: Generative models improve fairness of medical classifiers under distribution shifts. Nat. Med. **30**(4), 1166–1173 (2024)
13. Lee, S.I., Topol, E.J.: The clinical potential of counterfactual ai models. Lancet **403**(10428), 717 (2024)
14. Liang, K.,et al.: Pie: simulating disease progression via progressive image editing. arXiv preprint arXiv:2309.11745 (2023)
15. Lipman, Y., Chen, R.T., Ben-Hamu, H., Nickel, M., Le, M.: Flow matching for generative modeling. arXiv preprint arXiv:2210.02747 (2022)
16. Luo, X., et al.: Editscore: unlocking online RL for image editing via high-fidelity reward modeling. arXiv preprint arXiv:2509.23909 (2025)
17. Ma, C., et al.: Towards interpretable counterfactual generation via multimodal autoregression. In: International Conference on Medical Image Computing and Computer-Assisted Intervention, pp. 611–620. Springer (2025)
18. Min, H., You, T., Lee, H., Cho, Y., Cho, S.: Instructx2x: an interpretable local editing model for counterfactual medical image generation. In: International Conference on Medical Image Computing and Computer-Assisted Intervention, pp. 279–288. Springer (2025)
19. Ning, J., et al.: Unimedvl: unifying medical multimodal understanding and generation through observation-knowledge-analysis. arXiv preprint arXiv:2510.15710 (2025)
20. Ouyang, L.: Training language models to follow instructions with human feedback. In: Advances in Neural Information Processing Systems, vol. 35, pp. 27730–27744 (2022)
21. Pérez-García, F., et al.: Radedit: stress-testing biomedical vision models via diffusion image editing. In: European Conference on Computer Vision, pp. 358–376. Springer (2024)
22. Rombach, R., Blattmann, A., Lorenz, D., Esser, P., Ommer, B.: High-resolution image synthesis with latent diffusion models. In: Proceedings of the IEEE/CVF Conference on Computer Vision and Pattern Recognition, pp. 10684–10695 (2022)
23. Schulman, J., Wolski, F., Dhariwal, P., Radford, A., Klimov, O.: Proximal policy optimization algorithms. arXiv preprint arXiv:1707.06347 (2017)
24. Shao, Z., et al.: Deepseekmath: pushing the limits of mathematical reasoning in open language models. arXiv preprint arXiv:2402.03300 (2024)
25. Shentu, J., Al Moubayed, N.: Cxr-irgen: an integrated vision and language model for the generation of clinically accurate chest x-ray image-report pairs. In: Proceedings of the IEEE/CVF Winter Conference on Applications of Computer Vision, pp. 5212–5221 (2024)
26. Singla, S., Eslami, M., Pollack, B., Wallace, S., Batmanghelich, K.: Explaining the black-box smoothly–a counterfactual approach. Med. Image Anal. **84**, 102721 (2023)
27. Wang, Y., Zang, Y., Li, H., Jin, C., Wang, J.: Unified reward model for multimodal understanding and generation. arXiv preprint arXiv:2503.05236 (2025)
28. Wu, C., et al.: Qwen-image technical report. arXiv preprint arXiv:2508.02324 (2025)
29. Zhang, S., et al.: Biomedclip: a multimodal biomedical foundation model pretrained from fifteen million scientific image-text pairs. arXiv preprint arXiv:2303.00915 (2023)
30. Zhang, S., et al.: Hive: Harnessing human feedback for instructional visual editing. In: Proceedings of the IEEE/CVF Conference on Computer Vision and Pattern Recognition, pp. 9026–9036 (2024)

MINT: Molecularly Informed Training with Spatial Transcriptomics Supervision for Pathology Foundation Models

Minsoo Lee, Jonghyun Kim, Juseung Yun, Sunwoo Yu, and Jongseong Jang(✉)

LG AI Research, Seoul, South Korea
{minsoo.lee,kimjh0107,js.yun,ysw0920,j.jang}@lgresearch.ai

Abstract. Pathology foundation models learn morphological representations through self-supervised pretraining on large-scale whole-slide images, yet they do not explicitly capture the molecular state of the tissue. Spatial transcriptomics technologies bridge this gap by measuring gene expression in situ, offering a natural cross-modal supervisory signal. We propose MINT (Molecularly Informed Training), a fine-tuning framework that uses spatial transcriptomics as molecular supervision for pretrained pathology Vision Transformers. MINT appends a learnable ST token to the ViT input to encode transcriptomic information separately from the morphological CLS token, preventing catastrophic forgetting through DINO self-distillation and explicit feature anchoring to the frozen pretrained encoder. Gene expression regression at both spot-level (Visium) and patch-level (Xenium) resolutions provides complementary supervision across spatial scales. Trained on 577 publicly available HEST samples, MINT achieves the best overall performance on both HEST-Bench for gene expression prediction (mean Pearson r=0.440) and EVA for general pathology tasks (0.803), demonstrating that spatial transcriptomics supervision complements morphology-centric self-supervised pretraining.

Keywords: Pathology Foundation Model · Spatial Transcriptomics

1 Introduction

Foundation models pretrained on large-scale whole-slide image (WSI) collections have become central to computational pathology [1,2,4,5]. Through self-supervised objectives such as DINO [6] and DINOv2 [7], these models learn morphological representations from millions of histopathology tiles without manual annotation, enabling effective transfer to diverse downstream tasks including tissue classification, biomarker prediction, and survival analysis. Recent scaling efforts—H-optimus-0 [1] (ViT-G on 500K slides), UNI2-h [3] (ViT-H on 350K slides), and Virchow2 [20] (ViT-H on 3.1M slides)—have progressively improved performance across standardized benchmarks.

G. Yang et al. (Eds.): MICCAI 2026, LNCS 16878, pp. 398–407, 2027.
https://doi.org/10.1007/978-3-032-38059-3_38

However, these models are trained *exclusively* on visual patterns. Histopathology images implicitly encode the molecular state of the tissue microenvironment: cellular composition, signaling pathway activity, and gene expression programs all manifest in tissue morphology. Spatial transcriptomics (ST) technologies provide direct measurements of this link. Spot-level platforms [9], from the original Spatial Transcriptomics to its commercial successor Visium (10x Genomics), profile transcriptomes at spot-level resolution ($\sim$55 μm, roughly 10–50 cells per spot), while Xenium [10] provides single-molecule transcript detection at sub-cellular resolution. Recently, the HEST dataset [8] has aggregated over 1,100 such paired histology–transcriptomics samples spanning 131 organs, making large-scale training data for this modality publicly available.

A natural question arises: *can spatially-resolved gene expression serve as supervisory signal to improve the representations of pretrained pathology foundation models?* Prior work has primarily explored histology-to-gene-expression prediction as a task-specific supervised problem [11–14], but has not used spatial transcriptomics as molecular supervision to enhance pathology foundation model representations. The key challenge is that fine-tuning on gene expression objectives risks catastrophic forgetting of the morphological representations acquired during large-scale pretraining.

We introduce MINT (Molecularly Informed Training with Spatial Transcriptomics Supervision), a multi-task fine-tuning framework that addresses this challenge through three design principles. First, rather than modifying the CLS token that encodes morphological features, we append a dedicated learnable *ST token* to the ViT input sequence, providing the model with a separate channel for molecular information. Second, we combine DINO self-distillation for continued visual learning with explicit feature distillation from the frozen pretrained model, creating a dual mechanism against catastrophic forgetting. Third, we exploit both spot-level and Xenium patch-level gene expression as complementary supervision at different spatial resolutions.

Our contributions are as follows:

1. We propose MINT that uses spatial transcriptomics as molecular supervision for pretrained pathology ViTs while preventing catastrophic forgetting through a dedicated ST token and dual distillation mechanisms.
2. We show that the ST and CLS tokens capture complementary information—the ST token specializes for molecular signals while the CLS token retains morphological transferability—and that combining both yields consistent improvements in a backbone-agnostic manner.
3. Under the official HEST-Bench and EVA evaluation framework, MINT achieves the best overall performance on both benchmarks (mean Pearson r=0.440 and EVA average 0.803).

2 Method

2.1 Overview

Given a pretrained ViT encoder f_θ with embedding dimension D, MINT constructs a student–teacher framework and fine-tunes with four complementary objectives (Fig. 1). The student and teacher are initialized from f_θ; the student is updated via gradient descent while the teacher tracks the student through exponential moving average (EMA). A frozen copy of f_θ serves as a distillation anchor. The key idea is to augment the ViT with a learnable *ST token* that encodes transcriptomic information, while the original CLS and patch tokens retain their morphological roles.

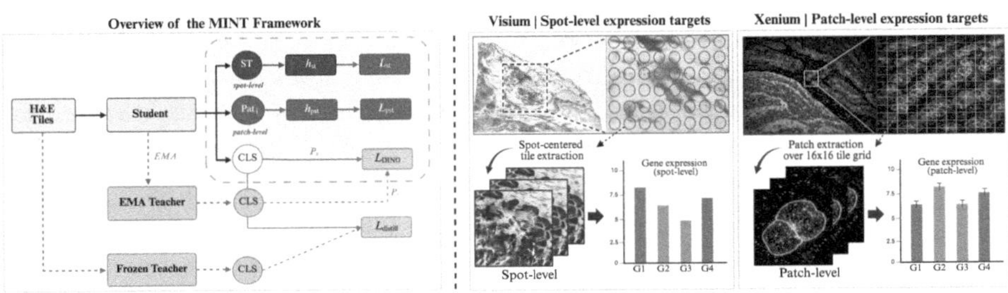

Fig. 1. MINT framework. The student ViT, augmented with a learnable ST token, outputs CLS, ST, and patch token representations. Gene expression regression from the ST token ($\mathcal{L}_{\mathrm{ST}}$) and patch tokens ($\mathcal{L}_{\mathrm{pST}}$) provides transcriptomic supervision, while DINO self-distillation ($\mathcal{L}_{\mathrm{DINO}}$) and feature anchoring ($\mathcal{L}_{\mathrm{distill}}$) preserve morphological representations. Only the student is updated via backpropagation.

2.2 ST Token Design

We augment the ViT input with a learnable *ST token* $\mathbf{t}^{\mathrm{st}} \in \mathbb{R}^D$, appended alongside the CLS token and N patch tokens (with N=256 for ViT-g/14 at 224×224). After processing through all transformer layers, the model produces three output representations: the CLS token $\mathbf{z}^{\mathrm{cls}} \in \mathbb{R}^D$ retaining morphological features, the ST token $\mathbf{z}^{\mathrm{st}} \in \mathbb{R}^D$ encoding transcriptomic features, and patch tokens $\{\mathbf{z}_i^{\mathrm{pat}}\}_{i=1}^N$ providing spatially-localized representations.

The motivation for a separate token, rather than supervising the CLS token itself for gene expression, is twofold. First, it preserves the CLS token's morphological representations acquired through large-scale WSI pretraining. Directly supervising the CLS token for gene expression risks overwriting these features; the separate ST token instead specializes for gene expression while the CLS token continues to serve its original role in the self-supervised and feature anchoring objectives. Second, the ST token participates in self-attention with CLS and patch tokens across all transformer layers, enabling it to learn molecular representations from the full spatial context. At inference, concatenating both tokens

($[\mathbf{z}^{\text{cls}} \| \mathbf{z}^{\text{st}}] \in \mathbb{R}^{2D}$) combines both types of information, while using CLS alone recovers the original feature space.

2.3 Training Objectives

DINO Self-distillation. Following DINO [6], we apply multi-crop augmentation: 2 global crops (224×224) and 8 local crops (96×96). The student processes all 10 views while the teacher sees only global views. Both project their CLS tokens through a shared MLP head to a 65,536-dimensional probability space with centering and sharpening:

$$\mathcal{L}_{\text{DINO}} = -\sum_{x \in \mathcal{V}_g} \sum_{\substack{x' \in \mathcal{V} \\ x' \neq x}} P_t(x) \log P_s(x') \tag{1}$$

where $\mathcal{V}_g$ and $\mathcal{V}$ denote global and all crops, and P_t, P_s are teacher and student distributions. This loss maintains the self-supervised learning dynamics of the original pretraining, enabling continued visual representation learning during fine-tuning.

Feature Distillation. To explicitly anchor the student's CLS representation near the pretrained feature space, we minimize the L_2 distance to the frozen model's output on all crops:

$$\mathcal{L}_{\text{distill}} = \frac{1}{|\mathcal{V}|} \sum_{x \in \mathcal{V}} \left\| \mathbf{z}^{\text{cls}}_{\text{student}}(x) - \mathbf{z}^{\text{cls}}_{\text{frozen}}(x) \right\|_2^2 \tag{2}$$

This provides a direct regularization against catastrophic forgetting, complementing the implicit regularization from DINO's EMA teacher. The combination of both mechanisms—EMA-based self-distillation and explicit feature anchoring—provides redundant safeguards for representation stability.

Spot-Level ST Regression. Each training tile corresponds to a Visium spot. We define a shared vocabulary of G=38,710 genes as the union across all slides; since each slide measures only a subset, the loss is computed only over genes present in the current slide. The ST token is passed through an MLP head $h_{\text{st}} : \mathbb{R}^D \to \mathbb{R}^G$ to predict $\log(1+x)$-transformed expression values. To further focus on informative genes, we apply stochastic highly variable gene (HVG) selection [17]: at each iteration, with probability p_{hvg} we restrict the loss to the top-k HVGs, and otherwise use all measured genes:

$$\mathcal{L}_{\text{ST}} = \frac{1}{|\mathcal{G}_s|} \sum_{g \in \mathcal{G}_s} \left(h_{\text{st}}(\mathbf{z}^{\text{st}})_g - y_g \right)^2 \tag{3}$$

where $\mathcal{G}_s$ is the (stochastically) selected gene subset and y_g the target expression.

Patch-Level Xenium Regression. Spot-level platforms provide one expression profile per spot, averaging over 10–50 cells. Xenium technology [10], in contrast, detects individual transcripts at sub-cellular resolution, enabling construction of patch-level targets. For each 16×16 grid of ViT patches within a tile, we aggregate Xenium transcripts falling within each patch's spatial extent to create per-patch expression vectors.

Each patch token $\mathbf{z}_i^{\text{pat}}$ is passed through a dedicated MLP head $h_{\text{pst}} : \mathbb{R}^D \to \mathbb{R}^{G_{\text{xen}}}$, where $G_{\text{xen}} = 5{,}783$ is the union of genes across available Xenium panels. Since each panel covers a different gene subset, the loss for a given sample is restricted to its panel's genes. We further apply *all-zero patch filtering*: the loss is computed only over patches containing at least one detected transcript across the sample's measured Xenium genes, while zero-valued gene entries within these non-empty patches are retained in the regression target, avoiding penalization for regions without detected transcripts:

$$\mathcal{L}_{\text{pST}} = \frac{1}{|\mathcal{P}^+|} \sum_{p \in \mathcal{P}^+} \frac{1}{|\mathcal{G}|} \sum_{g \in \mathcal{G}} \left(h_{\text{pst}}(\mathbf{z}_p^{\text{pat}})_g - y_{p,g}\right)^2 \tag{4}$$

where $\mathcal{P}^+$ is the set of patches containing at least one detected transcript and $\mathcal{G}$ the set of genes measured by the sample's Xenium panel.

Total Objective. The four losses are combined as:

$$\mathcal{L} = \mathcal{L}_{\text{DINO}} + \lambda_{\text{distill}} \mathcal{L}_{\text{distill}} + \lambda_{\text{ST}} \mathcal{L}_{\text{ST}} + \lambda_{\text{pST}} \mathcal{L}_{\text{pST}} \tag{5}$$

where λ_{distill}, λ_{ST}, and λ_{pST} are scalar weights for each auxiliary objective.

3 Experiments

3.1 Training Setup

We fine-tune H-optimus-0 [1] (ViT-g/14, 1.1B parameters) on the HEST dataset [8], which provides paired histology tiles and spot-level spatial transcriptomics measurements (Visium and earlier ST platforms) across diverse tissue types. From the 649 human samples in HEST, we exclude 72 samples included in HEST-Bench to prevent data leakage, yielding 577 training samples. For patch-level Xenium supervision (Sect. 2.3), we use 44 Xenium-profiled samples from HEST, oversampled 5× to balance against the more abundant spot-level data. We train for 90K iterations using AdamW with learning rate 5×10^{-5}, cosine schedule, batch size 64 per GPU across 8 GPUs, and bf16 mixed precision. The loss weights λ_{distill}, λ_{ST}, and λ_{pST} in Eq. 5 are each set to 100. For HVG selection, we use k=200 and p_{hvg}=0.5. We denote the resulting model as **MINT**. At inference, we concatenate the CLS and ST tokens to form $[\mathbf{z}^{\text{cls}} \| \mathbf{z}^{\text{st}}] \in \mathbb{R}^{3072}$ as the default feature representation.

Table 1. HEST-Bench results (PCA+Ridge with 256 PCA components; Pearson correlation; higher is better). Best in **bold**, second-best underlined.

Model	IDC	PRAD	PAAD	SKCM	COAD	READ	ccRCC	LUAD	LYMPH	Avg
CTransPath [15]	0.511	0.343	0.438	0.511	0.229	0.110	0.228	0.499	0.235	0.345
Phikon [19]	0.533	0.342	0.443	0.536	0.259	0.152	0.242	0.547	0.237	0.366
CONCH [5]	0.536	0.355	0.448	0.579	0.253	0.167	0.218	0.531	0.251	0.371
REMEDIS [22]	0.529	0.347	0.464	0.582	0.286	0.115	0.265	0.534	0.247	0.374
CONCH 1.5 [5]	0.544	0.381	0.457	0.552	0.280	0.160	0.218	0.551	0.270	0.379
GigaPath [16]	0.551	0.371	0.477	0.554	0.301	0.186	0.239	0.540	0.249	0.385
UNI [2]	0.570	0.314	0.476	0.625	0.263	0.176	0.243	0.551	0.257	0.386
H0-mini [23]	0.586	0.368	0.492	0.601	0.249	0.186	0.267	0.548	0.263	0.396
Virchow [4]	0.570	0.331	0.488	0.609	0.311	0.202	0.264	0.546	0.259	0.398
Virchow2 [20]	0.592	0.347	0.466	0.617	0.258	0.208	0.279	0.561	0.258	0.398
UNI2-h [3]	0.590	0.357	0.500	0.660	0.301	0.223	0.264	0.558	0.273	0.414
H-optimus-0 [1]	0.598	0.385	0.493	0.643	0.299	0.229	0.265	0.558	0.260	0.415
MINT	**0.655**	**0.389**	**0.509**	**0.695**	**0.314**	**0.244**	**0.283**	**0.597**	**0.275**	**0.440**

3.2 Benchmarks and Protocols

We evaluate on two complementary benchmarks, strictly following each benchmark's official protocol. First, we report results on HEST-Bench, which assesses how well histology features support gene expression prediction across 9 cancer types. Following the benchmark specification, we use the standard PCA+Ridge evaluation pipeline with 256 PCA components and report Pearson correlation (higher is better). Second, we evaluate general-purpose pathology transferability using the EVA evaluation framework, which consists of nine pathology benchmarks spanning classification (BreakHis, CRC, Gleason, MHIST, PCam), weakly-supervised whole-slide tasks (Cam16, PANDA), and nuclei segmentation (CoNSeP, MoNuSAC); we refer to [18] for full protocol details. For classification and whole-slide tasks, we use the [CLS || ST] concat feature. For segmentation tasks, we follow the EVA protocol using spatially arranged patch features concatenated with the input image under online linear evaluation, without using the ST token. For fair comparison, all baseline numbers reported in Tables 1 and 2 (except MINT) are taken from the official benchmark leaderboards. We additionally run the official evaluation pipelines to verify that these reported baseline results are reproducible under the benchmark settings. All MINT results are obtained using the same official evaluation pipelines without any modifications.

3.3 Comparison with Existing Models

Table 1 presents HEST-Bench results. MINT achieves the best overall mean Pearson correlation (0.440), surpassing all compared models including H-optimus-0 (0.415) and UNI2-h (0.414), and ranking first across all 9 cancer types. On EVA

Table 2. EVA evaluation framework results (higher is better). Best in **bold**, second-best underlined. Average is the arithmetic mean over 9 benchmarks.

Model	BHis	CRC	Glea	MHIST	PCam	Cam16	PANDA	CoNS	MoNuS	Average
Phikon-v2 [24]	0.713	0.939	0.757	0.777	0.894	0.805	0.625	0.627	0.644	0.753
Phikon [19]	0.717	0.940	0.729	0.803	0.920	0.798	0.644	0.627	0.635	0.757
CONCH [5]	0.681	0.953	0.729	0.832	0.922	0.841	0.628	0.618	0.632	0.760
hibou-L [25]	0.735	0.932	0.764	0.839	**0.955**	0.823	0.634	**0.646**	0.669	0.777
UNI [2]	0.785	0.944	0.750	0.843	0.937	0.833	0.659	0.628	0.659	0.782
Pv-GigaPath [16]	0.827	0.951	0.724	0.829	0.945	0.815	0.653	0.626	0.680	0.783
Midnight-12k [26]	0.819	0.966	**0.800**	0.804	0.929	0.849	0.649	0.623	0.659	0.789
H-optimus-0 [1]	0.801	0.955	0.770	0.843	0.943	0.827	0.671	0.644	**0.685**	0.793
UNI2-h [3]	0.859	0.965	0.775	0.824	0.950	0.849	0.657	0.630	0.642	0.795
Virchow2 [20]	0.821	**0.967**	0.783	**0.861**	0.938	**0.861**	0.646	0.640	0.669	0.798
MINT	**0.864**	0.960	0.769	0.834	0.943	0.851	**0.673**	**0.646**	**0.685**	**0.803**

(Table 2), MINT also attains the highest average (0.803), exceeding Virchow2 (0.798) and H-optimus-0 (0.793). Compared to the pretrained H-optimus-0 backbone, MINT maintains comparable or improved performance across all EVA tasks except MHIST, where a marginal decrease is observed.

Notably, MINT ranks first on both benchmarks simultaneously. HEST-Bench measures molecular prediction capability while EVA assesses general-purpose pathology transferability—two axes that are not inherently aligned. The fact that MINT improves on both indicates that spatial transcriptomics supervision provides complementary information to morphology-centric pretraining, rather than introducing a trade-off between molecular and morphological performance.

These gains are achieved not by scaling up image data, but by introducing a new supervisory modality: MINT fine-tunes on only 577 paired histology–transcriptomics samples from the publicly available HEST dataset, yet improves both molecular and morphological benchmarks. This suggests that cross-modal supervision offers a distinct and complementary axis for improving pathology foundation models beyond image-only self-supervised scaling, and that further growth of paired histology–transcriptomics datasets may yield additional gains.

3.4 Representation Analysis

Table 3 examines how the choice of feature representation affects downstream performance across two backbone architectures. Since the EVA segmentation benchmarks (CoNSeP, MoNuSAC) rely on spatially arranged patch embeddings rather than token-level representations, the EVA average here is computed over classification and whole-slide tasks only. The CLS and ST tokens exhibit clear specialization: ST-only outperforms CLS-only on HEST-Bench (0.428 vs. 0.413 for H-optimus-0), while CLS-only outperforms ST-only on EVA (0.828 vs. 0.823), indicating that each token captures distinct information—molecular and

Table 3. Representation variants across backbones, including pretrained baselines and ablations. HEST-Bench reports mean Pearson correlation; EVA-CLS averages over classification and whole-slide tasks (7 benchmarks). Best in **bold**, second-best underlined.

Backbone	Training	Representation	HEST-Bench	EVA-CLS
H-opt-0	No training	CLS-only	0.415	0.830
	L_{ST} on [CLS] w/o $L_{distill}$	CLS-only	0.424	0.811
	L_{ST} on [CLS] + $L_{distill}$	CLS-only	0.426	0.827
	MINT	CLS-only	0.413	0.828
		ST-only	0.428	0.823
		CLS ⊕ ST (sum)	0.431	0.837
		[CLS ‖ ST] (concat)	**0.440**	**0.842**
UNI2-h	No training	CLS-only	0.414	0.840
	L_{ST} on [CLS] w/o $L_{distill}$	CLS-only	0.420	0.824
	L_{ST} on [CLS] + $L_{distill}$	CLS-only	0.426	0.836
	MINT	CLS-only	0.410	0.839
		ST-only	0.426	0.834
		CLS ⊕ ST (sum)	0.433	0.844
		[CLS ‖ ST] (concat)	**0.435**	**0.848**

morphological, respectively. Combining both tokens improves performance on both benchmarks. Notably, element-wise summation (CLS ⊕ ST) uses the same dimensionality as CLS-only (1536-d) yet substantially outperforms it (e.g., 0.431 vs. 0.413 on HEST-Bench), confirming that the gains arise from genuinely complementary information rather than increased feature dimensionality. Concatenation ([CLS ‖ ST]) achieves the best results across both benchmarks. These patterns are consistently observed across both H-optimus-0 and UNI2-h backbones, confirming that the benefit of the ST token is backbone-agnostic.

To validate the necessity of the dedicated ST token, we ablate it by applying $\mathcal{L}_{ST}$ directly to the CLS token, with and without feature distillation ($\mathcal{L}_{distill}$). Both variants achieve higher HEST-Bench scores than the pretrained baseline, indicating that the CLS token successfully learns molecular information regardless of distillation. However, without distillation, EVA-CLS drops substantially (0.830→0.811 for H-optimus-0; 0.840→0.824 for UNI2-h), revealing catastrophic forgetting of morphological representations. Adding distillation largely recovers EVA performance (to 0.827 and 0.836, respectively), confirming its role in preventing forgetting. Nevertheless, both $\mathcal{L}_{ST}$ on [CLS] variants fall short of MINT's token-separated design on both benchmarks for both backbones: even CLS ⊕ ST summation outperforms the best $\mathcal{L}_{ST}$ on [CLS] variant (0.431/0.837 vs. 0.426 / 0.827 for H-optimus-0; 0.433/0.844 vs. 0.426/0.836 for UNI2-h), demonstrating that decoupling molecular and morphological learning into separate tokens is more effective than encoding both in a shared representation. MINT's dedicated ST token avoids this trade-off entirely: the CLS-only variant of MINT closely matches the pretrained baseline on EVA (0.828 vs. 0.830 for H-optimus-0; 0.839 vs. 0.840 for UNI2-h), confirming that the ST token successfully decouples molecular learning from the pretrained feature space.

4 Conclusion

We presented MINT, which incorporates spatial transcriptomics supervision into pretrained pathology ViTs through a dedicated ST token and dual distillation. By decoupling molecular and morphological learning, MINT achieves the highest scores on both HEST-Bench and EVA in a backbone-agnostic manner, suggesting that cross-modal supervision offers a complementary axis beyond image-only self-supervised scaling.

Disclosure of Interests. LG AI Research has filed patent application(s) related to the method described in this paper. The authors are employees of LG AI Research. The authors declare no other competing interests.

References

1. Bioptimus: H-optimus-0. GitHub. https://github.com/bioptimus/releases/tree/main/models/h-optimus/v0. Accessed 24 Jun 2026
2. Chen, R.J., et al.: Towards a general-purpose foundation model for computational pathology. Nat. Med. **30**(3), 850–862 (2024)
3. MahmoodLab: UNI2-h model. Hugging Face. https://huggingface.co/MahmoodLab/UNI2-h. Accessed 24 Jun 2026
4. Vorontsov, E., et al.: A foundation model for clinical-grade computational pathology and rare cancers detection. Nat. Med. **30**(10), 2924–2935 (2024)
5. Lu, M.Y., et al.: A visual-language foundation model for computational pathology. Nat. Med. **30**(3), 863–874 (2024)
6. Caron, M.: Emerging properties in self-supervised vision transformers. In: Proceedings of the IEEE/CVF International Conference on Computer Vision, pp. 9650–9660 (2021)
7. Oquab, M.: DINOv2: Learning robust visual features without supervision. Trans. Mach. Learn. Res. (2024)
8. Jaume, G.: HEST-1k: A dataset for spatial transcriptomics and histology image analysis. In: Advances in Neural Information Processing Systems, Datasets and Benchmarks Track (2024)
9. Ståhl, P.L., et al.: Visualization and analysis of gene expression in tissue sections by spatial transcriptomics. Science **353**(6294), 78–82 (2016)
10. Janesick, A., et al.: High resolution mapping of the tumor microenvironment using integrated single-cell, spatial and in situ analysis. Nat. Commun. **14**(1), 8353 (2023)
11. He, B., et al.: Integrating spatial gene expression and breast tumour morphology via deep learning. Nat. Biomed. Eng. **4**(8), 827–834 (2020)
12. Pang, M.: Leveraging information in spatial transcriptomics to predict super-resolution gene expression from histology images in tumors. bioRxiv preprint (2021)

13. Xie, R., et al.: Spatially resolved gene expression prediction from H&E histology images via bi-modal contrastive learning. In: Advances in Neural Information Processing Systems (2023)
14. Jia, Y.: THItoGene: A deep learning method for predicting spatial transcriptomics from histological images. Brief. Bioinform. **25**(1) (2024)
15. Wang, X., et al.: Transformer-based unsupervised contrastive learning for histopathological image classification. Med. Image Anal. **81**, 102559 (2022)
16. Xu, H., et al.: A whole-slide foundation model for digital pathology from real-world data. Nature **630**, 181–188 (2024)
17. Luecken, M.D., Theis, F.J.: Current best practices in single-cell RNA-seq analysis: a tutorial. Mol. Syst. Biol. **15**(6), e8746 (2019)
18. Gatopoulos, I.: EVA: Evaluation framework for pathology foundation models. Med. Imaging Deep Learn. (2024)
19. Filiot, A.: Scaling self-supervised learning for histopathology with masked image modeling. medRxiv preprint 2023.07.21.23292757 (2023)
20. Zimmermann, E.: Virchow2: scaling self-supervised mixed magnification models in pathology. arXiv preprint arXiv:2408.00738 (2024)
21. He, K.: Deep residual learning for image recognition. In: Proceedings of the IEEE Conference on Computer Vision and Pattern Recognition, pp. 770–778 (2016)
22. Azizi, S., et al.: Robust and data-efficient generalization of self-supervised machine learning for diagnostic imaging. Nat. Biomed. Eng. **7**(6), 756–779 (2023)
23. Filiot, A.: Distilling foundation models for robust and efficient models in digital pathology. arXiv preprint arXiv:2501.16239 (2025)
24. Filiot, A.: Phikon-v2, a large and public feature extractor for biomarker prediction. arXiv preprint arXiv:2409.09173 (2024)
25. Nechaev, D.: Hibou: A family of foundational vision transformers for pathology. arXiv preprint arXiv:2406.05074 (2024)
26. Karasikov, M.: Training state-of-the-art pathology foundation models with orders of magnitude less data. arXiv preprint arXiv:2504.05186 (2025)

Ontology-Grounded Structured Prediction for Dental CBCT Reporting

Luca Lumetti[1], Mattia Di Bartolomeo[2], Arrigo Pellacani[3], Alexandre Anesi[1], Costantino Grana[1], and Federico Bolelli[1(✉)]

[1] University of Modena and Reggio Emilia, Modena, Italy
federico.bolelli@unimore.it
[2] Independent Researcher, Roma, Italy
mattiadibartolomeo@unimore.it
[3] University Hospital of Modena, Modena, Italy

Abstract. We present a dataset and baseline for ontology-grounded structured prediction from dental Cone-Beam Computed Tomography (CBCT) volumes. Building on the public ToothFairy3 benchmark (532 volumes with expert-level segmentations), we contribute *(i)* a total of 893 free-text clinical reports for 529 publicly available CBCT volumes, *(ii)* their conversion into validated RDF/Turtle (Resource Description Framework) instances aligned with a clinician-designed OWL (Web Ontology Language) ontology spanning 13 finding types and multiple qualifier axes, and *(iii)* a strong baseline demonstrating the effectiveness of our setup and establishing a foundation for future work. We formulate CBCT reporting as a *three-stage* structured prediction problem—i.e., finding *detection*, anatomical *slot allocation*, and *property* prediction—and introduce a hierarchical evaluation suite of six clinically interpretable metrics that decouple detection, localization, and characterization. A baseline model using frozen multi-scale VoxTell features, a structure-indexed encoder, and ontology-driven prediction heads achieves strong results under 5-fold cross-validation, with stage-decoupled analysis identifying presence detection as the primary deployment bottleneck. Dataset, ontology, and code are publicly released at https://github.com/AImageLab-zip/CBCT-Report.

Keywords: Cone-Beam Computed Tomography · Report Generation · Medical Ontology · ToothFairy

1 Introduction

Cone-Beam Computed Tomography (CBCT) is central to dental and maxillofacial practice, supporting implant planning, surgery, and pathology assessment by providing volumetric visualization of teeth, jawbones, canals, and adjacent structures. In routine workflows, clinicians translate CBCT observations into reports that specify *what* is present, *where* it is located (often tooth-specific),

G. Yang et al. (Eds.): MICCAI 2026, LNCS 16878, pp. 408–417, 2027.
https://doi.org/10.1007/978-3-032-38059-3_39

and *how* it should be characterized (e.g., severity, morphology, treatment quality). Automating reporting could reduce clinical workload and enable screening, but requires models that jointly handle multiple anatomical structures and compositional clinical findings.

Most CBCT benchmarks emphasize segmentation, focusing on structures such as teeth, jaws, inferior alveolar canals, and pulp [8,11,12,16,18,20,23]. The ToothFairy challenge series [5–7,17] has advanced multi-structure maxillofacial segmentation, culminating in ToothFairy3 with 532 public CBCT scans and 50 private test scans for long-term benchmarking [19], each accompanied by 77 different accurate segmentation classes. However, moving from segmentation to finding detection and reporting introduces challenges: findings are heterogeneous and imbalanced, must be grounded to an anatomical region or structure (e.g., ISO 3950/FDI tooth identifiers [9]), and require clinically relevant qualifiers.

A key bottleneck for generative reporting is evaluation. Standard free-text captioning metrics (e.g., BLEU [21], ROUGE [13], METEOR [2]) measure surface similarity and can correlate poorly with clinical correctness, missing hallucinations, omissions, or wrong localization [1,3,4,14,26]. Clinically informed alternatives such as entity/relation extraction (e.g., RadGraph [10]) or LLM-based factuality scoring (e.g., RadFact [3]) remain domain- and schema-dependent; for dental CBCT, tooth-indexed multi-instance findings and property axes are rarely standardized.

We therefore formulate CBCT reporting as ontology-grounded[1] structured prediction rather than free-text generation. We introduce a resource derived from the publicly available ToothFairy3 cases, consisting of *(i)* free-form clinical reports authored by clinical staff, and *(ii)* their conversion into validated ontology instances. The central representation is a clinician-authored OWL/RDF (Web Ontology Language/Resource Description Framework) ontology defining 13 finding types and orthogonal qualifier axes (e.g., severity, morphology, orientation), serialized in Turtle (Terse RDF Triple Language) [24,25]. A companion SHACL (Shapes Constraint Language) specification constrains applicable properties, cardinalities, and value ranges, enabling automated validation and schema-driven model head configuration.

Contributions. In summary, the paper provides *(i)* 893 free-text reports for 529 ToothFairy3 cases and validated ontology instances, enabling learning and evaluation beyond segmentation; *(ii)* an OWL/RDF schema spanning 13 finding types and multiple qualifier axes, with ad-hoc SHACL shapes for validation and model head auto-configuration; and *(iii)* a multi-stage framework predicting finding *presence*, *anatomical slots*, and *properties*, with code for reproducibility.

2 The Proposed Dataset and Ground-Truth Construction

We build on the ToothFairy3 benchmark, which provides 532 public CBCT volumes of the maxillofacial region, each paired with expert ground-truth multi-

[1] An ontology is a clinician-defined, machine-readable model of the reporting domain. It lists relevant concepts (e.g., findings and anatomical entities) and restricts how they can be combined using relations and attributes (e.g., severity and morphology).

structure segmentation masks (77 segmentation classes) and tooth-level localization consistent with ISO 3950/FDI identifiers.
Our reports were authored by 2 radiologists and 6 maxillofacial surgeons with 5–20 years of experience in CBCT interpretation, who examined the anonymized CBCT volumes. No standardized reporting template was imposed; reports were primarily free-text narratives that followed routine clinical practice, typically including systematic evaluation of the dentition, periapical regions, and bone structures. In total, the generated dataset comprises *893 free-text reports* across 529 of the 532 cases (one or two per patient); three patients were excluded due to image-quality issues, making report authoring unreliable.

We convert free-text CBCT reports into structured supervision targets (ontology instances) that are directly learnable and automatically checkable. Concretely, each report becomes a patient-level RDF graph composed of atomic *finding* instances that encode *what* was observed (a *finding type*) and *where* it applies (i.e., tooth, side, or quadrant). Each finding can also include standardized *qualifier properties* that describe *how* it should be characterized (e.g., severity and morphology). To ensure consistency, we validate all graphs against SHACL constraints and perform clinician-guided iterative corrections. An example of a finding extracted from an ontology instance is provided in Algorithm 1. We suggest the reader explore a complete ontology instance for a broader understanding.[2]

Target Representation: Ontology-Grounded Findings (*what* and *where*). Findings are the smallest reportable CBCT observations in our target space. In our scenario, each finding has *exactly one* finding type from a *closed set of 13* clinically validated categories. Anatomical grounding is encoded via the corresponding slot (*where*). Bilateral observations are always split into two independent finding instances, i.e., left and right. Tooth identifiers support multi-tooth entities, e.g., a bridge spanning teeth 14–16 carries three `:hasToothId` assertions, each activating its own slot bit independently.

Algorithm 1: Instance excerpt for `Patient_3F_004__Finding_1`

rdfs:comment

```
Mandibular canal with a predominantly lingual course, in close
relationship with the roots of dental elements 37 and 38.
```

rdf:properties

```
:findingType → :MandibularCanalCourse
:onSide → :Left
:hasCanalTrajectory → :PrevalentlyLingual
:hasCanalToothContact → :IntimateContact
:hasToothId → "37", "38"
```

[2] https://github.com/AImageLab-zip/CBCT-Report/blob/main/data/ttl_reports/ToothFairy3F_001.ttl.

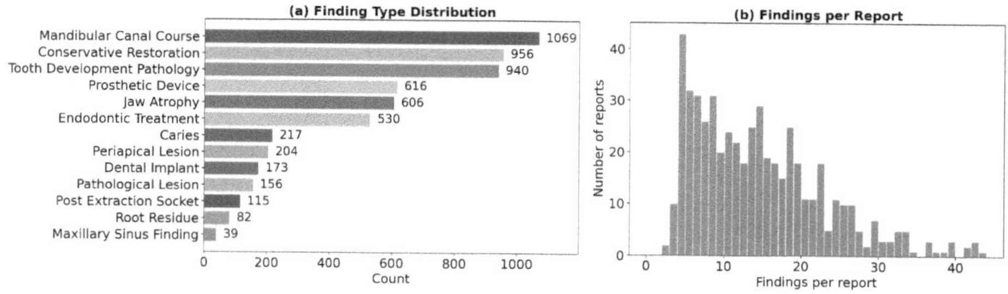

Fig. 1. Distribution of the number of extracted findings per patient and the distribution of each finding type in the dataset.

Anatomical Properties (*how*). They enforce single-valued semantics (e.g., `:hasSeverity` ∈ {`Mild`, `Moderate`, `Severe`}; `:hasRadiodensity` ∈ {`Radiolucent`, `Radiopaque`, `MixedDensity`}), or multi-label assertions (e.g., `:hasMorphology`, `:hasPeriodontalFeature`).

2.1 SHACL Validation and Automated Extraction

SHACL (Shapes Constraint Language) is a standard used to validate RDF graph data against a set of conditions or constraints. Our SHACL core shape requires exactly one `:findingType`, constrains qualifier value ranges, and validates tooth IDs using an FDI regex. Additional shapes restrict qualifiers by finding type and specific conditional requirements.

In our context, ground-truth instances are produced from free-text reports using GPT-5.2. For each patient, the prompt includes: *(i)* the OWL/RDF ontology schema defined by clinicians, *(ii)* the raw report text (both reports when available) with instructions to consolidate duplicates, *(iii)* three few-shot examples authored by clinicians, and *(iv)* explicit extraction rules. The model is tasked to output pure Turtle code, i.e., a syntax for defining RDF graphs. Outputs are parsed with `rdflib` and validated with `pySHACL`. Each extracted finding includes an `rdfs:comment` that quotes the source phrase, enabling targeted auditing. The extracted ontology instances (our labels) are released alongside free-text reports and the final validated ontology instances, enabling future researchers to use the labels without re-running the LLM extraction pipeline.

Supervised Iterative Refinement. We employ an iterative and interactive extraction process: each SHACL violation (i.e., an output that is non-compliant with the defined rules/shapes) was reviewed by a dental clinician who either corrected a report typo, updated the ontology constraint, or manually authored the instance for that patient. When systematic errors were identified, the prompt and few-shot examples were updated, and affected cases were re-extracted. Iterations continued until no extraction errors remained.

After the SHACL-based structural validation process, all generated ontology instances were further manually verified by clinicians, leveraging the free-text

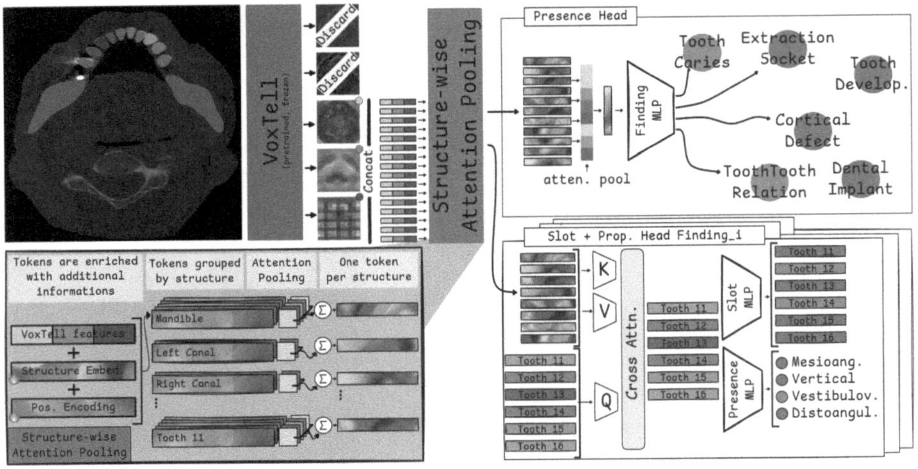

Fig. 2. Overview of the proposed pipeline (Sect. 3). Frozen VoxTell features are tokenized per anatomical structure and pooled into one embedding per structure. A presence head detects which finding types are active; per-finding-type slot heads localize each finding to its anatomical index (tooth/side/quadrant) and predict its qualifiers.

report excerpt from which the finding was extracted. Notably, the final structured graphs represent *clinician-validated annotations rather than raw LLM outputs*.

Dataset Statistics and Label Distribution. Across successfully parsed cases, the extraction pipeline yields a total of *5,703* finding instances over the *13* finding types, exhibiting strong class imbalance consistent with clinical prevalence. Figure 1 shows all selected types and their distribution.

Clinical Agreement. To quantify human agreement, we separately applied the above-described free-text-to-ontology automatic conversion pipeline to the subset of 358 patients for whom at least two independent reports were available. We refer the reader to Sect. 4, where the evaluation metrics and the corresponding agreement scores are reported.

3 Methods

The proposed ontology allows us to reframe the free-text report generation task into a "simpler" classification problem. To this aim, we develop and propose a model that maps a CBCT volume and its multi-structure segmentation to an ontology-grounded structured report through four pipeline stages: *(i)* frozen multi-scale feature extraction, *(ii)* segmentation-guided regional tokenization, *(iii)* a trainable encoder producing one embedding per anatomical structure, and *(iv)* ontology-driven prediction heads for finding presence (*what*), slot allocation (*where*), and property prediction (*how*). The pipeline is depicted in Fig. 2.

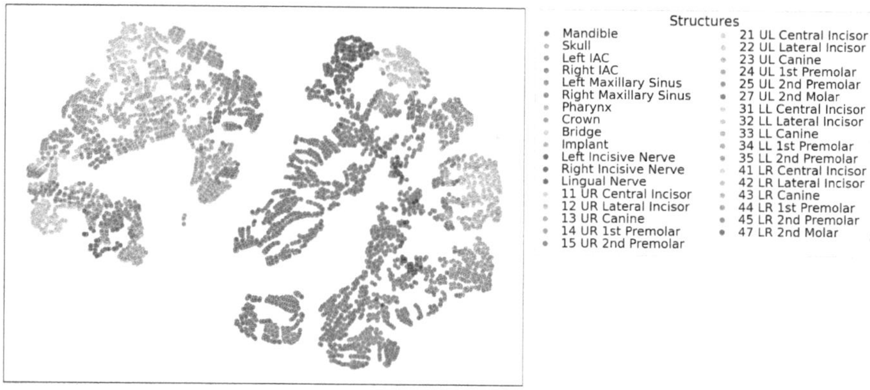

Fig. 3. t-SNE applied to extracted tokens after applying max-pooling to the largest structures. Features used are the concatenation of VoxTell Stages 3, 4, and 5.

Feature Extraction and Tokenization. We adopt VoxTell [22] as a pretrained and frozen 3D encoder. Features from encoder levels 3, 4, and 5 are upsampled to the level-3 spatial grid and concatenated along the channel dimension (C=896), combining coarse semantic content (level 5) with finer spatial detail (level 3). The ground-truth segmentation mask from the ToothFairy3 dataset is downsampled to this grid via adaptive max-pooling to preserve structure labels.

This provides typically ~10,000 structure-aligned tokens per patient. Figure 3 shows a t-SNE visualization of extracted structure-level tokens, confirming that embeddings form semantically coherent clusters by anatomical category.

Structure Encoder. Tokens are then projected to d=256, enriched with a learned structure-type embedding (8 coarse anatomical categories) and a spatial positional encoding, then grouped by structure ID. Within each group, additive attention pooling collapses the tokens into a single structure embedding:

$$\mathbf{s}_\ell = \sum_{n \in \mathcal{I}_\ell} \alpha_{\ell,n} \, \tilde{\mathbf{x}}_n, \quad \alpha_{\ell,n} = \mathrm{softmax}_{n \in \mathcal{I}_\ell}\big(\mathbf{v}^\top \tanh(\mathbf{W}\tilde{\mathbf{x}}_n)\big). \tag{1}$$

Here, ℓ indexes an anatomical structure and $\mathcal{I}_\ell$ denotes the set of token indices belonging to structure ℓ; $\tilde{\mathbf{x}}_n \in \mathbb{R}^d$ is the embedded feature vector of token n. The pooled structure representation $\mathbf{s}_\ell \in \mathbb{R}^d$ is obtained as an attention-weighted sum of its tokens, where $\alpha_{\ell,n}$ is the normalized attention weight computed via additive attention: $\mathbf{W}$ is a learnable projection matrix, $\tanh(\cdot)$ is applied element-wise, and $\mathbf{v}$ is a learnable vector producing a scalar score per token. The softmax is taken over $n \in \mathcal{I}_\ell$ so that $\sum_{n \in \mathcal{I}_\ell} \alpha_{\ell,n} = 1$. The resulting set of tokens is defined as $\mathbf{H} \in \mathbb{R}^{B \times S \times 256}$.

Ontology-Driven Prediction Heads. Head configuration is derived automatically from our SHACL schema at each stage.

Stage 1—finding presence (what). A global attention pool over $\mathbf{H}$ produces a patient-level embedding, from which a small MLP predicts 13 sigmoid logits.

Stage 2—slot allocation (where). For each finding type f with K_f ontology-defined slots, slot embeddings are computed via cross-attention to $\mathbf{H}$ and a binary activation is predicted per slot:

$$\mathbf{Z}_f = \mathrm{CrossAttn}(\mathbf{Q}_f, \mathbf{H}, \mathbf{H}) \in \mathbb{R}^{B \times K_f \times 256}, \qquad \hat{y}^{\mathrm{slot}}_{f,k} = \sigma(\mathbf{w}_f^\top \mathbf{Z}_{f,k} + b_f). \quad (2)$$

The slot queries $\mathbf{Q}_f$ follow the ontology indexing scheme: tooth-indexed finding types use K_f=32 slots (one per FDI tooth 11–48) with queries initialized from the corresponding tooth embeddings in $\mathbf{H}$. Side-indexed types (K_f=2) and quadrant-indexed types (K_f=4) use fully learned queries.

Stage 3—property prediction (how). For each active (f, k) slot, a per-property MLP maps $\mathbf{Z}_{f,k}$ to qualifier predictions: softmax for single-choice categorical properties, sigmoid for multi-label. During training, losses are applied only to ground-truth active slots.

Training Objective. We minimize a weighted sum of per-stage losses:

$$\mathcal{L} = \mathcal{L}_{\mathrm{pres}} + \mathcal{L}_{\mathrm{slot}} + \mathcal{L}_{\mathrm{prop}}. \quad (3)$$

$\mathcal{L}_{\mathrm{pres}}$ and $\mathcal{L}_{\mathrm{slot}}$ use focal-BCE (γ=0.5) with per-class positive weights; $\mathcal{L}_{\mathrm{slot}}$ additionally applies label smoothing (ϵ=0.1). $\mathcal{L}_{\mathrm{prop}}$ uses cross-entropy for categorical and BCE for multi-label properties, both with label smoothing (ϵ=0.1).

4 Experiments

Evaluation Protocol. A clinically useful structured report must simultaneously satisfy *detection*, *localization*, and *characterization* axes. To this aim, we evaluate performance at four levels, aligned with the staged task:
Finding-Type Detection (pres-F1). Macro F1 over the 13 finding types, ignoring localization and properties.

Instance Localization (inst-F1). Macro F1 over ⟨finding type, slot⟩ pairs. A predicted slot activation is a true positive only if both the finding type and the anatomical slot match a ground-truth target exactly.

Characterization Accuracy (char-acc). Mean per-property accuracy over true-positive localized instances, isolating characterization from upstream errors.

Patient-Level Report Quality (rep-F1). Per-patient F1 between the predicted and ground-truth sets of ⟨finding type, slot⟩ pairs, averaged across patients.

To decouple pipeline stages for ablation, we also report *slot-F1* (slot-activation F1 conditioned on ground-truth present finding types) and *prop-F1* (property F1 conditioned on ground-truth active slots). These GT-conditioned metrics isolate localization and characterization quality independently of upstream errors.

Training Setup. We use all 529 ToothFairy3 volumes with successfully extracted ground-truth. A fixed held-out test set (20%, ~107 volumes) is isolated via stratified sampling over finding types, slots, and properties; the remaining

Table 1. Effect of segmentation source on full-model performance (5-fold CV means ± std, fixed held-out test set). GT segmentations use ToothFairy3 masks; predicted segmentations are obtained from an nnU-Net trained on the same split.

Segmentation	pres-F1	slot-F1	prop-F1	inst-F1	char-acc	rep-F1
nnU-Net	71.2 ± 0.6	51.7 ± 2.2	50.1 ± 0.6	43.3 ± 0.9	51.0 ± 0.6	49.8 ± 1.7
GT	72.0 ± 1.0	53.4 ± 1.1	61.2 ± 1.5	46.7 ± 0.8	85.4 ± 0.7	53.8 ± 0.9

80% (~425 volumes) is partitioned into 5 folds, stratified by finding-type prevalence. Each fold trains on four-fifths of the pool (~340 samples). We report means over folds. Training uses AdamW [15] with cosine LR schedule ($\eta_{peak}=10^{-3}$, $\eta_{\min}=10^{-5}$), batch size 32, dropout $p=0.2$, and early stopping on validation pres-F1 (patience 50, max 150 epochs). Post-hoc per-class threshold calibration is applied on the validation fold before reporting test metrics.

Segmentation Sensitivity. Table 1 reports model performance when leveraging automatic segmentation produced by an nnU-Net trained on the same split *vs* the GT segmentation masks. The small gap between these two variations (pres-F1: 71.2 *vs.* 72.0; rep-F1: 49.8 *vs.* 53.8) indicates that the pipeline is robust to imperfect structure delineation and quantifies the contribution of segmentation quality to reporting performance.

Stage-Decoupling Analysis. In the same table (Table 1), distinct behaviors at each prediction stage are reported. Under GT-conditioned evaluation, slot-F1 = 53.4 and prop-F1 = 61.2 confirm that, given correct upstream inputs, both localization and qualifier prediction perform reliably. In a fully-predicted evaluation (segmentation predicted), inst-F1 drops to 43.3 due to cascading detection errors. Notably, with GT segmentation, char-acc is high (85.4): if a finding is correctly detected and localized, its clinical properties are predicted with high fidelity. This identifies *presence detection as the primary deployment bottleneck*, not characterization.

Multiscale Feature Ablation. Table 2 compares individual VoxTell encoder levels against the full multi-scale concatenation. Level 5 alone performs worst: at $8\times16\times16$ resolution, structure-masked pooling collapses the localization signal. Levels 3 and 4 retain finer spatial detail and perform substantially better. Concatenating all three levels yields the best results, confirming that high-level semantic content and fine spatial resolution are complementary.

Clinical Agreement. Human agreement is high overall for finding detection (pres-F1 0.81±0.16) and characterization (char-acc. 0.83±0.13) and remains substantial for structured report matching (rep-F1 0.67±0.21) and instance localization (inst-F1 0.70±0.20).

5 Conclusion

We introduce the first ontology-grounded CBCT report resource, with 893 clinical reports and clinician-validated structured labels spanning 13 finding types.

Table 2. Effect of VoxTell encoder level on full-model performance (5-fold CV means). Single-level inputs use features of that level only; concat uses levels 3+4+5.

Features	pres-F1	slot-F1	prop-F1	inst-F1	char-acc	rep-F1
Lv. 3	69.4 ± 1.8	52.4 ± 1.4	58.6 ± 1.7	44.1 ± 1.3	84.2 ± 0.9	51.1 ± 2.2
Lv. 4	70.1 ± 1.5	52.8 ± 1.2	59.4 ± 1.6	44.9 ± 1.2	84.6 ± 0.8	51.9 ± 1.9
Lv. 5	60.1 ± 3.1	48.7 ± 2.8	53.6 ± 2.2	37.2 ± 1.9	80.7 ± 1.5	43.1 ± 4.1
Concat	$\mathbf{72.0 \pm 1.3}$	$\mathbf{53.4 \pm 1.1}$	$\mathbf{61.2 \pm 1.5}$	$\mathbf{46.7 \pm 0.8}$	$\mathbf{85.4 \pm 0.7}$	$\mathbf{53.8 \pm 0.9}$

We also propose a hierarchical metric suite that decouples detection, localization, and characterization, replacing surface-level text similarity with clinically interpretable scores, and a lightweight ontology-driven baseline. Overall, the released dataset, ontology, metrics, and baseline provide a rigorous foundation for moving dental CBCT AI beyond segmentation toward clinically actionable interpretation.

Acknowledgments. This project has received funding from Fondazione di Modena, through FAR 2024 (E93C24002080007).

Disclosure of Interests. The authors have no competing interests to declare that are relevant to the content of this article.

References

1. Babar, Z., van Laarhoven, T., Zanzotto, F.M., Marchiori, E.: Evaluating diagnostic content of AI-generated radiology reports of chest X-rays. Artif. Intell. Med. **116** (2021)
2. Banerjee, S., Lavie, A.: METEOR: an automatic metric for MT evaluation with improved correlation with human judgments. In: Proceedings of the ACL Workshop on Intrinsic and Extrinsic Evaluation Measures for Machine Translation and/or Summarization, (2005)
3. Bannur, S., et al.: MAIRA-2: grounded Radiology Report Generation. arXiv preprint arXiv:2406.04449 (2024)
4. Boag, W., Kané, H., Rawat, S., Wei, J., Goehler, A.: A pilot study in surveying clinical judgments to evaluate radiology report generation. In: Proceedings of the 2021 ACM Conference on Fairness, Accountability, and Transparency (2021)
5. Bolelli, F., et al.: Multi-structure segmentation in CBCT volumes: the toothfairy2 challenge. Med. Image Anal. (2026)
6. Bolelli, F., et al.: Segmenting the inferior alveolar canal in CBCT volumes: the toothfairy challenge. IEEE Trans. Med. Imaging (2024)
7. Bolelli, F., et al.: Segmenting maxillofacial structures in CBCT volumes. In: Proceedings of the IEEE/CVF Conference on Computer Vision and Pattern Recognition (2025)
8. Gamal, M., Baraka, M., Torki, M.: Automatic mandibular semantic segmentation of teeth pulp cavity and root canals, and inferior alveolar nerve on pulpy3D dataset. In: Medical Image Computing and Computer Assisted Intervention – MICCAI 2024 (2024)

9. ISO: Dentistry - Designation system for teeth and areas of the oral cavity (2016). https://www.iso.org/standard/68292.html. Accessed 11 Nov 2024
10. Jain, S., et al.: RadGraph: Extracting Clinical Entities and Relations from Radiology Reports. arXiv preprint arXiv:2106.14463 (2021)
11. Jaskari, J., et al.: Deep learning method for mandibular canal segmentation in dental cone beam computed tomography volumes. Sci. Rep. **10**(1) (2020)
12. Lahoud, P., et al.: Artificial intelligence for fast and accurate 3-dimensional tooth segmentation on cone-beam computed tomography. J. Endodontics **47**(5) (2021)
13. Lin, C.Y.: ROUGE: a package for automatic evaluation of summaries. In: Text Summarization Branches Out (2004)
14. Liu, Y., et al.: MRScore: evaluating medical report with LLM-based reward system. In: Medical Image Computing and Computer Assisted Intervention – MICCAI 2024 (2024)
15. Loshchilov, I., Hutter, F.: Decoupled Weight Decay Regularization. arXiv preprint arXiv:1711.05101 (2017)
16. Lumetti, L., et al.: Taming Mambas for 3D Medical Image Segmentation, IEEE Access (2025)
17. Lumetti, L., Pipoli, V., Bolelli, F., Ficarra, E., Grana, C.: Enhancing patch-based learning for the segmentation of the mandibular canal. IEEE Access **12** (2024)
18. Lumetti, L., Pipoli, V., Bolelli, F., Ficarra, E., Grana, C.: Location matters: harnessing spatial information to enhance the segmentation of the inferior alveolar canal in CBCTs. In: 27th International Conference on Pattern Recognition (ICPR) (2024)
19. Lumetti, L., et al.: ToothFairy3: scaling cbct maxillofacial segmentation to 77 classes with U-Mamba2. In: Medical Image Computing and Computer Assisted Intervention – MICCAI 2026 (2026)
20. van Nistelrooij, N., et al.: ToothSeg: robust tooth instance segmentation and numbering in CBCT using deep learning and self-correction. IEEE J. Biomed. Health Inform. (2025)
21. Papineni, K., Roukos, S., Ward, T., Zhu, W.J.: Bleu: a method for automatic evaluation of machine translation. In: Proceedings of the 40th Annual Meeting of the Association for Computational Linguistics (2002)
22. Rokuss, M., et al.: VoxTell: free-text promptable universal 3D medical image segmentation. In: Proceedings of the IEEE/CVF Conference on Computer Vision and Pattern Recognition (2026)
23. Usman, M., et al.: Dual-stage deeply supervised attention-based convolutional neural networks for mandibular canal segmentation in CBCT scans. Sensors **22**(24) (2022)
24. W3C OWL Working Group: OWL 2 Web Ontology Language Document Overview (2009)
25. World Wide Web Consortium: RDF 1.1 Turtle: terse RDF triple language (2014)
26. Yu, F., et al.: Evaluating progress in automatic chest X-ray radiology report generation. Patterns **4**(9) (2023)

OPVD: On-Policy Chain-of-Visual-Thought Distillation for Medical Reasoning

Ningyue Zhang[1] and Qiushi Yang[2(✉)]

[1] Department of Engineering, The Hong Kong University of Science and Technology, Hong Kong SAR, China

[2] Department of Electrical Engineering, City University of Hong Kong, Hong Kong SAR, China

qsyang2-c@my.cityu.edu.hk

Abstract. Recent advancements in enhancing the medical reasoning capabilities of Vision-Language Models (VLMs) have predominantly relied on rule-based reinforcement learning (RL) paradigms. However, such approaches are often constrained by their dependence on *sparse answer-level rewards*, which fail to provide dense supervisory signals for intermediate reasoning steps, and by the *computational inefficiency inherent* in requiring extensive sampling for policy exploration. To address these limitations, we propose *On-Policy Chain-of-Visual-Thought Distillation* (OPVD) for vision-language medical reasoning, a concise yet efficient framework designed to imbue VLMs with robust medical reasoning abilities. OPVD introduces an interleaved vision-language Chain-of-Visual-Thought (CoVT) mechanism to construct reliable and granular reasoning trajectories. Unlike previous approaches, our OPVD leverages these trajectories to perform token-level dense supervision, effectively distilling multi-modal CoVT knowledge into the model via an on-policy self-distillation strategy where the teacher and student share the same policy model. This eliminates the need for external teachers or costly sampling procedures while maximizing data efficiency. Extensive experiments conducted on the public multi-modal medical reasoning dataset demonstrate that OPVD remarkably outperforms existing models trained via standard supervised fine-tuning and RL methods, establishing a new state-of-the-art in efficient and accurate medical visual reasoning.

Keywords: On-Policy Distillation · Medical Reasoning · Visual-Language Model · Chain-of-Thought

1 Introduction

Vision-language medical reasoning [6,10,16] represents a cornerstone of modern clinical decision support systems, bridging the gap between complex multi-modal patient data and actionable diagnostic insights. The ability to simultaneously interpret medical imagery, such as MRI, CT or X-ray, and correlate these

G. Yang et al. (Eds.): MICCAI 2026, LNCS 16878, pp. 418–427, 2027.
https://doi.org/10.1007/978-3-032-38059-3_40

visual findings with textual clinical information is critical for accurate diagnosis, treatment planning, and prognosis. As healthcare increasingly relies on artificial intelligence to augment human expertise, developing Vision-Language Models (VLMs) that possess robust, step-by-step reasoning capabilities akin to expert clinicians has become an urgent priority [2,10,12,16]. Such models must not only identify abnormalities but also articulate the logical deduction process that links visual evidence to medical conclusions, thereby enhancing trustworthiness and interpretability in high-stakes clinical environments.

Recent research in vision-language medical reasoning [3,6–10,14–16,19] has largely gravitated towards rule-based Reinforcement Learning (RL) paradigms to unlock the latent reasoning potential of VLMs. Prominent among these approaches is the application of Group Relative Policy Optimization (GRPO) and its variants [4,17], which have demonstrated remarkable success in stimulating complex reasoning chains. For instance, MedVLM-R1 [16] adapts GRPO to medical Visual Question Answering (VQA) tasks, encouraging the model to explore diverse Chain-of-Thought (CoT) trajectories to derive correct answers. Similarly, Med-R1 [10] systematically analyzes the impact of incorporating explicit CoT processes across various medical tasks, confirming that structured reasoning significantly boosts diagnostic accuracy. ViTAR [3] extends the Deep-Eyes [23] paradigm by encouraging the model to explore subtle local visual regions for reasoning, optimized via GRPO. While these RL-driven methods have set new benchmarks, they predominantly rely on trial-and-error exploration. Concurrently, although a nascent body of work in general Large Language Models (LLMs) has begun to explore distillation strategies [20,22] for efficiently enhancing reasoning, such methodologies remain largely unexplored within the specialized domain of medical reasoning, leaving a significant gap in efficient training frameworks for medical VLMs.

Despite the successes of rule-based RL methods, they suffer from two fundamental limitations that hinder their scalability and efficiency. First, these approaches typically optimize models using sparse rewards derived solely from the final answer correctness and format adherence. This lack of dense supervisory signals for intermediate reasoning steps often leads to inefficient optimization, where the model struggles to learn how to reason correctly even when it occasionally arrives at the right answer. Second, RL-based training necessitates extensive multiple sampling per instance to explore the policy space adequately. This requirement incurs prohibitive computational costs and can introduce training instability due to the high variance in reward estimates. Motivated by these challenges, we propose a paradigm shift: instead of relying on sparse feedback and exhaustive exploration, we enable the model to engage in selective self-learning during optimization. By injecting medical diagnosis-specific priors to construct reliable vision-language CoT trajectories and leveraging these trajectories for token-level dense supervision, we aim to harness the model's own capabilities more efficiently to elevate its reasoning proficiency.

To realize this vision, we present On-Policy Chain-of-Visual-Thought Distillation (OPVD), a novel framework that operationalizes dense supervision through an interleaved vision-language reasoning process. Our approach proceeds in a parallel branches within a shared-parameter policy model. First, acting as a teacher, the policy model is provided with the ground-truth answer alongside the input query and image. Guided by this prior, it generates a preliminary reasoning path and identifies critical local visual regions (e.g., subtle lesions). We then employ automated "crop-and-zoom" tools to magnify these regions, feeding them back as enhanced visual cues to generate a comprehensive, high-quality text-CoT trajectory and the final solution. Second, acting as a student, the same policy model processes the original input without answer priors. We impose an on-policy Chain-of-Visual-Thought (CoVT) distillation constraint that forces the student's generated reasoning tokens to align with the teacher's dense, step-by-step output at the token level. This strategy requires only two rounds of interleaved vision-text sampling per instance, yet effectively transfers reasoning knowledge via dense supervision, bypassing the need for costly multi-group exploration.

In summary, our main contributions are threefold:

- We propose OPVD, a novel on-policy self-distillation framework tailored for medical VLMs, which replaces sparse reward-based RL with token-level dense supervision derived from interleaved CoVT trajectories.
- We design a lightweight "crop-and-zoom" iterative reasoning mechanism that leverages ground-truth priors to construct reliable teacher trajectories, enabling efficient knowledge transfer with minimal sampling overhead compared to traditional RL methods.
- Extensive experiments on the public dataset demonstrate that OPVD outperforms both standard supervised fine-tuning and advanced RL-based competitors, establishing a new benchmark for efficiency and accuracy in multi-modal medical reasoning.

2 Approach

2.1 Overview

As illustrated in Fig. 1, we develop On-Policy chain-of-Visual-thought Distillation (OPVD) framework, which consists of two parallel branches instantiated from the same Vision-Language Models (VLMs): a *teacher policy model* π_T and a *student policy model* π_S. Crucially, both policies share the same model parameters θ, differing only in their conditioning contexts during training. The teacher branch acts as a guide with access to privileged information. Its inputs include the query Q, the original image I, and the ground-truth answer A. For cases requiring fine-grained visual understanding, a cropped and zoomed-in local visual region L is also generated and fed into the model. The teacher generates a comprehensive reasoning trajectory with the final answer prediction $\{\text{CoT}_{\text{com}}, \pi_T^a\} = \pi_T(Q, I, A, L)$. Simultaneously, the student branch operates under standard inference conditions, taking only the query Q and image I as input to produce its own trajectory with the predicted answer

$\{\mathrm{CoT_S}, \pi_S^a\} = \pi_S(Q, I)$. The core of our method lies in aligning the student's output with the high-quality trajectories generated by the teacher through token-level dense supervision, thereby achieving efficient knowledge transfer.

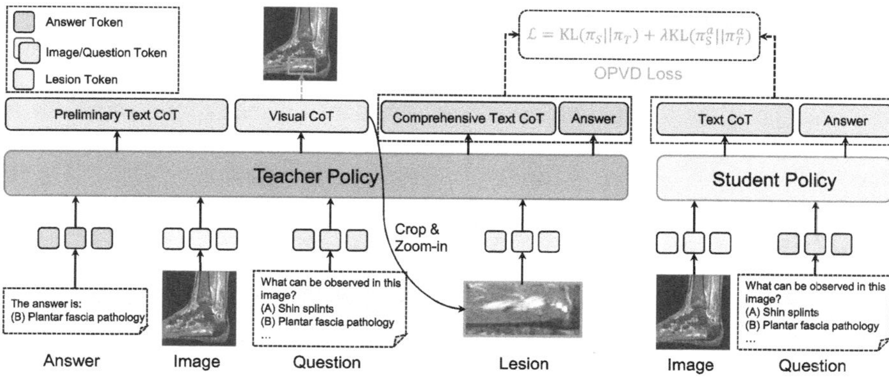

Fig. 1. Overview of OPVD, which employs a shared-parameter model acting as both teacher and student policy models. The teacher utilizes privileged answers and iterative visual zooming to construct reliable Chain-of-Visual-Thought (CoVT) trajectories. The student learns via on-policy distillation to match these trajectories without requiring external sampling or sparse rewards.

2.2 Chain-of-Visual-Thought (CoVT) Construction

To obtain reliable reasoning trajectories for distillation, we design a mechanism where the teacher model leverages both textual and visual priors to construct high-quality, interleaved vision-language CoT information. This process involves one or two forward passes depending on the complexity of the visual evidence. Specifically, in the first pass, the teacher receives the query Q, image I, and ground-truth answer A. The prompt includes a system instruction guiding the model to determine if specific local regions require re-examination. Based on this, the model generates a preliminary text CoT. If the model identifies difficult-to-discriminate regions (e.g., small lesions), it outputs the bounding box coordinates $\mathcal{B}$ for these areas. We then invoke an external tool to crop and zoom in on these regions from the original image I, resizing them to match the original input dimensions to form the local image L.

In the second pass (if triggered), L is concatenated with Q, I, and A and fed back into the teacher. This allows the model to refine its reasoning based on enhanced visual details, ultimately producing a comprehensive text CoT trajectory $\mathrm{CoT_{com}}$ and the predicted answer π_T^a. By interleaving text reasoning with adaptive visual refinement, the teacher constructs a robust trajectory that serves as the dense supervision target for the student.

2.3 On-Policy CoVT Distillation

To enable the model to continuously self-correct and improve its reasoning capabilities, we adopt an on-policy self-distillation strategy. Unlike traditional methods that rely on static datasets or multiple stochastic samples, our approach dynamically generates targets based on the current model state.

Let the student generate a response trajectory $\mathrm{CoT_S} \sim \pi_S(\cdot \mid Q, I)$. Both the teacher and student policies then evaluate this student-generated trajectory. At each token position n, they induce next-token distributions conditioned on the same prefix $\mathrm{CoT_S^{<n}} = (\mathrm{CoT_S^1}, \ldots, \mathrm{CoT_S^{n-1}})$. However, their conditioning contexts differ:

$$\pi_S(\mathrm{CoT_S^n} \mid Q, I, \mathrm{CoT_S^{<n}}) \quad \text{and} \quad \pi_T(\mathrm{CoT_S^n} \mid Q, I, A, L, \mathrm{CoT_S^{<n}}), \tag{1}$$

where $\pi_S(\cdot|\cdot)$ and $\pi_T(\cdot|\cdot)$ refers to the probability distributions over the vocabulary $\mathcal{V}$ for the student and teacher, respectively. Note that while the teacher has access to the privileged answer A and local image L to guide the generation, it evaluates the *student's* partial generation $\mathrm{CoT_S^{<n}}$ rather than its own optimal path. This setup encourages the teacher to provide informed guidance on how to correct or continue the student's current reasoning step, effectively creating a dynamic curriculum. The learning objective minimizes the divergence between these two distributions along the student's rollout, allowing gradients to backpropagate only through the student's logits.

2.4 Optimization and Inference

During the optimization phase, we calculate the full-vocabulary divergence distillation loss for both the reasoning trajectory and the final answer prediction. To ensure the model prioritizes diagnostic accuracy, we apply a weighting coefficient to the loss term associated with the final answer. The total loss function $\mathcal{L}_{total}$ is defined as:

$$\mathcal{L}_{\mathrm{OPVD}} = \mathrm{KL}(\pi_S(\cdot \mid Q, I) \parallel \pi_T(\cdot \mid Q, I, A, L)) + \lambda \mathrm{KL}(\pi_S^a \parallel \pi_T^a), \tag{2}$$

where $\mathrm{KL}(\cdot||\cdot)$ denotes the Kullback-Leibler divergence, π_S^a and π_T^a represent the distributions over the final answer tokens, and λ is a hyperparameter controlling the weight of the answer alignment. Note that gradients are backpropagated only to the student policy for update, while the teacher policy shares parameters with the student initially but remains frozen throughout training.

At the inference stage, only the student policy is utilized. Given an input image and query, the model performs a single forward pass to generate the chain-of-thought reasoning and the final diagnosis. Since the teacher branch and the iterative zooming mechanism are only active during training, our method introduces no additional computational overhead or latency during deployment compared to standard VLM inference.

Table 1. Results of VQA-VLMs on MRI (in-domain), and CT and X-Ray (out-of-domain) modalities.

Methods	Sampling Num.	In-Domain/Out-of-Domain			Average
		MRI	CT	X-ray	
Zero-Shot VLM					
Qwen2-VL-2B [18]	–	61.67	50.67	53.00	55.11
Qwen3-VL-2B [1]	–	67.28	55.80	58.62	60.57
Qwen2-VL-7B [18]	–	72.33	68.67	66.63	69.21
Qwen2-VL-72B [18]	–	68.67	60.67	72.33	67.22
Huatuo-GPT-vision-7B [2]	–	71.00	63.00	73.66	69.22
MRI RL trained VLM					
MedVLM-R1-2B [16]	6	96.83	72.03	70.30	79.72
Med-R1-2B [10]	4	96.11	71.26	70.09	79.15
ViTAR [3]	16	97.30	72.48	70.78	80.19
MRI Distillation trained VLM					
Qwen2-VL-2B-SFT	-	95.05	55.26	35.10	61.80
Qwen2-VL-2B-OPVD	2	**97.84**	**73.61**	**71.44**	**80.96**
w/o CoVT	1	97.25	72.33	70.72	80.10
Qwen3-VL-2B-SFT	–	96.80	56.76	36.74	63.43
Qwen3-VL-2B-OPVD	2	**98.54**	**73.77**	**72.13**	**81.48**
w/o CoVT	1	98.32	73.10	71.56	80.99

3 Experiments

3.1 Datasets and Setup

Dataset. Following the experimental protocol established by MedVLM-R1 [16], we conduct our evaluations on the publicly available HuatuoGPT-Vision dataset [2]. This benchmark aggregates data from several prominent medical VQA sources, including VQA-RAD [11], SLAKE [13], PathVQA [5], OmniMedVQA [6], and PMC-VQA [21], comprising over 17,300 multiple-choice questions across diverse imaging modalities. Since MedVLM-R1 did not release its specific training splits, we construct our own subsets focusing on radiology modalities (CT, MRI, and X-ray). Specifically, we selected 2,000 MRI image-question pairs for training. For evaluation, we reserved a total of 900 pairs: 300 MRI, 300 CT, and 300 X-ray samples. In this setup, the MRI test set serves as the in-domain evaluation benchmark, while the CT and X-ray sets are utilized to assess out-of-distribution (OOD) generalization capabilities.

Implementation Details. We employ Qwen2-VL-2B [1] and Qwen3-VL-2B as the VLM backbones for all experiments. The models are trained using the PyTorch framework on NVIDIA A100 GPUs for 3,000 iterations. Key hyper-parameters include an initial learning rate of $1e-6$, a global batch size of 64, and a distillation weight λ set to 5.0. To optimize computational efficiency during the calculation of KL divergence, we restrict the computation to the top 128 vocabulary tokens with the highest probabilities predicted by the student model.

Prompt Template and Evaluation Metrics. In the CoVT construction phase, the teacher policy model receives a structured text prompt (illustrated in Fig. 3) containing the input image, the query, and the ground-truth answer as a prior. For instance, given an image and the query "What can be observed in this image? (A) Shin splints, (B) Tennis elbow, (C) Plantar fascia pathology, (D) Meniscal tear" along with the ground truth "The answer is: (C) Plantar fascia pathology" the model is guided to generate the intermediate reasoning trajectory. For evaluation, we use standard multiple-choice accuracy: 1 point for the correct option letter, 0 otherwise; results are reported as correct ratio.

3.2 Comparison Results

We evaluate the in-domain and out-of-domain (OOD) generalization performance by training on MRI data and testing on MRI, CT, and X-ray datasets. As shown in Table 1, our OPVD framework outperforms larger zero-shot models. Specifically, when fine-tuning Qwen2-VL-2B, our method achieves accuracies of 97.84%, 73.61%, and 71.44% on the MRI, CT, and X-ray test sets, respectively. This surpasses the second-best comparison, ViTAR, by margins of 0.54%, 1.13%, and 0.66%. Notably, while ViTAR requires 16 sampling iterations per step, incurring eight times our computational cost, OPVD maintains superior efficiency with minimal sampling. Furthermore, consistent improvements over standard SFT are observed across both Qwen2-VL-2B and Qwen3-VL-2B backbones, validating the model-agnostic nature of our approach. These results confirm that our on-policy strategy for structuring visual-language thought processes effectively enhances reasoning capabilities with high data efficiency. From Fig. 2, we show two examples of medical VQA generated by our OPVD, which demonstrates that our method can produce reliable, interpretable reasoning processes and yield correct answers.

3.3 Ablation Studies

Effectiveness of Chain-of-Visual-Thought. To investigate the contribution of local visual cues within the Chain-of-Visual-Thought (CoVT), we conduct an ablation study by removing the visual zoom-in mechanism. In this variant, the teacher policy generates a preliminary text-only CoT based solely on the answer context for distillation. As detailed in Table 1 (w/o CoVT), upon Qwen2-VL-2B, eliminating the visual component resulted in performance drops of 0.59%, 1.28%, and 0.72% on the MRI, CT, and X-ray test sets, respectively. This degradation underscores the critical role of interleaved visual information in constructing reliable reasoning trajectories, thereby boosting overall inference accuracy.

Question:
What can be observed in this image? (A) Shin splints, (B) Tennis elbow, (C) Plantar fascia pathology, (D) Meniscal tear

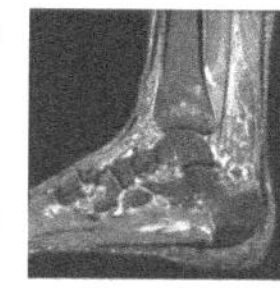

Response:
Reasoning Process: The image is a sagittal MRI of the ankle/foot region, showing the calcaneus, talus, and plantar soft tissues. There is increased signal intensity along the plantar fascia origin at the medial calcaneal tuberosity, consistent with plantar fasciitis or fascial pathology.
Answer: (C) Plantar fascia pathology

Question:
What is the designation of the anomaly shown in the visual representation? (A) Nodule formation, (B) Pulmonary embolism, (C) Airspace opacity, (D) Pleural effusion

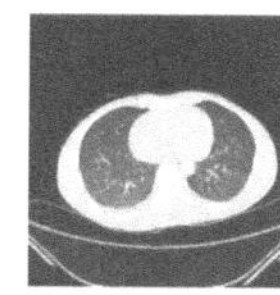

Response:
Reasoning Process: The CT image shows bilateral, patchy, ill-defined areas of increased attenuation in the lung parenchyma—consistent with airspace (alveolar) opacities, such as in pneumonia or pulmonary edema.
Answer: (C) Airspace opacity

Fig. 2. Example of medical VQA generated by OPVD, highlighting its ability to generate reliable, interpretable reasoning and accurate answers.

Influence of Answer Re-Weighting Term λ. We examine the impact of the hyperparameter λ, which balances the distillation loss between the reasoning process and the final answer. Experiments were conducted with $\lambda \in \{1.0, 2.0, 5.0, 10.0\}$, as summarized in Table 2. Optimal performance is achieved at $\lambda = 5.0$. Deviating from this value proved detrimental: insufficient weighting ($\lambda < 5.0$) causes objective misalignment, while excessive emphasis on the answer ($\lambda > 5.0$) impairs the model's ability to learn robust intermediate reasoning steps. This suggests that a balanced focus on both process and outcome is essential for effective knowledge distillation.

Text Prompt Template

Question:
What can be observed in this image? (A) Shin splints, (B) Tennis elbow, (C) Plantar fascia pathology, (D) Meniscal tear

The ground truth answer is: (C) Plantar fascia pathology

Analyze the provided image, question, and ground truth answer. First, deduce the logical reasoning steps required to bridge the visual evidence with the answer.

If specific target regions in the image are ambiguous, difficult to identify, or critical for understanding the reasoning, output the bounding boxes in JSON format.

Output Format:
1. Reasoning Process: [Your step-by-step deduction]
2. Critical Regions (if any): [JSON list of bboxes or "None"]

Fig. 3. Text prompt template of the teacher policy model at the first round. Similarly, the template at the second round incorporates additional cropped and zoomed-in local regions to generate final reasoning process and the answer.

Table 2. Ablations on distillation part re-weighting term λ

λ	MRI	CT	X-ray	Average
1.0	98.06	73.25	71.54	80.95
2.0	98.21	73.37	71.66	81.08
5.0	98.54	73.77	72.13	81.48
10.0	98.09	73.27	71.57	80.98

4 Conclusion

In this work, we propose OPVD, an efficient framework that addresses sparse rewards and high computation in RL-based medical VLMs. By using interleaved vision-language trajectories for token-level dense supervision via on-policy self-distillation, OPVD enables robust reasoning—without external teachers or heavy sampling. Experiments on public benchmarks show OPVD significantly surpasses SFT and RL baselines in efficient, accurate medical visual reasoning.

Disclosure of Interests. The authors have no competing interests to declare that are relevant to the content of this article.

References

1. Bai, J., et al.: Qwen-vl: a versatile vision-language model for understanding, localization, text reading, and beyond. arXiv preprint arXiv:2308.12966 (2023)
2. Chen, J., et al.: Huatuogpt-vision, towards injecting medical visual knowledge into multimodal LLMs at scale (2024)
3. Chen, K., et al.: Think twice to see more: iterative visual reasoning in medical VLMs. arXiv preprint arXiv:2510.10052 (2025)
4. Guo, D., et al.: Deepseek-r1: incentivizing reasoning capability in LLMs via reinforcement learning. arXiv preprint arXiv:2501.12948 (2025)
5. He, X., Zhang, Y., Mou, L., Xing, E., Xie, P.: Pathvqa: 30000+ questions for medical visual question answering. arXiv preprint arXiv:2003.10286 (2020)
6. Hu, Y., et al.: Omnimedvqa: a new large-scale comprehensive evaluation benchmark for medical LVLM. In: CVPR, pp. 22170–22183 (2024)
7. Huang, X., Wu, J., Liu, H., Tang, X., Zhou, Y.: Medvlthinker: simple baselines for multimodal medical reasoning. arXiv preprint arXiv:2508.02669 (2025)
8. Huang, Z., Mu, L., Zhu, Y., Zhao, X., Zhang, S., Zhang, X.: Elicit and enhance: advancing multimodal reasoning in medical scenarios. arXiv preprint arXiv:2505.23118 (2025)
9. Kumar, A., et al.: Training language models to self-correct via reinforcement learning. arXiv preprint arXiv:2409.12917 (2024)
10. Lai, Y., et al.: Med-r1: reinforcement learning for generalizable medical reasoning in vision-language models. In: IEEE TMI. IEEE (2026)
11. Lau, J.J., Gayen, S., Ben Abacha, A., et al.: A dataset of clinically generated visual questions and answers about radiology images. Sci. Data **5**(1), 1–10 (2018)

12. Li, C., et al.: Llava-med: training a large language-and-vision assistant for biomedicine in one day. Adv. Neural. Inf. Process. Syst. **36**, 28541–28564 (2023)
13. Liu, B., Zhan, L.M., Xu, L., Ma, L., Yang, Y., Wu, X.M.: Slake: a semantically-labeled knowledge-enhanced dataset for medical visual question answering. In: ISBI, pp. 1650–1654 (2021)
14. Liu, C., et al.: Beyond distillation: pushing the limits of medical LLM reasoning with minimalist rule-based RL. arXiv preprint arXiv:2505.17952 (2025)
15. Ouyang, L., et al.: Training language models to follow instructions with human feedback. NeurIPS **35**, 27730–27744 (2022)
16. Pan, J., et al.: Medvlm-r1: incentivizing medical reasoning capability of vision-language models (VLMs) via reinforcement learning. In: MICCAI, pp. 337–347. Springer (2025)
17. Shao, Z., et al.: Deepseekmath: pushing the limits of mathematical reasoning in open language models. arXiv preprint arXiv:2402.03300 (2024)
18. Wang, P., et al.: Qwen2-vl: enhancing vision-language model's perception of the world at any resolution. arXiv preprint arXiv:2409.12191 (2024)
19. Xu, W., et al.: Lingshu: a generalist foundation model for unified multimodal medical understanding and reasoning. arXiv preprint arXiv:2506.07044 (2025)
20. Ye, T., Dong, L., Wu, X., Huang, S., Wei, F.: On-policy context distillation for language models. arXiv preprint arXiv:2602.12275 (2026)
21. Zhang, X., et al.: PMC-VQA: visual instruction tuning for medical visual question answering. arXiv preprint arXiv:2305.10415 (2023)
22. Zhao, S., et al.: Self-distilled reasoner: on-policy self-distillation for large language models. arXiv preprint arXiv:2601.18734 (2026)
23. Zheng, Z., et al.: Deepeyes: incentivizing "thinking with images" via reinforcement learning. In: arXiv preprint arXiv:2505.14362 (2025)

Precision Recall Controllable Radiology Report Generation via Hybrid Natural Language and Clinical Reward Learning

Ling Chen[1], Ruinan Jin[1], Jun Luo[1], Hanliang Chen[1], Quirin Strotzer[2], Rongkai Yan[1], Yuan Xue[1], Luciano Prevedello[1], and Dufan Wu[1(✉)]

[1] The Ohio State University, Columbus, OH 43221, USA
{ling.chen,dufan.wu}@osumc.edu
[2] University Medical Center Regensburg, Regensburg, Germany

Abstract. Automated radiology report generation (RRG) has gained increasing attention because it can reduce the heavy workload of clinical report writing. However, most existing methods mainly optimize for natural language generation (NLG) metrics that focus on language fluency, while providing little control over clinically important factors such as precision and recall. As consequence, generated reports may be fluent but not well aligned with different clinical needs. To address this challenge, we propose a reinforcement learning framework for precision recall controllable RRG, where a control parameter explicitly adjusts the trade-off between clinical precision and recall during inference. This design allows the model to flexibly generate reports according to different clinical requirements. To ensure clinical correctness, we introduce a clinical reward into the training objective, which helps improve clinical efficacy (CE) beyond standard language-based optimization. In addition, we apply a group-relative training strategy that normalizes rewards within each training group, reducing reward variance and improving training stability. Extensive experiments on the MIMIC-CXR dataset show that our method consistently outperforms state-of-the-art approaches in both NLG and CE evaluation metrics, while providing reliable control over the CE precision recall trade-off.

Keywords: Radiology Report Generation · Receiver Operation Curve · Reinforcement Learning · Group-Relative Training · Chest X-ray

1 Introduction

Automated radiology report generation (RRG) aims to produce free-text clinical reports directly from medical images, such as chest X-rays, and has received growing attention in recent years. By reducing the time and effort required for manual report writing, these systems have the potential to improve clinical workflow efficiency and support large-scale medical imaging analysis. Most existing

G. Yang et al. (Eds.): MICCAI 2026, LNCS 16878, pp. 428–438, 2027.
https://doi.org/10.1007/978-3-032-38059-3_41

approaches treat report generation as a sequence-to-sequence learning problem and are commonly trained using maximum likelihood estimation (MLE) with word-level supervision [1,2].

While MLE-based methods can generate fluent and grammatically correct reports, they often suffer from exposure bias and show limited alignment with clinical goals [3] [4]. To address these issues, recent studies have introduced reinforcement learning (RL)–based optimization strategies that directly optimize sentence-level evaluation metrics [5–9]. However, most prior work focuses on improving natural language generation (NLG) scores, such as BLEU [10], METEOR [11], and ROUGE-L [12], which mainly measure text similarity and linguistic quality. As a result, generated reports may appear fluent but still contain clinically incorrect, incomplete, or unbalanced findings. More importantly, existing methods provide little control over clinically critical trade-offs during report generation. In real-world practice, different clinical scenarios may require different emphasis on precision and recall [13]. For example, screening exams often favor high recall to avoid missing abnormalities, while diagnostic tests may prioritize high precision to reduce false positives. Current report generation models lack explicit mechanisms to adjust such trade-offs, limiting their practical usability in diverse clinical settings.

In this work, we propose a reinforcement learning framework for precision recall controllable RRG for chest X-rays, in which a continuous control parameter explicitly adjusts the balance between clinical precision and recall at inference time. This design allows the same model to flexibly generate reports that align with different clinical requirements, without retraining or changing decoding strategies. To improve clinical performance, we further introduce a clinical reward explicitly supervises the presence and absence of clinically relevant findings, enhancing clinical efficacy beyond language-only optimization. In addition, reinforcement learning for report generation often suffers from high reward variance and unstable training [14]. To address this challenge, we adopt a group-relative training strategy [15] that normalizes rewards within each sample group, effectively reducing variance and stabilizing policy optimization. This strategy improves learning efficiency and leads to more consistent performance gains across datasets. We evaluate the proposed method on MIMIC-CXR [16] dataset. Experimental results show that our approach consistently outperforms state-of-the-art methods in both language quality and clinically oriented evaluation metrics, while providing reliable and interpretable control over the precision recall trade-off.

2 Method

The overall framework is illustrated in Fig. 1, which consists of the overall architecture (Fig. 1(a)), the reinforcement learning process (Fig. 1(b)), and the transformer decoder details (Fig. 1(c)).

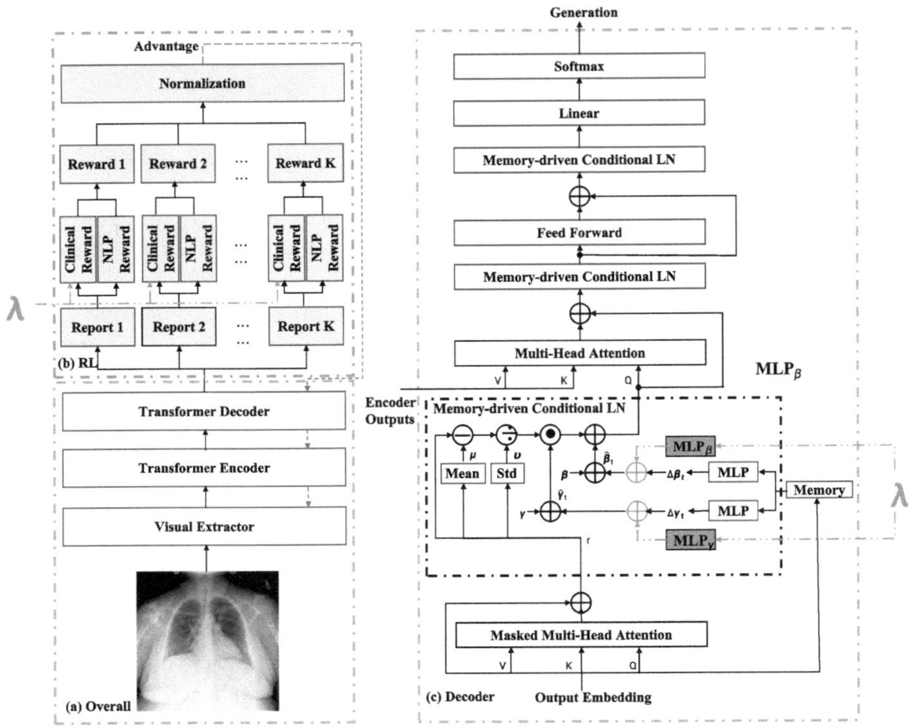

Fig. 1. Overall framework of the proposed precision recall controllable RRG model. (a) Overall architecture. (b) Reinforcement learning process. (c) Transformer decoder. The overall framework and decoder are shown in gray dashed boxes with details omitted. Red dotted lines denote gradient back-propagation during training. Blue dotted lines indicate the precision recall control parameter λ applied at both the representation and reward levels. The group-relative training component is highlighted in blue dashed boxes. (Color figure online)

2.1 Precision Recall Controllable Reinforcement Learning

We formulate RRG as a reinforcement learning problem, where the model generates a report sequence $Y = (y_1, \ldots, y_T)$ conditioned on an input image X.

To explicitly control the precision recall trade-off, we introduce a continuous control parameter $\lambda \in [0, 1]$ that is consistently applied at both the representation level and the reward level. This unified design ensures coherent controllability throughout the generation and optimization process.

At the representation level, inspired by conditional normalization techniques such as FiLM [17], we introduce a Precision Recall Conditioned Adaptive Layer Normalization (PRC-AdaLN) module to incorporate the clinical control parameter λ into the decoder representations. Specifically,

$$\text{PRC-AdaLN}(x,\lambda) = \big(\gamma + \Delta\gamma + \text{MLP}_\gamma(\lambda)\big) \odot \frac{x-\mu(x)}{\sigma(x)} + \big(\beta + \Delta\beta + \text{MLP}_\beta(\lambda)\big), \tag{1}$$

where x denotes the decoder feature, $\mu(x)$ and $\sigma(x)$ are the mean and standard deviation computed along the feature dimension, and γ and β are the learnable affine parameters in standard LayerNorm. The terms $\Delta\gamma$ and $\Delta\beta$ denote additional residual modulation parameters, while two lightweight MLPs, MLP_γ and MLP_β, map the clinical control parameter λ to separate conditioning offsets for the normalization scale and bias, respectively. As illustrated in Fig. 1(c), this representation level modulation dynamically adjusts feature statistics, enabling the decoder to explicitly steer generation along the clinically meaningful precision recall axis.

At the reward level, λ further controls the optimization objective by balancing precision- and recall-oriented clinical efficacy (CE) rewards:

$$R_{\text{CE}}(\lambda) = \lambda\, R_{\text{prec}} + (1-\lambda)\, R_{\text{rec}}. \tag{2}$$

During training, for each training sample, the representation modulation and reward optimization shared a random λ sampled uniformly from $[0,1]$. Thus the model learns a family of conditional policies with consistent precision recall characteristics. At inference time, adjusting λ allows a single trained model to flexibly generate reports that emphasize either higher precision or higher recall, without retraining or modifying the decoding strategy.

2.2 Hybrid Natural Language and Clinical Reward Learning

We optimize the report generation model using a hybrid training objective that combines reinforcement learning with standard cross-entropy loss:

$$\mathcal{L}_{\text{total}} = \alpha\mathcal{L}_{\text{rl}} + (1-\alpha)\mathcal{L}_{\text{ce}}, \tag{3}$$

where α balances reinforcement learning loss and cross-entropy loss.

For the reinforcement learning component, we minimize the negative expected reward:

$$\mathcal{L}_{\text{rl}} = -\mathbb{E}_{Y\sim p_\theta}[R(Y)], \tag{4}$$

where θ refers to the model parameters and the reward function $R(Y)$ integrates both natural language generation (NLG) quality and CE:

$$R(Y) = \rho R_{\text{NLG}}(Y) + (1-\rho) R_{\text{CE}}(Y). \tag{5}$$

Here, $\rho \in [0,1]$ is a fixed hyperparameter that controls the trade-off between natural language quality and clinical efficacy. The clinical reward is based on the clinical efficacy metric, with $R_{\text{CE}}(Y)$ computed according to Eq. 2. The natural language generation reward $R_{\text{NLG}}(Y)$ is defined as:

$$R_{\text{NLG}}(Y) = \sum_i \omega_i r_i(Y), \tag{6}$$

where $r_i(Y)$ represents multiple NLG evaluation metrics, including BLEU-4, METEOR, and ROUGE-L [7], and ω_i are non-negative weighting coefficients with $\sum_i \omega_i = 1$.

2.3 Group-Relative Training

Reinforcement learning for report generation often suffers from high reward variance, which can lead to unstable training, especially when multiple reward components are involved. To alleviate this issue, we adopt a group-relative self-critical sequence training strategy [15].

For each input image x, we perform group sampling by drawing K candidate reports $\{Y^{(1)}, Y^{(2)}, \ldots, Y^{(K)}\}$ from the current policy. Each sampled report is evaluated using the task-specific reward function, yielding a set of rewards $\{R^{(k)}\}_{k=1}^{K}$.

Instead of relying on absolute reward values, we compute a group-relative baseline by normalizing rewards within each group:

$$\hat{A}^{(k)} = \frac{R^{(k)} - \text{mean}(\{R^{(i)}\}_{i=1}^{K})}{\text{std}(\{R^{(i)}\}_{i=1}^{K}) + \epsilon}, \tag{7}$$

where $\hat{A}^{(k)}$ denotes the standardized advantage of the k-th sampled report and ϵ is a small constant for numerical stability. The reinforcement learning loss is computed according to Eq. 4 by replacing $R(Y)$ with $\hat{A}$ for each sample. We then use the REINFORCE algorithm [18] to compute policy gradients and update the model parameters.

By using the group-relative standardized advantage as the learning signal, the model focuses on the relative quality of candidate reports generated for the same image. This strategy effectively reduces reward variance and stabilizes policy optimization.

3 Experiments

Datasets and Evaluation Metrics. We evaluate our model on the MIMIC-CXR dataset [16], a large-scale chest X-ray benchmark containing 473,057 radiographs paired with 206,563 reports. Following the preprocessing protocol in [2], we adopt the official patient-level split with 270,790 training, 2,130 validation, and 3,858 testing samples. Each study is paired with its corresponding report as ground truth. We only used the findings section for each report.

Following prior work [7,19], we evaluate model performance using a combination of natural language generation (NLG) metrics and CE metrics. The NLG metrics include BLEU [10], METEOR [11], and ROUGE-L [12]. For CE evaluation, generated and ground-truth reports are first converted into 14 CheXbert labels [20]. Following our training setup, uncertain findings are considered positive, while negative and blank labels are considered negative. We then report micro-averaged precision, recall, and F1 score across all labels and test samples.

Implementation Details. We used $\alpha = 0.99$ in Eq. 3 that balances RL and cross-entropy loss, $\rho = 0.1$ in Eq. 5 that balances NLG and clinical efficacy. Following [7], the weighting coefficients for the NLG reward in Eq. 6 were set to $\omega_1 = 5/11$, $\omega_2 = 5/11$, and $\omega_3 = 1/11$ for BLEU-4, METEOR, and ROUGE-L, respectively. For the group-relative training, we set the group size $K = 5$. All remaining hyperparameters are kept consistent with [7]. A pretrained model [2] was fine-tuned on the MIMIC-CXR dataset for 5 epochs with a batch size of 4 using 16 NVIDIA A100 GPUs provided by the Ohio Supercomputer Center [21]. We further applied post-processing to all generated reports to remove repeating sentences and sentences that were shorter than three words. Our code is available at: https://github.com/98lingchen/MICCAI2026.

4 Results

Parameter Performance. As shown in Fig. 2, the precision recall control parameter λ provides effective and continuous regulation over clinical precision and recall. We performed inference using different λ values from 0 to 1 with an interval of 0.1. As λ increases, clinical precision consistently improves while recall gradually decreases, demonstrating a clear trade-off between precision and recall. The F1-score exhibits a unimodal trend and reaches its maximum at $\lambda = 0.3$. These results validate that λ enables flexible adjustment of report generation behavior to meet different clinical requirements, such as high-recall screening and high-precision confirmation, without requiring model retraining. For the results reported below, we use $\lambda = 0.3$, which achieved the best F1 scores.

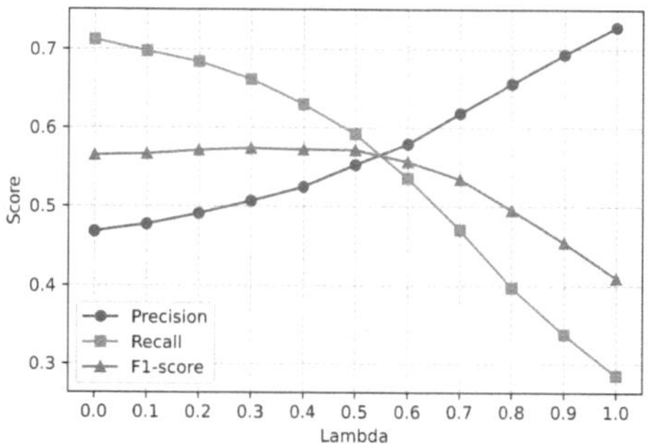

Fig. 2. Effect of the precision recall control parameter λ on clinical precision, recall, and F1-score.

Performance Comparison. We compared our method with several state-of-the-art methods on the MIMIC-CXR dataset, including R2Gen [2], METrans [22], R2GenGPT [23], BoostRRG [24], Diff-RRG [25], MLRG [26], and MedGemma 1.5 4B [27]. Results of the methods except MedGemma were

Table 1. Performance comparisons between the proposed method and existing approaches on the test set of the MIMIC-CXR dataset using both NLG and CE metrics.

Method	NLG Metrics						CE Metrics		
	B-1	B-2	B-3	B-4	RG-L	MET.	PREC.	REC.	F1
R2Gen	0.353	0.218	0.145	0.103	0.277	0.142	0.333	0.273	0.276
METrans	0.386	0.250	0.169	0.124	0.291	0.152	0.364	0.309	0.334
R2GenGPT	0.411	0.267	0.186	0.134	0.297	0.160	0.392	0.387	0.389
BoostRRG	0.402	0.262	0.180	0.128	0.291	0.175	0.465	0.482	0.473
Diff-RRG	0.405	0.251	0.169	0.120	0.276	0.164	0.528	0.430	0.474
MLRG	0.411	0.277	0.204	0.158	0.320	0.176	**0.549**	0.468	0.505
MedGemma	0.200	0.114	0.071	0.045	0.160	0.125	0.520	0.383	0.441
Ours	**0.451**	**0.310**	**0.219**	**0.159**	**0.333**	**0.177**	0.505	**0.656**	**0.571**

reported as provided in their original papers. For MedGemma, the reports were generated as follows:

The prompt for system is "You are a professional radiologist", followed by the user prompt: "Please provide a radiology report of the following chest X-ray image. Keep only the 'Findings' and 'Impression' sections in your report" and the CXR image. The findings section was then retrieved by extracting the part between "findings" and "impression" via simple text search.

As shown in Table 1, our method achieved state-of-the-art performance on the MIMIC-CXR dataset across both NLG and CE metrics. In particular, our model obtains notable improvements on NLG metrics, indicating stronger capability in modeling long-range dependencies. Moreover, our approach yields substantial gains in CE metrics, demonstrating its effectiveness in generating clinically accurate reports.

Ablation Study. Table 2 presents an ablation study on the MIMIC-CXR dataset to evaluate the individual and combined effects of the clinical reward and group-relative training. We did not separately ablate the two λ-related designs because they are coupled: λ conditioned decoder enables test-time control, while λ weighted rewards provide the supervision during training that makes this control meaningful. Without either component, the model achieves relatively limited performance on both NLG and CE metrics. Introducing the clinical reward alone leads to substantial improvements in CE metrics, demonstrating its effectiveness in enhancing clinical correctness. Applying group-relative training alone also yields notable gains across both NLG and CE metrics. When both components are jointly applied, the model achieves the best overall performance across all metrics, highlighting their complementary roles in improving both linguistic quality and clinical efficacy.

Table 2. Ablation study of clinical reward and group-relative training on the MIMIC-CXR dataset.

Method		NLG Metrics						CE Metrics		
Clinical	Group-rel.	B-1	B-2	B-3	B-4	RG-L	MET.	PREC.	REC.	F1
×	×	0.400	0.253	0.171	0.120	0.276	0.154	0.392	0.335	0.342
✓	×	0.404	0.260	0.180	0.129	0.312	0.153	**0.590**	0.330	0.423
×	✓	0.433	0.298	0.212	0.153	**0.339**	0.170	0.546	0.396	0.459
✓	✓	**0.451**	**0.310**	**0.219**	**0.159**	0.333	**0.177**	0.505	**0.656**	**0.571**

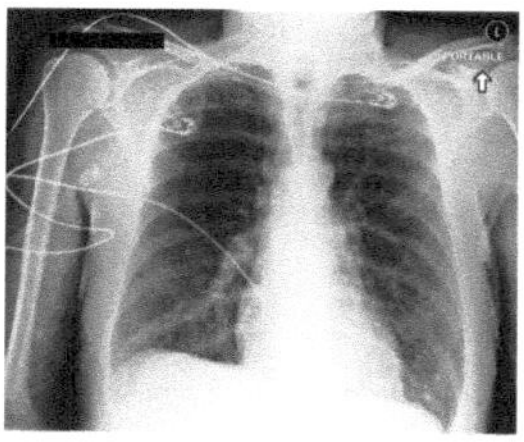

Ground Truth: ap portable upright chest radiograph was provided . the lungs are hyperinflated with upper lobe lucency compatible with emphysema . no focal consolidation effusion or pneumothorax seen . cardiomediastinal silhouette is normal . bony structures are intact .

Baseline: ap portable upright view of the chest . overlying ekg leads are present . there is no focal consolidation effusion or pneumothorax . the cardiomediastinal silhouette is normal . imaged osseous structures are intact .

Ours: ap view of the chest . the lungs are hyperinflated . the lungs are clear . there is no focal consolidation . no pleural effusion or pneumothorax . the cardiomediastinal silhouette is normal . the mediastinal and hilar contours are normal .

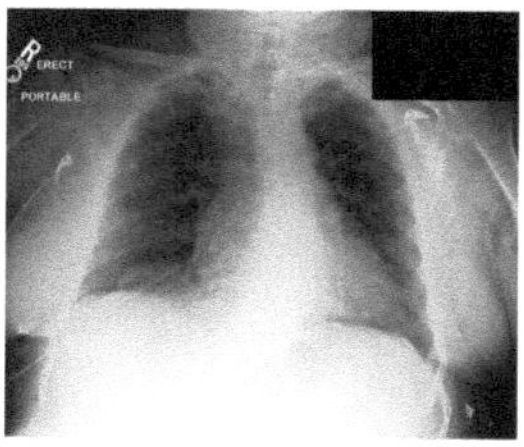

Ground Truth: cardiomegaly is stable . there is no focal consolidation concerning for pneumonia . there is no pleural effusion pneumothorax or pulmonary edema . scoliosis is again noted . an old left clavicular deformity is noted .

Baseline: as compared to the previous radiograph there is no relevant change . low lung volumes . borderline size of the cardiac silhouette without pulmonary edema . no pleural effusions . no pneumonia no pneumothorax .

Ours: ap view of the chest . the lungs are clear . there is no focal consolidation . no pleural effusion or pneumothorax . the cardiac silhouette is enlarged . the mediastinal and hilar contours are normal .

Fig. 3. Qualitative comparison of radiology reports generated by the baseline method and our method with ground truth references. Main improvements are highlighted in blue. (Color figure online)

Quantitative Analysis. We further present qualitative analysis of representative samples in Fig. 3 to illustrate the differences between our method and the baseline model [2]. The reports generated by our method demonstrated improved clinical accuracy and closer alignment with the ground truth. In the first case, our model correctly identified the hyperinflated lung which was missed by the baseline model. In the second case our model confirmed the enlarged cardiac silhouette whereas the baseline model described the size as "borderline". These examples demonstrate that the reports generated by our framework are more consistent with the ground truth and better reflect clinically meaningful findings.

5 Conclusion

This paper proposed a novel approach to enhance clinical consistency and controllability in RRG. Our method enables flexible precision recall control through a hybrid natural language and clinical efficacy reward mechanism, allowing the model to better capture clinically relevant findings while maintaining semantic coherence. Extensive experiments on the MIMIC-CXR dataset demonstrated the effectiveness and robustness of our approach compared with existing methods. However, the current model is limited by the small size of the model as well as uncleaned training data, e.g., some data points were lateral CXR only, or the reports mentioned previous exam but the input had only the current CXR. In future works, we plan to extend Group reward-decoupled normalization (GDPO) [28] to our method and extend the framework to LLMs such as MedGemma and finetune it with a cleaner dataset. We also plan to perform more comprehensive evaluations of the methods with more advanced metrics [3], human reader study, and institutional data, and compare our method with recent approaches such as CURE [29] for curriculum-guided anatomy grounding.

Acknowledgments. This work was supported in part by NIBIB under Award Number R01EB035394. The content is solely the responsibility of the authors and does not necessarily represent the official views of NIH.

Disclosure of Interests. The authors have no competing interests in the paper.

References

1. Jing, B., Xie, P., Xing, E.: On the automatic generation of medical imaging reports. In: Proceedings of the 56th Annual Meeting of the Association for Computational Linguistics (volume 1: long papers), pp. 2577–2586 (2018)
2. Chen, Z., Song, Y., Chang, T.-H., Wan, X.: Generating radiology reports via memory-driven transformer arXiv preprint arXiv:2010.16056 (2020)
3. Yu, F., et al.: Evaluating progress in automatic chest x-ray radiology report generation. Patterns **4**(9) (2023)
4. Sloan, P., Clatworthy, P., Simpson, E., Mirmehdi, M.: Automated radiology report generation: a review of recent advances. IEEE Rev. Biomed. Eng. **18**, 368–387 (2024)
5. Zhang, Y., Merck, D., Tsai, E., Manning, C.D., Langlotz, C.: Optimizing the factual correctness of a summary: a study of summarizing radiology reports. In: Proceedings of the 58th Annual Meeting of the Association for Computational Linguistics, pp. 5108–5120 (2020)
6. Li, Y., Liang, X., Hu, Z., Xing, E.P.: Hybrid retrieval-generation reinforced agent for medical image report generation. Adv. Neural Inf. Process. Syst. **31**(2018)
7. Yi, X., Fu, Y., Yu, J., Liu, R., Zhang, H., Hua, R..: LHR-RFL: linear hybrid-reward-based reinforced focal learning for automatic radiology report generation. IEEE Trans. Med. Imaging **44**(3), 1494–1504 (2025)
8. Qin, H., Song, Y.: Reinforced cross-modal alignment for radiology report generation. In: Findings of the Association for Computational Linguistics: ACL 2022, pp. 448–458 (2022)

9. Celikyilmaz, A., Bosselut, A., He, X., Choi, Y.: Deep communicating agents for abstractive summarization. arXiv preprint arXiv:1803.10357 (2018)
10. Papineni, K., Roukos, S., Ward, T., Zhu, W.-J.: Bleu: a method for automatic evaluation of machine translation. In: Proceedings of the 40th annual meeting of the Association for Computational Linguistics, pp. 311–318 (2002)
11. Banerjee, S., Lavie, A.: Meteor: an automatic metric for mt evaluation with improved correlation with human judgments. In: Proceedings of the ACL Workshop on Intrinsic and Extrinsic Evaluation Measures for Machine Translation And/or Summarization, pp. 65–72 (2005)
12. Lin, C.-Y.: Rouge: a package for automatic evaluation of summaries. In: Text Summarization Branches Out, pp. 74–81 (2004)
13. Topol, E.J.: High-performance medicine: the convergence of human and artificial intelligence. Nat. Med. **25**(1), 44–56 (2019)
14. Paulus, R., Xiong, C., Socher, R.: A deep reinforced model for abstractive summarization arXiv preprint arXiv:1705.04304 (2017)
15. Shao, Z., et al. Deepseekmath: pushing the limits of mathematical reasoning in open language models. arXiv preprint arXiv:2402.03300 (2024)
16. Johnson, A.E.W., et al.: Mimic-CXR-Jpg, a large publicly available database of labeled chest radiographs. arXiv preprint arXiv:1901.07042 (2019)
17. Perez, E., Strub, F., De Vries, H., Dumoulin, V., Courville, A.: Film: visual reasoning with a general conditioning layer. In: Proceedings of the AAAI Conference on Artificial Intelligence, vol. 32 (2018)
18. Williams, R.J.: Simple statistical gradient-following algorithms for connectionist reinforcement learning. Mach. Learn. **8**(3), 229–256 (1992)
19. Liu, H., Wei, D., Xu, Z., Wu, X., Zheng, Y., Wang, L.: RRG-DPO: direct preference optimization for clinically accurate radiology report generation. In: International Conference on Medical Image Computing and Computer-Assisted Intervention, pp. 552–562. Springer (2025)
20. Smit, A., Jain, S., Rajpurkar, P., Pareek, A., Ng, A.Y.,Lungren, M.: Chexbert: combining automatic labelers and expert annotations for accurate radiology report labeling using Bert. arXiv preprint arXiv:2004.09167 (2020)
21. Ohio Supercomputer Center. Ohio supercomputer center (1987)
22. Wang, Z., Liu, L., Wang, L., Zhou, L.: Metransformer: radiology report generation by transformer with multiple learnable expert tokens. In: Proceedings of the IEEE/CVF Conference on Computer Vision and Pattern Recognition, pp. 11558–11567 (2023)
23. Wang, Z., Liu, L., Wang, L., Zhou, L.: R2gengpt: radiology report generation with frozen LLMs. Meta-Radiol. **1**(3), 100033 (2023)
24. Sha, Y., Pan, H., Meng, W., Li, K.: Contrastive knowledge-guided large language models for medical report generation. In: International Conference on Medical Image Computing and Computer-Assisted Intervention, pp. 111–120. Springer (2025)
25. Yun, H., Maeng, J., Kang, E., Suk, H.: Diff-RRG: longitudinal disease-wise patch difference as guidance for LLM-based radiology report generation. In: International Conference on Medical Image Computing and Computer-Assisted Intervention, pp. 152–161. Springer (2025)
26. Liu, K., et al.: Enhanced contrastive learning with multi-view longitudinal data for chest x-ray report generation. In: Proceedings of the Computer Vision and Pattern Recognition Conference, pp. 10348–10359 (2025)
27. Sellergren, A., et al.: Medgemma technical report. arXiv preprint arXiv:2507.05201 (2025)

28. Liu, S.Y., et al. GDPO: group reward-decoupled normalization policy optimization for multi-reward RL optimization. arXiv preprint arXiv:2601.05242 (2026)
29. Messina, P., et al: Cure: curriculum-guided multi-task training for reliable anatomy grounded report generation. arXiv preprint arXiv:2601.15408 (2026)

PriLoRA: Prior-Conditioned Low-Rank Adapters for Medical Vision Models

Amirhossein Kazerouni[1,2,3](✉), Sanaz Karimijafarbigloo[4], Ali Aghayari[5], Ali Ghasemzadeh Khorzoughi[5], Reza Azad[4], Wojciech Samek[6,7], Dorit Merhof[4], Michael Brudno[1,2,3], and Babak Taati[1,2,3]

[1] University of Toronto, Toronto, Canada
[2] Vector Institute, Toronto, Canada
[3] University Health Network, Toronto, Canada
amirhossein@cs.toronto.edu
[4] University of Regensburg, Regensburg, Germany
[5] Berlin, Germany
[6] Technische Universität Berlin, Berlin, Germany
[7] Fraunhofer Heinrich Hertz Institute, Berlin, Germany

Abstract. Adapting large foundation models to medical imaging is costly due to limited curated data, dense expert annotation requirements, and the expense of full fine-tuning. Parameter-efficient fine-tuning (PEFT) methods such as LoRA mitigate these costs, but linear low-rank updates can be expressivity-limited. Recent sinusoidal variants such as SineLoRA improve capacity, yet depend on a fixed frequency hyperparameter that is difficult to transfer across modalities and tasks. We propose **Prior-Conditioned Low-Rank Adapters (PriLoRA)**, an input-conditioned sinusoidal adapter that leverages input-derived prior features to modulate the adapter response per block and per sample while preserving LoRA's modularity and parameter efficiency. PriLoRA predicts lightweight modulation parameters that control the sinusoidal update, enabling adaptive expressivity without increasing the adapter rank or requiring task-specific frequency search. We evaluate PriLoRA across multiple medical backbones, modalities, and tasks, including generation, classification, segmentation, and CT reconstruction, and observe consistent improvements in robustness and transfer over standard LoRA and representative PEFT baselines across datasets and modalities. **Code:** https://github.com/xmindflow/PriLoRA.

Keywords: PEFT · LoRA · Medical imaging · Prior-conditioned adapters

A. Kazerouni, S. Karimijafarbigloo, A. Aghayari and A. Ghasemzadeh Khorzoughi—Equal contribution.

A. Aghayari and G. Khorzoughi—Independent Researcher.

G. Yang et al. (Eds.): MICCAI 2026, LNCS 16878, pp. 439–449, 2027.
https://doi.org/10.1007/978-3-032-38059-3_42

1 Introduction

Foundation models [18,19,23] are increasingly adopted in medical imaging due to their strong transferability. However, full fine-tuning is often expensive in memory, compute, and data, a challenge amplified by limited access to curated datasets and expert annotations [12]. This has motivated parameter-efficient fine-tuning (PEFT) methods that reuse pretrained backbones while updating only a small set of parameters [8]. LoRA [10] instantiates this idea by introducing lightweight trainable updates to selected layers while freezing the backbone, substantially reducing training cost—an advantage that is particularly important in medical imaging with limited annotations and constrained resources [6,29].

Recent work in PEFT has moved beyond *vanilla* LoRA by modifying *where* and *how* lightweight updates are parameterized. Following [16], existing extensions can be broadly grouped into: (i) **adaptation-improving** designs that redistribute adapter capacity across layers or refine the update parameterization [14,15,17,30]; (ii) **operator-aware** designs that generalize LoRA to non-linear-layer structures such as convolutional kernels, enabling parameter-efficient tuning of CNN backbones for medical tasks [21]; and (iii) **efficiency-focused** designs that combine LoRA with quantization to enable low-resource fine-tuning and deployment, including healthcare-oriented decision support pipelines [1,4]. Beyond these design choices, LoRA also faces a *rank bottleneck*: purely linear low-rank updates are constrained in representational capacity by construction. SineLoRA [11] addresses this bottleneck by injecting a sinusoidal non-linearity into a low-rank parameterization, increasing the effective rank of the induced transformation. However, its performance is sensitive to the frequency choice, which is treated as a hyperparameter and can impact optimization stability [11].

In this work, we aim to retain the rank-enhancing benefit of sinusoidal low-rank parameterizations while avoiding reliance on a globally fixed frequency choice. This is particularly important in medical imaging, where different modalities and tasks demand different levels of expressivity, making a single fixed setting difficult to transfer across datasets and use cases. More broadly, this reflects a central challenge in parameter-efficient adaptation: achieving high accuracy with sufficient expressivity while preserving the desirable low-rank structure of adapter updates. We propose **Prior-Conditioned Low-Rank Adapters (PriLoRA)**, a parameter-efficient adapter that conditions a sinusoidal low-rank *weight update* on input-derived prior knowledge. PriLoRA introduces a lightweight **modulator** that takes prior features extracted from the input and predicts a small set of parameters that specialize the adapter for each block and each sample. For a linear layer with frozen weight W_0 and LoRA factors B, A, we first form the low-rank update BA, and then apply an input-conditioned sinusoidal transform: $\Delta W_{\mathrm{Pri}} = \alpha \sin\Big(\beta\,(\omega_0\, BA) + \gamma\Big) + \delta$, where α controls the amplitude, β scales the effective frequency, γ sets the phase, and δ provides an additive shift (all broadcast to the shape of BA). This turns sinusoidal LoRA into an input-conditioned adapter that adjusts its non-linear response per block and per sample, while preserving LoRA-style modularity and parameter efficiency.

2 Method

Background. LoRA [10] enables parameter-efficient fine-tuning by freezing pre-trained weights and learning a low-rank residual update. For a linear layer with weight $W_0 \in \mathbb{R}^{d \times k}$, LoRA parameterizes the update as

$$\Delta W = BA, \qquad B \in \mathbb{R}^{d \times r},\ A \in \mathbb{R}^{r \times k},\ r \ll \min(d, k), \tag{1}$$

and applies the adapted layer as $y = (W_0 + \Delta W)x = W_0 x + BAx$. Since only A and B are trained, LoRA reduces the number of trainable parameters from $\mathcal{O}(dk)$ to $\mathcal{O}(r(d+k))$, enabling efficient adaptation of large backbones. A central limitation of this parameterization is a *rank bottleneck*: by construction, $\text{rank}(\Delta W) \leq r$. Increasing r improves expressivity and can better approximate the generally full-rank update required for optimal adaptation, but it also increases parameter count and compute. This trade-off motivates techniques that enhance the representational capacity of low-rank updates without sacrificing parameter efficiency. SineLoRA [11] alleviates this limitation by injecting a sinusoidal non-linearity into the low-rank update. Concretely, it applies a non-linearity to a low-rank matrix, and specializes to LoRA via: $\Delta W = \frac{1}{g} \sin(\omega_0 \cdot BA)$, where $g > 0$ is a fixed gain (scaling) factor used to control the magnitude of the update and stabilize optimization. This formulation increases the effective rank and improves approximation quality without adding trainable parameters [11]. However, performance is sensitive to the hyperparameters, specifically frequency ω_0, which is a hyperparameter that can affect optimization stability and transfer across medical tasks and modalities. This motivates an adaptive alternative that conditions the sinusoidal update on input-derived priors (Fig. 1).

2.1 Prior-Conditioned Low-Rank Adapters (PriLoRA)

We propose **PriLoRA**, a parameter-efficient adapter that preserves the expressivity gains of sinusoidal low-rank updates while avoiding a single globally fixed effective sinusoidal behavior. SineLoRA can be viewed as applying a *fixed* sinusoidal transform to a low-rank update, where the primary controllable knob is the frequency ω_0. Varying ω_0 exposes a rank–capacity trade-off: increasing frequency reduces redundancy and diversifies directions, leading to higher *effective rank*, but overly large values can push the update toward an unnecessarily high-rank regime, resulting in over-parameterization and degraded generalization. In medical imaging, the appropriate operating point is not universal; different modalities and tasks induce different distribution shifts and require different adaptation capacity, making a single fixed ω_0 brittle to tune and hard to transfer.

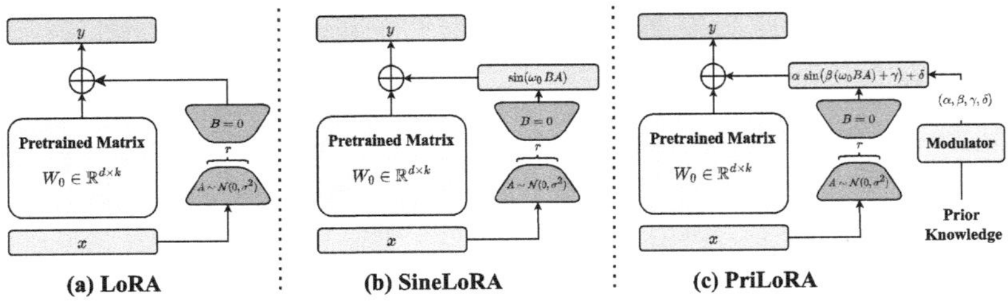

Fig. 1. Comparison of LoRA, SineLoRA, and PriLoRA. (a) *LoRA* adapts a frozen pretrained weight W_0 using a low-rank update $\Delta W = BA$ (with rank r), producing the output y via the residual path. (b) *SineLoRA* increases expressivity by applying an element-wise sinusoidal transform to the low-rank update, e.g., $\sin(\omega_0 BA)$. (c) *PriLoRA* conditions the sinusoidal update on input-derived *prior knowledge* using a lightweight modulator that predicts $(\alpha, \beta, \gamma, \delta)$, where β rescales a fixed base frequency ω_0 to obtain an input-conditioned effective frequency.

SineLoRA fixes remaining degrees of freedom through a shared gain $\frac{1}{g}$ and zero phase/offset, enforcing one global sinusoid for all inputs. However, stable transfer depends not only on frequency but also amplitude and where the nonlinearity acts (phase and shift), which vary across medical tasks. In segmentation, a fixed sinusoid may behave nearly linear in some layers or overly saturate others, harming boundary refinement. In image generation/restoration it can underfit fine textures or introduce oscillatory artifacts, and in classification excessive nonlinearity may amplify spurious high-frequency features and reduce robustness. These task-dependent effects motivate conditioning amplitude, frequency, phase, and shift on input-derived priors instead of using a single global sinusoid.

Formulation. Using the LoRA low-rank update BA from Eq. (1), PriLoRA replaces the globally fixed sinusoid with a prior-conditioned sinusoidal transform whose amplitude, frequency scaling, phase, and offset are predicted from input-derived prior knowledge. Let $z = p(x) \in \mathbb{R}^r$ be a vector of prior features extracted from the current medical input x. A lightweight modulator $g(\cdot;\theta) : \mathbb{R}^r \to \mathbb{R}^4$ maps z to four scalars: $\big(\alpha, \beta, \gamma, \delta\big) = g(z;\theta)$, which parameterize the sinusoidal update for that input. PriLoRA then defines the weight update as

$$\Delta W_{\text{Pri}} = \alpha \sin\Big(\beta\,(\omega_0\, BA) + \gamma\Big) + \delta, \tag{2}$$

where $\sin(\cdot)$ is applied element-wise and the scalar parameters are broadcast over the entries of BA. The adapted layer becomes

$$y = \big(W_0 + \Delta W_{\text{Pri}}\big)\,x = W_0 x + \Delta W_{\text{Pri}}\,x. \tag{3}$$

Effect of Parameters. The parameters in Eq. (2) control complementary aspects of the update. The amplitude term α gates the overall strength of the adapter, scaling the deviation from W_0 and allowing PriLoRA to attenuate or amplify the update based on the difficulty or reliability of the input-derived prior (e.g., down-weighting noisy priors). The frequency-scaling term β rescales the argument of the sinusoid, effectively stretching or compressing $\sin(\cdot)$ along the BA directions: larger β induces more oscillatory behavior and higher effective rank (capturing fine-grained variations), while smaller β keeps the update closer to a low-rank, near-linear regime, which can improve stability and reduce overfitting. The phase term γ shifts where BA lies on the sinusoid, allowing PriLoRA to reposition the operating point into regions with favorable slope (e.g., avoiding saturation for certain modalities or intensity ranges) without changing the magnitude of BA itself. Finally, the offset term δ adds a learned bias to the sinusoidal output, enabling modest input-dependent baseline corrections (e.g., domain-specific shifts) without requiring a separate bias adapter. Together, these four coefficients let PriLoRA adapt the shape and strength of the sinusoidal low-rank update to the medical input, and SineLoRA is recovered as a special case by fixing $(\alpha, \beta, \gamma, \delta)$ to global constants.

Table 1. Comparison across Generation, Segmentation, Classification, and CT Reconstruction tasks. For generation, lower is better for FID and higher is better for CLIP Score, reported as mean ± std. For segmentation and classification, higher is better for all metrics. For CT reconstruction, higher PSNR/SSIM and lower LPIPS indicate better quality; results are reported at projection angles $\{50, 100, 150, 200\}$. Best performance is shown in **bold**, and second-best performance is underlined.

(a) Comparison on generation and segmentation tasks.

Methods	Generation				Segmentation		
	FID_{Total} ↓	FID_{Test} ↓	CLIPScore ↑	# **Params (M)**	IoU ↑	Dice ↑	# **Params (M)**
AdaLoRA [30]	151.26 ± 1.82	158.11 ± 1.65	**28.08 ± 0.04**	4.948	0.8209	0.8930	16.911
DoRA [14]	114.72 ± 1.03	120.57 ± 1.15	28.00 ± 0.05	5.021	0.8110	0.8864	16.975
LoRA [9]	112.32 ± 0.63	118.52 ± 0.63	28.01 ± 0.07	4.706	0.8388	0.9051	16.911
PiSSA [17]	119.01 ± 1.56	125.19 ± 1.52	27.26 ± 0.08	4.706	0.8404	0.9057	16.911
SineDoRA [11]	79.07 ± 0.49	90.74 ± 0.44	27.55 ± 0.13	5.021	0.7841	0.8680	16.975
SineLoRA [11]	85.05 ± 1.43	93.84 ± 1.50	27.47 ± 0.04	4.706	0.8349	0.9029	16.911
PriLoRA (Ours)	**70.15 ± 1.23**	**80.73 ± 0.99**	27.65 ± 0.07	4.949	**0.8533**	**0.9156**	17.856

(b) Comparison on Classification and CT Reconstruction tasks.

Methods	Classification					CT Reconstruction				
	AUC ROC ↑	Recall ↑	Precision ↑	F1-score ↑	# Trainable (M)	PSNR ↑/LPIPS ↓/SSIM ↑				# Trainable
						50 views	100 views	150 views	200 views	
Baseline	0.9872	0.8900	0.9131	0.8883	17.927	25.53/0.249/0.548	27.41/0.192/0.692	27.54/0.200/0.685	28.55/0.178/0.747	198,401
AdaLoRA [30]	0.9891	0.8660	0.9033	0.8624	3.815	25.22/0.266/0.543	26.50/0.229/0.648	26.47/0.227/0.638	27.38/0.197/0.700	49,248
DoRA [14]	**0.9929**	0.8850	0.9151	0.8851	3.819	25.36/0.257/0.547	26.82/0.221/0.663	27.17/0.209/0.669	27.64/0.193/0.716	49,920
LoRA [9]	0.9803	0.8160	0.8872	0.8019	3.814	25.36/0.258/0.546	26.83/0.220/0.662	27.09/0.209/0.667	27.66/0.193/0.716	49,152
PiSSA [17]	0.9768	0.8050	0.8826	0.8072	3.814	25.32/0.259/0.545	26.74/0.221/0.659	26.97/0.212/0.661	27.55/0.193/0.711	49,152
SineDoRA [11]	0.9941	0.8780	0.9118	0.8810	3.819	25.40/0.249/0.554	26.69/0.222/0.663	27.12/0.207/0.677	27.24/0.202/0.696	49,920
SineLoRA [11]	0.9849	0.8210	0.8780	0.8157	3.814	25.42/0.255/0.552	26.82/0.224/0.661	27.18/0.206/0.671	27.34/0.198/0.703	49,152
PriLoRA (Ours)	0.9823	**0.8980**	**0.9161**	**0.8989**	3.876	**25.48/0.247/0.567**	**27.53/0.190/0.693**	**27.83/0.194/0.693**	**27.85/0.190/0.720**	49,740

3 Experimental Results

We comprehensively evaluate PriLoRA across four medical vision tasks: image generation, segmentation, classification, and CT reconstruction. We report both quantitative and qualitative results across diverse medical datasets, backbone architectures, and imaging modalities, showing that conditioning on input-derived priors consistently outperforms existing low-rank adapters. Following prior work, we use a rank of 16 by default. AdaLoRA halves the rank mid-training, and we use ranks of 32 for CT reconstruction.

3.1 Generation

Dataset. We use the IU X-Ray dataset [27], which contains 3,955 radiology reports paired with frontal and lateral chest X-ray images. The data are split into 80%/16%/4% for training, validation, and testing, respectively.

Architecture. We build on Stable Diffusion v1.5 [22] using the diffusers framework, freezing the VAE and pretrained backbones while applying parameter-efficient adaptation to the UNet and CLIP text encoder [20]. PriLoRA partitions the UNet into ten blocks, each equipped with a lightweight modulation module. For each block, a prior feature vector is formed by concatenating channel-wise statistics (mean, standard deviation, min, max) of the preceding block's hidden states with mean-pooled text embeddings to keep it efficient. A two-layer MLP with LayerNorm and SiLU predicts four block-specific parameters that condition the subsequent block. Training follows the standard diffusion noise-prediction objective with MSE loss, using AdamW (learning rate (1×10^{-4})), batch size 8, and 8 epochs. Images are resized and center-cropped to (224×224). Sampling uses 50 diffusion steps with classifier-free guidance (scale 7.5).

Analysis. We evaluate generation using $\mathrm{FID}_{\mathrm{Total}}$, $\mathrm{FID}_{\mathrm{Test}}$, and CLIPScore, where $\mathrm{FID}_{\mathrm{Total}}$ compares generated samples to the full dataset and $\mathrm{FID}_{\mathrm{Test}}$ to the held-out test set. Figure 2 presents qualitative results for two representative prompts. While all methods produce visually plausible radiographs, PriLoRA more consistently preserves anatomical structure and spatial coherence, yielding images closer to the ground truth. This trend is reflected in Table 1, where PriLoRA achieves the lowest FID while maintaining competitive CLIP scores, indicating improved image fidelity without sacrificing textâĂŞimage alignment.

3.2 Segmentation

Dataset. We use ISIC 2016 [7] for 2D melanoma segmentation, comprising 900 training images with expert lesion masks and 379 test dermoscopic images.

Architecture. We follow [26] for 2D medical segmentation and adapt Segment Anything (SAM) [13] to the medical domain while preserving its architecture. As in the generation setting, modulation uses prior features extracted via adaptive average pooling that halves the dimension of each preceding layer. The pooled features are passed through a lightweight MLP with a single 32-unit hidden layer, LayerNorm, and SiLU activation to predict the modulation parameters.

Analysis. As shown in Table 1, PriLoRA achieves the highest IoU and Dice, indicating improved region overlap and spatial structure preservation. Qualitative examples in Fig. 2 further show that PriLoRA produces smoother, more contiguous lesion boundaries that closely match the ground truth, while other

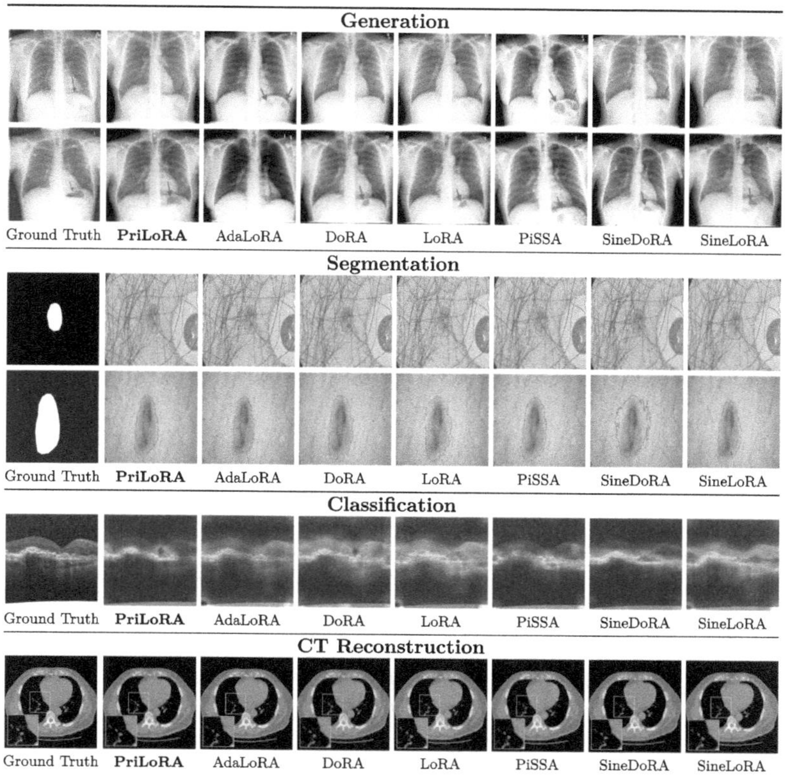

Fig. 2. Qualitative comparison across tasks. The first two rows show generation results for two prompts: (1) *There are no focal areas of consolidation. No pleural effusions. No pneumothorax. Heart size within normal limits. Osseous structures intact.* and (2) *The heart is normal in size and contour. The lungs are clear, without evidence of infiltrate. There is no pneumothorax or effusion.* The third and fourth rows present segmentation results. The fifth row shows Grad-CAM++ [2] results for Choroidal Neovascularization (ChN). The last row reports CT reconstruction results.

adapters often exhibit boundary fragmentation or spurious activations. These results indicate that conditioning low-rank updates on input-derived priors leads to more stable and precise segmentation without increasing model capacity.

3.3 Classification

Dataset. We use OCTMNISTv2 [28], a four-class retinal OCT dataset (ChN, DME, Drusen, Normal) commonly used for retinal disease diagnosis [5]. We follow the official split: the training set contains 33,484 ChN, 10,213 DME, 7,754 Drusen, and 46,026 Normal images, while the validation and test sets include 2,708 and 250 samples per class, respectively.

Architecture. We extend layer-wise adaptation to the convolutional domain by attaching a lightweight modulator to each of VGG-16 [25]'s 13 convolutional layers while jointly training the two classification heads. The modulator pools global spatial and channel context and uses a single-layer MLP to predict depth-specific modulation parameters.

Analysis. We analyze attention using Grad-CAM++ [2] on the final VGG-16 convolutional layer. Compared to baseline and other PEFT methods, PriLoRA yields more localized, anatomically relevant activations, consistent with its improved precision, recall, and F1-score in Table 1.

3.4 CT Reconstruction

Dataset. We evaluate CT reconstruction using 256×256 lung CT images from the publicly available Kaggle Lung Nodule Analysis dataset [3].

Architecture. We adopt SIREN [24] as the base implicit neural representation for CT reconstruction, implemented as a 3-hidden-layer MLP mapping 2D coordinates (x, y) to CT intensities. Given a specified number of projection views, sinograms are generated via the Radon transform, and training minimizes the discrepancy between predicted and ground-truth sinograms by applying the Radon transform to the reconstructed output. As a baseline, SIREN is fully trained for 1,000 iterations, with an intermediate checkpoint saved at iteration 500; starting from this checkpoint, PEFT is performed for the remaining 500 iterations under the same objective. We apply PriLoRA, which modulates the hidden linear layers of SIREN while keeping the base weights frozen, using a lightweight, zero-initialized modulator (LayerNormâĂŞSiLUâĂŞLinear) that takes a 32-dimensional pooled feature representation and predicts four modulation parameters per hidden layer to control the sinusoidal low-rank update.

Analysis. For CT reconstruction, PriLoRA achieves improved quantitative (Table 1) performance across PSNR, LPIPS, and SSIM at all projection counts. Qualitatively (Fig. 2), while differences are subtle, PriLoRA preserves slightly finer structural details without introducing additional artifacts.

3.5 Ablation Studies

Parameter-Wise Contribution Analysis. We ablate the modulation parameters by selectively disabling each component. As shown in Table 2 (left), removing any parameter degrades generation quality, indicating that the components play complementary roles. Activating all four yields the best FID on both training and test sets, while partial configurations may improve individual metrics (e.g., CLIP) at the cost of overall generative fidelity.

Rank Ablation on Classification. We compare LoRA, SineLoRA, and PriLoRA across adapter ranks to assess classification performance. As shown in Table 2 (right), PriLoRA consistently matches or outperforms baselines, with particularly strong performance at low and mid ranks. Notably, PriLoRA remains competitive even at rank 2 and achieves peak accuracy and F1 at modest ranks, demonstrating efficient and stable adaptation without large adapter capacity.

Table 2. Ablation studies on PriLoRA. *(Left)* parameter ablation over modulation parameters α–δ on the generation task. *(Right)* Rank ablation comparing LoRA, SineLoRA, and PriLoRA on classification, reporting AUC-ROC/accuracy (top) and precision/F1-score (bottom) for each rank.

α	β	γ	δ	FID_T	FID_{Te}	CLIP
✓	✗	✗	✗	85.51	94.88	27.45
✗	✓	✗	✗	82.20	92.21	27.56
✗	✗	✓	✗	89.18	99.05	**27.96**
✗	✗	✗	✓	85.57	95.32	27.50
✓	✓	✗	✗	97.48	106.79	28.16
✓	✓	✓	✗	94.47	104.96	26.78
✓	✗	✓	✓	98.02	109.25	27.50
✗	✓	✓	✗	90.03	100.86	27.90
✓	✓	✓	✓	**70.15**	**80.73**	27.65

Rank	LoRA	SineLoRA	PriLoRA
2	0.9833 / 0.851	**0.9858** / 0.8470	0.9754 / **0.8600**
	0.8989 / 0.8480	0.8985 / 0.8454	0.8986 / **0.8586**
4	0.9863 / 0.868	0.9891 / 0.835	**0.9903** / **0.8770**
	0.9078 / 0.8644	0.8935 / 0.8344	0.9059 / **0.8745**
8	0.9847 / 0.842	0.9782 / 0.810	**0.9864** / **0.880**
	0.8912 / 0.8363	0.8820 / 0.8061	**0.9101** / **0.8799**
16	0.9803 / 0.8160	**0.9849** / 0.8210	0.9823 / **0.8980**
	0.8872 / 0.8019	0.8780 / 0.8157	**0.9161** / **0.8989**
32	0.9808 / 0.870	**0.9889** / 0.865	0.9803 / **0.878**
	0.9079 / 0.8682	0.9058 / 0.8629	**0.9109** / **0.8751**

Model Efficiency. Using CUDA event timing over 50 samples, LoRA achieves 2.40 ms inference latency, while SineLoRA, SineDoRA, and PriLoRA incur slightly higher latencies of 2.84 ms, 3.05 ms, and 2.88 ms, respectively, indicating only a modest overhead relative to LoRA while preserving parameter-efficient inference. Compared to LoRA, PriLoRA increases the number of trainable parameters by 1–6%. Compared to SineDoRA, PriLoRA uses fewer parameters for generation (−1.4%) and CT reconstruction (-0.36%), and incurs only a modest increase for segmentation (+5.2%, 0.88M) and classification (+1.5%, 0.06M).

4 Discussion and Conclusion

We introduce PriLoRA, a prior-conditioned sinusoidal low-rank adapter that improves the expressivity of parameter-efficient fine-tuning while preserving LoRA's efficiency. By conditioning sinusoidal low-rank updates on input-derived priors, PriLoRA adapts its non-linear behavior across medical tasks without increasing adapter rank or requiring task-specific tuning. Experiments across generation, classification, segmentation, and CT reconstruction show consistent gains in robustness and transfer over standard LoRA and representative PEFT baselines.

Limitations. PriLoRA is particularly effective in settings with heterogeneous adaptation demands, limited adapter rank, and distribution shifts across modalities, where fixed hyperparameter adapters can be brittle. While the method adaptively selects its modulation parameters, performance may be sensitive to to extreme initialization of the base weights W_0, which can reduce the effective adaptation capacity during training. Although we evaluate across tasks and modalities, each task is evaluated on one main dataset; multi-institution or scanner robustness remains future work. Overall, these results suggest that input-conditioned modulation is a promising direction for scalable and reliable adaptation of medical foundation models.

Acknowledgment. This work was supported by the German Research Foundation (DFG) as research unit DeSBi [KI-FOR 5363] (459422098)

Disclosure of Interests. The authors have no competing interests to declare that are relevant to the content of this article.

References

1. Ansari, M.S., Khan, M.S.A., Revankar, S., Varma, A., Mokhade, A.S.: Lightweight clinical decision support system using qlora-fine-tuned llms and retrieval-augmented generation, (2025). arXiv:2505.03406 arXiv preprint
2. Bany Muhammad, M., Yeasin, M.: Eigen-cam: Visual explanations for deep convolutional neural networks. SN Computer Science **2**(1), 47 (2021)
3. Brag, Y.E.: Lung nodules detection dataset annotations. https://www.kaggle.com/datasets/younesselbrag/lung-nodules-detection-dataset-annotations (2025), accessed: 2026-02-12
4. Dettmers, T., Pagnoni, A., Holtzman, A., Zettlemoyer, L.: Qlora: efficient finetuning of quantized LLMs. Adv. Neural. Inf. Process. Syst. **36**, 10088–10115 (2023)
5. Ferrara, N.: Vascular endothelial growth factor and age-related macular degeneration: from basic science to therapy. Nat. Med. **16**(10), 1107–1111 (2010)
6. Guan, H., Liu, M.: Domain adaptation for medical image analysis: a survey. IEEE Trans. Biomed. Eng. **69**(3), 1173–1185 (2021)
7. Gutman, D., et al.: Skin lesion analysis toward melanoma detection: a challenge at the international symposium on biomedical imaging (ISBI) 2016, hosted by the international skin imaging collaboration (ISIC). arXiv preprint arXiv:1605.01397 (2016)
8. Han, Z., Gao, C., Liu, J., Zhang, J., Zhang, S.Q.: Parameter-efficient fine-tuning for large models: a comprehensive survey. arXiv preprint arXiv:2403.14608 (2024)
9. Hu, E.J., et al.: Lora: low-rank adaptation of large language models. arXiv preprint arXiv:2106.09685 (2021)
10. Hu, E.J., et al.: Lora: low-rank adaptation of large language models. ICLR **1**(2), 3 (2022)
11. Ji, Y., Saratchandran, H., Gordon, C., Zhang, Z., Lucey, S.: Efficient learning with sine-activated low-rank matrices. In: The Thirteenth International Conference on Learning Representations (2025). https://openreview.net/forum?id=cWGCkd7mCp
12. Kazerouni, A., et al.: Diffusion models in medical imaging: a comprehensive survey. Med. Image Anal. **88**, 102846 (2023)
13. Kirillov, A., et al.: Segment anything. In: Proceedings of the IEEE/CVF International Conference on Computer Vision, pp. 4015–4026 (2023)
14. Liu, S.Y., et al.: Dora: weight-decomposed low-rank adaptation. In: Forty-first International Conference on Machine Learning (2024)

15. Mao, Y., Li, H., Pang, W., Papanastasiou, G., Yang, G., Wang, C.: Selora: self-expanding low-rank adaptation of latent diffusion model for medical image synthesis. arXiv preprint arXiv:2408.07196 (2024)
16. Mao, Y., et al.: A survey on lora of large language models. Front. Comp. Sci. **19**(7), 197605 (2025)
17. Meng, F., Wang, Z., Zhang, M.: Pissa: principal singular values and singular vectors adaptation of large language models. Adv. Neural. Inf. Process. Syst. **37**, 121038–121072 (2024)
18. Moor, M., et al.: Foundation models for generalist medical artificial intelligence. Nature **616**(7956), 259–265 (2023)
19. Podell, D., et al.: SDXL: improving latent diffusion models for high-resolution image synthesis. arXiv preprintarXiv:2307.01952 (2023)
20. Radford, A., et al.: Learning transferable visual models from natural language supervision. In: International Conference on Machine Learning, pp. 8748–8763. PMLR (2021)
21. Rivera, M., Hoyos, A.: Colora: efficient fine-tuning for convolutional models with a study case on optical coherence tomography image classification. arXiv preprint arXiv:2505.18315 (2025)
22. Rombach, R., Blattmann, A., Lorenz, D., Esser, P., Ommer, B.: High-resolution image synthesis with latent diffusion models. In: Proceedings of the IEEE/CVF Conference on Computer Vision and Pattern Recognition, pp. 10684–10695 (2022)
23. Siméoni, O., et al.: Dinov3. arXiv preprint arXiv:2508.10104 (2025)
24. Sitzmann, V., Martel, J., Bergman, A., Lindell, D., Wetzstein, G.: Implicit neural representations with periodic activation functions. Adv. Neural. Inf. Process. Syst. **33**, 7462–7473 (2020)
25. Vedaldi, A., Zisserman, A.: VGG Convolutional Neural Networks Practical, vol. 66, p. 27. Department of Engineering Science, University of Oxford (2016)
26. Wu, J., et al.: Medical sam adapter: adapting segment anything model for medical image segmentation. Med. Image Anal. **102**, 103547 (2025)
27. Yang, F., et al.: Annotations of lung abnormalities in the shenzhen chest x-ray dataset for computer-aided screening of pulmonary diseases. Data **7**(7), 95 (2022)
28. Yang, J., et al.: Medmnist v2-a large-scale lightweight benchmark for 2d and 3d biomedical image classification. Sci. Data **10**(1), 41 (2023)
29. Zech, J.R., Badgeley, M.A., Liu, M., Costa, A.B., Titano, J.J., Oermann, E.K.: Variable generalization performance of a deep learning model to detect pneumonia in chest radiographs: a cross-sectional study. PLoS Med. **15**(11), e1002683 (2018)
30. Zhang, Q., et al.: Adaptive budget allocation for parameter-efficient fine-tuning. In: International Conference on Learning Representations. Openreview (2023)

Proto-CAP: Prototype Memory Fusion and Curriculum-Adaptive Loss for Pediatric Myopia Visual Question Answering

Xu Yan[1,2], Yanlin Wu[1,2], Zixun Wang[3], Jingtao Yu[3], Ruihua Wei[3], and Tao Li[1,2](✉)

[1] Nankai University, Tianjin, China
litao@nankai.edu.cn
[2] Haihe Laboratory of ITAI, Tianjin, China
[3] Tianjin Medical University Eye Hospital, Tianjin, China

Abstract. Medical Visual Question Answering (Med-VQA) is pivotal for assisting clinical decision-making and automated diagnosis. While deep learning has achieved remarkable success in general VQA tasks, Med-VQA research targeting specific diseases—particularly pediatric myopia involving multimodal data—remains largely underexplored, primarily due to the scarcity of publicly available, expert-annotated datasets. In this paper, we present the Pediatric Myopia VQA dataset (PM-VQA). To the best of our knowledge, this is the first publicly available Med-VQA dataset dedicated to pediatric myopia. It comprises **2,741** multimodal images, totaling **14,781** expert-verified question-answer pairs. Furthermore, we propose an enhanced framework named Proto-CAP, designed to address the semantic gap and overfitting issues inherent in medical small-scale data. Within Proto-CAP, we introduce a Question-Guided Prototype Fusion mechanism and a Curriculum-Adaptive Poly-2 Loss to enhance prior knowledge injection and dynamically balance training difficulty. Experiments on PM-VQA demonstrate that our proposed method achieves state-of-the-art performance, reaching an accuracy of **79.19%** and an F1-score of **78.74%**, which outperforms existing methods by **2.10%** in accuracy and **2.14%** in F1-score.Code and dataset will be available at https://github.com/PixelOptic-Dev/Proto-CAP.

Keywords: Pediatric Myopia · Medical Visual Question Answering · Multimodal Fusion

1 Introduction

Pediatric myopia has emerged as an increasingly severe public health challenge globally, projected to affect approximately 50% of the world's population by

X. Yan, Y. Wu—Equal contribution.

G. Yang et al. (Eds.): MICCAI 2026, LNCS 16878, pp. 450–459, 2027.
https://doi.org/10.1007/978-3-032-38059-3_43

2050 [4]. Its accurate diagnosis and intervention rely heavily on the comprehensive multimodal analysis of Color Fundus Photography(CFP) and Optical Coherence Tomography (OCT). Medical Visual Question Answering, capable of interpreting medical imagery and providing interactive diagnostic evidence, presents a promising solution to mitigate the shortage of ophthalmic medical resources [2, 11]. However, the direct deployment of current Med-VQA in this field is constrained by a multitude of complex factors.

One major hurdle is the scarcity of domain-specific data. Existing public datasets (e.g., VQA-RAD [7] and SLAKE [12]) primarily target general radiology scenarios and lack fine-grained annotations specific to ophthalmic lesions. Such general-purpose data fails to support models in learning the complex correlations between subtle fundus changes and refractive status in pediatric myopia [21], thereby severely impeding the development of professional-grade diagnostic tools.Another is the challenge of optimizing deep models on small-scale medical data. In the absence of large-scale pre-training data, directly training deep models is highly prone to overfitting [18]. Although strategies such as sample re-weighting [17] and static polynomial loss [8] have alleviated this issue to some extent, they are either computationally expensive or lack adaptability to training dynamics, failing to strike an optimal balance between gradient boosting and hard example mining.

To address these challenges, we propose Proto-CAP, an enhanced framework designed specifically for high-performance Med-VQA. We first construct PM-VQA—the first large-scale, expert-verified dataset dedicated to pediatric myopia. To effectively utilize this data, we design a prototype fusion module that explicitly models medical concepts via a learnable memory bank, injecting question-guided prior knowledge into visual representations. Simultaneously, we introduce a Curriculum-Adaptive Poly-2 Loss, which dynamically adjusts curriculum difficulty based on the model's convergence status.

The main contributions are summarized as follows:

1. We create a pediatric myopia medical visual question answering dataset called PM-VQA, which is the first expert-verified dataset specifically for pediatric myopia.
2. We propose the Proto-CAP framework, featuring two core innovations: a Prototype Memory Fusion module that bridges semantic gaps via learnable medical concepts, while a Curriculum-Adaptive Poly-2 Loss dynamically balances gradient boosting and hard example mining based on real-time proficiency.
3. Extensive experiments demonstrate that Proto-CAP significantly outperforms state-of-the-art Med-VQA methods across all metrics.

2 Dataset

2.1 Comparison with Existing Datasets

We performed a detailed comparison between PM-VQA and existing mainstream medical VQA datasets. As shown in Table 1, current datasets exhibit certain limitations. While SLAKE [12] and VQA-RAD [7] contain multimodal data and have undergone expert validation, they focus primarily on radiology rather than ophthalmology. Conversely, ophthalmology-related datasets such as OphthalWeChat [22], DME-VQA [19], and INSEGCAT-VQA [20] mostly lack multimodal support. Furthermore, they are largely constructed from public datasets without rigorous QA validation, making it difficult to meet the requirements for multimodal data fusion analysis in clinical practice. To this end, we create PM-VQA, a rigorously expert-verified, multimodal dataset derived from clinical practice. With **2,741** images and **14,781** QA pairs, it fills the void for high-quality clinical ophthalmic VQA data.

Table 1. Comparison of existing medical VQA datasets in terms of scale, domain, modality, data source and QA verification.

Reference	Image Number	QA Pairs	Domain	Multi-modal	Data Source	QA Verification
SLAKE [12]	642	14028	Radiology	✓	Public datasets	✓
VQA-RAD [7]	315	3515	Radiology	✓	Public datasets	✓
OphthalWeChat [22]	3469	30120	Eye	✓	Public datasets	✗
Holland et al. [5]	41926	479710	Eye	✗	Clinical	✓
DME-VQA [19]	679	13470	Eye	✗	Public datasets	✗
INSEGCAT-VQA [20]	4647	39008	Eye	✗	Public datasets	✗
PM-VQA(ours)	2741	14781	Eye	✓	Clinical	✓

2.2 Data Collection and Construction Process

As illustrated in Fig. 1, the PM-VQA dataset construction involves three phases. First, we collected multimodal data—comprising Color Fundus Photography (CFP), Optical Coherence Tomography (OCT), and clinical records—from a collaborating hospital. Following rigorous quality screening and data anonymization, two ophthalmologists formulated clinical QA pairs, which were subsequently validated by a senior expert to ensure medical accuracy. Finally, the processed data was formatted into JSON and partitioned into **10,271** training, **2,287** validation, and **2,223** testing QA pairs. This study and the use of all data were approved by the ethics committee of the Tianjin Medical University Eye Hospital [No. 2024KY-67].

3 Method

We propose a Proto-CAP enhanced framework for Medical Visual Question Answering. As illustrated in Fig. 1, this framework is designed to mimic the clinical cognitive pattern where physicians simultaneously consider multiple correlated questions during diagnosis. The model comprises three core components: a multi-modal backbone based on the Medical Multi-question Learning (MMQL) [1], a Prototype Memory Fusion module, and the Curriculum-Adaptive Poly-2 Loss function (Fig. 2).

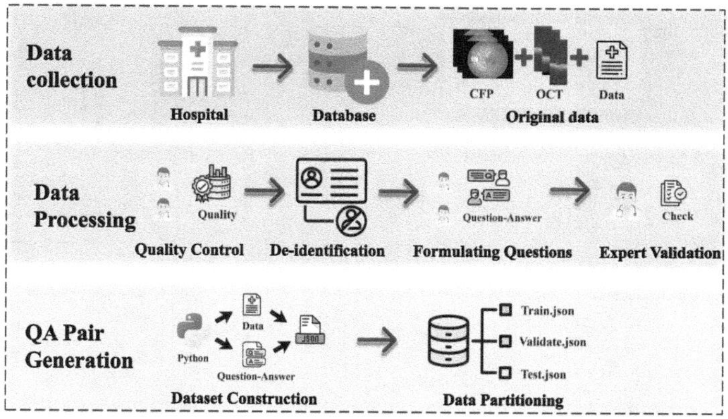

Fig. 1. Construction pipeline of the PM-VQA dataset. The process consists of three stages: Data Collection, Data Processing, and QA Pair Generation.

3.1 Medical Multi-question Learning Backbone

To capture correlations among questions, we adopt the Medical Multi-question Learning (MMQL) [1], consolidating all questions Q_k related to an image v_k into a unified sequence T for joint reasoning. The concatenation is formulated as follows:

$$T = q_1 \circ q_2 \circ \cdots \circ q_i \circ a_i \circ \cdots \circ q_{|Q_k|} \tag{1}$$

where $\circ$ denotes the concatenation operation. During training, answers are randomly replaced with [MASK] tokens based on a probability p. Remaining ground-truth answers serve as context prompts, guiding the model to infer diagnoses from known information.

The sequence T is encoded by a pre-trained BERT-Base [6] to extract textual embeddings $Q_{emb} \in \mathbb{R}^{L \times D}$. Concurrently, the image I is processed by a Swin Transformer [14] to obtain visual features $V \in \mathbb{R}^{49 \times D}$. Before multimodal fusion, these features are enhanced by the Prototype Memory Fusion module to improve the representation of clinical concepts.

3.2 Prototype Memory Fusion

While MMQL provides inter-question context, the model may still lack domain-specific medical knowledge across samples. To address this limitation, we design a **Prototype Fusion** module positioned prior to the Transformer fusion stage, serving to enhance the visual features.

We maintain a learnable memory matrix $M \in \mathbb{R}^{K \times D}$, where K denotes the number of medical concept prototypes. To retrieve the medical knowledge most relevant to the current multi-question sequence T, we first apply global average pooling to the textual embeddings Q_{emb} to obtain a global vector q_{global}.

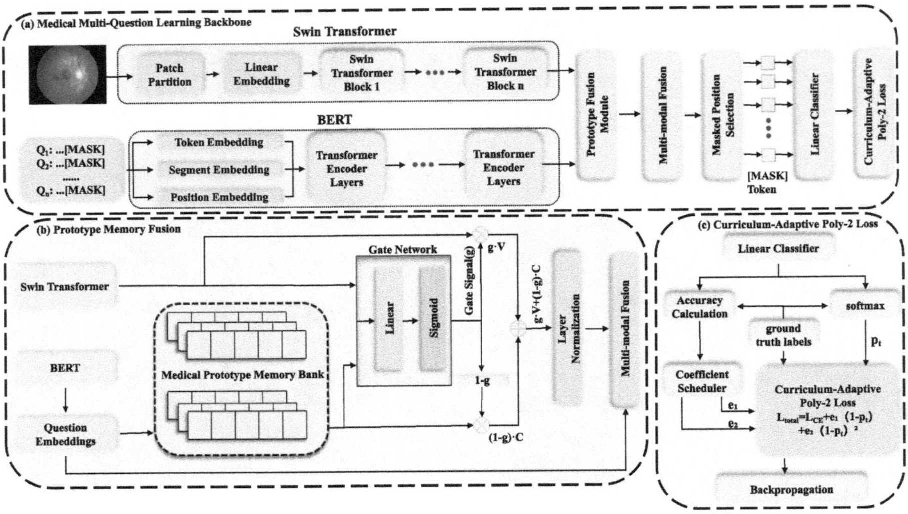

Fig. 2. (a) The MMQL backbone encodes images and masked question sequences via Swin Transformer and BERT for joint prediction. (b) Prototype Memory Fusion injects retrieved concept features C into visual features V via a gating mechanism defined as $g \cdot V + (1-g) \cdot C$. (c) Curriculum-Adaptive Poly-2 Loss dynamically adjusts coefficients e_1 and e_2 based on training accuracy to balance sample difficulty.

This vector serves as the query to compute attention scores against the memory matrix:

$$Attn_{proto} = \text{Softmax}((q_{global} W_q) \cdot M^T) \tag{2}$$

Based on these attention weights, we aggregate the prototypes to yield the concept features C that best align with the current context.

The retrieved concept features are subsequently broadcasted and injected into the original visual features V. We employ a Gating Mechanism to dynamically regulate the injection intensity, ensuring the preservation of intrinsic image features while incorporating external knowledge:

$$g = \sigma(W_g \cdot [V; C]) \tag{3}$$

$$V_{enhanced} = \text{LayerNorm}(g \cdot V + (1-g) \cdot C) \tag{4}$$

Finally, the enhanced visual features $V_{enhanced}$ replace the original V and are fed into the main Transformer alongside the textual embeddings Q_{emb} for final interaction and prediction.

3.3 Curriculum-Adaptive Poly-2 Loss

During the training process of Medical VQA, the learning state of the model evolves dynamically. Traditional loss functions typically rely on fixed parameters, lacking sensitivity to the model's actual current capabilities. To address this limitation, we refine the polynomial loss framework and propose a **Curriculum-Adaptive Poly-2 Loss.** This loss function eliminates the dependency on static parameters by dynamically modulating polynomial coefficients based on current prediction accuracy, thereby facilitating self-paced learning. The total loss function is defined as a polynomial form based on Taylor expansion:

$$L_{total} = L_{CE} + e_1 \cdot (1 - P_t) + e_2 \cdot (1 - P_t)^2 \quad (5)$$

L_{CE} is the Cross-Entropy loss and P_t is the target prediction probability. The coefficients e_1 and e_2 dynamically adapt to the model's current proficiency: e_1 follows a cosine decay to accelerate initial convergence, while e_2 increases linearly to emphasize hard samples (low P_t) during later training stages.

Through this mechanism, when the model exhibits low accuracy, the loss function prioritizes the correction dominated by the e_1 term to facilitate rapid convergence. Conversely, as the model achieves high accuracy, the influence of the e_2 term is amplified, compelling the model to focus on hard samples with lower prediction probabilities, thereby further enhancing generalization capability.

4 Experiment

4.1 Implementation Details

In the experiments, the input images are resized to 224 × 224 to match the input requirements of the Swin Transformer. We adopt Swin Transformer [14] and BERT [6] pre-trained on ImageNet-1K and Wikipedia, respectively, as the visual and textual backbones. We use the AdamW [15] optimizer with differential learning rates to fine-tune different components: the learning rate is set to 5×10^{-6} for the BERT encoder, 1×10^{-5} for the Swin Transformer and the fusion Transformer, and initialized at 1×10^{-4} for the classifier and other randomly initialized parameters. The proposed Curriculum-Adaptive Poly-2 Loss is employed as the objective function. Training is conducted for 80 epochs with a batch size of 16. Our framework was implemented using PyTorch, and all experiments were conducted on a single Nvidia RTX A6000 GPU.

4.2 Comparisons With Other Methods

To evaluate the effectiveness of our proposed Proto-CAP, we conducted comprehensive experiments and compared it with several methods in the medical visual question answering domain. The baseline models include CMSA [3], M-Mixup [13], M2I2 [10], MUMC [9], MEVF-BAN [23], Surgical-VQA [16], and MMQL [1].

We evaluate the performance using four standard metrics: Accuracy, Precision, Recall, and F1-score. The quantitative comparison results on the PM-VQA dataset are presented in Table 2 Our Proto-CAP achieves the state-of-the-art performance with an Accuracy of 79.19% and an F1-score of 78.74%. Specifically, it surpasses the second-best method, MMQL, by a significant margin of 2.10% in Accuracy and 2.14% in F1-score (Table 2).

Table 2. Comparison of performance metrics on the PM-VQA.

Method	Accuracy(%)	Precision(%)	Recall(%)	F1-Score(%)
CMSA [3]	50.07	39.49	50.07	41.15
M-Mixup [13]	53.58	30.65	53.58	38.68
M2I2 [10]	57.76	48.79	57.76	51.76
MUMC [9]	63.52	62.06	63.52	61.86
MEVF-BAN [23]	65.76	65.82	65.76	65.48
Surgical-VQA [16]	69.09	71.77	69.09	67.48
MMQL [1]	77.09	78.35	77.09	76.60
Proto-CAP(ours)	**79.19**	**79.41**	**79.19**	**78.74**

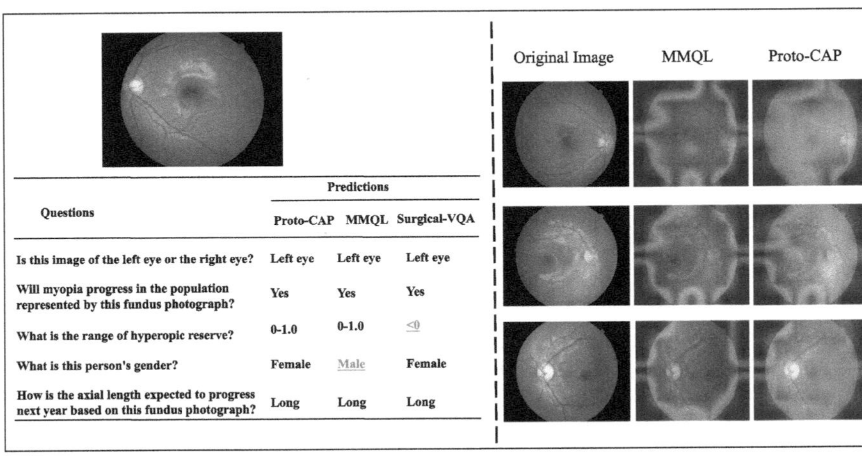

Fig. 3. The left panel shows the prediction results of different models on pediatric myopia-related questions. The right panel displays the visualization of attention heatmaps, comparing the interpretability of MMQL and Proto-CAP.

Furthermore, we provide a qualitative comparison to intuitively demonstrate the superiority of our model. As illustrated in Fig. 3, the left panel shows specific question-answering cases on a pediatric myopia fundus image. Proto-CAP successfully answers all questions correctly, whereas MMQL struggles with demographic features (misclassifying the patient's gender as "Male") and Surgical-VQA fails on fine-grained clinical assessments (predicting the hyperopic reserve as "<0" instead of "0–1.0"). The right panel of Fig. 3 visualizes the attention heatmaps. Compared to the original MMQL, Proto-CAP generates more concentrated and precise focal areas around the key anatomical structures.

The remarkable reasoning capability of Proto-CAP stems fundamentally from its mechanistic breakthroughs. While the traditional MMQL excels at capturing inter-question dependencies, its lack of medical priors hinders the alignment of localized visual anomalies with deep clinical semantics. Addressing this bottleneck, our Prototype Memory Fusion module successfully bridges the vision-language semantic gap by explicitly injecting parameterized ophthalmic concepts into the visual space. This innovation forces the network to move beyond fitting superficial image features and pivot towards authentic pathology-aware reasoning. Consequently, the model's attention dynamically converges on the most diagnostically valuable clinical regions, ultimately generating predictions that are not only accurate but also fully consistent with ophthalmic clinical logic.

4.3 Ablation Studies

To evaluate each component's contribution, we conducted ablation studies on the PM-VQA dataset using the MMQL backbone as the baseline. The quantitative results are summarized in Table 3. Integrating the Prototype Memory Fusion (+Proto) increases Accuracy by 1.18% and F1-score by 0.98%, demonstrating that dynamically injecting medical concepts enhances feature representation. Applying the Curriculum-Adaptive Poly-2 Loss (+CAP) alone yields a 0.65% Accuracy and 0.74% F1-score improvement, validating its ability to dynamically balance optimization between hard and easy samples. Finally, combining both modules (+Proto-CAP) achieves the best overall performance (79.19% Accuracy, 78.74% F1-score), outperforming the baseline by 2.10% and 2.14%, respectively. This highlights their strong complementarity: the Proto module provides enriched semantic features, while the Curriculum-Adaptive Poly-2 Loss ensures they are optimally learned.

Table 3. Ablation study on the proposed Proto and CAP modules.

Method	Accuracy(%)	Precision(%)	Recall(%)	F1-Score(%)
baseline	77.09	78.35	77.09	76.60
+Proto	78.27	77.63	78.27	77.58
+CAP	77.74	77.50	77.74	77.34
+Proto-CAP	**79.19**	**79.41**	**79.19**	**78.74**

5 Conclusion

In this paper, we propose Proto-CAP, a novel Med-VQA framework for pediatric myopia assessment. Built on a Medical Multi-question Learning backbone, it simultaneously processes multiple related clinical inquiries. Proto-CAP introduces two key innovations: a Prototype Memory Fusion module that integrates learnable medical concepts into visual features, and a Curriculum-Adaptive Poly-2 loss that dynamically adapts to sample difficulty during training. Extensive experiments on the PM-VQA dataset demonstrate that Proto-CAP significantly outperforms state-of-the-art models, establishing a robust baseline for pediatric ophthalmology and broader computer-aided diagnosis.

Acknowledgments. This work is supported by the National Natural Science Foundation of China (No. 62272248), the Natural Science Foundation of Tianjin (No. 23JCZDJC01010), the Tianjin Key Medical Discipline Construction Project (No. TJYXZDXK-3-004A-2) and Haihe Lab of ITAI Science and Technology Project (No. 25HHXCSS00007).

Disclosure of Interests. The authors have no competing interests to declare that are relevant to the content of this article.

References

1. Chen, Q., Bian, M., Xu, H.: MMQL: multi-question learning for medical visual question answering. In: International Conference on Medical Image Computing and Computer-Assisted Intervention, pp. 480–489. Springer (2024)
2. Eslami, S., De Melo, G., Meinel, C.: Does clip benefit visual question answering in the medical domain as much as it does in the general domain? arXiv preprint arXiv:2112.13906 (2021)
3. Gong, H., Chen, G., Liu, S., Yu, Y., Li, G.: Cross-modal self-attention with multi-task pre-training for medical visual question answering. In: Proceedings of the 2021 International Conference on Multimedia Retrieval, pp. 456–460 (2021)
4. Holden, B.A., et al.: Global prevalence of myopia and high myopia and temporal trends from 2000 through 2050. Ophthalmology **123**(5), 1036–1042 (2016)
5. Holland, R., et al.: Specialized curricula for training vision language models in retinal image analysis. NPJ Digit. Med. **8**(1), 532 (2025)
6. Kenton, J.D.M.W.C., Toutanova, L.K.: Bert: pre-training of deep bidirectional transformers for language understanding. In: Proceedings of NAACL-HLT, vol. 1, pp. 4171–4186, Minneapolis, Minnesota (2019)
7. Lau, J.J., Gayen, S., Ben Abacha, A., Demner-Fushman, D.: A dataset of clinically generated visual questions and answers about radiology images. Sci. Data **5**(1), 1–10 (2018)
8. Leng, Z., et al.: Polyloss: a polynomial expansion perspective of classification loss functions. arXiv preprint arXiv:2204.12511 (2022)
9. Li, P., Liu, G., He, J., Zhao, Z., Zhong, S.: Masked vision and language pre-training with unimodal and multimodal contrastive losses for medical visual question answering. In: International Conference on Medical Image Computing and Computer-Assisted Intervention, pp. 374–383. Springer (2023)
10. Li, P., Liu, G., Tan, L., Liao, J., Zhong, S.: Self-supervised vision-language pre-training for medial visual question answering. In: 2023 IEEE 20th International Symposium on Biomedical Imaging (ISBI), pp. 1–5. IEEE (2023)
11. Lin, Z., et al.: Medical visual question answering: a survey. Artif. Intell. Med. **143**, 102611 (2023)
12. Liu, B., Zhan, L.M., Xu, L., Ma, L., Yang, Y., Wu, X.M.: Slake: a semantically-labeled knowledge-enhanced dataset for medical visual question answering. In: 2021 IEEE 18th International Symposium on Biomedical Imaging (ISBI), pp. 1650–1654. IEEE (2021)
13. Liu, L., Su, X.: How well apply multimodal MIXUP and simple MLPs backbone to medical visual question answering? In: 2022 IEEE International Conference on Bioinformatics and Biomedicine (BIBM), pp. 2648–2655. IEEE (2022)

14. Liu, Z., et al.: Swin transformer: hierarchical vision transformer using shifted windows. In: Proceedings of the IEEE/CVF International Conference on Computer Vision, pp. 10012–10022 (2021)
15. Loshchilov, I., Hutter, F.: Decoupled weight decay regularization. arXiv preprint arxiv:1711.05101 (2017) arXiv:1711.05101 (2017)
16. Seenivasan, L., Islam, M., Krishna, A.K., Ren, H.: Surgical-VQA: visual question answering in surgical scenes using transformer. In: International Conference on Medical Image Computing and Computer-Assisted Intervention, pp. 33–43. Springer (2022)
17. Shu, J., et al.: Meta-weight-net: learning an explicit mapping for sample weighting. Adv. Neural. Inf. Process. Syst. **32** (2019)
18. Tajbakhsh, N., Jeyaseelan, L., Li, Q., Chiang, J.N., Wu, Z., Ding, X.: Embracing imperfect datasets: a review of deep learning solutions for medical image segmentation. Med. Image Anal. **63**, 101693 (2020)
19. Tascon-Morales, S., Márquez-Neila, P., Sznitman, R.: Consistency-preserving visual question answering in medical imaging. In: International Conference on Medical Image Computing and Computer-Assisted Intervention, pp. 386–395. Springer (2022)
20. Tascon-Morales, S., Márquez-Neila, P., Sznitman, R.: Localized questions in medical visual question answering. In: International Conference on Medical Image Computing and Computer-Assisted Intervention, pp. 361–370. Springer (2023)
21. Ting, D.S., et al.: Deep learning in ophthalmology: the technical and clinical considerations. Prog. Retin. Eye Res. **72**, 100759 (2019)
22. Xu, P., et al.: Benchmarking large multimodal models for ophthalmic visual question answering with ophthalwechat. arXiv preprint arXiv:2505.19624 (2025)
23. Zhan, L.M., Liu, B., Fan, L., Chen, J., Wu, X.M.: Medical visual question answering via conditional reasoning. In: Proceedings of the 28th ACM International Conference on Multimedia, pp. 2345–2354 (2020)

QCAgent: An Agentic Framework for Quality-Controllable Pathology Report Generation from Whole Slide Image

Rundong Wang[1,2], Wei Ba[5], Yingtai Li[1,2], Ying Zhou[1,2], Bowen Liu[1], Baizhi Wang[1,2], Yuhao Wang[1,2], Zhidong Yang[6], Kun Zhang[1,2], Rui Yan[1,2,3,4(✉)], and Shaohua Kevin Zhou[1,2,3,4(✉)]

[1] School of Biomedical Engineering, Division of Life Sciences and Medicine, University of Science and Technology of China (USTC), Hefei, Anhui, China
yanrui@ustc.edu.cn, s.kevin.zhou@gmail.com

[2] Center for Medical Imaging, Robotics, Analytic Computing and Learning (MIRACLE), Suzhou Institute for Advanced Research, USTC, Suzhou, Jiangsu, China

[3] Jiangsu Provincial Key Laboratory of Multimodal Digital Twin Technology, Suzhou, Jiangsu, China

[4] Biomedical Basic Research Center (BBRC) of Jiangsu, Suzhou, Jiangsu, China

[5] Chinese PLA General Hospital Ninth Medical Center, Beijing, China

[6] Hong Kong University of Science and Technology (HKUST), Hong Kong, China

Abstract. Recent methods for pathology report generation from whole-slide images (WSIs) are capable of producing slide-level diagnostic descriptions but fail to ground fine-grained statements in localized visual evidence. Furthermore, they lack control over which diagnostic details to include and how to verify them. Inspired by emerging agentic analysis paradigms and the diagnostic workflow of pathologists, who selectively examine multiple fields of view, we propose **QCAgent**, an agentic framework for quality-controllable WSI report generation. The core innovations of this framework are as follows: (i) it incorporates a customized critique mechanism guided by a user-defined checklist specifying required diagnostic details and constraints; (ii) it re-identifies informative regions in the WSI based on the critique feedback and text-patch semantic retrieval, a process that iteratively enriches and reconciles the report. Experiments demonstrate that by making report requirements explicitly prompt-defined, constraint-aware, and verifiable through evidence-grounded refinement, QCAgent enables controllable generation of clinically meaningful and high-coverage pathology reports from WSI. Code is available at https://github.com/cs-w0r1d/qcagent.

Keywords: Whole Slide Image · Pathology Report Generation · Agent · Quality-controllability

R. Wang and W. Ba—Equal contribution.

G. Yang et al. (Eds.): MICCAI 2026, LNCS 16878, pp. 460–470, 2027.
https://doi.org/10.1007/978-3-032-38059-3_44

1 Introduction

Pathology foundation models [12,17,21,25,26] have significantly advanced whole slide image (WSI) analysis and its applications. Among its applications, WSI-to-Text generation promises to reduce the pathologist's burden and enhance diagnostic consistency by generating descriptions [4,10,15,20]. Unlike general image captioning, however, a clinical standard report must not only provide a final diagnosis but also rigorously cover specimen context, microscopic morphological descriptors (e.g., nuclear pleomorphism), and critical predictors such as lymphovascular invasion (LVI).

Due to the gigapixel scale of WSIs, current generative models (e.g., HistGen [10], SGMT [18] and PRISM [20]) predominantly follow a single-pass paradigm, leading to two intractable issues. First, feature collapse occurs: to maintain computational feasibility, the model compresses thousands of image patches into compact global vectors. This drastic dimensionality reduction, while capturing general lesions, inevitably erases spatially sparse but diagnostically decisive features, such as micro-lesions or evidence of LVI [5]. Second, linguistic prior bias arises: lacking a "visual look-back" mechanism, Large Language Models (LLMs) tend to hallucinate generic clinical commons (e.g., "cellular atypia") based on diagnostic labels (e.g., "gastric cancer") when global features are insufficient for detailed synthesis. Such details lack correlation with specific visual evidence from the current slide and do not address missing information required for clinical quality control, severely undermining the clinical credibility of reports [7].

To bridge the gap between global context and local evidence, we draw inspiration from how pathologists iteratively navigate between low-power screening and high-power verification. This workflow aligns with emerging agentic paradigms [13,19,23,24,29]. Although recent work has explored agents in computational pathology [2,3,9,11,16,22], agentic *on-demand back-tracking* for controllable, evidence-grounded report generation remains underexplored.

Motivated by this, we propose QCAgent, an agent-based, quality-controllable pathology report generation framework. First, the system employs a WSI foundation model, PRISM [20], to extract global semantic features, supplemented by representative patch information via a cluster-based random sampling strategy. Subsequently, it invokes a pathology-expert reasoning model, Patho-R1 [28], to generate an initial draft. Crucially, we introduce a quality control module that utilizes advanced LLMs [1,6,27] to simulate a pathologist's auditing behavior. It identifies critical missing information, converts them into targeted retrieval instructions, and triggers Patho-R1 to revise the report. This process iterates until the report passes the QC criteria. This mechanism enables evidence-based report generation by explicitly aligning visual findings with clinical narratives. The main contributions of this paper are three-fold:

- **Iterative QC architecture.** We reformulate report generation as a multi-round *audit–retrieve–revise* process. A structured QCAgent automatically identifies non-compliant statements and missing clinical information, enabling closed-loop refinement.

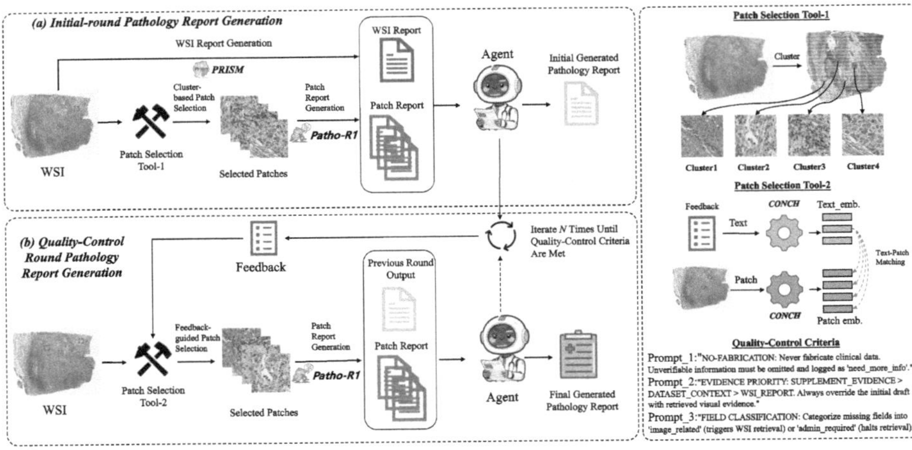

Fig. 1. Pipeline of QCAgent: (a) initial-round report generation using PRISM and cluster-based patch selection (Tool-1) with Patho-R1 patch reports; (b) QC-round iterative refinement using feedback-guided CONCH retrieval (Tool-2) and Patho-R1 evidence until QC criteria are met.

- **Demand-driven evidence retrieval.** We introduce a *text-guided evidence retrieval* mechanism. Instead of random sampling, the QCAgent converts missing fields into morphology-oriented queries to retrieve the most relevant visual patches from the WSI, grounding the report completion in targeted evidence.
- **Strict anti-hallucination design.** We adopt an evidence-priority scheme (supplemental visual evidence > initial draft) to reduce hallucinations. By explicitly separating verifiable from non-verifiable fields, only evidence-supported claims are retained in the final report.

2 Method

This section elaborates on the proposed QCAgent, an agent-based, quality-controllable framework for pathology report generation from whole-slide images (WSIs), as illustrated in Fig. 1. The overall computational pipeline of this framework consists of two core stages: (a) **Initial-round generation**, which aims to rapidly construct a WSI-level draft and extracts local visual evidence with global coverage; and (b) **Quality-Control (QC) rounds of iterative generation**, which translates information gaps within the report into explicit "evidence demands". Through a closed-loop mechanism of "feedback-guided evidence retrieval" and "evidence-driven revision", this stage progressively fulfills stringent quality control criteria.

Problem Formulation. Given a whole-slide image (WSI) S and a set of pre-extracted patch features $\mathcal{F} = \{f_1, \ldots, f_N\}$ obtained by CONCH [14], together with an initial draft report r_0 produced by a foundation model, our goal is to generate an informative and accurate pathology report r^* through T rounds of iterative refinement. We formulate the process as a multi-round optimization with a maximum of T iterations:

$$r^* = \text{QC-Iterate}(r_0, \mathcal{F}, S; T), \tag{1}$$

where each iteration comprises four consecutive stages: (i) quality-control (QC) assessment, (ii) evidence retrieval, (iii) visual diagnosis, and (iv) report revision.

We define *quality-controllability* as the capability of the system to explicitly follow and enforce a predefined set of quality control criteria $\mathcal{K}$ during inference. These criteria include, but are not limited to, a report schema, required fields, anti-hallucination rules, evidence-priority principles, and standardized writing templates. At each iteration, the system outputs structured QC results, enabling users to dynamically configure $(\mathcal{K}, T)$ to achieve a controllable trade-off between generation quality and computational cost. Under this formulation, "controllability" is not merely post-hoc scoring; instead, hard constraints are embedded into an iterative "audit–evidence seeking–revision" closed loop, yielding a transparent and auditable reasoning trace.

Concretely, our QC prompts implement three core rules:

NO-FABRICATION: Never fabricate clinical data. Any unverifiable information must be omitted from the report and explicitly logged as `need_more_info`.

EVIDENCE PRIORITY: When supplemental evidence conflicts with the initial draft, revisions follow the priority order SUPPLEMENT_EVIDENCE > DATASET_CONTEXT > WSI_REPORT.

FIELD CLASSIFICATION: Missing fields are categorized into `image_related` and `admin_required`.

Initial Round: Global Draft and Coverage-Oriented Evidence Extraction. As illustrated in Fig. 1(a), the initial round consists of two parallel streams:

1. **WSI-level report generation (PRISM).** We first employ a pretrained WSI-level foundation model to produce an initial draft report r_0, which serves as a global narrative backbone for subsequent refinement.
2. **Cluster-based patch selection (Patch Selection Tool-1).** To provide representative *coverage-oriented* evidence in the initial round, we perform clustering in the feature embedding space over candidate patches and uniformly sample from each cluster to construct a representative patch set $\mathcal{P}^{(1)}$. This strategy ensures comprehensive initial information selection.
3. **Patch-level report generation (Patho-R1).** The selected patches $\mathcal{P}^{(1)}$ are then fed into the pathology vision-language model (VLM) Patho-R1 to generate patch-level descriptions $\mathcal{E}^{(1)}$, which serve as detailed evidence. In practice, Patho-R1 is executed in batches and produces fine-grained diagnostic cues, providing direct visual evidence to complement the WSI-level draft.

Finally, the agent fuses the global draft and local evidence to produce the initial generated report r'_0, which is the starting point for the subsequent QC iterations.

Quality-Control Rounds: Feedback-Guided Evidence Seeking and Evidence-Driven Revision. As shown in Fig. 1(b), the QC rounds take the *previous-round report* together with accumulated evidence and the quality criteria as inputs, and perform closed-loop refinement with a maximum of T iterations.

At iteration t, the QCAgent conducts a fine-grained, structured audit of the previous report r_{t-1} and outputs: (i) a set of missing key fields $\mathcal{M}_t$, (ii) a set of morphology-driven retrieval queries $\mathcal{Q}_t$, and (iii) a revised draft r'_t. The query set $\mathcal{Q}_t$ follows a demand-driven principle ("retrieve what is missing"). For example, when lymphovascular invasion (LVI) is missing, the agent generates a morphology-oriented query such as *"Histopathological evidence for lymphovascular invasion in gastric adenocarcinoma."* Meanwhile, the QCAgent enforces the **NO-FABRICATION** rule: for any field that cannot be supported by reliable visual evidence from the WSI, the system refuses to hallucinate and explicitly logs it as `need_more_info`.

Tool-2 performs text-guided retrieval in the CONCH's embedding space:

1. **Text encoding:** Compute the normalized query embedding $q = \text{CONCH}_{\text{text}}(\mathcal{Q}_t)$;
2. **Similarity search:** Retrieve TopK patches via cosine similarity using a FAISS IndexFlatIP index, i.e., $\text{TopK}(\{q \cdot f_i\}_{i=1}^{N})$;
3. **De-duplication:** Exclude patches already retrieved in previous rounds to reduce redundancy;
4. **Patch cropping:** Return the Top-K patch coordinates and similarity scores, and crop the corresponding high-resolution patches from the original WSI.

Compared with random sampling or global clustering alone, this demand-driven retrieval paradigm pinpoints tissue regions most relevant to the current QC feedback, substantially improving the efficiency of evidence acquisition.

The retrieved patches are then re-input to Patho-R1 to generate fine-grained descriptions, yielding a new set of supplemental evidence $\mathcal{E}_t$. This evidence is fed back into the subsequent QC loop as key supporting material for revision. The agent produces the revised report r_t by integrating $(r_{t-1}, \mathcal{E}_t, \mathcal{K})$. The iterative process terminates when any of the following conditions holds: (i) the report satisfies all QC criteria; (ii) the remaining missing fields are all non-image-evidenceable; or (iii) the iteration budget reaches T.

3 Experiments and Results

3.1 Datasets and Evaluation Metrics

Datasets. To assess generalization across clinical settings and languages, we evaluate our framework on two independent, large-scale cohorts of gastric cancer:

Algorithm 1. Agent-Based Framework for Quality-Controllable

Input: WSI $\mathbf{S}$, patch features $\mathcal{F}$, coordinates $\mathcal{C}$, quality criteria $\mathcal{K}$, max rounds T
Output: Final generated report r^*

$r_0 \leftarrow \text{PRISM}(\mathbf{S})$
$\mathcal{P}^{(1)} \leftarrow \text{Tool-1\_ClusterSelect}(\mathbf{S})$
$\mathcal{E}^{(1)} \leftarrow \text{PathoR1}(\mathcal{P}^{(1)})$
$r'_0 \leftarrow \text{AgentFuse}(r_0, \mathcal{E}^{(1)}; \mathcal{K})$
for $t = 1, \ldots, T$ **do**
  $(\mathcal{M}_t, \mathcal{Q}_t, r'_t) \leftarrow \text{QC}(r_{t-1}, \mathcal{E}_{t-1}; \mathcal{K})$
  if $\text{Pass}(\mathcal{M}_t, \mathcal{K})$ **then**
    break
  end if
  $\mathcal{P}_t^{(2)} \leftarrow \text{Tool-2\_CONCH-Retrieve}(\mathcal{Q}_t, \mathcal{F}, \mathcal{C})$
  $\mathcal{E}_t \leftarrow \text{PathoR1}(\mathcal{P}_t^{(2)})$
  $r_t \leftarrow \text{Revise}(r'_t, \mathcal{E}_t; \mathcal{K})$
end for
return $r^* \leftarrow r_t$

TCGA-STAD (Public, English). This cohort contains high-resolution WSIs from stomach adenocarcinoma in The Cancer Genome Atlas (TCGA). For TCGA-STAD, we use the slide-level pathology reports from TITAN [8] as ground-truth.

In-house Dataset (Private, Chinese). This cohort consists of 524 H&E WSIs. Ground-truth is the corresponding Chinese clinical pathology reports, which are highly structured and concise, emphasizing factual statements.

Evaluation Metrics. We report both traditional text overlap metrics (BLEU-1/4, ROUGE-L, METEOR) and a semantic similarity metric (BERTScore). We also introduce a clinically motivated *Field Recall*, which measures the proportion of clinically critical pathology fields successfully covered by the generated report.

Baseline and Implementation Details. We adopt PRISM, a publicly available WSI-level foundation model, as the initial-report baseline. To validate the effectiveness of iterative quality control, we compare the one-pass draft generated by PRISM (without QC) against the final output of QCAgent.

Our setup employs the CONCH encoder for text retrieval (Tool-2), Patho-R1 for visual generation, and Qwen3-VL-30B-A3B-Thinking as the core QCAgent. The QC loop runs for up to $T = 3$ iterations. At each iteration, Tool-2 retrieves the top $K = 3$ patches for each QC query, with previously retrieved patches excluded to reduce redundancy.

3.2 Experimental Results

Results on TCGA-STAD (English). Compared with the PRISM baseline, our QCAgent achieves consistent and substantial improvements across all metrics on TCGA-STAD (Table 1). Notably, Field Recall increases from 32.08% to

Table 1. Quantitative results of WSI report generation on two datasets.

Dataset	Method	BLEU-1	BLEU-4	ROUGE-L	METEOR	FieldRecall	BERTScore
TCGA-STAD	PRISM (Baseline)	0.0986	0.0319	0.0947	0.0985	0.3208	0.1245
	QCAgent (Ours)	0.2857	0.0836	0.2784	0.4087	0.6335	0.2738
In-house	PRISM (Baseline)	0.0226	0.0083	0.0484	0.0333	0.1908	−0.0831
	QCAgent (Ours)	0.0924	0.0186	0.2490	0.2307	0.4863	0.2370

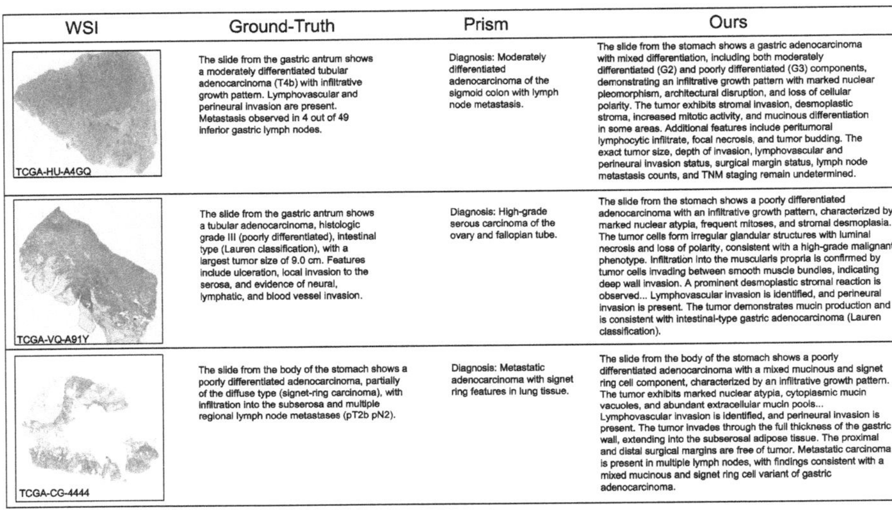

Fig. 2. Qualitative comparison of pathology reports on TCGA-STAD. For each case, we show the WSI thumbnail, the ground-truth clinical report, the one-pass draft generated by PRISM, and our QC-refined report.

63.35%, indicating that the iterative QC process successfully recovers clinically important missing findings. In addition, we observe a clear increase in report completeness: PRISM reports are typically brief, while QC-refined reports provide more detailed and comprehensive descriptions.

Results on In-house Cohort (Chinese). We further evaluate cross-lingual transfer by directly applying our QCAgent to the private Chinese cohort without any finetuning on Chinese pathology data. As shown in Table 1, the QCAgent yields marked improvements over PRISM across overlap-based metrics and semantic similarity. We note that the Chinese ground-truth reports are highly condensed, often in phrase-like formats, while our system generates more detailed narrative paragraphs. This stylistic mismatch can depress n-gram matching metrics. Nevertheless, the improvement in BERTScore-F1 suggests better alignment at the semantic and factual level.

Qualitative Comparison. QCAgent improves clinical completeness. Compared with PRISM draft, QCAgent adds morphology-grounded descriptions for previously missing fields and avoids unsupported claims by marking non-image-verifiable information as undetermined (Fig. 2); a representative step-by-step case study is shown in Fig. 3.

3.3 Ablation Study

To evaluate the contribution of key components (Clustering, Retrieval, and Patho-R1), we conducted an ablation study on 100 randomly selected WSIs from the TCGA-STAD dataset. During preliminary evaluations, we observed a pronounced "length bias". Specifically, the complete framework (Full) generates highly detailed reports. While clinically informative, this verbosity heavily penalizes precision-based metrics like F_1-score. Conversely, configurations lacking fine-grained details achieve spuriously high F_1-scores. To objectively measure the coverage of core pathological features, we employ recall-centric metrics for this analysis.

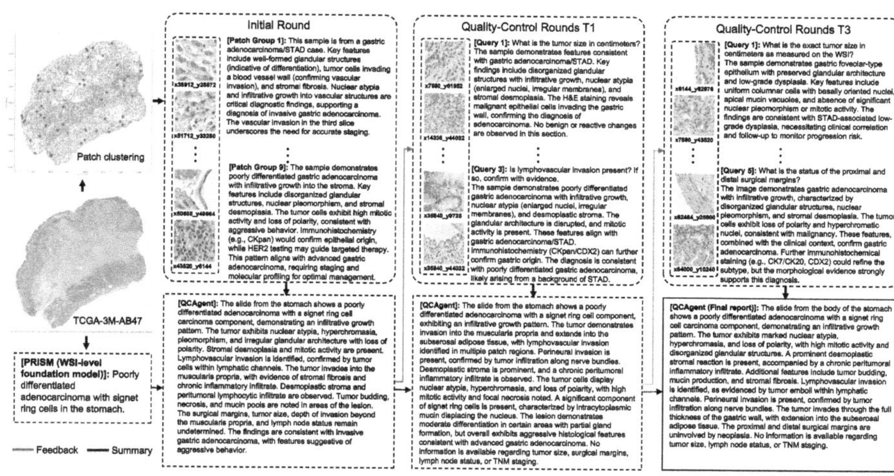

Fig. 3. Case study of the iterative audit–retrieve–revise workflow. Left: WSI thumbnail and patch clustering, with PRISM providing the initial global context. Middle/Right: across QC rounds, QCAgent issues missing-field queries, retrieves supporting patches, and revises the report, improving completeness while marking non-image-verifiable items as *undetermined*.

Table 2. Ablation study results using recall-centric metrics on 100 WSIs.

Configuration	AvgLen	R1-Recall	R2-Recall	RL-Recall	BS-Recall	FieldRecall
Full (Ours)	130	**0.5706**	**0.3061**	**0.5139**	0.3703	0.6690
w/o Retrieval	138	0.5561	0.2652	0.4809	0.3423	0.6440
w/o Patho-R1	98	0.5310	0.2696	0.4495	**0.3986**	**0.7306**

Ablation experiments confirm that each component is important (Table 2). (a) Removing Patho-R1 (visual feature extraction) module causes the most severe performance degradation alongside a significant reduction in report length. This demonstrates the critical role of Patho-R1 in extracting fine-grained histological features. (b) Removing the CONCH retrieval module reduces ROUGE-L Recall by 6.4%. This validates that dynamically retrieving local WSI regions based on missing fields effectively recovers omitted information.

4 Conclusion

We present an agentic, quality-controllable framework for WSI-based pathology report generation. Through an audit–retrieve–revise loop with cross-modal patch retrieval and evidence-driven revision, the QCAgent improves completeness while enforcing anti-hallucination constraints that distinguish verifiable findings from administrative items. On TCGA-STAD and an in-house gastric cancer cohort, our method consistently outperforms PRISM across overlap, semantic similarity, and Field Recall in a zero-shot, training-free setting. Future work will add expert evaluation and optimize clinical deployment.

Acknowledgments. This work was supported by the National Natural Science Foundation of China (No. 62402473, 62502490, 62271465), Basic Research Program of Jiangsu (No. BK20255001), Natural Science Foundation of Jiangsu Province (BK20250496, BK20240464), Beijing Natural Science Foundation (L252175), and Suzhou Basic Research Program (SYG202338).

Disclosure of Interests. The authors have no competing interests to declare that are relevant to the content of this article.

References

1. Bai, S., et al.: Qwen3-VL technical report. arXiv preprint arXiv:2511.21631 (2025). https://arxiv.org/abs/2511.21631
2. Buckley, T.A., et al.: Navigating gigapixel pathology images with large multimodal models. arXiv preprint arXiv:2511.19652 (2025). https://arxiv.org/abs/2511.19652
3. Chen, J., et al.: PathAgent: toward interpretable analysis of whole-slide pathology images via large language model-based agentic reasoning. arXiv preprint arXiv:2511.17052 (2025). https://arxiv.org/abs/2511.17052
4. Chen, P., et al.: WsiCaption: multiple instance generation of pathology reports for gigapixel whole-slide images (2024). https://arxiv.org/abs/2311.16480
5. Chen, P., Zhu, C., Zheng, S., Li, H., Yang, L.: WSI-VQA: interpreting whole slide images by generative visual question answering (2024). https://arxiv.org/abs/2407.05603
6. Chen, Q., et al.: Benchmarking large language models for biomedical natural language processing applications and recommendations. Nat. Commun. **16**(1), 3280 (2025). https://doi.org/10.1038/s41467-025-56989-2

7. Chen, Y., et al.: SlideChat: a large vision-language assistant for whole-slide pathology image understanding. In: Proceedings of the IEEE/CVF Conference on Computer Vision and Pattern Recognition (CVPR), pp. 5134–5143 (2025)
8. Ding, T., et al.: A multimodal whole-slide foundation model for pathology. Nat. Med. 1–13 (2025)
9. Ghezloo, F., et al.: PathFinder: a multi-modal multi-agent system for medical diagnostic decision-making applied to histopathology. In: Proceedings of the IEEE/CVF International Conference on Computer Vision, pp. 23431–23441 (2025)
10. Guo, Z., et al.: HistGen: Histopathology report generation via local-global feature encoding and cross-modal context interaction (2024). https://arxiv.org/abs/2403.05396
11. Huang, G., et al.: SurvAgent: hierarchical cot-enhanced case banking and dichotomy-based multi-agent system for multimodal survival prediction (2025). https://arxiv.org/abs/2511.16635
12. Jaume, G., et al.: Transcriptomics-guided slide representation learning in computational pathology. In: Proceedings of the IEEE/CVF Conference on Computer Vision and Pattern Recognition, pp. 9632–9644 (2024)
13. Liu, X., et al.: AgentBench: evaluating LLMs as agents. In: International Conference on Learning Representations, vol. 2024, pp. 52989–53046 (2024)
14. Lu, M.Y., et al.: A visual-language foundation model for computational pathology. Nat. Med. **30**(3), 863–874 (2024)
15. Lu, M.Y., et al.: A multimodal generative AI copilot for human pathology. Nature **634**(8033), 466–473 (2024)
16. Lyu, X., et al.: WSI-Agents: a collaborative multi-agent system for multi-modal whole slide image analysis. In: Proceedings of MICCAI 2025. LNCS, vol. 15964, Springer Nature, Switzerland (2025)
17. Ma, J., et al.: A generalizable pathology foundation model using a unified knowledge distillation pretraining framework. Nat. Biomed. Eng. 1–20 (2025)
18. Qin, W., et al.: What a whole slide image can tell? subtype-guided masked transformer for pathological image captioning (2023). https://arxiv.org/abs/2310.20607
19. Sapkota, R., Roumeliotis, K.I., Karkee, M.: AI agents vs. agentic AI: a conceptual taxonomy, applications and challenges. Inf. Fus. 103599 (2025)
20. Shaikovski, G., et al.: PRISM: a multi-modal generative foundation model for slide-level histopathology (2024). https://arxiv.org/abs/2405.10254
21. Song, A.H., et al.: Morphological prototyping for unsupervised slide representation learning in computational pathology. In: Proceedings of the IEEE/CVF Conference on Computer Vision and Pattern Recognition, pp. 11566–11578 (2024)
22. Sun, Y., et al.: CPathAgent: an agent-based foundation model for interpretable high-resolution pathology image analysis mimicking pathologists' diagnostic logic. Adv. Neural. Inf. Process. Syst. **38**, 101673–101731 (2026)
23. Truhn, D., et al.: Artificial intelligence agents in cancer research and oncology. Nat. Rev. Cancer 1–14 (2026)
24. Wang, H., et al.: Beyond direct diagnosis: LLM-based multi-specialist agent consultation for automatic diagnosis. arXiv preprint arXiv:2401.16107 (2024). https://arxiv.org/abs/2401.16107
25. Wang, X., et al.: A pathology foundation model for cancer diagnosis and prognosis prediction. Nature **634**(8035), 970–978 (2024)
26. Xu, H., et al.: A whole-slide foundation model for digital pathology from real-world data. Nature **630**(8015), 181–188 (2024)
27. Yang, A., et al.: Qwen3 technical report. arXiv preprint arXiv:2505.09388 (2025). https://arxiv.org/abs/2505.09388

28. Zhang, W., et al.: Patho-R1: a multimodal reinforcement learning-based pathology expert reasoner (2025). https://arxiv.org/abs/2505.11404
29. Zhao, W., et al.: An agentic system for rare disease diagnosis with traceable reasoning. Nature, 1–10 (2026)

RadHiera: Semantic Hierarchical Reinforcement Learning for Medical Report Generation

Bodong Du[1], Honglong Yang[1,2], and Xiaomeng Li[1,2](✉)

[1] The Hong Kong University of Science and Technology, Hong Kong SAR, China
eexmli@ust.hk
[2] Shenzhen Loop Area Institute, Shenzhen, China

Abstract. Vision-language models have shown promising results in radiology report generation. However, most existing methods generate reports as flat text and do not explicitly model the semantic dependency between the *Findings* and *Impression* sections, which can lead to inconsistencies between clinical observations and diagnostic conclusions. In this paper, we propose **RadHiera**, a semantic hierarchical reinforcement learning framework for radiology report generation. **RadHiera** follows the semantic organization of radiology reports by first optimizing overall report quality, then improving the diagnostic accuracy of the *Impression* section, and finally enforcing consistency between *Findings* and *Impression* so that diagnostic conclusions are supported by clinical evidence. Specifically, we begin with a base reward that combines linguistic quality and medical factuality to provide supervision on the whole report. On this basis, we introduce a severity-aware reward for the *Impression* section that places greater emphasis on errors involving clinically critical conditions, thereby reducing both missed diagnoses and overstatement. We further enforce cross-section consistency using Expert Model-derived label sets, with subset constraints and hallucination penalties to ensure that impressions remain faithful to the findings. Experiments on three public chest X-ray benchmarks show that **RadHiera** consistently improves diagnostic accuracy and inter-section consistency over state-of-the-art methods, while also demonstrating good adaptability to ultrasound report generation. The source code is available at https://github.com/xmed-lab/RadHiera.

Keywords: Medical Report Generation · Reinforcement Learning · Semantic Hierarchy

1 Introduction

Radiology reports adhere to a hierarchical structure where Findings describe observations and Impression synthesizes conclusions, ensuring clinical reliability [6,15]. However, existing MRG systems underexplore this dependency.

G. Yang et al. (Eds.): MICCAI 2026, LNCS 16878, pp. 471–481, 2027.
https://doi.org/10.1007/978-3-032-38059-3_45

Early frameworks [4,8,10,12,17,18,21] neglect the Impression section due to its higher semantic complexity and optimization difficulty. In recent years, MLLMs [16,19,26] generate whole reports including the Findings and Impression sections but fail to explicitly model the inter-section hierarchy. Concurrent RL-based work such as UniRG [13] improves report-level optimization, but does not specifically target this Findings–Impression dependency. As a result, generated reports may exhibit diagnostic inaccuracies or inconsistencies, such as missing critical pathologies or introducing unsupported conclusions (see the "MedVersa" results in Fig. 2).

To address this structural mismatch and improve clinical reliability, we propose **RadHiera**, a semantic hierarchical reinforcement learning framework that directly optimizes policy behavior toward clinically structured reporting. Central to this approach is a multi-level reward mechanism that explicitly models the semantic dependency between Findings and Impression sections. First, we apply a base reward that integrates linguistic fluency and medical factual correctness, ensuring coherent and clinically valid report generation. Building on this, we introduce a **severity-aware Impression reward** for the Impression section. It compares the generated Impression with the reference and assigns higher weights to clinically critical conditions (e.g., pneumothorax, pulmonary edema). This reward penalizes omissions of high-risk findings and discourages overstatements, thus prioritizing critical diagnoses and improving clinical fidelity. Finally, we introduce a **cross-sectional consistency reward** between the *Findings* and *Impression* sections, using CheXbert-derived label sets. The reward enforces consistency by applying subset constraints and hallucination penalties, reducing diagnostic contradictions and promoting structured alignment across sections.

Extensive experiments across three public chest X-ray benchmarks [5,11,24] demonstrate that RadHiera consistently enhances diagnostic precision and inter-section coherence relative to state-of-the-art baselines. Furthermore, the framework exhibits remarkable cross-modal generalizability, extending its performance gains to the domain of ultrasound report generation in an in-house CarotidUS-MRG dataset.

To summarize, this work makes the following contributions:

- We propose **RadHiera**, a semantic hierarchical reinforcement learning framework that incorporates clinical constraints into the policy optimization process, explicitly modeling the hierarchical structure of MRG.
- We design a **multi-level reward mechanism** composed of a base reward, a severity-aware Impression reward for diagnostic prioritization, and a cross-sectional consistency reward that enforces alignment between report sections.
- We conduct comprehensive experiments on diverse MRG datasets, demonstrating that our method achieves state-of-the-art performance on MRG both quantitatively and qualitatively.

2 Method

In this section, we first formulate the medical report generation task and introduce the reinforcement learning algorithm we build upon. Then we present the overall design of **RadHiera**, a semantic hierarchical reinforcement learning framework as shown in Fig. 1.

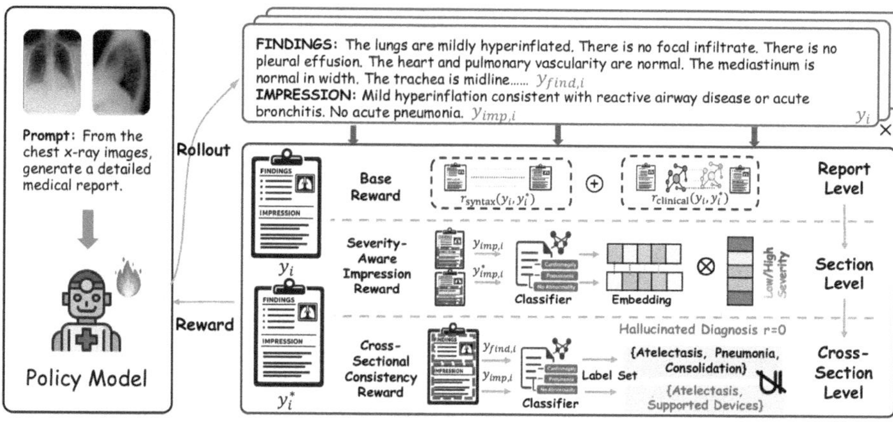

Fig. 1. Overall architecture of **RadHiera**. The policy model generates n rollouts from queries through GRPO optimization with the multi-level rewards: (1) base reward for policy optimization stability at the whole-report level, (2) Severity-Aware Impression Reward that leverages expert model embeddings to assess diagnostic severity levels, and (3) Cross-Sectional Consistency Reward that ensures alignment between Findings and Impression sections by detecting hallucinated diagnoses. This hierarchical reward enables generation of clinically effective and coherent radiology reports.

2.1 Problem Formulation and Preliminaries

Medical Report Generation (MRG) aims to generate structured and clinically coherent reports from medical images. A typical radiology report comprises two key sections: the *Findings* section, which describes observable visual cues, and the *Impression* section, which summarizes diagnostic conclusions.

Given an input medical image x and its reference report $y^* = \{y_1^*, \ldots, y_T^*\}$, a vision-language model π_θ models the conditional distribution of tokens:

$$p(y_t \mid x, y_{<t}),$$

where y_t denotes the t-th generated token. Standard supervised fine-tuning (SFT) maximizes the likelihood of the reference report:

$$\mathcal{L}_{\text{SFT}} = -\sum_{t=1}^{T} \log p(y_t^* \mid x, y_{<t}^*),$$

While SFT ensures fluency and domain alignment, it fails to explicitly enforce clinical consistency between Findings and Impression or optimize clinical metrics, necessitating structured reinforcement learning.

To align model behavior with clinically grounded signals, we adopt Group Relative Policy Optimization (GRPO), a reinforcement learning algorithm that evaluates groups of candidate completions without a value network. For input x, the policy π_{old} samples G candidate reports $\{s_i\}_{i=1}^{G}$, each receiving a scalar reward r_i from a reward function $R(s_i, x)$.

Within each group, rewards are normalized to compute relative advantage:

$$A_i = \frac{r_i - \text{mean}(\{r_j\}_{j=1}^{G})}{\text{std}(\{r_j\}_{j=1}^{G}) + \epsilon}, \tag{1}$$

where ϵ is a constant. For each token $s_{i,t}$ in candidate s_i, the policy ratio is:

$$r_{i,t}(\theta) = \frac{\pi_\theta(s_{i,t} \mid x, s_{i,<t})}{\pi_{\text{old}}(s_{i,t} \mid x, s_{i,<t})}. \tag{2}$$

The GRPO objective, incorporating clipping and KL regularization against a reference model π_{ref}, is:

$$\begin{aligned}\mathcal{J}(\theta) = \mathbb{E}\Bigg[\frac{1}{G}\sum_{i=1}^{G}\frac{1}{|s_i|}\sum_{t=1}^{|s_i|}\min\Big(r_{i,t}(\theta)A_i,\ \text{clip}(r_{i,t}(\theta), 1-\epsilon, 1+\epsilon)A_i\Big) \\ -\beta D_{KL}\big(\pi_\theta(\cdot \mid x) \,\|\, \pi_{\text{ref}}(\cdot \mid x)\big)\Bigg],\end{aligned} \tag{3}$$

where β weights the KL penalty. This formulation increases the probability of above-average reports while preserving pretrained behavior. To optimize policy behavior toward clinically structured reporting, we design a reward across different report semantic levels:

$$r(y, y^*) = r_{\text{base}}(y, y^*) + r_{\text{imp}}(y, y^*) + r_{\text{consist}}(y). \tag{4}$$

Each reward targets a specific semantic granularity, detailed as follows.

2.2 Report-Level Base Reward

The base reward maintains general language quality and medical validity during policy optimization, while the hierarchical rewards reshape model behavior toward workflow-aligned reporting. The base reward is defined as

$$r_{\text{base}}(y, y^*) = r_{\text{syntax}}(y, y^*) + r_{\text{clinical}}(y, y^*). \tag{5}$$

r_{syntax} evaluates surface-level fluency and semantic similarity and r_{clinical} measures factual correctness by extracting clinical entities from y and comparing them with y^*. This base reward ensures that policy updates preserve medical validity and prevents significant deviations from the initial policy model, avoiding drift in language style and medical terminology. While the base reward maintains general quality, it does not explicitly encode the report structure. We therefore introduce two additional semantic-level rewards that shape the policy toward clinically meaningful structured reports.

2.3 Severity-Aware Impression Reward

Clinically, the Impression section reflects a high-level diagnostic decision. Errors on critical conditions (e.g., pneumothorax) carry higher clinical risk than minor lexical deviations. To model this asymmetry, we design a severity-aware reward on the Impression section. Let y^{imp} denote the generated Impression and $y^{\text{imp}*}$ the reference Impression. A frozen classifier `fcls` extracts binarized clinical labels:

$$z^{\text{imp}} = \texttt{fcls}(y^{\text{imp}}), \quad z^{\text{imp}*} = \texttt{fcls}(y^{\text{imp}*}). \tag{6}$$

We compute a severity-weighted matching reward:

$$r_{\text{imp}}(y, y^*) = \sum_c w_c \cdot \left(\mathbb{I}_{\text{TP}}^{(c)} - \mathbb{I}_{\text{FN}}^{(c)} - \mathbb{I}_{\text{FP}}^{(c)}\right), \tag{7}$$

where $\mathbb{I}_{\text{TP}}^{(c)} = \mathbb{I}[z_c^{\text{imp}} = 1 \wedge z_c^{\text{imp}*} = 1]$ (and similarly for FN/FP), and w_c assigns larger weights to clinically critical conditions [2,7]. This reward shapes the policy toward risk-sensitive diagnostic alignment and improves diagnostic accuracy.

2.4 Cross-Sectional Consistency Reward

Radiology reports require *Impression* conclusions to be grounded in *Findings*. However, recent methods do not enforce this dependency, risking hallucinated or contradictory diagnoses. To address this, we extract clinical label sets from both sections:

$$z^{\text{find}} = \texttt{fcls}(y^{\text{find}}), \quad z^{\text{imp}} = \texttt{fcls}(y^{\text{imp}}). \tag{8}$$

Let $\mathcal{F}$ and $\mathcal{I}$ denote the sets of positive labels corresponding to z^{find} and z^{imp}, respectively. The consistency reward is defined as:

$$r_{\text{consist}}(y) = \begin{cases} 1, & \text{if } \mathcal{I} \subseteq \mathcal{F} \\ 0, & \text{otherwise} \end{cases} \tag{9}$$

Here, $\mathcal{I} \subseteq \mathcal{F}$ checks if the *Impression* set is a subset of the *Findings* set, and the reward is 0 for any *Impression* labels not found in the *Findings* section (hallucinated diagnoses). This reward term shapes the policy to reduce cross-sectional contradictions and discourage unsupported diagnoses, ensuring that diagnostic conclusions in the *Impression* are aligned with the *Findings* section. In summary, while the subset formulation in the consistency reward enforces rigorous consistency by grounding the *Impression* in the *Findings* to suppress unsupported conclusions, it may increase the risk of clinically significant omissions. The severity-aware reward addresses this trade-off by penalizing omissions of critical findings, thereby maintaining clinical sensitivity.

3 Experiments

Datasets. We evaluate our model on four datasets covering different imaging modalities: CarotidUS-MRG: A private dataset of 12,000 carotid ultrasound studies with expert-annotated reports, split by patient (70/15/15%) to prevent leakage. Reports follow a standardized format for structured clinical documentation. MIMIC-CXR [11]: A public chest X-ray dataset with 377,110 frontal images and 227,827 associated reports. IU-Xray [5]: A public chest X-ray dataset with 8,121 images and 7,470 reports from 3,955 patients. ReXGradient-160K [24]: The largest publicly available multi-site chest X-ray dataset, containing 273,004 images from 160,000 radiological studies across 109,487 patients from 79 medical sites. We adopt the official split for consistency with prior studies [25], enabling direct comparison with existing work.

Evaluation Metrics. We assess performance using natural language generation (NLG) and clinically grounded metrics. BLEU [14], BERTScore [23], and SembScore [17] evaluate report-level fluency and semantic similarity. For MIMIC-CXR, IU-Xray, and ReXGradient, we follow the RexRank [25] protocol with RadGraph [9], 1/RadCliQ-v1 [22], CheXbert-F1 [17], and consistency (CheXbert labeler subset check). For CarotidUS-MRG, where RadGraph and RadCliQ-v1 are inapplicable, KeyWordMatch uses a modality-specific labeler trained on privately annotated reports to extract and overlap key medical terms.

Baselines and Implementation. We evaluate our approach, which employs Supervised Fine-Tuning on Qwen2.5-VL-3B [1] followed by GRPO with our semantic hierarchical rewards, against five state-of-the-art methods: R2GenGPT [19], MedGemma-4B [16], MedVersa [26], CheXpertPlus [3], and RadFM [20]. All models are trained under identical settings: learning rate 1×10^{-5}, LoRA rank 8, four generations per case and a batch size of 2. Training took approximately 672 GPU-hours on 8× MetaX C500 GPUs (64 GB VRAM each). For reproducibility, r_{syntax} utilizes BLEU, ROUGE-L, BERTScore, and SembScore; r_{clinical} uses RadGraph micro-F1. r_{imp} and r_{consist} employ CheXbert (CXR) and a private ultrasound labeler. For fair comparison, MedGemma and MedVersa are fine-tuned on IU-Xray and ReXGradient.

3.1 Main Results

Cross-Dataset Performance Superiority. RadHiera consistently outperforms others across all four datasets in Tables 1 and 2, leading in six of seven MIMIC-CXR metrics and sweeping IU-Xray, ReXGradient, and the ultrasound CarotidUS-MRG. On MIMIC-CXR, while MedVersa slightly leads in BERTScore (0.430 vs. 0.421), RadHiera substantially improves semantic alignment with a SembScore of 0.442 vs. 0.315, alongside gains in clinical efficacy metrics like RadGraph (0.282 vs. 0.273) and Consistency (0.851 vs. 0.776). This gap widens on IU-Xray and ReXGradient, where RadHiera exceeds the runner-up by 0.473 in 1/RadCliQ-v1 (2.325 vs. 1.852), suggesting stronger generalization to other datasets. In CarotidUS-MRG, consistent improvements in all metrics,

Table 1. Performance comparison on MIMIC-CXR, IU-Xray, and ReXGradient. NLG metrics evaluate textual similarity at the report level. Clinical metrics assess factual and diagnostic correctness. Bold = best, underline = second-best.

Dataset	Model	NLG Metrics			CE Metrics			
		BLEU-2 ↑	BERTScore ↑	SembScore ↑	RadGraph ↑	1/RadCliQ-v1 ↑	F1 ↑	Consistency ↑
MIMIC-CXR	RadFM [20]	0.081	0.281	0.245	0.111	0.625	0.245	0.741
	CheXpertPlus [3]	0.196	0.389	0.429	0.166	0.838	0.322	0.729
	R2GenGPT [19]	0.203	0.407	0.305	0.243	0.909	0.348	0.753
	MedGemma-4B [16]	0.185	0.417	0.301	0.258	0.905	0.359	0.771
	MedVersa [26]	<u>0.193</u>	**0.430**	<u>0.315</u>	<u>0.273</u>	<u>0.919</u>	<u>0.376</u>	<u>0.776</u>
	RadHiera (Ours)	**0.206**	<u>0.421</u>	**0.442**	**0.282**	**0.939**	**0.394**	**0.851**
IU-Xray	RadFM [20]	0.228	0.491	0.589	0.254	1.320	0.421	0.841
	CheXpertPlus [3]	0.264	0.516	0.608	0.257	1.619	0.469	0.828
	R2GenGPT [19]	0.271	0.558	0.601	0.269	1.751	0.483	0.852
	MedGemma-4B [16]	<u>0.259</u>	0.571	0.603	0.268	1.821	<u>0.502</u>	0.891
	MedVersa [26]	0.251	<u>0.579</u>	<u>0.611</u>	<u>0.274</u>	<u>1.852</u>	0.494	<u>0.898</u>
	RadHiera (Ours)	**0.284**	**0.607**	**0.626**	**0.313**	**2.325**	**0.519**	**0.915**
ReXGradient	RadFM [20]	0.176	0.368	0.409	0.151	0.837	0.279	0.827
	CheXpertPlus [3]	0.238	0.437	0.463	0.208	1.038	0.313	0.809
	R2GenGPT [19]	0.251	0.478	0.479	0.231	1.050	0.334	0.812
	MedGemma-4B [16]	0.258	<u>0.486</u>	0.482	0.242	1.120	0.341	<u>0.882</u>
	MedVersa [26]	<u>0.266</u>	0.481	<u>0.488</u>	<u>0.249</u>	<u>1.170</u>	<u>0.349</u>	0.874
	RadHiera (Ours)	**0.298**	**0.544**	**0.492**	**0.299**	**1.450**	**0.376**	**0.891**

Table 2. Performance comparison on the CarotidUS-MRG dataset. F1 is computed on the Impression section over three observations (stenosis, plaque, and IMT thickness). RadGraph and RadCliQ-v1 are inapplicable, so we introduce KeyWordMatch, a metric based on a proprietary ultrasound labeler trained on expert-annotated data.

Model	BLEU-2 ↑	BERTScore ↑	SembScore ↑	Keyword ↑	F1 ↑	Consistency ↑
R2GenGPT [19]	0.453	0.844	0.815	0.589	0.449	0.823
MedGemma-4B [16]	0.458	0.846	0.812	0.571	0.461	0.873
MedVersa [26]	<u>0.473</u>	<u>0.868</u>	<u>0.837</u>	<u>0.593</u>	<u>0.469</u>	<u>0.885</u>
RadHiera (Ours)	**0.522**	**0.915**	**0.891**	**0.742**	**0.519**	**0.915**

including an increase in F1 from 0.469 to 0.519, demonstrate robustness to the modality shift.

Diagnostic Consistency and Cross-Sectional Alignment. RadHiera excels at maintaining agreement between Findings and Impression sections, as reflected by the Consistency metric. Across MIMIC-CXR, IU-Xray, and ReXGradient, RadHiera scores 0.851, 0.915, and 0.891, surpassing second-best methods by 0.075, 0.017, and 0.009. These gains correlate with higher CheXbert-F1 on the Impression section (0.394 on MIMIC-CXR), linking enhanced cross-sectional coherence to more accurate diagnostic labeling. This consistent improvement extends to CarotidUS-MRG, as evidenced by the Consistency metric (0.915 versus 0.885).

Case Study. Figure 2 compares MedVersa, RadHiera, and expert reports on MIMIC-CXR and CarotidUS-MRG. MedVersa often produces template-like descriptions focusing on superficial phrasing (e.g., pulmonary congestion, intima-media thickness, or plaque) while overlooking clinically critical indicators like ICD, stenosis severity, or plaque characteristics. Its Findings and Impression sections are sometimes weakly connected, limiting diagnostic coherence. In contrast, RadHiera generates clinically grounded reports with strong cross-sectional consistency. It links imaging evidence to the diagnostic process, quantifies stenosis severity and characterizes plaque morphology. The resulting Impression sections are concise yet logically traceable to Findings, closely resembling expert-level summaries in structure and clinical reliability.

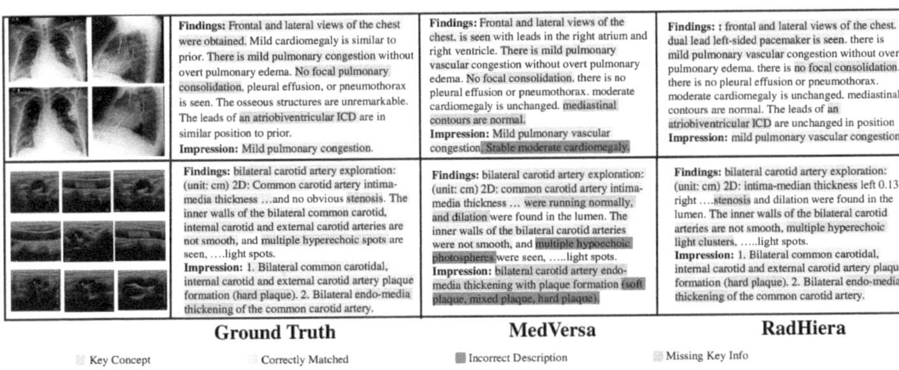

Fig. 2. Comparison of MedVersa and RadHiera outputs against ground truth. We observe that MedVersa sometimes misses clinical indicators and exhibits weak coherence between Findings and Impression sections. In contrast, RadHiera generates clinically grounded reports with higher cross-sectional consistency and diagnostic accuracy.

3.2 Ablation Study

Table 3 demonstrates that RadHiera consistently outperforms the SFT baseline and standard GRPO. GRPO improves SembScore from 0.402 to 0.411 and F1 from 0.332 to 0.368, whereas RadHiera further elevates these to 0.442 and 0.394. Ablation studies reveal synergistic effects: removing consistency rewards reduces Consistency from 0.851 to 0.821 and RadGraph from 0.282 to 0.274, while ablating Impression rewards causes a sharper Consistency drop to 0.789. These results confirm that joint hierarchical optimization effectively enhances both linguistic fluency and clinical alignment.

Table 3. Ablation study of reward components in RadHiera on MIMIC-CXR. We compare the full hierarchical reward against the SFT baseline, standard GRPO with base reward, and partial reward combinations.

Model Configuration	NLG Metrics			Clinical Efficacy (CE) Metrics			
	BLEU-2 ↑	BERTScore ↑	SembScore ↑	RadGraph ↑	1/RadCliQ-v1 ↑	F1 ↑	Cons. ↑
SFT (Qwen2.5-VL-3B)	0.175	0.395	0.402	0.241	0.901	0.332	0.747
GRPO (Base Reward Only)	0.188	0.402	0.411	0.255	0.915	0.368	0.772
RadHiera w/o Imp. Reward	0.202	0.417	0.426	0.261	0.912	0.368	0.789
RadHiera w/o Cons. Reward	0.204	0.419	0.433	0.274	0.927	0.359	0.821
RadHiera (Full)	**0.206**	**0.421**	**0.442**	**0.282**	**0.939**	**0.394**	**0.851**

4 Conclusion

In this paper, we introduce RadHiera, a semantic hierarchical reinforcement learning framework that improves medical report generation by enforcing structural and clinical consistency. By integrating severity-aware rewards and cross-sectional constraints, RadHiera ensures coherent and accurate diagnostics. Experiments across four datasets demonstrate that RadHiera consistently outperforms state-of-the-art methods in diagnostic accuracy and clinical alignment.

Acknowledgments. This work was supported by a research grant from the Joint Research Scheme (JRS) under the National Natural Science Foundation of China (NSFC) and the Research Grants Council (RGC) of Hong Kong (Project No. N_HKUST654/24); a research grant from the Innovation and Technology Commission (ITC) (Project No. GHP/124/22); and research grants from the RGC (Project Nos. 16500825, R6005-24, and C6088-25Y).

Disclosure of Interests. The authors have no competing interests to declare that are relevant to the content of this article.

References

1. Bai, S., et al.: Qwen2.5-vl technical report (2025). https://arxiv.org/abs/2502.13923
2. Cardinale, L., et al.: Revisiting signs, strengths and weaknesses of standard chest X-ray in acute dyspnoea. J. Thorac. Dis. **4**, 398–407 (2012). https://doi.org/10.3978/j.issn.2072-1439.2012.05.05
3. Chambon, P.: Chexpert plus: augmenting a large chest X-ray dataset with text radiology reports, patient demographics and additional image formats. arXiv preprint arXiv:2405.19538 (2024)
4. Chen, Z., Shen, Y., Song, Y., Wan, X.: Generating radiology reports via memory-driven transformer. In: Proceedings of the 59th Annual Meeting of the Association for Computational Linguistics and 11th International Joint Conference on Natural Language Processing (ACL-IJCNLP), pp. 1439–1449 (2021)
5. Demner-Fushman, D., et al.: Preparing a collection of radiology examinations for distribution and retrieval. J. Am. Med. Inform. Assoc. **23**(2), 304–310 (2016)

6. Ganeshan, D., et al.: Structured reporting in radiology. Acad. Radiol. **25**(1), 66–73 (2018)
7. Horng, S., Liao, R., Schwab, P., Dalal, S., Golland, P.: Deep learning to quantify pulmonary edema in chest radiographs (2020)
8. Hou, W., Cheng, Y., Xu, K., Li, W., Liu, J.: RECAP: towards precise radiology report generation via dynamic disease progression reasoning. In: Proceedings of Conference on Empirical Methods Natural Language Processing (EMNLP), pp. 2134–2147 (2023)
9. Jain, S., et al.: Radgraph: extracting clinical entities and relations from radiology reports. arXiv preprint arXiv:2106.14463 (2021)
10. Jing, B., Xie, P., Xing, E.: On the automatic generation of medical imaging reports. In: Proceedings of 56th Annual Meeting of the Association for Computational Linguistics (ACL), pp. 2577–2586 (2018)
11. Johnson, A.E., et al.: Mimic-cxr, a de-identified publicly available database of chest radiographs with free-text reports. Sci. Data **6**(1), 317 (2019)
12. Li, Y., Liang, X., Hu, Z., Xing, E.P.: Hybrid retrieval-generation reinforced agent for medical image report generation. Adv. Neural Inf. Process. Syst. **31** (2018)
13. Liu, Q.: Scaling medical imaging report generation with multimodal reinforcement learning. arXiv preprint arXiv:2601.17151 (2026)
14. Papineni, K., Roukos, S., Ward, T., Zhu, W.J.: Bleu: a method for automatic evaluation of machine translation. In: Proceedings of the 40th Annual Meeting of the Association for Computational Linguistics, pp. 311–318 (2002)
15. Schwartz, L.H., Panicek, D.M., Berk, A.R., Li, Y., Hricak, H.: Improving communication of diagnostic radiology findings through structured reporting. Radiology **260**(1), 174–181 (2011)
16. Sellergren, A., Kazemzadeh, S., Jaroensri, T., et al.: MedGemma technical report (2026). https://arxiv.org/abs/2507.05201
17. Smit, A., Jain, S., Rajpurkar, P., Pareek, A., Ng, A., Lungren, M.: Combining automatic labelers and expert annotations for accurate radiology report labeling using BERT. In: Proceedings of Empirical Methods Natural Language Processing (EMNLP), pp. 1500–1519 (2020)
18. Wang, Z., Han, H., Wang, L., Li, X., Zhou, L.: Automated radiographic report generation purely on transformer: a multicriteria supervised approach. IEEE Trans. Med. Imag. **41**(10), 2803–2813 (2022)
19. Wang, Z., Liu, L., Wang, L., Zhou, L.: R2gengpt: radiology report generation with frozen LLMs. Meta-Radiology **1**(3), 100033 (2023)
20. Wu, C., Zhang, X., Zhang, Y., Hui, H., Wang, Y., Xie, W.: Towards generalist foundation model for radiology by leveraging web-scale 2D&3D medical data. Nat. Commun. **16**(1), 7866 (2025)
21. Yan, S., Cheung, W.K., Chiu, K.W., Tong, T.M., Cheung, K.C., See, S.: Attributed abnormality graph embedding for clinically accurate x-ray report generation. IEEE Trans. Med. Imag. **42**(8), 2211–2222 (2023). https://doi.org/10.1109/tmi.2023.3245608
22. Yu, F., et al.: Evaluating Progress in Automatic Chest X-Ray Radiology Report Generation. Technical report, Radiology and Imaging (2022). https://doi.org/10.1101/2022.08.30.22279318
23. Zhang, T., Kishore, V., Wu, F., Weinberger, K.Q., Artzi, Y.: Bertscore: evaluating text generation with bert. In: International Conference on Learning Representations (2020)

24. Zhang, X., Acosta, J.N., Miller, J., Huang, O., Rajpurkar, P.: Rexgradient-160k: a large-scale publicly available dataset of chest radiographs with free-text reports. arXiv preprint arXiv:2505.00228 (2025)
25. Zhang, X., et al.: Rexrank: a public leaderboard for AI-powered radiology report generation. arXiv preprint arXiv:2411.15122 (2024)
26. Zhou, H.Y., Acosta, J.N., Adithan, S., Datta, S., Topol, E.J., Rajpurkar, P.: Medversa: a generalist foundation model for medical image interpretation. arXiv preprint arXiv:2405.07988 (2024)

RadSLDP: Selective Local Differential Privacy for Radiology Vision-Language Models

Konghao Zhao[1], Xinyang Xu[1], Runhui Xu[1], Yutaka Natsuaki[2], Wenjie Xiong[3], Sai Praneeth Karimireddy[1], and Ruishan Liu[1,2(✉)]

[1] Department of Computer Science, Viterbi School of Engineering, University of Southern California, Los Angeles, CA, USA
ruishanl@usc.edu

[2] Department of Radiation Oncology, Keck School of Medicine, University of Southern California, Los Angeles, CA, USA

[3] Department of Electrical and Computer Engineering, Virginia Polytechnic Institute and State University, Blacksburg, VA, USA

Abstract. Medical VisionLanguage Models (VLMs) for radiology report generation are often fine-tuned by external developers due to hospitals' limited computational infrastructure. It requires sensitive patient data to be transmitted to untrusted training environments, which creates privacy risks. Standard Differential Privacy protects the trained model but assumes trusted access to raw data, making it unsuitable for untrusted external training. Local Differential Privacy (LDP) enables privacy-preserving training in untrusted settings by perturbing all text token embeddings before transmission but significantly degrading clinical utility. We observe that clinical reports contain both image-inferable and non-inferable content and define the non-inferable portion as non-visual clinical context (NVCC), such as demographics and medical history, which constitutes the primary privacy risk. To preserve privacy while maintaining utility, we propose RadSLDP, a selective LDP mechanism that perturbs only expert-validated NVCC token embeddings, with each protected token embedding satisfying formal per-token (ε, δ)-LDP via the Gaussian mechanism. As a result, RadSLDP matches or outperforms uniform LDP in 94% of comparisons, including 29/36 improvements and 5/36 ties. It improves BLEU-4 by 24–77% and CheXbert Micro-F1 by 87–108% on ReXGradient dataset. RadSLDP also exhibits lower performance variability across privacy budgets, enabling practical privacy-preserving VLM fine-tuning in untrusted settings.

Keywords: Local Differential Privacy · Vision-Language Models · Radiology Report Generation

K. Zhao, X. Xu and R. Xu—Equal contribution.

G. Yang et al. (Eds.): MICCAI 2026, LNCS 16878, pp. 482–492, 2027.
https://doi.org/10.1007/978-3-032-38059-3_46

1 Introduction

Medical Vision-Language Models (VLMs) have demonstrated remarkable capability in generating radiology reports from chest X-ray images, offering the potential to reduce radiologists' workload and improve diagnostic consistency. However, modern VLMs comprise billions of parameters, making fine-tuning computationally prohibitive for most hospital systems with limited infrastructure. Hospitals must therefore rely on external cloud providers or model developers, requiring transmission of raw training data to untrusted environments. However, training datasets often contain sensitive patient information, including demographics, medical history, and clinical indications. Such outsourcing introduces fundamental privacy risks, as external trainers gain direct access to sensitive patient information, potentially violating regulatory frameworks (Fig. 1a).

Differential Privacy (DP) [6] is the standard approach for training neural networks with formal privacy guarantees by ensuring that model outputs are approximately invariant to the inclusion of any individual training sample. DP-SGD [1] achieves this in neural networks through per-sample gradient clipping and noise injection, and has been applied to vision and language tasks [15,32,37]. Recent methods further apply and improve this [27,31,36] in multi-modal training. However, DP-SGD assumes a trusted training environment and protects only the released model, making it unsuitable for untrusted trainer settings common in medical VLM development.

Local Differential Privacy (LDP) protects data in the untrusted trainer setting by perturbing inputs before they leave the trusted domain [5]. Prior work applies LDP to text through binary embedding perturbation [18], sentence-level mechanisms [4], and private inference methods [8]. However, existing approaches apply uniform noise to all tokens, resulting in substantial degradation in downstream task performance [11].

This limitation highlights an important structural property of radiology VLM training data. Clinical text accompanying chest X-rays contains two fundamentally different types of information. Image-inferable content, such as descriptions of cardiomegaly or device placement, reflects clinical findings visible in the radiograph. In contrast, image-uninferable content, including patient demographics and medical history, cannot be derived from the image alone. We define this portion as Non-Visual Clinical Context (NVCC) which constitutes the primary privacy risk.

This raises another question of whether de-identified chest radiographs carry residual privacy risk. In standard data-sharing workflows, hospitals remove all categories of identifiers specified by HIPAA Safe Harbor [28] before releasing imaging data, and public datasets such as MIMIC-CXR [13], CheXpert Plus [3], and ReXGradient [35] follow these protocols. Unlike head CTs or full-body MRIs, standard chest radiographs lack directly recognizable features such as facial structure. Prior work suggests that re-identification becomes plausible primarily in settings that enable cross-dataset linkage, for example when an adversary has access to identified reference images [22]. In typical outsourced training workflows, hospitals may share de-identified radiographs for model development but do not share institution-held identified image repositories with external parties; this limits the auxiliary data needed for cross-dataset linkage attacks. This motivates selectively perturbing NVCC tokens to protect patient-specific information while preserving diagnostic signal from the image.

Prior work has explored selective LDP to reduce the utility loss of uniform noise by restricting perturbation to sensitive content. Shi et al. [24] introduced this for RNN/LSTM architectures. Just Fine-tune Twice [25] extends it to Transformers via two-phase training. Policy-based embedding-level noise injection [30], adaptive token-weighted noise [33], split-and-denoise architectures [19], and gradient-based selective fine-tuning [34] further develop this direction. However, all existing selective DP methods target unimodal text-only models. We are the first to formalize and evaluate selective LDP for radiology VLM fine-tuning.

With this insight, we propose RadSLDP, a selective embedding-level LDP mechanism for privacy-preserving radiology VLM fine-tuning. To identify sensitive content, we develop a multi-stage annotation pipeline: clinical indications from three large-scale datasets are decomposed into 141,129 semantically unique units, clustered into 347 unique subcategories, and annotated for image-inferability using an LLM, followed by review from a certified medical physicist. RadSLDP then operates within the trusted hospital environment: for each training sample, a policy function identifies NVCC tokens. Their embeddings are clipped to bound ℓ_2-sensitivity and perturbed with calibrated Gaussian noise. In contrast, image-inferable token embeddings are transmitted unmodified. Each protected token embedding satisfies formal per-token (ε, δ)-LDP via the Gaussian mechanism [6], with guarantees that hold independently of the downstream model architecture.

We validate RadSLDP across three chest X-ray datasets and three VLM architectures. At per-token $\varepsilon = 8$, RadSLDP matches or outperforms uniform LDP in 94% of comparisons (34/36), with BLEU-4 improvements of 25–77% and CheXbert Micro-F1 gains of 87–108% on the ReXGradient dataset (Fig. 1c). RadSLDP also exhibits lower performance variability across privacy budgets than uniform LDP across all three architectures. Our main contributions are as follows:

- We study radiology VLM fine-tuning with an untrusted external trainer, and curate an expert-validated NVCC taxonomy to derive token-level sensitivity labels across three datasets.
- We propose RadSLDP, a model-agnostic selective LDP mechanism that perturbs only NVCC token embeddings with formal per-token (ε, δ)-LDP guarantees per protected token.
- We conduct extensive experiments across three large-scale datasets, demonstrating that selective perturbation achieves substantial utility gains over uniform LDP.

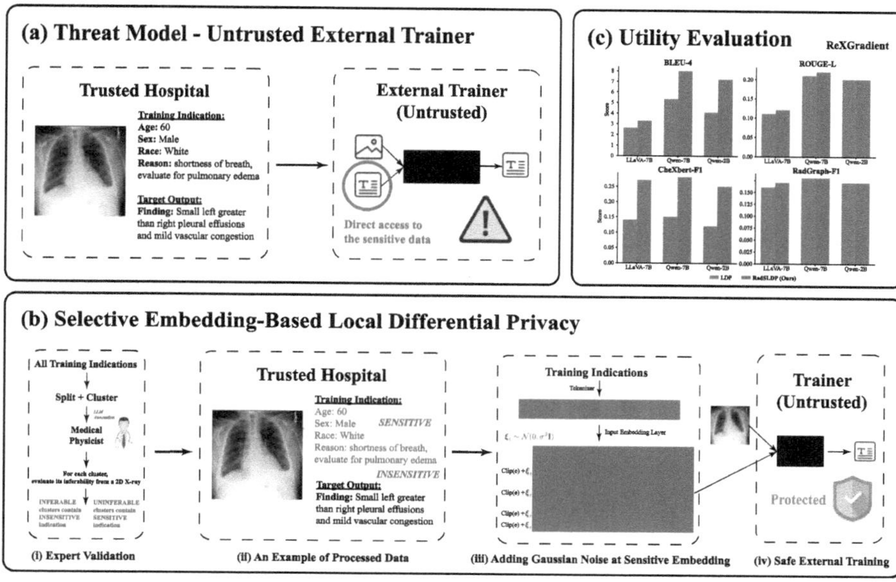

Fig. 1. Overview of RadSLDP for privacy-preserving clinical VLM fine-tuning.

2 Methodology

2.1 Problem Formulation

We consider a setting where a trusted hospital holds a dataset $\mathcal{D} = \{(I_i, X_i, Y_i)\}_{i=1}^{N}$, and I_i is a 2D chest X-ray image, X_i is the associated clinical context text, and Y_i is the target radiology report. The hospital lacks the computational resources to fine-tune a large VLM and must transmit training data to an untrusted external trainer. The goal is to protect sensitive patient attributes in X_i while preserving sufficient clinical signal to produce a high-quality report generation model.

2.2 Sensitive Patient Attribute Identification

By our observation, clinical context text X_i contains two fundamentally different types of information (NVCC and non-NVCC). To distinguish between these categories, we define a deterministic policy function $\pi : X \rightarrow \mathcal{W}$ that maps clinical context text to a set of sensitive words $\mathcal{W}$. The policy combines two complementary components: (1) a predefined rule that matches patient demographic attributes (e.g., age, sex, race, ethnicity), and (2) a data-driven annotation process.

We build our taxonomy using clinical indications pooled from three large, publicly available chest radiology datasets (see Sect. 3.1 for details), covering

Table 1. Taxonomy of clinical indications. For the example subcategory, "✓" denotes image-inferable and "×" denotes non-inferable. "# Sub." indicates the number of subcategories and "# Indi." indicates the number of unique granular indications under each summarized category.

Summarized Category (13)	Example Subcategory (347)	# Sub.	# Indi.
Thoracic Pathology Evaluation	Evaluation for pneumonia or pneumothorax (✓)	135	54,799
Medical Devices, Lines & Tubes	Status post chest tube placement or removal (✓)	41	19,527
Demographic & Administrative Context	Patient age (×)	36	17,603
Post-Procedural & Perioperative Context	Cesarean section delivery or birth (×)	27	10,151
Social History, Risk Factors & Exposures	History or presence of HIV/AIDS infection (×)	14	9,787
Non-Thoracic Conditions Prompting Chest Imaging	Nausea and vomiting symptoms (×)	16	5,865
Cardiovascular & Hemodynamic Conditions	Evaluation for mediastinal widening (✓)	22	5,196
Respiratory Support & Airway Management	Hypoxemic respiratory failure or distress (×)	12	4,695
Trauma, Injury & Hemorrhage History	Rib fracture(s) (✓)	11	4,338
Clinical Symptoms & Functional Status	Generalized or progressive weakness (×)	14	3,598
Neurologic & Cognitive Conditions	Altered mental status (×)	9	2,610
Laboratory & Physiologic Abnormalities	Evaluation or monitoring of leukocytosis (×)	8	2,542
Infectious, Inflammatory & Immune-Related	Evaluation or management of cellulitis (×)	2	418
Total		**347**	**141,129**

diverse institutions and clinical settings. We first normalize and split all clinical indications across three datasets into semantically coherent and unique units, yielding 141,129 unique parts. These parts are embedded using a domain-specific biomedical encoder based on PubMedBERT [9] which are then clustered using HDBSCAN [2] into 347 subcategories, each representing a group of semantically similar clinical expressions. We then label each subcategory for image-inferability using GPT-5.2 [21]. To further validate the labels, a certified medical physicist reviews all subcategories; 342 out of 347 labels are accepted and the remaining ones are corrected (Fig. 1b). The image-inferability label of each subcategory is propagated to all member phrases within its cluster.

These image non-inferable indications, together with predefined demographic attributes, constitute the sensitive set (Table 1). Across all datasets, an average of 10.6 out of 28.7 tokens per sample (37%) are identified as sensitive NVCC tokens. The resulting taxonomy captures recurring semantic patterns in radiology indications and supports incremental maintenance: new indications can be embedded and either assigned to existing subcategories or grouped into new clusters for review, without changing previously validated labels. For substantially out-of-distribution indications, additional expert annotation may be required before deployment.

2.3 Selective Embedding-Level Local Differential Privacy

Given the sensitive word set $\mathcal{W}_i = \pi(X_i)$ produced by the policy function, we tokenize the input as $\texttt{ids} = \text{Tok}(X_i)$ of length L_i and construct a binary mask $m_i = \text{Mask}(\mathcal{W}_i, \texttt{ids}) \in \{0,1\}^{L_i}$. For each token position $k \in \{1, \ldots, L_i\}$, $m_{i,k} = 1$ indicates that token position k corresponds to a sensitive NVCC attribute and

$m_{i,k} = 0$ indicates an image-inferable token that is left unperturbed under the stated outsourced-training threat model.

Rather than applying uniform noise to all token embeddings as in standard LDP, we selectively perturb only the embeddings at sensitive positions identified by the policy mask m_i. For each sample (I_i, X_i, Y_i) in a training batch, the mechanism proceeds as follows.

Encoding. The image and clinical text are independently encoded to produce visual embeddings $E_i^V = f_V(I_i; \theta_V)$ and text embeddings $E_i^X = f_X(X_i; \theta_X) \in \mathbb{R}^{L_i \times d}$, where d is the encoding dimension. Critically, f_X denotes the non-trainable token embedding layer, which maps each token independently to a fixed vector.

Selective Perturbation. For $k \in \{1, \ldots, L_i\}$, if $m_{i,k} = 1$ (sensitive), we clip the embedding to bound its ℓ_2-norm and add calibrated Gaussian noise:

$$\tilde{e}_{i,k} = E_{i,k}^X \cdot \min\left(1, \frac{C}{\|E_{i,k}^X\|_2}\right), \qquad E_{i,k}'^X = \tilde{e}_{i,k} + z, \quad z \sim \mathcal{N}(0, \sigma^2 I_d), \tag{1}$$

where C is the clipping norm. If $m_{i,k} = 0$ (non-sensitive), the embedding is passed through unmodified: $E_{i,k}'^X = E_{i,k}^X$ (Fig. 1b).

Fusion and Training. The privatized text embeddings $E_i'^X$ are fused with visual embeddings varied by model architecture, and the model is then trained with standard gradient descent on the report generation loss. The external trainer receives only the privatized embeddings and performs standard fine-tuning without access to the original sensitive information.

2.4 Privacy Analysis

Definition 1. (per-token (ε, δ)-Local Differential Privacy). *A randomized mechanism $\mathcal{M}$ operating on token embeddings satisfies per-token (ε, δ)-local differential privacy if, for any two possible embeddings $e, e' \in \mathbb{R}^d$ and any measurable set $\mathcal{S}$:*

$$\Pr[\mathcal{M}(e) \in \mathcal{S}] \leq e^{\varepsilon} \Pr[\mathcal{M}(e') \in \mathcal{S}] + \delta, \qquad \varepsilon > 0,\ \delta \in [0, 1). \tag{2}$$

Theorem 1. *Let $e \in \mathbb{R}^d$ denote the embedding of a sensitive token. The mechanism*

$$\mathcal{M}(e) = \tilde{e} + z, \qquad \tilde{e} = e \cdot \min\left(1, \frac{C}{\|e\|_2}\right), \quad z \sim \mathcal{N}(0, \sigma^2 I_d), \tag{3}$$

satisfies per-token (ε, δ)-local differential privacy for any $\varepsilon > 0$ and $\delta \in (0, 1)$ provided

$$\sigma \geq \frac{2C}{\varepsilon}\sqrt{2\ln\left(\frac{1.25}{\delta}\right)}. \tag{4}$$

Proof. After clipping, $\|\tilde{e}\|_2 \leq C$ for any input e. For any two possible token embeddings e, e' with clipped representations $\tilde{e}, \tilde{e}'$, the ℓ_2 sensitivity satisfies

$$\Delta_2 = \|\tilde{e} - \tilde{e}'\|_2 \leq \|\tilde{e}\|_2 + \|\tilde{e}'\|_2 \leq 2C. \tag{5}$$

The mechanism $\mathcal{M}(e) = \tilde{e} + z$ with $z \sim \mathcal{N}(0, \sigma^2 I_d)$ is therefore the Gaussian mechanism with sensitivity $\Delta_2 \leq 2C$. By the standard Gaussian mechanism bound [6], choosing σ as above ensures per-token (ε, δ)-differential privacy. Since the perturbation is applied locally to each embedding prior to any transmission or external computation, the mechanism satisfies per-token (ε, δ)-local differential privacy by definition.

Remark 1 (Composition and model independence). Theorem 1 establishes a per-token (ε, δ)-LDP guarantee for each protected (NVCC) token embedding. Under the record-level adjacency where unprotected content is fixed and only protected NVCC tokens may differ, releasing k independently privatized protected embeddings yields a conservative $(k\varepsilon, k\delta)$-LDP record-level bound by sequential composition [7] (with tighter bounds via advanced composition [14] or Rényi DP [20]). Accordingly, when experiments report utility at a fixed per-token (ε, δ), the induced conservative record-level bound varies across records through their realized k.

In this paper, we focus on per-token LDP because it matches the privacy-critical interface of the setting: NVCC embeddings are privatized locally before transmission, and leakage in clinical text arises from inference on localized sensitive spans. If record-level privacy is instead the target, our selectivity also improves record-level accounting relative to a uniform token-level LDP baseline. Under conservative sequential composition, perturbing all L tokens leads to a $(L\varepsilon, L\delta)$ record-level bound, whereas perturbing only the k protected tokens leads to $(k\varepsilon, k\delta)$. Since $k < L$, our method composes over fewer releases and therefore allows less noise per protected output under matched record-level privacy budgets. Concretely, under a matched record-level budget $(\varepsilon_{\text{rec}}, \delta_{\text{rec}})$, uniform LDP allocates $(\varepsilon_{\text{rec}}/L, \delta_{\text{rec}}/L)$ per token, while RadSLDP allocates $(\varepsilon_{\text{rec}}/k, \delta_{\text{rec}}/k)$ only to protected NVCC tokens. Since $k = 10.6 < L = 28.7$ on average, RadSLDP applies less noise per protected token under the same record-level bound, improving utility while preserving the matched record-level guarantee.

3 Experiments

3.1 Experiment Setup

Datasets and Evaluation. We evaluate on three public chest X-ray datasets. CheXpert Plus [3] contains 187,711 radiology studies with text radiology reports and patients' demographics; ReXGradient-160K [35] contains 160,000 radiology studies within 3 US health systems; MIMIC-CXR [13] contains 227,835 imaging studies presented to the Beth Israel Deaconess Medical Center Emergency Department. Together, these datasets comprise over 500,000 studies with paired radiographs and reports containing clinically rich indication text and demographic attributes. Model utility is evaluated using lexical overlap metrics BLEU-4 [23] and ROUGE-L [16], and clinically oriented metrics CheXbert Micro-F1 [26] and RadGraph F1 [12].

Implementation. We fine-tune LLaVA-1.5-7B [17], Qwen2-VL-7B and 2B [29] using LoRA [10] (rank 32) for 2 epochs with learning rate 2×10^{-4}, maximum sequence length 4096, clipping norm $C = 1.0$, and privacy parameter $\delta = 1/n$, where n is the training-set size of the corresponding dataset, on 8 NVIDIA A6000 GPUs. Training is implemented in PyTorch using the Hugging Face Trainer with the AdamW optimizer, warmup ratio 0.03, and bf16 mixed precision.

Table 2. Utility results across models and datasets. **DP-SGD and non-private training assume access to unperturbed training data and therefore do not apply to the external-training privacy setting considered in this work; we report them for reference only.*

			ReXGradient				CheXpert PLUS				MIMIC-CXR			
Model	**Training**	ε	BLEU↑	RG-L↑	Chex-F1↑	Rad-F1↑	BLEU↑	RG-L↑	Chex-F1↑	Rad-F1↑	BLEU↑	RG-L↑	Chex-F1↑	Rad-F1↑
LLaVA-7b	Non-private*	-	3.98	0.13	0.29	0.18	5.99	0.17	0.53	0.21	5.04	0.16	0.39	0.19
	DP-SGD*	8	1.95	0.12	0.18	0.15	1.05	0.13	0.30	0.08	3.17	0.15	0.18	0.17
	LDP	8	2.60	0.11	0.14	0.16	4.30	0.16	0.46	0.19	4.32	0.15	0.37	0.17
	RadSLDP	8	**3.26**	**0.12**	**0.27**	**0.17**	**4.54**	**0.17**	**0.51**	**0.20**	**4.52**	0.15	**0.40**	**0.19**
Qwen-7b	Non-private*	-	10.42	0.24	0.36	0.19	10.08	0.22	0.49	0.20	7.23	0.23	0.36	0.19
	DP-SGD*	8	4.62	0.19	0.18	0.15	1.51	0.16	0.30	0.09	3.27	0.19	0.16	0.14
	LDP	8	5.30	0.21	0.15	0.18	7.27	0.19	0.49	0.19	**6.50**	0.22	**0.35**	0.18
	RadSLDP	8	**7.99**	**0.22**	**0.28**	0.18	**7.86**	0.19	**0.51**	**0.20**	6.06	**0.23**	0.30	**0.19**
Qwen-2b	Non-private*	-	9.10	0.23	0.30	0.18	8.86	0.21	0.40	0.19	7.25	0.22	0.30	0.18
	DP-SGD*	8	4.47	0.18	0.17	0.14	2.31	0.19	0.25	0.12	3.81	0.19	0.11	0.14
	LDP	8	4.06	0.20	0.12	0.17	5.50	0.17	0.42	0.15	6.06	0.21	0.28	0.16
	RadSLDP	8	**7.20**	0.20	**0.25**	0.17	**5.88**	**0.18**	**0.43**	**0.18**	**6.72**	**0.22**	**0.32**	**0.18**

3.2 Results

Overall Utility Comparison. Table 2 reports utility at per-token $\varepsilon = 8$. We compare against LDP that adds uniform noise to all input embeddings as the primary baseline, as it is the only mechanism that satisfies our untrusted trainer threat model. To the best of our knowledge, no prior work has applied LDP to VLM fine-tuning; we therefore implement uniform embedding-level LDP using the same Gaussian mechanism and privacy budget as RadSLDP, applied to all input tokens rather than selectively. Across all 36 comparisons, RadSLDP outperforms LDP in 29 cases (80.5%) and matches its performance in 5 cases (14%), with improvements most pronounced on ReXGradient: 24–77% in BLEU-4 and 87–108% in Micro-F1 across all architectures. Notably, Qwen2-2B improves from 0.12 to 0.25 in Micro-F1 (+108%) and from 4.06 to 7.20 in BLEU-4 (+77%) on ReXGradient. On CheXpert Plus and MIMIC-CXR, RadSLDP consistently matches or exceeds LDP, indicating that selective perturbation preserves clinically relevant information that uniform noise can disrupt.

Table 3. Effect of per-token privacy budget ε on utility on the ReXGradient dataset. $\varepsilon = \infty$ represents clipping only. Std is computed over $\varepsilon \in \{4, 8, \infty\}$.

		$\varepsilon = 4$	$\varepsilon = 8$	$\varepsilon = \infty$	**Overall**
Model	**Training**	BLEU↑	BLEU↑	BLEU↑	BLEU Std_ε ↓
LLaVA-7B	LDP	2.36	2.60	2.91	0.23
	Ours	**3.32**	**3.26**	**3.48**	**0.09**
Qwen2-7B	LDP	5.09	5.30	9.52	2.04
	Ours	**7.99**	**7.99**	**9.60**	**0.76**
Qwen2-2B	LDP	3.40	4.06	7.40	1.75
	Ours	**7.10**	**7.20**	**8.05**	**0.43**

Stability Across Privacy Budgets. Table 3 examines the effect of varying $\varepsilon \in \{4, 8, \infty\}$ on ReXGradient. RadSLDP exhibits lower BLEU-4 standard deviation than LDP across all three architectures: LLaVA-7B (0.09 vs. 0.23), Qwen2-7B (0.76 vs. 2.04), and Qwen2-2B (0.43 vs. 1.75). Because RadSLDP perturbs only NVCC tokens, changes in the noise scale affect a smaller portion of the input, producing more stable performance as ε varies. In contrast, uniform LDP applies noise to all tokens, amplifying the effect of any change in privacy budget across the entire input. Both methods show a general increasing trend as ε grows, consistent with prior work.

4 Conclusion

We presented RadSLDP, a selective embedding-level local differential privacy mechanism for privacy-preserving fine-tuning of medical VLMs in untrusted training settings. By perturbing only Non-Visual Clinical Context (NVCC) token embeddings while leaving image-inferable content intact, RadSLDP provides formal per-token (ε, δ)-LDP guarantees independent of the downstream model architecture. Across three chest X-ray datasets and three VLM architectures, RadSLDP matches or outperforms uniform LDP in 94% of comparisons (34/36), achieving up to 108% Micro-F1 improvement on ReXGradient with lower performance variability across privacy budgets due to perturbing fewer tokens. To support reproducibility, the source code and our sensitivity annotations comprising 347 expert-validated subcategories derived from 141,129 clinical indications are available at: https://github.com/Epiphany625/RadSLDP.

Disclosure of Interests. The authors have no competing interests to declare that are relevant to the content of this article.

References

1. Abadi, M., Chu, A., Goodfellow, I.: Deep learning with differential privacy. In: Proceedings of the ACM SIGSAC Conference on Computer and Communications Security (CCS), pp. 308–318 (2016)
2. Campello, R.J., Moulavi, D., Sander, J.: Density-based clustering based on hierarchical density estimates. In: Pacific-Asia Conference on Knowledge Discovery and Data Mining (PAKDD) (2013)
3. Chambon, P., Delbrouck, J.B., Sounack, T.: Chexpert plus: augmenting a large chest X-ray dataset with text radiology reports, patient demographics and additional image formats (2024)
4. Du, M., Yue, X., Chow, S.S.M., Sun, H.: Sanitizing sentence embeddings (and labels) for local differential privacy. In: Proceedings of the ACM Web Conference, pp. 2349–2359 (2023)
5. Duchi, J.C., Jordan, M.I., Wainwright, M.J.: Local privacy and statistical minimax rates. In: IEEE 54th Annual Symposium on Foundations of Computer Science, pp. 429–438 (2013)
6. Dwork, C., McSherry, F., Nissim, K., Smith, A.: Calibrating noise to sensitivity in private data analysis. In: Theory of Cryptography Conference (TCC), pp. 265–284. Springer (2006)
7. Dwork, C., Roth, A.: The Algorithmic Foundations of Differential Privacy, vol. 9. Now Publishers (2014)
8. Flemings, J., Razaviyayn, M., Annavaram, M.: Differentially private next-token prediction of large language models. In: North American Chapter of the Association for Computational Linguistics (NAACL), pp. 4390–4404 (2024)
9. Gu, Y., et al.: Domain-specific language model pretraining for biomedical natural language processing. ACM Trans. Comput. Healthcare **3**(1) (2021)
10. Hu, E.J., Shen, Y., Wallis, P.: LoRA: low-rank adaptation of large language models. In: International Conference on Learning Representations (ICLR) (2022)
11. Hu, L.: Differentially private natural language models: recent advances and future directions, pp. 478–499 (2024)
12. Jain, S., Agrawal, A., Saporta, A.: RadGraph: extracting clinical entities and relations from radiology reports. In: Neural Information Processing Systems (NeurIPS) Datasets and Benchmarks Track (2021)
13. Johnson, A.E., Pollard, T.J., Berkowitz, S.J.: MIMIC-CXR, a de-identified publicly available database of chest radiographs with free-text reports. Sci. Data **6** (2019)
14. Kairouz, P., Oh, S., Viswanath, P.: The composition theorem for differential privacy. In: International Conference on Machine Learning (ICML), pp. 1376–1385. PMLR (2015)
15. Kaissis, G.A., Ziller, A., Passerat-Palmbach, J., et al.: End-to-end privacy preserving deep learning on multi-institutional medical imaging. Nat. Mach. Intell. **3**, 473–484 (2021)
16. Lin, C.Y.: ROUGE: a package for automatic evaluation of summaries. In: Text Summarization Branches Out, pp. 74–81 (2004)
17. Liu, H., Li, C., Wu, Q.: Visual instruction tuning. In: Advances in Neural Information Processing Systems (NeurIPS) (2024)
18. Lyu, L., He, X., Li, Y.: Differentially private representation for NLP: formal guarantee and an empirical study on privacy and fairness. In: Findings of the Association for Computational Linguistics: EMNLP, pp. 2355–2365 (2020)

19. Mai, P., Yan, R., Huang, Z.: Split-and-denoise: protect large language model inference with local differential privacy. In: Proceedings of the 41st International Conference on Machine Learning (ICML), pp. 34431–34461 (2024)
20. Mironov, I.: Rényi differential privacy. In: IEEE Computer Security Foundations Symposium (CSF), pp. 263–275. IEEE (2017)
21. OpenAI: GPT (version 5.2). Large Language Model (2025). https://platform.openai.com/
22. Packhäuser, K., Gündel, S., Münster, N., et al.: Deep learning-based patient re-identification is able to exploit the biometric nature of medical chest X-ray data. Sci. Rep. **12**, 14851 (2022)
23. Papineni, K., Roukos, S., Ward, T., Zhu, W.J.: BLEU: a method for automatic evaluation of machine translation. In: Annual Meeting of the Association for Computational Linguistics (ACL), pp. 311–318 (2002)
24. Shi, W., Cui, A., Li, E.: Selective differential privacy for language modeling. In: Proceedings of the 2022 Conference of the North American Chapter of the Association for Computational Linguistics (NAACL) (2022)
25. Shi, W., Shea, R., Chen, S.: Just fine-tune twice: selective differential privacy for large language models. In: Proceedings of the 2022 Conference on Empirical Methods in Natural Language Processing (EMNLP), pp. 6327–6340 (2022)
26. Smit, A., Jain, S., Rajpurkar, P.: CheXbert: combining automatic labelers and expert annotations for accurate radiology report labeling using BERT. In: Conference on Empirical Methods in Natural Language Processing (EMNLP), pp. 1500–1519 (2020)
27. Tran, L., Sun, W., Patterson, S., Milanova, A.: Privacy-preserving personalized federated prompt learning for multimodal large language models. In: The Thirteenth International Conference on Learning Representations (2025)
28. U.S. Department of Health and Human Services: Guidance regarding methods for de-identification of protected health information in accordance with the Health Insurance Portability and Accountability Act (HIPAA) privacy rule (2012)
29. Wang, P., Bai, S., Tan, S.: Qwen2-vl: enhancing vision-language model's perception of the world at any resolution. arXiv preprint arXiv:2409.12191 (2024)
30. Wang, T., Zhai, L., Yang, T., et al.: Selective privacy-preserving framework for large language models fine-tuning. Inf. Sci. **682**, 121000 (2024)
31. Wei, Q., Li, J., You, Z.: Dual-priv pruning: efficient differential private fine-tuning in multimodal large language models. arXiv preprint arXiv:2506.07077 (2025)
32. Yu, D., Naik, S., Backurs, A.: Differentially private fine-tuning of language models. arXiv preprint arXiv:2110.06500 (2021)
33. Yu, M., Singh, A., Li, S., Cao, Y.: Adaptive token-weighted differential privacy for LLMs: not all tokens require equal protection. arXiv preprint arXiv:2509.23246 (2025)
34. Zhang, R.: Privacy-oriented text generation in LLMs via selective fine-tuning and semantic attention masks. J. Comput. Technol. Softw. **4**(8) (2025)
35. Zhang, X., Acosta, J.N., Miller, J.: Rexgradient-160k: a large-scale publicly available dataset of chest radiographs with free-text reports (2025)
36. Zheng, L., Cao, Y., Yoshikawa, M., et al.: Sensitivity-aware differential privacy for federated medical imaging. Sensors **25**(9), 2847 (2025)
37. Ziller, A., Usynin, D., Braren, R., et al.: Medical imaging deep learning with differential privacy. Sci. Rep. **11**, 13524 (2021)

Reasoning Trace Divergence: An Empirical Signal for Trustworthy Black-Box MLLMs in Histopathology Classification

Anastasiia Kazmina[1], Constantin Venhoff[2], Bryan Wong[1], Sungrae Hong[1], Yutong Xie[3], and Mun Yong Yi[1(✉)]

[1] Korea Advanced Institute of Science and Technology, Daejeon, South Korea
{anastasiiaka,munyi}@kaist.ac.kr
[2] University of Oxford, Oxford, UK
[3] Mohamed bin Zayed University of Artificial Intelligence, Abu Dhabi, United Arab Emirates

Abstract. Multimodal Large Language Models (MLLMs) offer transformative potential for medical diagnostics, yet their tendency to generate plausible but ungrounded hallucinations remains a critical barrier. In black-box settings, where internal model states are inaccessible, existing uncertainty quantification methods like visual-induced agreement rely on final label consistency, which often fails to capture systematic overconfidence. In this work, we move from visual stability to reasoning stability, identifying a failure mode where models maintain consistent labels despite divergent and fabricated logic. We introduce Reasoning Trace Divergence (RTD), a novel empirical signal that exposes these latent hallucinations by probing the stability of the model's logical paths under stochastic visual perturbations. Evaluating this signal on two patch-level histopathology datasets using Gemini and MedGemma, we demonstrate that reasoning instability is a robust predictor of grounding failures. By filtering for high reasoning stability, we identify a safe operating zone that significantly improves classification reliability, providing a principled direction for trustworthy black-box medical AI. The code is available at https://github.com/anastlibr99/reasoning-trace-divergence-blackbox-mllms-histopath.

Keywords: Uncertainty Quantification · Reasoning Stability · Computational Pathology

1 Introduction

Multimodal Large Language Models (MLLMs) enable the analysis of multidimensional data [2,12], a capability essential in medicine where diagnostic insights are distributed across heterogeneous sources such as images and text. While these models offer transformative potential, they are prone to generating plausible but wrong content [17,18]. This raises a critical question for high-stakes clinical domains: how can we evaluate the reliability of a prediction? This challenge

G. Yang et al. (Eds.): MICCAI 2026, LNCS 16878, pp. 493–502, 2027.
https://doi.org/10.1007/978-3-032-38059-3_47

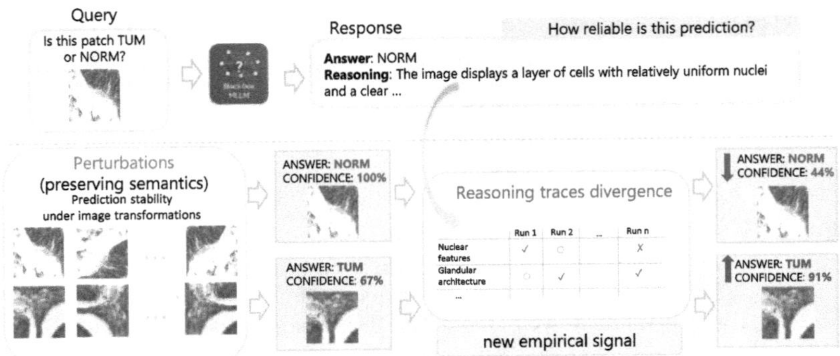

Fig. 1. Overview of the UVR Framework. UVR quantifies uncertainty by auditing reasoning equivariance across D_4 transformations. The pipeline calibrates MLLM confidence by lowering the scores for incorrect answers (Red) and increasing confidence for correct ones (Green). (Color figure online)

is exacerbated in black-box settings [1,16], where massive scale or proprietary APIs preclude access to internal model states, gradients, or logits.

Hallucinations were initially formalized in text-only Large Language Models (LLMs) [8], identifying root causes such as noisy training data, knowledge gaps, or data fabrication [11]. A prominent mitigation strategy is semantic entropy [5], which quantifies uncertainty by measuring consistency across semantically equivalent responses. This concept was further extended by introducing knowledge-preserving semantic transformations to the input. Adapting these techniques to MLLMs requires a more nuanced understanding of cross-modal interactions [13]. While defining semantic-preserving transformations for multimodal inputs has shown improvement over baseline entropy [19,20], clinical applications remain complex. In medical imaging, semantic-preserving distortions differ significantly from the general domain; in histopathology specifically, where small morphological features determine malignancy, the grounding of reasoning in specific visual tokens is critical.

While recent work has demonstrated that vision-amplified semantic entropy can effectively detect hallucinations in open-ended medical Visual Question Answering (VQA) [10,21], the efficacy of these techniques in structured classification tasks remains under-explored. Although classification is often treated as a subset of closed-set VQA, the transition to histopathology introduces a unique challenge: the reliance on few-shot In-Context Learning (ICL) to adapt generalist MLLMs to specialized domains [3,7]. This shift is critical because while few-shot prompting significantly improves top-line accuracy, it often induces a calibration-accuracy gap where models produce increasingly coherent reasoning traces that mask systematic overconfidence or latent hallucinations. Consequently, there is an urgent need to analyze the uncertainty signals inherent in these generated traces. Current black-box metrics, which rely solely on final label consistency, fail to capture the subtle failures in reasoning that few-shot settings can introduce, leaving a significant gap in our ability to detect failures in high-stakes diagnostic contexts.

To address this, we analyze the failure modes of semantic uncertainty in zero-shot and few-shot histopathology. We define a novel empirical signal that moves beyond token-level entropy to consider the divergence of the model's reasoning traces under stochastic visual perturbation (see Fig. 1). We evaluate this signal across both general-purpose (Gemini 2.5 Flash Lite [16]) and domain-specific (MedGemma 1.5 4b [14]) MLLMs using two histopathology patch datasets in both zero- and few-shot settings using ICL. Our main contributions are:

- We identify **systematic overconfidence** as a primary failure mode in black-box medical MLLMs, where standard semantic entropy fails to capture decoupled reasoning.
- We introduce **Reasoning Trace Divergence (RTD)** as an empirical signal that exposes these failure patterns by measuring the stability of the model's logic under visual pressure.
- We demonstrate the utility of this signal by identifying a **safe operating zone** for clinical decision support; filtering for high reasoning stability effectively isolates correct predictions and improves the reliability of the diagnostic pipeline.

2 Methodology

2.1 Task Formulation

In this section, we define a histopathology patch classification task with a discrete label set, under both zero-shot and few-shot settings. Let $x_i \in \mathcal{X}$ denote the i-th test image, for $i = 1, \ldots, N$. $\mathcal{Y}$ is a finite set of possible labels the image may belong to. $\mathcal{Y} = \{c_1, c_2, \ldots, c_C\}$ where C is the total number of classification labels. Let $y_i \in \mathcal{Y}$ be the ground truth label for x_i.

The complete model input is encapsulated in a multimodal classification query $Q_i = (x_i, s, u_E)$ where s denotes a system prompt defining the experimental constraints and output formatting requirements, and u_E represents the user prompt. The user prompt is an ordered sequence consisting of a set of E labeled demonstrations for every class from $\mathcal{Y}$ and a task instruction u^{inst}: $u_E = \left(\{(x_e, y_e)\}_{e=1}^{E}, u^{\text{inst}}\right)$. The setting $E = 0$ corresponds to zero-shot inference, while $E > 0$ denotes few-shot learning.

A MLLM parameterized by θ produces a class label prediction and reasoning text: $\hat{y}_i, r_i = f_\theta(Q_i; \tau)$ where $Q_i = (x_i, s, u_E)$ denotes the model input, $\hat{y}_i \in \mathcal{Y}$ is the predicted class label, r_i denotes the generated reasoning, and τ is the decoding temperature.

Our primary goal is to evaluate predictive certainty within black-box settings where internal model probabilities are inaccessible. To achieve this, we first establish a deterministic reference for each test image x_i. Given the input query $Q_i = (x_i, s, u_E)$ the model produces a reference prediction $\hat{y}_i = f_\theta(Q_i; \tau = 0)$. We then assess the reliability of this prediction by generating an empirical distribution $p_i(c)$ under various conditions.

2.2 Semantic-Preserving Perturbations

To more robustly probe the model's knowledge, we move beyond simple noise and consider perturbations to the input manifold. Not all perturbations of input can lead to meaningful uncertainty results that reflect the true model uncertainty about the specific task. Following the definition of knowledge-preserving perturbations for language tasks in [6] we define semantic-preserving perturbations for a query Q_i as follows.

Definition 1 (Semantic-Preserving Perturbation). *Let Q_i denote the model query corresponding to image x_i, including the image and associated prompts. A perturbation operator δ is said to be semantic-preserving if*

$$K(\delta(Q_i)) = K(Q_i), \tag{1}$$

where $K(\cdot)$ denotes the decision-relevant knowledge required to determine the true class label y_i.

In the analysis of histopathological patches, the orientation of morphological structures is entirely stochastic. Consequently, diagnostic features such as gland architecture or serration patterns can be observed at any angle. This inherent property ensures that the D_4 transformation group [15], consisting of eight rotations and reflections, functions as a set of strictly semantic-preserving perturbations. For any patch x_i, the underlying diagnostic knowledge $K(x_i, s, u_E)$ remains invariant regardless of the spatial transformation $\delta \in D_4$. We exploit this property to define Visual Entropy (H_{vis}), which quantifies the predictive instability across these eight symmetric views. Formally, $H_{vis} = -\sum_{c \in \mathcal{Y}} p(c) \log p(c)$, where $p(c)$ is the empirical label frequency across the 8 predictions. A high H_{vis} signifies a failure in visual equivariance, indicating that the model's internal representations are sensitive to non-semantic spatial shifts.

2.3 Reasoning Trace Divergence

A distinct advantage of MLLMs is their capacity to generate natural language justifications alongside categorical predictions, providing a dual-stream signal for uncertainty quantification. We leverage this to construct a higher-fidelity metric of model instability. However, as raw justifications r_i are unstructured, we utilize a secondary LLM to perform Semantic Feature Mapping, converting free-form text into structured Reasoning Traces ($\mathcal{T}$). For a given patch x and its D_4 transformations, a meta-analytical prompt first identifies m unique morphological features $F = \{f_1, \ldots, f_m\}$ present across all justifications. Note that F emerges from the model's own justifications, not a predefined ontology; alignment with descriptors like glandular architecture is observational, not expert-verified. Each r_i is then mapped to a feature vector $\boldsymbol{v}_i \in \mathcal{S}^m$, where the state space $\mathcal{S} = \{1, 0, -1\}$ represents the presence, omission, or explicit negation of a feature, respectively. This results in a structured logical state that allows

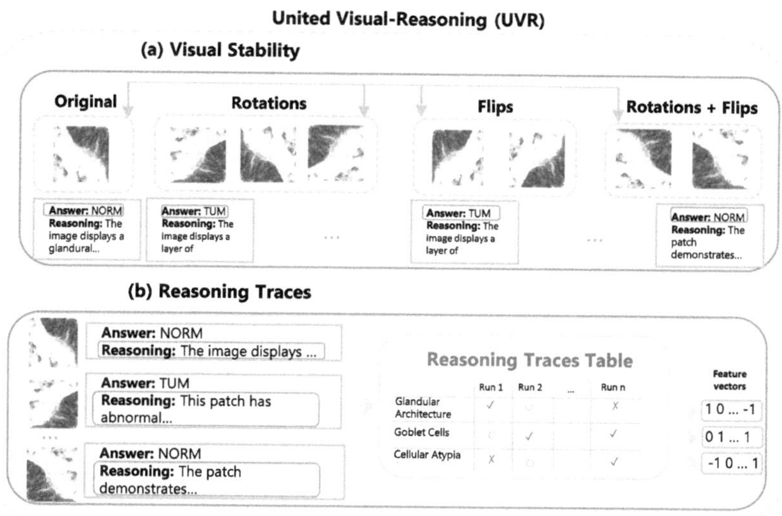

Fig. 2. UVR Framework. (a) Image-based D_4 perturbations evaluate model equivariance. (b) Multimodal reasoning traces are distilled into a Semantic Feature Matrix $(1, 0, -1)$ to compute logical uncertainty $\mathcal{H}_{RT}$.

for a precise mathematical comparison of the model's internal consistency. The Reasoning Trace is defined as the structured logical state:

$$\mathcal{T}_i = [v_{i,1}, v_{i,2}, \ldots, v_{i,m}] \tag{2}$$

We quantify logic-level instability using the Average Reasoning Entropy ($\mathcal{U}_{sem}$). For each feature f_j, we calculate the Shannon entropy $H(f_j)$ across the D_4 transformations. The average semantic uncertainty is defined as:

$$\mathcal{H}_{RT} = \frac{1}{m} \sum_{j=1}^{m} H(f_j), \tag{3}$$

where $H(f_j) = -\sum_{k \in \mathcal{S}} p_k \log p_k$. To capture the interaction between decision-level and logic-level instability, we employ a clipping operation to normalize entropy values into a strictly bounded $[0, 1]$ probability space, preventing extreme divergence from disproportionately skewing the joint metric. We propose the United Visual-Reasoning (UVR) certainty score:

$$\mathcal{C}_{UVR} = \text{clip}(1 - H_{vis}, 0, 1) \cdot \text{clip}(1 - \mathcal{H}_{RT}, 0, 1) \tag{4}$$

Figure 2 demonstrates this two-level approach.

3 Results

3.1 Experimental Setup

Datasets and Sampling. We evaluate the proposed uncertainty estimation methods on two benchmark histopathology patch classification datasets: Patch-

Camelyon (PCam) [4] and NCT-CRC-100K [9]. For PCam, which targets the detection of metastatic breast cancer in lymph node patches, we utilize a binary classification task consisting of two classes: METASTASIS and NO METASTASIS. We randomly sample 150 patches per class from the official test set for evaluation, while few-shot demonstration examples are drawn from the validation split. For NCT-CRC-100K, which comprises 100,000 H&E patches across nine tissue types, we constrain the task to a binary classification between tumor epithelium (TUM) and normal colon mucosa (NORM). We randomly sample 150 patches per class from the validation split for our evaluation cohort and draw few-shot demonstration examples from the training split. This sampling strategy ensures that the in-context examples and evaluation cases remain strictly separated across both datasets. To analyze the impact of context on reasoning stability, we evaluate the model in zero-shot and few-shot settings using $E \in \{0, 3, 6\}$ demonstration examples per class. Few-shot demonstration sets are sampled independently three times using random seeds {7, 13, 42} to account for variability introduced by in-context example selection. All reported few-shot results are averaged across these three samplings.

Model Selection and Prompting Strategy. To evaluate generalizability across architectures, we use two state-of-the-art MLLMs: the proprietary Gemini 2.5 Flash Lite (representing high-scale commercial APIs) and the open-weights MedGemma 1.5 4B (representing domain-specific medical models) [14]. For the generation and analysis of reasoning traces in both cases, we utilize Gemini 2.5 Flash Lite [16] by prompting it to extract and process the reasoning from the corresponding models. This approach allows us to consistently capture the logical divergence of the models' explanations for our reasoning stability metrics.

Table 1. AUROC (%) for Gemini 2.5 Flash Lite and MedGemma 1.5 4B. Mean ± std over three few-shot samplings (seeds 3, 13, 42). VASE is omitted for MedGemma due to safety-induced refusals under hard perturbations.

Model	Dataset	Score	0-Shot	3-Shot	6-Shot
Gemini 2.5 [16]	CRC-100K [9]	Sampling Entropy	58.9	56.8 ± 0.2	61.0 ± 1.0
		RadFlag [21]	58.7	56.7 ± 0.4	60.9 ± 1.0
		VASE [10]	56.8	60.0 ± 4.1	65.6 ± 7.5
		Ours (UVR)	**66.1**	**67.0 ± 1.5**	**67.9 ± 6.3**
	PatchCamelyon [4]	Sampling Entropy	64.0	57.1 ± 2.8	59.0 ± 5.1
		RadFlag [21]	63.8	57.1 ± 2.8	59.0 ± 5.1
		VASE [10]	56.8	46.2 ± 0.0	52.1 ± 6.3
		Ours (UVR)	**66.0**	**64.0 ± 8.9**	**66.3 ± 9.4**
MedGemma 4B [14]	CRC-100K [9]	Sampling Entropy	52.7	46.2 ± 5.9	46.3 ± 2.5
		RadFlag [21]	55.3	46.2 ± 5.9	46.0 ± 2.1
		Ours (UVR)	**70.2**	**72.0 ± 1.6**	**74.1 ± 2.3**
	PatchCamelyon [4]	Sampling Entropy	60.0	59.6 ± 2.9	59.6 ± 2.3
		RadFlag [21]	50.4	62.2 ± 3.4	58.0 ± 8.6
		Ours (UVR)	**60.1**	**63.8 ± 1.6**	53.0 ± 3.1

Evaluation Methods. We assess the reliability of a reference prediction $\hat{y}_i = f_\theta(Q_i; \tau = 0)$ using the Area Under the Receiver Operating Characteristic (AUROC) curve. In this framework, AUROC measures the probability that a correctly classified sample is assigned a higher certainty score than an incorrect one (hallucination). To ensure comparability, all entropy measures are normalized into certainty signals $\mathcal{C} \in [0, 1]$ as $\mathcal{C} = 1 - \text{clip}(H, 0, 1)$. Beyond global performance, we evaluate the utility of these signals for **reliable decision support**. Specifically, we define a safe operating zone by filtering for the **top 10% most confident predictions** according to each metric. We then calculate the classification accuracy within this high-certainty subset to determine which signal most effectively isolates the model's most trustworthy outputs. Crucially, this requires no ground-truth labels, so the safe operating zone applies directly to unseen samples. AUROC is threshold-independent; Top-10% accuracy reflects a clinically relevant operating point. Broader calibration curves are left to future work.

Table 2. Top-10% Accuracy (%) for Gemini 2.5 Flash Lite and MedGemma 1.5 4B. Mean ± std over three few-shot samplings (seeds 7, 13, 42).

Model	Dataset	Score	0-Shot	3-Shot	6-Shot
Gemini 2.5 [16]	CRC-100K [9]	Sampling	76.7	67.8 ± 5.1	68.9 ± 6.9
		RadFlag [21]	63.3	71.1 ± 5.1	74.4 ± 6.9
		VASE [10]	63.3	81.1 ± 1.9	92.2 ± 13.5
		Ours (UVR)	**86.2**	**89.7 ± 3.5**	<u>90.8</u> ± 7.2
	PatchCamelyon [4]	Sampling	64.0	57.1 ± 2.8	59.0 ± 5.1
		RadFlag [21]	68.4	56.9 ± 3.7	60.9 ± 4.9
		VASE [10]	56.8	46.2 ± 0.0	52.1 ± 6.3
		Ours (UVR)	<u>66.0</u>	**64.0 ± 8.9**	**66.3 ± 9.4**
MedGemma 4B [14]	CRC-100K [9]	Sampling	52.7	46.2 ± 5.9	46.3 ± 2.5
		RadFlag [21]	55.3	46.2 ± 5.9	46.0 ± 2.1
		Ours (UVR)	**70.2**	**72.0 ± 1.6**	**74.1 ± 2.3**
	PatchCamelyon [4]	Sampling	90.0	78.9 ± 9.6	70.0 ± 14.1
		RadFlag [21]	90.0	78.9 ± 9.6	70.0 ± 14.1
		Ours (UVR)	**93.1**	<u>73.6</u> ± 7.2	<u>68.0</u> ± 5.7

Table 3. Ablation study of visual and reasoning stability components on MedGemma 4B (CRC-100K, 6-Shot). Our final method (bottom row) combines both to achieve peak performance.

Visual Stability	Reasoning Stability	AUROC (%)	Acc@10% (%)
✗	✗	46.31	70.00
✓	✗	71.91	80.00
✗	✓	64.31	87.93
✓	✓	**74.07**	**93.10**

3.2 Experimental Results

We evaluate our proposed UVR framework against several established uncertainty and hallucination detection strategies, including: (1) Sampling Entropy, based on the principles of semantic entropy [5], which performs $N = 8$ independent forward passes at temperature $\tau = 1.0$ to measure label-level uncertainty through the diversity of predicted classes; (2) RadFlag [21], which estimates internal consensus by generating multiple responses and calculating the proportion of answers that align with the original diagnostic output, where lower agreement suggests a higher likelihood of error; and (3) Vision-Amplified Semantic Entropy (VASE) [10]: We adapted the VASE framework for medical classification by constructing a perturbation distribution that includes our weak D_4 transformations alongside three hard stressors: (i) high-sigma Gaussian blur ($\sigma = 15$) to preserve stain while destroying morphology, (ii) global mean pixel constant, and (iii) high-density salt-and-pepper noise. While this combined distribution exposes ungrounded logic in Gemini 2.5 Flash Lite, MedGemma 4B 1.5 frequently refused to answer when presented with these hard transformations. We therefore omit this metric for MedGemma to ensure a fair and consistent evaluation across the remaining benchmarks.

Comparison Analysis. As shown in Table 1, the proposed UVR metric demonstrates consistent improvement across all experimental settings, significantly outperforming standard sampling and existing baselines. The method reaches a peak AUROC of approximately 74.1% for the MedGemma 4B model on CRC-100K, marking a level of performance that is clinically relevant for computer-aided diagnosis. In Table 2, the Top-10% Accuracy results further confirm the metric's effectiveness as a reliable trust filter for medical decision-making. However, performance in MedGemma remains less stable in specific settings due to the model's inherent conservatism in histopathology tasks. This behavior results in minimal or repetitive reasoning chains that cause a performance drop compared to more expressive models.

Ablation Study. Ablation results in Table 3 confirm that while individual components provide significant benefits, combining visual and reasoning stability is essential for peak performance. Specifically, visual stability alone provides the largest single jump in AUROC (+25.60), while reasoning stability contributes an additional +18.00 in AUROC and +17.93 in Top-10% Accuracy. The two signals are thus complementary, capturing prediction and reasoning consistency respectively

4 Conclusion

In this work, we introduced a novel empirical signal for uncertainty quantification in black-box medical MLLMs based on RTD. By auditing the entropy of reasoning traces under visual, semantic-preserving perturbations, our framework moves beyond simple label consistency to a more robust representation of model certainty. Across two histopathology datasets, integrating reasoning traces improves both uncertainty estimation and classification accuracy for general-purpose and domain-specific models alike. Furthermore, our ablation study underscores the critical role of reasoning stability in identifying latent hallucinations that visual-only metrics often fail to capture.

Limitations and Future Work. While our framework enhances reliability, it introduces specific challenges. The extraction and analysis of reasoning traces by the MLLM itself can introduce a secondary layer of uncertainty and increased response variation. Expert validation of extracted concepts and broader calibration analysis remain important future directions. Future work will formalize RTD to better decouple diagnostic logic from linguistic stochasticity. Additionally, we aim to investigate the scaling of this signal across more diverse clinical modalities to further establish a safe operating zone for black-box AI in medicine.

Acknowledgments. This work was supported by the National Research Foundation of Korea (NRF) grant funded by the Korean government (MSIT) under grant number RS-2022-NR068758. We are also deeply grateful for the generous support provided by the Seegene Medical Foundation in South Korea. We thank Dr. Chris Abishek S, MBBS, currently pursuing residency in General Surgery (Chettinad Hospital and Research Institute), for helpful discussions on histopathological concepts and terminology.

Disclosure of Interests. The authors have no competing interests to declare that are relevant to the content of this article.

References

1. Achiam, J., et al.: GPT-4 technical report. arXiv preprint arXiv:2303.08774 (2023)
2. AlSaad, R., et al.: Multimodal large language models in health care: applications, challenges, and future outlook. J. Med. Internet Res. **26**, e59505 (2024)
3. Brown, T., et al.: Language models are few-shot learners. Adv. Neural. Inf. Process. Syst. **33**, 1877–1901 (2020)
4. Ehteshami Bejnordi, B., et al.: Diagnostic assessment of deep learning algorithms for detection of lymph node metastases in women with breast cancer. Jama **318**(22), 2199–2210 (2017)
5. Farquhar, S., Kossen, J., Kuhn, L., Gal, Y.: Detecting hallucinations in large language models using semantic entropy. Nature **630**(8017), 625–630 (2024)

6. Feng, Y., et al.: Rethinking LLM uncertainty: a multi-agent approach to estimating black-box model uncertainty. arXiv preprint arXiv:2412.09572 (2024)
7. Ferber, D., et al.: In-context learning enables multimodal large language models to classify cancer pathology images. Nat. Commun. **15**(1), 10104 (2024)
8. Huang, L., et al.: A survey on hallucination in large language models: principles, taxonomy, challenges, and open questions. ACM Trans. Inf. Syst. **43**(2), 1–55 (2025)
9. Kather, J.N., et al.: Predicting survival from colorectal cancer histology slides using deep learning: a retrospective multicenter study. PLoS Med. **16**(1), e1002730 (2019)
10. Liao, Z., Hu, S., Zou, K., Fu, H., Zhen, L., Xia, Y.: Vision-amplified semantic entropy for hallucination detection in medical visual question answering. In: International Conference on Medical Image Computing and Computer-Assisted Intervention, pp. 669–679. Springer (2025)
11. Ravichander, A., Ghela, S., Wadden, D., Choi, Y.: Halogen: fantastic LLM hallucinations and where to find them. In: Proceedings of the 63rd Annual Meeting of the Association for Computational Linguistics (Volume 1: Long Papers), pp. 1402–1425 (2025)
12. Rösler, W., et al.: An overview and a roadmap for artificial intelligence in hematology and oncology. J. Cancer Res. Clin. Oncol. **149**(10), 7997–8006 (2023)
13. Sahoo, P., Meharia, P., Ghosh, A., Saha, S., Jain, V., Chadha, A.: A comprehensive survey of hallucination in large language, image, video and audio foundation models. In: Findings of the Association for Computational Linguistics: EMNLP 2024, pp. 11709–11724 (2024)
14. Sellergren, A., et al.: Medgemma technical report. arXiv preprint arXiv:2507.05201 (2025)
15. Sharma, P.: Dihedral group D4–a new feature extraction algorithm. Symmetry **12**(4), 548 (2020)
16. Anil, R., et al.: Gemini: a family of highly capable multimodal models. arXiv preprint arXiv:2312.11805 (2023)
17. Yin, S., et al.: A survey on multimodal large language models. Natl. Sci. Rev. **11**(12), nwae403 (2024)
18. Yin, S., et al.: Woodpecker: hallucination correction for multimodal large language models. SCIENCE CHINA Inf. Sci. **67**(12), 220105 (2024)
19. Yu, Q., et al.: Hallucidoctor: mitigating hallucinatory toxicity in visual instruction data. In: Proceedings of the IEEE/CVF Conference on Computer Vision and Pattern Recognition, pp. 12944–12953 (2024)
20. Zhang, R., Zhang, H., Zheng, Z.: VL-uncertainty: detecting hallucination in large vision-language model via uncertainty estimation. arXiv preprint arXiv:2411.11919 (2024)
21. Zhang, S., Sambara, S., Banerjee, O., Acosta, J., Fahrner, L.J., Rajpurkar, P.: Radflag: a black-box hallucination detection method for medical vision language models. arXiv preprint arXiv:2411.00299 (2024)

Reformulating Chest X-Ray Report Generation as Fine-Grained Visual Question Answering with Structured Clinical Report Templates

Ziqin Xiao[1,2], Zehui Liao[1], Zilin Lu[1], Shishuai Hu[1], Shaoyi Leng[3], Jingfeng Zhang[3], and Yong Xia[1,2](✉)

[1] National Engineering Laboratory for Integrated Aero-Space-Ground-Ocean Big Data Application Technology, School of Computer Science and Engineering, Northwestern Polytechnical University, Xi'an 710072, China
yxia@nwpu.edu.cn
[2] Ningbo Institute of Northwestern Polytechnical University, Ningbo 315048, China
[3] Ningbo No. 2 Hospital, Wenzhou Medical University, Ningbo 315000, China

Abstract. Automated chest X-ray (CXR) report generation is essential for reducing radiologists' workload and enhancing diagnostic efficiency in pulmonary screening. While vision-language models (VLMs) have shown promise, most existing methods rely on holistic free-text supervision, which often provides highly concise, incomplete clinical signals and leads to factual hallucinations. To address this issue, we propose **TGST**, a **T**emplate-**G**uided **S**tructured **T**raining framework that reformulates report generation as a fine-grained, clinically-aligned visual question answering (VQA) task. Specifically, we first construct a tree-structured clinical template for pulmonary findings, decomposing complex reports into anatomical entities and their specific attributes. We then introduce a two-stage pipeline that maps original reports to the template and leverages a medical VLM to infer missing visual details, creating dense supervision signals. By supervising the model on these localized clinical queries, TGST enforces precise visual-textual alignment and penalizes reliance on linguistic priors. Extensive experiments on the IU-Xray dataset demonstrate that TGST significantly outperforms state-of-the-art methods. Notably, our framework achieves substantial gains in factual correctness (GREEN score) and node-level observation accuracy across multiple VLM backbones, highlighting its potential for reliable and systematic automated pulmonary screening.

Keywords: Chest X-ray report generation · Vision-language model · Structured template

Z. Xiao and Z. Liao—Contributed equally.

G. Yang et al. (Eds.): MICCAI 2026, LNCS 16878, pp. 503–513, 2027.
https://doi.org/10.1007/978-3-032-38059-3_48

1 Introduction

Medical image interpretation and standardized report generation are essential for clinical diagnosis and patient management. Chest X-ray (CXR), the most widely performed imaging examination worldwide, serves as a primary tool for pulmonary disease screening due to its low cost and broad accessibility. However, the rapid growth in CXR volume has imposed heavy workload pressure on radiologists [13]. Manual interpretation and reporting are still labor-intensive and error-prone, and subtle abnormalities are frequently missed under time constraints [4].

To address this clinical challenge, automated radiology report generation has emerged as a high-priority research direction, aiming to extract clinically relevant findings from medical images and generate reliable, standardized textual reports, thereby reducing radiologists' workload and improving diagnostic workflow efficiency [26]. Automated CXR report generation has evolved from CNN–RNN frameworks [16,17] to Transformer-based models [10,22,24], reinforcement learning approaches for enhancing factual correctness [21], and more recent vision-language models (VLMs) [2,3,7,12,18,20,27], achieving notable improvements in fluency and reasoning. However, most existing methods rely on holistic long-text supervision and one-step generation. In contrast, radiologists follow a structured, step-by-step process when reviewing CXR images, systematically examining each tissue or organ in a spatial sequence before summarizing their findings in a concise free-form report. As such, the report generated by radiologists for a given CXR does not serve as a comprehensive supervision signal. Using such reports to train VLMs inevitably limits the model's ability to fully understand the image, thereby hindering the accuracy and completeness of report generation.

To overcome this issue, we propose TGST, a Template-Guided Structured Training framework for VLM-based CXR report generation. TGST reformulates holistic report generation as a clinically guided, fine-grained VQA task. First, we construct a radiology textbook-aligned, tree-structured clinical template for pulmonary findings, where the root node represents the lung, the intermediate nodes correspond to organs or lesions, and each leaf node represents a clinically actionable visual attribute, such as specific observable features in the chest X-ray that are critical for clinical decision-making. Second, we introduce a two-stage pipeline: the free-text reports are mapped into the structured template, after which the missing parts within the template are filled in using a powerful medical VLM to generate explicit supervision signals with clear semantic targets. This transformation replaces one-step holistic generation with multi-step, detail-oriented reasoning, enforcing localized visual evidence learning and precise visual-text alignment. Finally, the structured reports generated from the template are used to train medical VLM. We conduct extensive experiments on the IU-Xray dataset [11]. The results show that TGST significantly outperforms conventional long-text supervision paradigms and recent report generation methods, with particularly notable improvements in factual correctness.

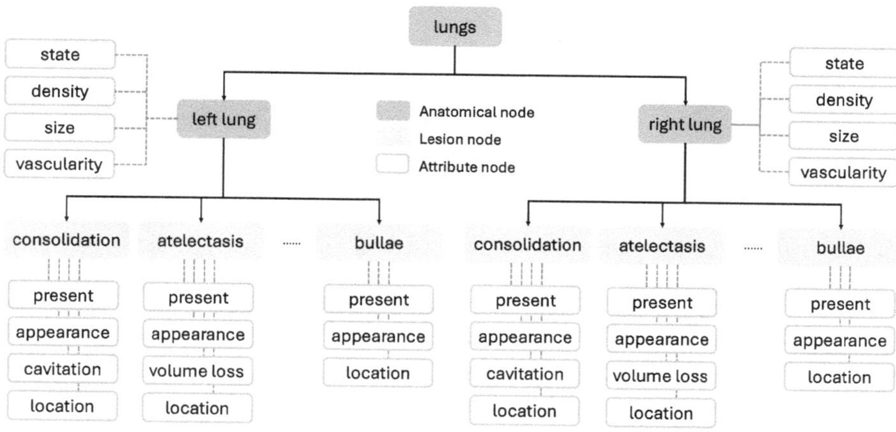

Fig. 1. Hierarchical clinical template for pulmonary CXR findings. The template consists of anatomical nodes, lesion nodes, and attribute nodes, forming a structured observation space for fine-grained VQA supervision.

The main contributions are three-fold: (1) We propose TGST, a new framework that reformulates CXR report generation as a fine-grained, clinically guided VQA task, introducing a two-stage pipeline for high-confidence supervision to reduce noise and hallucinations. (2) We develop a tree-structured template for CXR pulmonary findings, enabling systematic decomposition of free-text reports into semantically clear, clinically relevant supervision signals. (3) Extensive experiments on IU-Xray demonstrate that TGST achieves state-of-the-art factual correctness, highlighting its potential for reliable report generation.

2 Method

The proposed TGST framework reformulates holistic CXR report generation as a template-guided, fine-grained VQA task to mimic the systematic reviewing process of radiologists in routine clinical practice. To support this reformulation, we first construct a clinical knowledge-driven tree-structured template that formalizes the structured observation space for pulmonary CXR findings, as shown in Fig. 1. Based on this template, as illustrated in Fig. 2, TGST introduces a two-stage structuring pipeline that transforms sparse free-text reports into dense, high-confidence supervision signals by systematically mapping existing findings and inferring missing attributes. Finally, the model learns to generate reports via template-guided VQA, enforcing precise visual-text alignment through localized clinical queries across the hierarchical nodes.

2.1 Clinical Knowledge-Driven Structured Report Template

To mimic the systematic diagnostic process of radiologists and ensure clinical validity, we construct a tree-structured clinical template $\mathcal{T}$ that formalizes the observation space for CXR pulmonary findings. As illustrated in Fig. 1, $\mathcal{T}$ is designed as a hierarchical backbone enriched with node-specific attributes, encompassing three types of components: (1) **Anatomical Nodes:** The root node (*Lungs*) branches into *Left Lung* and *Right Lung*, establishing the spatial reference for subsequent findings. (2) **Lesion Nodes:** Under each lung node, we define seven clinically prevalent parenchymal abnormalities (e.g., *consolidation, atelectasis, bullae, granuloma, mass, nodule,* and *scarring*). These nodes represent the categorical identification of pathological entities within the anatomical context. (3) **Attribute Nodes:** Each anatomical or lesion node is linked to multiple fine-grained attribute nodes. Crucially, these attributes are heterogeneous and tailored to the specific node. For instance, a *consolidation* node triggers attributes like *appearance, cavitation,* and *location*, while the *left lung* node is associated with *state, density, size,* and *vascularity*.

Formally, each attribute node corresponds to a clinically actionable question $q \in \mathcal{Q}$, defining a closed-set VQA space. To handle out-of-template expressions, an *Other* option is provided for each attribute. This structured design decomposes holistic reports into deterministic, localized semantic targets, providing precise visual-textual alignment for model training. For example, consider an original report stating: *"There is minimal left basilar opacity compatible with scarring or atelectasis."* During the filling stage, the visual evidence may support the interpretation of *atelectasis* rather than *scarring*. Accordingly, the *present* attribute node under the *Left Lung* → *Atelectasis* node is set to *present*. Subsequently, the model proceeds to infer fine-grained attribute node such as *volume loss*. If, after both the text-guided initialization stage and the image-refined reasoning stage, none of the predefined options (e.g., *hilum displacement, rib approximation*) can be selected with high confidence, the corresponding *volume loss* attribute is assigned to *Other*. This mechanism ensures that the structured

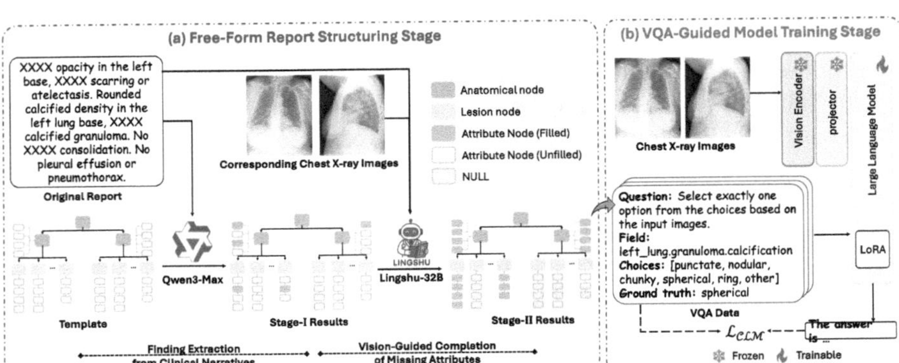

Fig. 2. Overview of the proposed TGST framework. We first construct a tree-structured clinical template for CXR pulmonary findings, then use a two-stage pipeline to structure free-form reports, finally fine-tune the VLM with the constructed VQA supervision.

template remains clinically faithful while preserving robustness to ambiguous or weak visual cues.

2.2 Two-Stage Pipeline for Free-Form Report Structuring

To transform sparse, holistic free-text reports into dense, structured supervision signals, we develop a two-stage pipeline. To prioritize pulmonary screening, we utilize Qwen3-Max [29] to extract lung-specific findings from the original free-form reports, effectively filtering out irrelevant clinical information.

Stage I: Finding Extraction from Clinical Narratives. In this stage, the lung-specific report is mapped onto the template $\mathcal{T}$ using Qwen3-Max with temperature set to 0.0. We impose a *strict-evidence constraint*: an attribute node is completed only if the report provides explicit and unambiguous evidence. Attributes omitted in the concise clinical narratives (e.g., laterality or specific morphology) remain unfilled to avoid erroneous assignments, with their inference deferred to the subsequent stage. Furthermore, *medical dependency rules* are incorporated to maintain logical coherence. For example, the identification of any lesion automatically triggers a state transition of the parent `lung.state` attribute node to *abnormal*.

Stage II: Vision-Guided Completion of Missing Attributes. To recover the visual details missing from the clinical narratives, we introduce a completion stage using Lingshu-32B [28], a domain-specific VLM, with temperature set to 0.0. Taking the CXR image and the partially completed template from Stage I as input, the VLM performs node-wise inference to populate the remaining unfilled attribute nodes. Specifically, each missing attribute node is reformulated as a visual query, enabling the model to supplement clinical details that were visually apparent but absent from the original textual narrative. To mitigate hallucination and redundancy, we introduce: (1) **Logical Pruning:** If a lesion (e.g., *atelectasis*) is confirmed absent, all its sub-attribute nodes are deterministically set to *null*, bypassing redundant VLM inference. (2) **Uncertainty Handling:** Attributes that remain clinically ambiguous after multimodal reasoning, or those that fall outside predefined categories, are assigned to the *Other* class. Finally, only node with valid assignments (excluding *null*) along with the CXR image is converted into image-question-answer triplet (x_i, q_i, a_i). All triplets together form our training data $\mathcal{D} = \{(x_i, q_i, a_i)\}_{i=1}^{N}$ for the next step, where N is the number of training samples. This filtering ensures that the VQA-based fine-tuning is guided by high-density, low-noise supervision signals, effectively reducing model hallucinations.

2.3 Learning to Report via Template-Guided VQA

After converting the lung-specific reports into corresponding fine-grained VQA samples, we leverage these samples $\mathcal{D} = \{(x_i, q_i, a_i)\}_{i=1}^{N}$ to fine-tune a VLM for CXR report generation to enhance pulmonary screening capabilities. Here, $x_i = \{x_i^f, x_i^l\}$ denotes the paired frontal and lateral CXR images as multi-view visual context, q_i is the node-specific clinical question serving as the textual instruction, and a_i is the corresponding high-confidence node value as the target output.

During training, we optimize the model parameters using the standard causal language modeling (CLM) loss. For sample (x_i, q_i, a_i), its loss $\mathcal{L}_i$ is calculated as follows:

$$\mathcal{L}_i = -\frac{1}{T}\sum_{t=1}^{T} \log p_\theta \left(y_t | y_{<t}, x_i, q_i\right), \tag{1}$$

where y_t is the t-th token of the target answer a_i, T is the total length of the target sequence, and θ denotes the learnable parameters of the VLM.

Compared with conventional long-form report supervision, this VQA-based training paradigm provides more explicit, localized training signals, forcing the VLM to attend to region visual evidence rather than relying primarily on linguistic priors. This fine-tuning strategy explicitly enforces fine-grained visual-text alignment, ensuring high consistency between the generated report and the underlying imaging evidence, and substantially improving the factual accuracy and descriptive granularity of the final report generation.

3 Experiments

3.1 Experimental Setup

Dataset. We conduct experiments on the IU-Xray dataset, a widely used benchmark for CXR report generation. The official release contains 7,470 CXR images paired with 3,955 radiology reports annotated by board-certified radiologists. To ensure view consistency, we select cases containing exactly one frontal view and one lateral view, resulting in 5,560 images and 2,780 reports. The subset is split into training, validation, and test sets with a ratio of 7:1:2. All methods are trained and evaluated on the lung-specific reports extracted using Qwen3-Max.

Evaluation Metrics. We evaluate the performance of the VLM fine-tuned via TGST based on two distinct output formats: (1) *Free-form Report Generation:* We assess the model's ability to generate descriptive reports by measuring the factual consistency between the generated text and ground-truth using GREEN [23]. (2) *Structured Template Completion:* We evaluate the model's fine-grained observation by treating each node in $\mathcal{T}$ as a VQA task. The accuracy of the filled attributes is measured via *Accuracy (Acc)*, *Precision (Pre)*, *Recall(Rec)*, and *F1-score*, aggregated through both macro and weighted averaging.

Implementation Details. For model fine-tuning, we evaluate three multimodal LLMs, including InternVL2.5-4B [8], Qwen2.5-VL-3B-Instruct [1], and HuatuoGPT-Vision-7B-Qwen2.5VL [5]. We fine-tune the language decoder using Low-Rank Adaptation, abbreviated as LoRA [14]. The rank is set to 8, and α is

Table 1. Performance of three VLMs fine-tuned using different supervision formats in free-form report generation (FRG) and structured template completion (STC). The best results are in **bold**.

Model	Training	Data	FRG	STC			
			GREEN	Acc.	F1 (M/W)	Prec. (M/W)	Rec. (M/W)
InternVL	Zero-shot	/	0.521	0.431	0.191/0.217	0.200/0.242	0.318/0.356
	LoRA	Report	0.502	0.607	0.247/0.311	0.300/0.338	0.340/0.387
		Ours	**0.756**	**0.817**	**0.339/0.416**	**0.401/0.497**	**0.370/0.410**
Qwen	Zero-shot	/	0.567	0.131	0.127/0.089	0.135/0.154	0.302/0.331
	LoRA	Report	0.629	0.250	0.184/0.154	0.218/0.309	0.325/0.352
		Ours	**0.828**	**0.830**	**0.329/0.400**	**0.351/0.468**	**0.356/0.399**
HuatuoGPT	Zero-shot	/	0.727	0.462	0.165/0.220	0.195/0.304	0.291/0.347
	LoRA	Report	0.774	0.302	0.164/0.151	0.167/0.209	0.307/0.328
		Ours	**0.815**	**0.831**	**0.306/0.341**	**0.336/0.386**	**0.342/0.354**

set to 16. Training is implemented using the TRL SFT framework based on HuggingFace Transformers. All models are trained for 10 epochs using the AdamW optimizer, with a per-device batch size of 16, gradient accumulation of 2 steps, and a learning rate of 2×10^{-5}. All experiments are conducted on two NVIDIA RTX 4090 GPUs.

3.2 Comparison Results of the Fine-Tuning Paradigm

Table 1 provides a comprehensive comparison of model performance in free-form report generation (FRG) and structured template completion (STC) across three distinct supervision paradigms and three backbone VLMs: (1) *Zero Shot*: the pre-trained VLMs are evaluated directly without any task-specific fine-tuning using the IU-Xray dataset, with their performance serving as the lower bound, (2) *Report-level Fine-tuning*: VLMs are fine-tuned using LoRA with conventional holistic free-text radiology reports; (3) *Ours*: VLMs are fine-tuned using the same LoRA strategy but under our proposed TGST supervision, reformulating the task into fine-grained VQA. The results demonstrate that our TGST significantly enhances VLMs' report generation capabilities, indicating that it offers a more comprehensive supervision signal. Across all three backbones, our TGST consistently improves factual correctness measured by GREEN. For example, by replacing traditional report supervision with our structured VQA data, GREEN increases from 0.629 to 0.828 on Qwen2.5-VL-3B-Instruct, resulting in an improvement of nearly 20%. Similar trends are observed on InternVL and HuatuoGPT, suggesting that the benefits of structured supervision generalize across architectures.

Beyond FRG evaluation, our TGST also yields notable improvements in STC, including Accuracy and Macro/Weighted F1. This demonstrates that decomposing long-form reports into semantically explicit supervision units provides clearer

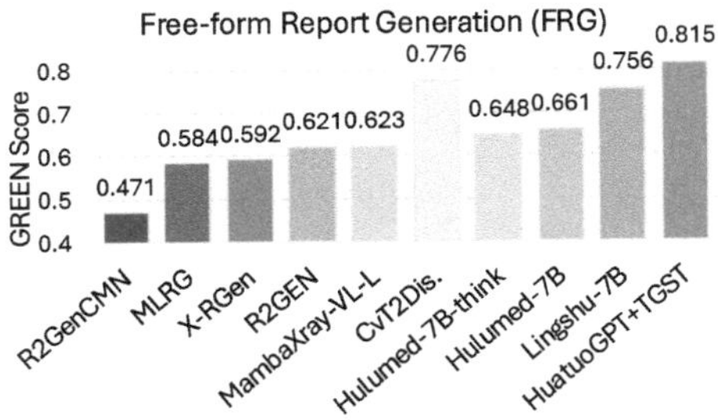

Fig. 3. FRG performance of our TGST+HuatuoGPT and other competing methods on the IU-Xray test set.

Table 2. Structured template completion performance of our proposed TGST using HuatuoGPT and other medical VLMs (Macro metrics) on the IU-Xray test set. The best results are in **bold**.

Model	Acc.	M-F1	M-Prec.	M-Rec.
Lingshu-7B	0.707	0.241	0.252	0.343
Hulumed-7B [15]	0.720	0.291	0.321	**0.350**
Hulumed-7B-think	0.713	0.273	0.304	0.335
Huatuo+TGST	**0.831**	**0.306**	**0.336**	0.342

optimization signals and facilitates more effective learning of structured clinical attributes. In contrast, report-level supervision requires the model to implicitly capture multiple anatomical findings from a single holistic target, which may dilute supervision signals. By explicitly modeling each clinical attribute as an independent learning objective, structured supervision promotes more stable optimization and leads to consistent gains in both report-level and fine-grained evaluation metrics.

3.3 Comparison with Other Report Generation Methods

Table 2 presents a comprehensive comparison between HuatuoGPT + TGST and several representative report generation approaches, including conventional encoder–decoder architectures [6,19,25] as well as recent medical VLMs. The GREEN scores for free-form report generation are further visualized in Fig. 3.

The results demonstrate that TGST substantially improves report-level factual correctness. Among traditional methods, CvT2Dis. [22] achieves the best FRG performance with a GREEN score of 0.776, outperforming earlier methods such as R2GEN [10] and R2GenCMN [9]. However, HuatuoGPT + TGST further increases the GREEN score to 0.815, establishing new state-of-the-art performance under the same evaluation protocol.

Beyond FRG evaluation, TGST also achieves superior performance in structured template completion. HuatuoGPT + TGST achieves the highest STC Accuracy of 0.831 and competitive Macro F1 scores, compared with recent medical VLMs such as Lingshu-7B and Hulumed-7B. This indicates that structured supervision not only improves holistic report generation but also enhances fine-grained attribute reasoning. In contrast, conventional report generation methods rely primarily on sequence-level optimization, which does not explicitly enforce alignment between localized visual evidence and structured clinical attributes. By modeling each attribute node as an independent supervision signal, TGST provides more precise optimization targets, leading to consistent gains across both report-level and fine-grained evaluation metrics.

4 Conclusion

In this paper, we present TGST, a new Template-Guided Structured Training framework that reformulates chest X-ray report generation as a fine-grained, clinically-aligned VQA task. By introducing a tree-structured clinical template and a two-stage supervision construction pipeline, TGST effectively addresses the limitations of sparse, holistic long-text supervision. This approach enables the model to bridge the gap between abstract clinical narratives and localized visual evidence, mimicking the systematic diagnostic reasoning of expert radiologists. Extensive experiments on the IU-Xray dataset demonstrate that TGST significantly outperforms existing methods across multiple VLMs, achieving superior performance in both free-form report generation and structured template completion. We believe that this structured training paradigm provides a scalable and reliable pathway for enhancing automated pulmonary screening, ultimately contributing to more accurate diagnostic workflows in clinical practice.

Acknowledgments. This work was supported in part by the National Natural Science Foundation of China under Grant 92470101 and Grant 62671517, in part by the "Pioneer" and "Leading Goose" R&D Program of Zhejiang, China, under Grant 2025C01201(SD2), in part by the Project of National Key Clinical Specialty (Department of Medical Imaging, No. 2024017), in part by the Ningbo Key Laboratory of Digital Imaging and Medical-Engineering Interdisciplinarity (No. 2024031). Zilin Lu was supported by the Innovation Foundation for Doctor Dissertation of Northwestern Polytechnical University under Grant CX2026XXX.

Disclosure of Interests. The authors have no competing interests to declare that are relevant to the content of this article.

References

1. Bai, S., et al.: Qwen2.5-vl technical report. arXiv preprint arXiv:2502.13923 (2025)
2. Bannur, S., et al.: Maira-2: grounded radiology report generation. arXiv preprint arXiv:2406.04449 (2024)
3. Bu, S., Li, T., Yang, Y., Dai, Z.: Instance-level expert knowledge and aggregate discriminative attention for radiology report generation. In: Proceedings of the IEEE/CVF Conference on Computer Vision and Pattern Recognition (CVPR), pp. 14194–14204 (2024)
4. Carter, A.J., Davis, K.A., Evans, L.V., Cone, D.C.: Information loss in emergency medical services handover of trauma patients. Prehosp. Emerg. Care **13**(3), 280–285 (2009)
5. Chen, J., et al.: Towards injecting medical visual knowledge into multimodal LLMs at scale. In: Proceedings of the 2024 Conference on Empirical Methods in Natural Language Processing, pp. 7346–7370 (2024)
6. Chen, Q., et al.: Act like a radiologist: radiology report generation across anatomical regions. In: Proceedings of the Asian Conference on Computer Vision, pp. 1–17 (2024)

7. Chen, W., Shen, L., Lin, J., Luo, J., Li, X., Yuan, Y.: Fine-grained image-text alignment in medical imaging enables explainable cyclic image-report generation. In: Proceedings of the 62nd Annual Meeting of the Association for Computational Linguistics (Volume 1: Long Papers), pp. 9494–9509 (2024)
8. Chen, Z., et al.: Expanding performance boundaries of open-source multimodal models with model, data, and test-time scaling. arXiv preprint arXiv:2412.05271 (2024)
9. Chen, Z., Shen, Y., Song, Y., Wan, X.: Cross-modal memory networks for radiology report generation. In: Proceedings of the 59th Annual Meeting of the Association for Computational Linguistics and the 11th International Joint Conference on Natural Language Processing (Volume 1: Long Papers), pp. 5904–5914 (2021)
10. Chen, Z., Song, Y., Chang, T.H., Wan, X.: Generating radiology reports via memory-driven transformer. In: Proceedings of the 2020 Conference on Empirical Methods in Natural Language Processing, pp. 1439–1449 (2020)
11. Demner-Fushman, D., et al.: Preparing a collection of radiology examinations for distribution and retrieval. J. Am. Med. Inform. Assoc. **23**(2), 304–310 (2016)
12. Gu, T., Yang, K., An, X., Feng, Z., Liu, D., Cai, W.: Orid: organ-regional information driven framework for radiology report generation. In: Proceedings of the Winter Conference on Applications of Computer Vision, pp. 378–387 (2025)
13. Hosny, A., Parmar, C., Quackenbush, J., Schwartz, L.H., Aerts, H.J.: Artificial intelligence in radiology. Nat. Rev. Cancer **18**(8), 500–510 (2018)
14. Hu, E.J., et al.: Lora: low-rank adaptation of large language models. In: Proceedings of the International Conference on Learning Representations (2022)
15. Jiang, S., et al.: Hulu-med: a transparent generalist model towards holistic medical vision-language understanding. arXiv preprint arXiv:2510.08668 (2025)
16. Jing, B., Wang, Z., Xing, E.: Show, describe and conclude: on exploiting the structure information of chest X-ray reports. In: Proceedings of the 57th Annual Meeting of the Association for Computational Linguistics, pp. 6570–6580 (2019)
17. Jing, B., Xie, P., Xing, E.: On the automatic generation of medical imaging reports. In: Proceedings of the 56th Annual Meeting of the Association for Computational Linguistics (Volume 1: Long Papers), pp. 2577–2586 (2018)
18. Liu, C., Tian, Y., Chen, W., Song, Y., Zhang, Y.: Bootstrapping large language models for radiology report generation. In: Proceedings of the AAAI Conference on Artificial Intelligence, vol. 38, pp. 18635–18643 (2024)
19. Liu, K., et al.: Enhanced contrastive learning with multi-view longitudinal data for chest X-ray report generation. In: Proceedings of the IEEE/CVF Conference on Computer Vision and Pattern Recognition, pp. 10348–10359 (2025)
20. Liu, T., et al.: HC-LLM: historical-constrained large language models for radiology report generation. In: Proceedings of the AAAI Conference on Artificial Intelligence, vol. 39, pp. 5595–5603 (2025)
21. Miura, Y., Zhang, Y., Tsai, E., Langlotz, C., Jurafsky, D.: Improving factual completeness and consistency of image-to-text radiology report generation. In: Proceedings of the 2021 Conference of the North American Chapter of the Association for Computational Linguistics: Human Language Technologies, pp. 5288–5304 (2021)
22. Nicolson, A., Dowling, J., Koopman, B.: Improving chest X-ray report generation by leveraging warm starting. Artif. Intell. Med. **144**, 102633 (2023)
23. Ostmeier, S., et al.: Green: generative radiology report evaluation and error notation. In: Findings of the Association for Computational Linguistics: EMNLP 2024, pp. 374–390 (2024)
24. Vaswani, A., et al.: Attention is all you need. Adv. Neural Inf. Process. Syst. **30** (2017)

25. Wang, X., et al.: Cxpmrg-bench: pre-training and benchmarking for X-ray medical report generation on chexpert plus dataset. In: Proceedings of the IEEE/CVF Conference on Computer Vision and Pattern Recognition, pp. 5123–5133 (2025)
26. Wang, X., Figueredo, G., Li, R., Zhang, W.E., Chen, W., Chen, X.: A survey of deep-learning-based radiology report generation using multimodal inputs. Med. Image Anal. **103**, 103627 (2025)
27. Wang, Z., Liu, L., Wang, L., Zhou, L.: R2gengpt: radiology report generation with frozen LLMs. Meta-Radiology **1**(3), 100033 (2023)
28. Xu, W., et al.: Lingshu: a generalist foundation model for unified multimodal medical understanding and reasoning. arXiv preprint arXiv:2506.07044 (2025)
29. Yang, A., et al.: Qwen3 technical report. arXiv preprint arXiv:2505.09388 (2025)

Residual-Aware Structured Pruning with Dual-Stream Alignment for Adapter-Tuned SAM

Xin Luo[1,2,3], Chunjiang Wang[1,2,3], and Shaohua Kevin Zhou[1,3,4,5,6](✉)

[1] School of Biomedical Engineering, Division of Life Sciences and Medicine, University of Science and Technology of China (USTC), Hefei 230026, Anhui, China
{luoxin_ustc,chunjiang_wang}@mail.ustc.edu.cn
[2] Suzhou Institute for Advanced Research, University of Science and Technology of China, Suzhou 215123, Jiangsu, China
[3] Medical Imaging, Robotics, Analytic Computing and Learning (MIRACLE) Lab, YRD-RIGHT, USTC Suzhou Institute for Advanced Research, Suzhou 215123, Jiangsu, China
skevinzhou@ustc.edu.cn
[4] Jiangsu Provincial Key Laboratory of Multimodal Digital Twin Technology, Suzhou 215123, Jiangsu, China
[5] Biomedical Basic Research Center (BBRC) of Jiangsu Province, Suzhou 215123, Jiangsu, China
[6] State Key Laboratory of Precision and Intelligent Chemistry, Hefei 230026, Anhui, China

Abstract. Adapter-tuned Segment Anything Models are promising for medical image segmentation, yet their backbone-heavy compute and adapter overhead hinder real-time deployment. Studying SAM-Med2D as a representative adapter-tuned SAM testbed, we observe that structured masking harms Dice far more when applied to the frozen backbone than to adapters, suggesting strong asymmetry in redundancy. We propose RASP-SAM, a structured pruning framework that exploits this decoupling. It defines dependency-safe pruning units under attention, FFN, adapter, and residual couplings, ranks units with Orthogonal Residual Scoring based on the orthogonal residual of downstream gradients relative to a merged pretraining and downstream gradient subspace with cross-fitting, and restores accuracy using Dual-Stream Compensation that distills adapters and backbone in two stages under a small calibration budget. With only 20K calibration images from SA-1B and SA-Med2D-20M, RASP-SAM achieves near-lossless segmentation on SAM-Med2D while reducing encoder compute from 65.15G to 15.62G MACs and parameters from 270.1M to 63.4M at 76% sparsity, outperforming a transferred SlimSAM baseline at comparable cost. These results validate RASP-SAM in the SAM-Med2D setting and motivate broader evaluation on adapter-style SAM variants.

Keywords: Structured pruning · Segment Anything Model · Adapter

G. Yang et al. (Eds.): MICCAI 2026, LNCS 16878, pp. 514–524, 2027.
https://doi.org/10.1007/978-3-032-38059-3_49

1 Introduction

Segment Anything (SAM) [10] has become a foundation for medical segmentation, typically adapted via full fine-tuning [15] or parameter-efficient adapters [3,7,8,17]. However, the heavy computational burden restricts real-time clinical deployment [3,9,22,23]; in SAM-Med2D, adapter parameters can even exceed those of the base model, forcing resolution compromises [3]. Because this model exposes the frozen-backbone/trainable-adapter split and residual coupling targeted by our method, we use SAM-Med2D as the primary validation setting for adapter-aware compression.

Existing SAM acceleration strategies—spanning architecture redesign [20, 24], quantization [12–14], and pruning [2,5,19]—typically apply uniform compression policies, neglecting the structural distinction between the frozen backbone and trainable adapters. This oversight obscures potential efficiency gains from treating these components differentially. To investigate their distinct redundancies, we conduct a pilot study on the SAM-Med2D setting [3].

We first run a pilot study on SAM-Med2D [3] by applying random channel-wise structured masks at the same sparsity to the backbone and adapters separately. As shown in Fig. 1, pruning the backbone degrades Dice much faster than pruning the adapters. This suggests that the backbone carries fragile shared representations, while adapters exhibit higher intrinsic redundancy, motivating asymmetric compression and dual-stream recovery.

Based on this observation, we study RASP-SAM in SAM-Med2D. The framework defines dependency-safe pruning units under backbone, adapter, and residual couplings, ranks them with Orthogonal Residual Scoring (ORS), and restores accuracy with Dual-Stream Compensation. Under a 20K-image calibration budget, as shown in Fig. 2, RASP-SAM achieves large parameter/MAC reductions with near-lossless Dice on SAM-Med2D. Our contributions are:

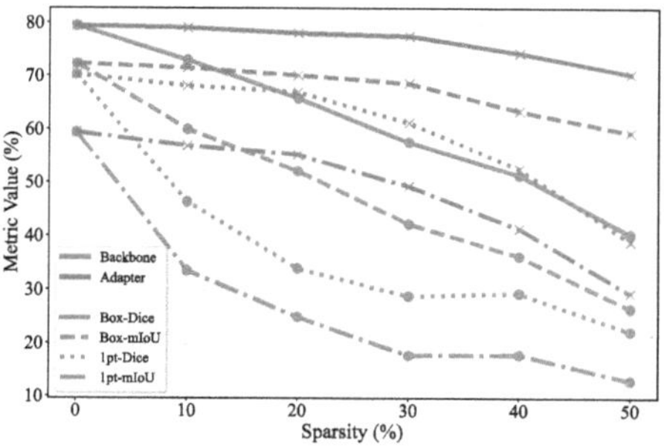

Fig. 1. Backbone vs. adapter random-mask sparsity sensitivity on SAM-Med2D.

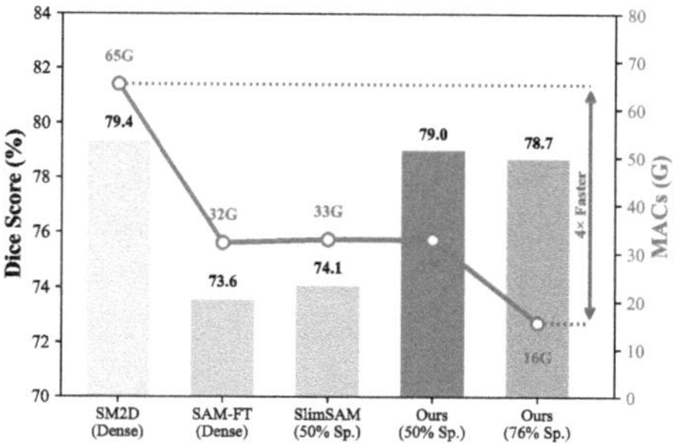

Fig. 2. RASP-SAM vs. Baselines: Dice score (left) vs. Computational cost (right).

- We analyze structural dependencies in the SAM-Med2D adapter-tuned encoder and derive practical pruning-safe units that preserve residual compatibility.
- We introduce *Orthogonal Residual Scoring*, which isolates non-substitutable gradient directions across pretraining and downstream streams to improve structured pruning decisions.
- We design *Dual-Stream Compensation*, a data-efficient recovery scheme that alternates supervision from both streams to preserve the shared backbone subspace and recover accuracy with only 20K images.

2 Related Work

Medical SAM Adaptation and Adapter Tuning. SAM [10] has been widely adapted to medical segmentation through both full/partial fine-tuning and adapter-based parameter-efficient tuning [3,15,17]. SAM-Med2D is closest to our setting because it inserts trainable adapters into the image encoder and exposes a clear backbone–adapter split. More broadly, adapter-style tuning is a standard PEFT paradigm in NLP and vision [7,18], alongside LoRA/QLoRA [4,8]. These methods improve adaptation efficiency, but not compression of adapter-tuned foundation models. Similar SAM adaptation trends in other vision tasks [1,16,21] further motivate adapter-aware compression.

Efficient and Compressed SAM. SAM acceleration mainly uses architecture redesign, quantization, and pruning. FastSAM [24] and MobileSAM [20] reduce cost by redesign or distillation; quantization approaches include general ViT methods [12,13] and PQ-SAM [14]; SlimSAM [2] shows effective pruning with distillation and limited calibration. However, these methods target generic SAM compression and do not explicitly exploit the backbone–adapter decoupling in adapter-tuned models such as SAM-Med2D.

Structured Pruning and Dependency-Aware Pruning. Pruning research spans classic saliency-based methods (e.g., OBD [11], OBS [6]) to modern structured pruning for transformers. Recent ViT studies show the importance of structured saliency and pruning granularity [19], while dependency-aware pruning stresses preserving structural couplings across layers and branches [5]. This is especially relevant to adapter-tuned SAM, where backbone and adapters are decoupled in parameterization but coupled through residual pathways. Our method targets this gap under limited calibration budgets.

3 Method

Problem Setup and Overview. We focus on structured pruning of the adapter-tuned SAM-Med2D image encoder. Given an encoder with n Transformer blocks (indexed by i) and embedding dimension $d_{\text{model}} = h \cdot d_h$ (where h

is the number of heads and d_h is the head dimension), our goal is to obtain a compressed model with target sparsity ρ that preserves both pretraining and medical knowledge. We utilize two small calibration streams: a pretraining stream $\mathcal{D}_{\text{pre}}$ and a downstream stream $\mathcal{D}_{\text{down}}$. The proposed RASP-SAM framework comprises three key stages: (1) Dependency-Aware Unit Definition to ensure structural consistency; (2) Orthogonal Residual Scoring (ORS) for importance ranking; and (3) Dual-Stream Compensation for accuracy recovery. We detail each stage below.

Dependency-Aware Pruning Units. To ensure shape consistency, we classify structural couplings into local and global dependencies, prioritizing the former to maximize pruning granularity.

Local dependencies, as shown in Fig. 3 except the residual-aligned group, define the smallest independent pruning units by coupling the output dimension of one layer with the input dimension of its successor within a block. In the backbone Multi-Head Attention, preserving the inner-product requires that pruning the output channels of the query/key/value projections $\{\mathbf{W}_{\text{Q}}, \mathbf{W}_{\text{K}}, \mathbf{W}_{\text{V}}\}$ be synchronized with pruning the input channels of the output projection $\mathbf{W}_{\text{O}}$. Similarly, in FFNs and paired adapters, the output channels of the expansion/first layer are coupled with the input channels of the reduction/second layer. Adhering to these intra-block couplings allows for layer-specific sparsity patterns, which minimizes structural perturbation compared to coarser constraints.

Global dependencies arise from residual connections, which enforce a shared mask across the residual stream. As illustrated by the residual-aligned group in Fig. 3, pruning an output channel of the first block's FFN necessitates synchronized pruning of the corresponding output channel of the parallel adapter, as well as the input and output channels of all subsequent blocks. This cascading constraint couples sensitive layers with redundant ones across the entire depth, leading to overly coarse granularity. We therefore adopt a hybrid strategy: we treat blocks as locally independent by default and enforce global constraints *only* at necessary residual summation points to ensure runtime validity.

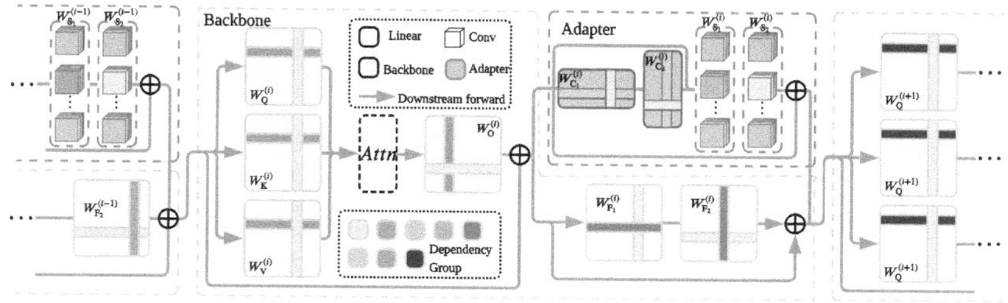

Fig. 3. Dependency schematic across encoder blocks $i-1$, i, and $i+1$. Colored rows/columns with the same color form one dependency group and must be retained or pruned jointly.

Orthogonal Residual Scoring. After adapter tuning, the backbone mainly preserves shared pretraining representations, while the adapters encode downstream corrections. We therefore score each dependency-safe unit with gradients from the pretraining and downstream calibration streams so that pruning preserves the shared backbone subspace and downstream task behavior. For a prunable module in block i, let n_u be the number of candidate units and d the flattened parameter dimension per unit. We collect unit-wise gradients averaged over calibration batches and stack them row-wise as $\mathbf{G}_{\text{pre}}^{(i)} \in \mathbb{R}^{n_u \times d}$ and $\mathbf{G}_{\text{down}}^{(i)} \in \mathbb{R}^{n_u \times d}$. During the pretraining pass, adapters are bypassed so the backbone receives only pretraining supervision, while during the downstream pass both backbone and adapters are active. Backbone units follow the dual-stream formulation below; adapters, lacking pretraining signals, default to standard Taylor scoring.

We extract dominant directions via SVD, retaining the top-r right singular vectors of the two gradient matrices:

$$\mathbf{G}_{\text{pre}}^{(i)} = \mathbf{U}\mathbf{\Sigma}\mathbf{V}^{\top}, \quad \mathbf{U}_{\text{pre}}^{(i)} = \mathbf{V}_{[:,1:r]}, \qquad \mathbf{G}_{\text{down}}^{(i)} = \hat{\mathbf{U}}\hat{\mathbf{\Sigma}}\hat{\mathbf{V}}^{\top}, \quad \mathbf{U}_{\text{down}}^{(i)} = \hat{\mathbf{V}}_{[:,1:r]}. \tag{1}$$

To form a joint pretraining-downstream subspace, we concatenate the two bases and orthonormalize them with a Gram construction:

$$\mathbf{C}^{(i)} = \left[\mathbf{U}_{\text{pre}}^{(i)} \ \mathbf{U}_{\text{down}}^{(i)}\right], \quad \mathbf{C}^{(i)\top}\mathbf{C}^{(i)} = \mathbf{V}_C\mathbf{\Lambda}_C\mathbf{V}_C^{\top}, \quad \mathbf{U}_{\cup}^{(i)} = \mathbf{C}^{(i)}\mathbf{V}_{C[:,1:k]}\mathbf{\Lambda}_{C,1:k}^{-1/2}, \tag{2}$$

where $k \leq 2r$ is chosen by a fixed rank. The basis $\mathbf{U}_{\cup}^{(i)}$ spans directions jointly explainable by the two streams.

Directly scoring a unit with a basis estimated from the same gradients introduces optimistic bias. We therefore use two-fold cross-fitting on units: split row indices of $\mathbf{G}_{\text{down}}^{(i)}$ into disjoint sets S_1 and S_2, build the merged basis on one fold, and score units on the other. Let $\mathbf{g}_u \in \mathbb{R}^d$ be the gradient of unit u, and let $\mathbf{U}_{\cup,[1]}^{(i)}$ and $\mathbf{U}_{\cup,[2]}^{(i)}$ be the two fold-specific merged bases. The residual energy is

$$R_u^2 = \begin{cases} \|\mathbf{g}_u\|_2^2 - \|\mathbf{U}_{\cup,[2]}^{(i)\top}\mathbf{g}_u\|_2^2, & u \in \mathsf{S}_1, \\ \|\mathbf{g}_u\|_2^2 - \|\mathbf{U}_{\cup,[1]}^{(i)\top}\mathbf{g}_u\|_2^2, & u \in \mathsf{S}_2, \end{cases} \tag{3}$$

and the ORS score is $s_u = \sqrt{\max(R_u^2, 0)}\ \|\boldsymbol{\theta}_u\|_2$, where $\boldsymbol{\theta}_u$ is the parameter vector of unit u. Crucially, if a unit's gradient direction is reconstructible by the subspace of other units, it implies redundancy. Therefore, ORS assigns higher scores to units with large orthogonal residuals, preserving components that contribute unique directional information not spanned by the rest.

Dual-Stream Compensation. After ORS selects dependency-safe units under target sparsity ρ, pruning still perturbs the shared backbone subspace and residual-aligned branch outputs. We therefore perform a two-stage recovery with the same two calibration streams used in ORS, namely the pretraining stream $\mathcal{D}_{\text{pre}}$ and the downstream stream $\mathcal{D}_{\text{down}}$, as summarized in Fig. 4. In each stage,

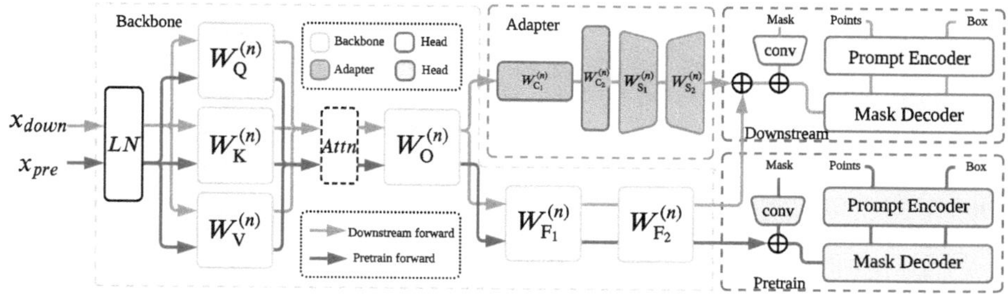

Fig. 4. Dual-Stream Compensation pipeline. Stage 1 prunes and recovers adapters with the backbone frozen using the downstream stream. Stage 2 prunes and recovers the backbone with adapters frozen using pretraining and downstream streams.

the teacher is the model before pruning that stage, and the student is the corresponding pruned model.

In Stage 1, we prune adapter units first and freeze the backbone. Taylor scores are computed on adapter dependency-safe units, and each adapter group is pruned to sparsity ρ. Recovery uses only $\mathcal{D}_{\text{down}}$ so that updates remain task-specific. For each encoder block i, we distill adapter outputs via MSE $(L_{\text{C}_2}^{(i)}, L_{\text{S}_2}^{(i)})$ and combine them with task loss which is a mixture of Focal, Dice, and IoU:

$$L_{\text{a}} = \lambda_{\text{a}} \sum_{i=1}^{n} \left(L_{\text{C}_2}^{(i)} + L_{\text{S}_2}^{(i)} \right) + L_{\text{task}},$$

where λ_{a} balances distillation and task supervision.

In Stage 2, we prune backbone units and freeze all adapters recovered in Stage 1. ORS is applied to backbone dependency-safe units, with the shared global mask only when residual alignment requires synchronized pruning. Recovery uses both streams to preserve the shared backbone subspace while restoring downstream performance. Each iteration processes two mini-batches, $\mathcal{B}_{\text{pre}} \sim \mathcal{D}_{\text{pre}}$ and $\mathcal{B}_{\text{down}} \sim \mathcal{D}_{\text{down}}$, in two separate passes. For each block i, we distill backbone activations at $\mathbf{W}_{\text{O}}^{(i)}$ and $\mathbf{W}_{\text{F}_2}^{(i)}$, defining MSE distillation losses $L_{\text{O}}^{(i)}$ and $L_{\text{F}_2}^{(i)}$. The objective is

$$L_{\text{b}} = \frac{\lambda_{\text{b}}}{2} \sum_{i=1}^{n} \left[\left(L_{\text{O}}^{(i)} + L_{\text{F}_2}^{(i)}\right)\Big|_{\mathcal{B}_{\text{pre}}} + \left(L_{\text{O}}^{(i)} + L_{\text{F}_2}^{(i)}\right)\Big|_{\mathcal{B}_{\text{down}}} \right] + L_{\text{task}}\Big|_{\mathcal{B}_{\text{down}}}, \quad (4)$$

where λ_{b} balances distillation and downstream supervision. The pretraining stream stabilizes backbone directions inherited from the foundation model, while the downstream stream restores task-specific behavior. This two-stage schedule matches the backbone and adapter roles used in ORS and enables recovery with a limited calibration budget.

4 Experimental Results

Experimental Settings. We build a 20K calibration set by sampling 10K images from SA-1B and 10K from SA-Med2D-20M and conduct experiments on NVIDIA H20 GPUs. To preserve foundation features during dual-stream compensation, inputs are resized to 1024×1024 for the pretraining stream and 256×256 for the downstream stream, while evaluation strictly follows the SAM-Med2D interactive protocol at 256×256. We prune the image encoder only in two stages, starting with adapters followed by the backbone, using one-shot pruning per stage with equal per-head channel pruning. ORS uses rank $r = k = 16$. Optimization uses Adam with learning rate 1×10^{-4} and batch size 16 for 40 epochs per stage with $\lambda_a = \lambda_b = 0.5$. We denote the pruning ratio for adapters and the backbone as Axx and Bxx respectively; for instance, A87.5 implies pruning adapters to 87.5% sparsity, and A75+B75 indicates 75% sparsity for both components. We report Dice under one box prompt and 1/3/5 point prompts.

Main Results. We benchmark RASP-SAM against the unpruned SAM-Med2D and a transferred SlimSAM baseline. To ensure a fair comparison, SlimSAM is calibrated using 20K downstream images to match our total budget. Table 1 shows that RASP-SAM achieves superior accuracy-efficiency trade-offs in this setting. At moderate compression, Dice remains nearly lossless. Even at high sparsity (A75+B75), where encoder MACs drop to approximately one-quarter of the original, the model retains strong performance across all prompts. Crucially, RASP-SAM outperforms the SlimSAM baseline at similar compute levels, supporting the effectiveness of our dependency-safe, dual-stream design on SAM-Med2D.

Inference Speed Analysis. To verify practical acceleration, Table 2 reports Mean latency (average ms/image) and TPS (throughput, images/s) at batch size 1. RASP-SAM (A75+B75) boosts encoder throughput by 13.7%. Crucially, this gain persists in the end-to-end pipeline with an 11.2% improvement, confirming that theoretical MAC reductions translate into effective and significant wall-clock acceleration.

Ablation on Pruning Order. Table 3 shows across both sparsity settings and all prompt types, pruning adapters before the backbone yields higher Dice than the reverse order. The gap becomes larger at higher sparsity. Pruning adapters first confines early changes to local dependencies and allows downstream-only compensation before backbone compression. Backbone pruning then uses dual-stream compensation to preserve the shared backbone subspace and reduce error accumulation. We therefore use adapter-first pruning in all other experiments.

Ablation on Compensation Data. We study compensation data sources under a fixed 20K-image budget with backbone pruning at B75. Table 4 shows, using only pretraining data preserves the foundation representation but underfits downstream prompts. Using only downstream data improves prompt adap-

Table 1. Main results on SAM-Med2D. **A**xx and **B**xx denote adapter and backbone pruning ratios; Sp. denotes overall encoder sparsity. Params and MACs are reported for the encoder at 256×256.

Model	Strat.	Sp.	Res.	BBox	1pt	3pt	5pt	Param.	MAC
SAM-B	N/A	0.00	256^2	61.63	18.94	28.28	37.47	89.67	32.07
SAM-B	N/A	0.00	1024^2	74.49	36.88	42.76	47.57	89.67	368.86
SM2D	N/A	0.00	256^2	79.35	70.01	76.35	78.68	270.08	65.15
SAM-B	Finetune	0.00	256^2	73.56	60.11	70.95	75.51	89.67	32.07
SM2D	SlimSAM	0.50	256^2	74.10	65.23	74.64	76.93	134.43	32.75
SM2D	B50	0.15	256^2	**79.23**	**69.92**	**76.27**	**78.94**	**227.58**	**49.29**
SM2D	A50	0.33	256^2	**79.32**	**69.43**	**76.26**	**78.84**	**179.88**	**48.63**
SM2D	B75	0.24	256^2	**79.11**	**69.67**	**75.06**	**78.82**	**202.30**	**41.07**
SM2D	A75	0.50	256^2	**79.39**	**68.81**	**75.78**	**78.98**	**134.76**	**40.36**
SM2D	A87.5	0.58	256^2	**79.07**	**66.39**	**75.05**	**77.65**	**112.20**	**36.22**
SM2D	A50+B50	0.50	256^2	**79.01**	**69.43**	**76.26**	**77.62**	**134.43**	**32.75**
SM2D	A75+B75	0.76	256^2	**78.65**	**68.13**	**74.26**	**76.85**	**63.43**	**15.62**

tation but weakens preservation of the shared backbone subspace. Mixing the two streams in equal proportion gives the best Dice across all prompt types. This supports the dual-stream compensation design used in the main experiments.

Table 2. Inference speed. Enc. denotes encoder only; E2E denotes end to end.

Strategy	Enc.		E2E	
	Mean	TPS	Mean	TPS
Dense	22.493	44.46	28.174	35.49
A50+B50	20.979	47.67	26.629	37.55
A75+B75	19.779	50.56	25.343	39.46

Table 3. Ablation on pruning order, $A \rightarrow B$: Prune adapter first, then backbone.

Strategy	Order	BBox	1pt	3pt	5pt
A50+B50	A→B	79.01	69.43	76.26	77.62
A50+B50	B→A	78.08	67.72	75.55	76.03
A75+B75	A→B	78.65	68.13	74.26	76.85
A75+B75	B→A	77.50	65.34	72.46	75.94

Table 4. Effect of compensation data source at B75. Pre: SA-1B; Down: SA-Med2D-20M; Metrics: Dice (%).

Data	#	BBox	1pt	3pt	5pt
Pre	20K	64.95	52.47	64.91	71.74
Down	20K	68.31	64.69	73.45	78.02
Pre+Down	10K+10K	**79.23**	**69.67**	**75.06**	**78.82**

Table 5. Importance scoring ablation at B75. Metrics: Dice (%).

Method	BBox	1pt	3pt	5pt
Magnitude	77.27	67.77	73.67	75.09
Hessian	77.92	66.26	73.84	77.08
Taylor	78.63	68.22	74.26	77.34
ORS	**79.11**	**69.67**	**75.06**	**77.82**

Ablation on Importance Scoring. We compare Magnitude, Hessian, Taylor, and ORS under the same prune-and-compensate schedule on SAM-Med2D at B75 with 20K mixed calibration. Figure 5 plots 1-point Dice during backbone compensation. ORS converges faster, reaches a higher plateau, and is more stable near convergence. Table 5 confirms the final Dice advantage across box and point prompts. ORS scores each unit by the orthogonal residual of its downstream gradient relative to the joint pretraining-downstream subspace, weighted by parameter scale. This suppresses substitutable directions and focuses compensation on directions that are harder to recover after pruning. Magnitude ignores direction, Taylor mixes substitutable and non-substitutable components, and Hessian is less stable under small calibration sets. These behaviors are consistent with the faster and stronger compensation observed for ORS.

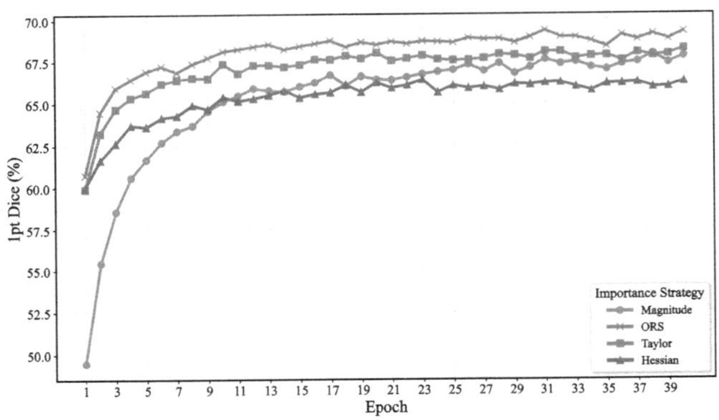

Fig. 5. 1-point Dice during backbone compensation after B75 pruning with four scoring criteria. ORS converges faster and reaches a higher plateau.

5 Conclusion

We presented RASP-SAM, a structured pruning framework for the adapter-tuned SAM-Med2D setting. By leveraging the structural decoupling between the backbone and adapters, our method integrates dependency-safe pruning, Orthogonal Residual Scoring, and Dual-Stream Compensation to maximize efficiency. Experiments on SAM-Med2D demonstrate that RASP-SAM achieves a 4$\times$ reduction in MACs with near-lossless performance. These results support the proposed dependency-safe and dual-stream design in a representative adapter-tuned SAM setting; extending the evaluation to other adapter-style SAM variants remains important future work.

Acknowledgments. Supported by Natural Science Foundation of China under Grant 62271465, National Key R&D Program of China under Grant 2025YFC3408300, Natural Science Foundation of Jiangsu Province (No. BK20255001), and Suzhou Basic Research Program under Grant SYG202338.

Disclosure of Interests. The authors have no competing interests to declare that are relevant to the content of this article.

References

1. Chen, T., et al.: Sam-adapter: adapting segment anything in underperformed scenes. In: Proceedings of the IEEE/CVF International Conference on Computer Vision, pp. 3367–3375 (2023)
2. Chen, Z., Fang, G., Ma, X., Wang, X.: Slimsam: 0.1% data makes segment anything slim. Adv. Neural Inf. Process. Syst. **37**, 39434–39461 (2024)
3. Cheng, J., et al.: Sam-med2d (2023). https://arxiv.org/abs/2308.16184
4. Dettmers, T., Pagnoni, A., Holtzman, A., Zettlemoyer, L.: Qlora: efficient finetuning of quantized LLMs. Adv. Neural. Inf. Process. Syst. **36**, 10088–10115 (2023)
5. Fang, G., Ma, X., Song, M., Mi, M.B., Wang, X.: Depgraph: towards any structural pruning. In: Proceedings of the IEEE/CVF Conference on Computer Vision and pattern Recognition, pp. 16091–16101 (2023)
6. Hassibi, B., Stork, D.: Second order derivatives for network pruning: optimal brain surgeon. Adv. Neural. Inf. Process. Syst. **5** (1992)
7. Houlsby, N., et al.: Parameter-efficient transfer learning for NLP. In: International Conference on Machine Learning, pp. 2790–2799. PMLR (2019)
8. Hu, E.J., et al.: Lora: low-rank adaptation of large language models. ICLR **1**(2), 3 (2022)
9. Jiang, Y., et al.: Foundation model-guided multi-view semi-supervised CT segmentation of liver tumors in resource-constrained settings. NPJ Digit. Med. **9**(1), 31 (2025). https://doi.org/10.1038/s41746-025-02190-0
10. Kirillov, A., et al.: Segment anything. In: Proceedings of the IEEE/CVF International Conference on Computer Vision, pp. 4015–4026 (2023)
11. LeCun, Y., Denker, J., Solla, S.: Optimal brain damage. Adv. Neural Inf. Process. Syst. **2** (1989)
12. Li, Z., Xiao, J., Yang, L., Gu, Q.: Repq-vit: scale reparameterization for post-training quantization of vision transformers. In: Proceedings of the IEEE/CVF International Conference on Computer Vision, pp. 17227–17236 (2023)
13. Lin, Y., Zhang, T., Sun, P., Li, Z., Zhou, S.: Fq-vit: post-training quantization for fully quantized vision transformer. arXiv preprint arXiv:2111.13824 (2021)
14. Liu, X., et al: PQ-SAM: post-training quantization for segment anything model. In: European Conference on Computer Vision, pp. 420–437. Springer (2024)
15. Ma, J., He, Y., Li, F., Han, L., You, C., Wang, B.: Segment anything in medical images. Nat. Commun. **15**(1), 654 (2024)
16. Ma, X., Wu, Q., Zhao, X., Zhang, X., Pun, M.O., Huang, B.: Sam-assisted remote sensing imagery semantic segmentation with object and boundary constraints. IEEE Trans. Geosci. Remote Sens. (2024)
17. Wu, J., et al.: Medical SAM adapter: adapting segment anything model for medical image segmentation. Med. Image Anal. **102**, 103547 (2025). https://doi.org/10.1016/j.media.2025.103547
18. Xia, Z., Pan, X., Song, S., Li, L.E., Huang, G.: Vision transformer with deformable attention. In: Proceedings of the IEEE/CVF Conference on Computer Vision and Pattern Recognition, pp. 4794–4803 (2022)
19. Yang, H., Yin, H., Shen, M., Molchanov, P., Li, H., Kautz, J.: Global vision transformer pruning with hessian-aware saliency. In: Proceedings of the IEEE/CVF Conference on Computer Vision and Pattern Recognition, pp. 18547–18557 (2023)

20. Zhang, C., et al.: Faster segment anything: towards lightweight SAM for mobile applications. arXiv preprint arXiv:2306.14289 (2023)
21. Zhang, R., et al.: Personalize segment anything model with one shot. arXiv preprint arXiv:2305.03048 (2023)
22. Zhang, X., et al.: Adapting vision foundation models for real-time ultrasound image segmentation. In: International Conference on Medical Image Computing and Computer-Assisted Intervention, pp. 24–34. Springer (2025)
23. Zhang, Y., et al.: Embedded framework for clinical medical image segment anything in resource limited healthcare regions. NPJ Digit. Med. **8**(1), 568 (2025)
24. Zhao, X., et al.: Fast segment anything. arXiv preprint arXiv:2306.12156 (2023)

REVEAL++: Differentiable Phenotypic Grouping for Vision–Language Retinal Modeling of Alzheimer's Disease Risk

Ethan Meidinger[1], Seowung Leem[2], Zeyun Zhao[2], and Ruogu Fang[2](✉)

[1] University of Virginia, Charlottesville, VA, USA
[2] J. Crayton Pruitt Family Department of Biomedical Engineering, Herbert Wertheim College of Engineering, University of Florida, Gainesville, FL, USA
ruogu.fang@vanderbilt.edu

Abstract. The retina offers a noninvasive window into neurodegenerative disease, capturing subtle structural patterns associated with a risk of future cognitive decline. Vision–language alignment frameworks such as REVEAL have shown that pairing retinal fundus images with structured clinical risk narratives improves early prediction of Alzheimer's disease (AD). A key design choice in these approaches is the use of phenotypic grouping, where individuals with similar risk profiles are treated as multi-positive pairs during contrastive learning. However, existing methods operationalize phenotypic similarity as a discrete construct, relying on hard group assignments that impose rigid supervision and decouple group formation from representation learning. We propose a continuous formulation of phenotypic structure within contrastive learning. Rather than assigning samples to fixed clusters, we model inter-subject similarity as a differentiable weighting function derived from intra-modality embedding similarities in both retinal images and risk profiles. These weights define soft multi-positive relationships through a continuous aggregation operator, enabling graded supervision that reflects the spectrum nature of disease risk. We further introduce a soft-target contrastive objective that jointly learns cross-modal alignment and phenotypic structure in an end-to-end manner. Evaluated on UK Biobank retinal imaging data for incident AD prediction, the proposed framework consistently outperforms discrete group-based contrastive learning and standard vision–language baselines. By treating phenotypic similarity as a learnable, continuous signal rather than a fixed grouping rule, our approach provides a principled and robust foundation for population-scale neurodegenerative risk modeling from multi-modal retinal and clinical data. Code is available at: https://github.com/REVEALPlusPlus.

Keywords: Neurodegenerative Disease · Multi-modal Alignment · Preclinical Prediction

G. Yang et al. (Eds.): MICCAI 2026, LNCS 16878, pp. 525–535, 2027.
https://doi.org/10.1007/978-3-032-38059-3_50

1 Introduction

Alzheimer's disease (AD) is a progressive neurodegenerative disorder characterized by a long preclinical phase during which pathological changes accumulate before clinical symptoms [6]. Advances in brain imaging and plasma-based biomarkers have substantially improved the ability to detect disease-related pathology. However, these approaches may remain costly, invasive, or impractical for large-scale population screening. Complementary modalities that are scalable and noninvasive therefore play an important role in early risk stratification. The retina has emerged as one such modality, as its structure and microvasculature share developmental and physiological links with the central nervous system and are associated with neurodegenerative and vascular processes relevant to AD [3]. In parallel, systemic and lifestyle-related risk factors, including cardiometabolic health and sleep patterns, capture longitudinal exposures that contribute to dementia risk decades before diagnosis [2,4]. Rather than serving as standalone diagnostic markers, these signals offer complementary population-level information that may help characterize early disease susceptibility (Fig. 1).

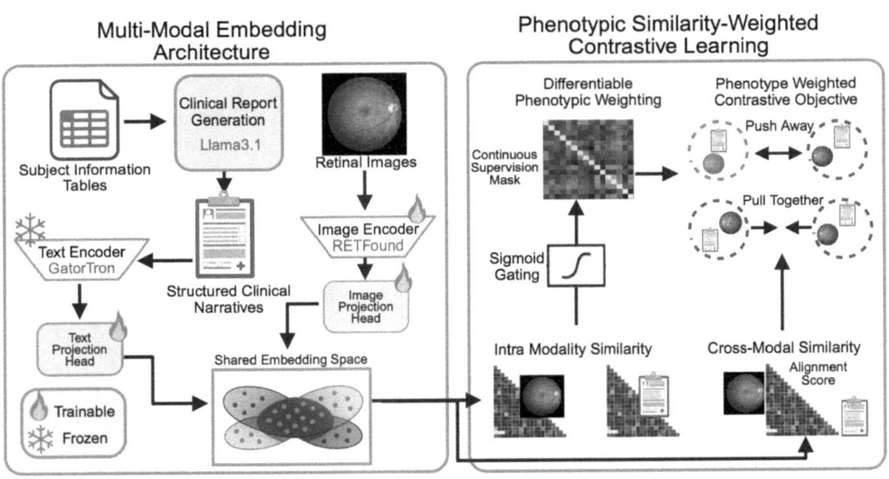

Fig. 1. Architecture of the proposed differentiable phenotypic weighting framework for group-aware contrastive learning. Image and text embeddings are aligned via a similarity-weighted multi-positive contrastive loss, where continuous phenotypic weights replace hard grouping to model the heterogeneous spectrum of Alzheimer's disease risk.

Recent advances in vision–language models (VLMs) have enabled joint representation learning across heterogeneous data modalities through contrastive alignment. Inspired by CLIP-style architectures, medical VLMs have increasingly been adapted to retinal imaging, leveraging large-scale pretraining to learn clinically meaningful visual representations [7,17,21]. Building on this paradigm, the

REVEAL framework aligned retinal fundus images with individualized clinical risk narratives derived from structured health data, enabling multi-modal modeling of neurodegenerative risk [13]. A central innovation of REVEAL was group-aware contrastive learning (GACL), which encouraged subjects with similar phenotypic profiles to act as multi-positive pairs during training. This strategy improved robustness to individual-level noise and promoted learning of shared disease-relevant structure, leading to improved downstream AD risk prediction compared with uni-modal and standard pairwise contrastive approaches.

Despite these advantages, phenotypic grouping in existing GACL formulations is constructed through discrete similarity thresholds, implicitly assuming that individuals belong to well-separated risk categories. From a biological perspective, however, neurodegenerative risk evolves along continuous and overlapping trajectories shaped by heterogeneous genetic, vascular, metabolic, and lifestyle factors. Individuals often exhibit partial similarity across multiple phenotypic axes rather than membership in a single homogeneous group. Hard group assignments may therefore introduce artificial boundaries that fail to reflect the graded and spectrum-like nature of disease vulnerability, while preventing the grouping process itself from adapting during representation learning.

In this work, we extend the original REVEAL framework by introducing a differentiable phenotypic weighting formulation for retinal image–clinical narrative alignment. REVEAL++ preserves the same multi-modal alignment setting and encoder backbone choices, while replacing REVEAL's hard threshold-based group assignments with continuous, differentiable phenotypic weights. Rather than treating subject pairs as strictly positive or negative, REVEAL++ allows phenotypic similarity to modulate contrastive supervision in a graded manner. These weights are computed from retinal image embeddings and clinical risk-profile embeddings, then combined through a soft aggregation operator to define a soft multi-positive contrastive objective. By modeling phenotypic relationships as a differentiable attention-like process, the proposed framework enables joint learning of representation alignment and population-level structure, more faithfully capturing the continuous and heterogeneous biological variability underlying neurodegenerative risk. Our contributions are three-fold:

- **Differentiable phenotypic weighting:** We replace hard threshold-based grouping in group-aware contrastive learning with continuous phenotypic similarity weights derived from retinal and clinical embeddings, enabling smooth data-driven cohort modeling that better captures heterogeneous Alzheimer's disease risk.
- **Soft multi-positive contrastive learning:** We introduce a soft-target contrastive objective that incorporates phenotypic similarity into cross-modal alignment, enabling graded multi-positive supervision instead of binary pair assignments.
- **State-of-the-art Alzheimer's risk prediction from retinal imaging:** We achieve new state-of-the-art performance on UK Biobank retinal imaging for incident Alzheimer's disease prediction, outperforming existing vision–language and group-aware contrastive learning methods.

2 Methods

2.1 Overview of REVEAL++

REVEAL++ learns joint image–text representations of retinal fundus images and structured clinical reports under a group-aware contrastive objective. Given a minibatch of N subjects, each subject p is associated with a retinal image and a clinical report. Image and text encoders produce modality-specific embeddings, which are projected into a shared latent space. To incorporate phenotypic structure into contrastive supervision, we compute intra-modality similarity matrices that capture retinal image embedding and risk-profile similarity between subjects. These similarities are transformed into a differentiable phenotypic weighting mask $W \in [0,1]^{N\times N}$, which acts as a soft pairwise target matrix in a multi-positive contrastive loss.

2.2 Clinical Report Generation

To enable alignment between retinal images and systemic risk factors within a vision–language framework, structured questionnaire data were converted into synthetic clinical narratives compatible with pretrained text encoders. Using the LLaMA-3.1 API as the text generation engine, each participant's tabular risk-factor profile was mapped into a standardized clinical-style summary [9]. For each subject, the LLM received a predefined documentation template, the subject's structured demographic, behavioral, cognitive, and lifestyle variables, and explicit instructions to generate a concise report without inferring missing values. The template was adapted from the "Patient Information" section of the CARE clinical case reporting guidelines, ensuring consistency with established medical documentation conventions [8]. To minimize variability and preserve numerical fidelity, the prompt enforced a one-to-one mapping between tabular entries and template fields, with unavailable values explicitly marked rather than imputed. This controlled translation process produces semantically enriched text representations that enable structured health information to be embedded within a shared multi-modal latent space.

2.3 Image and Text Encoders

Let $\mathbf{x}_p$ denote the retinal image and $\mathbf{t}_p$ the associated clinical report for subject p. The image encoder $E_I(\cdot)$ and text encoder $E_T(\cdot)$ produce modality-specific embeddings. In our implementation, we instantiate E_I as RETFound [21] and E_T as GatorTron [19]. Each encoder is followed by a projection head, denoted P_I and P_T, to map features into a shared embedding space of dimension d.

$$\mathbf{h}_p^I = E_I(\mathbf{x}_p), \qquad \mathbf{h}_p^T = E_T(\mathbf{t}_p). \tag{1}$$

The encoder outputs are projected into the shared embedding space and ℓ_2-normalized:

$$\hat{\mathbf{z}}_p^I = \frac{P_I(\mathbf{h}_p^I)}{\|P_I(\mathbf{h}_p^I)\|_2}, \qquad \hat{\mathbf{z}}_p^T = \frac{P_T(\mathbf{h}_p^T)}{\|P_T(\mathbf{h}_p^T)\|_2}. \tag{2}$$

2.4 Intra-modality Similarity for Phenotypic Grouping

To capture phenotypic similarity between subjects, we construct intra-modality similarity matrices based on cosine similarity between normalized projected representations. Let $\hat{\mathbf{z}}_p^I$ denote the normalized projected image embedding for subject p, and let $\hat{\mathbf{z}}_p^T$ denote the normalized projected text-derived risk-profile embedding for subject p.

$$S_{ii}(p,q) = \langle \hat{\mathbf{z}}_p^I, \hat{\mathbf{z}}_q^I \rangle \tag{3}$$

$$S_{tt}(p,q) = \langle \hat{\mathbf{z}}_p^T, \hat{\mathbf{z}}_q^T \rangle \tag{4}$$

Here, $S_{ii}(p,q)$ captures similarity between the learned retinal image embeddings of subjects p and q, while $S_{tt}(p,q)$ captures similarity between their clinical report embeddings.

2.5 Differentiable Phenotypic Weighting

We transform these similarities into soft membership signals using sigmoid gating with thresholds τ_F, τ_T and learnable sharpness parameters g_F, g_T:

$$a_F(p,q) = \sigma\left(\frac{S_{ii}(p,q) - \tau_F}{g_F}\right), \qquad a_T(p,q) = \sigma\left(\frac{S_{tt}(p,q) - \tau_T}{g_T}\right), \tag{5}$$

where $\sigma(\cdot)$ denotes the logistic sigmoid function. Finally, we combine the two signals using a differentiable probabilistic union operator to obtain the phenotypic weighting score:

$$W_{pq} = 1 - \big(1 - a_F(p,q)\big)\big(1 - a_T(p,q)\big), \qquad W_{pq} \in [0,1]. \tag{6}$$

Pairs with larger W_{pq} are treated as more strongly aligned in phenotype space and receive higher weight as positives in the multi-positive contrastive objective.

2.6 Phenotypic Similarity-Weighted Multi-positive Contrastive Loss

Cross-modal similarity between image and text embeddings is defined as:

$$S_{it}(p,q) = \langle \hat{\mathbf{z}}_p^I, \hat{\mathbf{z}}_q^T \rangle. \tag{7}$$

Logits are computed using temperature scaling, where the learnable temperature is parameterized as $T_c = \exp(-s)$, together with a learnable bias term β:

$$\ell_{pq} = \frac{S_{it}(p,q)}{T_c} - \beta, \qquad T_c = \exp(-s). \tag{8}$$

We optimize a soft-target multi-positive contrastive objective (Table 1):

$$\mathcal{L}_{\mathrm{MP}} = \frac{1}{N^2}\sum_{p=1}^{N}\sum_{q=1}^{N}\left[W_{pq}\,\log\!\big(1+\exp(-\ell_{pq})\big) + \big(1-W_{pq}\big)\,\log\!\big(1+\exp(\ell_{pq})\big)\right]. \tag{9}$$

Table 1. Cohort characteristics across data splits.

	Train (n = 30,462)	**Validation** (n = 3,384)	**Test** (n = 5,396)
Gender: (male %)	45.10	45.41	45.10
Age: mean (s.d)	55.53 (8.24)	55.78 (8.12)	55.52 (8.17)
Ethnicity: (British %)	84.08	83.51	88.51

When W_{pq} approaches 1, the pair (p, q) is treated as a positive match, when W_{pq} approaches 0, it is treated as a negative pair. Intermediate values allow soft supervision based on phenotypic similarity.

3 Experiments

3.1 Dataset and Preprocessing

A comprehensive set of demographic, behavioral, cognitive, and lifestyle variables was extracted from the UK Biobank [5] baseline assessment and compiled as candidate risk factors based on established epidemiological and biomarker evidence linking modifiable exposures to Alzheimer's disease and dementia risk [2,10,11,14,16,18]. These include factors associated with amyloid and tau pathology, sleep disturbance, cardiometabolic health, and other modifiable determinants of neurodegeneration.

Color fundus photographs (CFPs) from the initial UK Biobank assessment visit were used for image-based modeling. Images underwent automated quality control to exclude low-quality scans, retaining only high-quality CFPs for downstream analysis [22]. Preprocessed CFPs are input into a RETFound-initialized vision encoder, which is fine-tuned during training [21]. Each image was resized to match the input resolution of the pretrained RETFound encoder and normalized using standard channel-wise mean and standard deviation values consistent with its pretraining setup. To ensure consistent anatomical orientation across subjects, right-eye images were horizontally flipped prior to encoding.

Downstream incident AD prediction was formulated as a subject-level binary classification task. A total of 39,242 participants with high-quality CFPs were included and allocated into training (n = 30,462), validation (n = 3,384), and test (n = 5,396) sets. Participants free of prevalent and incident AD were split between each set, and each participant who later developed AD was put into the test pool. The training set was used for representation alignment, the validation set for hyperparameter optimization, and the test set for downstream incident AD prediction. To form the downstream evaluation cohort, controls were sampled from the non-AD test-set participants to obtain an approximate 8% AD prevalence while maintaining age- and gender-matched distributions. From this cohort (n = 1,163), 931 subjects (862 controls, 69 AD) were used to train SVM models with 5-fold cross-validation. The remaining 232 subjects (215 controls, 17 AD) were held out as an independent test set for AD prediction.

3.2 Implementation Details

RETFound and GatorTron were used as image and text encoders. The vision encoder is initialized with RETFound weights and fine-tuned end-to-end, while the text encoder is kept frozen. Linear projections map both modalities into a shared $d = 1024$-dimensional space. Embeddings were ℓ_2-normalized prior to similarity computation. The batch size was 128. Optimization used AdamW with hyperparameters selected via Optuna [1]. The final learning rate was 2.42×10^{-4}, $\epsilon = 8.61 \times 10^{-7}$, and weight decay 0.0232. Phenotypic similarity thresholds were initialized from empirical intra-modality cosine similarity distributions computed on the development set, restricting the search to the upper interquartile range.

4 Results

To evaluate the proposed framework, we compared our method against strong retinal and biomedical foundation models. We included RETFound [21], a large-scale retinal image foundation model; RET-CLIP [7], a retinal image–text contrastive pretraining framework; and KeepFIT-CFP [17], a knowledge-enhanced retinal vision–language model introduced with MM-Retinal. We additionally evaluated two general biomedical multi-modal foundation models: PMC-CLIP [15] and BiomedCLIP [20]. Because RETFound is an image-only encoder, we paired it with GatorTron [19] to construct multi-modal representations, concatenating image and text embeddings for downstream classification. In addition to vision–language baselines, we trained tabular SVM models using structured clinical variables and CFP-derived image features to assess whether performance gains stem from semantic narrative modeling or solely from image foundation representations. All methods were evaluated under an identical multi-modal SVM protocol. Each experiment was repeated across 10 random seeds, and we report mean $\pm$ standard deviation performance.

In the incident AD prediction task (Table 2), our phenotypic-weighted multi-positive contrastive framework consistently outperformed all comparison methods, indicating that soft, differentiable phenotypic alignment leads to more coherent multi-modal representations. Rather than relying on single positive pairs or hard grouping, the proposed formulation allows subjects with similar risk profiles to contribute proportionally during training, yielding stronger downstream discrimination. While pretrained vision–language baselines such as RETFound+GatorTron and RET-CLIP capture meaningful retinal–text correspondences, they do not explicitly model phenotypic structure, which may be important for long-horizon neurodegenerative risk prediction.

These findings suggest that modeling phenotypic similarity as a continuous, differentiable weighting mechanism enables smoother transitions between positive and negative supervision, leading to more coherent multi-modal embedding spaces and improved long-horizon neurodegenerative risk prediction.

Table 2. Comparison of multi-modal and baseline methods for incident AD prediction. Results are reported as mean ± standard deviation across 10 random seeds. **Bold** and <u>Underline</u> represent the best and the second best results.

	AUROC	Balanced Accuracy	F1-Score	MCC
Baseline SVM	0.593 ± 0.068	0.574 ± 0.083	0.140 ± 0.089	0.076 ± 0.099
KeepFIT-CFP	0.490 ± 0.063	0.505 ± 0.0412	0.099 ± 0.034	0.002 ± 0.046
BiomedCLIP	0.525 ± 0.064	0.522 ± 0.060	0.121 ± 0.052	0.023 ± 0.054
RETCLIP	0.558 ± 0.076	0.527 ± 0.042	0.106 ± 0.069	0.028 ± 0.051
PMC-CLIP	0.471 ± 0.049	0.484 ± 0.020	0.076 ± 0.023	−0.022 ± 0.023
RETFound + GatorTron	0.642 ± 0.052	0.581 ± 0.069	0.185 ± 0.099	0.119 ± 0.101
REVEAL (no GACL)	0.654 ± 0.092	0.602 ± 0.075	0.205 ± 0.096	0.144 ± 0.105
REVEAL (with GACL)	<u>0.658 ± 0.090</u>	<u>0.609 ± 0.079</u>	<u>0.207 ± 0.100</u>	<u>0.146 ± 0.111</u>
REVEAL++	**0.678 ± 0.061**	**0.613 ± 0.048**	**0.236 ± 0.079**	**0.168 ± 0.088**

5 Discussion

Alzheimer's disease is increasingly understood not as a binary condition but as a long-term neurodegenerative process that evolves over years prior to diagnosis. Pathological changes including amyloid deposition, tau accumulation, vascular dysfunction, and systemic metabolic dysregulation emerge progressively and interact across multiple biological scales before cognitive symptoms become apparent [6,16]. Retinal microvascular alterations and structural remodeling likewise develop along a continuum, reflecting cumulative exposure to systemic and neurodegenerative risk factors. Therefore, similarity between individuals along disease-relevant dimensions is continuous rather than discretely separable.

Hard similarity thresholds impose artificial boundaries on this biological continuum by assigning subjects to fixed phenotypic groups. While such grouping can strengthen contrastive supervision, it implicitly assumes well-defined subtype partitions that may not reflect the underlying progression of the disease. However, such discrete subtype boundaries may not exist during the preclinical stages of neurodegenerative disease, where pathological processes evolve gradually and heterogeneously across individuals. This mismatch can limit the ability of representation learning methods to capture subtle transitions in risk states. In contrast, the proposed differentiable phenotypic weighting mechanism allows phenotypic similarity to modulate supervision strength continuously. Participants with partially overlapping risk profiles or subtly similar retinal signatures contribute proportionally during training, enabling smoother organization of the shared embedding space while preserving meaningful inter-subject variation.

This formulation more closely reflects the pathophysiology of preclinical AD, where risk accumulates gradually and manifests heterogeneously across individuals [12]. By relaxing discrete grouping into continuous supervision, the model is able to represent intermediate phenotypic states that may correspond to early pathological changes. The resulting embedding geometry reflects a continuum of risk rather than rigid clusters, providing a representation that is both biologically plausible and better suited for early risk stratification.

Although REVEAL++ improves over REVEAL and other retinal vision–language baselines, the observed AUROC should not be interpreted as sufficient for clinical deployment. Rather, these results demonstrate improved representation learning for a difficult, low-prevalence, long-horizon incident AD prediction task.

6 Conclusion

We presented REVEAL++, a differentiable phenotypic alignment framework for multi-modal learning from retinal imaging and clinical risk narratives in preclinical Alzheimer's disease prediction. By replacing discrete threshold-based grouping with continuous similarity-driven supervision, the proposed approach enables phenotypic relationships to be learned jointly with representation alignment, allowing population structure to emerge directly from data. This formulation better captures the gradual and heterogeneous nature of neurodegenerative disease progression and leads to improved risk prediction performance. More broadly, differentiable phenotypic alignment offers a strategy for modeling structured variability in multi-modal biomedical data, with potential applications spanning chronic disease risk prediction, precision medicine, longitudinal health modeling, and large-scale population health analytics across diverse clinical domains.

Acknowledgments and Disclosure of Interests. This material is based upon work supported by the National Science Foundation under Grant No. (NSF 2123809). This research has been conducted using the UK Biobank Resource under Application Number 48388. The authors declare that they have no competing interests.

References

1. Akiba, T., Sano, S., Yanase, T., Ohta, T., Koyama, M.: Optuna: a next-generation hyperparameter optimization framework. In: Proceedings of the 25th ACM SIGKDD International Conference on Knowledge Discovery and Data Mining (KDD 2019), New York, NY, USA, pp. 2623–2631. ACM (2019). https://doi.org/10.1145/3292500.3330701
2. Aktan Süzgün, M., Tang, Q., Stefani, A.: Sleep abnormalities and risk of Alzheimer's disease. Curr. Neurol. Neurosci. Rep. **25**(1), 67 (2025). https://doi.org/10.1007/s11910-025-01451-5
3. Banna, H., Slayo, M., Armitage, J., Del Rosal, B., Vocale, L., Spencer, S.: Imaging the eye as a window to brain health: frontier approaches and future directions. J. Neuroinflammation **21**(1), 309 (2024). https://doi.org/10.1186/s12974-024-03304-3

4. Bueno Lopez, C., et al.: Cardiometabolic health and risk of dementia and brain atrophy: a community-based prospective cohort study of 0.5 million adults in China. Lancet Reg. Health West. Pacific **64**, 101743 (2025). https://doi.org/10.1016/j.lanwpc.2025.101743
5. Bycroft, C., et al.: The UK biobank resource with deep phenotyping and genomic data. Nature **562**(7726), 203–209 (2018). https://doi.org/10.1038/s41586-018-0579-z
6. Chow, K.H.M., Abel, T.: Neurodevelopmental origins of age-related neurodegenerative diseases. eBioMedicine **124**, 106151 (2026). https://doi.org/10.1016/j.ebiom.2026.106151
7. Du, J., et al.: RET-CLIP: a retinal image foundation model pre-trained with clinical diagnostic reports. arXiv preprint arXiv:2405.14137 (2024)
8. Gagnier, J.J., et al.: The care guidelines: consensus-based clinical case reporting guideline development. Global Adv. Health Med. **2**(5), 38–43 (2013). https://doi.org/10.7453/gahmj.2013.008
9. Grattafiori, A., Dubey, A., Jauhri, A., et al.: The Llama 3 herd of models. arXiv preprint arXiv:2407.21783 (2024). https://arxiv.org/abs/2407.21783
10. Hayden, K.M., et al.: Association between modifiable risk factors and levels of blood-based biomarkers of Alzheimer's and related dementias in the look ahead cohort. JAR Life **13**, 1–21 (2024). https://doi.org/10.14283/jarlife.2024.1
11. Huszár, Z., et al.: Association of modifiable risk factors with progression to dementia in relation to amyloid and tau pathology. Alzheimer's Res. Therapy **16**, 238 (2024). https://doi.org/10.1186/s13195-024-01602-9
12. Jack, C.R.J., et al.: NIA-AA research framework: toward a biological definition of Alzheimer's disease. Alzheimer's Dement. **14**(4), 535–562 (2018). https://doi.org/10.1016/j.jalz.2018.02.018
13. Leem, S., Gu, L., You, C., Gong, K., Fang, R.: REVEAL: multimodal vision–language alignment of retinal morphometry and clinical risks for incident AD and dementia prediction. In: Medical Imaging with Deep Learning (2026). https://openreview.net/pdf?id=aOKAXRHXVw, accepted by MIDL 2026. Proceedings of Machine Learning Research (PMLR)
14. Leshner, A.I., Landis, S., Stroud, C., Downey, A. (eds.): Preventing Cognitive Decline and Dementia: A Way Forward. National Academies Press, Washington, DC (2017). https://doi.org/10.17226/24782
15. Lin, W., et al.: PMC-CLIP: contrastive language-image pre-training using biomedical documents. arXiv preprint arXiv:2303.07240 (2023)
16. Livingston, G., et al.: Dementia prevention, intervention, and care: 2024 report of the lancet standing commission. Lancet **404**(10452), 572–628 (2024). https://doi.org/10.1016/S0140-6736(24)01296-0
17. Wu, R., Zhang, C., Zhang, J., Zhou, Y., Zhou, T., Fu, H.: MM-retinal: knowledge-enhanced foundational pretraining with fundus image-text expertise. arXiv preprint arXiv:2405.11793 (2024)
18. Xiong, J., Bhimani, R., Carney-Anderson, L.: Review of risk factors associated with biomarkers for Alzheimer disease. J. Neurosci. Nurs. **55**(3), 103–109 (2023). https://doi.org/10.1097/JNN.0000000000000705
19. Yang, X., et al.: A large language model for electronic health records. NPJ Digit. Med. **5**(1), 1–9 (2022). https://doi.org/10.1038/s41746-022-00742-2
20. Zhang, S., et al.: BiomedCLIP: a multimodal biomedical foundation model pre-trained from scientific image-text pairs. arXiv preprint arXiv:2303.00915 (2025)

21. Zhou, Y., et al.: A foundation model for generalizable disease detection from retinal images. Nature **622**(7981), 156–163 (2023). https://doi.org/10.1038/s41586-023-06555-x
22. Zhou, Y., et al.: AutoMorph: automated retinal vascular morphology quantification via a deep learning pipeline. Transl. Vis. Sci. Technol. **11**(7), 12 (2022). https://doi.org/10.1167/tvst.11.7.12

SAGEAgent: A Self-evolving Agent for Cost-Aware Modality Acquisition in Multimodal Survival Prediction

Chongyu Qu[1], Can Cui[1], Zhengyi Lu[1], Junchao Zhu[1], Tianyuan Yao[1], Junlin Guo[1], Juming Xiong[1], Yanfan Zhu[1], Yuechen Yang[1], Bennett A. Landman[1,2], and Yuankai Huo[1](✉)

[1] Vanderbilt University, Nashville, TN 37235, USA
yuankai.huo@vanderbilt.edu
[2] Vanderbilt University Medical Center, Nashville, TN 37232, USA

Abstract. Does every cancer patient truly need a complete diagnostic workup for accurate survival prediction? In multimodal clinical oncology, diagnostic modalities follow a clinically mandated order of escalating burden—from demographics collected at intake to genomic profiling requiring specialized tissue analysis. Current multimodal survival methods either assume all modalities are available or passively handle missing data, but none actively reason about whether acquiring the next modality is justified for a given patient along this ordered workflow. We formulate this as a sequential decision problem and propose SAGEAgent (**S**equential **A**cquisition **G**uided by **E**xperience), a self-evolving LLM-based clinical agent that decides which diagnostic modalities to acquire for each patient, balancing predictive accuracy against clinical invasiveness. SAGEAgent reasons about each patient's evolving diagnostic state through clinical tools that translate numerical predictions into text, an episodic memory that retrieves similar past cases, and a semantic memory that accumulates reusable decision patterns from experience. Experiments on a glioma cohort combining TCGA-LGG, TCGA-GBM, and BraTS with four diagnostic modalities demonstrate that SAGEAgent achieves competitive survival prediction accuracy while reducing average acquisition burden by 55%. Code is publicly available at: https://github.com/Chongyu1117/SAGEAgent.

Keywords: Modality Acquisition · LLM Agent · Survival Prediction

1 Introduction

Accurate survival prediction in oncology increasingly relies on integrating multiple diagnostic modalities [3,15], ranging from routine demographics and radiology to invasive biopsy and molecular profiling [16]. These modalities follow a strict clinical order: imaging precedes biopsy, which in turn provides the specimen for genomic analysis. Previous multimodal fusion methods [4,5,8] combine

G. Yang et al. (Eds.): MICCAI 2026, LNCS 16878, pp. 536–546, 2027.
https://doi.org/10.1007/978-3-032-38059-3_51

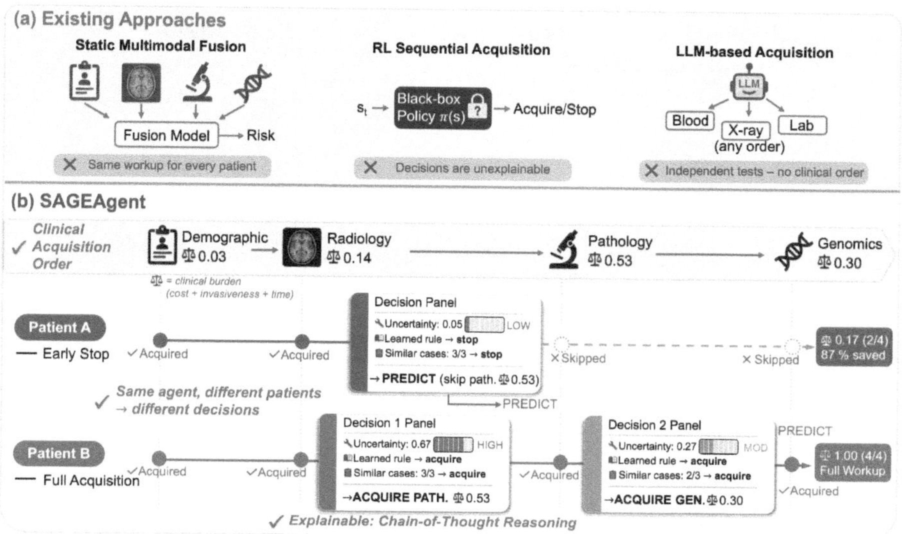

Fig. 1. Overview of modality acquisition strategies. (a) Existing approaches: static fusion applies all modalities uniformly, RL learns adaptive policies without explanation, and LLM-based methods reason over independent tests without clinical ordering. (b) SAGEAgent follows a clinically acquisition order, deciding at each stage whether to acquire or stop based on clinical tools, learned rules, and similar past cases. Patient A stops early (83% burden saved); Patient B proceeds through all modalities.

all available modalities through a fixed strategy or handle missing data at test time, but treat prediction as a static problem—the model outputs a risk score from whatever is observed, without reasoning about **whether the next diagnostic step is worth it for this specific patient**.

Sequential test acquisition has been studied from two directions. Reinforcement learning (RL) methods [2,20] train cost-aware acquisition policies but produce black-box decisions that limit clinical trust. Large language model (LLM) based approaches [1,13,18] offer transparent reasoning but operate on independent diagnostic tests without accounting for the sequential dependencies inherent in clinical modality acquisition. Neither direction models the clinically mandated ordering (see Fig. 1(a) for a comparison).

We propose SAGEAgent (**S**equential **A**cquisition **G**uided by **E**xperience), a training-free, self-evolving LLM-based framework for cost-aware sequential modality acquisition in multimodal survival prediction (Fig. 1(b)). At each stage of the clinical order (demographics → radiology → pathology → genomics), SAGEAgent decides whether to acquire or stop using three signal sources: clinical tools that translate numerical predictions into text, an episodic memory that retrieves similar past cases, and a semantic memory that accumulates decision rules through periodic self-reflection—all without gradient updates. Evaluated on a glioma cohort combining TCGA-LGG [19], TCGA-GBM [21], and BraTS [17]

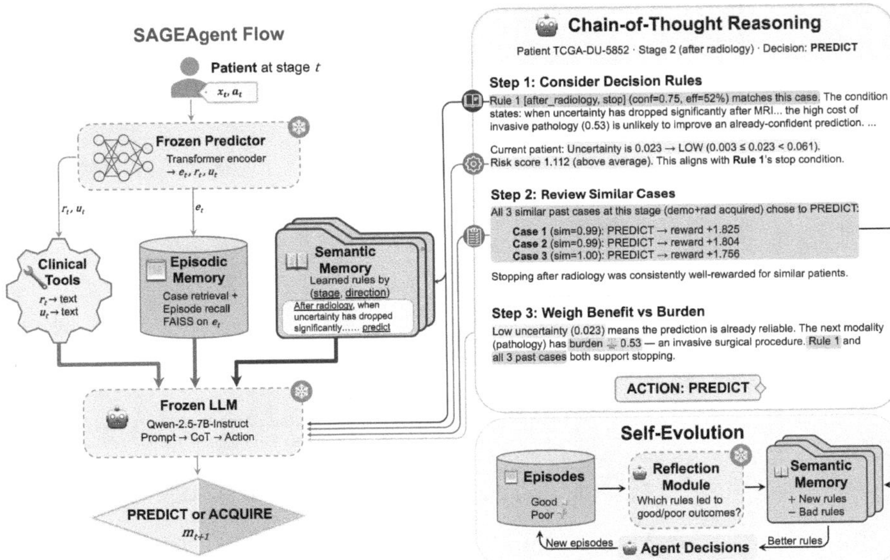

Fig. 2. SAGEAgent architecture and reasoning. SAGEAgent Flow: At each stage, the frozen predictor produces the patient embedding, risk score, and calibrated uncertainty. Three signal sources feed the frozen LLM: clinical tools convert numerical outputs to text, episodic memory retrieves similar patients and past decisions, and semantic memory provides learned decision rules. **Chain-of-Thought Reasoning**: A real reasoning trace after radiology, with highlights tracing each step to its source component. The agent synthesizes all three signals and decides to stop, skipping invasive pathology. **Self-Evolution**: Episodes accumulate with outcome labels; the reflection module analyzes which rules led to good or poor outcomes and updates semantic memory accordingly, forming a closed loop without gradient updates.

with nested 5×5 cross-validation, SAGEAgent achieves competitive survival prediction accuracy while reducing acquisition burden by 55% compared to full workup, providing transparent reasoning for each patient's acquisition pathway.

Our contributions are threefold:

1. **At the learning framework level**, we propose SAGEAgent, a novel framework that formulates cost-aware modality acquisition for multimodal survival prediction as a sequential decision-making problem, while explicitly respecting the clinical ordering of diagnostic procedures.
2. **At the training strategy level**, we introduce a dual-memory self-evolution mechanism that autonomously discovers decision rules from experience, eliminating the need for gradient-based updates or hand-crafted priors.
3. **At the interpretability level**, we provide fully auditable reasoning pathways in SAGEAgent, where each acquisition decision is grounded in explicit chain-of-thought reasoning, thereby bridging cost-aware modality acquisition with clinical transparency.

2 Method

2.1 Problem Formulation

Consider a patient with N diagnostic modalities $\{m_1, \ldots, m_N\}$ arranged in a clinically mandated acquisition order $m_1 \to m_2 \to \cdots \to m_N$. For glioma, this order is demographics $\to$ radiology $\to$ pathology $\to$ genomics, reflecting real clinical dependencies [16]. Each modality m_i carries a clinical burden $b(m_i) \in [0, 1]$, quantified via multi-criteria decision analysis (MCDA) [11] over monetary cost, turnaround time, patient invasiveness, and infrastructure requirements (details in Sect. 3.1). At decision stage $t \in \{1, \ldots, N-1\}$, the agent has already acquired the first t modalities. The decision is binary: acquire m_{t+1} (incurring burden $b(m_{t+1})$), or stop and predict. The goal is to maximize prediction accuracy while minimizing total acquisition burden $B_t = \sum_{i=1}^{t} b(m_i)$. Figure 2 provides an overview of the proposed framework.

2.2 Multimodal Predictor with Calibrated Uncertainty

Survival Predictor. The prediction backbone is a Transformer-based multimodal encoder that handles missing modalities via learnable mask tokens [8,9]. It maps each patient's available modalities to a fused embedding $\mathbf{e}_t \in \mathbb{R}^d$, from which a Cox proportional hazards head [7] produces a scalar risk score r_t. The predictor is trained with modality dropout and **frozen during agent operation**. Architecture and training details are in Sect. 3.2.

Calibrated Uncertainty Head. On top of the frozen predictor, we train a lightweight uncertainty head that estimates whether the current prediction would change if more modalities were acquired. Given the embedding $\mathbf{e}_t$ and acquisition mask $\mathbf{a}_t$, it outputs a calibrated probability:

$$u_t = \sigma\big(f_\phi([\mathbf{e}_t, \mathbf{a}_t])\big) \tag{1}$$

where f_ϕ is a lightweight MLP and σ is the sigmoid function. The training label is 1 when the absolute difference between the current and full-modality risk scores exceeds a threshold τ, and 0 otherwise. Training data is generated by enumerating clinical-order prefix masks on complete-modality patients. The resulting $u_t \in [0, 1]$ is patient-specific and mask-specific, giving the agent a calibrated signal of whether additional data is likely to matter.

2.3 SAGEAgent: LLM-Based Acquisition Agent

SAGEAgent is built on a frozen LLM that receives no gradient updates. At each stage, it gathers diagnostic signals, retrieves relevant experience, and outputs a binary decision (acquire or predict) with a chain-of-thought explanation. Its behavior improves solely through two memory systems following the episodic–semantic distinction in cognitive science [24].

Clinical Decision Tools. Two tools translate the frozen predictor's numerical outputs into text. The *Uncertainty Tool* reports u_t from Eq. 1 with a categorical level (very low to very high) based on quintile thresholds from the training distribution. The *Survival Predictor Tool* reports the risk score r_t and its position relative to the training cohort (below/around/above average). Both tools present outputs as natural-language sentences so that all reasoning occurs in text space.

Episodic Memory. The episodic memory provides case-based reasoning through two complementary retrieval mechanisms. *Case retrieval* draws from a FAISS-indexed [10] database of training patients, returning the k nearest neighbors by cosine similarity on ℓ_2-normalized embeddings together with their event status (survived or deceased) and survival time. This provides a static clinical reference: which training patients resemble the current one, and how long did they survive. *Episode recall* draws from the agent's own past decisions. When the agent acquires m_{t+1}, it receives a stage reward $R_{\text{stage}} = -b(m_{t+1}) + \max(0, u_t - u_{t+1})$ that penalizes cost and rewards uncertainty reduction. When the agent stops at stage t, a terminal reward $R_{\text{term}} = (1 - u_t) - B_t$ favors early stopping at low uncertainty. Retrieval is stage-matched and re-ranked by the sum of cosine similarity $\text{sim}(\mathbf{e}_t, \mathbf{e}_j)$ and min-max normalized episode reward $\tilde{r}_j \in [0, 1]$, returning the top-k episodes, including the action taken and its outcome.

Semantic Memory and Self-evolution. The semantic memory stores interpretable decision rules discovered through periodic self-reflection. Each rule is indexed by acquisition stage t and action direction $d \in \{\text{stop}, \text{acquire}\}$. After each episode, we record whether the agent followed or violated each relevant rule. Every N_r episodes (one reflection cycle), episodes are ranked by total reward, and those in the top and bottom quartiles are labeled good and poor respectively. Rule effectiveness is the fraction of followed episodes that led to good outcomes. The reflection module receives the effectiveness report and proposes new rules, which are verified for direction clarity and novelty. Rules that consistently lead to poor outcomes are deprecated, forming a closed loop where rules are discovered, validated, and either reinforced or removed.

3 Experiments and Results

3.1 Dataset and Evaluation Protocol

Dataset and Split. We evaluate on a glioma cohort of 962 patients (803 TCGA-LGG/GBM [19,21] and 159 BraTS [17]) spanning four diagnostic modalities in clinical order: demographics, radiology, pathology, and genomics, each encoded into a 32-dimensional feature vector by pretrained extractors [8]. Of the 962 patients, 170 possess all four modalities and form the evaluation set. We adopt nested 5×5 cross-validation. The outer loop splits 170 patients into 5 folds (136/34) for agent training and evaluation. Each inner 5-fold trains the predictor on the 136 patients plus 792 partial-modality patients as augmentation via modality dropout, yielding 25 independent pipelines. Per-patient decisions are aggregated by majority vote (≥3/5 inner models). We report mean ± std across 5 outer folds.

Clinical Burden. We quantify modality cost via MCDA [11] over monetary cost, turnaround time, invasiveness, and infrastructure requirements. After normalization, per-modality burdens are 0.03 (demographics), 0.14 (radiology), 0.53 (pathology), and 0.30 (genomics), totaling $B = 1.0$ for a full workup.

Metrics. We report C-index [14], average burden, and hypervolume $\text{HV} = (c - 0.5) \times (1.0 - b)$ which captures the accuracy–cost trade-off as a single scalar.

Table 1. Core results on the glioma survival cohort (170 complete-modality patients). Top (a) C-index in ascending order. RL-based methods form a lower cluster around 0.73, while the remaining six methods achieve C-index above 0.8. Top (b) C-index (solid) and burden (hatched) for the six methods with C-index > 0.8, where SAGEAgent saves the most burden. Bottom: per-method C-index and burden as the mean over 5 outer folds with 95% bootstrap confidence intervals (1000 stratified resamples). Bold marks the lowest burden among methods with C-index > 0.8. D, R, P, G denote demographics, radiology, pathology, and genomics, and ■ = 170 patients, □ = 5 patients, with shades varying linearly between these values. † = predictor backbone used by SAGEAgent.

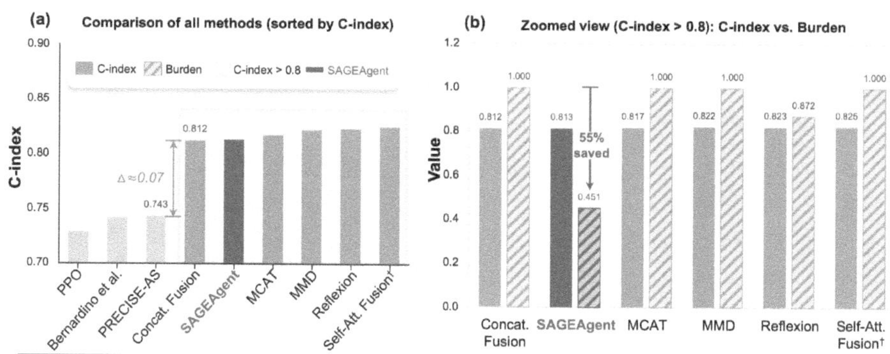

Category	Method	C-index↑	Burden↓	D	R	P	G
Static Fusion	Concat. Fusion	0.812 [.77, .86]	1.000 [1.00, 1.00]	■	■	■	■
	MCAT [6]	0.817 [.78, .86]	1.000 [1.00, 1.00]	■	■	■	■
	MMD [8]	0.822 [.78, .86]	1.000 [1.00, 1.00]	■	■	■	■
	Self-Att. Fusion†	0.825 [.78, .87]	1.000 [1.00, 1.00]	■	■	■	■
RL-based Acquisition	Bernardino et al. [2]	0.742 [.69, .79]	0.125 [.10, .15]	■	■		
	PRECISE-AS [20]	0.743 [.69, .79]	0.127 [.10, .16]	■	■		
	PPO [22]	0.729 [.68, .78]	0.124 [.10, .16]	■	■		
LLM-based Acquisition	Reflexion [23]	0.823 [.78, .87]	0.872 [.84, .90]	■	■	■	■
	SAGEAgent	0.813 [.77, .86]	**0.451** [.41, .50]	■	■	■	

3.2 Implementation Details

Predictor. The backbone is a 2-layer Transformer encoder ($d = 128$, 4 heads) with per-modality MLP projections from $\mathbb{R}^{32}$ to $\mathbb{R}^{128}$ and learnable mask tokens for missing modalities, trained with Cox partial likelihood, reconstruction, and alignment losses under 50% modality dropout (~458K parameters). The calibrated uncertainty head (Eq. 1) adds ~6.5K parameters trained with BCE loss, with $\tau = 0.3$ set by calibration quality (AUROC = 0.908, ECE = 0.110) on inner-fold validation after post-hoc temperature scaling [12].

SAGEAgent uses Qwen-2.5-7B-Instruct [25] with fully frozen weights, running on a single NVIDIA A6000 GPU. During experience accumulation, each training patient is processed for 3 episodes, yielding ~408 episodes per fold. Episodic memory retrieves $k = 3$ stage-matched neighbors via FAISS [10]. Semantic memory maintains up to 10 active rules with reflection triggered every 10 patients.

Table 2. Component ablation. Each row adds one component to the Base LLM. All configurations use frozen Qwen-2.5-7B-Instruct with majority-vote aggregation. C-index, burden, and HV are reported as the mean over 5 outer folds with 95% bootstrap confidence intervals, following the protocol of Table 1. Bold indicates best burden and HV. D, R, P, G denote demographics, radiology, pathology, and genomics, and ■ = 170 patients, □ = 5 patients, with shades varying linearly.

Configuration	C-index↑	Burden↓	HV↑	D	R	P	G
Base LLM	0.827 [.79, .87]	0.798 [.77, .82]	0.066 [.06, .08]	■	■	■	■
+ Tools	0.830 [.79, .87]	0.843 [.82, .87]	0.052 [.04, .06]	■	■	■	■
+ Tools + Episodic	0.814 [.77, .86]	0.611 [.56, .66]	0.122 [.10, .14]	■	■	■	■
+ Tools + Semantic	0.819 [.78, .86]	0.706 [.66, .75]	0.094 [.08, .11]	■	■	■	■
SAGEAgent	0.813 [.77, .86]	**0.451** [.41, .50]	**0.172** [.15, .20]	■	■	■	□

3.3 Core Results

Table 1 reports all methods under the nested 5×5 protocol of Sect. 3.1.

RL Methods Reduce Cost at the Expense of Accuracy. A C-index gap of approximately 0.07 separates the nine methods into two clusters (Table 1 top (a)). All three RL baselines reduce burden to ~0.12 but drop C-index to ~0.73, effectively discarding informative modalities. Static fusion methods require full workup by design, and Reflexion reduces burden by only 13%.

SAGEAgent Reduces Burden by 55% at No Significant Accuracy Cost. As shown in Table 1 top (b), SAGEAgent reaches a C-index of 0.813 while reducing burden to 0.451. Its C-index gap to every method with C-index above 0.8 is not statistically significant, with paired 95% bootstrap confidence intervals containing 0 and $p \geq 0.30$, whereas its burden reduction over those methods is significant at $p < 0.001$. SAGEAgent thus offers the strongest accuracy-to-cost balance among the methods compared, rather than the highest accuracy alone.

3.4 Ablation Study

Table 2 isolates the contribution of each component by progressively adding tools, episodic memory, and semantic memory to the frozen LLM. Adding tools alone raises burden from 0.798 to 0.843 because without past experience to contextualize uncertainty and risk signals, the LLM defaults to acquiring more. Episodic memory provides the largest individual burden reduction (0.843→0.611) by surfacing cases where early stopping succeeded, while semantic memory offers a complementary effect (0.843→0.706) through interpretable stopping criteria. Combined, SAGEAgent achieves the lowest burden (0.451) and highest HV (0.172), a 2.6× improvement over the Base LLM, whose marginally higher C-index is not significant. A naive policy that stops once u_t falls below a fixed τ reaches C-index 0.774, 0.769, and 0.750 at τ = 0.3, 0.5, and 0.7 with burden 0.347, 0.235, and 0.144, which is 0.04 to 0.06 below SAGEAgent. The tools and dual memory therefore add value beyond the uncertainty head.

Table 3. Rules discovered by semantic memory across five folds. Eff. = fraction of rule-following episodes with top-quartile reward (random baseline = 0.25). n = total episodes where the rule was followed across folds. Stability indicates whether the rule, once discovered, was ever deprecated and re-proposed during training.

Discovered Rule	Folds	n	Eff.	Stability
Stopping rules				
"After radiology, predict now ... risk error is small and pathology cost outweighs potential gain."	5/5	876	.46–.52	Stable
"After pathology, predict now ... prediction quality is high and genomics cost is not justified."	5/5	387	.39–.62	Stable
Acquisition rules				
"After demographics, acquire radiology ... uncertainty is high and imaging cost is low."	4/5	493	.38–.67	Stable
"After radiology, acquire pathology ... uncertainty remains high and risk error is large."	4/5	353	.44–.58	Churned

3.5 Self-evolution Analysis

The ablation in Sect. 3.4 shows that semantic memory reduces burden, but does not reveal what rules it learns. We now examine whether the discovered rules are consistent across folds. Each outer fold runs 10 reflection cycles from an empty rule set with no shared initialization.

Semantic Memory Discovers Consistent Rules Across Folds. Table 3 lists the four rules emerging in at least four of five folds, all achieving effectiveness well above the 0.25 random baseline and collectively covering over 2,100 decisions. Each fold generates 10.8 candidate rules on average, of which 4.6 survive pruning.

Stopping Rules are Easier to Learn Than Acquisition Rules. The two stopping rules emerge in the first two reflection cycles and are never deprecated. In contrast, the acquisition rule "after radiology, acquire pathology" accumulates 13 deprecated instances across folds before a stable version survives in four of five. Reflection logs reveals a two-phase pattern: cycles 1–2 establish core stopping rules that persist unchanged, while cycles 3–10 are dominated by acquire-rule exploration where most proposals are deprecated due to near-zero effectiveness.

4 Conclusion

We presented SAGEAgent, a training-free LLM agent for cost-aware sequential modality acquisition in survival prediction. On a glioma cohort it reduces acquisition burden by 55% with no statistically significant loss in C-index, grounding each decision in auditable reasoning. Notably, it learns to skip invasive pathology for half of the cohort, suggesting that many glioma patients may not require biopsy for reliable prognostication. Self-evolution analysis shows that stopping rules converge reliably while acquisition rules are harder to learn. Our evidence is limited to a single retrospective glioma cohort with complete four-modality data, so external multi-site validation, other cancer types, and prospective clinical study remain future work.

Acknowledgments. This research was supported by the National Institutes of Health (NIH) under grant R01DK135597, with $20000, representing 66% of the total project costs. The content is solely the responsibility of the authors and does not necessarily represent the official views of the National Institutes of Health. This material is based upon work supported by the U.S. National Science Foundation under Award Number 2434229. A total of $5000 representing 17% of the total project costs, was financed with this funding. Any opinions, findings, and conclusions or recommendations expressed in this material are those of the author(s) and do not necessarily reflect the views of the U.S. National Science Foundation. Non-governmental support was provided by the Vanderbilt Internal Grants. A total of $5000 representing 17% of the total project costs, was financed with non-governmental funds.

Disclosure of Interests. The authors have no competing interests to declare that are relevant to the content of this article.

References

1. Bani-Harouni, D., Pellegrini, C., Özsoy, E., Keicher, M., Navab, N.: Language agents for hypothesis-driven clinical decision making with reinforcement learning. arXiv preprint arXiv:2506.13474 (2025)

2. Bernardino, G., et al.: Reinforcement learning for active modality selection during diagnosis. In: Wang, L., Dou, Q., Fletcher, P.T., Speidel, S., Li, S. (eds.) Medical Image Computing and Computer Assisted Intervention – MICCAI 2022. LNCS, vol. 13431, pp. 592–601. Springer, Cham (2022). https://doi.org/10.1007/978-3-031-16431-6_56
3. Boehm, K.M., Khosravi, P., Vanguri, R., Gao, J., Shah, S.P.: Harnessing multimodal data integration to advance precision oncology. Nat. Rev. Cancer **22**(2), 114–126 (2022)
4. Braman, N., Gordon, J.W., Goossens, E.T., Willis, C., Stumpe, M.C., Venkataraman, J.: Deep orthogonal fusion: multimodal prognostic biomarker discovery integrating radiology, pathology, genomic, and clinical data. In: de Bruijne, M., et al. (eds.) Medical Image Computing and Computer Assisted Intervention – MICCAI 2021. LNCS, vol. 12905, pp. 667–677. Springer, Cham (2021). https://doi.org/10.1007/978-3-030-87240-3_64
5. Chen, R.J., et al.: Pathomic fusion: an integrated framework for fusing histopathology and genomic features for cancer diagnosis and prognosis. IEEE Trans. Med. Imaging **41**(4), 757–770 (2020)
6. Chen, R.J., et al.: Multimodal co-attention transformer for survival prediction in gigapixel whole slide images. In: Proceedings of the IEEE/CVF International Conference on Computer Vision, pp. 4015–4025 (2021)
7. Cox, D.R.: Regression models and life-tables. J. Roy. Stat. Soc.: Ser. B (Methodol.) **34**(2), 187–202 (1972)
8. Cui, C., et al.: Survival prediction of brain cancer with incomplete radiology, pathology, genomic, and demographic data. In: Wang, L., Dou, Q., Fletcher, P.T., Speidel, S., Li, S. (eds.) Medical Image Computing and Computer Assisted Intervention – MICCAI 2022. LNCS, vol. 13435, pp. 626–635. Springer, Cham (2022). https://doi.org/10.1007/978-3-031-16443-9_60
9. Devlin, J., Chang, M.W., Lee, K., Toutanova, K.: BERT: pre-training of deep bidirectional transformers for language understanding. In: Proceedings of the 2019 Conference of the North American Chapter of the Association for Computational Linguistics: Human Language Technologies, vol. 1 (Long and Short Papers), pp. 4171–4186 (2019)
10. Douze, M., et al.: The Faiss library. IEEE Trans. Big Data (2025)
11. Guitouni, A., Martel, J.M.: Tentative guidelines to help choosing an appropriate MCDA method. Eur. J. Oper. Res. **109**(2), 501–521 (1998)
12. Guo, C., Pleiss, G., Sun, Y., Weinberger, K.Q.: On calibration of modern neural networks. In: International Conference on Machine Learning, pp. 1321–1330. PMLR (2017)
13. Hager, P., et al.: Evaluation and mitigation of the limitations of large language models in clinical decision-making. Nat. Med. **30**(9), 2613–2622 (2024)
14. Harrell, F.E., Califf, R.M., Pryor, D.B., Lee, K.L., Rosati, R.A.: Evaluating the yield of medical tests. JAMA **247**(18), 2543–2546 (1982)
15. Lipkova, J., et al.: Artificial intelligence for multimodal data integration in oncology. Cancer Cell **40**(10), 1095–1110 (2022)
16. Louis, D.N., et al.: The 2021 who classification of tumors of the central nervous system: a summary. Neuro Oncol. **23**(8), 1231–1251 (2021)
17. Menze, B.H., et al.: The multimodal brain tumor image segmentation benchmark (BraTS). IEEE Trans. Med. Imaging **34**(10), 1993–2024 (2014)
18. Nori, H., et al.: Sequential diagnosis with language models. arXiv preprint arXiv:2506.22405 (2025)

19. Pedano, N., et al.: Radiology data from the cancer genome atlas low grade glioma (TCGA-LGG) collection. Cancer Imaging Arch. **2** (2016)
20. Saadat, A., et al.: PRECISE-AS: personalized reinforcement learning for efficient point-of-care echocardiography in aortic stenosis diagnosis. In: Gee, J.C., et al. (eds.) Medical Image Computing and Computer Assisted Intervention – MICCAI 2025. LNCS, vol. 15973, pp. 474–484. Springer, Cham (2025). https://doi.org/10.1007/978-3-032-05185-1_46
21. Scarpace, L., et al.: The cancer genome atlas glioblastoma multiforme collection (TCGA-GBM). Cancer Imaging Arch. (2016)
22. Schulman, J., Wolski, F., Dhariwal, P., Radford, A., Klimov, O.: Proximal policy optimization algorithms. arXiv preprint arXiv:1707.06347 (2017)
23. Shinn, N., Cassano, F., Gopinath, A., Narasimhan, K., Yao, S.: Reflexion: language agents with verbal reinforcement learning. In: Advances in Neural Information Processing Systems **36**, 8634–8652 (2023)
24. Tulving, E., et al.: Episodic and semantic memory. Organ. Memory **1**(381–403), 1 (1972)
25. Yang, A., et al.: Qwen2. 5 technical report. arXiv e-prints pp. arXiv–2412 (2024)

SAVer: Structure-Aware Validation of Standard Fetal Head Planes via Clinically Grounded Anatomical Completeness

Anusha Amba Prasanna[1(✉)], Vasanth Ambrose[1], Padmini Ramesh[2], Shyam Ayyasamy[2], Keerthi Ram[4], Suresh Seshadri[3], Manojkumar Lakshmanan[2], and Mohanasankar Sivaprakasam[1(✉)]

[1] Department of Electrical Engineering, Indian Institute of Technology, Madras (IITM), Chennai, India
anusha@htic.iitm.ac.in, mohan@ee.iitm.ac.in
[2] Healthcare Technology Innovation Centre (HTIC), IITM, Chennai, India
[3] Mediscan Systems, Chennai, India
[4] Sudha Gopalakrishnan Brain Centre, IITM, Chennai, India

Abstract. Standard fetal head plane identification in ultrasound is essential for accurate biometric measurements and early detection of neuro developmental abnormalities. Automating the identification of these planes can optimize diagnostic workflows. Existing approaches predominantly treat plane identification as a global image-level classification task, often relying on superficial visual features rather than clinically defined anatomical criteria. To address these limitations, we propose **SAVer**, structure-aware validation framework that couples anatomical reasoning with plane classification to ensure clinically consistent fetal ultrasound plane identification. SAVer integrates anatomical segmentation with plane identification through a unified architecture that combines a perception backbone with a Plane Classification Module along with an Anatomical Reasoning Engine. The reasoning module explicitly evaluates the presence, completeness, and structural consistency of clinically relevant landmarks to ensure alignment with diagnostic standards. In addition, we introduce a Adapter Feature Alignment Module (AFAM) that enables efficient feature alignment between anatomical segmentation representations and plane classification. We evaluate the proposed method on fetal ultrasound head plane data, achieving a structure-size weighted dice score of 95% for anatomical segmentation and demonstrating improved robustness. The proposed structure-aware framework enhances interpretability and reliability by embedding clinically meaningful constraints into the classification process.

Keywords: Fetal Brain Ultrasound · Standard Plane Identification · Structure Aware Learning · Anatomical Segmentation

G. Yang et al. (Eds.): MICCAI 2026, LNCS 16878, pp. 547–557, 2027.
https://doi.org/10.1007/978-3-032-38059-3_52

1 Introduction

Fetal Ultrasound is the most widely used modality for antenatal screening due to its safety and clinical accessibility. Precise acquisition of fetal standard head planes specifically Transventricular (TV), Transthalamic (TT) and Transcerebellar (TC) planes [14] are crucial for biometric measurements [1,2,5] and detection of neuro developmental anomalies [18]. Automating the identification of these planes can significantly optimize diagnostic workflows. However, reliable plane recognition remains challenging due to variability in fetal pose, probe orientation, and inter-operator acquisition differences, which frequently result in incomplete or ambiguous visualization of critical anatomical structures.

Existing methods for plane classification approaches this as an image level task by utilizing Convolutional Neural Networks (CNN) [4] and transformer [9,15] based architectures to learn discriminative representations in a latent feature space. In these frameworks, the structural information is progressively compressed into a low-dimensional global embedding from which the final prediction is derived. The abstract appearance statistics are encoded rather than explicitly modeling anatomically meaningful structures. Consequently, predictions are driven by learned correlations in feature space rather than verification of clinically required structural criteria. In contrast, anatomical structure segmentation operates directly in the input (pixel) space, producing spatially grounded masks corresponding to specific structures. A clinically valid standard plane is characterized by the presence, geometry and spatial relationships of multiple inner structures [14], for instance, the presence of Cerebellum and Cavum Septum Pellucidi (CSP) simultaneously represents the TC plane. Despite advances in anatomical structure segmentation for fetal ultrasound [8,12,16], segmentation is often treated as an independent task, with limited integration into standard plane classification frameworks. This separation between feature-level classification and pixel-level anatomical reasoning restricts interpretability and clinical reliability. Without explicit anatomical supervision, classification may be driven by global head morphology rather than the landmarks defining a standard plane. [19]. Consequently, there remains a critical need for frameworks that explicitly encode anatomical structure presence [13] into the plane classification process. To address the above limitations we propose structure-aware framework for fetal brain plane classification. Our contributions are four fold:

- **Comprehensive Multi Structure Segmentation:** We segment 32 clinically relevant inner fetal brain structures, which, to the best of our knowledge, represents the most comprehensive anatomical delineation for supporting downstream tasks such as biometric assessment, structural consistency analysis and anomaly identification.
- **Integrated SAVer Framework:** We develop a unified anatomy driven framework that leverages both global image features and localized anatomical structures to ensure classification aligns with clinical diagnostic criteria.
- **Anatomical Reasoning Engine:** We implement a reasoning module that explicitly evaluates the presence, visibility, and spatial relationships of key

anatomical landmarks, providing a "ultrasonographer-in-the-loop" logic to the automated classification process.
- **Adapter based feature alignment module:** We introduce a lightweight adapter module that enables efficient task-specific feature alignment without requiring full encoder retraining, preserving the generalization capability of the foundation model.

2 Dataset and Methodology

Dataset: We curated a dataset retrospectively from a tertiary clinical center, consisting of 400 TC, 325 TT, and 370 TV fetal ultrasound images, totaling 1,095 samples. Each image was annotated by two clinical experts and verified by two independent reviewers, covering 32 unique anatomical structures across all planes. As anatomical requirements are plane-specific, each plane includes a defined subset of structures according to clinical guidelines.

Architecture: The proposed SAVer framework, as shown in Fig. 1, processes an ultrasound(US) image I $\in \mathbb{R}^{3\times H\times W}$ as input and predicts whether it satisfies the criteria for a standard fetal head plane. Let $c \in \{TT, TC, TV\}$ denote the plane labels where C denotes the number of planes. This estimates $Y \in \{0,1\}^{K\times H\times W}$ segmentation masks where K represents the number of anatomical structures. The US image is processed by a fine-tuned MedSAM ViT-B [6,11] encoder ξ to extract hierarchical feature representations as shown in Eq. 1

$$f_{us} = \xi(I), \quad where \quad f_{us} \in \mathbb{R}^{256\times H\times W} \tag{1}$$

To enhance discriminative anatomical responses, the encoder features are refined using the Convolution Block Attention Module (CBAM) [17] which sequentially performs channel and spatial alignment to obtain f'_{us}. The refined features are processed by a lightweight decoder and segmentation heads to generate logits, which are then used to produce segmentation maps $\hat{Y}_k$ for localizing key anatomical structures in Eq. 2

$$\hat{\mathbf{Y}}_k = \mathcal{H}_k\left(f_{us'}\right), \quad \hat{\mathbf{Y}}_k \in \mathbb{R}^{H\times W}, \quad k = 1, 2, \ldots, 32. \tag{2}$$

While anatomical structures are localized in a bottom-up manner, structural validity is inherently plane-dependent. Different standard planes require distinct subsets of anatomical landmarks. Therefore, explicit plane prediction is necessary to provide contextual conditioning for structural reasoning. Thus, the encoder features f_{us} are then passed on to Global Average Pooling(GAP) module to obtain a pooled embedding feature vector $f_c \in \mathbb{R}^{256}$. The pooled embedding is projected into a higher dimensional latent space as depicted in Eq. 3 to align with the transformer embedding dimensionality.

$$z = W_p f_c + b_p, \quad W_p \in \mathbb{R}^{768\times 256}, z \in \mathbb{R}^{768} \tag{3}$$

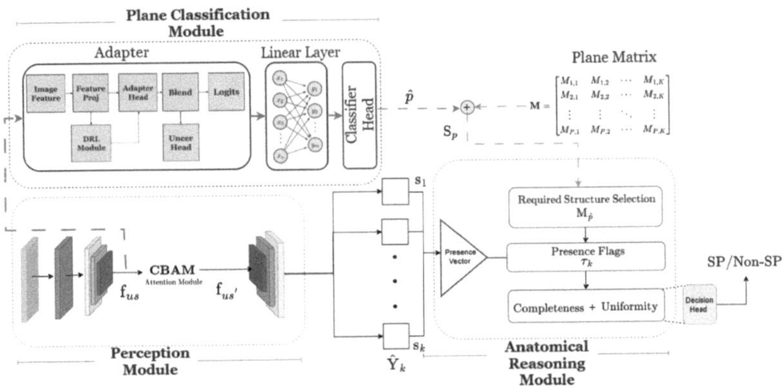

Fig. 1. SAVer framework: MedSAM-based multi-structure segmentation and adapter-based plane classification integrated with plane-aware anatomical reasoning for standard-plane validation.

The projected feature is passed through a bottleneck adapter [7], where the DRL module, uncertainty head, and blend operation enable lightweight task-specific adaptation, yielding a DRL-enhanced feature z_d, uncertainty estimate u, and adapted feature z_a:

$$z_a = \text{Blend}(z + W_u \sigma(W_d z_d), u), \quad z_d = \mathcal{H}_{DRL}(z), \; u = \mathcal{H}_{unc}(z_d), \tag{4}$$

The plane logits are then computed using a linear classifier head as in Eq. 5

$$\hat{p} = Softmax(W_c z_a + b_c), \quad \hat{p} \in \mathbb{R}^C \tag{5}$$

To explicitly model anatomical validity, we define a clinician-curated plane structure dependency matrix$M \in \{0,1\}^{C \times K}$, where $M_{c,k} = 1$ indicates that the structure k is required for the plane C. Anatomical evidence for the presence of a structure can be obtained from the predicted segmentation maps $\hat{Y}_k$, from which we compute a continuous structural presence score as denoted in Eq. 6

$$s_k = \frac{1}{HW} \sum_{i=1}^{H} \sum_{j=1}^{W} \hat{\mathbf{Y}}_k(i,j), \quad \mathbf{s} \in \mathbb{R}^K. \tag{6}$$

Here, s_k quantifies the normalized spatial extent of the Kth structure serving as a soft estimate of the anatomical presence. Given the predicted plane probabilities $\hat{p}$, the expected structure requirement vector that softly encodes which anatomical structures are expected under the predicted plane is given by Eq. 7

$$\mathbf{m} = \hat{p}^T M, \quad \mathbf{m} \in [0,1]^K \tag{7}$$

During training, we introduce a plane-aware anatomical loss that enforces minimum structural presence similar to [3] for extremely crucial anatomical landmarks. The required minimum structural presence is defined as shown in Eq. 8

$$\theta_k = \tau_k + \Delta_k, \tag{8}$$

where τ_k denotes the predefined minimal anatomical threshold and Δ_k is a learnable margin parameter. The resulting plane-aware anatomical loss in Eq. 10, together with the segmentation and classification terms, forms the overall training objective defined in Eq. 10.

$$\mathcal{L}_{seg} = \mathcal{L}_{Dice} + \mathcal{L}_{Focal} + \mathcal{L}_{SoftHD} + \mathcal{L}_{L2}, \qquad \mathcal{L}_{cls} = \mathcal{L}_{CCE} \tag{9}$$

$$\mathcal{L} = \mathcal{L}_{seg} + \lambda_{cls}\mathcal{L}_{cls} + \lambda_{anatomy}\left[\sum_{k=1}^{K} m_k \,\mathrm{Softplus}(\theta_k - s_k)\right] \tag{10}$$

Plane-Aware Structured Confidence Aggregation: Let $p_{\max} = \max(\hat{p})$ denote the predicted plane confidence. In addition, five complementary anatomical quality measures are derived from the segmentation maps: c_{struct}, the mean pixel-wise segmentation confidence within detected structures; c_{comp}, a completeness score that verifies the presence and sufficient extent of plane-specific mandatory anatomy; c_{uni}, the coefficient of variation of structure areas measuring anatomical balance; c_{dist}, a spatial plausibility score based on the distribution of structures across image quadrants; and c_{edge}, a framing quality score that penalizes structures intersecting image boundaries. Together, these metrics assess prediction certainty, anatomical completeness, balance, spatial consistency, and acquisition quality. All measures are normalized to $[0, 1]$ and aggregated into a unified structured confidence score defined in Eq. 11.

$$C_{std} = w_1 p_{\max} + w_2 c_{struct} + w_3 c_{comp} + w_4 c_{uni} + w_5 c_{dist} + w_6 c_{edge}, \tag{11}$$

where $\sum_{i=1}^{6} w_i = 1$. The final plane validity prediction is obtained by thresholding the aggregated confidence as defined in Eq. 12,

$$\hat{y}_{valid} = \mathbb{1}(C_{std} \geq \delta), \tag{12}$$

where δ is a predefined decision threshold.

3 Results and Discussion

Implementation Details: The model was trained on 1,095 TC, TT, and TV fetal US images, with 240 images (89 TC, 91 TT, and 60 TV) used for validation and testing. The model was trained for 60 epochs using the AdamW [10]optimizer with an initial learning rate of 10^{-4} freezing the encoder for the first 30 epochs before joint fine-tuning. All experiments were conducted on NVIDIA RTX 4090 GPU achieving an average inference time of 0.44 s per image.

Figure 2A depicts the predicted segmentations of inner key structures along with input images of the fetal head planes like TV, TC and TT. Figure 2B shows the radar plot for the weighted confidence along with overall SAVer score

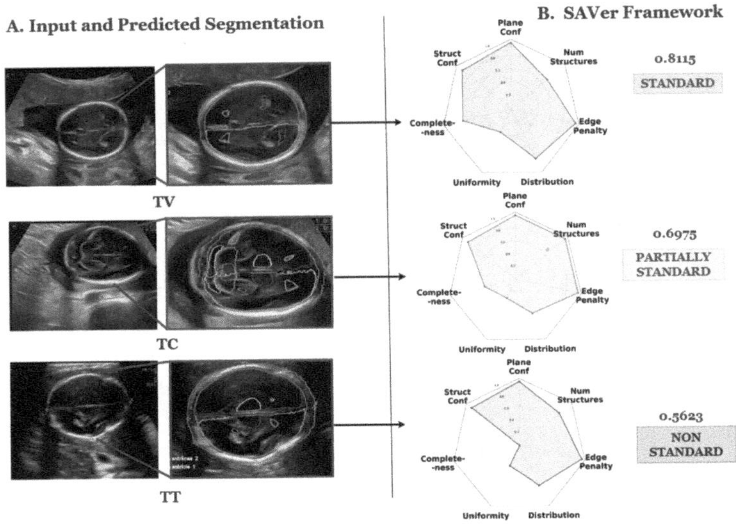

Fig. 2. SAVer framework (A) Input fetal ultrasound images and corresponding predicted multi-structure segmentations. (B) Radar-based confidence analysis for standard plane validity.

Standard Plane Validity Assessment: The proposed SAVer framework was evaluated for standard plane validity discrimination on a test cohort of 240 fetal brain ultrasound frames containing both standard and non-standard acquisitions as shown in Table 1. Performance was assessed using Accuracy, Precision, Recall, F1-score, and False Negative Rate (FNR), with emphasis on minimizing missed valid planes. Across all three planes, the model achieves high classification performance with F1-scores exceeding 91, indicating reliable discrimination between protocol-compliant and non-compliant frames. Notably, the FNR remains consistently low (≤ 0.05), demonstrating that valid diagnostic planes are rarely rejected. While the False Positive Rate (FPR) ranges from 0.20 to 0.36, SAVer demonstrates high sensitivity for standard plane detection, helping preserve diagnostically useful frames. At the same time, specificity remains critical for reliable biometry estimation, and future improvements will focus on further reducing false positives while maintaining strong detection performance.

Table 1. Standard Plane Classification Performance

Plane	Accuracy(%)↑	Recall(%)↑	Precision(%)↑	F1 Score(%)↑	FPR↓	FNR↓
TC	88.16	96.10	85.00	91.89	0.36	0.02
TT	88.57	94.44	91.07	92.73	0.31	0.05
TV	95.12	96.36	98.15	97.25	0.20	0.03

Anatomical Localisation and Structural Integrity: We evaluate anatomical segmentation performance across the three standard fetal brain planes for 32 key structures using Dice, area-weighted Dice, IoU, and HD95 as shown in Table 2. SAVer achieves macro Dice scores ranging from 68.76% to 72.58%, with the highest performance on the TV plane. The 32 structures vary considerably in size, from large regions to small fine-grained anatomies, and the lower macro Dice is mainly due to the inclusion of these smaller, inherently challenging structures as seen in Fig. 3. In contrast, area-weighted Dice consistently exceeds 95% across all planes, demonstrating robust performance on clinically dominant structures and effective mitigation of class imbalance. IoU values remain stable, indicating consistent regional overlap despite anatomical variability, while HD95 distances are confined to 17–18 pixels, reflecting stable boundary delineation with limited extreme contour deviations. Overall, this strong segmentation consistency supports the model's anatomical reasoning for standard plane validation, as accurate localization of required structures enables reliable completeness and presence verification.

Table 2. Anatomical Segmentation Performance

Plane	Dice(%)↑	Weighted Dice(%)↑	IOU↑	HD95(px)↓
TC	68.76	95.44	58.01	18.25
TT	69.67	95.52	58.88	17.39
TV	72.58	96.81	63.86	18.25

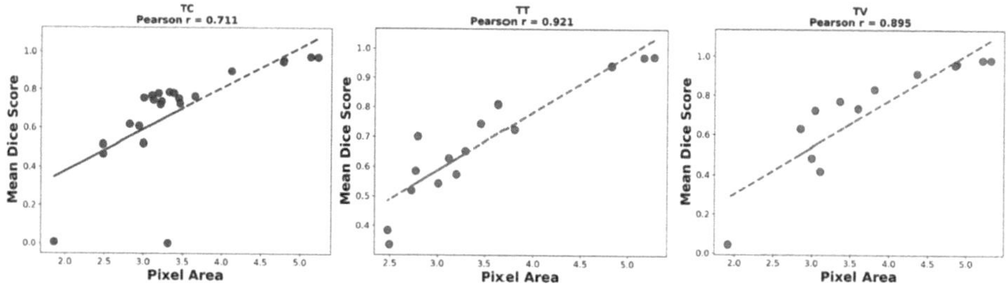

Fig. 3. Size–accuracy correlation across planes. Mean Dice score increases with anatomical structure size for TC, TT, and TV planes, with strong positive correlations.

Clinical Reliability Analysis. On the validation set of TV plane consisting of 60 US images, the model achieved 95% accuracy and 96.4% sensitivity for standard-plane detection, with a low unsafe acceptance rate of 1.7% false positives. Confidence analysis showed clear separation between correct and incorrect predictions, with misclassifications concentrated at lower confidence levels, indicating reliable uncertainty estimation. The risk–coverage curve further demonstrated that most errors occur in low-confidence cases; restricting acceptance to the top 80% most confident predictions increased accuracy to 97.9% as can be seen in Fig. 4. These results highlight high sensitivity, low clinical risk, and effective uncertainty awareness, supporting the model's use as a safety-oriented, real-time quality assurance tool for fetal neurosonography.

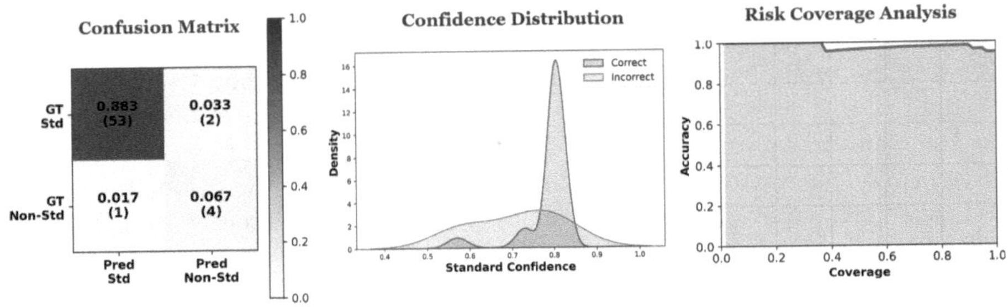

Fig. 4. Clinical Risk and Uncertainty Analysis. Confusion matrix (left) shows low unsafe acceptance. Confidence distributions (middle) separate correct and incorrect predictions, and the risk–coverage curve (right) demonstrates improved accuracy when low-confidence cases are deferred.

Table 3. Ablation study evaluating the contribution of each architectural component to standard-plane detection performance.

Model Variant	Accuracy(%)↑	Recall(%)↑	Precision(%)↑	F1 Score(%)↑	FPR↓	FNR↓
Only CLS	72.80	68.50	75.20	71.70	0.52	0.31
CLS + Seg	77.72	73.95	88.24	81.96	0.30	0.41
CLS + Seg + Struct	82.01	94.38	79.97	88.24	0.32	0.06
CLS + Seg + Struct + Uni	79.91	92.21	77.43	89.23	0.48	0.02
SAVer (Full Model)	**90.61**	**95.68**	**91.40**	**93.95**	**0.29**	**0.03**

Ablation Study. We conducted a systematic ablation study to quantify the utility of each architectural component as shown in Table 3. The baseline only classification model though gives an accuracy of 72.80%, is repressed by high FPR and FNR of 0.52 and 0.31 respectively. This implies that this would lead to false classification of non diagnostic planes. The integration of segmentation and the structure presence vector increased the performance improving the metrics while reducing the FNR upto 0.06. While addition of uniformity slightly reduces the accuracy, it also reduced the FNR to 0.02. The complete SAVer framework attained a peak accuracy of 90.61% and an F1-score of 93.95%. Clinically, this high precision (91.40%) ensures that captured images adhere to anatomical standards, while the concurrent minimization of FPR (0.29) and FNR (0.03) significantly reduces the cognitive load on sonographers and mitigates operator-dependent variability.

4 Conclusion and Future Works

The SAVer framework improves automated ultrasound quality assurance by combining anatomical segmentation with structure-aware reasoning. The ablation study demonstrates the importance of this multi-task design, achieving a peak accuracy of 90.61% and an F1-score of 93.95%. By explicitly verifying the presence and spatial arrangement of clinically relevant anatomical structures, SAVer provides an objective mechanism for standard-plane quality assessment that can support sonographers during image acquisition and reduce variability in plane selection. From a clinical perspective, such automated quality control has the potential to improve workflow efficiency, standardize image acquisition, and provide immediate feedback during fetal ultrasound examinations. Future work will focus on prospective multi-center validation across diverse scanners, operators, and patient populations to establish robustness in real-world settings. We also plan to extend the framework beyond binary standard/non-standard assessment to support graded plane quality evaluation, integrate anomaly-aware analysis of anatomical structures, and optimize the system for real-time deployment within routine clinical workflows.

Disclosure of Interests. The authors have no competing interests to declare that are relevant to the content of this article.

References

1. Alzubaidi, M., Shah, U., Agus, M., Househ, M.: Fetsam: advanced segmentation techniques for fetal head biometrics in ultrasound imagery. IEEE Open J. Eng. Med. Biol. **5**, 281–295 (2024). https://doi.org/10.1109/OJEMB.2024.3382487
2. Avisdris, N., et al.: Biometrynet: landmark-based fetal biometry estimation from standard ultrasound planes. In: Wang, L., Dou, Q., Fletcher, P.T., Speidel, S., Li, S. (eds.) Medical Image Computing and Computer Assisted Intervention - MICCAI 2022, pp. 279–289. Springer, Cham (2022)
3. Bach, S.H., Broecheler, M., Huang, B., Getoor, L.: Hinge-loss Markov random fields and probabilistic soft logic. J. Mach. Learn. Res. **18**(1), 3846–3912 (2017)
4. Baumgartner, C.F., et al.: Sononet: real-time detection and localisation of fetal standard scan planes in freehand ultrasound. IEEE Trans. Med. Imaging **36**(11), 2204–2215 (2017). https://doi.org/10.1109/TMI.2017.2712367

5. Chougule, A., Roy, P., Wagh, M.G., Goyal, D.: Artificial intelligence-based estimation of fetal head circumference with biparietal and occipitofrontal diameters using two-dimensional ultrasound images. J. Eng. Sci. Med. Diagn. Ther. **8**(2), 021103 (2024). https://doi.org/10.1115/1.4065799
6. Dosovitskiy, A., et al.: An image is worth 16x16 words: transformers for image recognition at scale. In: International Conference on Learning Representations (2021). https://openreview.net/forum?id=YicbFdNTTy
7. Gao, P., et al.: Clip-adapter: better vision-language models with feature adapters. Int. J. Comput. Vis. **132**(2), 581–595 (2024). https://doi.org/10.1007/s11263-023-01891-x
8. Kompella, G., et al.: Segmentation of femoral cartilage from knee ultrasound images using mask R-CNN. In: 2019 41st Annual International Conference of the IEEE Engineering in Medicine and Biology Society (EMBC), pp. 966–969 (2019). https://doi.org/10.1109/EMBC.2019.8857645
9. Krishna, T.B., Poreddy, A.K.R., Sindhu, K.G., Kokil, P.: Automated maternal fetal ultrasound image identification using a hybrid vision transformer model. In: Antonacopoulos, A., Chaudhuri, S., Chellappa, R., Liu, C.L., Bhattacharya, S., Pal, U. (eds.) Pattern Recognition, pp. 253–269. Springer, Cham (2025)
10. Loshchilov, I., Hutter, F.: Decoupled weight decay regularization. In: International Conference on Learning Representations (ICLR) (2019). https://arxiv.org/abs/1711.05101
11. Ma, J., He, Y., Li, F., Han, L., You, C., Wang, B.: Segment anything in medical images. Nat. Commun. **15**(1), 654 (2024). https://doi.org/10.1038/s41467-024-44824-z
12. Ma, L., Li, K., Xiao, Z., Tan, G., Wen, H., Li, S.: Structure-aware fine-grained instance segmentation for fetal brain ultrasound images. Neurocomputing 133093 (2026). https://doi.org/10.1016/j.neucom.2026.133093. https://www.sciencedirect.com/science/article/pii/S092523122600490X
13. Murugesan, B., Vijaya Raghavan, S., Sarveswaran, K., Ram, K., Sivaprakasam, M.: Recon-glgan: a global-local context based generative adversarial network for MRI reconstruction. In: International Workshop on Machine Learning for Medical Image Reconstruction, pp. 3–15. Springer (2019)
14. Salomon, L.J., et al.: Isuog practice guidelines (updated): performance of the routine mid-trimester fetal ultrasound scan. Ultrasound Obstet. Gynecol. **59**(6), 840–856 (2022). https://doi.org/10.1002/uog.24888. https://obgyn.onlinelibrary.wiley.com/doi/abs/10.1002/uog.24888
15. Tehsin, S., Alshaya, H., Bouchelligua, W., Nasir, I.M.: Hybrid state–space and vision transformer framework for fetal ultrasound plane classification in prenatal diagnostics. Diagnostics **15**(22) (2025). https://doi.org/10.3390/diagnostics15222879. https://www.mdpi.com/2075-4418/15/22/2879
16. Wang, Q., Zhao, D., Ma, H., Liu, B.: Fb-zwunet: a deep learning network for corpus callosum segmentation in fetal brain ultrasound images for prenatal diagnostics. Biomed. Signal Process. Control **104**, 107499 (2025). https://doi.org/10.1016/j.bspc.2025.107499. https://www.sciencedirect.com/science/article/pii/S1746809425000102
17. Yang, C., Zhang, C., Yang, X., Li, Y.: Performance study of CBAM attention mechanism in convolutional neural networks at different depths. In: 2023 IEEE 18th Conference on Industrial Electronics and Applications (ICIEA), pp. 1373–1377 (2023). https://doi.org/10.1109/ICIEA58696.2023.10241832

18. Yousefpour Shahrivar, R., Karami, F., Karami, E.: Enhancing fetal anomaly detection in ultrasonography images: a review of machine learning-based approaches. Biomimetics **8**(7) (2023). https://doi.org/10.3390/biomimetics8070519. https://www.mdpi.com/2313-7673/8/7/519
19. Zhao, L., Li, K., Pu, B., Chen, J., Li, S., Liao, X.: An ultrasound standard plane detection model of fetal head based on multi-task learning and hybrid knowledge graph. Future Gener. Comput. Syst. **135**, 234–243 (2022). https://doi.org/10.1016/j.future.2022.04.011. https://www.sciencedirect.com/science/article/pii/S0167739X22001340

Scalable Training of Spatially Grounded 2D Vision–Language Models for Radiology

Yusuf Salcan[1,4], Simon Ging[1,2(✉)], Robin Tibor Schirrmeister[3], Philipp Arnold[3], Elmar Kotter[3], Behzad Bozorgtabar[2], and Thomas Brox[1]

[1] Computer Vision Group, University of Freiburg, Freiburg im Breisgau, Germany
gings@cs.uni-freiburg.de
[2] Adaptive & Agentic AI (A3) Lab, Aarhus University, Aarhus, Denmark
[3] Department of Radiology, Medical Center – University of Freiburg, Freiburg im Breisgau, Germany
[4] CRIION-AI Lab, Freiburg im Breisgau, Germany

Abstract. We study how to train visually grounded vision–language models (VLMs) for radiology without manual spatial annotations. We introduce *RefRad2D*, a large-scale bilingual (German/English) dataset of 1.2M CT and MRI image–text pairs derived from clinical practice, with task-specific VQA and spatial grounding subsets generated automatically via LLM-based curation and automated segmentation. Trained on this data, our model *RadGrounder* jointly performs report generation, visual question answering, and spatial grounding via bounding-box detection or segmentation. On external VQA benchmarks (Slake, VQA-RAD), RadGrounder achieves competitive results with specialized medical VLMs. Adding our clinical data to the training mixture improves open-ended VQA over fine-tuning on the downstream datasets alone, showing the transferability of our dataset. Crucially, adding grounding supervision does not degrade language quality, enabling spatially verifiable outputs at no cost to VQA performance. Code and models are available at https://github.com/lmb-freiburg/radgrounder.

Keywords: Radiology VLM · Visual grounding · CT · MRI

1 Introduction

Vision–language models (VLMs) for radiology can generate coherent medical text, but a significant limitation remains: they cannot reliably ground their output in specific image regions. Without spatial grounding, predictions are difficult to verify against the underlying image, which limits clinical trust. Training such models for CT and MRI is further hindered by data scarcity. Approaches that consume 2D slices offer a scalable alternative to 3D volumetric methods, yet

Y. Salcan and S. Ging—Equal contribution.
B. Bozorgtabar and T. Brox—Equal supervision.

G. Yang et al. (Eds.): MICCAI 2026, LNCS 16878, pp. 558–568, 2027.
https://doi.org/10.1007/978-3-032-38059-3_53

prior efforts have struggled with scale; for instance, Med-Gemini authors [16,24] tailored a 2D CT slice dataset but their strict filtering yielded only 4,009 images, leaving 2D slice supervision at scale unexplored.

Recent medical VLMs achieve strong performance on radiology benchmarks, from lightweight generalists like BiomedGPT [26] and LLaVA-Med [11] to large-scale models like RadFM [22], which supports both 2D and 3D inputs. Most work on spatially grounded medical VLMs has focused on chest X-rays, including CheXagent [3] and MAIRA-2 [2]. VividMed [13] extended grounding to broader medical imaging, but relies on synthetic data from open datasets. Grounded VLMs trained on large-scale clinical CT and MRI data remain underexplored.

In this work, we introduce RadGrounder, a PaliGemma 2-based [19] multi-task VLM that performs report generation, VQA, and visual grounding on CT and MRI slices. To train it, we build RefRad2D, a bilingual (German/English) corpus of 1.2M image–text pairs derived from clinical routine via an LLM-driven curation pipeline, with spatial labels from TotalSegmentator [1,21].

Contributions:.

- We present RefRad2D, a large-scale bilingual (German/English) 2D CT/MRI dataset for VLM training with 1.2M image–text pairs and automatically derived spatial grounding annotations.
- We introduce RadGrounder, a multi-task model that jointly performs VQA and visual grounding. Built on standard VLM components, it shows that 2D slice-level supervision from routine reports is a scalable strategy for training radiology VLMs. Code and pretrained models are publicly available.
- We ablate training configurations and grounding strategies, finding that (1) token-based bounding box prediction provides effective localization without requiring an auxiliary segmentation head or additional loss terms, and (2) adding spatial grounding supervision does not degrade VQA or report generation quality.
- We validate our approach on external benchmarks (VQA-RAD [10], Slake [12]), showing competitive performance with other medical VLMs while enabling verifiable, spatially grounded interpretations.

2 Dataset

We introduce RefRad2D, a bilingual dataset with visual grounding annotations, derived from clinical logs at a university hospital. The dataset comprises 1.2 million unique image–caption pairs extracted from 945k CT and 321k MRI slices, representing a decade of clinical practice [18].

Preprocessing and Bilingual Expansion. Raw reports often contain longitudinal references (e.g., "lesion has grown") that induce hallucinations in single-frame VLMs. We used GPT-OSS (120B) [15] to rewrite captions, removing temporal context while preserving current findings [20]. To address Anglo-centric bias [14], we translated all German reports into English using Gemma 3 (27B) [6]. To refine translation quality, we sampled outputs and judged them with a stronger model, GPT-OSS (120B), iteratively improving the translation prompt based on identified errors [4]. We train on a mix of both languages.

Automated Anatomical Grounding. To generate dense, pixel-level annotations without manual labeling, we developed an automated pipeline using 181,362 CT and 36,026 MRI 3D volumes. This produced the RefRad2D-Grounded subset: 236,157 grounded slice–text pairs (217,692 CT, 18,465 MRI).

Anatomical Segmentation. We processed full 3D volumes using TotalSegmentator [1,21] to generate masks for 117 CT and 50 MRI classes. These were harmonized into a unified schema of $C = 121$ classes by merging shared anatomies. Volumes were sliced to match the VLM's 2D input format.

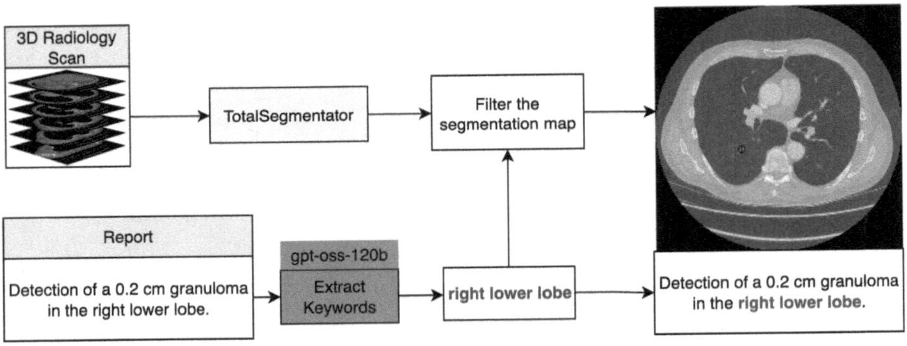

Fig. 1. Overview of the components to generate the data for visual grounding. We use TotalSegmentator [1,21] to create 3D segmentation masks. Via slicing and LLM-based keyword extraction we create dense 2D pixel-level annotations. A strict set-intersection filter ($C_T \cap C_I \neq \emptyset$) ensures high-quality matching between textual entities and visual regions.

Keyword-Mask Matching. To link visual regions to text (Fig. 1), we used GPT-OSS (120B) to extract anatomical mentions from each caption and map them to the same $C = 121$ class schema used by TotalSegmentator. Let $C_T \subseteq \{1, \ldots, C\}$ denote the classes mentioned in caption T, C_I the classes with masks in slice I, and M_I^T the segmentation mask for this class in this slice. We construct a valid training triplet (I, T, M_I^T) only if $C_T \cap C_I \neq \emptyset$. For detection, we convert the masks into bounding boxes by taking the box that covers the mask.

Synthetic VQA Generation. For VQA training, we generated the RefRad2D-VQA dataset (~9.6M pairs). Using Gemma 3 (27B), we derived 5 QA pairs per image directly from clinical findings (Open, Yes/No, Multiple Choice) and 3 QA pairs from the slice metadata [11].

3 Method

We propose RadGrounder, a multi-task Vision–Language Model (VLM) that generates radiology reports and localizes findings through bounding-box detection or segmentation.

Architecture. Our architecture builds upon the PaliGemma 2 (3B) framework [19], comprising a SigLIP-So400m [25] vision encoder and a Gemma-2B language decoder [5]. The model processes an image $I \in \mathbb{R}^{H \times W \times 3}$ into visual tokens Z_v, which are concatenated with text tokens Z_t and are processed autoregressively. We investigate two grounding strategies on top of this foundation: token-based bounding-box detection, which the model can perform natively via text generation, and an auxiliary segmentation head.

Bounding-Box Detection. For spatial grounding, we treat detection as a text-generation task. We extend the vocabulary with coordinate tokens discretized into 512 bins and class-identifying tokens. The model generates a structured sequence:

`<p bbox> [LOC] id=<segID> KEYWORD </p>` (1)

where `[LOC]` denotes the bounding box coordinates $\langle y_{min}, x_{min}, y_{max}, x_{max} \rangle$ and `id` maps to our unified schema of $C = 121$ anatomical classes. To resolve ambiguity in slices containing multiple instances of the same organ (e.g., multiple lymph nodes), we employ a class-wise merging strategy, predicting a single union bounding box for all instances of a class.

Segmentation Head. We also explore pixel-level grounding via a lightweight mask decoder following VividMed [13] and SAM [9]. When the model generates a `</seg>` token, its hidden state is projected into a 256-dim prompt that drives a Two-Way Transformer [9] over the image embeddings to produce a binary mask. Unlike detection, this approach requires an auxiliary decoder and an additional segmentation loss.

Training Objective. The model is trained end-to-end with auto-regressive cross-entropy $\mathcal{L}_{txt}$ for next-token prediction. Detection grounding is trained entirely through $\mathcal{L}_{txt}$, as bounding-box coordinates are generated as tokens. For the segmentation variant, we add an auxiliary loss:

$$\mathcal{L} = \mathcal{L}_{txt} + \lambda_{seg}(\mathcal{L}_{focal} + \mathcal{L}_{dice}) \quad (2)$$

where $\mathcal{L}_{focal}$ and $\mathcal{L}_{dice}$ handle the class imbalance inherent in medical segmentation. We employ a dynamic sampling strategy where batches alternate between report generation, VQA, and grounded tasks.

LLMScore. Standard n-gram metrics (e.g., CIDEr) struggle with semantic equivalence and the complexities of medical text. We developed LLMScore by adopting the "LLM-as-a-judge" paradigm [7,27]. We use Gemma 3 (27B) as the evaluator to compare the generated report against the ground truth. The judge assesses clinical factuality and semantic correctness on a 5-point scale normalized to [0, 1] and outputs a brief textual justification for its decision. To validate this metric, three radiologists independently scored 200 QA pairs from VQA-RAD. Inter-annotator agreement was high (ordinal Krippendorff's $\alpha = 0.958$) [8], and LLMScore achieved a Pearson correlation of $r = 0.977$ with the mean human scores, indicating strong alignment with expert judgment.

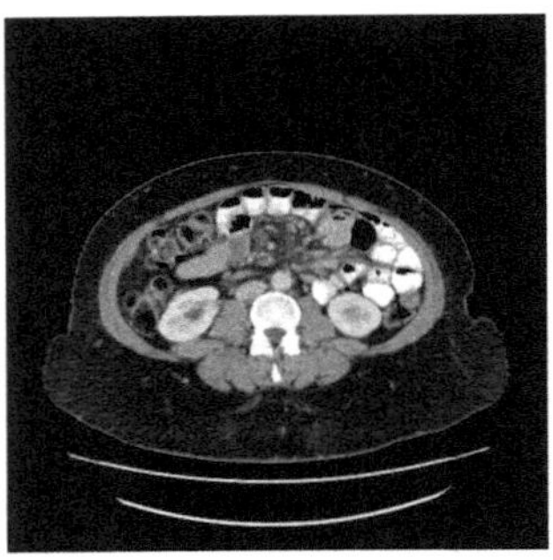

Prompt: <image> Clinical information: Status post sleeve gastrectomy, bypass, with increasing abdominal pain. Question: Internal hernia? Caption this image:
Ground Truth: Nonspecific increased mesenteric lymph nodes, up to 10mm.
Prediction: Increased mesenteric lymph nodes up to 1.2 cm.
Metrics:
CIDEr: 1.0877
LLM Score: 0.7500
LLM Reason: The candidate and reference both identify increased mesenteric lymph nodes. The sizes are very close (1.2cm vs 10mm = 1cm), indicating a mostly right answer.

Fig. 2. Qualitative example of RadGrounder's report generation on an abdominal CT slice (soft tissue window). The model successfully identifies increased mesenteric lymph nodes and accurately estimates their size. Despite a low traditional CIDEr score (1.08), the LLM-as-a-judge correctly assigns a high factual score (0.75) by recognizing the semantic equivalence of the measurements (1.2 cm vs. 10 mm).

Spatial-Semantic Evaluation (G-IoU). Standard Intersection over Union (IoU) assumes a fixed vocabulary and fails when a model generates a keyword that is semantically correct but lexically distinct from the ground truth. We propose Grounding-IoU (G-IoU) to jointly measure spatial and semantic fidelity. We first match predicted and ground-truth entities using the cosine similarity of their text embeddings. The G-IoU is then computed as the sum of their spatial IoUs weighted by this semantic similarity. The score is normalized by the maximum number of predicted or ground-truth entities, penalizing both spatial hallucinations (false positives) and diagnostic omissions (false negatives). We then average the per-sample G-IoU over all samples in the respective test set.

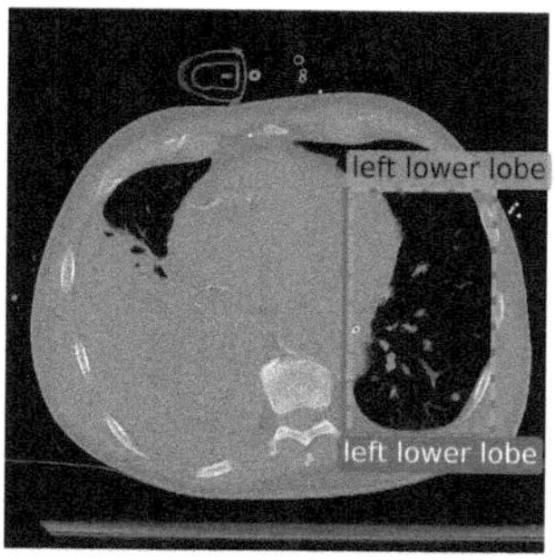

Prompt: <image> Clinical information: Severe pneumonia in a patient with COPD Gold 4. Question: Extent of the pneumonia? Caption this image and detect keywords:
Ground Truth: Approximately 6 mm measuring round lesion in the left lower lobe.
Prediction: Nodular densities in the left lung, exemplified in the left lower lobe with 5 mm, most likely postinflammatory.
Metrics:
CIDEr: 0.1373
LLM Score: 0.7500
LLM Reason: The candidate and reference both identify a lesion in the left lower lobe and provide similar size estimations (5mm vs 6mm). The candidate adds 'nodular densities' and 'postinflammatory' which aren't explicitly in the reference but are plausible given the clinical context.
G-IoU: 0.9497
mIoU: 0.9497

Fig. 3. Detection grounding performance on a chest CT (lung window). RadGrounder accurately localizes a nodular lesion in the left lower lobe, generating a bounding box that strongly aligns spatially and semantically with the ground truth (G-IoU: 0.9497).

Table 1. Quantitative results on the internal RefRad2D test set for Report Captioning, VQA, and spatial grounding tasks. The table compares our RadGrounder detection model (row 4) against PaliGemma 2 [19] (PaliG. 2), Gemma 3 [6], and MedGemma [17] (MedG.) (top), and ablates various training configurations (bottom). SL.: SigLIP vision encoder. We highlight **values within the confidence interval of the highest score** separately for the baseline comparison and the ablations.

Training data						Report		VQA			Grounded
Report	VQA	Slake	VQA-RAD	Detect.	Segment	CIDEr	LLM	F1 Open	LLM Open	Acc. Closed	G-IoU
PaliG. 2 (3B) [19]						0.6 0.1	2.6 0.5	12.7 1.9	22.8 2.3	51.5 2.7	–
Gemma 3 (4B) [6]						0.1 0.0	21.7 1.3	5.5 0.8	31.8 2.7	13.0 1.6	–
MedG. (4B) [17]						0.4 0.1	21.8 1.3	9.0 1.2	40.6 2.9	16.6 2.2	–
✓	✓	✓	✓	✓		**37.5** 3.3	**30.3** 1.4	**48.9** 2.8	**60.1** 2.6	**94.6** 1.4	**43.6** 1.6
+ Unfreeze finet. SL.						34.6 3.3	28.0 1.4	**47.8** 2.8	60.1 2.5	**95.1** 1.5	43.4 1.7
+ Unfreeze orig. SL.						**36.9** 3.2	**29.9** 1.4	**49.2** 2.8	60.8 2.6	94.0 1.5	**45.2** 1.7
✓	✓	✓	✓		✓	**36.9** 3.2	**30.1** 1.4	**50.4** 2.8	**64.1** 2.7	**95.6** 1.3	36.9 1.6
✓	✓					**39.0** 3.4	**30.7** 1.5	**49.2** 2.8	60.3 2.6	**94.5** 1.5	–
		✓	✓			0.1 0.0	1.5 0.5	15.7 2.2	21.8 2.5	32.4 2.2	–
✓	✓	✓	✓			**37.3** 3.3	**30.6** 1.4	**50.1** 2.8	61.2 2.6	**94.4** 1.5	–
✓	✓			✓		**39.2** 3.3	**30.6** 1.5	**49.1** 2.8	60.2 2.6	**94.5** 1.5	**43.6** 1.6

4 Experiments

Table 2. Performance on Slake [12] and VQA-RAD [10]. SL.: SigLIP vision encoder, PaliG.: PaliGemma, MedG.: MedGemma. We used the train/test split from Xu et al. [23] due to overlap between train and test images in the original VQA-RAD splits [24]. *Reported results are not available for that split. We highlight **values within the confidence interval of the highest score** separately for the baseline comparison and the ablations.

Training data						Slake				VQA-RAD		
Report	VQA	Slake	VQA-RAD	Detect.	Segment	F1	Recall Open	LLM Open	Acc. Closed	F1	LLM Open	Acc. Closed
PaliG. 2 (3B) [19]						24.5	24.3	34.2 3.5	58.1	37.9	26.0 4.1	55.6
Gemma 3 (4B) [6]						40.2	33.3	22.3 3.2	53.0	33.6	11.1 3.5	33.6
MedG. (4B) [17]						72.3	63.3	54.9 3.8	**87.6**	**49.9**	20.8 4.2	69.1
BiomedGPT-B [26]						85.2	–	–	**89.9**	*	–	*
LLaVA-Med [11]						–	**87.1**	–	86.8	*	–	*
RadFM (14B) [22]						84.4	–	–	–	*	–	–
Med-Gemini [24]						75.8	–	–	84.8	*	–	**78.8**
✓	✓	✓	✓	✓		**87.7** 2.4	**88.1** 2.4	**88.8** 2.3	**90.3** 2.7	**50.7** 3.5	**44.3** 4.8	64.7 4.6
Unfreeze finet. SL.						**88.4** 2.4	**88.9** 2.4	**89.9** 2.3	**90.8** 2.6	**48.7** 3.4	**44.3** 4.8	62.7 4.6
Unfreeze orig. SL.						**87.6** 2.3	**88.2** 2.3	**88.6** 2.2	**91.7** 2.7	**50.2** 3.4	**44.9** 4.8	**64.5** 4.6
✓	✓	✓	✓		✓	**86.4** 2.5	**86.9** 2.5	87.3 2.5	**89.6** 2.8	**50.4** 3.5	**46.6** 5.2	**65.9** 4.4
✓	✓					19.2 2.9	20.3 3.0	31.9 3.4	63.0 4.5	39.6 3.4	33.0 4.5	52.9 4.7
		✓	✓			**86.9** 2.5	**87.6** 2.5	**87.8** 2.4	88.4 3.2	**47.9** 3.4	28.6 4.2	**68.1** 4.4
✓	✓	✓	✓			**87.2** 2.5	**87.6** 2.4	**88.8** 2.3	**90.8** 2.7	**49.2** 3.4	**42.8** 4.7	63.2 4.5
✓	✓			✓		18.3 2.8	20.2 3.0	31.4 3.4	58.8 4.7	40.2 3.4	35.9 4.5	52.9 4.9

Implementation Details. All models were trained on a single NVIDIA H100 GPU for 6 epochs with a batch size of 24 and 12 gradient accumulation steps, using the Adafactor optimizer with a learning rate of 5e-5. We apply random scale cropping, contrast adjustment, and intensity shifts as data augmentation. Training with the frozen SigLIP encoder on the full dataset mixture took approximately 2.5 days.

Datasets and Protocols. The RefRad2D training split contains 760,409 CT and 256,197 MRI image–report pairs. The RefRad2D-VQA subset provides 8 QA pairs per image, and the RefRad2D-Grounded training subset contains 174,655 CT and 14,721 MRI spatially grounded triplets. All internal evaluation sets were standardized to 2,000 test samples. For external benchmarking, we use Slake [12] (9,849 training/2,070 testing QA pairs) and VQA-RAD [10] (3,064 training/451 testing QA pairs). For open-ended evaluation we use LLMScore (Sect. 3). We report 95% bootstrap confidence intervals (B=10,000) over test set samples.

Vision Encoder Strategy (Tables 1, 2). Before multi-task training, we adapt the SigLIP encoder to the medical domain via contrastive pretraining on RefRad2D. We then ablate whether to keep this adapted encoder frozen or unfrozen during the subsequent LLM training, and compare against unfreezing the original (non-adapted) SigLIP weights. All three strategies perform comparably within confidence intervals on both internal and external benchmarks. Since freezing the encoder reduces memory footprint and accelerates training, we adopt it as our standard configuration.

Dataset Mixture and Grounding (Tables 1, 2). Baseline models score near zero on report captioning CIDEr, though LLMScore is less affected (e.g., MedGemma: 0.4 vs. 21.8). CIDEr penalizes any n-gram mismatch with the reference, while LLMScore better captures semantic correctness (Fig. 2). The low baselines are expected: these models were not trained on our data, and the task requires generating hospital-specific text from a single slice without clinical context. Token-based detection (Fig. 3) achieved higher G-IoU (43.6 vs. 36.9) than the segmentation head while requiring no additional loss terms or decoder parameters. Neither grounding strategy degrades VQA or report generation quality compared to the no-grounding baseline. Mixture ablations show clear domain effects: training exclusively on internal data achieved the highest report generation score (CIDEr: 39.2) but failed to generalize to external VQA (Slake F1: 18.3). Conversely, adding our internal corpus to external fine-tuning improved open-ended VQA on most metrics (e.g., Slake F1: 87.7 vs. 86.9; VQA-RAD F1: 50.7 vs. 47.9). The exception was VQA-RAD closed accuracy, where the external-only model scored higher (68.1% vs. 64.7%).

Grounding Error Analysis. The 88 anatomical classes with test support are heavily long-tailed: the five most frequent cover half of all instances. Class frequency correlates only mildly with per-class G-IoU (Pearson r=0.34). The model misses 18.7% of ground-truth findings (false negatives) and produces 13.1% spurious predictions (false positives). Among matched boxes, 54.7% are fully correct. Another 28.6% localize correctly but name the wrong structure, 4.6% name the right structure but mislocalize, and 12.1% get both wrong. Spatial and textual errors correlate only weakly (r=0.29).

Comparison to State-of-the-Art (Tables 2). Using our standard configuration (frozen encoder, full multi-task mixture), RadGrounder achieves competitive performance on external benchmarks. On Slake, our model reaches F1 87.7

and Closed Accuracy 90.3, comparable to BiomedGPT-B [26] (85.2/89.9) and LLaVA-Med [11] (86.8 Acc), while surpassing Med-Gemini [16,24] (75.8/84.8) and MedGemma [17] (72.3/87.6). On VQA-RAD, our model achieves the highest Open F1 (50.7) among compared methods.

5 Conclusion

We presented RadGrounder, a visually grounded VLM for radiology trained without manual spatial annotations, and RefRad2D, a bilingual dataset of 1.2M CT and MRI image–text pairs with automatically derived spatial labels. On external VQA benchmarks, RadGrounder achieves competitive performance with specialized medical VLMs, and adding our clinical data improves over fine-tuning on downstream data alone. We explored two grounding strategies—token-based detection and auxiliary segmentation—finding that detection provides effective localization without additional model components or loss terms, and that neither strategy degrades VQA quality. RadGrounder thus enables spatially verifiable predictions at no cost to language performance.

Limitations. Raw images and reports cannot be shared for privacy reasons. The VQA and grounding annotations are generated automatically and require expert validation before clinical use. Our training data comes from a single hospital, and multi-center validation is needed to confirm generalization. The spatial grounding targets anatomical structures (via TotalSegmentator) rather than pathological findings, limiting clinical utility for lesion-level localization. Finally, comparing detection and segmentation grounding is inherently difficult, as the two approaches solve different geometric tasks (box overlap vs. pixel-level mask agreement), making direct comparison via a single metric nontrivial.

Future Work. Natural extensions include multi-center training to improve generalization and pathology-level grounding using lesion annotations, which would enable clinically actionable spatial predictions beyond anatomical localization. Since we operate on independent 2D slices, enforcing inter-slice spatial coherence for volumetric consistency is another natural direction.

Acknowledgments. This research was funded by the Deutsche Forschungsgemeinschaft (DFG, German Research Foundation) 417962828, 539134284, and 499552394 – SFB 1597 – Small Data, as well as through EFRE (FEIH_2698644) and the state of Baden-Württemberg.

Baden-Württemberg Co-funded by the European Union

References

1. Akinci D'Antonoli, T., Berger, L.K., Indrakanti, A.K., Vishwanathan, N., Weiss, J., et al.: TotalSegmentator MRI: robust sequence-independent segmentation of multiple anatomic structures in MRI. Radiology **314**(2) (2025). https://doi.org/10.1148/radiol.241613
2. Bannur, S., Bouzid, K., Castro, D.C., Schwaighofer, A., Thieme, A.: MAIRA-2: grounded radiology report generation. arXiv:2406.04449 (2024)
3. Chen, Z., Varma, M., Delbrouck, J.B., Paschali, M., Blankemeier, L.: CheXagent: towards a foundation model for chest X-ray interpretation. In: AAAI 2024 Spring Symposium Series on Clinical Foundation Models (2024)
4. Dubois, Y., Li, X., Taori, R., Zhang, T., Gulrajani, I.: AlpacaFarm: a simulation framework for methods that learn from human feedback. Adv. Neural Inf. Process. Syst. **36** (2023)
5. Team, G.: Gemma: open models based on Gemini research and technology. arXiv:2403.08295 (2024)
6. Team, G.: Gemma 3 technical report. arXiv:2503.19786 (2025)
7. Ging, S., Bravo, M.A., Brox, T.: Open-ended VQA benchmarking of vision-language models by exploiting classification datasets and their semantic hierarchy. In: The Twelfth International Conference on Learning Representations (2024)
8. Hayes, A.F., Krippendorff, K.: Answering the call for a standard reliability measure for coding data. Commun. Methods Meas. **1**(1), 77–89 (2007). https://doi.org/10.1080/19312450709336664
9. Kirillov, A., Mintun, E., Ravi, N., Mao, H., Rolland, C., et al.: Segment anything. In: Proceedings of the IEEE/CVF International Conference on Computer Vision, pp. 3992–4003. IEEE (2023). https://doi.org/10.1109/ICCV51070.2023.00371
10. Lau, J.J., Gayen, S., Ben Abacha, A., Demner-Fushman, D.: A dataset of clinically generated visual questions and answers about radiology images. Sci. Data **5**(1), 180251 (2018). https://doi.org/10.1038/sdata.2018.251
11. Li, C., Wong, C., Zhang, S., Usuyama, N., Liu, H., et al.: Llava-med: training a large language-and-vision assistant for biomedicine in one day. In: Advances in Neural Information Processing Systems, Datasets and Benchmarks Track, vol. 36 (2023)
12. Liu, B., Zhan, L.M., Xu, L., Ma, L., Yang, Y., Wu, X.M.: Slake: a semantically-labeled knowledge-enhanced dataset for medical visual question answering. In: 2021 IEEE 18th International Symposium on Biomedical Imaging (ISBI), pp. 1650–1654. IEEE (2021). https://doi.org/10.1109/ISBI48211.2021.9434010
13. Luo, L., Tang, B., Chen, X., Han, R., Chen, T.: VividMed: vision language model with versatile visual grounding for medicine. In: Proceedings of the 2025 Conference of the Nations of the Americas Chapter of the Association for Computational Linguistics: Human Language Technologies, vol. 1: Long Papers, pp. 1800–1821. Association for Computational Linguistics, Albuquerque (2025). https://doi.org/10.18653/v1/2025.naacl-long.89
14. Névéol, A., Dalianis, H., Velupillai, S., Savova, G., Zweigenbaum, P.: Clinical natural language processing in languages other than English: opportunities and challenges. J. Biomed. Semant. **9**(1), 12 (2018). https://doi.org/10.1186/s13326-018-0179-8
15. OpenAI: gpt-oss-120b & gpt-oss-20b model card. arXiv:2508.10925 (2025)
16. Saab, K., Tu, T., Weng, W.H., Tanno, R., Stutz, D.: Capabilities of Gemini models in medicine. arXiv:2404.18416 (2024)

17. Sellergren, A., Kazemzadeh, S., Jaroensri, T., Kiraly, A., Traverse, M.: MedGemma technical report. arXiv:2507.05201 (2025)
18. Smith-Bindman, R., Kwan, M.L., Marlow, E.C., Theis, M.K., Bolch, W., et al.: Trends in use of medical imaging in US health care systems and in Ontario, Canada, 2000–2016. JAMA **322**(9), 843–856 (2019). https://doi.org/10.1001/jama.2019.11456
19. Steiner, A., et al.: PaliGemma 2: a family of versatile VLMs for transfer. arXiv:2412.03555 (2024)
20. Thirunavukarasu, A.J., Ting, D.S.J., Elangovan, K., Gutierrez, L., Tan, T.F., Ting, D.S.W.: Large language models in medicine. Nat. Med. **29**(8), 1930–1940 (2023). https://doi.org/10.1038/s41591-023-02448-8
21. Wasserthal, J., Breit, H.C., Meyer, M.T., Pradella, M., Hinck, D., et al.: TotalSegmentator: robust segmentation of 104 anatomic structures in CT images. Radiol. Artif. Intell. **5**(5), e230024 (2023). https://doi.org/10.1148/ryai.230024
22. Wu, C., Zhang, X., Zhang, Y., Hui, H., Wang, Y., Xie, W.: Towards generalist foundation model for radiology by leveraging web-scale 2D&3D medical data. Nat. Commun. **16**(1), 7866 (2025). https://doi.org/10.1038/s41467-025-62385-7
23. Xu, S., Yang, L., Kelly, C., Sieniek, M., Kohlberger, T.: ELIXR: towards a general purpose X-ray artificial intelligence system through alignment of large language models and radiology vision encoders. arXiv:2308.01317 (2023)
24. Yang, L., Xu, S., Sellergren, A., Kohlberger, T., Zhou, Y.: Advancing multimodal medical capabilities of Gemini. arXiv:2405.03162 (2024)
25. Zhai, X., Mustafa, B., Kolesnikov, A., Beyer, L.: Sigmoid loss for language image pre-training. In: Proceedings of the IEEE/CVF International Conference on Computer Vision, pp. 11941–11952. IEEE (2023). https://doi.org/10.1109/ICCV51070.2023.01100
26. Zhang, K., Zhou, R., Adhikarla, E., Yan, Z., Liu, Y., et al.: A generalist vision-language foundation model for diverse biomedical tasks. Nat. Med. **30**(11), 3129–3141 (2024). https://doi.org/10.1038/s41591-024-03185-2
27. Zheng, L., Chiang, W.L., Sheng, Y., Zhuang, S., Wu, Z.: Judging LLM-as-a-judge with MT-Bench and chatbot arena. In: Advances in Neural Information Processing Systems, Datasets and Benchmarks Track, vol. 36 (2023)

Seeing Through Multiple Views: Parameter-Efficient Fine-Tuning via Selective Neurons for Consistent Radiology Report Generation

Yucheng Chen[1,2,3], Jinjing Zhu[4], Yang Yu[5], Yufei Shi[1,2], Hane Naghshbandi[1,2,3], Jinhua Liu[1,2], Angela S. Koh[6], Fang Fen[5], Kian Eng Ong[1], and Si Yong Yeo[1,2,3](✉)

[1] MedVisAI Lab, Singapore, Singapore
siyongyeo@ntu.edu.sg
[2] Lee Kong Chian School of Medicine, Nanyang Technological University, Singapore, Singapore
[3] Centre of AI in Medicine, Singapore, Singapore
[4] The Chinese University of Hong Kong, Sha Tin, New Territories, Hong Kong
[5] Institute for Infocomm Research (I^{2}R), A*STAR, Singapore, Singapore
[6] National Heart Centre Singapore (NHCS), Singapore, Singapore

Abstract. Recent years have seen substantial advances in radiology report generation (RRG), yet existing approaches predominantly adopt direct feature fusion when handling multi-view X-ray images. Such approaches overlook the potential clinical inconsistencies and inaccuracies arising when a single model processes different views, adversely impacting performance and clinical reliability. To this end, we introduce **View-PNDF** (**View**-specific **P**attern **N**euron **D**etection and **F**ine-tuning), a parameter-efficient framework that fosters view-consistent report generation from a neuronal perspective. Specifically, View-PNDF comprises: (i) a view-specific neuron detection module identifying neurons responsive to particular views, (ii) a verification module quantifying the existence of these neurons, and (iii) a selective fine-tuning strategy strengthening detected neurons while preserving view-agnostic representations. By updating only view-specific neurons, View-PNDF achieves consistent diagnoses across different views with reduced computational costs. Subsequently, we employ Large Language Models (LLMs) to consolidate the view-specific reports into a complete radiology report. Furthermore, we use traditional Natural Language Generation (NLG) metrics-based assessment on integrated reports for baseline comparison and employ LLM-based assessment (e.g., GPT-4o) on view-specific reports to capture clinical significance. Extensive experiments on two medical RRG benchmarks demonstrate that View-PNDF substantially improves view-specific chest X-ray report generation quality while maintaining robust general-view performance.

Keywords: Radiology Report Generation · Vision-Language Model · Multi-view Consistency

G. Yang et al. (Eds.): MICCAI 2026, LNCS 16878, pp. 569–579, 2027.
https://doi.org/10.1007/978-3-032-38059-3_54

1 Introduction

Radiology report generation (RRG) aims to automatically generate free-text descriptions from radiological images such as X-rays [19] and CT scans [11]. Current RRG approaches can be categorized into single-view and multi-view methods. Single-view models [7,12,15,21,23,28] focus on generating reports from individual radiographic projections but suffer from inherent limitations in capturing comprehensive anatomical information. Multi-view approaches [6,22,34] address these limitations by incorporating information from multiple perspectives - such as frontal, lateral, posteroanterior (PA), and anteroposterior (AP) views - demonstrating empirical improvements over their single-view counterparts. This multi-view paradigm aligns with clinical practice, where radiologists routinely examine each imaging view individually and then integrate these per-view observations into a comprehensive report.

Multi-view medical imaging is clinically essential for identifying posterior abnormalities, localizing lesions, and detecting conditions obscured in single projections [13,24,35,36]. However, existing multi-view RRG methods [6,22,26,34] primarily employ fusion strategies to enhance feature representation, overlooking a critical issue: different views exhibit intrinsic variations that challenge automated systems. Such variations may lead to inconsistent interpretations across views, compromising diagnostic reliability and report quality. Current systems like InternVL-DA-Medical-4B [5] generate reports from frontal views that generally align with GT reports, while lateral view outputs often contain contradictory findings. Such inconsistencies pose significant clinical risks, particularly when frontal images are unavailable due to patient positioning constraints, trauma conditions, or equipment limitations, making lateral views the primary diagnostic input. In settings where frontal and lateral views are provided separately, inconsistent outputs may further lead to confusion, misinterpretation, and unreliable clinical decision-making. This raises a fundamental question: *what underlying mechanisms cause these view-specific discrepancies, and how can we systematically address them?* Without understanding the root causes at the model's internal level, solutions may only address symptoms rather than the core problem. Inspired by [3,27], this motivates neuron-level analysis to identify which computational components are responsible for view-dependent variations, enabling targeted interventions to systematically improve cross-view consistency.

In this paper, we present a comprehensive neuron-level analysis of RRG models, namely **View**-specific **P**attern **N**euron **D**etection and **F**ine-tuning (**View-PNDF**). Rather than retraining the entire model, we first introduce a View-specific Neuron Detection (**VND**) module to identify a compact set of pattern neurons responsible for view-specific inconsistency. We then apply a View-specific Neuron Verification (**VNV**) module to quantitatively evaluate the contribution of these detected neurons by selectively deactivating them and measuring the impact on view-specific report generation. Building on these insights, we further propose a neuron-level adaptation strategy, View-specific Neuron Fine-tuning (**VNF**), which focuses training exclusively on the identified pattern neurons. The targeted fine-tuning enables the model to generate consistent view-specific findings from any available view, while findings from multiple views can be further consolidated into an integrated radiology report using LLMs. For evaluation, we use standard Natural Language Generation (NLG) metrics to assess the quality of the integrated reports. However, evaluating view-specific reports presents unique challenges as conventional NLG metrics fail to capture clinical validity: reports describing different anatomical views may receive similar scores even when they express contradictory clinical findings since NLG metrics measure textual overlap rather than medical semantics. To address this limitation, we employ a LLM–based evaluation method (e.g., GPT-4o) to assess the semantic fidelity and diagnostic accuracy of view-specific reports. By integrating view-aware methodology and appropriate evaluation strategies with RRG, our proposed framework offers a targeted and efficient solution for enhancing multi-view RRG performance.

2 Method

In this section, we introduce **View**-specific **P**attern **N**euron **D**etection and **F**ine-tuning (**View-PNDF**), a framework designed to enhance Radiology Report Generation (RRG) by identifying and optimizing neurons responsible for processing specific X-ray views (e.g., frontal vs. lateral). As illustrated in Fig. 1, our approach consists of three key components: View-specific Neuron Detection (**VND**), Neuron Verification (**VNV**), and Neuron Fine-tuning (**VNF**).

2.1 View-Specific Pattern Neuron Detector (VND)

We hypothesize that within the RRG model - specifically in the linear and self-attention layers of the Transformer decoder - distinct subsets of neurons are specialized for handling different X-ray views. We define these as *View-specific Pattern Neurons.*

Neuron Influence via Perturbation. Ideally, the importance of a neuron can be measured by the impact of its deactivation on the model's output. Formally, let $\mathcal{RRG}(x, \theta)$ denote the generated embedding for a view-specific input sample x (comprising visual features and prompt tokens) given model parameters θ. The influence I of the k-th neuron at the l-th layer, denoted as n_l^k, on a specific view pattern v is defined as the perturbation effect:

$$I(n_l^k|x^v) = \left\|\mathcal{RRG}(x^v, \theta) - \mathcal{RRG}(x^v, \theta \setminus w_l^k)\right\|_2, \quad (1)$$

where $\theta \backslash w_l^k$ represents the model weights with the specific neuron n_l^k deactivated (zeroed out).

Efficient Detection via Attribution Scores. Since calculating Eq. 1 for every neuron across the entire dataset is computationally prohibitive, we follow prior work [8], which suggests that neurons with high activation values are

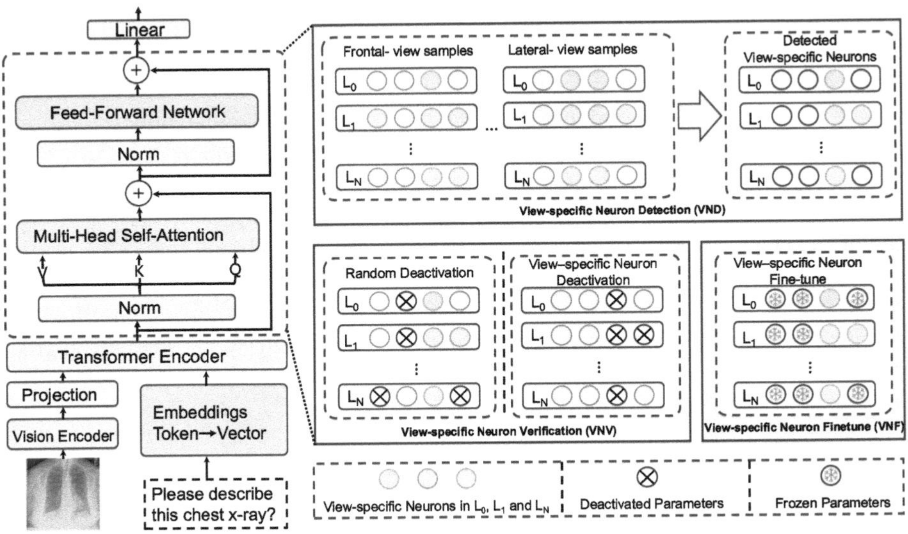

Fig. 1. The overall framework of View-PNDF, which consists of four components: the transformer structure in VLMs for RRG; View-specific Neuron Detection (VND) module identifies neurons associated with the specific views of chest X-rays; View-specific Neuron Verification (VNV) module verifies the detected pattern neurons; and View-specific Neuron Fine-tuning (VNF) module fine-tunes the detected view-specific neurons for RRG network.

strong indicators of pattern importance. We therefore employ an efficient *Attribution Score* as a proxy for detection. The attribution score for neuron n_l^k given input x^v is defined as the magnitude of its activation output:

$$\text{Attr}(n_l^k|x^v) = \left\| \sum_t f(x^v, w_l^k)_t \right\|_1, \tag{2}$$

where $f(\cdot)$ computes the activation values (e.g., the output of a specific linear projection or attention head) across the sequence dimension t.

Detection Strategy. To robustly identify neurons associated with a view v (e.g., lateral), we utilize a set of detection samples $\mathcal{D}_{detect} = \{x_1^v, \dots, x_B^v\}$. A neuron is classified as a *View-specific Pattern Neuron* if it consistently ranks among the most active neurons across all samples in the detection set. The set of pattern neurons $\mathcal{N}^v$ is formally defined as:

$$\mathcal{N}^v = \bigcap_{b=1}^{B} \{n_l^k \mid \text{rank}(\text{Attr}(n_l^k|x_b^v)) \leq \gamma N\}, \tag{3}$$

where $\text{rank}(\cdot)$ sorts neurons in descending order of attribution score, N is the total number of neurons in the layer, and γ is a selection ratio (e.g., top 5%). This intersection strategy ensures we select neurons that are consistently salient for the specific view, filtering out noise from instance-specific variations.

2.2 Verification and Fine-Tuning

View-Specific Neuron Verification (VNV). To validate that the detected set $\mathcal{N}^v$ genuinely controls view-specific generation, we employ the perturbation metric from Eq. 1 on a held-out verification set. We compare the performance drop when deactivating the detected pattern neurons versus deactivating a random set of neurons of the same size. The existence of view-specific neurons is confirmed if:

$$\sum_{d=1}^{D} \mathcal{L}_\Delta(\mathcal{N}_{deact}^{v}) \gg \sum_{d=1}^{D} \mathcal{L}_\Delta(\mathcal{N}_{deact}^{rand}), \tag{4}$$

where $\mathcal{L}_\Delta$ represents the degradation in generation quality (e.g., increase in loss or decrease in clinical accuracy metrics) when neurons are deactivated.

View-Specific Neuron Fine-tuning (VNF). Once identified and verified, we employ a targeted fine-tuning mechanism. Unlike standard fine-tuning methods which updates all parameters, VNF freezes the majority of the model and optimizes *only* the identified pattern neurons $\mathcal{N}^v$. This allows the model to enhance its representation of specific views without catastrophically forgetting general language generation capabilities. The optimization objective is the standard cross-entropy loss over the view-specific fine-tuning set $\mathcal{D}_{ft}$:

$$\mathcal{L}_{VNF} = - \sum_{(x,y)\in\mathcal{D}_{ft}} \sum_{t=1}^{T} \log P(y_t|y_{<t}, x; \theta_{\mathcal{N}^v}), \tag{5}$$

Table 1. NLG performance comparison of the integrated reports and reports generated by existing SOTA RRG methods on IU-XRay dataset. The subscripts MedGemma, InternVL, LLaVA, and Hulu denote that the VLM backbones used are MedGemma-4B, InternVL-DA-Medical-4B, LLaVA-Med-7B, and Hulu-Med-7B, respectively. BL: BLEU, MTR: METEOR, RG-L: ROUGE-L. Higher value indicates better performance.

METHOD	BL-1	BL-2	BL-3	BL-4	MTR	RG-L
R2GEN [7]	0.390	0.293	0.236	0.161	0.187	0.371
R2GENCMN [6]	0.402	0.310	0.225	0.175	0.190	0.365
XPRONet [32]	0.480	0.312	0.225	0.175	0.190	0.364
RGRG [31]	0.373	0.249	0.175	0.126	0.168	0.264
M2KT [34]	0.489	0.315	0.228	0.174	–	0.379
CAMANET [33]	0.478	0.314	0.229	0.179	0.202	0.361
PromptMRG [15]	0.488	0.253	0.237	0.189	0.192	0.374
View-PNDF$_{\text{MedGemma}}$	0.496	0.322	0.240	0.194	0.205	0.385
View-PNDF$_{\text{InternVL}}$	0.492	0.318	0.234	0.191	0.198	0.382
View-PNDF$_{\text{LLaVA}}$	0.505	0.330	0.248	0.201	0.212	0.392
View-PNDF$_{\text{Hulu}}$	0.510	0.335	0.252	0.205	0.216	0.396
Average	**0.501**	**0.326**	**0.244**	**0.198**	**0.208**	**0.389**

Table 2. NLG and CE performance comparison of the integrated reports and reports generated by existing SOTA methods on MIMIC-CXR dataset. The subscripts MedGemma, InternVL, LLaVA, and Hulu denote that the VLM backbones used are MedGemma-4B, InternVL-DA-Medical-4B, LLaVA-Med-7B, and Hulu-Med-7B, respectively. BL: BLEU, MTR: METEOR, RG-L: ROUGE-L, CE: Clinical Efficacy, Prec.: Precision, Rec.: Recall.

METHOD	NLG Metrics						CE Metrics		
	BL-1	BL-2	BL-3	BL-4	MTR	RG-L	Prec.	Rec.	F1
R2GEN [7]	0.353	0.218	0.145	0.103	0.142	0.270	0.406	0.203	0.204
R2GENCMN [6]	0.348	0.206	0.135	0.094	0.136	0.266	0.440	0.325	0.374
XPRONet [32]	0.344	0.215	0.146	0.105	0.138	0.279	0.463	0.285	0.353
M2KT [34]	0.368	0.220	0.149	0.107	–	0.259	0.420	0.339	0.352
CAMANET [33]	0.367	0.219	0.151	0.107	0.142	0.279	0.467	0.323	0.381
PromptMRG [15]	0.369	0.223	0.142	0.099	0.147	0.260	0.471	0.198	0.352
MLRG [22]	0.398	0.236	0.159	0.156	0.169	0.278	0.491	0.398	0.401
View-PNDF$_{\text{MedGemma}}$	0.408	0.245	0.166	0.164	0.174	0.286	0.502	0.415	0.408
View-PNDF$_{\text{InternVL}}$	0.402	0.240	0.162	0.160	0.168	0.282	0.495	0.405	0.398
View-PNDF$_{\text{LLaVA}}$	0.418	0.254	0.174	0.171	0.182	0.294	0.515	0.432	0.420
View-PNDF$_{\text{Hulu}}$	0.425	0.260	0.179	0.176	0.188	0.301	0.522	0.445	0.428
Average	**0.413**	**0.250**	**0.170**	**0.168**	**0.178**	**0.291**	**0.509**	**0.424**	**0.414**

where $\theta_{\mathcal{N}^v}$ denotes that only the parameters of the pattern neurons are updated.

3 Experiments and Results

3.1 Dataset and Evaluation Metrics

Datasets. We evaluate on **IU-XRay** [9] and **MIMIC-CXR** [17], both containing frontal and lateral views. For IU-XRay (2,955 samples), we adopt the standard 70%/10%/20% split [16]. For MIMIC-CXR, to ensure balanced view representation, we curate a subset of 10,000 training samples and 1,000 validation/test samples, instead of the standard split [7], which is dominated by single-view reports.

Evaluation Metrics. Integrated reports are evaluated using NLG metrics, including BLEU [29], METEOR [10], and ROUGE-L [20]. For view-specific reports, given the limited ability of n-gram metrics to assess clinical validity, we employ GPT-4o [1] to score semantic fidelity on a scale of 1–10, following recent LLM-based evaluation practices [37].

3.2 Implementation

The model is trained using a PyTorch implementation on two NVIDIA A6000 GPUs. The initial learning rate is set as 4×10^{-6} with a cosine learning rate

Table 3. Performance comparison of medical VLMs with and without View-PNDF. Scores are evaluated by GPT-4o.

Model	IU-XRay [9]				MIMIC-CXR [17]			
	Frontal		Lateral		Frontal		Lateral	
	w/o	w/	w/o	w/	w/o	w/	w/o	w/
MedGemma-4B [30]	7.1	7.3	2.9	6.8	6.2	6.3	3.6	5.9
InternVL-DA-Medical-4B [5]	8.5	8.6	3.7	7.9	7.4	7.6	2.9	7.1
LLaVA-Med-7B [18]	8.7	8.8	4.1	8.3	8.3	8.5	3.9	7.8
Hulu-Med-7B [14]	8.9	9.1	4.3	8.5	8.4	8.5	4.2	8.1

Table 4. NLG results of the InternVL-DA-Medical-4B on the IU-Xray dataset.

Before FT	BL-1	BL-4	MTR	RG-L
Frontal	0.398	0.156	0.163	0.265
Lateral	0.397	0.154	0.162	0.266
After FT	**BL-1**	**BL-4**	**MTR**	**RG-L**
Frontal	0.397	0.155	0.164	0.265
Lateral	0.397	0.154	0.163	0.264

Table 5. LLM evaluation scores of InternVL-DA-Medical-4B on the IU-Xray dataset.

Evaluator	Frontal	Lateral
GPT-4o	8.7	7.5
Qwen-72B	8.7	7.7
DeepSeek-V3	8.6	7.6

schedule. We use AdamW [25] optimizer to optimize the network with a weight decay of 0.5. We train the model for 1 epoch with a batch size of 8. The warm up ratio is set as 0.3.

3.3 Comparative Experiment

We evaluated our method on IU-XRay [9] and MIMIC-CXR [17] against state-of-the-art methods. As shown in Table 1, applying our method to four VLM backbones (MedGemma, InternVL, LLaVA, Hulu) on IU-XRay consistently outperformed baselines across all metrics, with View-PNDF$_{\text{Hulu}}$ achieving the best results (e.g., 0.510 BLEU-1). On the MIMIC-CXR dataset (Table 2), our approach maintained this advantage, surpassing the leading baseline MLRG in both linguistic quality (ROUGE-L 0.291 vs 0.278) and Clinical Efficacy (F1 0.414 vs 0.401). These results confirm that fine-tuning view-specific neurons significantly enhances both the linguistic fluency and clinical accuracy of generated reports.

3.4 Ablation Study

To verify the effectiveness of each module, we conduct ablation studies on the IU-XRay [9] and MIMIC-CXR [17] datasets using the InternVL medical model [5].

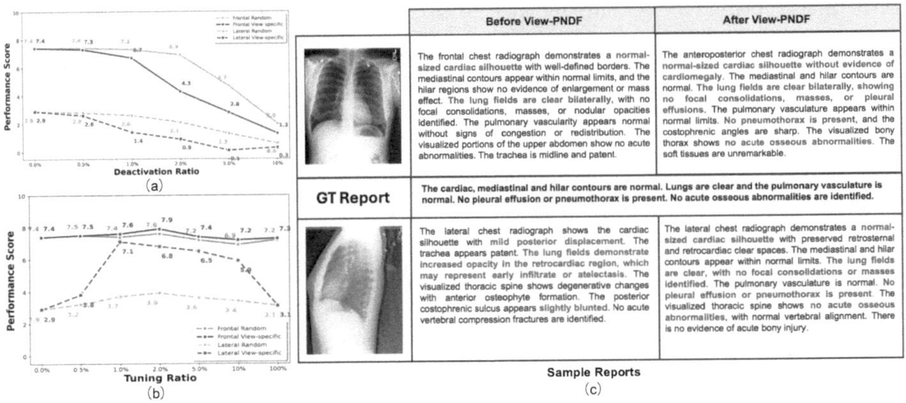

Fig. 2. Ablation studies of VND, VNV, and VNF on MIMIC-CXR dataset. (a) VND yields steeper performance drops than random deactivation, confirming neuron specificity. (b) VNF boosts lateral metrics while maintaining frontal quality. (c) Sample output from InternVL-DA-Medical-4B.

The Effectiveness of View-PNDF. Table 3 details the GPT-4o evaluation of four VLMs with and without View-PNDF. The method consistently improved performance across all models and datasets. Notably, lateral view scores saw significant gains (e.g., +4.2 points on MIMIC-CXR for InternVL), demonstrating that View-PNDF effectively captures view-specific patterns to generate more accurate descriptions for both views.

The Effectiveness of VND and VNV. To validate our detection mechanism, we compared view-specific versus random neuron deactivation (Fig. 2(a)). View-specific deactivation caused drastic performance drops (e.g., lateral score dropping from 2.9 to 0.3 on MIMIC-CXR), whereas random deactivation had minimal impact. This significant disparity confirms that VND successfully identifies neurons critical for specific views.

The Effectiveness of VNF. Figure 2(b) shows that view-specific fine-tuning significantly boosts lateral view performance (e.g., reaching 7.3 vs. 3.7 for random tuning on MIMIC-CXR) while preserving or enhancing frontal view quality. This proves VNF effectively bridges the performance gap between views without compromising general capabilities.

LLM Evaluators. Table 4 shows that standard NLG metrics yield nearly identical scores before and after fine-tuning, failing to capture clinical nuances. In contrast, evaluations using GPT-4o [1], Qwen [2], and DeepSeek [4] (Table 5) reveal clear performance differences, demonstrating that LLM-based evaluation aligns better with clinical validity for this task.

Qualitative Results. Figure 2(c) illustrates the qualitative impact of View-PNDF. While the baseline model misses critical findings (e.g., sternal wires, cardiomegaly) and lacks inter-view consistency, the fine-tuned model successfully captures these pathologies across both frontal and lateral views. This confirms that our method enhances the detection of view-specific pathological patterns, yielding more comprehensive reports.

4 Conclusion

We proposed View-PNDF to detect and fine-tune view-specific pattern neurons in RRG models. Experiments confirm the existence of these neurons and show that optimizing a small fraction ($< 5.0\%$) significantly enhances generation consistency and accuracy across views. While demonstrated on chest X-rays, our parameter-efficient method holds potential for other modalities (e.g., CT, MRI). Future work will focus on extending this approach to multi-modal data and developing model-agnostic implementations.

Acknowledgments. This project is supported by the Ministry of Education, Singapore, under its Academic Research Fund Tier 1 RS16/23. This Project is also partly supported by Centre of AI in Medicine, Singapore.

Disclosure of Interests. The authors have no competing interests to declare that are relevant to the content of this article.

References

1. Achiam, J., et al.: Gpt-4 technical report. arXiv preprint arXiv:2303.08774 (2023)
2. Bai, J., et al.: Qwen technical report. arXiv preprint arXiv:2309.16609 (2023)
3. Bau, D., Zhou, B., Khosla, A., Oliva, A., Torralba, A.: Network dissection: quantifying interpretability of deep visual representations. In: Proceedings of the IEEE Conference on Computer Vision and Pattern Recognition, pp. 6541–6549 (2017)
4. Bi, X., et al.: Deepseek llm: scaling open-source language models with longtermism. arXiv preprint arXiv:2401.02954 (2024)
5. Chen, Z., et al.: Internvl: scaling up vision foundation models and aligning for generic visual-linguistic tasks. In: Proceedings of the IEEE/CVF Conference on Computer Vision and Pattern Recognition, pp. 24185–24198 (2024)
6. Chen, Z., Shen, Y., Song, Y., Wan, X.: Cross-modal memory networks for radiology report generation. arXiv preprint arXiv:2204.13258 (2022)
7. Chen, Z., Song, Y., Chang, T.H., Wan, X.: Generating radiology reports via memory-driven transformer. arXiv preprint arXiv:2010.16056 (2020)
8. Dai, D., Dong, L., Hao, Y., Sui, Z., Chang, B., Wei, F.: Knowledge neurons in pretrained transformers. arXiv preprint arXiv:2104.08696 (2021)
9. Demner-Fushman, D., et al.: Preparing a collection of radiology examinations for distribution and retrieval. J. Am. Med. Inf. Assoc. **23**(2), 304–310 (2015)
10. Denkowski, M., Lavie, A.: Meteor 1.3: automatic metric for reliable optimization and evaluation of machine translation systems. In: Proceedings of the Sixth Workshop on Statistical Machine Translation, pp. 85–91 (2011)

11. Hamamci, I.E., Er, S., Menze, B.: Ct2rep: automated radiology report generation for 3d medical imaging. In: International Conference on Medical Image Computing and Computer-Assisted Intervention, pp. 476–486. Springer, Heidelberg (2024). https://doi.org/10.1007/978-3-031-72390-2_45
12. Huang, X., Chen, W., Liu, J., Lu, Q., Luo, X., Shen, L.: Damper: a dual-stage medical report generation framework with coarse-grained mesh alignment and fine-grained hypergraph matching. In: Proceedings of the AAAI Conference on Artificial Intelligence, vol. 39, pp. 3769–3778 (2025)
13. Ittyachen, A.M., Vijayan, A., Isac, M.: The forgotten view: chest x-ray-lateral view. Respir. Med. Case. Rep. **22**, 257–259 (2017)
14. Jiang, S., et al.: Hulu-med: a transparent generalist model towards holistic medical vision-language understanding. arXiv preprint arXiv:2510.08668 (2025)
15. Jin, H., Che, H., Lin, Y., Chen, H.: Promptmrg: diagnosis-driven prompts for medical report generation. In: Proceedings of the AAAI Conference on Artificial Intelligence, vol. 38, pp. 2607–2615 (2024)
16. Jing, B., Wang, Z., Xing, E.: Show, describe and conclude: on exploiting the structure information of chest x-ray reports. arXiv preprint arXiv:2004.12274 (2020)
17. Johnson, A.E., et al.: Mimic-cxr-jpg, a large publicly available database of labeled chest radiographs. arXiv preprint arXiv:1901.07042 (2019)
18. Li, C., et al.: Llava-med: training a large language-and-vision assistant for biomedicine in one day. Adv. Neural. Inf. Process. Syst. **36**, 28541–28564 (2023)
19. Liang, X., Zhang, Y., Wang, D., Zhong, H., Li, R., Wang, Q.: Divide and conquer: isolating normal-abnormal attributes in knowledge graph-enhanced radiology report generation. In: Proceedings of the 32nd ACM International Conference on Multimedia, pp. 4967–4975 (2024)
20. Lin, C.Y.: Rouge: a package for automatic evaluation of summaries. In: Text Summarization Branches Out, pp. 74–81 (2004)
21. Liu, C., Tian, Y., Chen, W., Song, Y., Zhang, Y.: Bootstrapping large language models for radiology report generation. In: Proceedings of the AAAI Conference on Artificial Intelligence, vol. 38, pp. 18635–18643 (2024)
22. Liu, K., et al.: Enhanced contrastive learning with multi-view longitudinal data for chest x-ray report generation. In: Proceedings of the Computer Vision and Pattern Recognition Conference, pp. 10348–10359 (2025)
23. Liu, Z., Zhu, Z., Zheng, S., Zhao, Y., He, K., Zhao, Y.: From observation to concept: a flexible multi-view paradigm for medical report generation. IEEE Trans. Multimedia **26**, 5987–5995 (2023)
24. Loh, D.R., Yeo, S.Y., Tan, R.S., Gao, F., Koh, A.S.: Explainable machine learning predictions to support personalized cardiology strategies. Eur. Heart J.-Digital Health **3**(1), 49–55 (2022)
25. Loshchilov, I., Hutter, F.: Decoupled weight decay regularization. arXiv preprint arXiv:1711.05101 (2017)
26. Miao, Q., et al.: Evoke: elevating chest x-ray report generation via multi-view contrastive learning and patient-specific knowledge. arXiv preprint arXiv:2411.10224 (2024)
27. Mu, J., Andreas, J.: Compositional explanations of neurons. Adv. Neural. Inf. Process. Syst. **33**, 17153–17163 (2020)
28. Nicolson, A., Dowling, J., Koopman, B.: Improving chest x-ray report generation by leveraging warm starting. Artif. Intell. Med. **144**, 102633 (2023)
29. Papineni, K., Roukos, S., Ward, T., Zhu, W.J.: Bleu: a method for automatic evaluation of machine translation. In: Proceedings of the 40th Annual Meeting of the Association for Computational Linguistics, pp. 311–318 (2002)

30. Sellergren, A., et al.: Medgemma technical report. arXiv preprint arXiv:2507.05201 (2025)
31. Tanida, T., Müller, P., Kaissis, G., Rueckert, D.: Interactive and explainable region-guided radiology report generation. In: Proceedings of the IEEE/CVF Conference on Computer Vision and Pattern Recognition, pp. 7433–7442 (2023)
32. Wang, J., Bhalerao, A., He, Y.: Cross-modal prototype driven network for radiology report generation. In: European Conference on Computer Vision, pp. 563–579. Springer, Heidelberg (2022). https://doi.org/10.1007/978-3-031-19833-5_33
33. Wang, J., Bhalerao, A., Yin, T., See, S., He, Y.: Camanet: class activation map guided attention network for radiology report generation. IEEE J. Biomed. Health Inf. **28**(4), 2199–2210 (2024)
34. Yang, S., Wu, X., Ge, S., Zheng, Z., Zhou, S.K., Xiao, L.: Radiology report generation with a learned knowledge base and multi-modal alignment. Med. Image Anal. **86**, 102798 (2023)
35. Yang, X., et al.: Cardiac image segmentation by random walks with dynamic shape constraint. IET Comput. Vision **10**(1), 79–86 (2016)
36. Yeo, S.Y., Xie, X., Sazonov, I., Nithiarasu, P.: Level set segmentation with robust image gradient energy and statistical shape prior. In: 18th IEEE International Conference on Image Processing, pp. 3397–3400. IEEE (2011)
37. Zheng, L., et al.: Judging llm-as-a-judge with mt-bench and chatbot arena. Adv. Neural. Inf. Process. Syst. **36**, 46595–46623 (2023)

Semantically Consistent Whole-Body PET/CT Report Generation with Local Lesion-Level Guidance

Mingyang Yu[1], Yaozong Gao[2], Yiran Shu[1,2], Jingyu Liu[2], Jiaming Wu[5], Kaicong Sun[1], Shuwei Bai[1], Yanbo Chen[2], Xiang Zhou[2], Yiqiang Zhan[2], Weifang Zhang[3], Shaonan Zhong[4], Xinlu Wang[4], Meixin Zhao[3(✉)], and Dinggang Shen[1,2,6(✉)]

[1] School of Biomedical Engineering and State Key Laboratory of Advanced Medical Materials and Devices, ShanghaiTech University, Shanghai 201210, China

[2] Shanghai United Imaging Intelligence Co., Ltd., Shanghai, China
Dinggang.Shen@gmail.com

[3] Department of Nuclear Medicine, Peking University Third Hospital, Haidian District, Beijing 100191, China
zhaomeixin.student@sina.com

[4] Department of Nuclear Medicine, The First Affiliated Hospital of Guangzhou Medical University, Guangzhou 510000, China

[5] Department of Health Technology and Informatics, The Hong Kong Polytechnic University, Hong Kong SAR, China

[6] Shanghai Clinical Research and Trial Center, Shanghai 201210, China

Abstract. Automated report generation for whole-body PET/CT imaging can substantially reduce radiologists' workload and improve clinical efficiency. Existing report generation approaches typically map the entire volumetric scan to a complete report. However, clinically meaningful findings in whole-body PET/CT are sparse and localized. Such global end-to-end modeling tends to omit lesion-specific semantic cues, limiting complete descriptions of lesions. To address this challenge, we propose a whole-body PET/CT report generation framework guided by local lesion-level cues. Our method is built upon three key strategies. First, we propose an aggregation module that compresses variable lesion candidates in PET/CT, enabling many-to-one lesion–description alignment that reflects real clinical reporting practice. Second, we propose local entity alignment and global relation alignment during pretraining to enforce cross-modal semantic consistency between report and image. Third, we use a two-stage report generation training pipeline, namely pretraining on structured reports and fine-tuning on free-text radiology reports, to reduce template bias. We have collected **3,015** whole-body PET/CT cases from **three** medical centers, including **120,549** annotated local lesions. Experimental results demonstrate that the proposed framework improves performance under clinically reliable evaluations of generated reports.

Keywords: Whole-body PET/CT · Report generation · Vision–language pretraining

G. Yang et al. (Eds.): MICCAI 2026, LNCS 16878, pp. 580–589, 2027.
https://doi.org/10.1007/978-3-032-38059-3_55

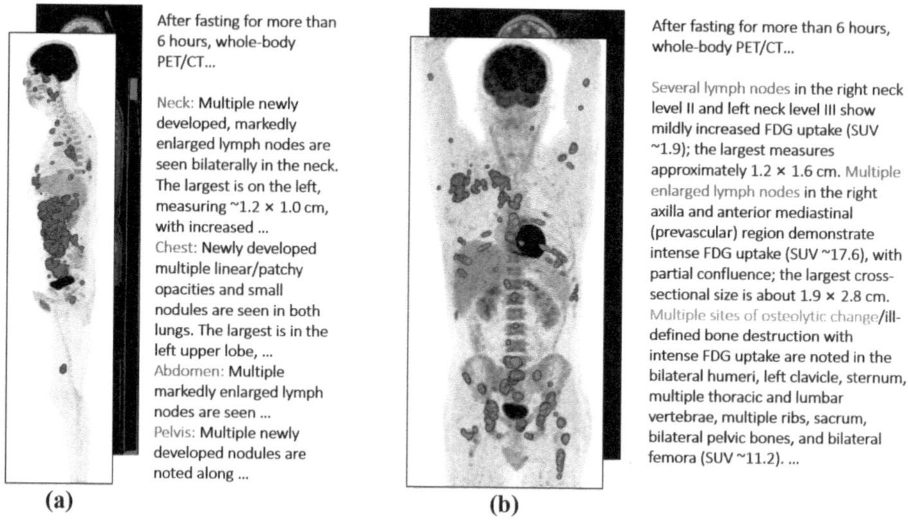

Fig. 1. Radiologists' reporting style in whole-body PET/CT often follows a many-to-one pattern, where multiple lesions are summarized in a single statement. Such summarization is commonly organized by anatomical regions (left, a), lesion types (right, b), or a hybrid of both.

1 Introduction

Automated report generation [12,16] has been extensively studied in recent years. However, most existing efforts focus on specific body parts, such as chest radiographs [10] and breast ultrasound [2], with limited anatomical field-of-view coverage, and the report content is largely driven by localized findings. In contrast, whole-body PET/CT spans diverse anatomical regions across the entire body, making report generation substantially more challenging due to more complex volumetric input and sparse localized abnormalities. Existing works [4,9] attempt to interpret whole-body PET/CT volumes using large models. However, this strategy is misaligned with clinical focus and computationally impractical. Moreover, the scarcity of large-scale paired whole-body PET/CT report datasets further limits effective training [9].

In contrast, studies [8,14] have explored mask-guided, local lesion-level captioning to produce more clinically useful lesion descriptions. In particular, PETAR [8] focuses on local finding generation with a relatively direct lesion–description correspondence, while PETRG [4] directly models the whole PET/CT volume for report generation. These studies demonstrate the potential of lesion-aware or full-scan PET/CT reporting, but local captioning does not fully address complete report generation, and whole-volume modeling may dilute small clinically relevant lesions within the large whole-body background. And other studies [1,13] show that structured reports can provide more efficient supervision and evaluation than free-text. However, transferring these ideas to whole-body, full-scan PET/CT report generation remains challenging. First, whole-body PET/CT scans often contain multiple metastatic lesions, so the report rarely follows a one-to-one image–text mapping. Instead, it frequently exhibits a many-to-one lesion–description pattern, where multiple lesions are summarized in a single statement. In practice (Fig. 1), such summarization is often organized by lesion types (e.g., bone lesions), anatomical regions (e.g., lymph nodes or soft-tissue masses), or a combination of both, reflecting reporting styles and clinical indications. Second, such report generation models rely heavily on a visual encoder with strong semantic understanding. Prior research often adopts CLIP-style [11] vision–language pretraining for image–text alignment. However, these mechanisms are typically designed for one-to-one matching and do not explicitly model the many-to-one alignment common in whole-body PET/CT reports. Finally, free-text reports, especially those from single-center cohorts, often contain substantial template-driven content. During early training, models may overfit these frequent patterns, which can hinder learning sparse yet clinically critical lesion-specific cues.

To address these challenges, we propose a local lesion-level guidance framework to generate whole-body PET/CT reports with guidance from specific local lesions. Specifically, our contributions are threefold. First, a SetTransformer-based aggregation module [5] is used to compress variable lesion candidates into a fixed-length entity set, enabling many-to-one lesion-description alignment that reflects real clinical reporting practice. Second, CLIP-style pretraining is built on one-to-one image–text matching and treats all other pairs in the minibatch as negatives, implicitly assuming that different reports (and different images) are unrelated. However, PET/CT images and reports are less diverse than natural images and texts and often share substantial anatomy- and template-driven content, making this fully discriminative objective suboptimal for PET/CT image–text alignment. To address this, we introduce local entity alignment and global relation alignment during pretraining. Local entity alignment preserves paired image-text alignment, while global relation alignment replaces the all-negatives assumption and encourages the image embedding space to align with the text embedding space. Third, to mitigate overfitting to template-driven contents, we adopt a two-stage report generation training pipeline, first training on structured positive findings to emphasize clinically salient cues, and then fine-tuning on free-text reports to reduce template bias.

2 Method

Our method extends the LLaVA [7] framework. As shown in Fig. 2b, the model takes a varying number of lesion-guided PET/CT volume patches, paired with lesion attributes (e.g., SUVmax and lesion size), as input to generate the final findings report. Specifically, given the grouped 3D lesion patch candidates (e.g., by anatomical region), patches within each group are first encoded into lesion features and then aggregated per group into a fixed-length set of visual entities. Subsequently, the resulting entity sets are projected into visual entity embed-

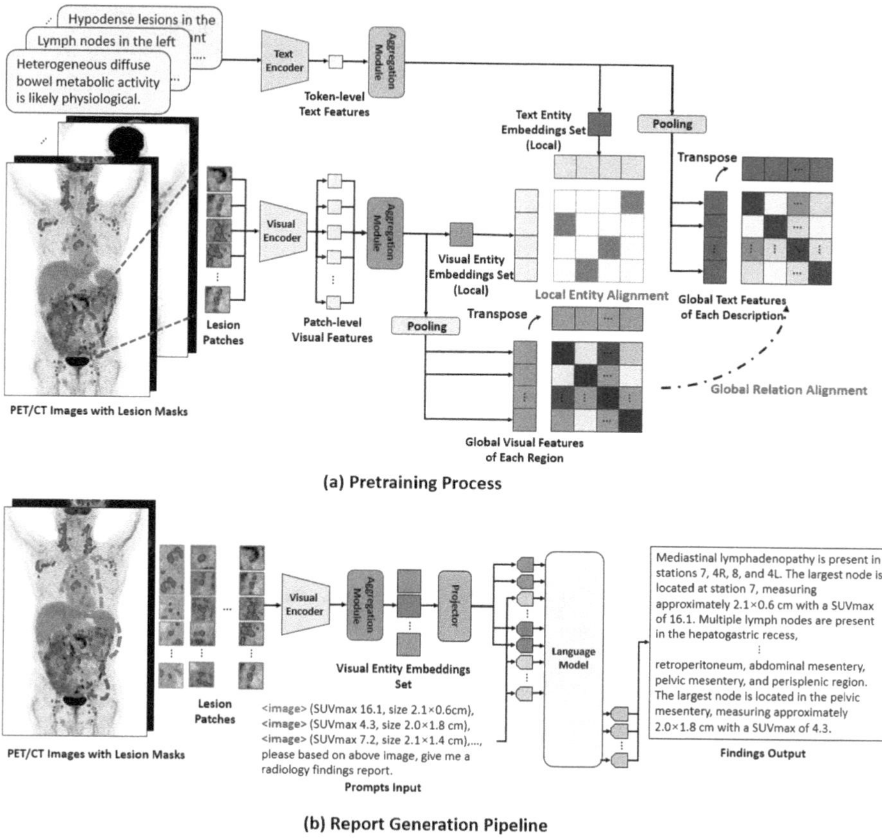

Fig. 2. Overview of the pretraining process and report generation pipeline. The pretraining pipeline (a, top) uses local entity alignment and global relation alignment to enforce semantic consistency both cross-modally (image–text) and within each modality. The model pipeline (b, bottom) shows how local lesion cues are encoded and aggregated, then combined with prompt embeddings to generate the final whole-body PET/CT report. All lesion volume patches are cropped as $64 \times 64 \times 64$ voxels with an isotropic spacing of 1.5 mm.

dings, fused with lesion-attribute prompt embeddings, and then fed to the language model to generate the whole-body PET/CT findings report.

2.1 SetTransformer-Based Aggregation Module

As illustrated in Fig. 1, whole-body PET/CT studies often contain a large and highly varying number of lesion ROIs, particularly in multifocal disease. In some patients, the scan may include hundreds of tracer-avid metastatic lesions. Directly feeding all lesion features into the LLM is inefficient and largely redundant, since many ROIs share similar uptake patterns and overlapping findings within the same anatomical region. In clinical practice, the corresponding report is typically concise, summarizing many lesion ROIs into a statement by lesion type or region, which naturally forms a many-to-one lesion-description relationship.

To better accommodate the many-to-one lesion–description relationship in whole-body PET/CT report generation, we introduce a SetTransformer-based aggregation module [5] into our framework. The SetTransformer uses attention-based set encoding to compress a variable-sized ROI set into a fixed number K of visual entities, and it is order-insensitive, making it naturally suitable for tasks that only care about lesion content rather than the input order. Accordingly, if all lesion ROIs in a case can be organized into n groups (e.g., by anatomical region), corresponding to n key summarization statements, then the ROIs can be compressed into $n \times K$ visual entities. This compact representation is designed to retain sufficient information for report generation and reflects the many-to-one nature of clinical reporting practice.

2.2 Local Entity Alignment and Global Relation Alignment

CLIP-style pretraining is widely used in vision–language learning. It uses a one-hot contrastive target. For each image, only the paired text is treated as positive, while all other texts are treated as negatives (and vice versa). This assumption is improper for whole-body PET/CT because PET/CT images and reports are less diverse than natural counterparts, and they often share substantial anatomy- and template-driven content, making the discriminative objective suboptimal for PET/CT image–text alignment.

To better support PET/CT alignment and our many-to-one lesion–description setting, we introduce two complementary objectives during the pre-training of the vision encoder and the aggregation module: local entity alignment and global relation alignment (Fig. 2a). For local entity alignment, we retain the paired positive supervision and model many-to-one correspondence by compressing paired lesion volumetric patches and their descriptions into fixed-length entity sets of size K. This design is motivated by the fact that a single report statement may mention multiple localized concepts, e.g., "Several lymph nodes are present in the right level II and left level III cervical nodal stations." which contains both "right neck level II" and "left neck level III". We then align K visual entities with K textual entities via one-to-one matching, so that each textual entity corresponds to a specific visual entity. For global relation alignment, we replace the absolute "all negatives" assumption with cross-sample relationships. We pool each entity set into a single embedding, build pair-to-pair similarity matrices in both modalities, and minimize their discrepancy so that visual representations align with the textual semantic space. The total loss is formulated as $\mathcal{L} = \mathcal{L}_{\text{align}} + \lambda_{\text{r}}\mathcal{L}_{\text{rel}}$, where $\lambda_{\text{r}} = 1/B$ balances the batch-level global relation loss with the local entity alignment loss; in our experiments, $B = 32$.

Local Entity Alignment. Given a mini-batch of B paired samples, we represent the b-th pair by K ℓ_2-normalized visual entities $E_b = \{e_{b,i}\}_{i=1}^{K}$ and K ℓ_2-normalized textual entities $T_b = \{t_{b,j}\}_{j=1}^{K}$, where $b \in \{1, \dots, B\}$. We compute the similarity matrix $S_b \in \mathbb{R}^{K \times K}$, obtain a one-to-one assignment π_b via Hungarian matching, and apply a temperature-scaled cross-entropy loss:

$$\mathcal{L}_{\text{align}} = -\frac{1}{BK} \sum_{b=1}^{B} \sum_{i=1}^{K} \log \frac{\exp\big(S_b(i, \pi_b(i))/\tau\big)}{\sum_{j=1}^{K} \exp\big(S_b(i,j)/\tau\big)}, \tag{1}$$

$$S_b(i,j) = e_{b,i}^{\top} t_{b,j}, \quad \pi_b = \text{Hungarian}(-S_b). \tag{2}$$

Global Relation Alignment. To align cross-sample semantic geometry across modalities, we first form sample-level embeddings by mean pooling the K entities: $u_b = \frac{1}{K}\sum_{i=1}^{K} e_{b,i}$ and $v_b = \frac{1}{K}\sum_{j=1}^{K} t_{b,j}$. We stack them as $U = [u_1; \dots; u_B]$ and $V = [v_1; \dots; v_B]$, compute relation distributions with temperature τ, and minimize a symmetric KL divergence:

$$P_I = \text{softmax}\big(UU^{\top}/\tau\big), P_T = \text{softmax}\big(VV^{\top}/\tau\big), \tag{3}$$

$$\mathcal{L}_{\text{rel}} = \frac{1}{2}\Big(\text{KL}(P_T \| P_I) + \text{KL}(P_I \| P_T)\Big). \tag{4}$$

2.3 Two-Stage Report Generation Training Pipeline

Free-text whole-body PET/CT reports often include many repeated template sentences (e.g., "No abnormal radiotracer uptake ..."). These sentences can dominate training, so the model learns common but low-information wording instead of rare yet clinically important abnormalities.

To address this issue, we employ a two-stage training pipeline. In Stage one, we train on structured reports that contain **positive lesions only**, covering **471** anatomical locations and **61** lesion-type categories, to keep the model focus on lesion-relevant cues and avoid learning template bias and normal findings. In Stage two, we then finetune this model on full free-text reports that include both positive and negative statements to learn realistic narrative style and institution-specific reporting conventions. These structured reports in Stage one are derived from structured labels rather than written as free text, i.e., they are constructed from each study via an automated lesion-centric pipeline. Specifically, we first apply a lesion segmentation model to extract 3D lesion volume patches. Then we use a set of lesion classification models to assign each lesion category (e.g., anatomical location, lesion type) together with computed quantitative attributes (e.g., SUVmax and lesion size). All results are manually reviewed by radiologists and mapped to corresponding sentence in free-text report to ensure consistency between structured report and free-text report.

3 Experiments and Results

3.1 Dataset

Our dataset is collected from three medical centers and including 3,015 whole-body PET/CT studies. We have identified 120,549 abnormal regions. Among which, 83,395 lesions are aligned with report-described content. The dataset contains **2,408** FDG studies and **607** PSMA studies. We use 2,811 studies for training and 204 studies for testing. We obtain lesion regions using an nnU-Net-based [3] lesion segmentation model. Each lesion has structured attributes, including anatomical location, CT findings, lesion size, and SUVmax. The structured labels cover 471 anatomical locations and 61 CT-finding categories. We normalize PET and CT images to $[0, 1]$ and apply standard data augmentation such as random rotation and scaling.

3.2 Training Details

Training was conducted on 8 NVIDIA H20 GPUs. During the pretraining stage, the CNN-based visual encoder was implemented as a 3D ResNet-50, and the Transformer-based visual encoder was implemented as a ViT-3D. The visual encoders were initialized with Kaiming initialization. Clinically similar lesion candidates were grouped by anatomical region or lesion category before Set-Transformer aggregation. We set the number of SetTransformer output entities to $K = 8$ to balance information preservation and computational efficiency. The temperature was set to $\tau = 0.1$. The batch size was $B = 32$, and $\lambda_r = 1/32$. The model was optimized using AdamW with a learning rate of 1×10^{-4}. During report generation fine-tuning, the model was initialized with the pretrained vision and text encoders. The vision encoder was frozen, and the language model was fine-tuned using LoRA adapters.

3.3 Comparison Study

We evaluate our method on our two-stage report generation training curriculum: 1) structured reports and 2) free-text reports. Structured reports contain positive lesion findings only, enabling a focused assessment of lesion-level semantic fidelity without interference from template-driven content. Free-text reports include both positive and negative statements and better reflect real-world reporting style. We use NLG (Natural Language Generation) metrics and CE (Clinical Efficacy) metrics as evaluation metrics. For NLG, we report BLEU and ROUGE-L (F1) scores computed on the Chinese report text. For CE metrics, we employ an LLM-based evaluator (ChatGPT-5.2) that matches each ground-truth (GT) statement to the most similar generated statement, since clinically correct findings may be expressed with different wording (e.g., "left lung" vs. "left upper lobe"). The evaluator outputs four accuracy scores, A-all, A-pos, F-all, and F-pos, which comprehensively quantify clinical efficacy by separately assessing anatomical and finding correctness on both positive and negative descriptions, highlighting the advantage of our method.

Table 1. Comparison study results on structured and free-text reports. All values are reported in $[0, 1]$. † denotes the language model that takes structured reports as input. B-3/B-4: BLEU-3/BLEU-4; R-L: ROUGE-L (F1). A-all/F-all: overall Anatomy/Finding accuracy (positive and negative descriptions); A-pos/F-pos: accuracy on positive lesions only. In PET2Rep [15], "Sep" and "Fus" denote the two PET/CT input strategies, "Separate" and "Fused," respectively. Best results are in bold. * indicates a statistically significant improvement of our method over all competing baselines (excluding our variants) on the CE metrics under the same setting ($p < 0.05$).

Method	LLMs/VLMs	NLG Metrics			CE Metrics			
		B-3	B-4	R-L	A-all	A-pos	F-all	F-pos
Structured Report								
LLaVA-Med	Llama-7B (ViT)	0.507	0.488	0.765	-	0.286	-	0.765
Ours	Llama-7B (CNN)	0.706	0.681	0.957	-	0.499	-	0.852
	Llama-13B (CNN)	0.719	0.694	0.962	-	0.502	-	0.855
	Llama-7B (ViT)	0.785	0.758	0.981	-	0.519	-	0.868
	Llama-13B (ViT)	**0.796**	**0.769**	**0.985**	-	**0.522**	-	**0.877**
Free-text Report								
S-to-F	QWen3-8B †	0.314	0.269	**0.316**	0.499	0.520	0.810	0.654
LLaVA-Med	Llama-13B (ViT)	0.143	0.105	0.273	0.296	0.300	0.757	0.608
PET2Rep (Sep)	Qwen3-VL-235B	0.201	0.179	0.255	0.491	0.413	0.695	0.635
PET2Rep (Fus)	Qwen3-VL-235B	0.216	0.196	0.258	0.504	0.420	0.695	0.521
Ours	Llama-13B (CNN)	0.274	0.232	0.299	0.646	0.571	0.720	0.611
	Llama-13B (ViT)	**0.326**	**0.282**	0.309	**0.649***	**0.594***	**0.817***	**0.670***

In Table 1, we can see that our method consistently outperforms LLaVA-Med [6] and PET2Rep [15] in both NLG and CE metrics, demonstrating clear advantages in generating clinically faithful whole-body PET/CT reports. Here, CNN/ViT denote 3D-CNN/3D-ViT vision encoders respectively, and LLaMA-7B/13B denote LLaMA-Chinese-7B/13B. It is shown that increasing the language model size from 7B to 13B further improves performance, while the choice of visual encoder has an even larger impact: 3D-ViT consistently achieves higher CE scores than 3D-CNN, suggesting that ViT-style visual representations are more compatible with LLM-based parsing and generation for whole-body PET/CT reporting. Finally, S-to-F denotes a structured-to-free strategy that uses structured-report text to predict free-text reports. The results indicate that structured reports provide helpful cues, but their information content remains limited.

3.4 Ablation Study

We conduct ablation study to validate the effectiveness of (i) local entity alignment with global relation alignment (*Alignment*), and (ii) the two-stage report generation training pipeline (*Structured Report*). The baseline is trained end-to-end on free-text reports using the same architecture as ours. As shown in Table 2, adding *Alignment* consistently improves performance, indicating that

Table 2. Ablation results for free-text report generation under CE metrics (all values in [0, 1]). All experiments use LLaMA-13B (ViT). CE metrics are defined the same as in the comparison study.

Method	CE Metrics			
	A-all	A-pos	F-all	F-pos
baseline	0.602	0.548	0.794	0.556
+*Alignment*	0.614	0.560	0.773	0.603
+*Structured Report*	0.630	0.581	0.809	0.651
+*Alignment* +*Structured Report*	**0.649**	**0.594**	**0.817**	**0.670**

local entity alignment and global relation alignment benefits the report generation task. Meanwhile, the *Structured Report* yields a larger gain than using Alignment alone, and combining them achieves the best results. This suggests that pretraining on structured, positive-only reports is a key factor for improving clinical report generation, as it helps the model focus on lesion-relevant semantics before learning free-text report templates.

4 Conclusion

In this paper, we present a local lesion-level guidance framework for whole-body PET/CT report generation. Inspired by the sparsity of clinically meaningful findings and the common many-to-one lesion-description pattern in clinical practice, we propose (1) a SetTransformer-based aggregation module to compress variable lesion candidates into a fixed entity set for many-to-one alignment, (2) local entity alignment and global relation alignment to enable image-text pretraining, and (3) a two-stage report generation training pipeline (structured reports to free-text reports) to mitigate template-driven bias. We also build a multi-center dataset containing **3,015** whole-body PET/CT image-report pairs from **three** institutions, with **120,549** annotated lesions. Experimental results demonstrate that the proposed framework improves performance under clinically reliable evaluations of generated reports.

Acknowledgments. This work was supported in part by National Natural Science Foundation of China (grant numbers 82441023, U23A20295, 62131015, 82394432), the China Ministry of Science and Technology (S20240085, STI2030-Major Projects-2022ZD0209000, STI2030-Major Projects-2022ZD0213100), Shanghai Municipal Central Guided Local Science and Technology Development Fund (No. YDZX20233100001001), The Key R&D Program of Guangdong Province, China (grant number 2023B0303040001), HPC Platform of ShanghaiTech University, the special fund of Beijing Clinical Key Specialty Construction Program, P. R. China (2022), China Medical Health Development Foundation, Peking University Third Hospital and United-Imaging Research Institution Intelligential Imaging Joint Research & Development Center Foundation, and Key Clinical Project of Peking University Third Hospital (BYSYZD2019038 and BYSYZD2023016). We also acknowledge the Department of Nuclear Medicine, The First Affiliated Hospital of Guangzhou Medical University, for providing validation data.

Disclosure of Interests. The authors have no competing interests to declare that are relevant to the content of this article.

References

1. Chen, K., Xu, W., Li, X.: The potential of Gemini and GPTs for structured report generation based on free-text 18F-FDG PET/CT breast cancer reports. Acad. Radiol. **32**(2), 624–633 (2025)
2. Huh, J., Park, H.J., Ye, J.C.: Breast Ultrasound Report Generation using LangChain. arXiv preprint arXiv:2312.03013 (2023)
3. Isensee, F., Jaeger, P.F., Kohl, S.A., Petersen, J., Maier-Hein, K.H.: nnU-Net: a self-configuring method for deep learning-based biomedical image segmentation. Nat. Methods **18**(2), 203–211 (2021)
4. Jiao, W., et al.: Vision-Language Models for Automated 3D PET/CT Report Generation. arXiv preprint arXiv:2511.20145 (2025)
5. Lee, J., Lee, Y., Kim, J., Kosiorek, A., Choi, S., Teh, Y.W.: Set transformer: a framework for attention-based permutation-invariant neural networks. In: International Conference on Machine Learning, pp. 3744–3753. PMLR (2019)
6. Li, C., et al.: LLaVA-Med: training a large language-and-vision assistant for biomedicine in one day. Adv. Neural. Inf. Process. Syst. **36**, 28541–28564 (2023)
7. Liu, H., Li, C., Wu, Q., Lee, Y.J.: Visual instruction tuning. Adv. Neural. Inf. Process. Syst. **36**, 34892–34916 (2023)
8. Maqbool, D., et al.: PETAR: Localized Findings Generation with Mask-Aware Vision-Language Modeling for PET Automated Reporting. arXiv preprint arXiv:2510.27680 (2025)
9. Nguyen, H.T., et al.: Toward a Vision-Language Foundation Model for Medical Data: Multimodal Dataset and Benchmarks for Vietnamese PET/CT Report Generation. arXiv preprint arXiv:2509.24739 (2025)
10. Ouis, M.Y., Akhloufi, M.A.: Deep learning for report generation on chest X-ray images. Comput. Med. Imaging Graph. **111**, 102320 (2024)
11. Radford, A., et al.: Learning transferable visual models from natural language supervision. In: International Conference on Machine Learning, pp. 8748–8763. PMLR (2021)
12. Wang, S., et al.: Interactive computer-aided diagnosis on medical image using large language models. Commun. Eng. **3**(1), 133 (2024)
13. Wray, R., Yeh, R.: Multimodal large language model based PET/CT report generation: capabilities and comparison of ChatGPT-4 versus ChatGPT-3.5. J. Nucl. Med. **65**(Suppl. 2), 242447 (2024)
14. Yu, M., et al.: Location-guided automated lesion captioning in whole-body PET/CT images. In: International Conference on Medical Image Computing and Computer-Assisted Intervention, pp. 348–357. Springer (2025)
15. Zhang, Y., et al.: PET2Rep: Towards Vision-Language Model-Drived Automated Radiology Report Generation for Positron Emission Tomography. arXiv preprint arXiv:2508.04062 (2025)
16. Zhong, T., et al.: ChatRadio-Valuer: A Chat Large Language Model for Generalizable Radiology Report Generation Based on Multi-institution and Multi-system Data. arXiv preprint arXiv:2310.05242 (2023)

Synergistic Perception-Reasoning Governance: Grounding Medical MLLMs with Verifiable Anatomical Evidence

Rui Hao[1], Qiankun Li[2,3](✉), Junyuan Mao[4], Linghao Meng[4], Dirui Xie[1], Dayu Tan[5], and Zhigang Zeng[1](✉)

[1] Huazhong University of Science and Technology, Wuhan, China
{ruihao,zgzeng}@hust.edu.cn

[2] Imperial Global Singapore, Imperial College London, London, UK
q.li2@imperial.ac.uk

[3] Nanyang Technological University, Singapore, Singapore

[4] National University of Singapore, Singapore, Singapore

[5] Anhui University, Hefei, China

Abstract. Multimodal large language models (MLLMs) show strong promise for clinical VQA and radiology report generation, yet inference-time hallucinations still undermine trustworthy use: models can produce fluent conclusions that conflict with imaging evidence. Existing mitigation strategies typically rely on additional training, external retrieval/-knowledge bases, or multi-stage post-hoc verification, which increases cost and pipeline complexity and often generalizes poorly across models and tasks. To address this, we propose a holistic, training-free evidence-injection framework that systematically mitigates hallucinations through dual-side evidence injection. By leveraging ROI priors acquired using MedSAM in our implementation, we recalibrate the visual perception trajectory via ROI-guided activation modulation while anchoring the textual reasoning trajectory by mapping anatomical coordinates into discrete semantic tokens as verifiable external memory. Then we introduce a task-aware dynamic router to select modality-specific interventions based on task semantics, balancing perceptual grounding and linguistic fluency. We conduct systematic evaluations on 2 tasks and 5 datasets using `LLaVA-1.5-7B`, `LLaVA-Med-1.5-7B`, `Qwen3-VL-8B/32B`, and `InternVL-3.5-8B/38B`. Controlled ablations and visualizations further validate the framework, which consistently outperforms baselines across medical benchmarks, improving close-ended accuracy by up to $\sim$ **6%** $\uparrow$ and reducing open-ended hallucinations by $\sim$ **35%** $\downarrow$. The code has been made available on GitHub: https://github.com/Henry991115/SPRG.

Keywords: Multimodal large language models · Visual misinterpretation hallucination · Hallucination mitigation · Evidence injection

G. Yang et al. (Eds.): MICCAI 2026, LNCS 16878, pp. 590–600, 2027.
https://doi.org/10.1007/978-3-032-38059-3_56

1 Introduction

MLLMs combine visual perception and language generation with emerging clinical reasoning, enabling medical VQA and radiology report generation [1–5]. However, hallucinations remain a key barrier to trustworthy deployment [6], as models may generate fluent outputs that are unsupported by or even contradict imaging evidence [7,8]. Recent benchmarks categorize medical hallucinations into visual misinterpretation, knowledge deficiency, and context misalignment [9]; in high-stakes settings, such evidence-inconsistent outputs can lead to incorrect lesion judgments and unsafe clinical attributions [10,11].

To mitigate these risks, many training-free, inference-time methods have emerged. VCD [12], DoLa [13], and OPERA [14]adjust decoding by contrasting perturbed inputs, contrasting layer logits, or penalizing over-trusted tokens. AVISC [15], DAMRO [16], and PAI [17]calibrate attention or visual tokens by correcting outliers, filtering vision-side noise, and reducing text-dominant bias. M3ID [18] strengthens image grounding through visually dominant constraints. However, most existing mitigations rely on internal statistical calibration without explicit anatomical grounding, which limits generalization across clinical tasks with different evidence requirements.

In this work, we propose a holistic, training-free inference-time evidence-injection framework to mitigate visual misinterpretation hallucinations in medical MLLMs. Using ROI priors as verifiable anatomical evidence, we improve visual grounding and clinical consistency through the perception-reasoning intervention mechanism, including text-side evidence injection and vision-side activation recalibration over visual tokens, with a joint strategy for complementary gains. To handle heterogeneous clinical tasks, we further introduce a lightweight task-aware dynamic router that dynamically selects modality-specific interventions based on task semantics. Extensive experiments across diverse medical benchmarks, supported by controlled ablations and visualizations, show superior robustness over existing baselines, with up to $\sim$ **6%**$\uparrow$ improvement in close-ended reasoning accuracy and up to $\sim$ **35%**$\downarrow$ reduction in open-ended hallucinations. Our key contributions can be summarized as follows:

- ❶ We introduce a novel training-free framework that mitigates hallucinations by injecting dual-side verifiable anatomical evidence, bridging the gap between raw visual perception and clinical reasoning.
- ❷ We develop a perception-reasoning intervention mechanism that recalibrates the visual perception trajectory via ROI-guided visual-side activation modulation, while mapping anatomical coordinates into discrete semantic tokens as verifiable external memory to anchor the textual reasoning trajectory.
- ❸ We propose a task-aware dynamic router that adjusts modality-specific intervention strategies according to task semantics, thereby achieving a better balance between perceptual grounding and linguistic fluency.
- ❹ We establish a rigorous cross-model, cross-task evaluation framework involving 2 tasks, 5 datasets and 6 MLLMs, providing a highly comprehensive study to date on inference-time medical hallucination mitigation.

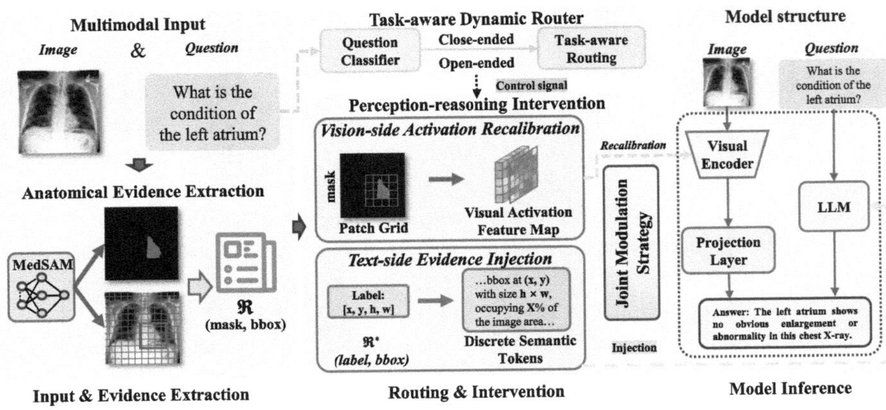

Fig. 1. The overall pipeline consists of all input, ROI prior extraction, intervention mechanism, task-aware routing, and model inference for the output generation.

2 Methods

Given an image I, a textual question q, and a pretrained MLLM parametrized by θ, the standard autoregressive generation is formulated as:

$$p_\theta(y \mid I, q) = \prod_{t=1}^{T} p_\theta(y_t \mid y_{<t}, \text{Enc}(I), q)\,, \tag{1}$$

where $\text{Enc}(I) \in \mathcal{R}^{N\times d}$ denotes the sequence of N visual patch tokens. To mitigate visual misinterpretation, we introduce an intervention function that conditionally modulates the visual representation $\tilde{V}$ and textual prompt $\tilde{q}$ based on verifiable anatomical priors $\mathcal{R}$, yielding $p_{\theta,\phi}(y \mid I, q, R) = \prod_{t=1}^{T} p_\theta(y_t \mid y_{<t}, \tilde{V}, \tilde{q})$.

As shown in Fig. 1, the pipeline comprises the anatomical evidence extraction, perception-reasoning intervention mechanism, and task-aware dynamic router.

2.1 Anatomical Evidence Extraction

We utilize MedSAM prompted by a candidate bounding box set $\mathcal{B} = \{b_k\}_{k=1}^{K}$ to extract verifiable anatomical evidence. For each valid box b_k, we obtain its corresponding segmentation mask m_k. After deduplication and invalid region filtering, we define the deterministic ROI prior set as $\mathcal{R} = \{(m_k, b_k, l_k)\}_{k=1}^{K'}$, where l_k is the anatomical semantic label derived from the clinical context.

2.2 Perception-Reasoning Intervention Mechanism

Text-Side Evidence Injection. We convert the ROI prior set $\mathcal{R}$ into a collection of verifiable region descriptors and inject them into the textual prompt in a structured form. Given $\mathcal{R} = \{(m_k, \hat{b}_k)\}_{k=1}^{K'}$, we represent each region as a detection pair (l_k, b_k), where l_k is the anatomical label and $b_k = [x, y, w, h]$ specifies its location and scale. We then sequentially construct the structured region set, the serialized evidence block, and the injected prompt as

$$\tilde{q} = [\mathcal{S}(\mathcal{R}) \parallel q \parallel \eta_{\text{task}}], \quad \tilde{V} = V, \tag{2}$$

where $\mathcal{S}(\cdot)$ is a deterministic serialization operator. The operator $\parallel$ denotes token-level concatenation, and $\eta_{\text{task}} =$ `"Answer Yes/No"` is optionally appended for binary QA to reduce format drift. The final prediction is generated by conditioning on $(I, \tilde{q})$.

Vision-Side Activation Recalibration. Let $\mathcal{P}_i \subset \mathbb{Z}^2$ denote the discrete spatial domain of the i-th visual patch token, where $i \in \{1, \ldots, N\}$. We compute the spatial coverage density γ_i by aggregating the global binary ROI mask $\mathbf{B}(u, v) = \bigvee_k m_k(u, v)$ over the patch, normalized by its cardinality $|\mathcal{P}_i|$:

$$\gamma_i = \frac{1}{|\mathcal{P}_i|} \sum_{(u,v)\in\mathcal{P}_i} \mathbf{B}(u, v). \tag{3}$$

To suppress out-of-distribution (OOD) structural collapse while extinguishing background noise, we formulate an activation soft-mask $\mathbf{m} \in \mathbb{R}^N$ governed by the threshold indicator $\mathbb{I}(\cdot)$:

$$m_i = \alpha \cdot \mathbb{I}(\gamma_i \geq \rho) + \beta \cdot (1 - \mathbb{I}(\gamma_i \geq \rho)), \tag{4}$$

where ρ is the density threshold, α serves as a gain factor about the ROI-aligned features, and $\beta \in (0, 1)$ acts as a severe L_2-norm attenuation factor. Implemented via a forward hook prior to the cross-modal projection layer, the recalibration is executed via broadcasting scalar multiplication:

$$\tilde{V} = \mathbf{Diag}(m)V, \quad \tilde{q} = q. \tag{5}$$

By shrinking the feature norm of non-ROI patches by a factor of β, their subsequent dot-product attention scores approach zero exponentially, effectively excising spurious visual pathways.

Joint Modulation Strategy. To achieve complementary regularization, the joint mapping simultaneously enforces L_2-norm visual filtering and discrete semantic anchoring:

$$\Phi_{\text{Joint}}: \quad \tilde{V} = \mathbf{Diag}(m)V, \quad \tilde{q} = [\mathcal{S}(\mathcal{R}) \parallel q \parallel \eta_{\text{task}}]. \tag{6}$$

2.3 Task-Aware Dynamic Router

Since distinct clinical question categories exhibit varying sensitivities to visual vs. textual constraints, statically applying a single strategy yields sub-optimal robustness. We therefore propose a zero-overhead dynamic router based on zero-shot LLM question classification at inference.

Given question q, a lightweight classifier yields its category $\hat{c} = h(q) \in \{\mathcal{C}_A, \mathcal{C}_M, \mathcal{C}_S, \mathcal{C}_R\}$, corresponding to Anatomy, Measurement, Symptom, and

Radiology features, respectively. We define the routing strategy dynamically based on the query category: $s(\hat{c})$ = **Vision-side** if $\hat{c} \in \{\mathcal{C}_A\}$, **Text-side** if $\hat{c} \in \{\mathcal{C}_M, \mathcal{C}_S\}$, and **Joint** if $\hat{c} \in \{\mathcal{C}_R\}$.

Rationale: Anatomy ($\mathcal{C}_A$) relies on spatial localization, naturally benefiting from vision-side L2-norm recalibration. Measurement ($\mathcal{C}_M$) and Symptoms ($\mathcal{C}_S$) demand precise semantic anchoring, making structured textual prompts superior. Radiology features ($\mathcal{C}_R$) typically require complex cross-modal reasoning, necessitating joint constraints on both visual activations and textual generation.

3 Experiments and Results

3.1 Datasets and Evaluation Metric

We evaluate hallucinations following the benchmark by Chang et al. [9]. For close-ended tasks, we report overall and per-type accuracy (anatomy, measurement, symptom, radiology) on MM-VisHal (comprising SLAKE [19] and VQA-RAD [20]) and CXR-VisHal (featuring IU-Xray [21] and MIMIC-CXR [22]). For open-ended evaluation, we assess hallucinations and text quality via CHAIR, CheXbert, RadGraph, RaTEScore, and Recall using 490 image-report pairs sampled from the MIMIC-CXR test set.

3.2 Results

On both close-ended and open-ended evaluation tasks, we systematically evaluate LLaVA-1.5-7B [23], LLaVA-Med-1.5-7B [24], Qwen3-VL-8B/32B [25], and InternVL-3.5-8B/38B [26]; the following sections present these results in turn.

Close-Ended Evaluation. Table 1 shows that our training-free evidence injection consistently improves accuracy across all MLLMs. Notably, text-side injection dominates, achieving the highest overall Acc in 9 of 12 settings (with 2 ties) and an average gain of +1.9%, peaking at +3.3% for Qwen3-VL-8B on MM-VisHal. Conversely, the vision-side constraint exhibits a strong bias toward anatomy-related questions, delivering the best Acc-A in 10 of 12 settings. Finally, the joint strategy excels in radiology-feature questions, attaining the best Acc-R in 9 of 12 settings. The domain-specific benefits of interventions make dynamic routing essential for superior robustness.

As shown in Fig. 2, our task-aware Router consistently outperforms less robust baselines. Each module exhibits selective advantages for different questions, By dynamically exploiting these complementary strengths, routing achieves the highest overall accuracy improvements (+3.2% on MM-VisHal; +2.4% on CXR-VisHal), demonstrating superior robustness over any fixed intervention.

Obs 1: Dynamic dual-side routing transcends the fragility of post-hoc corrections, consistently mitigating multi-source hallucinations.

Table 1. Close-ended evaluation results on MM-VisHal and CXR-VisHal datasets. A, M, S, and R denote Acc-A, Acc-M, Acc-S, and Acc-R, respectively. (**Note:** Due to space constraints, MM-VisHal includes SLAKE and VQA-RAD, while CXR-VisHal comprises IU-Xray and MIMIC-CXR.)

MLLM	Method	MM-VisHal					CXR-VisHal				
		A	M	S	R	Acc	A	M	S	R	Acc
LLaVA-1.5-7B	Original	58.5	42.3	55.7	50.8	52.3	73.2	43.0	52.7	58.2	55.7
	Text-side	65.6	**44.1**	**57.3**	51.3	**55.0**	73.6	**45.2**	**55.9**	56.9	**57.9**
	Vision-side	**69.8**	42.9	54.7	52.7	**55.0**	**74.7**	38.6	54.3	58.3	56.5
	Joint	61.7	42.8	56.2	**52.8**	53.6	72.4	41.7	54.9	**59.1**	56.9
LLaVA-Med-1.5-7B	Original	65.9	38.2	51.1	53.2	51.7	85.3	46.3	70.7	85.3	71.9
	Text-side	65.9	**40.1**	**57.3**	52.8	**54.4**	86.2	48.9	**73.3**	86.6	**74.1**
	Vision-side	**66.9**	39.1	53.6	51.8	52.8	**86.5**	48.1	71.3	85.6	72.7
	Joint	65.3	37.9	55.0	**54.6**	53.2	85.2	**49.2**	71.5	**87.6**	72.9
Qwen3-VL-8B	Original	73.1	61.7	66.0	72.0	67.6	77.0	46.8	78.5	80.7	75.0
	Text-side	76.9	**65.1**	70.2	**72.7**	**70.9**	77.8	**50.4**	**79.7**	81.7	**76.3**
	Vision-side	**77.9**	63.1	65.7	72.2	69.0	**78.3**	48.0	78.3	82.3	75.4
	Joint	72.9	63.0	**70.5**	**72.7**	69.6	78.1	50.0	78.0	**83.6**	75.5
InternVL3.5-8B	Original	76.6	65.0	70.4	84.4	72.9	88.1	73.5	85.1	93.7	85.1
	Text-side	**80.2**	**66.7**	**73.6**	84.4	**75.3**	**88.5**	74.5	85.4	92.9	85.4
	Vision-side	78.1	65.2	70.5	**84.8**	73.4	88.4	73.5	**85.7**	94.1	**85.6**
	Joint	78.6	66.0	73.0	84.1	74.5	87.8	**74.8**	85.3	**94.5**	85.4
Qwen3-VL-32B	Original	78.6	68.3	66.6	74.7	71.1	79.0	60.7	81.9	79.2	78.8
	Text-side	79.5	**69.9**	**70.6**	**76.7**	**73.5**	78.8	**62.5**	**83.1**	76.7	**79.5**
	Vision-side	**80.3**	68.0	69.5	74.8	72.5	**79.7**	62.2	81.8	79.8	79.1
	Joint	77.9	67.4	70.5	75.9	72.4	79.3	61.9	82.3	**81.5**	**79.5**
InternVL3.5-38B	Original	79.1	69.4	70.8	81.2	74.1	80.4	71.4	86.3	91.7	84.2
	Text-side	79.8	70.2	**74.8**	81.0	**75.9**	80.9	**75.6**	**87.0**	91.6	**85.1**
	Vision-side	**81.0**	**70.5**	71.9	81.4	75.3	**81.2**	72.8	86.4	91.7	84.5
	Joint	80.7	67.3	**74.8**	**82.6**	75.6	81.1	72.5	86.5	**93.1**	84.7

Open-Ended Evaluation. Figure 3 shows that ROI-based training-free interventions reduce hallucinations with metric-dependent trade-offs: vision-side constraints most consistently lower CHAIR and improve clinical consistency, text-side injection better preserves coverage and fluency (higher RaTEScore/Recall), and the joint strategy often provides the best balance, yielding strong Recall with competitive CHAIR (notably on InternVL3.5).

Figure 4 highlights the trade-offs of our ROI-based interventions against baselines. While OPERA yields the highest Recall (19.13%) and PAI excels in clinical metrics (CheXbert 23.13%, RadGraph 9.09%), both severely increase hal-

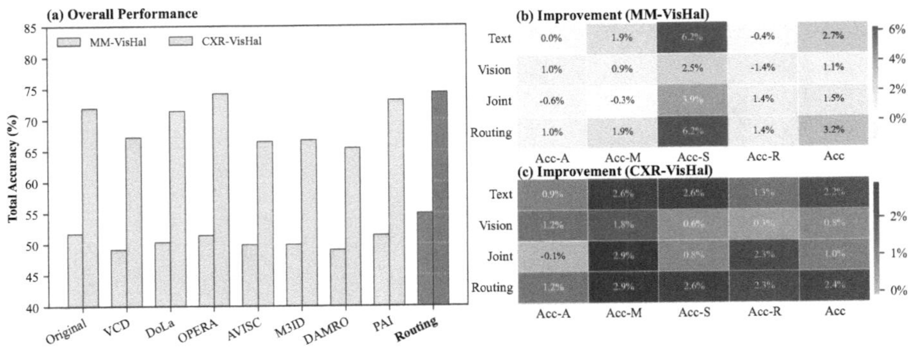

Fig. 2. Evaluation of mitigation strategies. (a) Overall accuracy comparison on MM-VisHal and CXR-VisHal datasets. (b)–(c) Absolute metric improvements of individual components and our Routing over the LLaVA-Med-1.5-7B baseline.

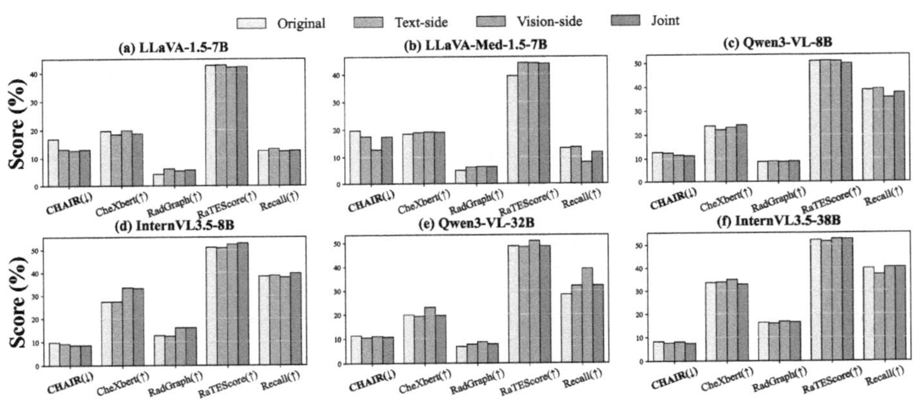

Fig. 3. Open-ended evaluation results on the MIMIC-CXR for report generation. Subfigures (a) to (f) illustrate the evaluation results across six different MLLMs.

lucinations. Our vision-side intervention achieves the lowest CHAIR (12.76%) and improves clinical consistency by curbing spurious visual findings, albeit at a slight cost to Recall. Meanwhile, our text-side intervention maintains Recall and boosts RaTEScore to 44.18%, proving structured textual injection enhances narrative quality and coherence while effectively mitigating hallucinations.

Obs 2: The central challenge in open-ended evaluation remains the difficulty of balancing report completeness and hallucination suppression.

3.3 Ablation Study and Qualitative Analysis

Ablation Study. Figure 5 illustrates the ablation studies for both interventions. **(1) Text-side:** Random bbox degrade performance across all benchmarks, proving that text-side improvements rely on the injected spatial grounding of ROI

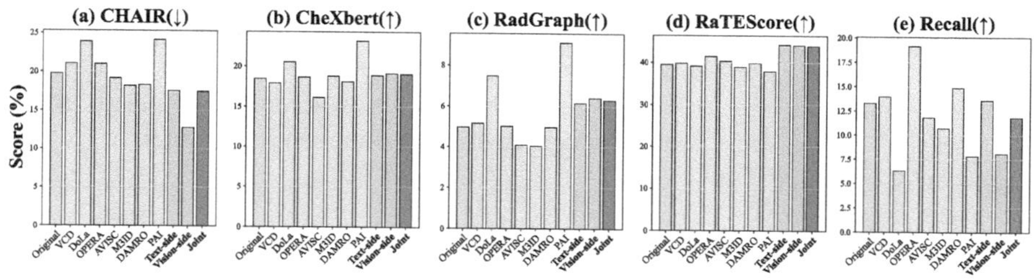

Fig. 4. Open-ended mitigation comparison on MIMIC-CXR using LLaVA-Med-1.5-7B.

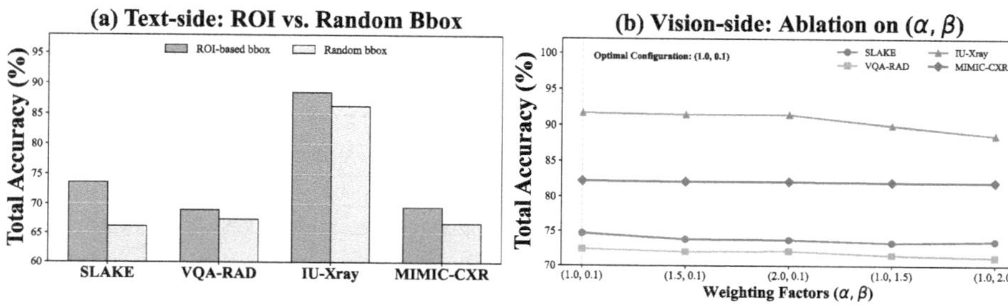

Fig. 5. Controlled comparisons for text-side and vision-side methods. (a) Performance comparison between ROI-based and random bounding boxes for Qwen3-VL-8B. (b) Ablation study of weighting factors (α, β) for InternVL3.5-8B.

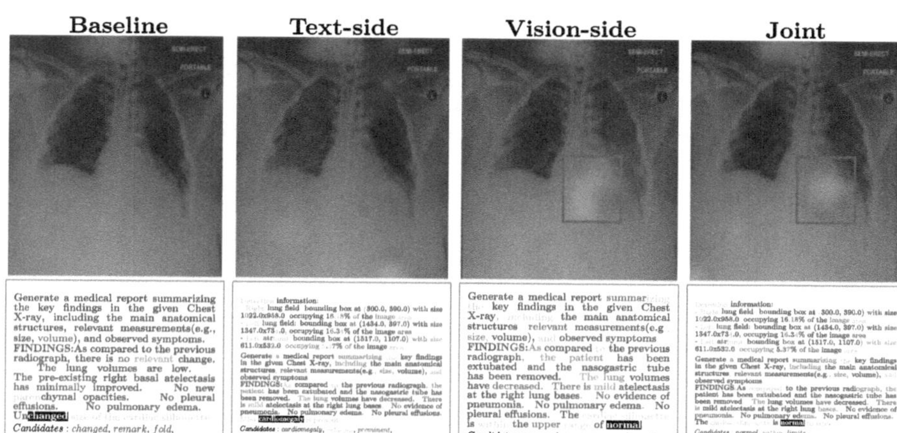

Fig. 6. Token Activation Map (TAM) for visualization presents high-quality localization results on MIMIC-CXR datasets under the InternVL3.5-8B.

coordinates. **(2) Vision-side:** In vision-side weighting, $(\alpha = 1.0, \beta = 0.1)$ is the optimal setting. Over-amplifying ROI features (α) provides no gain, while increasing non-ROI weights (β) consistently introduces distracting signals.

Qualitative Analysis. Figure 6 provide TAM visualizations to elucidate how our interventions steer the grounding mechanism. The baseline's diffuse attention during the generation of *"changed"* is sharpened by vision-side injection, which refocuses the model on the left atrial region—a key semantic anchor for the ground-truth *"normal"* heart size.

4 Conclusion

We address visual misinterpretation hallucinations in medical MLLMs with a training-free, evidence-injection framework. Using MedSAM-extracted ROI priors as verifiable anatomical evidence, our method improves visual grounding and clinical consistency through text-side evidence injection, vision-side activation recalibration, and their joint strategy. We further introduce a lightweight decoupled router that dynamically selects modality-specific interventions based on task semantics. Extensive evaluations on multiple MLLMs across close-ended and open-ended tasks, supported by controlled ablations and visualizations, show robust hallucination mitigation over existing baselines while improving reasoning accuracy and factual reliability. Overall, this work provides a practical path toward visually grounded and controllable generation in clinical workflows.

Acknowledgments. The work was supported by the National Key R&D Program of China under Grant 2025YFE0216500, the Major Program of National Natural Science Foundation of China under Grant 62595802, the Foundation for Outstanding Research Groups of Hubei Province of China under Grant 2025AFA012, the 111 Project on Computational Intelligence and Intelligent Control under Grant B18024, and the National Research Foundation, Prime Minister's Office, Singapore under its IN-CYPHER Campus for Research Excellence and Technological Enterprise (CREATE) Programme.

Disclosure of Interests. The authors have no competing interests in the paper.

References

1. Chen, Z., et al.: CheXagent: towards a foundation model for chest X-ray interpretation. In: AAAI 2024 Spring Symposium Series (Clinical Foundation Models). OpenReview.net (2024). https://openreview.net/forum?id=P3LOmrZWGR
2. Moor, M., et al.: Med-Flamingo: a multimodal medical few-shot learner. In: Proceedings of the 3rd Machine Learning for Health Symposium (ML4H 2023). Proceedings of Machine Learning Research, vol. 225, pp. 353–367. PMLR (2023)
3. Thawakar, O.C., et al.: XrayGPT: chest radiographs summarization using large medical vision-language models. In: Proceedings of the 23rd Workshop on Biomedical Natural Language Processing (BioNLP 2024), pp. 440–448. Association for Computational Linguistics (2024)
4. Wu, C., Zhang, X., Zhang, Y., Hui, H., Wang, Y., Xie, W.: Towards generalist foundation model for radiology by leveraging web-scale 2D&3D medical data. Nat. Commun. **16**, 7866 (2025)

5. Lee, S., Kim, W.J., Chang, J., Ye, J.C.: LLM-CXR: instruction-finetuned LLM for CXR image understanding and generation. In: International Conference on Learning Representations (ICLR) (Poster) (2024). OpenReview.net. https://openreview.net/forum?id=BqHaLnans2
6. Jing, L., Zhang, Y., Du, X.: Tutorial proposal: hallucinations in large language models and large vision-language models. In: Proceedings of the 2025 International Conference on Multimedia Retrieval (ICMR), pp. 2138–2139 (2025)
7. Lin, Z., Guan, S., Zhang, W., Zhang, H., Li, Y., Zhang, H.: Towards trustworthy LLMs: a review on debiasing and dehallucinating in large language models. Artif. Intell. Rev. **57**(9) (2024)
8. Tu, C., et al.: Attention reallocation: towards zero-cost and controllable hallucination mitigation of MLLMs. Int. J. Comput. Vision **134**(1) (2025)
9. Chang, A., Huang, L., Bhatia, P., Kass-Hout, T., Ma, F., Xiao, C.: MedHEval: benchmarking hallucinations and mitigation strategies in medical large vision-language models. arXiv:2503.02157 (2025)
10. Zhu, Z., et al.: Can we trust AI doctors? A survey of medical hallucination in large language and large vision-language models. In: Findings of the Association for Computational Linguistics: ACL 2025, pp. 6748–6769. Association for Computational Linguistics (2025)
11. Asgari, E., Montaña-Brown, N., Dubois, M., et al.: A framework to assess clinical safety and hallucination rates of LLMs for medical text summarisation. NPJ Digit. Med. **8**, 274 (2025)
12. Leng, S., et al.: Mitigating object hallucinations in large vision-language models through visual contrastive decoding. In: Proceedings of the IEEE/CVF Conference on Computer Vision and Pattern Recognition (CVPR), pp. 13872–13882 (2024)
13. Chuang, Y.-S., Xie, Y., Luo, H., Kim, Y., Glass, J., He, P.: DoLa: decoding by contrasting layers improves factuality in large language models. arXiv:2309.03883 (2023)
14. Huang, Q., et al.: OPERA: alleviating hallucination in multi-modal large language models via over-trust penalty and retrospection-allocation. In: Proceedings of the IEEE/CVF Conference on Computer Vision and Pattern Recognition (CVPR), pp. 13418–13427 (2024)
15. Woo, S., Kim, D., Jang, J., Choi, Y., Kim, C.: Don't miss the forest for the trees: attentional vision calibration for large vision language models. arXiv:2405.17820 (2024)
16. Gong, X., Ming, T., Wang, X., Wei, Z.: DAMRO: dive into the attention mechanism of LVLM to reduce object hallucination. In: Proceedings of the 2024 Conference on Empirical Methods in Natural Language Processing (EMNLP), pp. 7696–7712 (2024)
17. Liu, S., Zheng, K., Chen, W.: Paying more attention to image: a training-free method for alleviating hallucination in LVLMs. In: European Conference on Computer Vision (ECCV), pp. 125–140. Springer (2025)
18. Favero, A., et al.: Multi-modal hallucination control by visual information grounding. In: Proceedings of the IEEE/CVF Conference on Computer Vision and Pattern Recognition (CVPR), pp. 14303–14312 (2024)
19. Liu, B., Zhan, L.-M., Xu, L., Ma, L., Yang, Y., Wu, X.-M.: SLAKE: a semantically-labeled knowledge-enhanced dataset for medical visual question answering. In: 2021 IEEE 18th International Symposium on Biomedical Imaging (ISBI), pp. 1650–1654. IEEE (2021)

20. Lau, J.J., Gayen, S., Ben Abacha, A., Demner-Fushman, D.: A dataset of clinically generated visual questions and answers about radiology images. Sci. Data **5**(1), 1–10 (2018)
21. Demner-Fushman, D., et al.: Preparing a collection of radiology examinations for distribution and retrieval. J. Am. Med. Inform. Assoc. **23**(2), 304–310 (2016)
22. Johnson, A.E.W., et al.: MIMIC-CXR, a de-identified publicly available database of chest radiographs with free-text reports. Sci. Data **6**(1), 317 (2019)
23. Liu, H., Li, C., Li, Y., Lee, Y.J.: Improved baselines with visual instruction tuning. In: Proceedings of the IEEE/CVF Conference on Computer Vision and Pattern Recognition (CVPR), pp. 26296–26306 (2024)
24. Li, C., et al.: LLaVA-Med: training a large language-and-vision assistant for biomedicine in one day. In: Advances in Neural Information Processing Systems, vol. 36, pp. 28541–28564. Curran Associates, Inc. (2023)
25. Bai, S., et al.: Qwen3-VL technical report. arXiv preprint arXiv:2511.21631 (2025)
26. Wang, W., et al.: InternVL3.5: advancing open-source multimodal models in versatility, reasoning, and efficiency. arXiv preprint arXiv:2508.18265 (2025)

TAVR-VLM: Risk-Conditioned Causal Grounding for Hallucination-Resistant Report Generation

Zhixiang Lu[1,4](✉), Xiwei Liu[2], Sifan Song[1], Changkai Ji[3], Anh Nguyen[4], Jionglong Su[1], Imran Razzak[2], and Jinfeng Wang[1,5]

[1] Xi'an Jiaotong-Liverpool University, Suzhou, China
zhixiang@liverpool.ac.uk
[2] Mohamed bin Zayed University of Artificial Intelligence, Abu Dhabi, UAE
[3] Shanghai Jiao Tong University, Shanghai, China
[4] University of Liverpool, Liverpool, UK
[5] Kunming University of Science and Technology, Kunming, China

Abstract. Transcatheter Aortic Valve Replacement (TAVR) planning requires meticulous multimodal reasoning. However, adapting Multimodal Large Language Models (MLLMs) to this high-stakes domain is severely impeded by diagnostic hallucinations, where generated text lacks anatomical grounding. To address this, TAVR-VLM is introduced: a novel framework featuring Risk-Conditioned Causal Grounding Attention (R-CGA) that instantiates a model-internal "Risk → Region → Word" structural grounding pathway. R-CGA compresses multimodal inputs into a causal risk bottleneck, purifying dense visual features into a global risk mask. During autoregressive generation, a support-projected causal consistency objective constrains token-level grounding within the risk-defined support mask. Evaluated on M^{3}TAVR, TAVR-VLM demonstrates consistently strong performance across risk stratification, report generation, hallucination reduction and spatial grounding. The results indicate that R-CGA can enhance both the factual reliability and anatomical interpretability of vision-language models for TAVR preoperative planning.

Keywords: Vision-Language Models · Causal Grounding · Hallucination Mitigation · Transcatheter Aortic Valve Replacement

1 Introduction

Transcatheter Aortic Valve Replacement (TAVR) has established itself as the standard-of-care for severe aortic stenosis across the full spectrum of surgical risk [7,16]. Despite its clinical success, procedural outcomes remain highly sensitive to the precision of preoperative planning. Subtle inaccuracies in the assessment of 3D CT annular geometry or 2D echocardiographic hemodynamics can precipitate catastrophic complications, including annular rupture, coronary obstruction, or significant paravalvular leak [9,18]. To ensure safety and reproducibility,

G. Yang et al. (Eds.): MICCAI 2026, LNCS 16878, pp. 601–611, 2027.
https://doi.org/10.1007/978-3-032-38059-3_57

the Valve Academic Research Consortium (VARC-3) emphasizes the necessity of evidence-grounded evaluation [18]. Consequently, an ideal AI-driven TAVR system must transcend simple risk scoring to provide structured clinical reports and actionable recommendations explicitly grounded in multimodal imaging evidence.

The rapid evolution of multimodal large language models (MLLMs) has introduced the potential for generalist medical assistants capable of joint image-text interpretation [8]. While radiology report generation has progressed significantly using Transformer-based architectures on large-scale datasets [6,15], adapting these models to TAVR-grade clinical reasoning reveals a critical technical bottleneck. TAVR planning requires a rigorous coupling between global procedural risk states and localized anatomical findings. Contemporary medical MLLMs frequently exhibit diagnostic hallucinations, where the model generates plausible but non-existent clinical findings or recommendations unsupported by the underlying imagery [5,13]. In the context of structural heart interventions, such hallucinations are unacceptable as they directly compromise surgical decision-making. We posit that hallucinations persist in medical MLLMs due to the absence of explicit causal constraints [14,22]. Traditional models typically generate text by attending to dense visual tokens without ensuring that each reported clinical entity is anchored to risk-relevant evidence [13,15]. In professional TAVR workflows, clinicians utilize a top-down reasoning paradigm: they first establish a global risk hypothesis, then systematically interrogate specific anatomical regions to seek supporting evidence, and finally synthesize these insights into a structured plan [12,13]. This observation motivates our core design principle: Risk should serve as a causal bottleneck that governs the visual perception and linguistic output of the generator.

To operationalize this principle, we introduce **Risk-Conditioned Causal Grounding Attention (R-CGA)**, a top-down closed-loop module that integrates risk stratification with multi-granularity grounding for TAVR planning. R-CGA first predicts a global risk distribution and compresses it into a learnable, causal bottleneck. This bottleneck is then utilized to purify dense visual features into a global risk mask, ensuring the model encourages the model to focus on risk-relevant evidence relevant to the predicted risk state. Finally, we introduce a causal consistency supervision mechanism with a stop-gradient regularizer. This ensures that the token-level grounding maps for all generated clinical entities are spatially contained within the global risk-evidence region, thereby transforming risk prediction into a structural prior that mitigates hallucinations. Our main contributions are summarized as follows:

- We formalize TAVR assessment as a risk-conditioned structured generation task, identifying hallucination as a failure of the top-down causal link between global risk states and local anatomical evidence.
- We propose a three-stage architecture comprising a causal risk bottleneck, risk-to-region evidence selection, and region-to-word token grounding to enforce a strict "Risk → Region → Word" reasoning hierarchy.

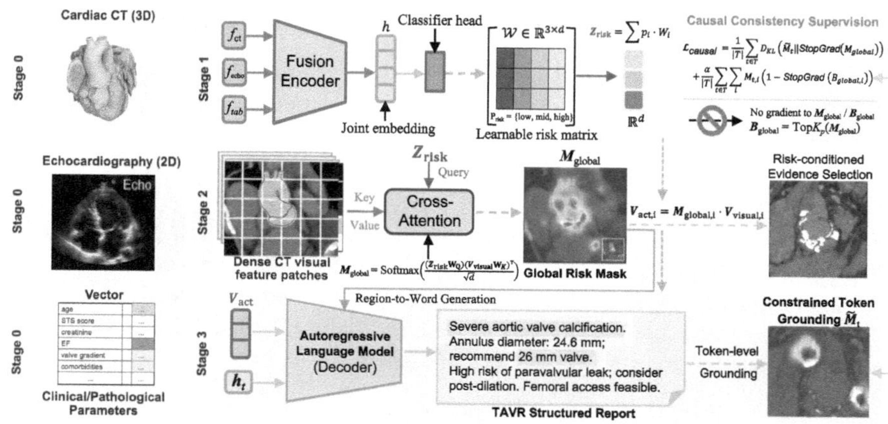

Fig. 1. Overview of the TAVR-VLM architecture. R-CGA implements a model-internal "Risk → Region → Word" structural grounding pathway. **Stage 1** encodes 3D CT, echocardiography clips, and clinical parameters into a joint embedding, predicts a risk distribution P_{risk}, and constructs a causal risk bottleneck $Z_{\text{risk}} = P_{\text{risk}}^{\top} W$. **Stage 2** uses Z_{risk} as the query and CT visual tokens as key/value features to obtain a global risk mask M_{global}, from which a risk-defined support mask $B_{\text{global}} = \text{TopK}\rho(M_{\text{global}})$ is derived. **Stage 3** generates the structured report from risk-activated visual features V_{act} while projecting raw token attention M_t into B_{global} to obtain the constrained token mask $\widetilde{M_t}$. The causal consistency loss aligns $\widetilde{M_t}$ with the global risk mask and penalizes attention outside the risk-defined support.

– We conduct comprehensive experiments on the M^{3}TAVR, including risk prediction, report generation, hallucination evaluation, spatial grounding, ablation analysis and qualitative grounding analysis.

2 Methodology

2.1 Problem Formulation and Causal Framework

We study TAVR pre-operative decision support as a risk-conditioned multimodal generation problem. Given multimodal inputs $\mathcal{X} = \{X_{\text{CT}}, X_{\text{Echo}}, X_{\text{tab}}\}$ (3D CT, 2D echocardiography, and clinical tables), our objective is to learn a joint distribution $P(Y, r \mid \mathcal{X})$. Here, $r \in \{l, m, h\}$ denotes the discrete procedural risk level, and $Y = (y_1, \ldots, y_T)$ represents the autoregressive structured report. To mitigate diagnostic hallucinations, we depart from standard multi-task learning by introducing a strict top-down causal constraint: Risk → Region → Word. We operationalize this via the **Risk-Conditioned Causal Grounding Attention (R-CGA)** module, which enforces that every clinically salient entity in Y is intrinsically grounded in risk-relevant anatomical evidence (Fig. 1).

2.2 Risk-Conditioned Causal Grounding Attention (R-CGA)

R-CGA implements the causal hierarchy through three interconnected stages, acting as an information bottleneck that filters irrelevant visual noise prior to text generation.

Stage 1: Causal Risk Bottleneck. Modality-specific encoders map inputs into a unified latent space, extracting dense visual tokens $V \in \mathbb{R}^{N \times d}$. A lightweight fusion network computes a joint patient embedding to predict the risk distribution $P_{\text{risk}} = [p_l, p_m, p_h]^\top$. Instead of passing raw embeddings to the generator, we introduce a learnable risk matrix $W \in \mathbb{R}^{3 \times d}$. The **Causal Risk Bottleneck** is computed as the expected prototype under P_{risk}:

$$Z_{\text{risk}} = P_{\text{risk}}^\top W \in \mathbb{R}^d. \tag{1}$$

Z_{risk} serves as a distribution-aware macro "commander" state, encapsulating the minimal sufficient clinical evidence required to justify the predicted risk profile.

Stage 2: Risk-to-Region Evidence Selection. Let $V_{\text{visual}} \in \mathbb{R}^{N \times d}$ denote the dense CT visual tokens used for spatial grounding. The risk bottleneck Z_{risk} acts as the query, while V_{visual} serves as key/value features. The global risk mask is obtained by cross-attention [4,11]:

$$M_{\text{global}} = \text{Softmax}\left(\frac{(Z_{\text{risk}} W_Q)(V_{\text{visual}} W_K)^\top}{\sqrt{d_k}}\right) \in \Delta^{N-1}. \tag{2}$$

The risk-activated visual features are then computed by token-wise gating:

$$V_{\text{act},i} = M_{\text{global},i} \cdot V_{\text{visual},i}, \qquad i = 1, \dots, N. \tag{3}$$

This separates the soft global risk mask used for visual activation from the support mask used for strict token containment.

Stage 3: Region-to-Word Token Grounding. During autoregressive generation, the decoder hidden state h_t computes a raw token-level attention map over the original CT visual tokens:

$$M_t = \text{Softmax}\left(\frac{(h_t W_Q^{(t)})(V_{\text{visual}} W_K^{(t)})^\top}{\sqrt{d_k}}\right) \in \Delta^{N-1}. \tag{4}$$

Let $\mathcal{T}$ denote the set of token indices corresponding to clinically salient entities, such as "calcification", "LVOT". For each $t \in \mathcal{T}$, M_t represents the raw visual evidence used by the decoder before causal support projection.

2.3 Support-Projected Causal Consistency and Optimization

The global risk mask M_{global} identifies the anatomical region that is relevant to the predicted risk state. To enforce that clinically salient tokens are grounded within this region, we derive a risk-defined support mask:

$$B_{\text{global}} = \text{TopK}_{\rho}(M_{\text{global}}), \tag{5}$$

where $\text{TopK}_{\rho}(\cdot)$ retains the top ρ proportion of visual tokens and returns a binary support mask. We then project the raw token grounding map M_t into this risk-defined support:

$$\widetilde{M}_t = \frac{M_t \odot \text{StopGrad}(B_{\text{global}})}{\sum_i M_{t,i}\text{StopGrad}(B_{\text{global},i}) + \epsilon}. \tag{6}$$

By construction,

$$\text{supp}(\widetilde{M}_t) \subseteq \text{supp}(B_{\text{global}}), \quad \forall t \in \mathcal{T}. \tag{7}$$

This support projection provides the formal containment property that the raw KL alignment alone cannot guarantee. The final **Support-Projected Causal Consistency Loss** $\mathcal{L}_{\text{causal}}$ contains two terms:

$$\begin{aligned}\mathcal{L}_{\text{causal}} = &\frac{1}{|\mathcal{T}|}\sum_{t\in\mathcal{T}} \mathbb{D}_{\text{KL}}\big(\widetilde{M}_t \parallel \text{StopGrad}(M_{\text{global}})\big) \\ &+ \frac{\alpha}{|\mathcal{T}|}\sum_{t\in\mathcal{T}}\sum_i M_{t,i}\big(1 - \text{StopGrad}(B_{\text{global},i})\big).\end{aligned} \tag{8}$$

The first term aligns the constrained token grounding distribution with the global risk mask. The second term penalizes raw token attention that leaks outside the risk-defined support. The StopGrad operator is applied only to the causal consistency branch, so that token grounding is optimized to obey the risk-defined region without allowing the token loss to expand or diffuse the global risk support. The end-to-end multi-task objective is jointly optimized:

$$\mathcal{L} = \mathcal{L}_{\text{risk}} + \mathcal{L}_{\text{LM}} + \lambda_{\text{rec}}\mathcal{L}_{\text{rec}} + \lambda_{\text{causal}}\mathcal{L}_{\text{causal}}, \tag{9}$$

where $\mathcal{L}_{\text{risk}}$ is the cross-entropy for risk prediction, $\mathcal{L}_{\text{LM}}$ is the autoregressive negative log-likelihood over V_{act}, and $\mathcal{L}_{\text{rec}}$ optimizes surgical recommendations. This design fundamentally transforms risk prediction from an isolated classification head into a structural prior that governs generation.

3 Experiments

3.1 Dataset

To rigorously evaluate our framework, we curate **M^{3}TAVR** (Multimodal, Multi-granularity, Multi-task cohort for TAVR), the largest structural heart disease benchmark to date, comprising 1,482 patients. Each patient record is strictly

Table 1. Comprehensive benchmarking on the $\mathbf{M^3TAVR}$ test set. Models are grouped into four paradigms. Best results are in **bold**, and second-best are <u>underlined</u>.

Model	Risk Prediction		Report Generation			Safety & Grounding	
	AUROC ↑	Macro-F1 ↑	ROUGE-L ↑	CIDEr ↑	ClinEnt-F1 ↑	Halluc. ↓	mIoU ↑
Risk-Only Baselines							
XGBoost [2]	0.785	0.742	–	–	–	–	–
TabNet [1]	0.791	0.750	–	–	–	–	–
Late-Fusion (CT+Echo+Tab)	0.832	0.796	–	–	–	–	–
Report Generation Baselines							
MCLN-Transformer [3]	–	–	0.316	0.662	0.432	37.9%	0.175
R2Gen-Mamba [17]	–	–	0.329	0.684	0.458	34.6%	0.210
METransformer [21]	–	–	0.341	0.703	0.470	33.8%	0.234
Open-Source Multimodal Models							
LLaVA-Med-1.5 [8]	0.804	0.773	0.388	0.821	0.614	24.2%	0.454
Qwen3-VL-8B [23]	0.849	0.823	0.401	0.868	0.695	17.9%	0.501
InternVL3.5-8B [20]	0.861	0.836	0.409	0.889	0.724	14.7%	0.516
Closed-Source Frontier Models							
GPT-4o	0.868	0.844	0.412	0.901	0.756	13.8%	–
Claude-4 Opus	0.875	0.852	0.415	0.908	0.764	12.9%	–
Gemini-3 Pro	0.883	0.860	0.418	0.914	0.772	11.5%	–
TAVR-VLM (Proposed Framework)							
R-CGA + Qwen3-VL-8B	<u>0.892</u>	<u>0.869</u>	<u>0.422</u>	<u>0.928</u>	<u>0.789</u>	<u>8.7%</u>	<u>0.603</u>
R-CGA + InternVL3.5-8B	**0.896**	**0.874**	**0.426**	**0.936**	**0.801**	**8.1%**	**0.624**

paired with three modalities: (i) 3D cardiac CT volumes, (ii) 2D echocardiography clips, and (iii) a comprehensive vector of clinical and pathological biomarkers X_{tab}. Furthermore, each case is annotated with a ground-truth procedural risk level $r \in \{l, m, h\}$ and a clinician-authored structured report Y containing diagnostic findings and surgical recommendations (e.g., valve sizing and access routes). To prevent data leakage and ensure robust evaluation, we employ a strict patient-level stratified split: 70% for training, 10% for validation, and 20% for testing. For a randomly sampled subset of the test set ($N = 200$), we additionally obtained expert-annotated region-of-interests (ROIs) for TAVR-critical structures to enable quantitative grounding evaluation.

3.2 Experimental Setup and Evaluation Metrics

To ensure a comprehensive evaluation, we benchmark our framework against four distinct paradigms: risk-only classifiers, medical report generators, open-source MLLMs, and closed-source frontier models, extracting standardized 2D slices for 2D-centric baselines to maintain domain fairness. Our architecture, driven by a 3D ViT for CT extraction and a temporal 2D-CNN for echocardiography, is optimized end-to-end via AdamW across four NVIDIA A100 GPUs with the causal consistency coefficient λ_{causal} empirically set to 0.5. We evaluate the joint modeling capabilities across three critical dimensions: clinical prediction performance (AUROC and Macro-F1), generation fidelity (ROUGE-L [10] and CIDEr [19]).

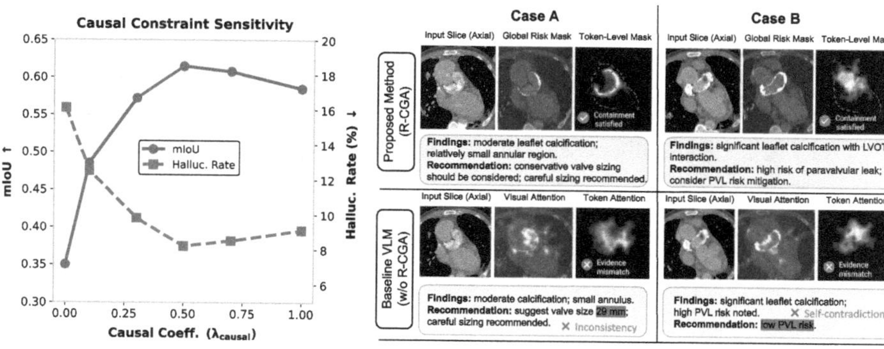

(a) Parameter Sensitivity (b) Qualitative Case Study

Fig. 2. Comprehensive Model Analysis. (a) Sensitivity of the causal consistency coefficient λ_{causal} on spatial grounding accuracy and hallucination rate. **(b)** Qualitative comparison between R-CGA and a baseline VLM without risk-conditioned grounding.

To rigorously penalize diagnostic fabrication, we define Hallucination Rate (HR) as the percentage of generated clinical entities lacking corresponding multimodal evidence, while spatial anchoring precision is quantified via mean Intersection over Union (mIoU) between the constrained token-level mask $\widetilde{M}_t$ and expert-annotated ROIs.

3.3 Main Results

Table 1 presents a comprehensive quantitative comparison on the **M^{3}TAVR** test set. By benchmarking against four distinct paradigms, we demonstrate that our proposed R-CGA framework establishes a new state-of-the-art across all evaluated clinical and generation dimensions.

Superiority in Clinical Prediction and Generation. While current open-source models and cutting-edge closed-source VLMs exhibit formidable general vision-language capabilities, they lack the explicit domain constraints required for high-stakes surgical planning. By integrating our plug-and-play R-CGA module into an 8B-parameter backbone (InternVL3.5-8B), we observe a transformative performance leap. Specifically, our model achieves an AUROC of 0.896 in risk prediction, outperforming the top-performing frontier model, Gemini-3 Pro (0.883). In report generation, our framework surpasses Gemini-3 Pro in CIDEr (0.936 vs. 0.914), indicating a higher consensus with clinical experts in utilizing precise TAVR-specific terminology rather than generic descriptions.

Breakthrough in Safety and Spatial Grounding. Diagnostic hallucination remains the most critical bottleneck for MLLM adoption in structural heart interventions. As shown in Table 1, traditional report generation baselines suffer from severe hallucinations ($> 33\%$) and diffuse visual attention (mIoU < 0.25). Even the

Table 2. Ablation of key R-CGA components on **M³TAVR**.

Variant	AUROC ↑	CIDEr ↑	Halluc. Rate ↓	mIoU ↑
Full TAVR-VLM	**0.896**	**0.936**	**8.1%**	**0.624**
w/o Risk Prototypes W	0.875	0.890	12.5%	0.542
w/o Purification ($V_{\text{act}} = V$)	0.881	0.865	18.4%	0.485
w/o Causal Loss ($\lambda_{\text{causal}} = 0$)	0.885	0.872	16.1%	0.350
w/o Stop-Gradient	0.888	0.895	14.3%	0.412

most advanced frontier model, Gemini-3 Pro, yields a hallucination rate of 11.5%, as its autoregressive generation relies heavily on generic memory priors rather than rigorous visual evidence.

In stark contrast, the top-down causal bottleneck in R-CGA reduces hallucinations through risk-conditioned grounding. Enforced by this structural prior, our hallucination rate decreases to 8.1%, while token-to-region grounding precision (mIoU) surges to 0.624. This empirically validates that our model does not merely memorize linguistic distributions; it accurately localizes pathological evidence (e.g., specific calcified leaflets) before emitting the corresponding clinical entities.

3.4 Ablation Study

Module Ablation. To disentangle the contributions of the R-CGA components, we conduct an ablation study (Table 2). Removing the learnable risk prototypes (W) degrades the macro bottleneck representation, lowering AUROC to 0.875. Crucially, bypassing the risk-driven visual purification (w/o V_{act}) exposes the language model to unconstrained tokens, triggering an immediate spike in hallucination rate (from 8.1% to 18.4%) as the generator exploits spurious correlations. Disabling the causal consistency loss ($\lambda_{\text{causal}} = 0$) severs the top-down alignment, causing a steep drop in spatial grounding precision (mIoU falls to 0.350). Finally, replacing the stop-gradient mechanism with standard backpropagation (w/o StopGrad) results in degenerate optimization; the model artificially inflates M_{global} to trivially minimize KL divergence, validating that risk must serve as a strict, unidirectional bottleneck.

Parameter Sensitivity. We analyze the sensitivity of the causal consistency coefficient λ_{causal} (Fig. 2a). A clear trade-off emerges: increasing λ_{causal} from 0 to 0.5 drastically improves mIoU (from 0.350 to 0.624) while curbing the diagnostic hallucination rate down to 8.2%. However, overly aggressive penalization ($\lambda_{\text{causal}} > 0.5$) marginally degrades performance, indicating that the language generator becomes too rigidly constrained by the global mask, sacrificing linguistic flexibility. Based on these dynamics, we adopt $\lambda_{\text{causal}} = 0.5$ and a prototype dimension of $d = 512$ for all experiments.

Qualitative Case Study. To intuitively understand how R-CGA suppresses hallucinations, we visualize representative cases comparing R-CGA with a baseline VLM without risk-conditioned grounding in Fig. 2b. The baseline model often covers diffuse or clinically irrelevant regions, leading to a mismatch of evidence and unsupported recommendations. In contrast, R-CGA first selects a global risk-relevant region and then constrains token-level grounding to this region through support projection. This mechanism reduces unsupported clinical entities and improves the anatomical consistency between the visual evidence and the generated report statements.

4 Conclusion

We presented TAVR-VLM[1], a pioneering risk-conditioned multimodal framework for hallucination-resistant TAVR report generation. The proposed R-CGA module constructs a model-internal "Risk → Region → Word" structural pathway by using a causal risk bottleneck to guide anatomical evidence selection and token-level grounding. Unlike unconstrained generation, R-CGA projects clinically significant token attention into a risk-defined support mask and penalizes grounding outside this support, thus mitigating evidence-mismatched clinical statements. Experiments on the M^3TAVR dataset demonstrate that TAVR-VLM outperforms both specialized medical models and cutting-edge frontier VLMs in risk prediction, report generation, hallucination reduction, and spatial grounding. Future work will focus on external validation, more detailed clinical endpoint modeling, and broader evaluation across structural heart interventions.

Disclosure of Interests. The authors have no competing interests to declare that are relevant to the content of this article.

References

1. Arik, S.Ö., Pfister, T.: Tabnet: attentive interpretable tabular learning. In: Proceedings of the AAAI Conference on Artificial Intelligence (2021). https://doi.org/10.1609/aaai.v35i8.16826
2. Chen, T., Guestrin, C.: Xgboost: a scalable tree boosting system. In: Proceedings of the 22nd ACM SIGKDD International Conference on Knowledge Discovery and Data Mining (2016). https://doi.org/10.1145/2939672.2939785
3. Chen, Z., Song, Y., Chang, T.H., Wan, X.: Generating radiology reports via memory-driven transformer. In: Proceedings of the 2020 Conference on Empirical Methods in Natural Language Processing (EMNLP), pp. 1439–1449. Association for Computational Linguistics (2020). https://doi.org/10.18653/v1/2020.emnlp-main.112
4. Gheini, M., Ren, X., May, J.: Cross-attention is all you need: adapting pretrained transformers for machine translation. In: Proceedings of the 2021 Conference on Empirical Methods in Natural Language Processing, pp. 1754–1765. Association for Computational Linguistics (2021). https://doi.org/10.18653/v1/2021.emnlp-main.132

[1] The code is available at: https://github.com/Leo1998-Lu/TAVR-VLM.

5. Huang, L., et al.: A survey on hallucination in large language models: principles, taxonomy, challenges, and open questions. ACM Trans. Inf. Syst. **43**(2) (2025). https://doi.org/10.1145/3703155
6. Johnson, A.E.W., et al.: MIMIC-CXR, a de-identified publicly available database of chest radiographs with free-text reports. Sci. Data (2019). https://doi.org/10.1038/s41597-019-0322-0
7. Leon, M.B., Smith, C.R., Mack, M., Miller, D.C., Moses, J.W., Svensson, L.G., et al.: Transcatheter aortic-valve implantation for aortic stenosis in patients who cannot undergo surgery. N. Engl. J. Med. (2010). https://doi.org/10.1056/NEJMoa1008232
8. Li, C., et al.: LLaVA-med: training a large language-and-vision assistant for biomedicine in one day. In: Thirty-Seventh Conference on Neural Information Processing Systems Datasets and Benchmarks Track (2023). https://openreview.net/forum?id=GSuP99u2kR
9. Lilly, S.M., Deshmukh, A.J., Epstein, A.E., Ricciardi, M.J., Shreenivas, S., Velagapudi, P., et al.: 2020 ACC expert consensus decision pathway on management of conduction disturbances in patients undergoing transcatheter aortic valve replacement. J. Am. Coll. Cardiol. (2020). https://doi.org/10.1016/j.jacc.2020.08.050
10. Lin, C.Y.: Rouge: a package for automatic evaluation of summaries. In: Workshop on Text Summarization Branches Out (WAS 2004) (2004)
11. Lu, Z., et al.: Deepgb-tb: a risk-balanced cross-attention gradient-boosted convolutional network for rapid, interpretable tuberculosis screening. In: Proceedings of the AAAI Conference on Artificial Intelligence, vol. 40, no. 46, pp. 38989–38997 (2026). https://doi.org/10.1609/aaai.v40i46.41245
12. Lu, Z., Su, J.: Hierrisk: a hierarchical framework for suicide risk prediction on social media. In: 2025 IEEE International Conference on Big Data (BigData), pp. 8169–8174 (2025). https://doi.org/10.1109/BigData66926.2025.11402629
13. Lu, Z., Su, J.: Dialectic-med: mitigating diagnostic hallucinations via counterfactual adversarial multi-agent debate (2026). https://arxiv.org/abs/2604.11258
14. Lu, Z., Xu, S., Li, Y., Su, J., Tang, T.: Causal-SAM-LLM: large language models as causal reasoners for robust medical segmentation. In: ICASSP 2026 - 2026 IEEE International Conference on Acoustics, Speech and Signal Processing (ICASSP), pp. 6246–6250 (2026). https://doi.org/10.1109/ICASSP55912.2026.11460869
15. Lu, Z., et al.: Skinclip-vl: consistency-aware vision-language learning for multimodal skin cancer diagnosis (2026). https://arxiv.org/abs/2603.21010
16. Mack, M.J., Leon, M.B., Thourani, V.H., Makkar, R.R., Kodali, S.K., Russo, M., et al.: Transcatheter aortic-valve replacement with a balloon-expandable valve in low-risk patients. N. Engl. J. Med. (2019). https://doi.org/10.1056/NEJMoa1814052
17. Sun, Y., Lee, Y.Z., Woodard, G.A., Zhu, H., Lian, C., Liu, M.: R2gen-mamba: a selective state space model for radiology report generation. In: 2025 IEEE 22nd International Symposium on Biomedical Imaging (ISBI), pp. 1–4 (2025). https://doi.org/10.1109/ISBI60581.2025.10980814
18. VARC-3 Writing Committee, et al.: Valve academic research consortium 3: updated endpoint definitions for aortic valve clinical research. J. Am. Coll. Cardiol. (2021). https://doi.org/10.1016/j.jacc.2021.02.038
19. Vedantam, R., Zitnick, C.L., Parikh, D.: Cider: consensus-based image description evaluation. In: 2015 IEEE Conference on Computer Vision and Pattern Recognition (CVPR), pp. 4566–4575 (2015). https://doi.org/10.1109/CVPR.2015.7299087
20. Wang, W., et al.: Internvl3.5: advancing open-source multimodal models in versatility, reasoning, and efficiency (2025). https://arxiv.org/abs/2508.18265

21. Wang, Z., Liu, L., Wang, L., Zhou, L.: Metransformer: radiology report generation by transformer with multiple learnable expert tokens. In: 2023 IEEE/CVF Conference on Computer Vision and Pattern Recognition (CVPR), pp. 11558–11567. IEEE Computer Society (2023). https://doi.org/10.1109/CVPR52729.2023.01112
22. Xue, C., Liu, Y., Zhou, M., Su, J., Lu, Z.: Semantic-topological graph reasoning for language-guided pulmonary screening (2026). https://arxiv.org/abs/2604.05620
23. Yang, A., et al.: Qwen3 technical report (2025). https://arxiv.org/abs/2505.09388

TC-SSA: Token Compression via Semantic Slot Aggregation for Gigapixel Pathology Reasoning

Zhuo Chen[1,2], Xiaoyu Yang[1], and Lijian Xu[1(✉)]

[1] Shenzhen University of Advanced Technology, Shenzhen, Guangdong, China
xulijian@suat-sz.edu.cn
[2] FoSE, University of Nottingham Ningbo China, Ningbo, Zhejiang, China

Abstract. The application of large vision-language models to computational pathology holds great promise for diagnostic assistants but faces a critical computational bottleneck: the gigapixel scale of Whole Slide Images (WSIs). A single WSI typically contains over 10^5 patches, creating sequence lengths that exceed the constraints of standard Transformer architectures. Existing solutions often resort to spatial sampling, which risks discarding diagnostically critical evidence. To address this, we propose TC-SSA (**T**oken **C**ompression via **S**emantic **S**lot **A**ggregation), a learnable token compression framework that aggregates patch features into a fixed number of semantic slots. A gated routing module assigns patches to slots using sparse Top-2 routing, followed by weighted aggregation, enabling global slide coverage under a strict token budget. The resulting representation retains diagnostically relevant information while reducing the number of visual tokens to 1.7% of the original sequence. On SlideBench (TCGA), our model achieves 78.34% overall accuracy and 77.14% on the diagnosis subset, outperforming sampling-based baselines under comparable token budgets. The method also generalizes to MIL classification, reaching AUCs of 95.83% on TCGA-BRCA and 98.27% on TCGA-NSCLC, and an ISUP grading accuracy of 79.80% on PANDA. These results suggest that learnable semantic aggregation provides an effective tradeoff between efficiency and diagnostic performance for gigapixel pathology reasoning. Our code is available at https://github.com/OzzyChen97/TC-SSA.git.

Keywords: Gigapixel pathology · Token compression · Vision-language models

1 Introduction

Vision-language models (VLMs) enable multimodal reasoning in computational pathology [12]. These models combine visual evidence with textual queries, offering an interface for slide-level question answering and clinical assistance. However, whole-slide images (WSIs) present a fundamental scalability challenge. A

Z. Chen and X. Yang—These authors contributed equally to this work.

G. Yang et al. (Eds.): MICCAI 2026, LNCS 16878, pp. 612–621, 2027.
https://doi.org/10.1007/978-3-032-38059-3_58

single WSI contains more than 10^5 patches, and sequence lengths exceed the memory and computational limits of standard Transformer architectures. Direct end-to-end processing of all patches is infeasible. Although recent advancements in VLMs have enabled highly capable pathology copilots through diverse pre-training and unified architectures [2,9,12,13,18,23,28], deploying these comprehensive models on gigapixel slides demands a critical balance between computational efficiency and diagnostic effectiveness. To address this challenge, existing approaches predominantly follow two directions. First, spatial sampling strategies, exemplified by LLaVA-Med [11] and Quilt-LLaVA [16], restrict the input to a fixed context window by discarding the majority of patches, which introduces a severe risk of omitting diagnostically critical regions. Second, sparse-attention frameworks such as SlideChat [3] retain broader visual evidence but incur substantially higher inference costs. Parallel to these primary directions, while foundational weakly supervised modeling is frequently formulated as Multiple Instance Learning [5,14,17,22], recent investigations extensively explore token compression, instruction tuning, and summarization techniques to facilitate efficient slide-level learning for pathology models [1,4,8,19,21,24,26]. To mitigate the efficiency bottleneck, we propose TC-SSA (Token Compression via Semantic Slot Aggregation), a learnable token-budgeting framework aggregating all patch features into fixed learnable semantic slots. To avoid systematically under-representing rare yet diagnostically critical evidence, a semantic slot regularization objective stabilizes slot utilization. Under a fixed token budget, this bottleneck maintains global slide coverage, eliminating dense attention's quadratic cost. Thus, our main contributions are: 1) **Semantic Slot-Based Token Compression:** We propose a semantic mechanism for WSI token compression, routing all visual tokens into a fixed set of learned semantic slots based on shared contextual relevance instead of spatial proximity. It aggregates sparse, diagnostically critical evidence while suppressing redundant background noise, thereby preserving the global context of the image under a rigorous token budget. 2) **Robust Regularization for Semantic Slots:** We propose semantic slot regularization to mitigate slot collapse and ensure routing stability during training. By jointly optimizing a soft load-balancing loss, an entropy regularizer, and a z-loss, the framework discourages degenerate slot flooding without forcing equal slot utilization. 3) **Superior efficiency-performance tradeoff:** Under a strict visual-token budget (only 1.7% of WSI), TC-SSA achieves 78.34% overall accuracy on SlideBench (TCGA) and improves the Diagnosis subset to 77.14%. Furthermore, it also demonstrates superior generalization with 55.94% on SlideBench (BCNB).

2 Methodology

2.1 Problem Formulation and Method Overview

We consider a WSI as a massive collection of patches. Let the visual features of these patches, extracted by the pre-trained vision encoder, be denoted as an input sequence $X \in \mathbb{R}^{B \times N \times D}$. Here, N represents the total number of patches, routinely exceeding 10^5 for a single gigapixel slide, and D denotes the feature dimension. The problem is defined as establishing a highly efficient compression function

$f : \mathbb{R}^{B\times N\times D} \rightarrow \mathbb{R}^{B\times K\times D}$. This function projects the vast input sequence X into a strictly bounded representation X', where $K \ll N$ is a predefined visual token budget. The fundamental objective is achieving this extensive token reduction while meticulously preserving the original image's global semantic context, thereby fitting dense diagnostic evidence into the restricted context window of the VLM. To resolve this bottleneck, we introduce a learnable token-budgeting framework,

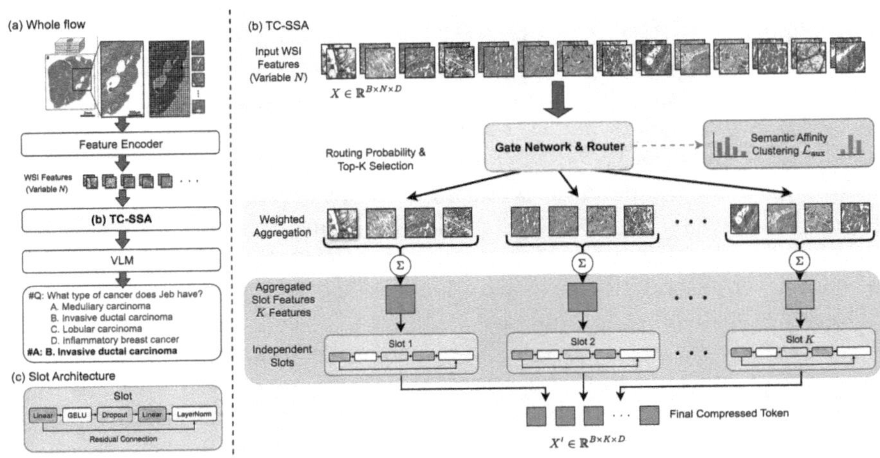

Fig. 1. Overview of the Token Compression via Semantic Slot Aggregation (TC-SSA) framework. (a) Whole flow: a feature encoder extracts a variable-length feature sequence (N) from partitioned gigapixel WSI patches. The TC-SSA module compresses these dense features into a fixed token budget before feeding them into the VLM for downstream diagnostic reasoning. **(b) TC-SSA Module:** a gate network processes input features $X \in \mathbb{R}^{B\times N\times D}$ to determine routing probabilities, followed by Top-2 routing. The auxiliary semantic slot regularization objective ($\mathcal{L}_{aux}$) stabilizes this routing distribution. Routed patch features undergo weighted aggregation into K distinct slots, yielding the final compressed representation $X' \in \mathbb{R}^{B\times K\times D}$. **(c) Slot Architecture:** the internal configuration of each independent slot.

Token Compression via Semantic Slot Aggregation (TC-SSA). As shown in Fig. 1, it comprises a gated routing mechanism and a slot-centric aggregation module. Initially, a lightweight gate computes a probability distribution over K predefined semantic slots for each input patch. To enforce sparse assignment and manage computational costs, the system applies Top-2 routing, where each patch contributes to at most two optimal slots. Subsequently, routed patches are aggregated via weighted pooling to construct compact slot embeddings. Each aggregated slot is independently refined by a multilayer perceptron to produce the final compressed sequence X'. This sequence is then projected into the downstream VLM embedding space. To maintain routing stability and prevent slot collapse during training, the framework incorporates auxiliary semantic slot regularization.

2.2 Semantic Slot Aggregation

For the j-th patch across a batch, we denote $x_j = X_{:,j,:} \in \mathbb{R}^{B \times D}$; equivalently, $x_{b,j} \in \mathbb{R}^D$ denotes the j-th patch feature of the b-th slide. A lightweight gating module computes a probability distribution over K distinct slots using shared parameters $W_g \in \mathbb{R}^{D \times K}$ and $b_g \in \mathbb{R}^K$. The resulting gate logits are $g_{b,j} = x_{b,j} W_g + b_g \in \mathbb{R}^K$, and stacking all patch logits gives $G \in \mathbb{R}^{B \times N \times K}$. The routing probability is then $P_{b,j,:} = \mathrm{Softmax}(G_{b,j,:})$. To enforce sparse assignment and control computational overhead, the system applies Top-2 routing. Rather than distributing patch information across all available slots, the framework isolates the two highest-probability slots. Let $m_{b,j,k} \in \{0, 1\}$ denote a binary mask indicating whether the k-th slot is among the Top-2 selections for the j-th patch of slide b. Consequently, truncated routing weights are defined as $\tilde{P}_{b,j,k} = m_{b,j,k} P_{b,j,k}$. This hard truncation guarantees that every patch contributes to at most two semantic concepts, thereby stabilizing the routing process and strictly controlling the subsequent aggregation cost. Following the determination of truncated routing probabilities, the framework executes a slot-centric feature aggregation step. TC-SSA does not perform explicit clustering; instead, patches with similar semantics naturally aggregate into the same slot as an emergent behavior of routing. For each slide b and semantic slot k, the aggregated feature vector $c_{b,k}$ is computed via a weighted pooling operation over all routed patches. The aggregation is formulated as

$$X'_{b,k,:} = c_{b,k} = \frac{\sum_{j=1}^{N} \tilde{P}_{b,j,k} \, x_{b,j}}{\sum_{j=1}^{N} \tilde{P}_{b,j,k} + \delta} \tag{1}$$

where $X' \in \mathbb{R}^{B \times K \times D}$ and $\delta = 10^{-9}$ is a minor numerical constant preventing division by zero if a slot receives no assignments. By normalizing the sum of the features by the sum of the routed weights, the magnitude of the resulting token remains stable regardless of the number of patches assigned to the specific slot. This mechanism effectively maps the high-dimensional visual evidence from the slide's physical coordinate space into a highly compressed, fixed-budget semantic space [25]. Through this procedure, critical diagnostic cues physically scattered across the tissue are successfully distilled into compact visual tokens without sacrificing global context.

2.3 Robust Semantic Slot Regularization

To prevent slot collapse, where a disproportionate number of patches are routed to a single semantic slot, the framework incorporates auxiliary semantic slot regularization during training. Without proper regularization, the gating module tends to heavily favor a few specific slots, which severely limits the representational capacity of the compressed sequence. To counteract this phenomenon, the primary component of the auxiliary objective is a soft load-balancing loss, denoted as

$$\mathcal{L}_{\text{switch}} = K \sum_{k=1}^{K} P_k f_k \tag{2}$$

where $P_k = \frac{1}{BN}\sum_{b=1}^{B}\sum_{j=1}^{N} P_{b,j,k}$ is the average routing probability to slot k and $f_k = \frac{1}{BN}\sum_{b=1}^{B}\sum_{j=1}^{N} m_{b,j,k}$ is the fraction of patches assigned to slot k under Top-2 routing. This loss formulation, widely adopted in sparse routing architectures, is used as a soft regularizer that penalizes degenerate cases where patches flood into one slot. It does not force equal distribution across slots: the gate can still assign background patches to a few slots while diagnostic regions occupy others. Furthermore, the auxiliary objective integrates two additional regularization terms to enhance routing stability. An entropy regularizer is applied to the gate's probability distribution to discourage overly confident but incorrect routing decisions early in training, denoted as:

$$\mathcal{L}_{\text{ent}} = 1 - \frac{-\sum_{k=1}^{K} P_k \log(P_k + \epsilon)}{\log K} \tag{3}$$

with $\epsilon = 10^{-8}$ for numerical stability. Additionally, a z-loss term $\mathcal{L}_z$ penalizes excessively large logit magnitudes produced by the gating network, thereby preventing numerical instability:

$$\mathcal{L}_z = \alpha \frac{1}{BN} \sum_{b=1}^{B} \sum_{j=1}^{N} \left(\log \sum_{k=1}^{K} \exp(g_{b,j,k}) \right)^2 \tag{4}$$

where $g_{b,j,k}$ denotes the pre-softmax gate logits and $\alpha = 10^{-4}$. The auxiliary routing regularization is optimized jointly with the primary task loss, denoted as $\mathcal{L}_{task}$, of the downstream VLM. Specifically, the overall training objective is formulated as:

$$\mathcal{L}_{total} = \mathcal{L}_{task} + \lambda(\mathcal{L}_{\text{switch}} + 0.5\,\mathcal{L}_{\text{ent}} + \mathcal{L}_z) \tag{5}$$

where λ is a hyperparameter balancing the contribution of these auxiliary terms. This regularization encourages diverse and stable semantic slots, yielding compact and informative patch representations.

3 Experiments and Results

Our model is trained on SlideBench following [3], while using 2×A6000 GPUs instead of 8×A100 of [3]. The benchmark defines **Microscopy** as low-level morphology and staining, and **Diagnosis** as clinical reasoning for grading and subtyping. **Overall Accuracy** is computed across all multiple-choice VQA pairs. We use a frozen CONCH encoder [12] to extract patch features. The model uses $K = 32$ slots and auxiliary routing loss weight $\lambda = 0.1$ with Top-2 routing.

1) Efficiency-Performance Comparative Studies. As shown in Table 1, TC-SSA achieves the best overall accuracy of **78.34%** on SlideBench (TCGA). Despite operating under the same token budget, TC-SSA surpasses SlideChat on the Diagnosis subset, highlighting the effectiveness of the routing-and-pooling strategy, which removes redundant patches while preserving critical evidence.

Table 1. Comparison of performance on SlideBench (TCGA) and zero-shot results on SlideBench (BCNB) and WSI-VQA*. SlideChat [3] is treated as the uncompressed upper bound for full WSI inference. We report FLOPs and accuracy as metrics. Best results among token-compression methods are shown in bold.

Methods	FLOPs	SlideBench (TCGA)			SlideBench	WSI
		Microscopy	Diagnosis	Overall	(BCNB)	(VQA*)
(Uncompressed Upper Bound)						
SlideChat [3]	133.3T	87.64	73.27	81.17	54.14	60.18
(Token Compression $\sim 60\times$)						
Random Baseline		24.44	24.91	25.02	24.40	24.14
GPT-4o [15]		62.89	46.69	57.91	41.43	30.41
LLaVA-Med [11]	1.70T	47.34	32.78	42.00	30.10	26.31
Quilt-LLaVA [16]	1.70T	57.76	35.96	48.07	32.19	44.43
MedDr [7]	1.70T	73.30	57.78	67.70	33.67	54.36
TC-SSA (Ours)	1.72T	**81.94**	**77.14**	**78.34**	**55.94**	**56.62**

Moreover, the superiority of TC-SSA further generalises to zero-shot settings, where it consistently outperforms competitors on both SlideBench (BCNB) and WSI-VQA*, demonstrating strong generalization of our method.

Figure 2 demonstrates superior performance on SlideBench (TCGA) of the proposed method. SlideChat serves as the uncompressed upper bound, whereas full-resolution WSI inference is impractical due to OOM constraints. By condensing global evidence into K tokens rather than discarding patches (e.g., LLaVA-

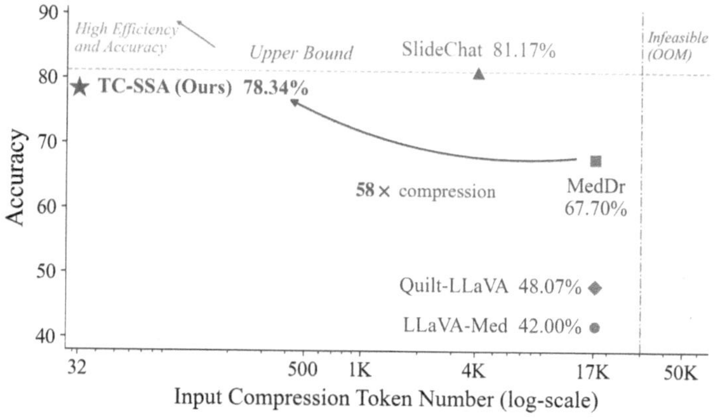

Fig. 2. Using only 32 visual tokens, TC-SSA achieves 78.34% overall accuracy on the SlideBench (TCGA) benchmark, yielding a 58× compression ratio compared to original patch features. SlideChat is reported as the uncompressed upper bound, while full-WSI inference is infeasible due to out-of-memory (OOM) constraints.

Med, Quilt-LLaVA), our method outperforms sampling baselines by 10.64% in accuracy with 98.3% fewer input tokens. Its linear-time complexity $O(N \cdot K)$ enables practical deployment in memory and latency-constrained clinical settings. We further visualize slot assignments across multiple TCGA cohorts. Figure 3 shows that each color denotes one semantic slot, and patches routed to the same slot often form coherent groups in embedding space, suggesting consistent grouping of related tissue patterns.

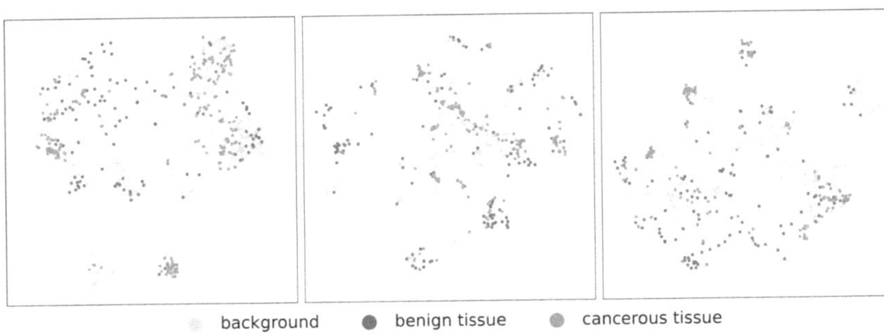

Fig. 3. Multi-dataset t-SNE visualization of semantic slot assignments. Each subplot corresponds to a different TCGA cohort. Each point is a patch embedding projected via t-SNE, with each color representing one semantic slot. The visualization shows that patches assigned to the same slot often form coherent groups, suggesting that routing captures related tissue patterns without using explicit tissue labels.

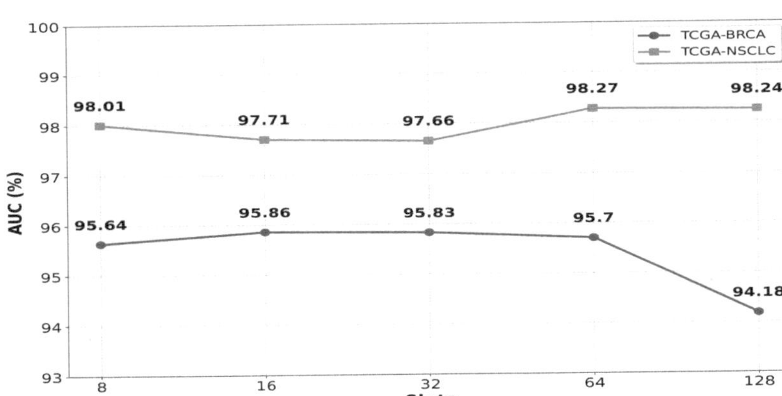

Fig. 4. Ablation experiments of the number of semantic slots K on TCGA-BRCA and TCGA-NSCLC. We set $K = 32$ by default.

2) Ablation Studies. We first ablate the number of semantic slots. Figure 4 reports MIL AUC as a function of the slot budget K. For TCGA-BRCA, performance is stable for $K \in \{16, 32, 64\}$ and drops at overly large budgets (e.g.,

$K = 128$), suggesting diminishing returns and possible over-fragmentation of semantic evidence. For TCGA-NSCLC, larger K brings only marginal gains beyond $K = 32$. Employing Top-2 routing with an auxiliary routing loss weight $\lambda = 0.1$ stabilizes slot utilization. The auxiliary terms are used as soft semantic slot regularizers: they discourage degenerate routing collapse without forcing all slots to receive equal numbers of patches. Without sufficient $\mathcal{L}_{\text{switch}}$ regularization, reducing λ can cause inactive slots and weakened performance on heterogeneous cases. Increasing K can over-fragment semantic evidence, allowing some slots to aggregate task-irrelevant artifacts and background noise that dilute the discriminative signals passed to the final classifier. Table 2 compares ResNet-50 [6], UNI [2], and GigaPath [23] encoders on TCGA-BRCA, TCGA-NSCLC, and PANDA. TCGA-BRCA and TCGA-NSCLC report cancer-subtyping AUC, while PANDA reports ISUP grading accuracy. All baseline results are reported as means, and the best results are shown in bold. Compared with the ImageNet-pretrained ResNet-50 baseline, pathology foundation encoders substantially improve performance across all datasets. Using the same UNI encoder for a controlled aggregation ablation, TC-SSA achieves the best results on TCGA-BRCA, TCGA-NSCLC, and PANDA, with gains of 1.01, 0.39, and 3.43 percentage points over the strongest baseline in each column. These results suggest that the improvement comes not only from stronger visual features, but also from the proposed semantic slot aggregation mechanism.

Table 2. Encoder comparison with public MIL baselines. BRCA and NSCLC are reported by AUC, while PANDA is reported by ISUP grading accuracy. Baseline results are reported as means. TC-SSA uses the UNI encoder. Best results are shown in bold.

Encoder	Method	TCGA-BRCA	TCGA-NSCLC	PANDA
ResNet-50 [6]	ABMIL [10]	83.80	92.32	58.89
	TransMIL [17]	88.52	92.49	56.42
	RRTMIL [20]	89.35	94.43	61.97
	2DMamba [27]	87.22	95.21	61.56
GigaPath [23]	ABMIL [10]	94.39	96.54	71.85
	TransMIL [17]	93.97	97.61	65.45
	GigaPath [23]	93.72	97.53	65.86
	RRTMIL [20]	94.82	97.63	72.46
	2DMamba [27]	93.84	96.87	75.72
UNI [2]	ABMIL [10]	94.05	97.04	74.69
	TransMIL [17]	93.33	97.27	68.06
	RRTMIL [20]	94.61	97.88	74.93
	2DMamba [27]	93.08	97.14	76.37
	TC-SSA (Ours)	**95.83**	**98.27**	**79.80**

4 Conclusion

We presented TC-SSA, a token compression module for gigapixel pathology VLMs utilizing semantic slot aggregation. Under a strict visual-token budget, TC-SSA achieves 78.34% overall accuracy on SlideBench (TCGA), improving the *Diagnosis* subset to 77.14%, while attaining AUCs of 95.83% on TCGA-BRCA and 98.27% on TCGA-NSCLC, and 79.80% ISUP grading accuracy on PANDA. This demonstrates learnable aggregation's strong performance under constrained budgets. Currently, a fixed slide-level slot budget K makes compression quality dependent on the patch encoder, and trading fine-grained spatial geometry for semantic structure may affect localization-heavy tasks.

Acknowledgments. Zhuo Chen and Xiaoyu Yang contributed equally to this work. Zhuo Chen conducted this work during an internship at Shenzhen University of Advanced Technology.

Disclosure of Interests. The authors have no competing interests to declare.

References

1. Chen, K., et al.: Cost-effective instruction learning for pathology vision and language analysis. Nat. Comput. Sci. (2025)
2. Chen, R.J., et al.: Towards a general-purpose foundation model for computational pathology. Nat. Med. (2024)
3. Chen, Y., et al.: Slidechat: a large vision-language assistant for whole-slide pathology image understanding. In: CVPR (2025)
4. Guo, Z., et al.: Focus: knowledge-enhanced adaptive visual compression for few-shot whole slide image classification. In: CVPR (2025)
5. Hashimoto, N., et al.: Multimodal gated mixture of experts using whole slide image and flow cytometry for multiple instance learning classification of lymphoma. J. Pathol. Inform. (2024)
6. He, K., Zhang, X., Ren, S., Sun, J.: Deep residual learning for image recognition. In: CVPR, pp. 770–778 (2016)
7. He, S., et al.: Meddr: diagnosis-guided bootstrapping for large-scale medical vision-language learning. CoRR (2024)
8. Hu, Q., et al.: Loc-path: learning to compress for pathology multimodal large language models. arXiv preprint arXiv:2512.05391 (2025)
9. Huang, Z., Bianchi, F., Yuksekgonul, M., Montine, T.J., Zou, J.: A visual-language foundation model for pathology image analysis using medical twitter. Nat. Med. (2023)
10. Ilse, M., Tomczak, J., Welling, M.: Attention-based deep multiple instance learning. In: ICML (2018)
11. Li, C., et al.: Llava-med: training a large language-and-vision assistant for biomedicine in one day. In: NeurIPS (2023)
12. Lu, M.Y., et al.: A visual-language foundation model for computational pathology. Nat. Med. (2024)
13. Lu, M.Y., et al.: A multimodal generative ai copilot for human pathology. Nature (2024)

14. Lu, M.Y., Williamson, D.F.K., Chen, T.Y., Chen, R.J., Barbieri, M., Mahmood, F.: Data-efficient and weakly supervised computational pathology on whole-slide images. Nat. Biomed. Eng. (2021)
15. OpenAI: GPT-4o system card. arXiv preprint arXiv:2410.21276 (2024)
16. Seyfioglu, M.S., Ikezogwo, W.O., Ghezloo, F., Krishna, R., Shapiro, L.G.: Quilt-llava: visual instruction tuning by extracting localized narratives from open-source histopathology videos. In: CVPR (2024)
17. Shao, Z., et al.: Transmil: transformer based correlated multiple instance learning for whole slide image classification. In: NeurIPS (2021)
18. Sun, Y., et al.: Cpath-omni: a unified multimodal foundation model for patch and whole slide image analysis in computational pathology. In: CVPR (2025)
19. Tang, W., et al.: Revisiting end-to-end learning with slide-level supervision in computational pathology. In: NeurIPS (2025)
20. Tang, W., Zhou, F., Huang, S., Zhu, X., Zhang, Y., Liu, B.: Feature re-embedding: towards foundation model-level performance in computational pathology. In: CVPR (2024)
21. Wang, B., et al.: Wsisum: WSI summarization via dual-level semantic reconstruction. Med. Image Anal. (2026)
22. Wu, J., et al.: Learning heterogeneous tissues with mixture of experts for gigapixel whole slide images. In: CVPR (2025)
23. Xu, H., et al.: A whole-slide foundation model for digital pathology from real-world data. Nature (2024)
24. Yang, X., Xu, L., Li, H., Zhang, S.: One leaf reveals the season: occlusion-based contrastive learning with semantic-aware views for efficient visual representation. In: ICML (2025)
25. Yang, X., Xu, L., Zeng, X., Wang, X., Li, H., Zhang, S.: Scalar: spatial-concept alignment for robust vision in harsh open world. Pattern Recogn. (2026)
26. Young, S., Zeng, X., Xu, L.: Fewer tokens, greater scaling: self-adaptive visual bases for efficient and expansive representation learning. arXiv preprint arXiv:2511.19515 (2026)
27. Zhang, J., et al.: 2dmamba: efficient state space model for image representation with applications on giga-pixel whole slide image classification. In: CVPR (2024)
28. Zhang, Y., et al.: Text-guided foundation model adaptation for pathological image classification. In: MICCAI (2023)

Text-Deficient Multimodal Stroke Segmentation with Lesion-Grounded Self-Retrieval-Augmented Generation

Heeseong Eum[1], Junhyeok Lee[1], Joon Jang[2], Han Jang[3], Songsoo Kim[4], and Kyu Sung Choi[4,5,6](✉)

[1] Interdisciplinary Program in Cancer Biology, Seoul National University College of Medicine, Seoul, South Korea
seong6466@snu.ac.kr

[2] Department of Biomedical Sciences, Seoul National University, Seoul, South Korea

[3] Interdisciplinary Program in Bioengineering, Seoul National University, Seoul, South Korea

[4] Department of Radiology, Seoul National University Hospital, Seoul, South Korea
ent1127@snu.ac.kr

[5] Department of Radiology, Seoul National University College of Medicine, Seoul, South Korea

[6] Healthcare AI Research Institute, Seoul National University Hospital, Seoul, South Korea

Abstract. Acute ischemic stroke requires rapid delineation of infarct core on diffusion-weighted imaging (DWI) and apparent diffusion coefficient (ADC) to guide time-critical decisions. Clinical urgency yields preliminary reporting as an early workflow summary centered on lesion location and extent, which enables report-conditioned multimodal segmentation. Paradoxically, the same urgency also makes preliminary reports frequently missing or inconsistently archived in retrospective cohorts, creating a key bottleneck for multimodal stroke segmentation. To address this gap, we present (i) Lesion-Grounded Self-Retrieval-Augmented Generation (LeG-RAG), a training pipeline that retrieves lesion-concordant reports using atlas-normalized lesion masks and synthesizes missing reports with a multimodal large language model without additional training, and (ii) LLMSwin, a report-conditioned multimodal segmentation architecture for brain MRI that extracts text features by combining the report with learnable prompts and feeding them into a frozen Llama-3-8B-Instruct model, and fuses these representations with multi-scale features from a SwinUNETR 3D backbone via two-way transformers. Experiments on a multi-institutional 3D stroke MRI dataset show that LeG-RAG achieves superior retrieval performance compared with state-of-the-art image-embedding methods across text-deficiency regimes, while remaining diagnostically consistent under ASPECTS-based reliability checks. LLMSwin achieves state-of-the-art segmentation across text-availability settings, outperforming image-only and prior mul-

H. Eum—First author.

G. Yang et al. (Eds.): MICCAI 2026, LNCS 16878, pp. 622–631, 2027.
https://doi.org/10.1007/978-3-032-38059-3_59

timodal segmentation baselines. Code is available at https://github.com/HeeseongEom/LeG-RAG.

Keywords: Multimodal Segmentation · Stroke MRI · Retrieval-Augmented Generation · Synthesized Report · Large Multimodal Model

1 Introduction

Acute ischemic stroke management hinges on rapid localization of infarct core on DWI/ADC, where minutes can influence treatment eligibility and outcome [3,11]. In routine neuroradiology, early communication often takes the form of preliminary reporting: a short impression emphasizing *where* the lesion is (laterality, standardized regions, vascular territories) and *how extensive* it appears [6,8]. However, preliminary texts are often absent or heterogeneously stored in retrospective imaging cohorts, despite availability of images and lesion masks.

We mitigate this mismatch by synthesizing missing reports during training via lesion-grounded retrieval. Since the multimodal LLM can ingest the *query image* at synthesis time, the information most likely to require external grounding is fine-grained spatial semantics (location/extent), which are explicitly encoded in lesion masks and dominate preliminary-style descriptions. After spatial normalization to a common atlas space (e.g., MNI), lesion masks become directly comparable across patients and align with atlas-defined regions used in clinical reasoning. We therefore propose *LeG-RAG*, which retrieves report exemplars from the same institutional archive using a coarse-to-fine, mask-driven strategy and generates a concise *synthesized report* via in-context learning [2,7,12]. We evaluate clinical reliability using the *Alberta Stroke Program Early CT Score (ASPECTS)*, a standardized regional scoring scheme [1,10].

In parallel, we introduce a dedicated report-conditioned multimodal segmentation model tailored to volumetric brain MRI. Inspired by LLM-based multimodal segmentation for target contouring [9], we extend the paradigm to stroke MRI and integrate it with a strong 3D backbone. Our *LLMSwin* encodes reports with a frozen pretrained LLM using learnable prompts and fuses the resulting representation with multi-scale SwinUNETR features [4] through two-way transformers at each stage. We also instantiate a UNETR-based variant (*LLMUNETR*) [5] using the same report pathway; LLMSwin consistently performs best, aligning with the empirical suitability of SwinUNETR-style representations for brain MRI [4].

In summary, we propose (i) **LeG-RAG**, a lesion-grounded retrieval-and-synthesis framework for text-deficient stroke training; (ii) **LLMSwin**, a report-conditioned volumetric multimodal segmentation architecture; and (iii) comprehensive evaluation showing improved synthesized-report quality and robust segmentation under text deficiency (Fig. 1).

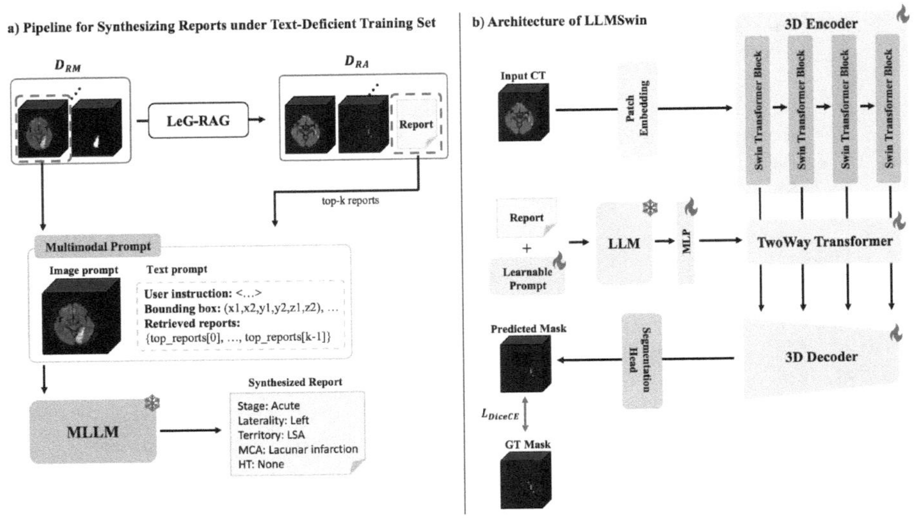

Fig. 1. (a) LeG-RAG overview: for each report-missing case in D_{RM}, retrieve lesion-concordant reports from D_{RA} in atlas space and synthesize a preliminary-style report for multimodal training. (b) Overview of the proposed **LLMSwin** report-conditioned volumetric segmentation architecture.

2 Method

2.1 Problem Setup

We study training under a text availability ratio p (in %), defined as the fraction of cases with available preliminary reports. The dataset is partitioned into a report-available subset

$$D_{RA} = \{(v_j, r_j, m_j)\}_{j=1}^{R}, \tag{1}$$

and a report-missing subset

$$D_{RM} = \{(v_i, m_i)\}_{i=1}^{S}, \tag{2}$$

where $N = R + S$ and $p = \frac{R}{N} \times 100$. Each v is a 3D MRI input formed by paired DWI and ADC volumes (two channels in $\mathbb{R}^{H\times W\times D\times 2}$), each $m \in \{0,1\}^{H\times W\times D}$ is a binary infarct mask, and each r is a short preliminary report. All volumes and masks are spatially normalized to a common atlas space (e.g., MNI). Our goal is to generate a synthesized report $\hat{r}_i$ for each $(v_i, m_i) \in D_{RM}$, yielding an augmented multimodal training set

$$D' = D_{RA} \cup \{(v_i, \hat{r}_i, m_i)\}_{i=1}^{S}, \tag{3}$$

which is then used to train report-conditioned multimodal segmentation models.

2.2 Lesion-Grounded Self-Retrieval-Augmented Generation

Spatial Key Retriever (SKR). For each query $(v_i, m_i) \in D_{RM}$, SKR computes an axial lesion profile $S_i \in \mathbb{R}^D$ by counting lesion voxels per axial slice and normalizing to unit length. For each candidate $(v_j, r_j, m_j) \in D_{RA}$, we compute S_j likewise. SKR retains candidates whose cosine similarity exceeds a threshold δ:

$$S_i = SK(m_i), \quad S_j = SK(m_j), \quad FD_{RA}(i) = \{(v_j, r_j, m_j) \in D_{RA} \mid \cos(S_i, S_j) > \delta\}. \quad (4)$$

We use $\delta = 0$ by default.

Clinical Context Retriever (CCR). CCR ranks candidates in $FD_{RA}(i)$ using a clinical context score combining overlap and boundary consistency of lesion masks. Let $\mathrm{DSC}(\cdot,\cdot)$ be Dice similarity and $\mathrm{HD95}(\cdot,\cdot)$ the 95th percentile Hausdorff distance. We define:

$$\mathrm{CCS}(m_i, m_j) = \lambda\, \mathrm{DSC}(m_i, m_j) + (1-\lambda)\, \frac{1}{1+\mathrm{HD95}(m_i, m_j)}, \quad (5)$$

with $\lambda = \frac{1}{2}$. The top-k exemplars are:

$$\mathcal{T}_i = \text{Top-}k_{(v_j, r_j, m_j) \in FD_{RA}(i)} \big[\, \mathrm{CCS}(m_i, m_j) \,\big]. \quad (6)$$

Synthesized Report Generation. Given $\mathcal{T}_i$, we provide the MLLM with the query image v_i and the retrieved reports $\{r_j\}$ as in-context demonstrations, and generate a concise synthesized report:

$$\hat{r}_i = \mathrm{MLLM}\big(\text{instruction};\ v_i, m_i, \{(m_j, r_j)\}_{(v_j, r_j, m_j) \in \mathcal{T}_i}\big). \quad (7)$$

Applying this to all $(v_i, m_i) \in D_{RM}$ yields D' for multimodal segmentation training.

2.3 LLMSwin: Report-Conditioned Multimodal Segmentation

LLMSwin integrates report semantics into a SwinUNETR-style volumetric backbone [4] using a frozen pretrained LLM and two-way transformer fusion. Let the Swin encoder produce multi-scale feature maps $\{F_\ell\}_{\ell=1}^{L}$, where $F_\ell \in \mathbb{R}^{H_\ell \times W_\ell \times D_\ell \times C_\ell}$. For a report r, we prepend learnable prompt tokens P and a special token $\langle\text{seg}\rangle$, and feed the token sequence into a frozen pretrained LLM:

$$H = \mathrm{LLM}\big([P;\ \langle\text{seg}\rangle;\ \mathrm{Tok}(r)]\big), \qquad h_{\text{seg}} = H_{\langle\text{seg}\rangle} \in \mathbb{R}^{d_L}. \quad (8)$$

For each encoder scale ℓ, we project h_{seg} to match the channel dimension:

$$z_\ell = \mathrm{MLP}_\ell(h_{\text{seg}}) \in \mathbb{R}^{C_\ell}. \quad (9)$$

Table 1. Stroke segmentation DSC on Test A/B under text-deficient training (text availability p). Img: image-only. MM-P: multimodal with available real reports only. MM-SR: multimodal with LeG-RAG synthesized reports. MM-Full: multimodal with all real reports (upper bound). Best results among Img, MM-P, and MM-SR at each text availability are in **bold**.

Model	Text avail.	Test A				Test B			
		Img	MM-P	MM-SR (ours)	MM-Full	Img	MM-P	MM-SR (ours)	MM-Full
LLMSeg	20%	0.708	0.684	**0.719**	0.729	0.615	0.595	**0.649**	0.645
	40%	0.708	0.700	**0.722**	0.729	0.615	0.597	**0.659**	0.645
	60%	0.708	0.715	**0.725**	0.729	0.615	0.631	**0.645**	0.645
	80%	0.708	0.719	**0.723**	0.729	0.615	0.612	**0.652**	0.645
LLMUNETR	20%	0.628	0.652	**0.664**	0.668	0.551	0.591	**0.601**	0.612
	40%	0.628	0.671	**0.689**	0.668	0.551	0.568	**0.601**	0.612
	60%	0.628	**0.676**	0.674	0.668	0.551	**0.617**	0.586	0.612
	80%	0.628	0.670	**0.696**	0.668	0.551	0.564	**0.624**	0.612
LLMSwin	20%	0.705	0.722	**0.755**	0.755	0.623	0.589	**0.685**	0.671
	40%	0.705	0.739	**0.755**	0.755	0.623	0.652	**0.661**	0.671
	60%	0.705	0.746	**0.752**	0.755	0.623	0.666	**0.664**	0.671
	80%	0.705	0.743	**0.749**	0.755	0.623	0.667	**0.673**	0.671

We fuse z_ℓ with visual tokens of F_ℓ using a two-way transformer TWT_ℓ. Let $\phi(\cdot)$ flatten spatial dimensions into tokens:

$$(\tilde{X}_\ell, \tilde{z}_\ell) = \mathrm{TWT}_\ell(\phi(F_\ell),\ z_\ell), \qquad \tilde{F}_\ell = \phi^{-1}(\tilde{X}_\ell). \tag{10}$$

The fused features $\{\tilde{F}_\ell\}$ are forwarded to a 3D decoder with skip connections to predict the lesion mask $\hat{m}$:

$$\hat{m} = \mathrm{Dec}(\{\tilde{F}_\ell\}_{\ell=1}^{L}). \tag{11}$$

Training minimizes a combined Dice and cross-entropy loss:

$$\mathcal{L}_{\mathrm{DiceCE}} = \mathcal{L}_{\mathrm{Dice}}(\hat{m}, m) + \mathcal{L}_{\mathrm{CE}}(\hat{m}, m). \tag{12}$$

Only prompts, projection MLPs, fusion modules, and the segmentation backbone/decoder are optimized; the pretrained LLM remains frozen.

3 Experiments and Results

3.1 Dataset and Evaluation

Dataset and Text Availability Protocol. We evaluate on a multi-institutional 3D stroke MRI cohort from three hospitals (anonymized). The training split contains 441 cases, and evaluation is performed on two held-out test sets (Test A: 112 cases; Test B: 151 cases). All images underwent skull stripping, resampling to 2 mm isotropic resolution, and Z-score normalization, followed by

Table 2. Evaluation of synthesized report under varying text availability p (aggregated F1 and accuracy) across embedding-based retrieval baselines. Best/second-best are in **bold**/<u>underline</u>.

Model	Text availability							
	20%		40%		60%		80%	
	F1↑	Acc↑	F1↑	Acc↑	F1↑	Acc↑	F1↑	Acc↑
Rad_DenseNet121	0.500	0.487	0.507	0.486	0.475	0.458	0.492	0.473
OpenCLIP	0.526	0.514	0.533	0.517	<u>0.537</u>	<u>0.522</u>	0.549	0.530
Rad_InceptionV3	<u>0.527</u>	0.509	0.534	0.517	0.528	0.516	0.535	0.523
BiomedCLIP	0.526	0.507	<u>0.535</u>	0.514	0.523	0.507	<u>0.560</u>	<u>0.543</u>
UNI-2	0.524	<u>0.519</u>	0.524	<u>0.520</u>	0.517	0.512	0.518	0.509
LeG-RAG (ours)	**0.715**	**0.671**	**0.733**	**0.713**	**0.754**	**0.728**	**0.750**	**0.711**

spatial normalization to a common atlas space. We simulate text-deficient training by varying text availability $p \in \{20, 40, 60, 80\}\%$ on the training set; for each p, D_{RA} contains the text-available subset and D_{RM} contains the remaining report-missing subset.

Evaluation Metrics. We evaluate segmentation using DSC. For synthesized report quality, we extract clinical attributes from real and synthesized reports using the same structured pipeline and report aggregated F1 and accuracy. For diagnostic consistency, we compute ASPECTS region-involvement agreement (F1/Acc) and total-score MAE from atlas-space lesion involvement [1,10].

3.2 Quantitative Analysis

Segmentation Under Text Deficiency. Table 1 shows that MM-SR improves over MM-P, particularly at low p, suggesting that LeG-RAG synthesized reports provide useful supervision when real reports are scarce. LLMSwin achieves the strongest performance across all p, outperforming image-only training and multimodal baselines, consistent with the strong volumetric representations of SwinUNETR-style backbones [4].

ASPECTS Reliability and Diagnostic Consistency. Table 4 summarizes ASPECTS-based agreement derived from atlas-space lesion involvement [1,10]. Across models and test sets, MM-SR improves or matches region-wise agreement while reducing MAE relative to text-deficient baselines, supporting diagnostic consistency under synthesized-report-driven training.

Synthesized Report Quality. Table 2 shows that LeG-RAG retrieval yields higher attribute agreement than image-embedding retrieval baselines across all p, consistent with the benefit of explicit spatial grounding over generic embedding

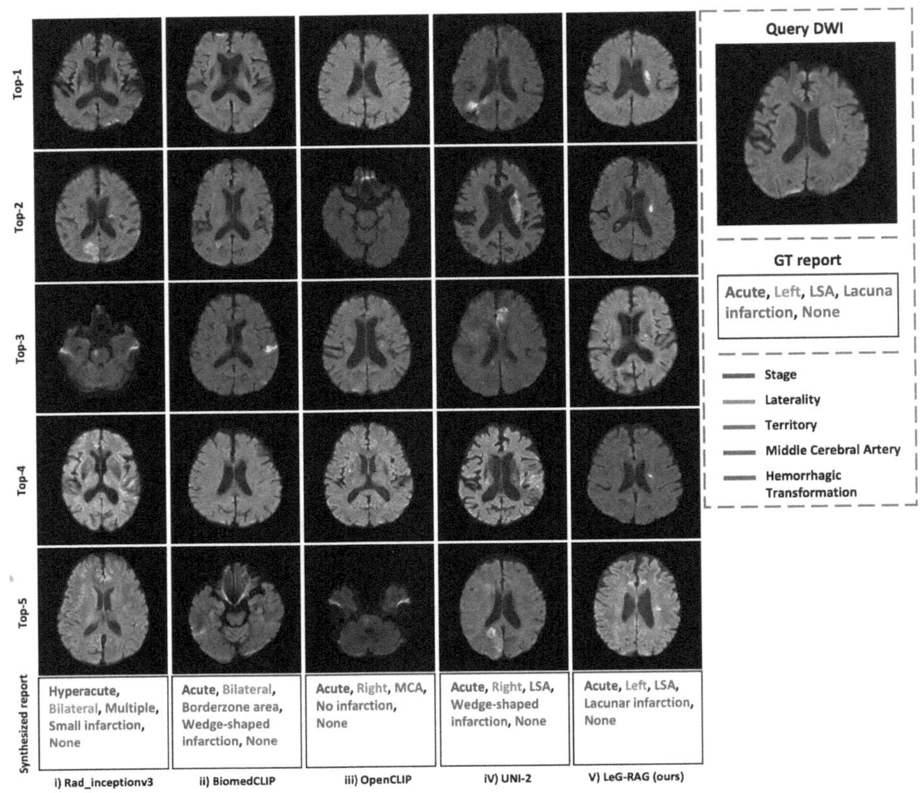

Fig. 2. Example retrievals and synthesized reports. LeG-RAG retrieves lesion-concordant exemplars in atlas space, yielding reports that better match laterality and territory-related attributes than embedding-based retrieval.

Table 3. MLLM comparison for report synthesis under fixed **LeG-RAG** retrieval (aggregated F1 and accuracy across text availability p). Best/second-best are in **bold**/underline.

Model	Text availability							
	20%		40%		60%		80%	
	F1↑	Acc↑	F1↑	Acc↑	F1↑	Acc↑	F1↑	Acc↑
Qwen2.5	0.691	0.795	0.725	0.808	0.739	0.815	0.750	0.825
Qwen2.5 (multi-slice)	0.693	**0.816**	0.725	**0.823**	0.736	**0.815**	0.749	**0.834**
VideoLLaVA	0.670	0.680	0.729	0.741	0.742	0.756	**0.774**	0.789
GPT-o1	0.730	0.763	0.749	0.801	0.770	0.800	0.773	0.806
LLaVA-OneVision	0.741	0.757	0.756	0.773	0.776	0.798	0.760	0.783
LLaVA-OneVision (multi-slice)	**0.749**	0.767	**0.765**	0.783	**0.777**	0.800	0.763	0.780

Table 4. ASPECTS-based diagnostic consistency on Test A/B (region involvement F1/Acc, total-score MAE). Img: image-only. MM-P: multimodal with available real reports only. **MM-SR**: multimodal with LeG-RAG synthesized reports. MM-Full: multimodal with all real reports (upper bound). Bold: best among Img/MM-P/MM-SR.

Model	Setting	Test A			Test B		
		F1↑	Acc↑	MAE↓	F1↑	Acc↑	MAE↓
LLMSeg	Img	0.879	0.929	0.411	0.909	0.919	0.748
	MM-P	0.916	0.951	0.393	0.912	0.920	0.737
	MM-SR (ours)	**0.944**	**0.967**	**0.315**	**0.934**	**0.939**	**0.657**
	MM-Full	0.954	0.973	0.339	0.927	0.933	0.726
LLMUNETR	Img	0.829	0.893	0.509	0.883	0.896	0.807
	MM-P	0.858	0.908	0.496	0.885	0.894	0.828
	MM-SR (ours)	**0.862**	**0.911**	**0.451**	**0.901**	**0.909**	**0.817**
	MM-Full	0.857	0.911	0.491	0.893	0.904	0.852
LLMSwin	Img	0.886	0.929	0.393	0.927	0.933	0.711
	MM-P	0.879	0.922	0.344	0.923	0.930	0.654
	MM-SR (ours)	**0.910**	**0.942**	**0.283**	**0.937**	**0.941**	**0.604**
	MM-Full	0.907	0.938	0.295	0.929	0.933	0.622

Table 5. Ablation of CCS components in **LeG-RAG** retrieval (top-5 average attribute F1 across text availability p).

Setting	Text availability			
	20%	40%	60%	80%
LeG-RAG + DSC only	0.547	0.558	0.616	0.635
LeG-RAG + HD95 only	<u>0.633</u>	<u>0.660</u>	<u>0.686</u>	<u>0.702</u>
LeG-RAG + CCS (DSC+HD95)	**0.635**	**0.663**	**0.694**	**0.713**

similarity [2,7]. Under fixed LeG-RAG retrieval, Table 3 indicates that synthesis quality depends on the chosen MLLM, while retrieval provides a strong, model-agnostic foundation.

3.3 Qualitative Analysis

Figure 2 provides representative examples, where LeG-RAG selects exemplars with closer lesion distribution in atlas space and yields synthesized reports that better reflect laterality and territory-relevant descriptions.

3.4 Ablation Analysis

Table 5 shows that using only DSC degrades retrieval quality, while the HD95-derived term is more robust; the combined CCS yields the best performance, supporting complementary overlap and boundary cues.

4 Conclusion

We presented **LeG-RAG**, a lesion-grounded retrieval and report-synthesis pipeline for text-deficient stroke training that leverages atlas-aligned lesion masks to retrieve spatially relevant exemplars and generate preliminary-style reports via in-context learning [2,7,12]. We also introduced **LLMSwin**, which fuses prompt-conditioned report representations with multi-scale SwinUNETR volumetric features [4] via two-way transformer fusion. Across multi-institutional stroke MRI, LeG-RAG improves attribute agreement and ASPECTS-based consistency [1,10], while LLMSwin achieves state-of-the-art segmentation across text availability settings.

Acknowledgments. This work was supported by the National Research Foundation of Korea (NRF) grant funded by the Korea government (MSIT) (No. RS-2026-25479661) (K.S.C.); the SNUH Research Fund (No. 04-2024-0600; No. 04-2025-2060) (K.S.C.); and the Korea Health Technology R&D Project through the Korea Health Industry Development Institute (KHIDI) grant funded by the Ministry of HealthWelfare (No. RS-2024-00439549) (K.S.C.).

Disclosure of Interests. The authors have no competing interests to declare that are relevant to the content of this article.

References

1. Barber, P., Demchuk, A., Zhang, J., Buchan, A.: Validity and reliability of a quantitative computed tomography score in predicting outcome of hyperacute stroke before thrombolytic therapy. Lancet **355**, 1670–1674 (2000)
2. Chen, W., Hu, H., Chen, X., Verga, P., Cohen, W.: Murag: multimodal retrieval-augmented generator for open question answering over images and text. In: EMNLP 2022, pp. 5558–5570. Association for Computational Linguistics, Abu Dhabi, United Arab Emirates (2022). https://doi.org/10.18653/v1/2022.emnlp-main.375
3. Goyal, M., et al.: Endovascular thrombectomy after large-vessel ischaemic stroke: a meta-analysis of individual patient data from five randomised trials. Lancet **387**, 1723–1731 (2016)
4. Hatamizadeh, A., Nath, V., Tang, Y., Yang, D., Roth, H., Xu, D.: Swin unetr: swin transformers for semantic segmentation of brain tumors in MRI images. In: BrainLes Workshop, MICCAI. LNCS, pp. 272–284. Springer (2022)
5. Hatamizadeh, A., et al.: Unetr: transformers for 3D medical image segmentation. In: WACV 2022, pp. 574–584 (2022)

6. Issa, G., et al.: The discrepancy rate between preliminary and official reports of emergency radiology studies: a performance indicator and quality improvement method. Acta Radiol. **56**(5), 598–604 (2015). https://doi.org/10.1177/0284185114532922
7. Lewis, P., et al.: Retrieval-augmented generation for knowledge-intensive NLP tasks. In: NeurIPS 2020, pp. 9459–9474 (2020)
8. Mates, J., Branstetter, B.F., Morgan, M.B., Lionetti, D.M., Chang, P.J.: 'wet reads' in the age of PACs: technical and workflow considerations for a preliminary report system. J. Digit. Imaging **20**(3), 296–306 (2007). https://doi.org/10.1007/s10278-006-1049-y
9. Oh, Y., et al.: LLM-driven multimodal target volume contouring in radiation oncology. Nat. Commun. **15**, 9186 (2024)
10. Pexman, J., et al.: Use of the alberta stroke program early CT score (aspects) for assessing CT scans in patients with acute stroke. AJNR Am. J. Neuroradiol. **22**, 1534–1542 (2001)
11. Saver, J.L.: Time is brain–quantified. Stroke **37**(1), 263–266 (2006). https://doi.org/10.1161/01.STR.0000196957.55928.ab
12. Wei, J., et al.: Chain-of-thought prompting elicits reasoning in large language models. NeurIPS **35**, 24824–24837 (2022)

Towards Unified Image–Video Modeling Across B-Mode and Contrast-Enhanced Ultrasound for Multi-task Analysis

Qiang Huang[1], Dengqiang Jia[1], Xiangmei Chen[2], Zehui Lin[1], Zhizhou Wang[1], Ligang Cui[3], Zhiyuan Wang[4(✉)], and Tao Tan[1(✉)]

[1] Faculty of Applied Sciences, Macao Polytechnic University, Macao, China
taotan@mpu.edu.mo

[2] Department of Ultrasound, Peking University Shenzhen Hospital, Shenzhen, China

[3] Department of Ultrasound, Peking University Third Hospital, Beijing, China

[4] Department of Ultrasound, The Affiliated Cancer Hospital of Xiangya School of Medicine, Central South University, Changsha, China
wangzhiyuan@hnca.org.cn

Abstract. Ultrasound imaging is widely used for disease diagnosis in clinical practice due to its non-invasive and portable advantages. However, current deep learning-based models are predominantly designed for specific organs or tasks, leading to performance with limited cross domain generalization and insufficient exploitation of available ultrasound data such as ultrasound images and videos. To address this limitation, we propose UIV-US, a unified multi-organ and multi-task framework for ultrasound image and video understanding across B-mode and contrast-enhanced ultrasound (CEUS). UIV-US operates within a shared parameter space to jointly support segmentation and classification tasks across both images and videos, enabling consistent representation learning across modalities and temporal settings. We design a Structured Semantic Conditioning Prompt (SSCP) that comprises five semantic attributes grouped into structural and spatial priors for adaptive conditioning of shared features, forming the foundation for scalable learning from large-scale ultrasound datasets. To further bridge image and video reasoning, a Memory-Reinforced Temporal Attention (MRTA) is introduced to ensure temporal consistency for video segmentation, while a segmentation-aware Feature-Level Attention (FLA) strategy performs reliable temporal aggregation for video classification. Extensive experiments on a large-scale multi-organ benchmark comprising 27,804 segmentation images, 7,320 classification images, and over 1,400 ultrasound videos demonstrate the effectiveness of UIV-US. The proposed framework achieves 89.81% Dice Similarity Coefficient for image segmentation, 96.28% for video segmentation, and 89.26% for CEUS segmentation. For classification, it reaches 85.76% Area Under the ROC Curve on image, 80.32% on video and 72.16% on CEUS, demonstrating strong performance across tasks. The code is available at https://github.com/miccai26-1426/UIV-US.

Q. Huang, D. Jia and X. Chen—These authors contributed equally to this work.

G. Yang et al. (Eds.): MICCAI 2026, LNCS 16878, pp. 632–642, 2027.
https://doi.org/10.1007/978-3-032-38059-3_60

Keywords: Universal Framework · Image & Video Analysis · Multi-Task · Multi-Organ

1 Introduction

Ultrasound imaging is one of the most widely used diagnostic tools in clinical practice and produces large-scale heterogeneous data across organs, modalities, and temporal settings [27]. A routine examination may simultaneously involve structural delineation, perfusion assessment, and diagnostic classification within a single session.

Despite the rapid progress of deep learning, most existing ultrasound analysis methods remain task- and organ-specific. Dedicated segmentation [14] and classification [7] networks dominate conventional B-mode applications, while CEUS approaches exploit contrast dynamics via cross-modality fusion [21] and temporal modeling [3]. For ultrasound videos, spatiotemporal aggregation [13] and lightweight temporal modules [23] are introduced for classification and segmentation. Although these methods achieve strong performance within restricted settings, image and video modeling are typically developed as separate pipelines, limiting parameter sharing and scalability.

To improve generalization, recent studies have explored multi-task and generalizable representation learning for ultrasound analysis. Large-scale pretraining frameworks such as USFM [10] and TinyUSFM [18] aim to learn transferable representations, and multi-organ segmentation strategies with shared representations and anatomical priors [4] have been investigated. In parallel, prompt-based paradigms establish flexible task conditioning mechanisms. Specifically, SAM [11] and SAM2 [25] introduce prompt-conditioned segmentation for images and videos, and medical adaptations including MedSAM [19], SAM-Med2D [5], and SAMUS [15] extend this strategy to clinical domains. More recent promptable frameworks such as UniUSNet [16] and PerceptGuide [17] support joint segmentation and classification for ultrasound images. However, most existing approaches remain predominantly image-oriented or segmentation-focused, and explicit unification of image and video reasoning across multiple ultrasound modalities within a single multi-task framework remains limited.

Motivated by these observations, we propose UIV-US, a unified multi-task image–video modeling framework for B-mode and contrast-enhanced ultrasound (CEUS). UIV-US operates within a shared parameter space to jointly support image and video segmentation and classification. To adapt shared representations to diverse clinical scenarios without increasing model complexity, we employ structured semantic conditioning and lightweight temporal reasoning within a single architecture.

Contributions: (1) We propose UIV-US, a shared-parameter ultrasound framework that unifies image and video segmentation and classification across B-mode and CEUS under a single architecture. (2) We design a Structured Semantic Conditioning Prompt (SSCP) mechanism that organizes five semantic attributes into structural and spatial priors to adaptively condition shared intermediate features, providing a scalable foundation for learning from large-scale ultrasound datasets. (3) We introduce a temporal modeling scheme within the shared architecture: a Memory-Reinforced Temporal Attention (MRTA) module that enforces temporally consistent segmentation through dynamic memory encoding and cross-attention, and a Feature-Level Attention (FLA) strategy that enables segmentation-aware frame aggregation for video classification.

2 Method

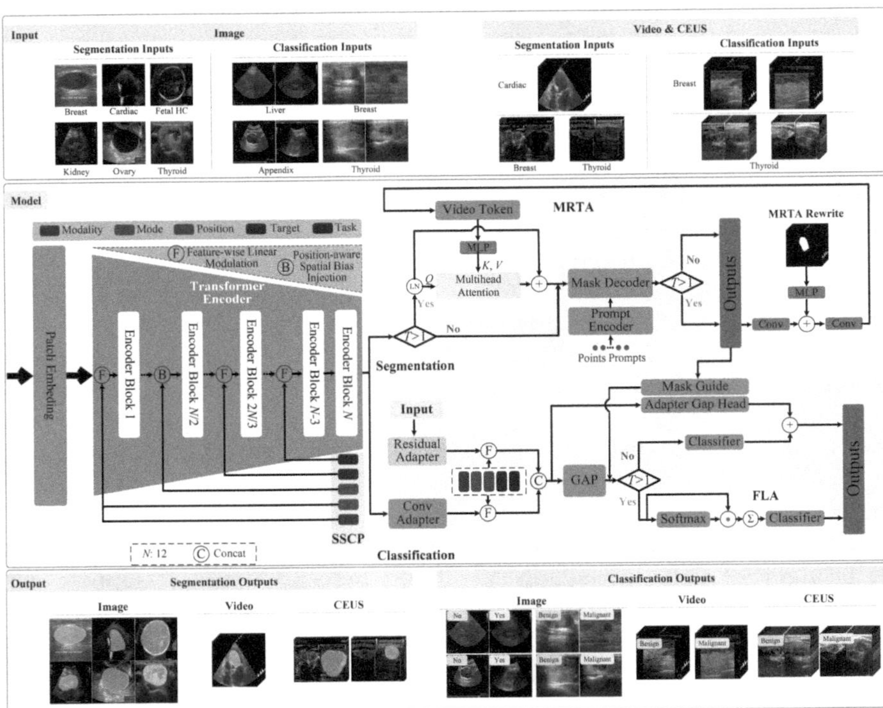

Fig. 1. The overview of UIV-US. The framework shares a Transformer encoder across segmentation and classification tasks, with SSCP for adaptive modulation. MRTA ensures temporal consistency for video segmentation, while FLA performs feature-level aggregation for video classification.

2.1 Overall Architecture

UIV-US is a universal multi-organ and multi-task model for unified ultrasound image and video understanding across B-mode and CEUS, as illustrated in Fig. 1. A single shared encoder is jointly optimized for segmentation and classification, while SSCP inject structural and spatial priors to adaptively modulate the shared representation space. For video inputs, temporal consistency is modeled through MRTA for segmentation and FLA for classification, enabling unified spatiotemporal reasoning within the same parameter space.

2.2 Structured Semantic Conditioning Prompt (SSCP)

We introduce an SSCP mechanism that dynamically adapts the shared representation space without duplicating model parameters.

Given a sample, a structured semantic prompt is built from five attributes, which are organized into two types of priors. The structural priors (*Task*, *Target*, *Mode*, and *Modality*) are injected via learnable feature modulation, while *Position* is modeled as a spatial prior through spatial bias. These attributes are predefined categorical metadata obtained from task settings and dataset/examination labels and are available before model input. For each attribute $a \in \mathcal{A}$, a one-hot vector $\mathbf{p}_a$ is mapped through a learnable prompt bank as $\mathbf{e}_a = \mathbf{p}_a \mathbf{W}_a$. For the four structural attributes, $(\boldsymbol{\gamma}_a, \boldsymbol{\beta}_a) = \text{MLP}_a(\mathbf{e}_a)$ modulates an intermediate feature as $\mathbf{h}' = \mathbf{h} \odot (1 + \tanh(\boldsymbol{\gamma}_a)) + \boldsymbol{\beta}_a$. For *Position*, we compute $\mathbf{c}_{\text{pos}} = \text{MLP}_{\text{pos}}(\mathbf{e}_{\text{pos}})$ and apply the spatial bias $\mathbf{h}'_{xy} = \mathbf{h}_{xy} + \mathbf{c}_{\text{pos}} B_{xy}$, where $B_{xy} = (x + y)/2$ and $x, y \in [-1, 1]$.

Instead of concatenating prompt information to the input, SSCP modulates intermediate feature representations within the shared backbone. Different semantic components are injected at different depths of the network. Coarse-grained attributes such as modality and mode influence early layers, while task- and target-specific priors refine higher-level semantic features.

Importantly, SSCP introduces only lightweight conditioning layers, while backbone parameters remain shared across tasks and modalities. This design maintains a unified parameter space and enables flexible adaptation to diverse clinical scenarios without requiring task-specific model duplication. Since all attributes are predefined metadata, SSCP is deterministic and reproducible and requires no additional manual annotation at test time.

2.3 Spatiotemporal Reasoning and Unified Multi-task Learning

For video inputs, segmentation adopts explicit memory-based temporal reasoning via Memory-Reinforced Temporal Attention (MRTA), whereas classification performs feature-level temporal aggregation using Feature-Level Attention (FLA) within the shared architecture.

Memory-Reinforced Temporal Attention (MRTA) for Video Segmentation. For video segmentation ($T > 1$), the standard encoder-decoder pipeline [11] is augmented with an MRTA to enforce temporal consistency across frames.

After predicting the segmentation mask $\hat{Y}_t$ for frame t, MRTA encodes a spatial memory representation

$$\mathbf{M}_t = \mathcal{E}_{\text{mrta}}(\mathbf{F}_t, \hat{Y}_t), \tag{1}$$

where $\mathbf{F}_t$ denotes backbone features and $\mathcal{E}_{\text{mrta}}$ represents the MRTA encoder. The encoded memory is stored in a dynamic MRTA bank with K frames per video, containing conditioned reference frames and recent historical frames.

Before decoding frame $t+1$, the feature $\mathbf{F}_{t+1}$ is refined by MRTA:

$$\tilde{\mathbf{F}}_{t+1} = \mathcal{A}_{\text{mrta}}(\mathbf{F}_{t+1}, \{\mathbf{M}_{\leq t}\}), \tag{2}$$

where $\mathcal{A}_{\text{mrta}}$ denotes multi-head cross-attention over aggregated memory tokens derived from the encoded reference and recent frames. This refinement enhances inter-frame interaction and improves temporal stability without introducing recurrent networks or heavy video transformer architectures.

Feature-Level Attention (FLA) for Classification. Let $P = \{\mathbf{e}_a\}_{a\in\mathcal{A}}$ denote the SSCP embeddings. For image inputs ($T = 1$), classification operates on a segmentation-aware representation

$$\mathbf{z} = \mathcal{H}_{\text{feat}}(I, \tilde{\mathbf{F}}, \hat{Y}, P), \tag{3}$$

where global and mask-guided ROI representations are adaptively fused based on mask reliability. The final prediction is obtained via the shared classification head, i.e., $\hat{C} = \mathcal{H}_{\text{cls}}(\mathbf{z})$.

For video inputs ($T > 1$), frame-level representations $\{\mathbf{z}_t\}_{t=1}^{T}$ are first extracted using the same head:

$$\mathbf{z}_t = \mathcal{H}_{\text{feat}}(I_t, \tilde{\mathbf{F}}_t, \hat{Y}_t, P). \tag{4}$$

Feature-Level Attention (FLA) aggregates them into a video-level representation, where s_t denotes a learnable scalar importance score for frame t.

$$\mathbf{z}_{\text{video}} = \sum_{t=1}^{T} \alpha_t \mathbf{z}_t, \quad \alpha_t = \frac{\exp(s_t)}{\sum_{k=1}^{T} \exp(s_k)}, \tag{5}$$

which is then fed into the same classification head to produce video-level prediction, i.e., $\hat{C} = \mathcal{H}_{\text{cls}}(\mathbf{z}_{\text{video}})$.

3 Experiments

3.1 Experimental Setup

Experiments are conducted on a large-scale multi-organ ultrasound benchmark (UM^2-US) covering both B-mode and CEUS modalities. For image segmentation, we include BUS-BRA [8], TN3K [24], MMOTU [29], Fetal_HC [9], kidneyUS_capsule [26], and EchoNet-Dynamic [22], totaling 27,804 images. For image classification, BUS-BRA, TN3K, Fatty-Liver [2], and Appendix [20] are used, comprising 7,320 labeled images. Video segmentation is evaluated on CAMUS [12], while video classification uses a private breast ultrasound dataset Breast* (300 videos).

Table 1. Validation comparison of UIV-US with SOTA pretrained models on UM^2-US. We present the mean DSC over all segmentation datasets, and mean AUC/Accuracy over all classification datasets.

	Model	Task	Mode	Prompt	Seg(DSC↑)	Cls(AUC↑/Acc↑)
Image	SAM [11]	Seg	zero-shot	Box	71.23%	-
	SAM [11]		zero-shot	Point	37.13%	-
	SAM-Med2D [5]		zero-shot	Box	69.00%	-
	SAM-Med2D [5]		zero-shot	Point	31.24%	-
	SAM-Med2D [5]		fine-tune	Box	69.56%	-
	SAM-Med2D [5]		fine-tune	Point	72.95%	-
	MedSAM [19]		zero-shot	Box	78.21%	-
	MedSAM [19]		fine-tune	Box	67.25%	-
	SAMUS [15]		fine-tune	Point	85.91%	-
	UniUSNet [16]	Seg & Cls	-	One-hot	82.62%	78.64%/79.95%
	PerceptGuide [17]		-		87.62%	79.33%/80.69%
	Ours	All	-	Dual-Priors	**89.81%**	**85.76%/82.23%**
Video	SAMUS [15]	Seg	fine-tune	Point	92.35%	-
	MemSAM [6]	Seg	fine-tune	Point	95.23%	-
	Ours	All	-	Dual-Priors	**96.28%**	**80.32%/86.33%**
CEUS	**Ours**	All	-	Dual-Priors	**89.26%**	**72.16%/77.33%**

CEUS segmentation and classification are further evaluated on two private datasets, CEBreast* (300 videos) and CEThyroid* (300 videos).

All data are split at the patient level into 70%/10%/20% training, validation, and testing sets. The model is implemented in PyTorch and optimized using AdamW with an initial learning rate of 5×10^{-4} and weight decay of 0.001.

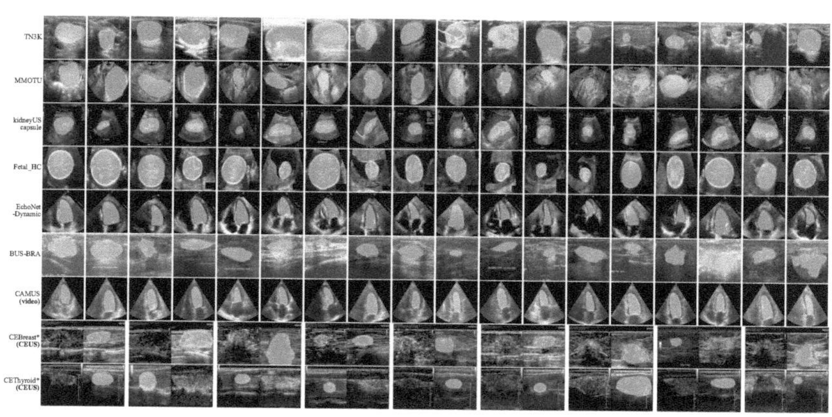

Fig. 2. Qualitative results of UIV-US across multiple datasets, with more accurate predictions on the left and slightly less precise results on the right. Yellow regions denote the overlap between prediction and ground truth, green and red regions denote false positives and false negatives, respectively. (Color figure online)

Table 2. Ablation results of SSCP, MRTA, and FLA. DSC is reported for segmentation and AUC for classification. Inter-frame Dice (IF-D, ↑) and Area Variation (AV, ↓) measure temporal stability.

Task	Variant/Method	Image		Video		CEUS		Stability (Video)	
		DSC↑	AUC ↑	DSC↑	AUC↑	DSC↑	AUC ↑	IF-D ↑	AV ↓
All Task	W/o SSCP	87.34%	82.94%	94.72%	76.24%	86.51%	68.27%	-	-
	W/o FiLM	87.29%	84.81%	94.79%	77.26%	88.29%	71.39%	-	-
	W/o SpatialBias	89.15%	83.27%	95.50%	79.83%	89.02%	69.91%	-	-
	Ours	**89.81%**	**85.76%**	**96.28%**	**80.32%**	**89.26%**	**72.16%**	-	-
Video & CEUS Seg.	W/o MRTA	-	-	93.12%	-	83.81%	-	86.47%	0.2003
	MRTA(K=1)	-	-	94.49%	-	87.23%	-	95.84%	0.0529
	MRTA(K=3)	-	-	95.00%	-	86.62%	-	97.33%	0.0451
	MRTA(K=6)	-	-	95.43%	-	87.12%	-	97.49%	0.0410
	Ours(K=10)	-	-	**96.28%**	-	**89.26%**	-	**97.59%**	**0.0382**
Video & CEUS Cls.	Mean pooling	-	-	-	76.15%	-	69.83%	-	-
	Max pooling	-	-	-	74.51%	-	66.37%	-	-
	Transformer pooling	-	-	-	79.83%	-	71.24%	-	-
	FLA	-	-	-	**80.32%**	-	**72.16%**	-	-

Training is performed in two stages: segmentation is first optimized with the classification branch frozen, followed by classification training with the encoder and segmentation branch frozen. We employ Dice and cross-entropy losses for segmentation, and cross-entropy loss for classification.

3.2 Comparison with State-of-the-Art Methods

We conduct systematic comparisons across multiple ultrasound tasks to evaluate the overall performance of the proposed framework. As existing models are primarily designed for individual tasks, UIV-US is compared against task-specific state-of-the-art (SOTA) methods for each corresponding setting.

For image segmentation, we compare with representative SAM-based baselines, including SAM [11], MedSAM [19], SAM-Med2D [5], and SAMUS [15]. For video segmentation, we compare with video-adapted SAMUS and MedSAM, as well as the memory-based method MemSAM [6]. For image classification and multi-task learning, we further compare with unified frameworks such as UniUSNet [16] and PerceptGuide [17].

Table 1 presents quantitative comparisons with representative SOTA methods across image and video segmentation as well as classification tasks. For image segmentation, zero-shot SAM-based models achieve moderate performance. Fine-tuning improves performance but remains sensitive to prompt type. Domain-specific models such as SAMUS achieve stronger results, demonstrating the importance of task adaptation in ultrasound settings. For video segmentation, MemSAM achieve competitive DSC scores. In unified multi-task settings, existing approaches (UniUSNet and PerceptGuide) demonstrate promising joint learning capability for segmentation and classification.

In contrast, UIV-US consistently achieves superior performance across all tasks, reaching 89.81% DSC for image segmentation, 96.28% for video segmentation, and 89.26% for CEUS segmentation. Classification performance also improves to 85.76%/82.23% (AUC/Acc) for image tasks and 72.16%/77.33% for CEUS. Qualitative results are illustrated in Fig. 2. These results demonstrate the effectiveness of structured semantic prompting and unified optimization.

3.3 Ablation Study

To validate the effectiveness of SSCP, MRTA, and FLA, we remove each module individually under the same experimental setting. The results are presented in Table 2.

For SSCP, we evaluate three variants: w/o SSCP, w/o FiLM, and w/o SpatialBias. Removing SSCP degrades performance across all tasks, confirming the effectiveness of prompt-guided modulation. Both FiLM and SpatialBias contribute complementary improvements, and the full model achieves the best overall performance. For MRTA, we compare w/o MRTA and different memory lengths ($K = 1, 3, 6, 10$), and evaluate temporal stability using Inter-frame Dice (IF-D) and Area Variation (AV). Removing MRTA significantly reduces video DSC (93.12%) and temporal stability (IF-D: 86.47%, AV: 0.2003). Increasing memory length consistently improves both accuracy and consistency, with the best performance achieved at $K = 10$. For FLA, we replace it with mean pooling, max pooling, and Transformer pooling. FLA achieves the best results, demonstrating the advantage of adaptive feature aggregation.

3.4 External Validation

To evaluate cross-domain generalization, we further conduct segmentation experiments on external ultrasound datasets, including BUSI [1], BUSIS [28], and DDTI [24], which are not used during training. UIV-US achieves DSC scores of 73.99%, 89.76%, and 67.83% on BUSI, BUSIS, and DDTI, respectively, maintaining competitive performance across domains. These results demonstrate the

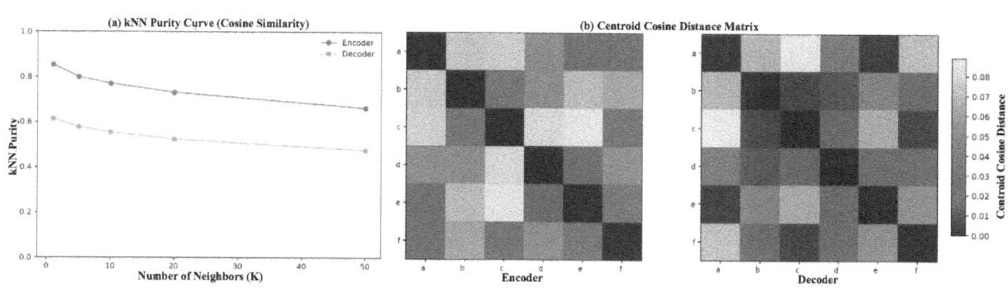

Fig. 3. Encoder-to-decoder feature evolution across all segmentation datasets. (a) kNN purity under cosine similarity for encoder and decoder representations across different neighborhood sizes K. (b) Inter-dataset centroid cosine distance matrices.

robustness and transferability of the proposed framework beyond the training datasets.

To examine the hierarchical evolution of feature representations, we analyze encoder and decoder features using kNN purity and centroid cosine distance. As shown in Fig. 3, encoder features exhibit stronger domain clustering, whereas decoder features show enhanced cross-domain alignment. This progressive change indicates a transition from domain-aware encoding to task-aligned decoding, suggesting reduced domain discrepancy in the decoder representation in the proposed framework.

4 Conclusion

In this work, we present UIV-US, a unified multi-organ and multi-task framework for B-mode and CEUS ultrasound image and video understanding. Built upon a shared parameter space with structured and spatial semantic conditioning, the proposed model jointly supports segmentation and classification tasks across image and video within a single architecture. Through memory-reinforced temporal attention for video segmentation and feature-level attention for classification, UIV-US effectively exploits both spatial and temporal information without introducing task-specific model duplication. Extensive experiments on large-scale multi-organ benchmarks demonstrate consistent improvements over task-specific baselines, as well as generalization to three unseen external datasets. These results suggest that unified semantic-conditioned modeling provides an effective and scalable paradigm for comprehensive ultrasound analysis.

Acknowledgments. This work was supported by Shenzhen Medical Research Fund (D2501013), Macao Polytechnic University Grant (RP/FCA-13/2026), the Science and Technology Development Fund, Macau SAR (File no. 0004/2025/ASJ) under the FDCT-FAPESP Joint Funding Scheme, and the Science and Technology Development Fund of Macao under Grant 0041/2023/RIB2.

Disclosure of Interests. The authors have no competing interests to declare that are relevant to the content of this article.

References

1. Al-Dhabyani, W., Gomaa, M., Khaled, H., Fahmy, A.: Dataset of breast ultrasound images. Data Brief **28**, 104863 (2020)
2. Byra, M., et al.: Transfer learning with deep convolutional neural network for liver steatosis assessment in ultrasound images. Int. J. Comput. Assist. Radiol. Surg. **13**(12), 1895–1903 (2018). https://doi.org/10.1007/s11548-018-1843-2
3. Chen, C., Wang, Y., Niu, J., Liu, X., Li, Q., Gong, X.: Domain knowledge powered deep learning for breast cancer diagnosis based on contrast-enhanced ultrasound videos. IEEE Trans. Med. Imaging **40**(9), 2439–2451 (2021)
4. Chen, H., et al.: Multi-organ foundation model for universal ultrasound image segmentation with task prompt and anatomical prior. IEEE Trans. Med. Imaging **44**(2), 1005–1018 (2025)

5. Cheng, J., et al.: Sam-med2d. arXiv preprint arXiv:2308.16184 (2023)
6. Deng, X., Wu, H., Zeng, R., Qin, J.: Memsam: taming segment anything model for echocardiography video segmentation. In: Proceedings of the IEEE/CVF Conference on Computer Vision and Pattern Recognition (CVPR), pp. 9622–9631 (2024)
7. Ding, W., Wang, J., Zhou, W., Zhou, S., Chang, C., Shi, J.: Joint localization and classification of breast cancer in b-mode ultrasound imaging via collaborative learning with elastography. IEEE J. Biomed. Health Inform. **26**(9), 4474–4485 (2022)
8. GómezâĂŘFlores, W., GregorioâĂŘCalas, M.J., Coelho de Albuquerque Pereira, W.: Bus-bra: a breast ultrasound dataset for assessing computer-aided diagnosis systems. Med. Phys. (2023)
9. van den Heuvel, T.L., de Bruijn, D., de Korte, C.L., van Ginneken, B.: Automated measurement of fetal head circumference using 2D ultrasound images. PLoS ONE **13**(8), e0200412 (2018)
10. Jiao, J., et al.: USFM: a universal ultrasound foundation model generalized to tasks and organs towards label efficient image analysis. Med. Image Anal. **96**, 103202 (2024)
11. Kirillov, A., et al.: Segment anything. In: Proceedings of the IEEE/CVF International Conference on Computer Vision, pp. 4015–4026. IEEE (2023)
12. Leclerc, S., et al.: Deep learning for segmentation using an open large-scale dataset in 2D echocardiography. IEEE Trans. Med. Imaging **38**(9), 2198–2210 (2019)
13. Li, M., et al.: Joint lesion detection and classification of breast ultrasound video via a clinical knowledge-aware framework. IEEE Trans. Circuits Syst. Video Technol. **35**(1), 45–61 (2025)
14. Li, R., et al.: Gaze-assisted human-centric domain adaptation for cardiac ultrasound image segmentation. In: ICASSP 2025 - 2025 IEEE International Conference on Acoustics, Speech and Signal Processing (ICASSP), pp. 1–5 (2025)
15. Lin, X., Xiang, Y., Zhang, L., Yang, X., Yan, Z., Yu, L.: Samus: adapting segment anything model for clinically-friendly and generalizable ultrasound image segmentation. arXiv preprint arXiv:2309.06824 (2023)
16. Lin, Z., Zhang, Z., Hu, X.: Uniusnet: a promptable framework for universal ultrasound disease prediction and tissue segmentation. In: 2024 IEEE International Conference on Bioinformatics and Biomedicine (BIBM), pp. 3501–3504. IEEE (2024)
17. Lin, Z., et al.: An orchestration learning framework for ultrasound imaging: prompt-guided hyper-perception and attention-matching downstream synchronization. Med. Image Anal. **104**, 103639 (2025)
18. Ma, C., et al.: Tinyusfm: towards compact and efficient ultrasound foundation models (2025). https://api.semanticscholar.org/CorpusID:282272257
19. Ma, J., He, Y., Li, F., Han, L., You, C., Wang, B.: Segment anything in medical images. Nat. Commun. **15**(1), 654 (2024)
20. Marcinkevičs, R., et al.: Regensburg pediatric appendicitis dataset (2023)
21. Meng, Z., Zhu, Y., Fan, X., Tian, J., Nie, F., Wang, K.: Ceusegnet: a cross-modality lesion segmentation network for contrast-enhanced ultrasound. In: 2022 IEEE 19th International Symposium on Biomedical Imaging (ISBI), pp. 1–5 (2022)
22. Ouyang, D., et al.: Echonet-dynamic: a large new cardiac motion video data resource for medical machine learning. In: NeurIPS ML4H Workshop (2019)
23. Pang, X., et al.: Blenet: a bio-inspired lightweight and efficient network for left ventricle segmentation in echocardiography. IEEE Trans. Circ. Syst. Video Technol. **35**(9), 9218–9233 (2025)

24. Pedraza, L., Vargas, C., Narváez, F., Durán, O., Muñoz, E., Romero, E.: An open access thyroid ultrasound image database. In: 10th International Symposium on Medical Information Processing and Analysis, vol. 9287, pp. 188–193. SPIE (2015)
25. Ravi, N., et al.: Sam 2: segment anything in images and videos. arXiv preprint arXiv:2408.00714 (2024)
26. Singla, R., et al.: The open kidney ultrasound data set. In: International Workshop on Advances in Simplifying Medical Ultrasound, pp. 155–164. Springer (2023)
27. Zhang, T., Tan, T., Samperna, R., et al.: Radiomics and artificial intelligence in breast imaging: a survey. Artif. Intell. Rev. **56**(Suppl 1), 857–892 (2023)
28. Zhang, Y., et al.: Busis: a benchmark for breast ultrasound image segmentation. Healthcare **10**(4), 729 (2022)
29. Zhao, Q., et al.: Mmotu: a multi-modality ovarian tumor ultrasound image dataset for unsupervised cross-domain semantic segmentation. arXiv preprint arXiv:2207.06799 (2022)

VIHD: Visual Intervention-Based Hallucination Detection for Medical Visual Question Answering

Jiayi Chen[1(✉)], Benteng Ma[4], Zehui Liao[1], Winston Chong[2,3], Yasmeen George[1], and Jianfei Cai[1]

[1] Department of Data Science and AI, Faculty of Information Technology, Monash University, Melbourne, VIC 3800, Australia
jiayi.chen@monash.edu
[2] Alfred Health Radiology, Alfred Health, Melbourne, VIC 3004, Australia
[3] School of Translational Medicine, Faculty of Medicine, Nursing and Health Sciences, Monash University, Melbourne, VIC 3800, Australia
[4] Hong Kong Polytechnic University, Hong Kong SAR, China

Abstract. While medical Multimodal Large Language Models (MLLMs) have shown promise in assisting diagnosis, they still frequently generate hallucinated responses that appear linguistically plausible but lack visual evidence. Such hallucinations pose risks to clinical decision-making and necessitate effective detection. Existing introspective detection methods primarily perform uncertainty estimation or logical verification by analyzing model responses conditioned on original or perturbed inputs. However, such external perturbations are often heuristic and context-agnostic, which overlooks the internal cross-modal dependency between generated tokens and related visual tokens during decoding. To address this issue, we propose VIHD, a **V**isual **I**ntervention-based **H**allucination **D**etection method that leverages targeted visual token masking to calibrate semantic entropy for more effective hallucination detection. VIHD locates visually dominant decoder layers via *Visual Dependency Probing (VDP)*, executes *Visual Intervention Decoding (VID)* via token masking to calibrate the semantic distribution, and quantifies the resulting *Calibrated Semantic Entropy (CSE)* as a reliable hallucination signal. Extensive experiments on three medical VQA benchmarks with two medical MLLMs demonstrate that VIHD consistently outperforms state-of-the-art methods, underscoring the importance of fine-grained visual dependency for hallucination detection. The code will be available at https://github.com/Jiayi-Chen-AU/VIHD.

Keywords: Vision-Language Model · Hallucination Detection · Visual Question Answering

G. Yang et al. (Eds.): MICCAI 2026, LNCS 16878, pp. 643–653, 2027.
https://doi.org/10.1007/978-3-032-38059-3_61

1 Introduction

Medical multimodal large language models (MLLMs) have shown remarkable efficacy in assisting medical diagnosis, supporting visual question answering (VQA) and automated report generation [11,16,23,27]. However, these models are susceptible to *hallucinated responses* that are linguistically coherent yet contradict visual evidence [9,12,18]. This issue poses risks to clinical decision-making and underscores the need for effective hallucination detection.

Existing hallucination detection methods can be broadly categorized into three paradigms: supervised detection, external verification, and introspective detection. **Supervised methods** [2,6,22,26] typically train auxiliary detectors on annotated hallucination datasets. While effective in-distribution, they hinge on labor-intensive annotations and generalize poorly to novel domains or architectures. **External verification** [3,24,28,29] leverages auxiliary LLMs, MLLMs, or knowledge bases to validate model outputs, but introduces substantial computational overhead and relies heavily on external tools. In contrast, **introspective approaches** [5,12,17,18,25] are entirely self-contained, requiring neither extra training nor external resources, making them efficient and deployment-friendly. Existing introspective approaches typically estimate token uncertainty [17,20], semantic uncertainty [5,18,30], or logical consistency [12,25] by analyzing model responses to original or perturbed inputs. However, such external perturbations are inherently heuristic and context-agnostic [10]. They treat MLLMs as black boxes, failing to probe the internal causal dependencies between generated tokens and visual evidence during decoding.

To mitigate this issue, we propose VIHD, a training-free **V**isual **I**ntervention-based **H**allucination **D**etection method, that probes the fidelity of cross-modal reasoning through targeted token-level interventions. VIHD operates in three stages. First, we identify visually dominant layers in the LLM decoder where cross-modal attention is most prominent. Secondly, we introduce a visual intervention decoding scheme that masks high-attention visual tokens to produce intervened responses, thereby revealing the model's dependency on specific visual evidence. Third, we design an adaptive strategy that integrates original and intervened semantic distributions via complementary or contrastive fusion, yielding a calibrated semantic entropy metric for hallucination detection.

Our main contributions are threefold: (1) We introduce VIHD, a training-free framework that harnesses fine-grained visual interventions to probe internal causal dependencies for hallucination detection. (2) We design an adaptive pipeline that identifies critical decoder layers, performs attention-guided visual token masking, and derives calibrated semantic entropy to quantify hallucination severity. (3) Extensive experiments on three medical VQA benchmarks and two medical MLLMs demonstrate the superiority of VIHD in hallucination detection.

2 Methodology

As illustrated in Fig. 1, VIHD detects hallucinations via fine-grained and targeted visual interventions. It comprises three components: (1) *Visual Dependency*

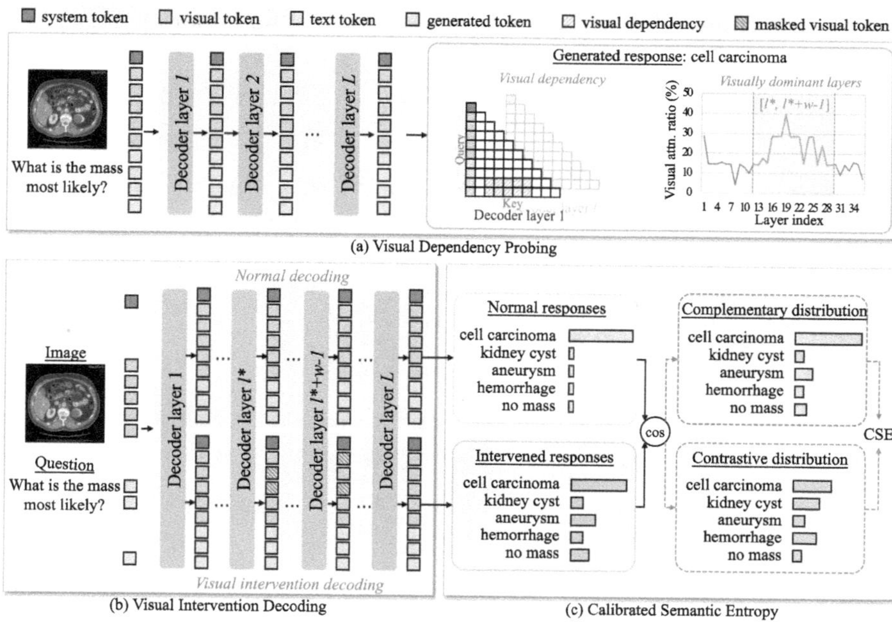

Fig. 1. Framework of VIHD. VIHD comprises (a) visual dependency probing, (b) visual intervention decoding, and (c) calibrated semantic entropy.

Probing, which pinpoints visually dominant decoder layers for targeted intervention; (2) *Visual Intervention Decoding*, which performs attention-guided token masking to produce intervened responses; (3) *Calibrated Semantic Entropy*, which calibrates semantic distribution based on normal and intervened responses and quantifies semantic entropy as a hallucination signal.

2.1 Visual Dependency Probing (VDP)

Given an input image x_v and a textual query x_q, a medical MLLM parametrized with θ generates a response $r = (r_1, \ldots, r_T)$ autoregressively, where T is the sequence length. The probability of generating response r [15] is factorized as:

$$p_\theta(r|x_v, x_q) = \prod_{t=1}^{T} p_\theta(r_t \mid x_v, x_q, r_{<t}). \quad (1)$$

where r_t is the t-th generated token and $r_{<t}$ is the preceding sequence.

Hallucinations often arise when tokens are generated primarily from linguistic priors without sufficient visual evidence. To quantify visual dependency during decoding, we first analyze the attention interactions between generated tokens and the input sequence. Let $\mathbf{Q}_r^{(l)} \in \mathbb{R}^{T \times d}$ denote the query matrix of generated tokens and $\mathbf{K}^{(l)} \in \mathbb{R}^{N \times d}$ represent the key matrix of the input sequence. The attention matrix $\mathbf{A}^{(l,h)} \in \mathbb{R}^{T\times N}$ for the h-th head is computed as:

$$\mathbf{A}^{(l,h)} = \text{softmax}\left(\frac{\mathbf{Q}_r^{(l,h)}\mathbf{K}^{(l,h)\top}}{\sqrt{d_h}}\right), \quad h \in \{1,\ldots,H\}, \tag{2}$$

where $d_h = \frac{d}{H}$ denotes the head dimension. We then quantify the layer-wise visual dependency $s^{(l)}$ by aggregating the attention weights assigned to visual tokens (indexed by $\mathcal{V}$) across all heads and generation tokens:

$$s^{(l)} = \frac{1}{H\cdot T}\sum_{h=1}^{H}\sum_{t=1}^{T}\sum_{v\in\mathcal{V}} A_{t,v}^{(l,h)}. \tag{3}$$

Finally, we localize a window of w consecutive layers with the strongest visual dependency $[l^*,\ l^*{+}w{-}1]$. The optimal starting layer l^* is defined as:

$$l^* = \underset{l}{\text{argmax}}\left(\frac{1}{w}\sum_{i=0}^{w-1} s^{(l+i)}\right). \tag{4}$$

2.2 Visual Intervention Decoding (VID)

After identifing visually dominant layers via VDP, we propose the visual intervention decoding algorithm to generate the intervened response r'. Given a visual input x_v and a textual query x_q, VID dynamically intervens the decoding process by applying fine-grained and targeted visual token masking. For each visually dominant layer l, we calculate the attention weights between the generated token and visual tokens at each generation step t.

$$\mathbf{A}_{t,v}^{(l)} = \frac{1}{H}\sum_{h=1}^{H}\mathbf{A}_{t,v}^{(l,h)}, \quad (l \in [l^*, l^*{+}w{-}1]). \tag{5}$$

We then mask the top-ρ fraction of visual tokens with the highest attention weight $\mathbf{A}_{t,v}^{(l)}$ to suppress their contribution. The intervened response r' is then generated conditioned on the perturbed visual representation. The probability of generating r' is formulated as:

$$p_\theta(r') = \prod_{t=1}^{T} p_\theta(r'_t | x_v, x_q, r'_{<t}). \tag{6}$$

2.3 Calibrated Semantic Entropy (CSE)

To enhance hallucination detection, we compute semantic entropy [5] over a calibrated semantic distribution. Specifically, we perform M independent sampling runs to obtain two sets of responses: M normal responses $\{r^{(i)}\}_{i=1}^{M}$ and M visual-intervened responses $\{r'^{(i)}\}_{i=1}^{M}$. Then, we utilize DeBERTa-Large-MNLI [7] to cluster them into N semantic-equivalence classes [13], yielding a normal semantic distribution $P(s|x_v, x_q)$ and a visual-intervened one $P(s'|x_v, x_q)$.

Crucially, visual intervention impacts vary by query granularity. General queries (*e.g.*, imaging modalities) remain robust to token masking, while detailed ones are relatively susceptible to semantic shifts. Accordingly, we propose an adaptive fusion strategy based on distribution alignment. When the intervened distribution aligns closely with the original, we treat token masking as a weak perturbation and apply *complementary fusion* to reinforce shared semantics. Conversely, we adopt *contrastive fusion* to suppress masking-induced hallucinations. The calibrated semantic distribution $P_c(s, s')$ is formulated as:

$$P_c(s,s') = \begin{cases} \sigma\left(\frac{1}{1+\alpha}P(s|x_v,x_q) + \frac{\alpha}{1+\alpha}P(s'|x_v,x_q)\right), & \text{if } \cos(P(s),P(s'))\geq\tau, \\ \sigma\left((1+\alpha)P(s|x_v,x_q) - \alpha P(s'|x_v,x_q)\right), & \text{otherwise.} \end{cases} \tag{7}$$

where α denotes the fusion factor, $\sigma(\cdot)$ is the softmax function, and τ is the similarity threshold. Finally, we calculate the calibrated semantic entropy CSE to quantify hallucination severity.

$$CSE = -\sum P_c(s,s') \cdot \log P_c(s,s') \tag{8}$$

A higher calibrated semantic entropy indicates significant semantic dispersion, serving as a robust indicator of hallucination.

3 Experiments

3.1 Experimental Settings

Datasets. We conduct hallucination detection on the test split of three medical VQA datasets covering multiple imaging modalities (CT, MRI, and X-ray): VQA-RAD [14], SLAKE [19], and VQA-Med-2019 [1]. **VQA-RAD** contains 451 test samples with 200 open-ended and 251 close-ended questions. **SLAKE** contains 1061 test samples in English with 645 open-ended and 416 close-ended questions. **VQA-Med-2019** comprises 500 test samples with 436 open-ended and 64 close-ended questions.

Backbones. We perform hallucination detection on two advanced medical MLLMs, Hulu-4B [11] and Lingshu-7B [27], for a robust assessment across diverse architectures.

Evaluation Metrics. Following [18], we employ the GREEN model [21] to generate hallucination labels. The GREEN model identifies matched findings and clinical errors in generated responses and calculates the GREEN score as $\frac{\#matched}{\#matched+\#errors}$. Samples with GREEN < 1.0 are labeled as hallucinated, while others (GREEN $= 1.0$) are non-hallucinated. Performance is quantified using two metrics: (1) *Area Under the ROC Curve (AUC)*, which measures the ranking capability to differentiate hallucinated from non-hallucinated responses; and

Table 1. Performance comparison of VIHD and seven competing methods.

Method	VQA-RAD				SLAKE				VQA-Med-2019			
	Open-Ended		All		Open-Ended		All		Open-Ended		All	
	AUC	AUG	AUC	AUG	AUC	AUG	AUC	AUG	AUC	AUG	AUC	AUG
Hulu-4B												
AvgProb	62.01	53.68	60.13	67.02	60.30	68.24	59.13	72.59	70.48	51.56	67.90	55.00
MaxProb	62.52	53.54	61.17	66.98	60.37	68.21	59.26	72.62	71.45	51.86	69.45	55.39
AvgEnt	61.15	53.56	59.32	66.88	60.39	68.26	59.03	72.56	70.43	51.68	67.72	55.00
MaxEnt	62.49	53.71	61.12	66.98	60.45	68.22	59.24	72.61	71.67	52.74	69.64	55.45
RadFlag	50.00	49.30	50.00	61.26	50.00	59.02	50.00	68.12	50.00	21.74	50.00	22.84
SE	50.89	51.77	49.48	62.24	49.29	59.02	49.30	66.89	47.53	19.58	47.42	21.43
VASE	76.98	69.09	66.23	71.51	74.41	79.12	70.31	81.27	75.09	54.30	72.99	57.41
VIHD	**83.13**	**72.01**	**78.23**	**80.82**	**79.94**	**81.11**	**78.91**	**84.96**	**78.65**	**59.64**	**78.10**	**60.39**
Δ	+6.15	+2.92	+12.00	+9.31	+5.53	+1.99	+8.60	+3.69	+3.56	+5.34	+5.11	+2.98
Lingshu-7B												
AvgProb	56.23	39.63	54.72	59.29	53.94	75.76	53.30	77.39	55.61	35.54	55.11	36.81
MaxProb	56.21	39.67	54.71	59.30	53.93	75.75	53.30	77.38	55.59	35.54	55.12	36.78
AvgEnt	56.28	39.73	54.80	59.34	53.97	75.82	53.34	77.42	55.68	35.63	55.11	36.90
MaxEnt	56.27	39.74	54.74	59.33	53.99	75.82	53.37	77.44	55.64	35.59	55.15	36.82
RadFlag	53.27	36.33	52.94	58.94	57.07	77.14	55.46	78.25	57.30	35.91	56.04	37.85
SE	50.37	35.93	50.92	55.27	52.72	74.13	52.45	76.44	54.27	34.69	53.96	35.96
VASE	71.45	48.02	69.62	69.35	72.37	85.06	70.56	84.52	69.57	56.77	68.85	60.80
VIHD	**73.13**	**55.84**	**71.58**	**70.86**	**75.36**	**88.61**	**72.95**	**86.49**	**74.71**	**64.11**	**74.18**	**64.54**
Δ	+1.68	+7.82	+1.96	+1.51	+2.99	+3.55	+2.39	+1.97	+5.14	+7.34	+5.33	+3.74

(2) *Area under the GREEN Curve (AUG)* [18], which evaluates the alignment between estimated hallucination scores and factual correctness.

Implementation Details. During hallucination label construction, we set the LLM sampling temperature to 0.1 to produce stable primary responses. During hallucination detection, we increase the temperature to 1.0 and perform $M = 10$ independent sampling runs following [18]. We adopt Nucleus sampling [8], employ a sliding window of width $w = \frac{L}{2}$ for visual dependency probing, and mask $\rho = 10\%$ of high-attention visual tokens in visual intervention decoding. For semantic distribution calibration, we set the cosine similarity threshold to $\tau = 0.95$ and the calibration coefficient to $\alpha = 1.0$. All experiments are conducted on a single NVIDIA A100 GPU with 80 GB of memory.

Comparison Methods. We compared VIHD with seven SOTA methods: (1) four token uncertainty-based methods: AvgProb, MaxProb, AvgEnt, and MaxEnt [17], (2) one consistency-based method: RadFlag [31], and (3) two semantic uncertainty-based methods: SE [5] and VASE [18]. VCD [15] and SID [10] are excluded for focusing on hallucination mitigation rather than detection.

3.2 Results

Table 1 presents the hallucination detection performance of VIHD against seven competing methods across three medical VQA benchmarks. The best performances are highlighted in **bold** and improvements over the second-best one are shown in blue. As reported in Table 1, VIHD consistently outperforms competing methods by a notable margin across all VQA datasets and medical MLLMs. On **Hulu-4B**, it improves the average AUC by 8.57% for all questions and 5.08% for open-ended questions. Notably, the most substantial gain is observed on VQA-RAD, where VIHD achieves a 12% AUC increase across all questions. On the more capable **Lingshu-7B** backbone, VIHD maintains its advantage with average AUC improvements of 3.22% overall and 3.27% on open-ended questions, with a substantial 5.33% gain on VQA-Med-2019. These results validate VIHD's effectiveness in detecting hallucinations across diverse medical imaging modalities and question types. The consistent improvements over VASE across both model capacities and dataset characteristics underscore the efficacy of internal fine-grained visual intervention in hallucination detection. Furthermore, VIHD requires 11.59 s per sample at inference, comparable to VASE (11.55 s), demonstrating performance gains without extra computational cost.

Table 2. Ablation study on VQA-RAD and VQA-Med-2019 using Hulu-4B.

Module			VQA-RAD				VQA-Med-2019			
VDP	VID	CSE	Open-Ended		All		Open-Ended		All	
			AUC	AUG	AUC	AUG	AUC	AUG	AUC	AUG
✗	✗	✗	50.89	51.77	49.48	62.24	47.53	19.58	47.42	21.43
✗	✓	✗	79.66	71.79	67.85	73.96	76.23	56.54	74.77	59.23
✗	✓	✓	79.97	71.65	76.30	80.35	77.64	58.80	76.56	59.73
✓	✓	✓	**83.13**	**72.01**	**78.23**	**80.82**	**78.65**	**59.64**	**78.10**	**60.39**

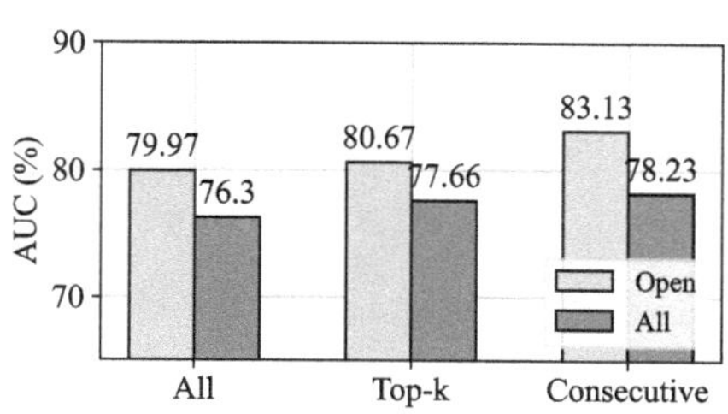

Fig. 2. Variants of VDP.

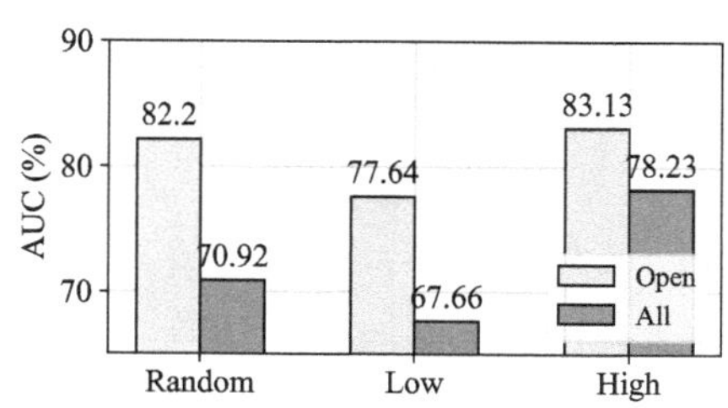

Fig. 3. Variants of visual token masking.

3.3 Ablation Study and Further Analysis

Ablation Study. Table 2 presents the component-wise ablation study on VQA-RAD and VQA-Med-2019 using Hulu-4B. Without any component, the baseline, Semantic Entropy [5], yields overall AUC scores of 49.48% and 47.42% on VQA-RAD and VQA-Med-2019, respectively. The introduction of visual intervention decoding (VID) substantially improves AUC by 18.37% and 27.35%, demonstrating the effectiveness of targeted visual intervention. Augmenting VID with calibrated semantic entropy (CSE) further boosts performance by 8.45% and 1.79%. Finally, incorporating visual dependency probing (VDP) to focus interventions at visually dominant layers leads to the best overall performance. These results validate the effectiveness of each proposed component.

Variants of Visual Dependency Probing (VDP). As shown in Fig. 2, we evaluate three layer selection strategies for visual dependency probing: (1) *all-layer selection*, which applies token masking across all decoder layers; (2) *attention-based selection*, which identifies isolated layers with peak visual dependency; and (3) *consecutive attention-based selection*, which selects a sequential span of high cumulative dependency to enforce continuity. Compared to the all-layer selection, the attention-based strategy yields a 1.36% AUC gain on VQA-RAD, highlighting the necessity of targeting visually dominant layers. Enforcing layer continuity further improves AUC by 0.57%. Therefore, we adopt the consecutive attention-based strategy as our default.

Variants of Visual Token Masking. We evaluate three token masking strategies within visual intervention decoding: (1) *random masking*, (2) *low-attention masking*, and (3) *high-attention masking*. As shown in Fig. 3, high-attention masking outperforms random masking by 7.41% AUC on VQA-RAD, which

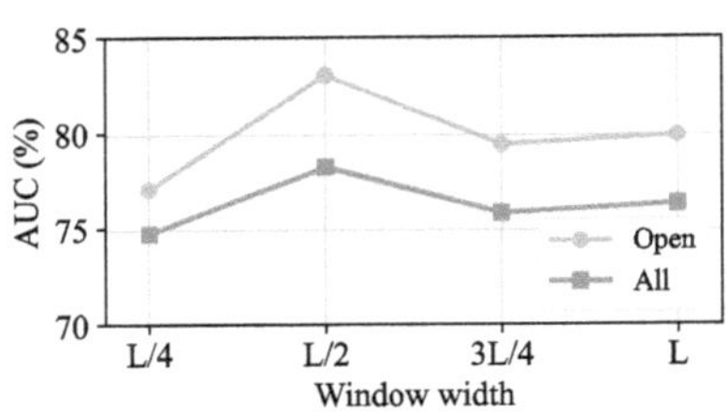

Fig. 4. Effect of sliding window width.

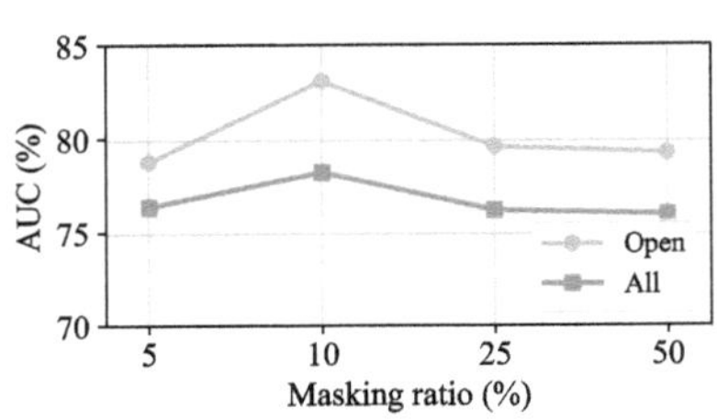

Fig. 5. Effect of masking ratio.

Table 3. Performance comparison on VQA-RAD using Hulu-4B with Top-K sampling.

Method	VQA-RAD			
	Open-Ended		All	
	AUC	AUG	AUC	AUG
AvgProb	65.06	53.51	60.68	69.58
MaxProb	65.89	53.24	61.73	69.79
AvgEnt	63.61	51.83	59.88	69.30
MaxEnt	65.06	53.41	61.73	69.85
RadFlag	50.00	44.77	50.00	62.54
SE	48.21	42.30	48.56	58.64
VASE	76.84	67.50	70.40	75.48
VIHD (Ours)	80.56	71.85	77.07	80.25
Δ	+3.72	+4.35	+6.67	+4.77

demonstrates that targeted perturbations of salient visual tokens facilitate effective semantic calibration. Conversely, low-attention masking acts merely as a subtle augmentation with negligible impact, thus leading to limited detection. Therefore, we adopt high-attention masking in visual intervention decoding.

Effect of Sliding Window Width. We also analyze the impact of sliding window width $w = \{\frac{L}{4}, \frac{L}{2}, \frac{3L}{4}, L\}$ on intervention efficacy. As illustrated in Fig. 4, performance peaks at $w = \frac{L}{2}$. Narrower windows $\frac{L}{4}$ restrict contextual propagation of perturbations, limiting their discriminative power, while wider windows $\frac{3L}{4}$ introduce excessive noise by intervening semantically heterogeneous layers. The $\frac{L}{2}$ configuration strikes an optimal balance, enabling localized yet contextually coherent perturbation propagation. We adopt $w = \frac{L}{2}$ for all experiments.

Effect of Masking Ratio. We investigate the impact of masking ratio $\rho \in \{5\%, 10\%, 25\%, 50\%\}$ on VQA-RAD. As shown in Fig. 5, insufficient masking ($\rho = 5\%$) yields weak perturbations that fail to calibrate semantic distribution, while aggressive masking ($\rho > 25\%$) corrupts visual semantics and degrades calibration fidelity. In contrast, the ratio of 10% achieves the optimal detection performance by introducing discriminative yet non-destructive interventions. Therefore, we fix $\rho = 10\%$ throughout experiments.

Effect of Sampling Strategy. To assess robustness under diverse decoding strategies, we evaluate all methods using Top-K sampling (K = 50) [4] on VQA-RAD. As shown in Table 3, VIHD maintains superior performance across both AUC and AUG metrics, surpassing the strongest baseline (VASE) by 6.67% in AUC and 4.77% in AUG overall. These results validate the effectiveness of visual intervention for hallucination detection across multiple sampling paradigms.

4 Conclusion

We have proposed **V**isual **I**ntervention-based **H**allucination **D**etection (VIHD), a training-free method that exploits fine-grained visual intervention to detect hallucinations. VIHD identifies visually dominant decoder layers, performs visual intervention decoding to derive calibrated semantic entropy as a hallucination score. Extensive experiments on three medical VQA datasets and two MLLMs have demonstrated the superiority of VIHD in hallucination detection.

Acknowledgments. This work was supported by the Medical Research Future Fund (MRFF) under Grant NCRI000074.

Disclosure of Interests. The authors declare no relevant competing interests.

References

1. Ben Abacha, A., Hasan, S.A., Datla, V.V., Liu, J., Demner-Fushman, D., Müller, H.: VQA-Med: overview of the medical visual question answering task at imageclef. In: CLEF, vol. 2380 (2019)

2. Chen, J., et al.: Detecting and evaluating medical hallucinations in large vision language models. arXiv preprint arXiv:2406.10185 (2024)
3. Cohen, R., Hamri, M., Geva, M., Globerson, A.: LM vs LM: detecting factual errors via cross examination. In: Proceedings of the 2023 Conference on Empirical Methods in Natural Language Processing, pp. 12621–12640 (2023)
4. Fan, A., Lewis, M., Dauphin, Y.: Hierarchical neural story generation. In: ACL, pp. 889–898 (2018)
5. Farquhar, S., Kossen, J., Kuhn, L., Gal, Y.: Detecting hallucinations in large language models using semantic entropy. Nature **630**(8017), 625–630 (2024)
6. Gunjal, A., Yin, J., Bas, E.: Detecting and preventing hallucinations in large vision language models. In: AAAI, vol. 38, pp. 18135–18143 (2024)
7. He, P., Liu, X., Gao, J., Chen, W.: DeBERTa: decoding-enhanced bert with disentangled attention. In: ICLR (2021)
8. Holtzman, A., Buys, J., Du, L., Forbes, M., Choi, Y.: The curious case of neural text degeneration. In: ICLR (2020)
9. Huang, Q., et al.: OPERA: alleviating hallucination in multi-modal large language models via over-trust penalty and retrospection-allocation. In: CVPR, pp. 13418–13427 (2024)
10. Huo, F., Xu, W., Zhang, Z., Wang, H., Chen, Z., Zhao, P.: Self-introspective decoding: alleviating hallucinations for large vision-language models. In: ICLR (2024)
11. Jiang, S., et al.: Hulu-Med: a transparent generalist model towards holistic medical vision-language understanding. arXiv preprint arXiv:2510.08668 (2025)
12. Jin, M., Liao, Z., Xia, Y.: V-Loop: visual logical loop verification for hallucination detection in medical visual question answering. arXiv preprint arXiv:2601.18240 (2026)
13. Kuhn, L., Gal, Y., Farquhar, S.: Semantic uncertainty: linguistic invariances for uncertainty estimation in natural language generation. In: ICLR (2023)
14. Lau, J.J., Gayen, S., Ben Abacha, A., Demner-Fushman, D.: A dataset of clinically generated visual questions and answers about radiology images. Sci. Data **5**(1), 1–10 (2018)
15. Leng, S., et al.: Mitigating object hallucinations in large vision-language models through visual contrastive decoding. In: CVPR, pp. 13872–13882 (2024)
16. Li, C., et al.: LLaVA-Med: training a large language-and-vision assistant for biomedicine in one day. NeurIPS **36**, 28541–28564 (2023)
17. Li, Q., Geng, J., Lyu, C., Zhu, D., Panov, M., Karray, F.: Reference-free hallucination detection for large vision-language models. In: EMNLP, pp. 4542–4551 (2024)
18. Liao, Z., Hu, S., Zou, K., Fu, H., Zhen, L., Xia, Y.: Vision-amplified semantic entropy for hallucination detection in medical visual question answering. In: MICCAI, pp. 669–679. Springer (2025)
19. Liu, B., Zhan, L.M., Xu, L., Ma, L., Yang, Y., Wu, X.M.: SLAKE: a semantically-labeled knowledge-enhanced dataset for medical visual question answering. In: ISBI, pp. 1650–1654. IEEE (2021)
20. Ma, H., Chen, J., Zhou, J.T., Wang, G., Zhang, C.: Estimating LLM uncertainty with evidence. arXiv preprint arXiv:2502.00290 (2025)
21. Ostmeier, S., et al.: GREEN: generative radiology report evaluation and error notation. In: EMNLP, pp. 374–390 (2024)
22. Prabhakaran, V., Aggarwal, P., Verma, V.K., Swamy, G., Saladi, A.: Vade: visual attention guided hallucination detection and elimination. In: ACL, pp. 14949–14965 (2025)

23. Sellergren, A., et al.: MedGemma technical report. arXiv preprint arXiv:2507.05201 (2025)
24. Sun, Z., et al.: ReDeEP: detecting hallucination in retrieval-augmented generation via mechanistic interpretability. In: ICLR (2025)
25. Wu, J., et al.: Logical closed loop: uncovering object hallucinations in large vision-language models. In: ACL, pp. 6944–6962 (2024)
26. Xiao, W., et al.: Detecting and mitigating hallucination in large vision language models via fine-grained AI feedback. In: AAAI, vol. 39, pp. 25543–25551 (2025)
27. Xu, W., et al.: Lingshu: a generalist foundation model for unified multimodal medical understanding and reasoning. arXiv preprint arXiv:2506.07044 (2025)
28. Yu, W., et al.: MM-Vet: evaluating large multimodal models for integrated capabilities. In: ICML, pp. 57730–57754. PMLR (2024)
29. Zhang, J., Li, Z., Das, K., Malin, B., Kumar, S.: SAC3: reliable hallucination detection in black-box language models via semantic-aware cross-check consistency. In: EMNLP, pp. 15445–15458 (2023)
30. Zhang, R., Zhang, H., Zheng, Z.: VL-Uncertainty: detecting hallucination in large vision-language model via uncertainty estimation. arXiv preprint arXiv:2411.11919 (2024)
31. Zhang, S., Sambara, S., Banerjee, O., Acosta, J., Fahrner, L.J., Rajpurkar, P.: RadFlag: a black-box hallucination detection method for medical vision language models. arXiv preprint arXiv:2411.00299 (2024)

Wasserstein-Aligned Localisation for VLM-Based Distributional OOD Detection in Medical Imaging

Bernhard Kainz[1,2](✉), Johanna P. Mueller[1], Matthew M. G. Baugh[2], and Cosmin Bercea[3,4]

[1] Department of AIBE, FAU Erlangen–Nürnberg, Erlangen, Germany
bernhard.kainz@fau.de
[2] Department of Computing, Imperial College London, London, UK
[3] Technical University Munich, Munich, Germany
[4] Munich Center for Machine Learning (MCML), Munich, Germany

Abstract. Zero-shot anomaly localisation via vision-language models (VLMs) offers a compelling approach for rare pathology detection, yet current methods suffer from: lack of healthy reference context, high prediction variance, and suboptimal reference selection. We introduce **WALDO**, a training-free framework grounded in optimal transport theory that addresses these limitations through: (i) *entropy-weighted Sliced Wasserstein distance* for anatomically-aware reference selection from DINOv3 patch distributions, (ii) *Goldilocks zone sampling* exploiting the non-monotonic relationship between reference similarity and localisation accuracy, and (iii) *self-consistency aggregation* via weighted non-maximum suppression. On the NOVA brain MRI benchmark, WALDO with Qwen2.5-VL-72B achieves $43.5_{\pm 1.6}\%$ mAP@30 (95% CI: [40.4, 46.7]), representing +19% relative improvement over zero-shot baselines. Cross-model evaluation shows consistent gains: GPT-4o achieves $32.0_{\pm 6.5}\%$ and Qwen3-VL-32B achieves $32.0_{\pm 6.6}\%$ mAP@30. Paired McNemar tests confirm statistical significance for the primary pair ($p = 1.8 \times 10^{-6}$, hit@30). We provide theoretical analysis of the Goldilocks effect through distributional divergence, demonstrating why moderate-similarity references optimise the bias-variance trade-off in comparative visual reasoning. Code: https://github.com/bkainz/WALDO_MICCAI26_demo.

Keywords: Anomaly localisation · Optimal transport · Vision-language models · Self-consistency · MRI · Chest X-ray

1 Introduction

Localising pathological regions in medical images is fundamental to clinical diagnosis. While supervised object detection has achieved remarkable success [15], it fundamentally cannot generalise to pathologies absent from training data. This is

G. Yang et al. (Eds.): MICCAI 2026, LNCS 16878, pp. 654–664, 2027.
https://doi.org/10.1007/978-3-032-38059-3_62

a critical limitation given that rare diseases comprise over 7,000 conditions affecting 300 million people worldwide. Vision-language models (VLMs) [13,20,28] offer zero-shot reasoning capabilities, but the NOVA benchmark [5] reveals that state-of-the-art VLMs achieve only 24.5–37.7% mAP@30 on brain MRI rare anomaly localisation, indicating substantial performance gaps.

Limitations of Current Approaches. We identify three structural limitations in VLM-based localisation: **(L1)** VLMs must identify anomalies without reference to what constitutes "normal" anatomy. This is analogous to asking a radiologist to find pathology without ever seeing healthy cases. **(L2)** When references are provided, performance depends on their anatomical correspondence; random sampling introduces mismatched views and inconsistent comparisons. **(L3)** Stochastic sampling causes high prediction variance across runs, with standard deviations exceeding 5% mAP; Existing self-consistency methods [29] reduce variance in language tasks but does not extend to spatial localisation.

Contributions. We propose WALDO (**W**asserstein-**A**ligned **L**ocalisation for **D**istributional **OOD**), addressing each limitation: **(1)** We formulate reference selection as optimal transport over DINOv3 [26] patch distributions, using entropy-weighted Sliced Wasserstein distance to preserve spatial structure while achieving $O(n \log n)$ computation. **(2)** We identify and theoretically analyse the non-monotonic relationship between reference similarity and localisation accuracy ("Goldilocks effect"), showing that references with moderate similarity optimise the bias-variance tradeoff. **(3)** We extend self-consistency to spatial localisation via confidence-weighted NMS, reducing prediction variance by ~40%. **(4)** We demonstrate consistent improvements across VLM architectures and imaging modalities. On NOVA [5], WALDO achieves $43.5_{\pm 1.6}\%$ mAP@30 with Qwen2.5-VL-72B (+19% relative; $p < 0.01$ vs. baseline) with similar gains for GPT-4o and Qwen3-VL-32B [2]. Evaluation on VinDr-CXR [19] chest X-rays reveals imaging modality-specific challenges, but equally promising results.

Related Work. *Self-supervised anomaly detection* has driven progress in medical imaging. Reconstruction-based methods like f-AnoGAN [24] detect anomalies via generative reconstruction error. Knowledge distillation approaches [6] exploit student-teacher discrepancy. Recent work on synthetic anomaly generation [25,27] enables self-supervised training without real anomaly labels. The Many Tasks framework [3] learns to localise anomalies from multiple synthetic proxy tasks, while cold-diffusion restorations [18] provide ensemble-based unsupervised detection. MOOD [31] and nnOOD [4] establish benchmarks for these training-based approaches. Industrial methods like PatchCore [23] and PaDiM [10] use patch-level memory banks. *Unlike these training-based approaches, WALDO operates training-free*, representing a new paradigm for OOD detection that complements benchmarks like NOVA by enabling zero-shot evaluation on previously unseen pathologies. *In-context learning (ICL)* [8,11] enables foundation models to perform tasks from examples without parameter updates. While ICL has proven effective for language tasks [29,30], its application to medical anomaly localisation in images remains underexplored. *Optimal*

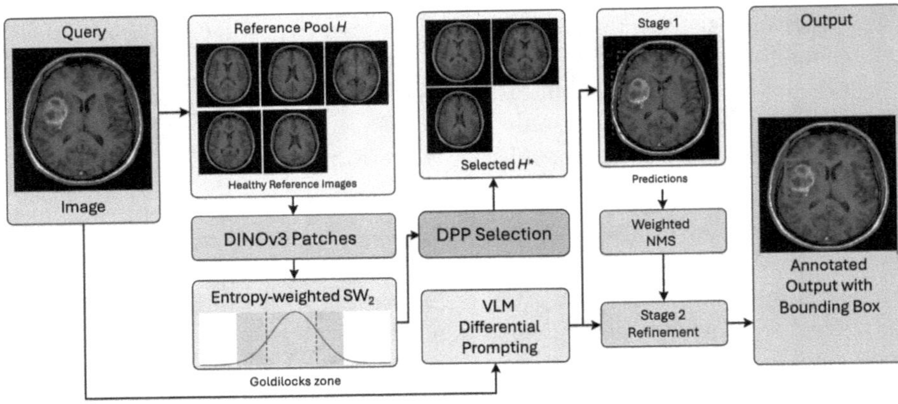

Fig. 1. WALDO idea. The query image and healthy reference pool are processed through DINOv3 to extract patch embeddings. Entropy-weighted Sliced Wasserstein distance ($SW_2^{(w)}$) measures distributional similarity, with references from the Goldilocks zone (30–70th percentile similarity) selected via DPP for diversity. The VLM performs differential prompting across selected references, with Stage 1 producing initial bounding boxes that are aggregated via weighted NMS. Stage 2 refinement produces the final localisation $\hat{\mathcal{B}}$.

transport has proven effective for domain adaptation [9] and generative modelling [1]; Sliced Wasserstein [7] enables efficient computation. VLMs [13,22,28] show strong medical VQA performance. The NOVA benchmark [5] evaluates VLM localisation on 281 rare pathologies, finding GPT-4o achieves 24.49% and Qwen2.5-VL-72B achieves 37.7% mAP@30, motivating a training-free approach.

2 Method

Problem Formulation and Overview. Given query image $\mathbf{x}_q \in \mathbb{R}^{H\times W\times 3}$, healthy reference pool $\mathcal{H} = \{\mathbf{h}_1, \ldots, \mathbf{h}_N\}$, and VLM $\mathcal{V}$, we seek bounding boxes $\hat{\mathcal{B}} = \{(b_i, c_i)\}_{i=1}^M$ where $b_i = (x_1, y_1, x_2, y_2)$ and $c_i \in [0, 1]$ is confidence. The goal is to predict bounding boxes that accurately localise pathological regions. Our method is outlined in Fig. 1. Given a query image and a pool of healthy references, WALDO proceeds in three stages. First, it extracts DINOv3 patch features and computes entropy-weighted SW distances. Second, it selects K diverse references from the Goldilocks zone via DPP. Third, it generates predictions with differential prompting and aggregates them via weighted NMS. The total complexity is $O(NMT \log T + K \cdot C_{\text{VLM}})$ where C_{VLM} is VLM inference cost.

Similarity via Entropy-Weighted Sliced Wasserstein. We extract DINOv3-ViT-B/16 [26] features, obtaining patch tokens $\mathbf{P}_\mathbf{x} = \{\phi_1, \ldots, \phi_T\} \subset \mathbb{R}^{768}$ where $T = (H/16) \times (W/16)$. Unlike CLS-based global similarity, patch distributions preserve spatial structure essential for localisation. DINOv3's self-supervised pretraining yields semantically rich patch representations without task-specific fine-tuning. The 2-Wasserstein distance between empirical distributions requires $O(T^3)$ computation via linear programming. We use Sliced Wasserstein [7,21]: $SW_2^2(P,Q) = \mathbb{E}_{\theta \sim \mathcal{U}(\mathbb{S}^{d-1})}[W_2^2(\theta^\top P, \theta^\top Q)]$ where 1D Wasserstein reduces to sorted differences: $W_2^2(P_\theta, Q_\theta) = \frac{1}{T}\sum_{i=1}^{T}(p_{(i)} - q_{(i)})^2$ for sorted projections. We approximate with $M = 100$ random projections, achieving $O(TM \log T)$ complexity.

Not all patches carry equal information. We weight by local entropy: $w_i = H(\phi_i)/\sum_j H(\phi_j)$ where $H(\phi) = -\sum_k \sigma_k(\phi) \log \sigma_k(\phi)$ and $\sigma(\cdot)$ is softmax over feature dimensions. High-entropy patches (textured regions) receive higher weight than homogeneous background. The weighted SW distance becomes:

$$SW_2^{(w)}(P,Q) = \left(\sum_{i=1}^{T} w_i \cdot (p_{(i)} - q_{(i)})^2\right)^{1/2}. \tag{1}$$

Goldilocks Zone Reference Selection. Counter-intuitively, the most similar reference ($\arg\min SW_2$) does not yield best localisation. We observe that mAP@30 follows an inverted-U with respect to reference similarity, peaking at moderate similarity (30–70th percentile). Formally, let $d(q,h)$ denote reference similarity. The VLM's localisation error decomposes as $\mathbb{E}[\epsilon^2] = \text{Bias}^2(d) + \text{Var}(d)$. *High similarity* ($d \to 0$): Low bias (anatomically matched) but high variance, *i.e.*, subtle pathology-reference differences approach a VLM's discrimination threshold, causing noisy predictions. *Low similarity* ($d \to \infty$): Low variance (obvious differences) but high bias, *i.e.*, anatomical misalignment, causes spurious "anomaly" detections at normal variations. The *Goldilocks zone* ($d \in [d_l, d_u]$) provides optimal tradeoff where references are similar enough for anatomical correspondence yet different enough for reliable pathology contrast. We rank references by $SW_2^{(w)}(P_q, P_h)$ and sample from percentile range $[\alpha, 1-\alpha]$ with $\alpha = 0.3$. For diversity, we apply Determinantal Point Process (DPP)-like repulsive sampling [14]: $\mathcal{H}^* = \arg\max_{|\mathcal{S}|=K} \det(\mathbf{L}_\mathcal{S})$ where $L_{ij} = q_i q_j \cdot \exp(-\beta \cdot SW_2(h_i, h_j))$, $q_i = \mathbf{1}[h_i \in \text{Goldilocks zone}]$, and β controls diversity.

Differential Prompting and Aggregation. For each selected reference $h_k \in \mathcal{H}^*$, we prompt the VLM: *"Compare the patient scan (left) to the healthy reference (right). Identify and localise any regions that appear abnormal, different, or pathological. Return bounding box coordinates [x1, y1, x2, y2] normalised to [0, 1000]."* This differential framing leverages VLMs' comparative reasoning capabilities rather than requiring absolute anomaly detection.

Given predictions $\{(b_k^{(j)}, c_k^{(j)})\}$ from K references with J boxes each, we aggregate via weighted NMS: $\hat{\mathcal{B}} = \text{NMS}(\bigcup_{k,j}\{(b_k^{(j)}, \tilde{c}_k^{(j)})\}, \theta_{\text{IoU}} = 0.5)$ where confidence $\tilde{c}_k^{(j)} = c_k^{(j)} \cdot \exp(-\lambda \cdot SW_2(P_q, P_{h_k}))$ downweights predictions from less similar references. We use $\lambda = 0.1$.

Implementation. We use N = 30-50 healthy reference images (30 IXI brain MRI for NOVA; 50 "No Finding" cases for VinDr-CXR), $K = 5$ references per query ($K = 3$ for Stage 1, $K = 2$ for Stage 2 refinement), $M = 100$ random

Table 1. NOVA brain MRI localisation results (n = 906, seed = 42). Best per-model in **bold**. †: $p < 0.05$, ††: $p < 0.01$ (paired McNemar on hit@30, full NOVA).

Model	Method	Samples	mAP@30 (%)	95% CI@30	mAP@50 (%)	95% CI@50	Avg IoU (%)	s/img
Qwen2.5-VL-72B [28]	Zero-shot [5]	906	37.7	–	24.5	–	–	∼2
	Zero-shot (rep.)	906	$36.4_{\pm1.5}$	[33.3, 39.4]	$23.4_{\pm1.4}$	[20.6, 26.2]	23.6	
	WALDO (ours)	906	$\mathbf{43.5}_{\pm1.6}$	[40.4, 46.7]	$\mathbf{26.3}_{\pm1.4}$	[23.5, 29.0]	**29.6**††	
GPT-4o [20]	Zero-shot	2×50	$19.0_{\pm5.5}$	[8.0, 30.0]	$3.0_{\pm2.4}$	[0.0, 8.0]	14.2	∼3
	WALDO (ours)	2×50	$\mathbf{32.0}_{\pm6.5}$	[20.0, 44.0]	$\mathbf{14.0}_{\pm4.9}$	[4.0, 24.0]	**21.7**†	
Qwen3-VL-32B	Zero-shot	906	$20.4_{\pm1.3}$	[17.7, 23.0]	$13.8_{\pm1.1}$	[11.5, 16.1]	13.8	∼3
	WALDO (ours)	50	$\mathbf{32.0}_{\pm6.6}$	[20.0, 46.0]	$\mathbf{18.0}_{\pm5.4}$	[8.0, 28.0]	**22.7**††	
Qwen3-VL-235B (MoE)	**Zero-shot**	906	$\mathbf{36.3}_{\pm1.6}$	[33.3, 39.5]	$\mathbf{20.1}_{\pm1.3}$	[17.5, 22.6]	**25.1**††	∼5
	WALDO (ours)	906	$31.8_{\pm1.6}$	[28.7, 34.9]	$15.2_{\pm1.2}$	[12.8, 17.5]	21.7	
Gemini-2.0-Flash [13]	Zero-shot [5]	906	20.16	–	7.37	–	–	∼4
	Zero-shot (rep.)	906	$18.1_{\pm1.3}$	[15.6, 20.6]	$6.4_{\pm0.8}$	[4.8, 8.0]	15.2	
	WALDO (ours)	50	$\mathbf{38.0}_{\pm7.2}$	[24.0, 52.0]	$\mathbf{10.0}_{\pm4.2}$	[2.0, 18.0]	**24.5**††	

projections for SW computation, Goldilocks percentile range $\alpha = 0.3$, NMS IoU threshold 0.5, and confidence weight $\lambda = 0.1$ (selected by grid search; see Sensitivity Analysis). NOVA images are 480 × 480 px; VinDr-CXR DICOMs are up to 1 024 × 1 024 px (resized from variable source resolution). Both are resized to 512 × 512 for feature extraction and to 1 024 × 1 024 for VLM inference. VLM inference uses temperature 0.7 with nucleus sampling (top-p = 0.95). We report results on the full NOVA test set (n = 906) and all VinDr-CXR images with ≥1 annotated finding (n = 949).

3 Experiments

Datasets. *NOVA* [5]: 906 brain MRI samples with 281 rare pathology types and expert-annotated bounding boxes. *VinDr-CXR* [19]: 18,000 chest X-rays with radiologist annotations. Following NOVA, we report mAP@30, mAP@50 (IoU thresholds 0.3, 0.5), average IoU. We report mean with standard error as subscript (e.g., $43.5_{\pm1.6}\%$) and 95% bootstrap confidence intervals (1000 resamples) [12]. Statistical significance is assessed using paired McNemar tests [17] on the binary hit@30 (IoU ≥ 0.3) indicator, with $\alpha = 0.05$.

Models. Primary VLM: Qwen2.5-VL-72B [28]. Cross-model: GPT-4o [20], Qwen3-VL-32B [2], Gemini-2.0-Flash [13]. As reference features, we use DINOv3-ViT-B/16 [26].

Table 1 presents results across VLM architectures on NOVA. With Qwen2.5-VL-72B, WALDO achieves $43.5_{\pm1.6}\%$ mAP@30, a +19.5% relative improvement over our zero-shot reproduction (36.4%). A paired McNemar test (on hit@30) confirms this improvement is statistically significant ($p = 1.8 \times 10^{-6}$). Using API-based inference, zero-shot prediction requires ∼2–5 s/image, while WALDO adds ∼3× overhead due to reference embedding and multi-image prompts. Smaller models exhibit substantially lower zero-shot accuracy (Qwen3-VL-32B achieves $20.4_{\pm1.3}\%$ mAP@30) than larger models (Qwen2.5-VL-72B: 37.7%, Qwen3-VL-235B: 36.3%), confirming that model scale strongly affects zero-shot medical

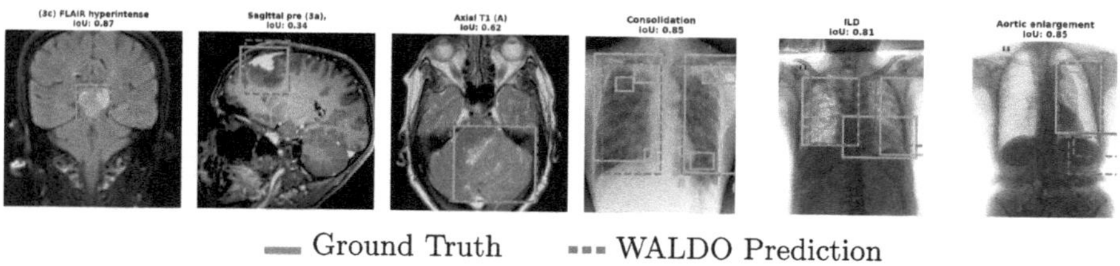

Fig. 2. Qualitative results. Left three: NOVA brain MRI; Right three: VinDr-CXR. Lesion type and IoU shown above each image.

Table 2. VinDr-CXR (n=949 with ≥1 annotated finding; all lesion types included). "no findings" used as healthy references. ††: $p < 0.01$.

Model	Method	mAP@30	mAP@50	IoU
Qwen2.5-72B	Zero-shot	$18.7_{\pm 2.5}$	$4.0_{\pm 1.2}$	$14.4_{\pm 1.2}$
	WALDO	$\mathbf{22.3}_{\pm 2.7}$	$\mathbf{5.7}_{\pm 1.5}$	$\mathbf{18.2}^{\dagger\dagger}_{\pm 1.1}$
Qwen3-32B	Zero-shot	$12.8_{\pm 2.1}$	$4.3_{\pm 1.3}$	$8.9_{\pm 1.1}$
	WALDO	$\mathbf{34.1}_{\pm 3.0}$	$\mathbf{10.7}_{\pm 2.0}$	$\mathbf{22.2}^{\dagger\dagger}_{\pm 1.3}$
GPT-4o	Zero-shot	$3.3_{\pm 1.1}$	$0.4_{\pm 0.4}$	$2.3_{\pm 0.6}$
	WALDO	$\mathbf{10.9}_{\pm 2.0}$	$\mathbf{1.5}_{\pm 0.8}$	$\mathbf{9.4}^{\dagger\dagger}_{\pm 0.9}$

Table 3. Ablation on NOVA: Component contributions.

Configuration	mAP@30	mAP@50	IoU
Baseline (zero-shot)	$36.4_{\pm 1.5}$	$23.4_{\pm 1.4}$	23.6
+ Random reference	$35.8_{\pm 1.6}$	$21.2_{\pm 1.4}$	24.1
+ Cosine selection	$37.2_{\pm 1.5}$	$22.8_{\pm 1.4}$	24.8
+ SW_2 selection	$39.8_{\pm 1.6}$	$24.1_{\pm 1.4}$	26.2
+ Entropy weighting	$40.9_{\pm 1.6}$	$24.8_{\pm 1.4}$	27.1
+ Goldilocks zone	$41.8_{\pm 1.6}$	$25.4_{\pm 1.5}$	28.2
+ Self-consistency	$42.6_{\pm 1.6}$	$25.9_{\pm 1.5}$	28.9
+ DPP diversity	$\mathbf{43.5}_{\pm 1.6}$	$\mathbf{26.3}_{\pm 1.5}$	**29.6**

localisation. WALDO improves Qwen3-VL-32B to 32.0% mAP@30 (+57% relative), demonstrating that reference-augmented reasoning can compensate for limited model capacity. Similarly, Gemini-2.0-Flash improves from 18.1% to 38.0% mAP@30 (+110% relative), confirming cross-architecture generalisability. Qwen3-VL-235B (MoE) shows that WALDO achieves 31.8% versus 36.3% zero-shot (−12% relative). This cannot be attributed to model scale alone: Qwen2.5-VL-72B achieves comparable zero-shot performance (36.4%) despite being ~3× smaller, yet benefits strongly from WALDO (+19.5%). The degradation appears specific to the MoE architecture, suggesting its expert routing mechanism may conflate reference and query contexts during differential reasoning.

Figure 2 shows representative examples with WALDO predictions. On NOVA brain MRI, WALDO accurately localises focal lesions with IoU up to 0.87. The differential prompting helps the VLM identify subtle intensity differences between query and reference. Failure cases typically involve very small lesions (< 5% image area) or diffuse pathologies lacking clear boundaries.

Table 2 shows lower performance on chest X-rays (10.9–34.1% mAP@30) vs. 43% on brain MRI. This domain gap likely reflects: (i) high inter-patient cardiac and mediastinal variability confounding healthy references, *i.e.*, heart size and shape vary substantially among healthy individuals; brain MRI variability exists but focal pathologies produce contrast changes absent in healthy tissue, making healthy references more discriminative; (ii) diffuse pathologies (pulmonary oedema, fibrosis, infiltrates) lacking focal boundaries amenable to bounding box localisation; (iii) overlapping anatomical structures (ribs, vessels) creating com-

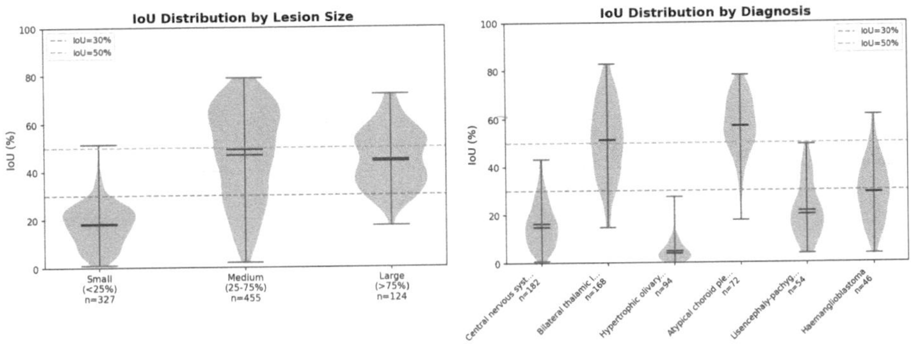

Fig. 3. Error stratification analysis on NOVA. IoU distributions grouped by lesion size for NOVA (left) and disease category (right). Small lesions exhibit markedly lower accuracy, whereas performance varies across diagnostic groups.

plex spatial patterns. Notably, WALDO still consistently improves over zero-shot across all models, with the largest relative gain for GPT-4o (+230%).

Table 3 shows contributions. The largest gain comes from replacing cosine similarity with Sliced Wasserstein (+6% mAP@30), validating the distributional formulation. Entropy weighting adds +2%, Goldilocks sampling +2%, and self-consistency with DPP diversity adds +4%. Each component addresses a distinct failure mode: SW distance enforces anatomical alignment, Goldilocks avoids over/under-similar references, and self-consistency reduces prediction variance.

Selection Metric Comparison. Entropy-weighted SW_2 achieves highest performance (41% mAP@30 for selection alone, 43.5% with full pipeline). CLS-based cosine similarity (37%) underperforms mean-patch L2 (39%) and unweighted SW_2 (40%), confirming that (i) distributional formulation outperforms point estimates and (ii) entropy weighting provides meaningful gains.

Goldilocks Effect Analysis. Stratifying performance by reference similarity percentile reveals a non-monotonic relationship: most-similar references (lowest decile) achieve 35% mAP@30, while references from the median percentile reach 43.5%. The differences between these regimes is statistically significant ($p = 0.02$, paired t-test), supporting an inverted-U similarity-accuracy relationship.

Differential vs. Direct Prompting. Differential prompting ("compare and find differences") outperforms direct detection ("find anomalies") by +7% mAP@30 (43.5% vs. 36.4%), indicating that comparative reasoning leverages VLM's relational capabilities more effectively than absolute anomaly detection.

Number of References. Performance saturates at $K = 5$ references: $K = 1$: 39%, $K = 3$: 42%, $K = 5$: 43.5%, $K = 7$: 43%, indicating sufficient Goldilocks zone coverage with marginal benefit from additional VLM calls.

Sensitivity Analysis. We analyse sensitivity to hyperparameters. Goldilocks percentile range $[\alpha, 1-\alpha]$: $\alpha = 0.2$ achieves 42%, $\alpha = 0.3$ achieves 43.5%, $\alpha = 0.4$

achieves 43%, indicating moderate sensitivity with a broad optimum. Number of SW projections M: 50 projections achieve 42.5%, 100 achieve 43.5%, 200 achieve 43.5%, indicating saturation at $M = 100$. Confidence weight $\lambda = 0.05$ achieves 42.5%, $\lambda = 0.1$ achieves 43.5% (optimal), and $\lambda = 0.2$ achieves 42%.

Error Analysis. Figure 3 shows performance stratified by lesion size and diagnosis on NOVA. WALDO failure modes can be categorised as: (i) *Small lesions* (35% of errors): Lesions <5% of image area are frequently missed, likely due to VLM resolution limits; (ii) *Diffuse pathologies* (28%): White matter changes and oedema lack clear boundaries; (iii) *Reference mismatch* (22%): Unusual slice orientations or acquisition artefacts reduce reference pool quality; (iv) *VLM hallucination* (15%): False positive detections at normal anatomical variants. The first two categories represent fundamental localisation challenges, while the latter two are addressable through reference curation and confidence calibration. Quantitatively, 27–47% of WALDO predictions on VinDr-CXR are false positives (IoU = 0 vs. all GT); however, WALDO reduces the fraction of images with no correct localisation by 22–55 percentage points compared to zero-shot across all models, confirming that the gain in detection recall outweighs the increase in false-positive volume.

Computational Cost. DINOv3 feature extraction requires ~50ms per image on an A100 GPU. Sliced Wasserstein distance computation for 30 references requires ~20ms. VLM inference dominates runtime at ~5 s per call via API; with $K = 5$ references, this corresponds to ~25 s per query. The reference selection overhead (< 1s) is negligible relative to VLM inference time.

Discussion. WALDO's design parallels radiological practice in which clinicians compare patient scans to mental models of normal anatomy and identify deviations. The observed Goldilocks effect reflects this intuition that comparison cases should be "similar enough to be relevant, different enough to be informative." The MRI-CXR performance gap (43.5% vs. 10.9-34.1% mAP@30) likely arises from modality differences, as many brain pathologies manifest as focal intensity changes against homogeneous tissue, while chest X-rays present overlapping structures with high inter-patient variability even among healthy individuals.

Limitations: CXR performance remains lower due to diffuse pathologies and anatomical overlap, and the method requires curated healthy reference pools with clinical expertise. VLM latency limits real-time deployment. At 43.5% mAP@30, and with small lesions (<5% area) frequently missed, WALDO suits triage and attention guidance rather than primary diagnosis. Future work could distil selection strategies into lighter models to improve efficiency. We do not compare directly with supervised detection methods, as these require annotated training data for each pathology type; WALDO specifically targets the setting where labelled training data is absent, which is the norm for rare diseases.

Synthetic Anomaly ICL: We explored an alternative approach inspired by the Many Tasks framework [3], in which synthetic anomalies with known bounding boxes are introduced into healthy references and used as in-context learning exemplars. This approach achieved 15% mAP@30. The domain shift between synthetic manipulations (Poisson patches, texture perturbations) and real pathologies appears to confuse spatial reasoning in VLMs, suggesting that training-based synthetic anomaly approaches [3,25] may not transfer effectively yet.

Broader Implications: A key insight from our results is that providing healthy reference images enables VLMs to perform differential diagnosis without disease-specific training. The reference-based comparison paradigm effectively provides in-context anatomical priors, allowing the model to identify deviations from healthy structure rather than relying on learned disease representations.

4 Conclusion

We introduced WALDO, a training-free approach for VLM anomaly localisation grounded in optimal transport theory. Key contributions include: (1) entropy-weighted Sliced Wasserstein reference selection using DINOv3 patch distributions to preserve spatial structure; (2) a theoretical characterisation of the Goldilocks effect explaining why moderate-similarity references optimise bias-variance tradeoff; (3) self-consistency aggregation via weighted NMS for robust bounding box prediction. WALDO achieves $43.5_{\pm 1.6}$% mAP@30 on NOVA with Qwen2.5-VL-72B (+19.5% relative improvement, $p < 0.01$), with consistent gains across GPT-4o (+68%), Qwen3-VL-32B (+57%), and Gemini-2.0-Flash (+110%). Evaluation on VinDr-CXR also showed promising zero-shot capabilities. Our results show that providing healthy reference images enables VLMs to perform differential diagnosis via in-context anatomical priors, advancing zero-shot anomaly detection towards clinical utility. *Future directions* include learned end-to-end reference selection, multi-scale patch analysis, uncertainty calibration for clinical deployment, and cross-modal transfer of selection strategies.

Acknowledgments. We acknowledge HPC resources from NHR@FAU (projects b143dc, b180dc), funded by federal and Bavarian state authorities and Gerhard Wellein's HPC approach. NHR@FAU hardware is partially funded by DFG 440719683. Additional support was received from ERC projects MIA-NORMAL 101083647 and CHARMS 101246053, DFG 513220538 and 512819079, and the state of Bavaria (HTA and the Bavarian Foundation Model Initiative). We further acknowledge resources provided by the Isambard-AI National AI Research Resource (AIRR), operated by the University of Bristol and funded by DSIT via UKRI and STFC [ST/AIRR/I-A-I/1023] [16]. We used coding agents and LLMs from Anthropic, OpenAI, Google, and Mistral AI, for text polishing, coding, experiment orchestration, and cluster monitoring. Most remarkable was the models' rigorous support with falsification of weaker hypotheses through extensive testing and mathematical scrutiny.

Disclosure of Interests. The authors have no relevant competing interests.

References

1. Arjovsky, M., Chintala, S., Bottou, L.: Wasserstein generative adversarial networks. In: International Conference on Machine Learning, pp. 214–223. ICML - PMLR (2017)
2. Bai, S., Chen, K., Liu, X., Wang, P.: Qwen3-VL technical report. arXiv preprint arXiv:2511.21631 (2025)
3. Baugh, M., Tan, J., Müller, J.P., Dombrowski, M., Batten, J., Kainz, B.: Many tasks make light work: learning to localise medical anomalies from multiple synthetic tasks. In: MICCAI, pp. 162–172. Springer (2023)
4. Baugh, M., Tan, J., Vlontzos, A., Müller, J.P., Kainz, B.: nnood: a framework for benchmarking self-supervised anomaly localisation methods. In: International Workshop on Uncertainty for Safe Utilization of Machine Learning in Medical Imaging, pp. 103–112. Springer (2022)
5. Bercea, C.I., et al.: NOVA: a benchmark for anomaly localization and clinical reasoning in brain MRI. arXiv preprint arXiv:2505.14064 (2025), NeurIPS 2025
6. Bergmann, P., Fauser, M., Sattlegger, D., Steger, C.: Uninformed students: student-teacher anomaly detection with discriminative latent embeddings. In: Proceedings of the IEEE/CVF Conference on Computer Vision and Pattern Recognition, pp. 4183–4192 (2020)
7. Bonneel, N., Rabin, J., Peyré, G., Pfister, H.: Sliced and radon wasserstein barycenters of measures. J. Math. Imaging Vision **51**(1), 22–45 (2015)
8. Brown, T., et al.: Language models are few-shot learners. In: NeurIPS 2020, vol. 33, pp. 1877–1901 (2020)
9. Courty, N., Flamary, R., Tuia, D., Rakotomamonjy, A.: Optimal transport for domain adaptation. IEEE TPAMI **39**(9), 1853–1865 (2017)
10. Defard, T., Setkov, A., Loesch, A., Audigier, R.: Padim: a patch distribution modeling framework for anomaly detection and localization. In: ICPR 2021, pp. 475–489. Springer (2021)
11. Dong, Q., et al.: A survey on in-context learning. In: ELNLP'24 - arXiv:2301.00234, pp. 1107–1128 (2024)
12. Efron, B., Tibshirani, R.: Bootstrap methods for standard errors, confidence intervals, and other measures of statistical accuracy. Stat. Sci. 54–75 (1986)
13. Team, G.: Gemini: a family of highly capable multimodal models. arXiv:2312.11805 (2023)
14. Kulesza, A., Taskar, B., et al.: Determinantal point processes for machine learning. Found. Trends Mach. Learn. **5**(2–3), 123–286 (2012)
15. Litjens, G., et al.: A survey on deep learning in medical image analysis. Med. Image Anal. **42**, 60–88 (2017)
16. McIntosh-Smith, S., Alam, S.R., Woods, C.: Isambard-AI: a leadership class supercomputer optimised specifically for artificial intelligence (2024)
17. McNemar, Q.: Note on the sampling error of the difference between correlated proportions. Psychometrika **12**, 153–157 (1947)
18. Naval Marimont, S., Siomos, V., Baugh, M., Tzelepis, C., Kainz, B., Tarroni, G.: Ensembled cold-diffusion restorations for unsupervised anomaly detection. In: MICCAI 2024. LNCS, vol. 15011, pp. 243–253. Springer (2024)
19. Nguyen, H.Q., et al.: VinDr-CXR: an open dataset of chest X-rays with radiologist's annotations. Sci. Data **9**(1), 429 (2022)
20. Hurst, A., Lerer, A., Goucher, A.P., et al.: Gpt-4o system card. arXiv:2410.21276 (2024). https://openai.com/index/gpt-4o-system-card/
21. Peyré, G., Cuturi, M., et al.: Computational optimal transport: with applications to data science. Found. Trends® Mach. Learn. **11**(5-6), 355–607 (2019)

22. Radford, A., et al.: Learning transferable visual models from natural language supervision. In: International Conference on Machine Learning, pp. 8748–8763. ICML - PmLR (2021)
23. Roth, K., Pemula, L., Zepeda, J., Schölkopf, B., Brox, T., Gehler, P.: Towards total recall in industrial anomaly detection. In: Proceedings of the IEEE/CVF Conference on Computer Vision and Pattern Recognition, pp. 14318–14328 (2022)
24. Schlegl, T., Seeböck, P., Waldstein, S.M., Langs, G., Schmidt-Erfurth, U.: f-AnoGAN: fast unsupervised anomaly detection with generative adversarial networks. Med. Image Anal. **54**, 30–44 (2019)
25. Schlüter, H.M., Tan, J., Hou, B., Kainz, B.: Natural synthetic anomalies for self-supervised anomaly detection and localization. In: ECCV, pp. 474–489. Springer (2022)
26. Siméoni, O., Vo, H.V., Seitzer, M., Baldassarre, F., Oquab, M.: DINOv3: self-supervised learning with gram anchoring. arXiv preprint arXiv:2508.10104 (2025)
27. Tan, J., Hou, B., Batten, J., Qiu, H., Kainz, B., et al.: Detecting outliers with foreign patch interpolation. Mach. Learn. Biomed. Imaging arXiv:2011.04197 **1**(April 2022 issue), 1–27 (2022)
28. Wang, P., Bai, S.: Qwen2-VL: enhancing vision-language model's perception at any resolution. arXiv:2409.12191 (2024)
29. Wang, X., Wei, J., Schuurmans, D., et al.: Self-consistency improves chain of thought reasoning. In: ICLR - arXiv:2203.11171 (2023)
30. Wei, J., et al.: Chain-of-thought prompting elicits reasoning in large language models. In: NeurIPS 2022, vol. 35, pp. 24824–24837 (2022)
31. Zimmerer, D., et al.: Mood 2020: a public benchmark for out-of-distribution detection and localization on medical images. IEEE Trans. Med. Imaging **41**(10), 2728–2738 (2022)

Author Index

G. Yang et al. (Eds.): MICCAI 2026, LNCS 16878, pp. 665–669, 2027.
https://doi.org/10.1007/978-3-032-38059-3

M

N

O

P

Q

R

S

T

V

W

X

FSC
www.fsc.org
MIX
Papier | Fördert
gute Waldnutzung
FSC® C083411

Kolibri 360 GmbH
Ferdinand-Jühlke-Straße 7
99095 Erfurt, Deutschland
produktsicherheit@kolibri360.de